AF541595

OILSEED CROPS

a division of

NIPA GENX ELECTRONIC RESOURCES & SOLUTIONS P. LTD.

New Delhi-110 034

About the Author

Dr M.V.R. Prasad represents the category of scientists, who are not confined exclusively to the narrow walls of their fields of specialization; but have grown to look at agricultural science as the integrated whole of diverse disciplines for sustainable progress of research. With his specialization in Genetics and Plant Breeding, Dr Prasad strived to work with a multidisciplinary approach to achieving solutions to crop production particularly in dryland-land Agriculture.

Possessing over fifty years of experience at the national and international levels in the improvement of oilseed crops, grain legumes and tree species, he has contributed to the amelioration of dryland agriculture through development of crop varieties of groundnut, castor, mung bean, moth bean, guar (cluster beans), cashew nut and Pongamia.Among his research contributions, a particular mention may be made of the development and use of aerial pod-bearing attribute, a novel leaf mutant genotype with higher dry matter production and higher oleic acid content, and selection criteria for efficient groundnut breeding in addition tobreeding productive castor genotypes possessing wilt resistance for the first time in the country.

At the international level, Dr Prasad has worked as the International Crop Improvement Specialist and Visiting Professor of IICA (OAS) with EMBRAPA in Brazil. He also worked as the International Cashew Specialist and Technical Adviser of Af.D.B. and World Bank Projects on Cashew Development in Africa. He also participated at the FAO Regional Conference at Bangkok in 1991 and International Safflower Conference held in Beijing in 1993. He was the Leader of the Team constituted by the European Commission to assist the Mozambican Cashew Institute in the preparation of the plans for cashew development in 2007.

As the Director of the Directorate of Oilseeds Research (Currently Indian Institute of Oilseeds Research), his rich contributions to strengthening of oilseeds research and production are well known.

Dr Prasad has trained a number of scientists and post-graduate students ingenetics and breeding of oilseed crops in India and abroad. He possesses over 130 publications on oilseeds, grain legumes, and problems of dryland agriculture.

OILSEED CROPS

M.V.R. Prasad

a division of

NIPA GENX ELECTRONIC RESOURCES & SOLUTIONS P. LTD.

New Delhi-110 034

a division of
NIPA GENX ELECTRONIC RESOURCES & SOLUTIONS P. LTD.
101,103, Vikas Surya Plaza, CU Block
L.S.C.Market, Pitam Pura, New Delhi-110 034
Ph : +91 11 27341616, 27341717, 27341718
E-mail:newindiapublishingagency@gmail.com
www: www.nipabooks.com

For customer assistance, please contact
Phone: + 91-11-27 34 17 17 Fax: + 91-11- 27 34 16 16
E-Mail: feedbacks@nipabooks.com

© 2021, Publisher

ISBN: 978-93-90175-970

All rights reserved. No part of this publication may be reproduced, stored in a retrieval system or transmitted in any form or by any means, including electronic, mechanical, photocopying recording or otherwise without the prior written permission of the publisher or the copyright holder.

This book contains information obtained from authentic and highly reliable sources. Reasonable efforts have been made to publish reliable data and information, but the author/s, editor/s and publisher cannot assume responsibility for the validity, accuracy or completeness of all materials or information published herein or the consequences of their use. The work is published with the understanding that the publisher and author/s are not attempting to render any professional services. The author/s, editor/s and publisher have attempted to trace and acknowledge the copyright holders of all material reproduced in this publication and apologize to copyright holders if permission and/or acknowledgements to publish in this form have not been taken. If any copyrighted material has not been acknowledged, please write to us and let us know so that we may rectify the error, in subsequent reprints.

Trademark Notice: NIPA, the NIPA logos and their presentations (the way they are written/ presented) in this book are the trademarks of the publisher and hence may not be used without written permission, if copied or used without authorization, the infringer will be prosecuted as per law.

NIPA also publishes books in a variety of electronic formats. Some content that appears in print may not be available in electronic books, and vice versa.

Composed and Designed by NIPA.

M.S. SWAMINATHAN RESEARCH FOUNDATION

M.S. Swaminathan
Founder Chairman
Ex-Member of Parliament (Rajya Sabha)

Foreword

Fats and oils are important components of food, apart from their industrial uses. Food applications account for a major share to the order of 75% worldwide consumption of fats and oils. Oilseed bearing plants, which are energy rich, form an important source of edible fats and oils. Paradoxically, these energy-rich crops are being grown largely under energy-starved conditions of dryland agriculture in India with their productivity levels remaining rather low.

India is the second largest producer of groundnut after China and the third largest producer of rapeseed-mustard group after China and Canada. India is the largest producer of castor, sesame, safflower and niger.

In the decade of 1950s, India produced a meagre 5 million tonnes of oilseeds. This number went up to 9 million tonnes in 1970 and subsequently there has been remarkable progress in research and production of oilseed crops, which triggered what is popularly termed the "Yellow Revolution" from late 1980s to the 1990s.

Over the last few years, domestic consumption of edible oils increased and touched the level of 18.90 million tonnes in 2011-12 and is likely to increase further. With per capita consumption of vegetable oils at the rate of 16 kg/year/person for a projected population of 1276 million, the total vegetable oil demand has touched more than 20.4 million tonnes by 2017. It may however be observed, that the higher rates of consumption of vegetable fats is restricted to the rich and upper middle class populations of India.

On the other hand, India has one of the lowest per capita daily supply of calories, protein and fat, according to the Organisation for Economic Co-operation and Development (OECD). Considering the calorie needs of Indian populations, particularly in respect of persons putting in heavy work, such as manual labourers, factory workers, construction labourers and farmers, energy allowance prescribed by the Indian Council of Medical Research (ICMR) is 3,800 k.cals., and protein requirement of 60 g/day; the actual availability of calories and proteins in these sections of population is far below the recommendations. Hence there is need to ensure availability of calorie-protein rich food to needy sections of populations to ensure their nutrition security. Increasing further the production and productivity of oilseed crops is needed in this context, since edible oilseeds are not only rich sources of calories but offer considerable proteins too, paving the way for protein-rich product development as in the case of soybean, linseed, sesame, groundnut and sunflower.

With the Indian acreage of oilseeds fluctuating between 22 to 29 million hectares, the production of major oil seeds remains 33 million tons, with the current productivity levels fluctuating at 1.0 to 1.5 t/ha. Currently there is limited scope to enhance the cultivable area under oilseed crops. Hence, research has to be intensified on oilseed crops to enhance their productivity levels through application of modern tools of plant genetics and physiology including biotechnology and crop management technologies.

In marching towards achieving higher oilseed production, tree borne oilseeds (TBOs], like karanj, sal, mahua, simarouba, kokum, jatropha, neem, jojoba, cheura, wild apricot, walnut, tung etc. assume considerable importance in addition to annual oilseed crops. This group of oilseed bearing plants can be grown under different agro-climatic conditions on waste land /deserts/hilly areas. These TBOs are also good source of vegetable oil which need to be supported for their improvement and cultivation. Apart from their yields of oilseed, these trees are valuable in cutting down emissions, building up soil organic matter and ensuring carbon sequestration, thereby bringing about environmental amelioration. The oilseed bearing trees have not yet received adequate research support to augment their production and productivity. It is therefore necessary to intensify research on this group of oil-yielders.

There has been a continued shift of vegetable oils particularly of soybean and rapeseed-mustard from food to industrial consumption, particularly to biodiesel, which should be discouraged, owing to the increasing demands of edible oilseeds in Asia and Africa. Alternatively, non-edible oilseed crops like castor and oil bearing tree species should be exploited for this purpose, avoiding the use of edible oilseed crops.

Oilseeds as a whole form a unique and important group of crops which demand a comprehensive and thorough understanding of the potentials and intricacies involved. In the course of unravelling the problems of this group, scientists have to face certain intractable challenges, which need thorough understanding of their peculiarities. Hence, there is a felt need for a comprehensive book on oilseed crops for the benefit of scientists, students and those engaged in oilseeds research and production.

I hope that the book on Oilseed Crops by Dr. M.V.R. Prasad, former Director of the ICAR Directorate of Oilseeds Research, who possesses vast experience in oilseeds through his work in India and abroad, fulfils the above need. The New India Publishing Agency deserves compliments to have undertaken to publishing the book on Oilseed Crops.

M S Swaminatha

3rd Cross Road, Taramani Institutional Area, Chennai (Madras) - 600 113, INDIA
Phone:+91-44-2254 2790, 2254 1698, Fax:+91-44-2254 1319
E-mail: founder@mssrf.res.in, swami@mssrf.res.in

Acknowledgement

In the task of writing the book on 'Oilseed Crops' which took me almost one year, I must acknowledge the help and encouragement extended by my well-wishers, friends and colleagues. Their support rendered my job pleasant and enjoyable.

My sincere thanks go to Drs. V. Ranga Rao, M. Ramachandram, A.R.G. Ranganatha, K. Ajnani, Prakash Ghorpade, A. L. Rathnakumar, M. Sujatha, and K.H. Singh who had gone through certain chapters of the draft book and extended valuable suggestions.

I am grateful to Dr M. Murali Krishna who stood by me constantly encouraging me at every step apart from offering very valuable suggestions after going through some chapters of the draft book.

Thanks are due to Dr A. Vishnuvardhan Reddy, Director, Indian Institute of Oilseeds Research (IIOR) Hyderabad to have put the IIOR library facilities at my disposal in this context. I acknowledge ready help from Drs Harvir Singh, Senthilvel and I.Y.L.N. Murthy in finding some relevant references. I must thank Shri V.V. Siva Prasad for his help in framing and uploading the final document in the form of Google drive.

I sincerely thank New India Publishing Agency who prevailed upon me to write the book and bearing with my intricacies and complexities arising out of the current difficult situation in the country. I shall be failing in my duty if I don't acknowledge the support that I enjoyed from my family members at every step of this task.

Last but not the least, I am extremely grateful to my guru, Prof. (Dr) M.S. Swaminathan to have given his invaluable foreword to the book.

It shall be failing on my part, if I don't acknowledge the abundant Grace and Love that I received form Shree Hari-Sai, without which it wouldn't have been possible for me to accomplish this task.

M.V.R. Prasad

Preface

Oil bearing plant species that contribute to the vegetable oil kitty of India are varied and diverse. There has been a long and continuous effort both from research side and on the production front, to enhance the production and productivity of these diverse plant species. There was a breakthrough in the stagnant Indian oilseed production during the decades of nineteen eighties to nineteen nineties, which gave confidence to achieve self-sufficiency in the production of vegetable oils. Nevertheless, one could see the inescapable escalation of the demand for vegetable oils not only in India; but world over. While the area expansion of oilseed crops is out of question in a country like India, the only option that we have is to enhance the recoverable genetic potential of oilseed crops at the same time strengthening their management at farmers' level with emphasis on enhancement of productivity per unit of water and nutrients utilized. The era of biotechnology genetic engineering and genomics should take oilseed crops to their new heights of production potentials. A research worker or a student of oilseeds naturally should be interested to gather the relevant details with regard to each of the oilseed bearing plant species that would not only initiate him into the study of oilseeds but also generate keen interest to probe more into the intricacies of the subject. An effort is made in this book to meet the above requirement.

A considerable chunk of edible vegetable oils like those of soybean and rapeseed-mustard are being diverted for production of biodiesel, which should not be encouraged, since we see increasing number of low income populations are in need of the edible oils for their consumption added to the prevailing wide gap between the demand and production of vegetable oils. The nonedible oils from vast number of productive oil bearing trees should be used for the production of biodiesel. Steps should be taken to enhance the research on this group of oil bearing trees to reap benefits out of them. At present there is no tangible effort to enhance the genetic potential of this group of trees.

Inclusion of a perennial oil-bearing trees in the cropping systems of farmers would not only contribute to the overall production of oilseeds at the national level; but would also augment the income levels of farmers particularly the small and marginal dryland farmers imparting stability to overall farm production. It may be recalled that during the worst drought years of India viz., 1972, 1974 and 1979, the 500 and odd villages in the country escaped the adverse impact drought largely due to the presence of *Pongamia pinnata* tree populations on the farmers' holdings. The perennial oil bearing trees like Pongamia offer a silver lining to the otherwise dark picture of loss of soil fertility and complex environmental deterioration through their proven capacity to ameliorate soil and enhance the agro-ecosystems.

Another aspect that deserves more attention is about the protein component of oilseeds. Apart from soybean known for its protein content, every oilseed bearing plant has the potential to contribute to the protein kitty of the nation, provided we pay more attention to this aspect.

For example, it has been conclusively demonstrated by CFTRI that protein of Pongamia cake is comparable to that of soybean in terms of quality following removal of some irritants and anti-nutritional substances such as karnjin, saponins and phytates, although not being toxic. It should become the mandate of oilseed scientists and research institutes to initiate concerted efforts towards the cost effective recovery of the protein component of oilseeds in effort to augment the nutritional security of India consisting of vast masses suffering from protein malnutrition.

M.V.R. Prasad

Contents

1

Introduction

Oilseed Crops consist of those groups of plants cultivated for substantial quantities of the oil, vegetable fat obtained from the seeds (endosperm) of the plants.

The major oilseeds grown world over are soybeans, sunflowers, rapeseed – mustard (or canola in USA & Canada), groundnuts (peanuts) and linseed. Seed oils from flax (linseed) and castor beans are used for industrial purposes. These Oilseed crops including oilseed bearing trees are the basic biological systems that produce oils, contributing to renewable energy production, help stabilize greenhouse gases, and mitigate the risk of climate change (Jaradat, 2016)

The oilseed crops are of heterogeneous group of plant species not having any common morphological and genetic features except production of seed oil. Crops like groundnut, sesame, castor bean, soybean and niger are grown during the rainy season and the others like rapeseed and mustard and safflower are grown in the post rainy season. Sunflower, however, may be grown in both the seasons. Hence the individual crops forming the conglomerate of Oilseed crops have to be treated differently for their improvement and production. Nevertheless, all of them produce a common end product viz., oil or vegetable fat, the importance of which is growing in product development, commerce and human and farm animal nutrition. The oilseed bearing crops have to be treated as a unique group in terms of research and development to come up with strategies for their judicious exploitation in the national farm economy. From the scientific point of view, the oilseed crops are energy rich and offer impressive potentials in terms of realizable production levels and valuable product development. In India good progress has been achieved so far in research and development of some oilseed crops; but there exists a need to scale to newer heights in terms of realising the hidden potentials of oilseed crops through basic and strategic research, with an integrated approach.

In the case of groundnut, recent work carried out at ICRISAT has resulted in the development of genotypes with higher levels of oleic acid to the tune of 80%, which is a commendable achievement for it opens up new avenues in human nutrition and product development. Nevertheless, there is a need to come up with new strategies in terms of selection criteria for kernel productivity and new information on physiological parameters to be employed to enhance the efficiency of groundnut breeding and production. Considering the rapeseed-mustard group of crops, it should be observed that very good progress has been achieved in the context of development of productive varieties and hybrids of *Brassica juncea* (mustard). The development of hybrid in Indian mustard by ICAR National Research Centre on Rapeseed-Mustard viz., NRCHB-506, using *moricandia* cytoplasmic genetic male sterility system is a major advancement in the field of varietal technology. However, the crop of taramira (*Eruca sativa* Mills) of *Brassica* group grown in the post rainy season on poor soils without irrigation in the north and north-west India has not received adequate attention in terms of research and development. This crop has potential of becoming an important industrial crop. This aspect is being highlighted indicating the key agronomic inputs that could enhance the productivity of the crop. Similarly, niger (*Guizotica abyssinica* L.) is another oilseed crop that remains neglected despite its impressive nutritional qualities and the potential for development of hybrids. Development of hybrids in Sesame (*Sesamum indicum)* is lagging behind despite the possibilities and potential that can open-up in this direction.

A wide variety of oil seeds are produced in India as indicated above. India holds a significant share in world oil seed production. It is the second largest producer of groundnut after China and third largest producer of rapeseed after China and Canada. India is the largest producer of castor, sesame, safflower and niger.

Oilseeds constitute the second largest agricultural commodity after cereals in India, sharing 14% of the gross cropped area and accounting for about 5% of India's gross national product and about 10% of the value of all the agricultural products (Gupta and Gupta, 2016). The oil bearing plant species could become important feed stocks for production of biodiesel. Nearly14 millon people are involved in the production of oilseed crops and one million in the processing of oilseeds (Gupta and Gupta, 2016).

Agronomic approaches to oilseed production require adjustments to account for physiological differences among these crops. Some of the important production challenges for the oilseed crops grown on drylands of India are (1) plant establishment, (2) confronting diverse biotic stresses, and (3) deficient water and nutrient availability under conditions of adverse temperatures. By coordinating production practices with growth stages and growing conditions, the impacts of moisture and temperature stress can be minimized. This requires crops to be more reliant on stored water and nutrients in the subsoils for sustaining growth and development due to greater water and heat stress during reproductive stages.

Oilseeds are recognized as potential rotational crops due to their ability to extract deep soil moisture in water-limited environments more effectively than wheat or peas (Merrill *et al.,* 2004). Opportunities for integrating oilseeds into traditional cropping patterns consist of (i) including them as non-competing intercrops in the stands of principal crop zones and (ii) restructuring the crop sequences including the oilseed crops to make the whole cropping system more productive and profitable.

Yield under environmental stress is a state-of-the-art reference that investigates the effect of environmental stress on oil seed crops and outlines effective ways to reduce stress and improve crop yield (Klepper *et al.,* 1984)

Vegetable fats are mixtures of triglycerides. A triglyceride is an ester derived from glycerol and three fatty acids. Triglycerides are the main constituents of body fat in humans and other animals, as well as vegetable fat.

Fats and oils are concentrated forms of energy. The energy yield from the complete oxidation of fatty acids is about 9 kcal per gram, in comparison with about 4 kcal per gram for carbohydrates and proteins. Triglycerides are molecules made of fatty acids (chain-like molecules of carbon, hydrogen, and oxygen) linked in groups of three to a backbone of glycerol. When foods containing fats are consumed, the fatty acids are separated from their glycerol backbone during the process of digestion. Fats and oils in the diet are thus available to the body as fatty acids.

The traditional food crops such as wheat and rice too have oil in their grains; but it is of the order of 2 to 3%. On the other hand, the plant species grown for oil contain seed oil of 20% and above. Edible fats and oils are similar in molecular structure; however, fats are solid at room temperature, while oils are liquid.

Fats and oils are essential nutrients, form a range of 25 to 40% of calories of human diet world over. Edible vegetable oils are used as salad or cooking oils, or may be solidified (by a process called hydrogenation) to make margarine and shortening. These products supplement or replace animal products such as butter and lard, supplies of which are inadequate to meet the needs of an increasing world population.

Dietary fats including vegetable oils are essential to give human body energy and to support cell growth. Additionally, they help protect body organs and help keep the body warm. They also help the body absorb some nutrients and produce important hormones.

There has been considerable debate regarding type and optimal amounts of fats / oils used in the diet, their role in regulating body weight and their importance in the aetiology of chronic disease (Caballero and Allen, 2012).

The recommendation of the joint group of Food and Agricultural Organization (FAO) and World Health Organization (WHO) on the dietary role of fats and oils is as follows:

"Adequate amounts of dietary fats are essential for health. In addition to contributing to meeting energy needs, intakes of dietary fat must be sufficient to meet requirements for essential fatty acids and fat-soluble vitamins. The minimum intake consistent with health varies throughout a person's life and among individuals. Adequate intake of dietary fat is particularly important prior to and

during pregnancy and lactation. Increasing the availability and consumption of dietary fats is often a priority for overcoming the problems of protein-energy malnutrition. Recommendations to populations concerning desirable ranges of fat intakes may vary according to prevailing conditions, especially dietary patterns and the prevalence of diet-related non-communicable diseases."

Despite the contentious issues surrounding dietary fats, they are considered essential nutrients because they are required to perform critical functions in the body including serving as a carrier of preformed fat-soluble vitamins, enhancing the bioavailability of fat-soluble micronutrients and providing essential substrate for the synthesis of metabolically active compounds (such as the steroid hormones, testosterone, oestrogen and progesterone) among other useful functions.

India's vegetable oil economy is world's fourth largest after USA, China & Brazil. The oilseed accounts for 13% of the Gross Cropped Area, 3% of the Gross National Product and 10% value of all agricultural commodities. This sector has recorded annual growth rate of area, production and yield @ 2.44%, 5.47% and 2.96% respectively during last decade (ISOR, 2020).

The diverse agro-ecological conditions in the country are favourable for growing 9 annual oilseed crops, which include 7 edible oilseeds (groundnut, rapeseed & mustard, soybean, sunflower, sesame, safflower and niger) and two non-edible oilseeds (castor and linseed). Oilseeds cultivation is undertaken across the country in about 27 million hectares mainly on marginal lands, of which 72% is confined to rainfed farming posing challenges in achieving sustainability in their production (Sharma, 2014).

In 1951, India produced a meagre 5 million tonnes of oil seeds. This number went up to 9 million tonnes in 1970 and there has been remarkable progress. Currently, oil seeds share 14% of the area under major crops. At present, India's largest oilseed producing state is Gujarat due to the state's highest groundnut production. Rajasthan is India's top rapeseed and mustard producing state, followed by Madhya Pradesh and Haryana. Almost half (48.12%) of rapeseed and mustard is produced in Rajasthan. India's top Soybean producing state is Madhya Pradesh with a share of 44% in India's total production of this protein rich crop. Karnataka enjoys premier position in Sunflower production.

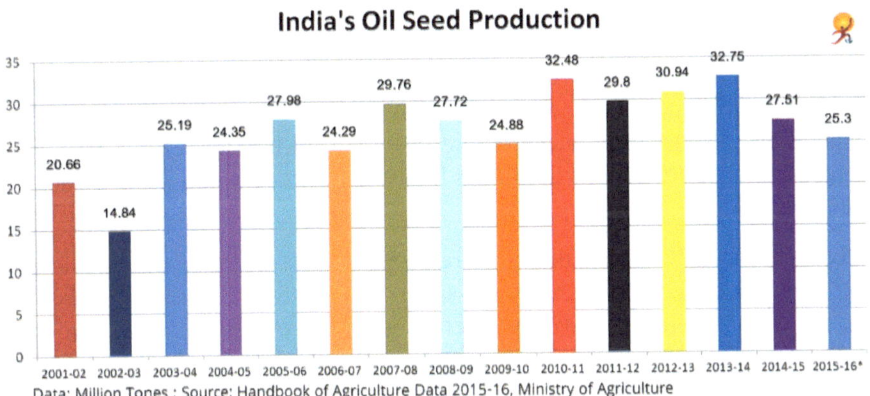

Fig. 1: Indian Oilseeds Production from the year 2000 to 2016
(Source: Ministry of Agriculture Govt. of India)

The domestic sources of vegetable oil are of two types, viz., (1) Primary and (2) Secondary.

(1) The Primary Source consists of the production from traditional oilseed crops viz., groundnut, rapeseed-mustard, soybean, sunflower, Sesame, Niger, Safflower, Linseed and castor.

(2) The Secondary Source comprise Coconut, Cotton seed, Rice Bran and Oilseed baring tree species.

Primary sources cover around 60MT of the total demand, while secondary cover 29 MT, thus making domestic supply to meet around 89MT of the total domestic requirement.

Of this, around 5MT is exported or used in industries. The net-domestic availability of Edible Oils thus remains around 86.37 MT and rest is met through imports. In 2015-16, India imported 148.20 million tonnes of edible oils.

During the years starting from 1986 to 1995 India achieved a tremendous progress in oilseed crop production under the Technology Mission on Oilseeds sponsored by the Government of India, which has been acclaimed as "Yellow Revolution" (Sharma, 2001) The productivity of oilseeds got enhanced from 650 kg/ha to 900kg/ha during this period, with major jump in total oilseed production. This achievement was possible with introduction of high yielding varieties, productive management technologies and expansion of the area under oilseeds. Nevertheless, with the increasing standard of living of population and changes in the lifestyle, there is ever increasing demand of the tune of 6% annually for vegetable oils with in the country, while the domestic production has been increasing at the rate of 2% per year. The demand for vegetable oil both in quantity and quality is increasing.

During the last few years, the domestic consumption of edible oils has increased substantially and has touched the level of 18.90 million tonnes in 2011-12 and is likely to increase further. With per capita consumption of vegetable oils at the rate of 16 kg/year/person for a projected population of 1276 million, the total vegetable oils demand has touched more than 20.4 million tonnes by 2017. The consumption of vegetable oils in India is likely to increase to 34 million tonnes by 2030 (Agribusiness, 2018).

A substantial portion of the Indian requirement of edible oil is met through import of palm oil from Indonesia and Malaysia. It is, therefore, necessary to exploit domestic resources to maximize production to ensure edible oil security for the country.

Tree Borne Oilseeds (TBOs), like sal, mahua, simarouba, kokum, olive, karanja, jatropha, neem, jojoba, cheura, wild apricot, walnut, tung etc. are growing in the country under different agro-climatic conditions in a scattered form in forest and non-forest areas as well as on waste land /deserts/hilly areas. These TBOs are also good source of vegetable oil and therefore need to be supported for their improvement and cultivation. The oilseed bearing trees have not yet received adequate research support to augment their production and productivity. There is every need to intensify research on this group of oil-yielders (Prasad, 1994c).

Area under oil seeds in India remains between 22 to 29 million hectares while production of the major oil seeds remains 33 million tons. Currently there is no scope for enhancing the cultivable area under oilseed crops. The continuous and ever increasing shortage of vegetable oil suggests that the introduction of oil palm had little impact on the vegetable oil production.

The major constraint in increasing the oilseed production in India is due to the cultivation of most of the oilseed crops on drylands solely depending on rainfall without any source of irrigation facility. These energy-rich crops are being grown under poor agro-ecological conditions leading to energy starvation, additionally being subjected to onslaught of biotic stresses. In the race against receding soil moisture, the oilseed plants despite their drought resistance face the necessity to grow their roots faster to tap the moisture. Hence there is urgent need to intensify research on incorporating genetic and physiological attributes that promote faster rate of root and shoot growth combing with earliness of flowering to achieve a major break-through in the productivity of oil bearing plant species and to come up with sustainable dryland production technologies relevant to promotion of oilseed crops.

According to the statement of National Mission On Oilseeds and Palm (NMOOP) its short-term targets (IFY 2017/18 to IFY 2021/22) are specified below.

Table 1: Oilseeds and Vegetable Oil Production Targets by IFY 2021/22 (Million Metric Tons)

Crops	Short-term target for 2021-22	
	Oilseeds	Oil
Groundnut	10.75	2.47
Soybean	16.50	2.64
Rapeseed & Mustard	9.75	3.02
Sesame	1.10	0.33
Others	4.00	1.50
Total	42.1	9.96

Table 2: Area, Production and Yield of Oilseed Crops in India 2018-19 (Vishnuvardhan Reddy et al., 2020)

Crop	Area(million ha)	Production (million t)	Productivity (kg/ha)
Soybean	11.33	13.79	1217
Rapeseed-Mustard	6.23	9.34	1499
Groundnut	4.81	6. 69	1393
Castor	0.76	1.20	1565
Sesame	1.60	0.76	473
Sunflower	0.25	0.22	886
Linseed	0.30	0.16	533
Niger	0.19	0.06	334
Safflower	0.04	0.02	592
Total	25.51	32.24	1264

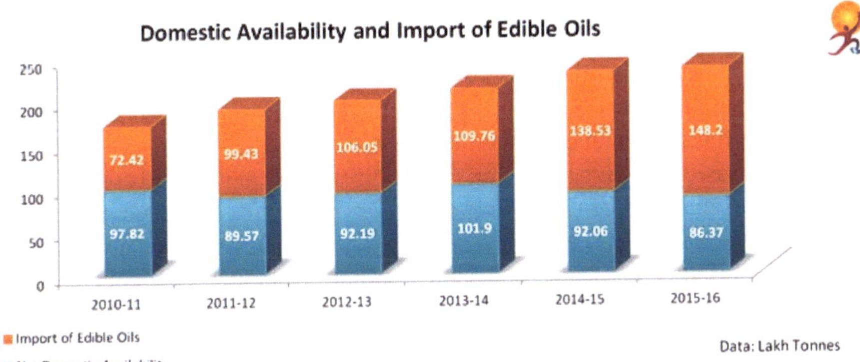

Fig. 2: Domestic Availability and Import of Edible Oils in India

In recent years world over, oilseed crops are being increasingly exploited for production of pharmaceuticals, surfactants, plasticisers, emulsifiers, detergents, lubricants, adhesives, cosmetics, oleo-chemicals, biofuels apart from their demand for their healthy vegetable oils and livestock feeds (Metzger and Bornscheuer, 2006), which resulted in an 82% expansion of oilseed crop cultivation areas and about a 240% increase in total world production over the last 30 years (Rahman,

M. and Jimenez (2016)). Notwithstanding the overall enhancement in oilseeds production in the past three decades, the growth and production of oilseed crops exhibit fluctuation without any consistency. Although many of the lipid based products are used for industrial chemicals, they often are not of the most favourable composition for non-food applications. The impending challenge confronting the vegetable oil industry is to arise to the needs of growing demand for oil by producing adequate amounts of plant derived oils (Lu *et al.,* 2011). A search for new oil-seed crops with more advantageous oil composition has led to the development of excellent candidates that are now close to commercial acceptance. Among them are *Crambe*, *Limnanthes*, *Vernonia*, *Sapium* and *Simmondsia*. Other crops are at a much lower stage of development but also have excellent potential. They include *Cuphea*, *Foeniculum*, *Stokesia,Lesquerella* and *Lunaria*. In this age of searching for renewable resources to replace petrochemicals and imported strategic materials, a well-organized research and development programme on new oilseed crops could soon result in self-sufficiency for industrial oils and fatty acids.

As compared to cereal food crops like rice, wheat maize and millets, the productivity of annual oilseed crops in general is far lower. This is because the oilseeds are energy rich crops in which plant's physiological system expends more energy towards the biosynthesis and storage of fat/ oil. This, however, does not mean that there is no scope for yield enhancement of oilseed crops. It has been shown as to how to enhance the profitability of oilseeds production systems through resource conservation technologies and efficient cropping systems (Hegde, 2020). The genetic modification of plant types with a thorough understanding of interrelationships among canopy structure and its development at different stages of plant's life cycle in relation to the source-sink relationships are to be undertaken (Cahener and Ashri, 1974; Prasad, 1988). For such an effort, considerable basic data on genetics and plant physiological mechanisms are to be worked out. Since most of the oilseed crops are grown largely under conditions of dryland agriculture, the relevant research strategies involving combination of breeding for drought resistance and yield attributes together with enhancement of their physiological efficiency for photosynthesis, dry matter enhancement, its partitioning and moisture and nutrient uptake should be in place in order to sustain the oilseeds productivity at higher levels. Such a complex task would necessitate a thorough understanding of each of the oilseed crops with regard to their genetic and physiological idiosyncrasies to bring in appropriate modification in the plant architecture conducive to the above mentioned attributes. In order to instil sustainability and consistency in oilseeds production and safe guard the income levels of oilseed farmers, most of who are engaged in dryland agriculture, there is an urgent need for augmenting research efforts to open new avenues for higher levels of oilseeds production and productivity. In the current era biotechnology and genetic engineering ushering into genomics, a bright future for oilseeds research on their production and productivity is not far off.

2

Groundnut (*Arachis hypogaea* L.)

Introduction

Groundnut (*Arachis hypogaea* L.) is an important traditional oilseed crop of India contributing to 3.9 million tons of global vegetable oil ranking the sixth in the sequence of oilseed crops. Added to its importance as an oilseed, groundnut is also an important food legume cultivated in an area of about 22 million hectares worldwide leading to a production of 38.6 million tons as per the FAO statistics of 2011 (ICRISAT, 2017). China and India are the leading groundnut producers followed by USA and Nigeria. Africa with 12.40 m ha area and 11.54 m tons of production, and Asia with 11.87 m ha and 29.95 m tons, together account for 95% global groundnut area and 91% of global groundnut production. There was a substantial increase in global groundnut production by about 5 m tons in 2013, from 40 m tons in 2012. In the last decade, 2004–2013, global groundnut production increased by 24%, contributed by 7% increase in groundnut area, and 16% increase in yield. The projected global demand for groundnut and its related products is expected to increase and so there is a need to further increase production and productivity to meet the demand.

Among the different oilseed crops grown in India, groundnut is the major crops grown in kharif / monsoon season. Among different states if India, Gujarat is the prominent state growing groundnut.

Groundnut has a long history of cultivation since the plant was domesticated and cultivated by the indigenous tribes such as Incas of Peru in South America for more than 1500 years. It is believed that the cultivated groundnut was introduced into India by the Spanish colonizers (Krapovickas, 1968). Groundnut is believed to have been introduced into the USA by the African slaves. Groundnut crop has found an ecological niche in the tropical and warm temperate zones of world, due its impressive adaptability.

Although the crop is cultivated in over 100 countries of the world, China and India remain the major producers of groundnut contributing to over 55% of world's production. India has over 40% of world's area of groundnut and contributes to 30% of the global production. The United States on the other hand has only 3% of world's area under groundnut, but significantly produces 9% of the world's production.

In spite of the instability in its production (Hammons 1976, Gregory *et al.,* 1951) groundnut plant has exhibited a very high degree of yield potential under favourable conditions (Prasad 1994d). It has been observed by plant physiologists (Sinclair and dewit, 1975) that unlike soybean, groundnut plant has not yet reached the peak evolutionary stage in terms of its reproductive efficiency; but at the same time it retains the most appropriate pattern of nitrogen metabolism conducive to attain new heights in terms of seed yield.

Edible oil is the major product for which groundnut is grown in many parts of the world. With 3.9 million tons of production, groundnut oil ranks sixth in the sequence of vegetable oils produced in the world.

About two thirds of world's groundnuts produced are used for extracting vegetable oil. However, in the USA on the other hand, 70% of the peanuts produced are used to produce edible products such as peanut butter, snack food, confectionaries etc. Peanut oil is obtained from kernels which won't fit into the confectionary industry. The oil content of groundnut varies from 42 to 52%. Most of the recommended varieties of groundnut contain around 50% oil. Groundnut oil is considered good cooking oil with good storability largely due to high content of Oleic Acid (55 to 60%).

Ecology of Groundnut: Groundnut plant is native of eastern South America bisected by 20°S line of latitude, where a large number of wild species of *Arachis* are known to exist. As a domesticated and cultivated plant groundnut's range is between 35°S and 40°N in Central Asia and North America with hot summer ecosystem.

Groundnut plant grows best in light textured sandy loam soils with neutral pH. Optimum temperature for its growth and development ranges from 28 to 30 °C and the crop requires about 500-600 mm of well distributed rainfall (Varaprasad *et al.,* 2011) However, Groundnut crop has shown itself to get acclimatised to wide range of agro-ecological situations and is considered drought resistant, since the crop can grow satisfactorily and produce even with a rainfall of 400 to 450 mm (Prasad and Reddy, 1992).

Groundnut is grown commercially below 1250 meters from sea level and is day-neutral. Nevertheless, the plant could be affected by very low light intensity during its early growth phase. Cloudy weather at flowering phase could result in decline in flower number. There is a differential response of the varieties to the photperiod-temperture interactions (Bell and Wright, 1998).

Low temperature (22/18*C) affected the reproductive development by reducing the proportion of reproductive nodes in relation to total (vegetative + reproductive) nodes. The conversion of pegs into pods, as indicated by pod to peg ratio (PPR), was lower in long day than in short day conditions, thereby indicating that the PPR could be used as an indicator of genotypic sensitivity to photoperiod in groundnut. (Nigam *et al.*, 1998).Bunch (*sapnish* and *valencia*) types are stated to be more adversely affected by climatic variation than the runner (*virginia)* type of groundnuts (Weiss, 2000), which probably refers to their differential adaptability to climate change.

Origin, Diversity and Cytogenetics: The family Leguminosae or Fabaceae to which groundnut belongs is one of the three largest of the families of flowering plants with around 700 genera containing 18000 species of all habitats starting from herbs to large trees. The most important attribute of this group of plants is their characteristic to maintain sustained symbiotic relationship with specific bacteria known as *Rhizobia* to fix atmospheric nitrogen. The sub family of Papilionoideae to which groundnut pertains, contains about 480 generea and 12,000 species.Groundnut plant is especially fascinating because it develops pods below the ground, a process known as geocarpy.

The genus *Arachis* is a member of family Fabaceae (synonym: Leguminosae) subfamily Papilionoidae, tribe Aeschynomeneae and subtribe Stylosanthinae (Polhill *et al.*,1981). At the time Linnaeus first named the cultivated groundnut *Arachis hypogaea* L., it was the only member of the genus.

The genus was in a state of chaos until 1980 when Gregory, Krapovickas and Gregory (1980) published a taxonomic revision of the genus. This work took them nearly thirty-five years, and involved re-visiting and collecting specimens at the type locations of each known species. This taxonomic revision recognises 69 species in 9 sections. Distinctions are made on the basis of morphological characters and life cycle attributes, although eco-geographic distribution, crossability evidence, cytological information, plant form, as well as chromatographic and antigenic reactions were all considered in the groupings.

All species, except the cultivated species *Arachis hypogaea* and its close relative *A. monticola*, in Section *Arachis*, and certain species in Section *Rhizomatosae*, are diploid (2n=2x=20).

Cultivated groundnut also known as peanut, has originated in South America, specifically at the foothills of Andes around South Bolivia to North-western Argentina (Krapovickas 1968). South America is considered the primary centre of genetic variability of groundnut in view of its enormous genetic diversity there. However, Africa is considered the secondary centre of diversity of groundnut. Groundnut plant (*Arachis hypogaea* L.) is self-pollinated, amphidiploid ($2n = 4x = 40$) with a genome size of 2891 Mbp. The genus *Arachis* belongs to the family *Fabaceae*, subfamily *Faboideae*. Based on morphology, geographical distribution, and cross compatibility, the genus is divided into nine taxonomic sections. The genus consists of 80 described species (Valls and Simpson, 2005), that include diploids and tetraploids, of which *A. hypogaea* L. is the most important and widely cultivated. *A. hypogaea* is a segmental amphidiploid but cytologically behaves like a diploid.

Cytological and molecular studies suggest that two diploid wild species *Arachis duranensis* (AA) and *Arachis ipaensis* (BB) are progenitors of cultivated groundnut (Favero *et al.*, 2006 ; Seijo *et al.*, 2007; Bertioli and Cannon, 2016). The significant laboratory data developed by Seijo *et al.*, (2007) using double GISH (genomic *in situ* hybridization) revealed that *A. duranensis* and *A. ipaensis* appeared to be the most likely source genomes for cultivated groundnut.It was based on the hybridization of these species' chromosomes with groundnut. Additionally, earlier Fávero *et al.*, (2006) successfully hybridized A. ipaensis and *A. duranensis* and doubled the chromosomes of the resulting hybrid, to recover the "amphidiploid". Fertile crosses were earlier effected successfullybetween the synthetic amphidiploid and different kinds of cultivated groundnut (Fávero *et al.*, 2006). These successful hybridizations amply supported that *A. ipaensis* and *A. duranensis* are progenitors of cultivated groundnut (Fávero *et al.*, 2006). Archaeological excavations on the coast of Peru have also indicated the genesis of groundnut confirming the molecular evidence (Simpson *et al.*, 2001). A single hybridization event between progenitors followed by chromosome duplication about 3500 years ago led to origin of cultivated groundnut. It was observed that these genomes are similar to cultivated groundnut's A and B sub-genomes and can be used to identify candidate disease resistance genes, to guide tetraploid transcript assemblies and to detect genetic exchange between cultivated goundnut's sub-genomes (Bertioli and Cannon, 2016). Based on cross-compatibility among the members of the genus *Arachis*, four gene pools were identified (Singh and Simpson, 1994).There are vast number of wild species of *Arachis* with different degrees of genetic relationship with the cultivated groundnut *Arachis hypogea*, L. All wild species of the genus *Arachis* are found only in South America and are distributed from north-eastern Brazil to north-western Argentina and from south coast of Uruguay to north western Moto Grosso south of Amazon and from east base of the Andes to the Atlantic. (Gregory and Gregory, 1976; Gregory *et al.*, 1980).

The genus *Arachis* is divided into nine different sections viz., Arachis, Erectoides, Heteranthae, Caulorrhizae, Rhizomatosae, Extranervosae, Triseminatae, Procumbentes and Trierectoides based on morphology, geographic distribution and cross compatibility relationships.

The Species present in sections *Erectoides*, *Extranervosae* and *Triseminatae* and diploid species of section *Rhizomatosae* are believed to be basal in their divergence when compared to the species in other sections. Most of the wild species of *Arachis* possess impressively wider genetic variability and genes for many important characters which can be used in groundnut breeding. However, many of these wild species are not easily crossable with the cultivated groundnut. The barriers of crossability may however, be overcome employing modern genetic and cytogenetic techniques. Singh and Simpson (1994) classified the genetic diversity of *Arachis* into four gene pools consisting of (i) the tetraploid species *Arachis hypogaea* and *A. monticola* as primary genepool. (ii)The secondary gene pool consists of diploid species from the section *Arachis* that are cross compatible with *A. hypogaea. (iii)* The third or tertiary genepool includes *Procumbentes* which are feebly crossable with *A. hypogaea* with a very low frequency. (iv) The fourth category is represented by the rest of the sections of wild species of *Arachis* forming around seven sections. These are mostly perennials not crossable with cultivated *A. hypogaea* or *A. monticola*.

The perennial species possess very small stigmas surrounded by a dense ring of hairs hindering out-crossing. The annuals on the other hand have larger stigmas and are least hairy and as such amenable for cross pollination. The products of hybridization involving the annuals are recoverable through the application of growth regulators to enhance pollen tube growth and fertilization, followed up with the techniques of embryo rescue (Stalker *et al*., 1987*)*. The pollen tube growth could be achieved through bud pollination in the crosses involving *A. spegazzini* (Lu *et al.,*1990). Most of the intersectional hybrids exhibited sterility (Stalker and Simpson, 1995). However, the subsequent investigations carried out indicate that back-crossing of the intersectional hybrids to the *A. hypogaea* would enhance the fertility and seed formation in the following BC generations.

- **Section *Arachis*** with**Section *Rhizomatosae:*** The crosses involving perennial *Arachis glabrata* with *A. hypogaea* resulted in hybrids that were perennial devoid of underground rhizome, which is the main diagnostic feature of the section. The chromosome associations in the hybrid was much variable (Mallikarjuna and Sastry, 2002). The back cross of the hybrid to *A. hypogaea* was more stable with more ring bivalents. Higher mean ring bivalents indicated homology between the genomes of *A. hypogaea* and *A. glabrata*. The back cross hybrid was more stable than the original hybrid since the mean number of bivalents ranged from 10 to 20 with more number of ring bivalents of eleven indicating homology between the chromosomes of the two species. The presence of univalent in the BC1 hybrid is an indication of the preferential pairing of A genome of triploids and tetraploids (*A. hypogaea*) and lack of homology between the B gnome of triploids and the R genome of *A. glabrata* (Mallikarjuna, 2005). It is also suggested that the fertility status of the triploid hybrid between *Arachis duranensis* and *A. glabrata* could be enhanced by back crossing with *A. hypogaea* (Mallikarjuna, 2005). Another hybrid involving *Arachis diogoi* and *A. glabrata* exhibited more number of rod bivalents instead of ring bivalents of the 9 bivalents formed along withn12 univalents (Mallikarjuna, 2005). The studies of Stalker and Simpson (1995) and Smartt *et al.,* (1978) indicated that one of the genomes of *Arachis hypogaea* is related to the genome of *A. glbrata.*

- **Section *Arachis*** with **Section *Procumbentes:*** The triploid hybrid of *Arachis hypogaea* and *A. chiquitana* exhibited pollen fertility in the range of 5 to 24%. Following the back crosses effected to *A. hypogaea* used as the male parent, some of the BC1 plants didn't attain the flowering stage despite exhibiting a normal vegetative growth. Nevertheless, the rest of the plants showed normal flowering, with perceptible enhancement in pollen fertility (Mallikarjuna, 2005). The intersectional hybrids of the parental combination of *A. hypogaea* and *A. kretschmeri* were successfully recovered with the application of growth hormones to the pollinated pistils followed by embryo rescue. The F1 exhibited a slow and retarded growth and as such were back crossed to *A. hypogaea.* The BC1 plants produced single seeded pods with prominent beaks. Following the subsequent second back cross to the recurrent parent, the resultant BC2 plants produced two-seeded pods exhibiting the features of *A. hypogaea,* the recurrent parent (Mallikarjuna and Hoisington, 2009).

- **Section *Arachis*** with **Section *Erectoides:*** Stem cuttings of the hybrids of *A. rigonii* and *A.* spp (PI 262278) were treated with colchicine for doubling the chromosomes. The seeds obtained from the induced polyploids were crossed with two diploid species viz., *A. stenosperma* and *A. duranensis.* It was shown that the introgressions of genes from section *Erectiodes* would be possible by using diploids as the bridging species (Stalker, 1981). The hybirds of *A. hypogaea* and *A. paraguariensis* (diploid) exhibited poor vigour; but produced 95 to 98% fertile pollen with 19 bivalents at meiosis. The plant was categorised as nullisomic which exhibited normal meiosis (Bera *et al.,* 2002).

- **Genomes of Section *Arachis:*** The Section *Arachis* has 31 species, which are cultivated for seeds and pods such as, *Arachis hypogaea,* forage species *Arachis pintoi*, *Arachis glabrata* and *A. sylvestris* and ornamental species *Arachis repens.*

  The section *Arachis* consists of both annuals and perennial species.

The cluster analysis (Chandan and Pandya, 2009) showed unique clustering of accessions of most of the species under the section *Arachis*.

The clustering pattern of *Arachis duranensis* and *A. kempff-mercadoi,* however indicated the wide morphological variability within the species. The perennial species *A. cardenasi, A. correntina* (ICG 8132 and 8918) and *A. kempff-mercadoi* (ICG 8164) showed very high similarity of morphological traits. Though *A. khulmannii* is a perennial species, it was found clustered with *A. duranensis* accessions of the fourth cluster. In the agglomeration schedule, the last two accessions which combined to form a cluster were ICG 8124 (*A. batizocoi*) and ICG 5813 (*A. hypogaea*), with a distance of 174.4, thus showing wide morphological dissimilarity. Though *A. batizocoi* was reported to be intermediate between *A. duranensis* and *A. monticola*, cluster analysis showed that there exists a wide variation between the morphological characters of these species. *Arachis batizocoi* formed a separate group with high Euclidian distance with ICGs 8123 (*A. duranensis*), 8197 (*A. monticola*), 8216 (*A. cardenasi*), 11548 (*A. valida*) and 5813 (*A. hypogaea*).

The cross involving the tetraploid *A. hypogaea* and diploid *A. cardenasii* gave rise to the triploid progeny, which showed 3.45 univalents,10.42 bivalents, 1.14 trivalents, 0.26 quadrivalents and 0.04 pentavalents (Bera *et al.*, 2002). The chromosomes of the sterile triploid were doubled to produce hexaploid. The hexaploids consisted of the chromosomal configurations like 26.29 univalents,14.49 bivalents, 0.02 trivalents, and 0.06 quadrivalents. The larger number of univalents in the above hexaploids could be due to preferential pairing (Riley and Law, 1965). The triploids also showed autosyndetic pairing in the form of tetravalents (Bera *et al.*, 2002). Bera et al., (2002) came up with the suggestion that the basic chromosome number of Arachis might be 9 (x = 9) and stated that the karyotype with x = 10 originated during the course of evolution as a result of selective duplication of chromosomes.

The triploid of the cross *A. hypogaea* and *A. cardenasii* was back crossed to *Arachis hypogaea* and the resulting BC1 progeny exhibited profuse growth with dark green foliage and prolific flowering with rhizomatous roots (Vindhiyavarman *et al.,* 2004). The BC1 progeny yielded multiple cuttings of the vegetative parts and as such recommended as a forage legume.

A case of triploid hybrid of *A. hypogaea* and *A. kempff-mercadoi* was reported by Mallikarjuna *et al.*, (2005), with a mean pollen fertility of the hybrid 0-5%. The back cross of the triploid hybrid as female with *A. hypogaea* resulted in normal seed set in the BC1 generation. The reciprocal cross on the other hand failed.

The triploid F1 of another cross of *A. hypogaea* and *A. diogoi* effected by Pompeu (1983) exhibited a pollen fertility of 10%, while thc F1 of the cross of *A. hypogaea* and an amphidiploid of (*A. stenosperma* x *A. cardenasii*) reported by Vindhiyavarman *et al.*, (2000) resulted in low (5%) pollen fertility.

Involving *Arachis hypogaea* as the parent for back cross for hybrids of *Arachis cardenasii x A. villosa or* that of the reciprocal cross of the same set of parents, the pollen fertility of the resultant back-cross progeny could be enhanced to the level of 30.5 to 38.5 % (Vindhiyavarman *et al.,*2000). However, there were a differences in fertility levels of the back-cross generations of the reciprocal crosses when *Arachis stenosperma* was involved in the cross combinations in the place of *A. villosa* cited above. Not withstanding the problems posed by the cross-incompatibilities of the species chosen for hybridisation for gene-transfer and lack of fertility in the hybrids realised, it has been proven by the investigations presented above that the intricacies of sterility may be overcome by back-crossing the hybrids to the genotypes of *Arachis hypogaea*

Many wild species possess durable resistance to many diseases of cultivated groundnut and hence form the valuable genetic materials particularly for disease resistance. However, most of the wild species of *Arachis* being diploid, posing crossability barriers with *A. hypogaea* L., Simpson (1991) and Simpson *et al.,* (2001) have suggested four different pathways viz., (a) triploid pathway, (b) hexaploid pathway, (c) amphidiploid pathway and (d) autotetraploid pathway.

(a) Triploid pathway: The cultivated groundnut *A. hypogaea* L. (AABB 2n=4x= 40) a when crossed with a diploid species (AA or BB), the triploid recovered, may produce unreduced gametes (2n) leading to sporadic seed set. The resulting hexaploids are either selfed or back crossed to *A. hypogaea* L.

(b) Hexaploid pathway: *A. hypogaea* L. (2n=4x=40) with AABB genome complex is crossed with a diploid (2n=2x=20; AA or BB). The resultant F1, the triploid (2n=3x=30; AAB or ABB), may be treated with colchicine to obtain a hexaploid (2n=6x= 60; AAAABB or AABBBB) through chromosome doubling. The hexaploid may be back-crossed with *A. hypogaea* L (2n=4x=40). The resultant hybrid may be selfed or back crossed with *A. hypogaea* L. in order to recover a stable tetraploid which may be selected from the progenies segregating for variable chromosome numbers.

(c) Amphidiploid pathway: This pathway is the simulation of the evolution of *Arachis hypogaea* L., involving two wild diploid species of varying genome constitution

(i) Two diverse species of *Arachis* of A genome are crossed and the resulting F1 the diploid (2n=2x=20) is crossed to another species of B genome (2n=2x=20). The resulting F1 hybrid (2n=2x=20) is treated with colchicine to obtain the amphidiploid (2n=4x=40), which may be crossed with cultivated groundnut through repeated back-crosses to recover a stable amphidiploid.

(ii) Two wild diploid species of differing genome constitution are crossed viz., (AA) & (BB) after thorough cytological screening for relative compatibility and the resulting F1 which may be sterile is subjected to colchicine treatment to recover an amphidiploid. The synthesised amphidiploid may be subjected to repeated back crosses with cultivated groundnut to recover a stable and usable amphidiploid derivative.

(d) Autotetraploid pathway: The diploid wild species of differing genomic constitution AA and BB are subject to colchicine treatment to obtain autotetraploids of each type (AAAA and BBBB) These autotetraploids are crossed with the cultivated *A. hypogaea* L. (2n=4x=40). The backcrosses with the cultivated groundnut are repeated till the stabilized derivative tetraploids with the desired gene combinations are recovered.

Classification of the intra specific variability of groundnut

Even in the cultivated groundnut there exists substantial variation for growth habit, branching, flowering and pod and seed characters.

Krapovickas and Rigoni (1960) and Krapovickas (1968,1973) studied extensive collections of cultivated groundnut from diverse zones of South America and proposed a system of sub-specific nomenclature of *A. hypogea*. They related the 5 subspecies to 5 geographical regions of South America viz., Gurananian, Bolivian, Peruvian, Amazonian and the region of Goias & Minas Gerais. Gregory and Gregory (1976) extended the number of regions to 6 and included north-eastern Brazil. The sub-specific classification as proposed by Gregory and Gregory (1976) is described as followed by Gibbons *et al*., (1972)

Arachis hypogaea, subspecies *hypogea*., var. *hypogaea* : Type *virgnia*

A. hypogea sub species *fastigiata* var. *fastigiata:* Type *valencia*

A. hypogaea subspecies *fastigiata* var. *peruviana* – Type *peruvian*

A. hypogaea sub species *fastigiata* var. *aequatoriana* –Type *aequatorian*

A.hypogaea sub species *fastigiata* var. *vulgaris* –Type *spanish*

The principal diagnostic characters of the sub species and botanical varieties of *Arachis hypogaea* are described below.

(A) **Sub species *hypogaea***: habit procumbent or erect; branching alternate, inflorescence simple and never borne directly on the main axis. First branch on the cotyledonery lateral always vegetative. Two or two to four seeds per pod; pod beak pronounced, moderate or absent; pod constriction prominent, moderate or absent; pods very large (>20mm) or small (<10mm); seed testa colour commonly tan but red, white, purple and variegated. Seed dormancy usually present in all the varietal forms, foliage dark green.

1. **Var. *hypogaea:*** habit procumbent, decumbent or erect; main axis in procumbent forms short (less than 50cm) stem not hairy; usually two seeded, medium-late maturing.
2. **Var. *hirsuta***: habit procumbent, main axis may exceed one-meter length; stem hairy; pods strongly beaked with 2 to 4 seeds; very late in maturity.

(B). **Sub species *fastigiata***: habit procumbent; branching sequential; inflorescence simple or compound, inflorescence always borne on the main axis; first branches of the cotyledonary laterals always reproductive; seed dormancy absent; foliage usually lighter green as compared to that of sub species *hypogaea*

1. **Var. *fastigiata***: Vegetative branches on the primaries absent or regularly placed at the distal nodes; inflorescence usually simple; pods with usually with 2 to 4 or rarely 5 seeds; beak absent, slight or prominent; seed size medium to small; testa colour tan, red, white, yellow, purple or variegated. Commonly known as *Valencia* type
2. **Var.*peruviana:*** Main axis has both vegetative and reproductive branches. The vegetative branches of the main axis appear near the base while the reproductive flowering shoots of the main stem are borne in the upper part. The lateral reproductive branches are mostly flower bearing with multiple flowers. The leaves are larger, glabrous and thicker. Foliage is dense. The leaflets are hairy on the margins, on the lower midrib. The pods are three to four seeded. The pod surface is dense with prominent linear ribs. The seed coat is dark brownish.
3. **Var.*aequatoriana***: Main axisis erect and branched. The reproductive branches are shorter and the lateral branches decumbent. The leaves are glabrous and hairy on both sides. Pods are three to four seeded. Pod surface is prominently reticulated with pronounced longitudinal ribs. The seed coat is brownish.
4. **Var.*vulgris***: Vegetative branches occasional and irregularly placed; inflorescence compound; pods usually two seeded; beak present or absent, slight or prominent; size medium to small; testa colour tan, red, white or purple. Popularly known as *Spanish* type.

"One can speculate that in the evolutionary history of peanut species the rate of partitioning of fruit was related to survival in different environments. Where conditions were such that insects, animals and diseases destroyed leaves rapidly, the more successful plant might have been the one that filled its few seeds only by expending most of its assimilates to maintain its leaf canopy. Under less severe conditions the more successful plant would be the one that made more seed by partitioning more of its energy to fruit. Selection for a crop plant would be towards a type that would make more seed by a higher partitioning factor." (Duncan *et al.*,1978)

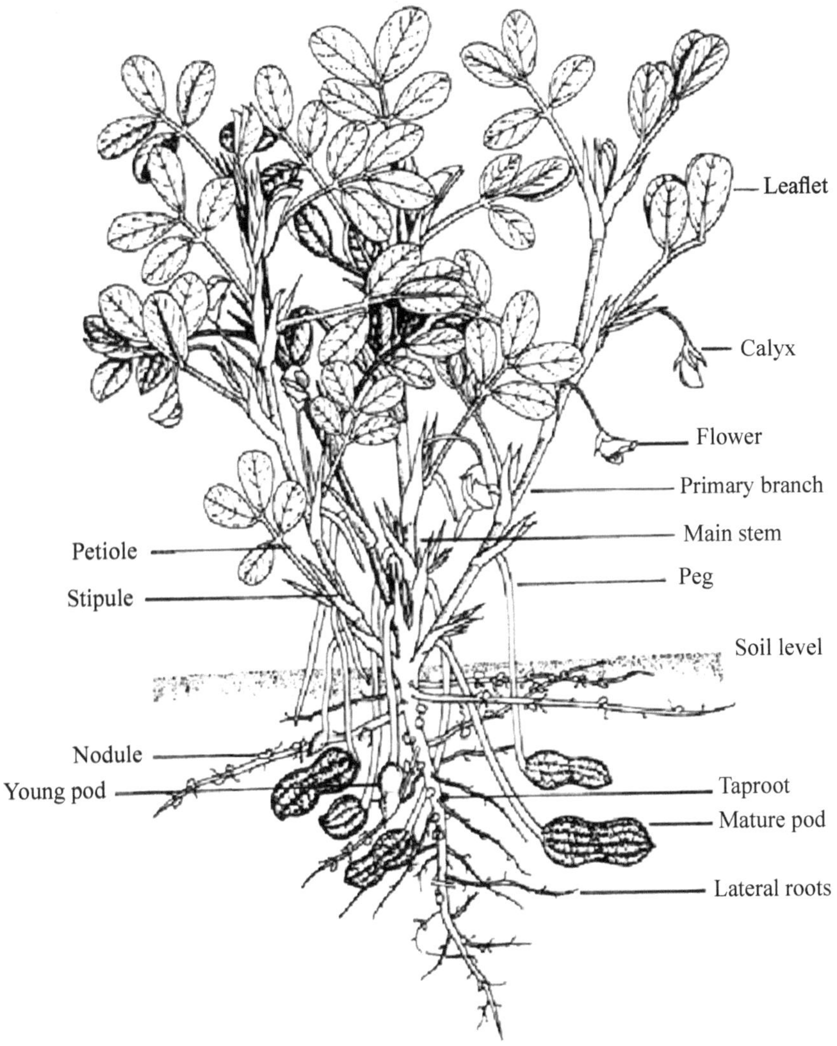

Fig. 1: Stylized Groundnut Plant with its parts (Rao, V.R., 1988).

Morphology of groundnut plant

Groundnut is a leguminous plant; but differs from other legumes in growth, branching and reproductive biology. Most of the other legumes exhibit epigeal germination, while groundnut has hypogeal germination. Groundnut also differs from other legumes in the sense that it has aerial flowering and underground pod development unlike all other legumes. Floral biology of groundnut reflected in its inflorescence is rather complex as compared to several other legumes. The variation that is noted in the branching pattern and the pod and seed characters including the seed dormancy in certain types of groundnut make groundnut a special kind of legume. The essential morphological attributes that are unique to groundnut are described below.

1. Seed

The seeds of domesticated groundnut plant exhibit a wide range of variation in size ranging from 0.17 to 1.24g/seed (Rao 1988). In general, the varieties of var. *hypogaea* have larger and heavier seeds than those of var. *fastigiata.* The seed testa colour include different shades such as variegated, dark, pink, light rose, red, purple white-blotched, white and yellow. Seed colour is an important diagnostic character of groundnut varieties. Each seed contains two cotyledons, upper stem axis and young leaf primordia, lower stem axis which is the hypocotyl and primary root. The embryo of groundnut unlike those of other legumes is straight rather than curved and contains all the leaves and other above ground parts that appear during the first two weeks of growth (Gregory *et al.,* 1951 and Maeda 1972)

2. Root System

Groundnut plant exhibits a fairly well-developed root system with a pronounced tap root (Fig.1). It may not be out of context to mention that in some wild species of *Arachis* the roots and hypocotyl have been found modified to form tubers in the section *Erectoides* and tuberiods in the sections *Erectoides* and *Extranervosae.* The diameter of the tap root may vary from few millimetres in annual species to 10 cm in some perennial species of *Arachis* (Gregory *et al.,* 1980). A well-developed tap root may penetrate to a soil depth of 90 to 130 cm.

In the case of groundnut, tap root of 50 cm long has been reported in 37 days old plants (Yarbrough 1949). The cells originating in the lateral root cap region and in the outer cortex form the surface layers of the root, the outermost of which split and slough off' (Yarbrough 1949).

2 (i). **Rhizobia induced root nodules in groundnut:** Even though root nodules are not part of root system in the botanical sense, their close association with the root system of groundnut makes it essential to study them in association with the root system. Root nodules make their first appearance on the roots of 15 days' old plants. The nodules are caused by the nitrogen fixing bacteria of the genus *Rhizobium.* The genotypic differences of groundnut have been observed for root nodulation (Iselib *et al* 1978; Nambiar and Dart 1980).

Differences in nodulation in the hypocotyl region have also been observed. The strains of var. *hypogaea* are reported to exhibit higher density of nodules than the strains of var. *fastigiata* (Nambiar and Dart 1980)

According to Chandler (1978) the infection process in *Arachis hypogaea* by rhizobia differs from that normally found in *Trifolium* spp. in that no infection threads are formed. The root hairs, which are long (up to 4 mm), septate, and often with large basal cells, occur only at the sites of emerging lateral roots. Infection occurs only where the root hairs have large basal cells. Rhizobia cause curling and deformation of the root hairs (as in *Trifolium* spp.) but enter the root at the junction of the root hair and the epidermal and cortical cells. The bacteria are distributed intracellularly via the middle lamellae and enter the cortical cells through the structurally altered cell wall, often close to the host cell nucleus. The root hairs and large basal cells become infected in the same way. Within the cortical cells of the emerging lateral root the rhizobia multiply rapidly and the invaded cells divide repeatedly to form the nodule tissue. Bacteriod formation occurs only when the host cell ceases to divide (Chandler 1978).The nitrogen-fixing activity of root nodules is closely related with their size. In the root nodules of the medium size group (1.5-2.0 mm in diameter), nitrogen-fixing activity per unit fresh weight of nodule was the highest at the flowering stage and rapidly decreased thereafter. The nitrogen-fixing activity of root nodules larger than 2.0 mm in diameter did not vary significantly with their size. Colours of rhizobium-infected zones varied with their size such as white in small nodules; red in medium-sized nodules; and greenish in larger nodules, which suggests that the concentration of leghaemoglobin is the highest in the medium-sized nodules. Nitrogen-fixing activities of the medium-sized nodules might determine the amount of nitrogen fixation in the whole root system during pod-filling because medium-sized nodules had high activity and were large in number (Tajima *et al* 2007). Classification of root nodule size based on the circle template is a simple, rapid, and useful method to evaluate nitrogen-fixing activity of root nodules (Tajima *et al.,* 2007).

3. Shoot system and growth habit

The leaves of groundnut plant which appear initially are those in the embryo and new leaves are formed much later. The leaves of the genus *Arachis* are tetrafoliolate, except in the case of series *Trifoliate* of the section *Erectoides* that have trifoliate leaves. The leaves of groundnut plant are paripinnate with two pairs of opposite leaflets.

The groundnut plant has the main stem and secondary and tertiary branches, depending upon the kind of subspecies and corresponding botanical variety. There are two specific types of growth habit viz., spreading and erect. Under the spreading category there are runner, trailing, procumbent and prostrate types, while the erect types show the variation such as upright, erect bunch and bunch types.

Runner types are generally observed in the sub specific var. *hypogaea* and erect and bunch forms are in larger frequency in var. *fastigiata*.

Nevertheless, spreading forms of sequential branching pattern of Var. *hypogaea* and bunch types in the var. *hypogaea* have been observed. In addition, there are intermediate forms between spreading and bunch types in both the botanical varietal forms.

The classification has been carried out based on the number and arrangement of vegetative and reproductive branches of the shoot system. According to Gregory *et al* (1951), Bunting (1955) and Krapovickas (1968), the forms of var. *hypogeaea* always possess a clear vegetative main axis and numerous secondary (n+2) and tertiary (n+3) branches with alternate flowering character on their nodes.

The strains of var. *fastigiata* on the other hand exhibit reproductive main axis and sequential branching as is described in greater detail earlier.

The groundnut haulms contain 8–15% protein, 1–3% lipids, 9–17% minerals, and 38–45% carbohydrates. It is used as cattle feed either in fresh or dried stage, or by preparing hay or silage. The digestibility of nutrients in groundnut haulm is around 53% and that of crude protein is 88% when fed to cattle. Haulms release energy up to 2.337 cal kg^{-1} of dry matter. Being a legume crop, groundnut has unique ability to fix atmospheric nitrogen through symbiotic association with *Rhizobium*. This ability is greatly useful in improving the fertility of degraded soils, and economically beneficial to the small-holder farmers as they can save money on resources such as nitrogen fertilizers (Janila *et al*., 2016b).

4. Flowering and pod development

The inflorescence of groundnut varies with the subspecific pattern. It could be solitary with a sngle flower as in the case of sub species *hypogea* or a raceme (compound inflorescence) as found in the sub-species *vulgaris*. Groundnut plant starts flowering from 20 to 30 days flowing germination.

The flowers are paplinaceous with one standard, two wing and one pair of keel petals. The corolla is mounted on a pedicel like elongated structure known as 'hypanthium' which is nothing but a calyx-tube (Fig.2). The standard petal is normally yellow in colour with a slight occasional variation towards orange or dark orange. Coloured crescent lines radiating from the base of the standard petal could be observed. The wing petals enclose the keel,

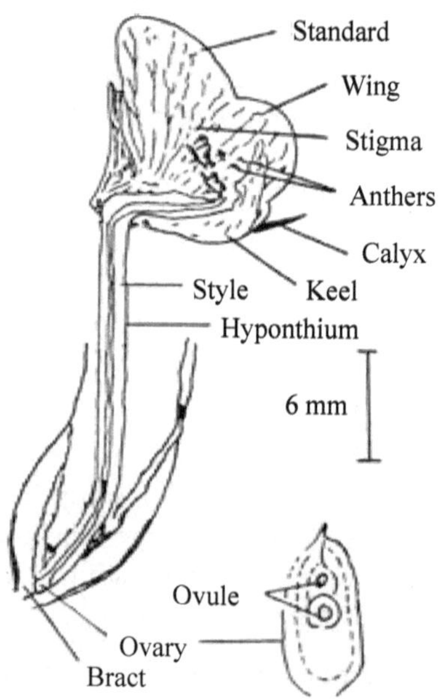

Fig. 2: Parts of a groundnut flower (Smith, 1950)

which totally wraps the reproductive parts. The staminal column is composed of ten monodelphous stamens of which two possess sterile anthers. The style is long, slanting and hairy distally with a club-shaped stigma. The ovary is superior located at the base of the hypanthium. It consists of two to four ovules. Occasionally the ovary may contain more than 4 ovules too (Rao, 1988).

The flowering is generally influenced by temperature. The temperature of the order of 20° -30°C influence precocious flowering. Flowering is spread in two phases. The first peak of flowering occurs from 25 days following emergence and proceeds for a period of approximately 20 to 22 days. The second peak occurs later while the first instalment of gynophores starts growing and lasts for 60 to 75 days from the emergence. Flowering declines after 70days, depending upon the genotype. Spanish types in general are precocious with a duration of flowering of 50 days, while the duration of flowering lasts for more than 65 days. The forms of var. *fastigiata* are generally more precocious than those of var. *hypogaea.* The anthesis characterised by flower opening during the early fore-noon starting from 6.00 to 9.00 AM

Also a definite correlation exists between flowering and maturity of plant.

The flowers are cleistogamous and self-pollination is the rule (Lim *et al.*, 1984). Nevertheless, some cross-pollination is detected (Fig.3) at the levels ranging from 1.0 to 3.9% (Bolhuis 1951, Srinivasulu and Chandrasekharan 1958, Hammons 1963, Gibbons and Tattersfield 1969) involving bee pollinators.

Fertilization occurring by mid-day is completed in six hours following pollination. During the early development of the young embryo the ovary at the base of the calyx tube gets mobilized for growth within a week. The intercalary meristem below the ovary gets activated and the green ovary turns purplish from the tip downwards. The developing ovary pierces through the other floral parts by the activity of the meristem to reveal the elongating peg or gynophore, which bears the fertilized ovules in the ovary at the tip. The tip then becomes diageotropic and gets positioned in the soil to get enlarged into fruit known as pod. Occasionally under humid conditions some pegs in var. *fastigiata* develop into aerial pods without seeds.

However, fully fertile aerial pod bearing mutants of the Brazilian groundnut variety 'Tatu" have been reported by Prasad (1985). The seed number and size vary in pods of groundnut based on the subspecies and corresponding varietal forms as already described earlier in the preceding pages.

Fig. 3: Bee pollinator on a groundnut flower.

5. Pod & Seeds

Pods usually contain four ovoid seeds. Each seed is enclosed in a papaery testa. Colour of seed varies from creamy white or light yellw to dark red or purple. Seeds vary in shape from rougly cylindrical to almost spherical. Seeds comprise 55 to 75% of unhulled pods.

Enyi (1977) defined seed yield of a given area of land in groundnut as number of pods, number of seeds per pod and size of individual seed. He also reported that seed yield depends on the number of effective pegs (gynophores) that turn into mature pods at harvest.

A positive correlation between number of effective pegs turning into mature pods and seed yield was also reported by several workers (Badwal and Gupta 1968, Jaswal and Gupta 1967, Rahman and Rahman1976, Dorairaj1962, Chandra Mohan *et al* 1967, Sangha 1971, Sangha and Sandhu 1974, Labana *et a*l 1980).

On the other hand, Lawrence (1974) found that pod number per plant was most variable component bearing inverse relation to plant density. Nevertheless, pod yield was highly and positively related to the number of secondary branches, number of pods and 100 kernel weight (Ishag1970)

Seeds are often diagonally flattened at the end, where they are in proximity in the pod. Each seed consists of two diagonally elongated cream coloured cotyledons with a short radical and plumule.

Regarding the oil content of the seed, it is the cotyledon or kernel that contains the fat or oil. The normal range of values over may types of seeds is around 45 - 50% oil, with high oleic acid and around 25 to 35% protein. It has been observed that in some genotypes the seeds of the same pod exhibit significant differences in oil content, the basal one usually showing the highest (Weiss, 2000). Grosso and Guzman (1995) found that the seed composition of a wide range of wild *Arachis hypogaea* L.in Peru had significant variation in all components of fat.

It has been observed that in general the total fatty acid content was greatest at physiological maturity which is very close to harvest. During seed development, oleic acid content increased and linoleic acid decreased, while palmitic and eiosenoic acids decreased slowly from seed formation to maturity (Weiss, 2000).

6. Crop Nutrition

Low levels of productivity of groundnut on drylands may be attributed to delegation of the crop to marginal lands at the level of small and marginal farmers, apart from factors imposed by the uncertain rainfall and its distribution (Prasad and Reddy,1992; Prasad and SudhakarBabu, 1997). A major proportion of the soils on which the crop is grown are not fertile enough to support a successful crop of groundnut. Hence, right kind of nutrient management of the crop assumes relevance in this context, notwithstanding the common adage that groundnut is an 'unpredictable legume' as it does not respond normally to the application of fertilizers (Arant *et al.,* 1951).

Groundnut plant has the ability to extract the hidden soil nutrients from deeper layers successfully, which could be reflected in plant growth and that is one of the reasons for the lack response for applied nutrients in terms of yield (Weiss,2000). According to Weiss (2000), in Nigeria the proportion of mean nutrient uptake by plants and which remained in kernels was 63% N, 68% P, 23%K, 24% Mg and 4% Ca. Also in the case of East Africa (Weiss, 2000), total N and 80% P was found in kernels at harvest and it is stated that similar trends were observed in other countries too. However, the proportion of nutrients absorbed at different growth phases of groundnut seem to vary as reported by several workers. It is stated that generally plants absorb approximately 10% of their total requirements of N, P, Ca and Mg during the vegetative phase, 40 to 50% during the reproductive phase and the rest of the remainder during the phase of pod maturity (Weiss, 2000). A combination of farm yard manure (FYM) @ 2 tons/ha and single superphosphate (SSP)@ 100kg/ha applied basally before planting has given significantly higher yields over 2 tons/ha in Africa (Weiss, 2000). Nutrients applied to the preceding crop are reported to have positive effect on groundnut yields. Hence, the crop rotation is an important factor to be considered with regard to the nutrition management of groundnut crop. While fertilizers might promote seed development; they are not found to influence oil content and its composition (Weiss, 2000), since environmental factors have little effect on the genes controlling these characters.

Harris (1949) found root-hair-like growths on the pegs or gynophores of peanuts. Also some varieties are known to have similar or root-like formations on the hull of the pod, showing that gynophores absorb water and nutrients. The evidence (Harris 1949) is that calcium is beneficial in the pegging zone as the element is selectively absorbed by the gynophores to promote pod development and that merely supplying the roots with adequate nutrients will not result in proper nut development.

Role of certain major nutrients in groundnut production are discussed below.

Nitrogen: Groundnut being a legume has the added advantage of harnessing symbiotic nitrogen fixation through rhizobial root nodules.

Well nodulated plants also tend to be more drought resistant and produce higher yields (Weiss, 2000). In case groundnut is to be grown on a land where no legume was grown earlier, inoculation seeds with rhizobium helps in promoting root nodulation.

It has been found that any application of nitrogen might suppress the root nodulation of groundnut, while application of phospahtic fertilizers particularly single super phosphate is beneficial for root nodulation of groundnut. Excess nitrogen also depresses oil content and increases protein content. Nevertheless, on soils of poor fertility, a small quantity of nitrogen applied as starter dose in seed bed might assist seedling growth before root nodulation sets in.

The nodule nitrogen added to the soil during the growth cycle of groundnut is reported to be beneficial to the succeeding crop. According to Weiss (2000), in Malaysia, an equivalent of 77 to 105 kg N/ha is through the residues of the previous crop of groundnut.

Phosphorous: Phosphorous is one of the key elements in groundnut crop nutrition since it is essential for root and kernel development, and promoting the effective root nodulation. Phosphorous also is known to increase the uptake of many nutrients. In order for Phosphate to be effective, there should be adequate level of soil moisture when crop is grown on drylands. Lack of response to P in many cases might be due to inadequate soil moisture, which restricts the uptake of P. The requirement of P in respect of groundnut is fairly high over 60 kg P2O5/ha under normal conditions (Prasad and Reddy, 1992). However, it has been observed that even small doses of the order of 6 to 15 kg P/ha have resulted in significantly higher yields on small holders' farms in Africa and India (Weiss, 2000). Choice of single super phosphate apart from providing P, also helps in alleviating the deficiency of Sulphur, which is needed for groundnut.

Phosphate deficiency is indicated in young plants, which presents a stunted appearance with a reddish or purple tinge on leaves and stems, stunted growth and small bluish green leaves which may disappear as the root growth expands (Weiss, 2000; Harris, 1949).

Potassium: Groundnut crop's response to Potassium depends upon the availability of other crop nutrients particularly magnesium and calcium. Potassium is also related to the uptake of nitrogen. It was considered that Indian soils are rich in potassium and hence hitherto not much emphasis was given to this nutrient. Of late deficiency of potassium is noticed particularly in rice fields. Potassium deficiency may reflect as stunted growth, dark green leaves and some reddening and dying of the tissue near the tips of the branches (Harris, 1949).

Potassium uptake is most active in the first six weeks of growth and gradually decreases with age. The nutrient also moves from older to younger parts of plant and developing pods. Potassium uptake can enhance the number of pods; but the calcium and magnesium are needed to increase the number and size of kernels (Weiss, 2000).

Potassium may prolong the vegetative phase. The nutrient must be applied as basal dose; preferably placed in the root zone before sowing, if the soil is deficient. High levels of K in the pegging zone might inhibit calcium uptake by developing pods. When Potassium is needed, lime or gypsum may be required. Potassium may be applied only based on soil analysis.

Calcium

Groundnut plant needs calcium in right quantities for proper pod development as the gynophores penetrate the soil. High level o of soil calcium may reduce the incidence of pod and root rots. If the soil is deficient in calcium, gypsum may be applied to facilitate good pod and seed development.Deficiency of calcium may cause stunted plant, breaking down of the bud area, and numerous fine brown spots on the leaves, producing a bronze colour (Harris, 1949). Balanced supply of nutrients in the root zone as well as calcium and sulphur in the pegging zone are necessary for the best production of nuts (Harris, 1949). Calcium can increase the number of kernels per pod, not by raising the number of available ovule cavities, but by enhancing the number of kernels reaching the maturity. Calcium absorbed by roots went to stem and foliage and little to developing pods. Calcium is relatively immobile within the groundnut plant and its concentration at a particular part is dependent upon availability at the time the particular part is formed. Calcium must be fairly close to the zone of formation of pods. It cannot be translocated to the developing pod. Hence, the developing gynophores or pegs have to absorb calcium directly from the soil. Hence, the application of gypsum @ 500 kg /ha in pegging zone is recommended at peak flowering and pegging stages (Prasad and Reddy, 1992).

Sulphur: Sulphur is found to be deficient in many tropical soils. Sulphur deficiency is indicated as chlorosis of terminal leaves, which can be alleviated easily by applying sulphur containing fertilizers.

The crop receiving sulphur produced higher quality and bolder kernels. One of the less expensive sources of sulphur together with Phosphorous is single super phosphate. If soil application of sulphur is not possible due to deficient moisture conditions, foliar sprays containing S could be useful in alleviating the crop deficiency. Sulphur deficiency may reduce the root nodulation; but excess may inhibit it. Sulphur also influences the oil content (Pathak and Pathak, 1972) To assess the effect of sulphur, elemental sulphur was added to fertilisers at equivalent rate to that contained in the sulphate compound.

Table. 1: Effect of Sulphur on nodules and oil percentage in India (Pathak and Pathak, 1972.)

Treatment (mean of levels)	Mean root nodule number	Seed oil (%)
No fertilizer	306	48.2
Ammonium sulphate	376	50.4
Ammonium chloride + S	380	50.7
Ammonium chlorie	389	47.8
Calcium sulphate	386	51.5
Calcium sulphate + S	360	51.2
Calcium chloride	319	48.6

Micro nutrients: The most important micro nutrients in that order are

Boron.> Suphur>Zinc>Manganese > Copper

- Magnesium (Mg) deficiency is indicated by interveinal chlorosis of terminal leaves and stunted plant growth. In case of severe shortage of Mg, plants become discoloured and eventually die.
- Boron deficiency adversely affects pod production and kernel quality. Visually it is difficult to perceive the deficiency symptoms.
- Molybdenum (Mo) deficiency can affect flowering and peg development.
- Zinc toxicity symptoms become apparent in the second month of the crop emergence. Affected plant presents a stunted look with stem splitting. Groundnut should not be planted in the areas of excess zinc. In the case of zinc deficiency, application of zinc sulphate and gypsum gave very good results.

Genetic Resources and Groundnut Breeding

Groundnut breeding started as early as 1910 in India; but was not pursued with any degree of persistence. Gregory *et al* (1951) reviewing the previous work on the subject summed up the position as under: "In reporting the published work on peanut breeding and genetics, we have been unable to unfold a sequence of events or to build a coordinated body of knowledge, culminating in recent advanced studies logically derived from the results of previous experiments. Instead we have attempted to preserve the mosaic of unrelated patterns........."

Limited progress in groundnut breeding for productivity has been attributed to subterranean pod bearing habit of the plant (Wynne and Gregory 1991) and consequent lack of information on inter-relationship between canopy characters and pod or kernel yield.

According to Gregory *et al.*, (1973) despite a long history of its cultivation, broad and sub-specific variability and geographical distribution, difficulties in its composition offer limited remedial genetic resources among its varietal forms. Therefore, groundnut breeders were constrained with restricted genetic variability among the varietal forms to achieve major advance in productivity.

The major constraint of the groundnut breeding in India has been that the breeding work was revolving around the parental choices from the locally available varietal forms. Since groundnut is an introduced crop to India, the genetic variability among the available varietal forms is not diverse enough to warrant any significant recombinational variability for selection of superior progenies (Prasad 1988; Prasad 1992). The wealth of genetic diversity of the cultivated groundnut is embedded in the land-races of *Arachis hypogaea* in south America. As described by Krapovickas and Rigoni (1960) and Camacho-Villa *et al.*, (2005), land races are dynamic populations of *Arachis* distinct in their genetic attributes, variability and diversity bestowed with local adaptations associated with traditional local farming systems. Land races offer rich sources of genes not only for resistance against biotic stresses and adverse environmental conditions; but also for plant type attributes of physiological efficiency. The diverse species of *Arachis* were managed under the local production systems by the native tribes of south America over several thousands of years and as such got differentiated into diverse forms in respect of various attributes including nutritional parameters (Adriana *et al.*, 2014). The above lacunae of lack of genetic diversity in the cultivated groundnut in India, has been filled to a considerable extent by the International Crops Research Institute for Semi-Arid Tropics (ICRISAT) through augmentation of the germplasm resources. The gene bank at ICRISAT, India holds the largest collection of groundnut accessions. The other gene banks with groundnut collection are, National Bureau of Plant Genetic Resources (NBPGR) and Directorate of Groundnut Research (DGR) in India; Oil Crops Research Institute (OCRI) of Chinese Academy of Agricultural Sciences (CAAS) and Crops Research Institute of Guangdong Academy of Agricultural Sciences in China; U. S. Department of Agriculture—Agricultural Research Service, Tifton Georgia; Texas A&M University; and North Carolina State University in the USA; EMBRAPA—CENARGEN and the Instituto Agronômico de Campinas in Brazil; and Instituto Nacional de Tecnología Agropecuaria (INTA) and Instituto de Botánica del Nordeste (IBONE) in Argentina. The number of groundnut accessions available in the gene banks has been reviewed by Pandey *et al.,* (2012). For the ease in identification of trait-specific genetic resources, core and mini core collections that represent 10 and 1% of the entire germplasm collection, respectively, were developed at ICRISAT, USDA/ARS, and China. The gene banks are also repositories of wild species which are source of novel alleles (Janila *et al.*, 2016)

Fig.4: Some degree of variability for kernel characters in groundnut. (www.motherearthnews.com)

The selection criteria for enhanced groundnut productivity in India revolved around pod size (Reddy 1988). Although pod weight is a yield component positively correlated with productivity, an emphasis on selection for pod weight would ultimately result in low pod number per plant without much enhancement in productivity. This would also cause bolder seed size which would indirectly enhance the cost of seed material thereby forcing farmer to spend more on seed which is already a costly input. This, however, has led the groundnut breeders to lay emphasis on pod number per plant as the selection criterial for groundnut breeding (Prasad *et al.*, 1984). That is to say that the selection criterion for groundnut yield is based on pod number and pod weight which are highly unstable (Nagabhushanam and Prasad 1992b).

Difficulties in hybridization, poor realization of crossed seed (Venkateswaran 1980) restricted recombination and hereditary instability (Hammons 1976), poor pod-set after effecting crosses (Thangavelu *et al.,* 1980), genetic problems arising out of segmental allopolyploid nature, adherence to an unstable character system viz., pod number and weight per plant as a selection criteria and absence of clear-cut relationship between canopy characters and yield components (Prasad and Kaul 1980 and Prasad *et al.,* 1984) have been attributed for slow progress in groundnut breeding.

The major objectives of groundnut breeding

Indian groundnut production is largely confined to the rain-fed unirrigated drylands in the monsoon season and is subjected to the abiotic and biotic stresses, which is the characteristic features of kharif season.

- The yield losses of groundnut due to the incidence of pests and diseases is estimated at 57% without pesticidal control; but the yield losses of the order of 27 to 37% occur in spite of pesticidal applications (Tanzubil and Yahaya, 2017), which means development of pest and disease resistance should be the priority objective.
- The groundnut pod yield decline due to drought stress was 72% at high temperature and 55% at moderate temperature (Hamidou, *et al.,*2012). So the other important objective is breeding for drought tolerance.
- Difference between the potential yield and realised yield in groundnut on average is 47% (Bhatia *et al.*,2006), which means genetic enhancement of the yielding ability of groundnut plant should be an important objective.
- The major objective of peanut breeding should be to achieve a character combination conducive to the attainment of the above objectives.

Many groundnut varieties have been released through the adoption of traditional breeding methods. Nevertheless, the yield differences between the developed varieties and local types has not been significant. Induced mutations also played a considerable role in generating useful new genetic variability that has strengthened groundnut breeding (Prasad *et al* 1984).

ICAR-ICRISAT collaborative research programmes in groundnut production technology in India have shown recoverable production potentials of about 5 to 7 tonnes even in the decade of 1990s. These studies have also shown that the improved cultivars contributed as much as 25 to 28% while improved agronomy contributed 30 to 32% increase in groundnut productivity. A combination of improved genotype and agronomy resulted in synergistic enhancement in yield from 50 to 150% (Nageswara Rao and Singh 1994) particularly under irrigated conditions.

During the decades of 1970s and 1980s significant progress was achieved in developing groundnut varieties resistant to biotic stresses of groundnut (Reddy 1988) under the aegis of ICRISAT. In 2016, the first near-isogenic lines (NIL) trial was conducted for 12 marker-assisted backcrossing (MABC)-derived lines combining foliar fungal disease resistance with early maturity. Moreover, three drought-tolerant lines were identified for national release and they gave higher yields than the best check and zonal check varieties. Of the three drought-tolerant lines, two are also high-oil lines and have consistently recorded over 50% oil across locations during the last three years of evaluation under AICRP-Groundnut. Absence of trade-off between oil content and yield (Janila *et al* 2014) enabled development of high oil yielding breeding lines. Back cross breeding was rarely used because most of the desired economic traits in groundnut are quantitatively inherited. However, with the advent of molecular markers for the traits of interest, back cross breeding has been used to transfer major effect QTLs governing resistance to nematode and foliar fungal diseases, and mutant FAD (Fatty acid dehydrogenase) alleles for enhanced fatty acid profile. Wide hybridization techniques were used to tap the potential of wild species, particularly for disease resistance, while mutation breeding played an important role for release of several groundnut varieties in India and China.

Significant advances have also been made in groundnut genomics including genome sequencing, marker development and genetic and trait mapping. These advances have led to a better understanding of the groundnut genome, discovery of genes/variants for traits of interest and integration of marker□assisted breeding for selected traits. The integration of genomic tools into the breeding process accompanied with increased precision of yield tailoring and phenotyping will increase the efficiency and enhance the genetic gain for release of improved groundnut varieties (Varshney, 2016).

Development of linkage maps of groundnut during the last decade was followed by identification of markers and quantitative trait loci for the target traits. Consequently, the last decade has witnessed the deployment of molecular breeding approaches to complement the ongoing groundnut improvement programs in USA, China, India, and Japan. The other potential advantages of molecular breeding are the feasibility to target multiple traits for improvement and provide tools to tap new alleles from wild species. The first groundnut variety developed through marker-assisted back-crossing is a root-knot nematode-resistant variety, ***Nema TAM*** in USA (Simpson *et al*., 2003).

Availability of draft genome sequence for diploid (AA and BB) and tetraploid, AABB genome species of *Arachis* in coming years is expected to bring low-cost genotyping to the groundnut community that will facilitate use of modern genetics and breeding approaches such as genome-wide association studies for trait mapping and genomic selection for crop improvement (Janila *et al* 2013a).

Spurt in genetic information of groundnut was facilitated by development of molecular markers, genetic, and physical maps, generation of expressed sequence tags (EST), discovery of genes, and identification of quantitative trait loci (QTL) for some important biotic and abiotic stresses and quality traits. Since then, USA, China, Japan, and India have begun to use genomic tools in routine groundnut improvement programs. Introgression lines that combine foliar fungal disease resistance and early maturity were developed using MAB. Establishment of marker-trait associations (MTA) paved way to integrate genomic tools in groundnut breeding for accelerated genetic gain. Genomic Selection (GS) tools are employed to improve drought tolerance and pod yield, governed by several minor effect QTLs. Also, an improved version of the *Arachis hypogaea cv Tifrunner* genome was released. The improvements were made on the basis of HiC assessments, contributed by the Aiden lab at Baylor College of Medicine (Peanutbase, 2019).

Nevertheless, the productivity in terms of yield /plant or yield per unit area or yield per unit of moisture / nutrients remains static and whatever progress that has been attained is only marginal (Siddiq *et al*., 2004). Between the decades of 60s and 70s, there is practically little difference in productivity (700-800 kg ha-1) and the increase in production was largely due to the expansion in the area. But during 80s, particularly during 1988-89 the productivity crossed one tonne (1132 kg /ha), which is due to the transfer of available technologies to farmers' fields through Technology Mission on Oilseeds (Prasad 1991).

Groundnut Canopy and Productivity: The growth and development of groundnut plant appears to be characterized by excessive vegetative growth for its pod yield. It has been suggested by Swaminathan and Jain (1973) that restructuring of morphological frame of groundnut plant towards compaction would bring about considerable enhancement of yield. Cahener and Ashri (1974) suggested using efficient selection criteria involving certain canopy attributes together with yield components in groundnut breeding.

However, little progress was made in terms of identification and exploitation of genetic traits for plant physiological attributes influencing yield components (Prasad 1994d). In other words, selection criteria in groundnut breeding all these years have been based on unstable characters viz., pod number and pod weight

(Nagabhushanam and Prasad 1992b), resulting in little progress with regard to yield per se.

Prasad *et al.,* (1984) indicated that genetic restructuring of peanut plant to combine compact canopy frame and higher pod yield was possible in the case of *Virginia* genotypes, while *Spanish* bunch on the other hand did not tolerate any reduction in vegetative growth which reflected in low pod production probably because they attained a critical genetic balance for canopy attributes.

Simple plant physiological models involving only a few parameters derived for the potential greatly improve the efficiency of breeding process without major changes in data capture and processing capabilities (Duncan *et al.,* 1978). It must be recognized that these improved models integrate many complex processing into a single parameter (Prasad 1994d)

According to Duncan *et al*., (1978) the basic model is

YR=C*DR*p

Where YR is reproductive yield

C is the mean crop growth rate

DR is the duration of the reproductive growth

P is the mean fraction of crop growth rate partitioned towards reproductive organs.

In the above model C provides a measure of 'source' (an integration of the effects of the radiation intercepted by the crop and photosynthetic result of this).

The duration (DR) is relatively simple matter, while partitioning matter (p) provides the integration of reproductive initiation process and translocation. P and DR describe the sink.

The above model by Duncan *et al.,* (1978) is probably the most easily applied to breeding and the concept of portioning is better than that of harvest index, particularly in an indeterminate crop such as groundnut (Prasad 1994d).

Since groundnut pod development occurs below the soil surface, it was suggested that the canopy or vegetative growth of groundnut plant didn't exhibit any positive relationship with its pod yield (Gregory *et al* 1951). It was also suggested that the plant's dry matter production contributing to the canopy development might be diverted towards pod yield by appropriate genetic mechanisms to achieve major advance in pod and kernel yield.

On the other hand, Sinclair and de Wit (1975) found that Nitrogen requirements of legumes including soybean and groundnut were so great that sustained seed growth demanded continued nitrogen translocation from vegetative tissues and such translocation must eventually induce senescence in these tissues, restrict the duration of the seed-fill period, and limit seed yield. According to Sinclair and de Wit (1975) groundnut plant has not yet reached the desired portioning efficiency conducive for higher grain productivity levels corresponding to its dry-matter production and that groundnut has higher dry matter potential than soybean.

Further, the pod yield had positive correlation with total dry matter at 60 DAE, leaf area index, net assimilation rate, nitrogenase and leaf nitrogen. The peak leaf area index (LAI) and crop growth rate of groundnut plant occurred in the mid pod-setting phase (78-81 days after sowing,) and about 80% of the total dry mass was accumulated during pod formation; but total dry matter at harvest had negative relationship with harvest index (Singh, 2004). Chhonkar and Kumar (1987) reported that crop growth rate (CGR), relative growth rate (RGR) and net assimilation rate (NAR) between 60-90 DAS were significantly correlated with pod yield. During the early stages, the partitioning of dry matter is into the leaf and stem but in later stages it accumulates simultaneously in the vegetative parts and fruits as these two phases overlap in groundnut.

Quicker cessation of the vegetative growth makes more photosynthates available for pods and is a desirable character (Singh, 2004). According to Duncan *et al.,* (1978) yield improvement was due to partitioning of assimilates between vegetative and reproductive parts, length of grain-filling period and rate of fruit development. Duncan *et al.*, (1978) reported that partitioning factor is a better estimate than harvest-index.

The ideal type of groundnut, therefore, would be one that sets all its young fruits within 4-5 days and spends the rest of the growing season filling them. Same time, there are two peaks of flowering in normal sown crop, causing higher reproductive efficiency than the late sown crop which has no peak. More numbers of flowers produced up to 45 DAS cause a greater reproductive efficiency and higher pod yield. (Singh, 2004).

Early seeding of spreading-type cultivars (e.g., Florunner), planted in approximately square patterns, should maximize both vegetative and reproductive growth and development, capture more light, and produce more pod yield without sacrificing market quality (Gardner and Auma1989)

In the case of groundnut, it was observed that the plant suffers the disadvantage in terms of poor canopy development at the peak flowering phase. At the peak flowering and initial pegging phases there is a translocation of the photosynthates towards the developing reproductive parts, which makes it imperative for an adequate canopy development at the crucial reproductive stage is important. Groundnut pod development suffers if the canopy can't support the developing pods with the supply of photosynthates. It is observed that this is one of the major limiting factors for groundnut productivity. It is worth maintaining high leaf area index during peak flowering and pod filling period since the yields were not limited by the size of the photosynthetic sink but probably by the low leaf area and less efficient leaves during the final filling period (Singh, 2004)

Spanish genotypes of groundnut show high percentage senescence of leaves than *valencia* and both these types show higher rate of senescence than *virginia* (Narayanan and Chand, 1986).

The attributes of (i) more efficient carbon fixation through efficient leaves, better canopy geometry or greater leaf area duration, (ii) the partitioning factor and (iii) duration of fruit growth (filling period) are the crucial for higher productivity of groundnut plant (Duncan *et al.*,1978)

The partitioning of photosynthate to fruit during pod filling stage is the most influential physiological factor in yield determination of groundnut. The high yield is associated with rapid increase in pod number and near cessation of vegetative growth during pod filling. The partitioning coefficient ranges from 40-98 % and the partitioning of a higher percentage of photosynthate to pods result in higher pod growth rates and higher yields (Singh 2004).

Duncan *et al.,* (1978) made the first attempt to analyse the physiological factors accounting for increase in yield potential in USA and reported that the partitioning of assimilate had the greatest effect on fruit yield and the estimates of partitioning to fruit ranged from 41% in the `Dixie Runner' released in 1943 to 98% in `Early Bunch' released in 1977.

The *Virginia* Groundnut genotypes on the other hand, produce excessive canopy at the pod maturity, which is unwanted in terms of pod yield.

The limitations in canopy structure of groundnut in nutshell are as follows:

1. Inadequate initial canopy development that continues till the peak flowering and fails to support reproductive development. This limitation may be overcome by adequate nutritional management to enhance initial canopy growth and sustaining the same till peak flowering and gynophore penetration for pod development.
2. Excessive canopy growth in terms of new shoots in *virgina* types at the pod development stage causing inadequate pod development and maturity. This defect arising in terms of production of more of tertiary and fourth order branches at pod development has to be curtailed by selecting suitable genotypes such as *virginia* bunch types and avoiding *virginia* runner / spreading types. In other words, the intermediate canopy types with higher degree of initial canopy development and curtailed canopy growth at the later pod development phase are the most desirable plant types in terms of pod and kernel yield (Prasad and Nagabhushananm, 1991).

3. Hence selection for higher degree of canopy development at peak flowering and sustenance of the same at pod development stage and curtailment of new vegetative growth in post flowering phase should be important guidelines for efficient groundnut breeding (Prasad 2019b).

Canopy categories of groundnut in relation to yield: Groundnut is different from other grain legumes since it has aerial flowering and subterranean pod development. In view of the positive geotropism of gynophores, which develop into sink in the soil, it becomes difficult for the plant breeder to exercise section for yield based on above ground characters, which are not considered indicative of the reproductive efficiency (Gregory *et al.,* 1973). While correlations have been developed among seed and pod characters on one hand and yield of groundnuts on the other (Coffelt and Hammons, 1974; Singh *et al.,* 1979; Yadava *et al.*, 1984 and Kataria *et al.*, 1984), the knowledge of inter-relationship between canopy characters and yield components remains fragmentary.

An examination of canopy frames of diverse genotypes of groundnut revealed (Nagabhushanam and Prasad, 1992a) that the most compact canopy category (canopy category 1) (Ex. MH2 variety) was restricted to the sequential branching types, while the most vegetative canopy forms (canopy category 4) (Ex. M-13 variety) were essentially the alternate branching *viginia* types and both these extreme categories were not characterised by high degree of stability for kernel yield. The next canopy frame which falls in intermediate category (canopy category 2) (PGN 1 and MH2 BC28), however, of the sequential branching *spanish* and *valencia* forms respectively. The next canopy type (canopy category 3) (Ex. TMV2NLM and 32-2-5) which is of more compact type than canopy category 4, falls in the alternate branching *virginia* type. Both the canopy categories viz., 2 and 3 were characterised by high degrees of yield stability unlike the canopy categories 1 and 4. In other words, the canopy compaction in groundnut from canopy category 4 to canopy category 1 marked a shift from *virginia* forms to sequential branching types and *vice-versa,* with the attainment of the highest levels of yield stability in canopy categories 2 and 3, consisting of both sequential and alternate branching types. These findings are supported by the findings of Prasad *et al., (*1984), based on induced mutations that higher levels of kernel yield could be achieved by genetic manipulation of canopy characters leading to enhancement of canopy in sequential branching *spanish* and *valencia* types and canopy compaction in alternate branching *virginia* types. In the above study an effort was made to identify the groundnut genotypes on the basis of the canopy frame so as to relate it to their agronomic potential and to work out appropriate parental combinations based on canopy types for use in groundnut breeding. A similar exercise carried out by Metz *et al.,* (1984) in soybean was useful in relating maturity patterns to canopy development and also choosing the appropriate parents for plant breeding work.

Root nodules in relation to other characters of groundnut: Although root nodules do not form the root system in the botanical sense as stated earlier, their close association with the root system of groundnut makes it essential to study them in association with the other parts of groundnut plant. Study of root nodulation and its relation with kernel yield indicated that total number of nodules and dry weight of nodules per plant exhibited a strong positively significant phenotypic correlation with kernel yield in all the varieties studied (Nagabhushanam, 1989). However, when the nodulation at different zones of the root system viz., 0 -5 cm and 5 -10 cm exhibited a significant positive correlation in *virginia* genotypes and positive but nonsignificant correlation with kernel yield in sequential branching *spanish* and *valencia* genotypes.

Investigations carried out by Elkan *et al.,* (1980) also brought out the positive role of root nodules in groundnut physiology with relative differences among the genotypes with reference to their branching pattern.

Based on heritability estimates, there would be a possibility to improve nitrogen fixation and agronomic traits by selecting families for root nodule mass which would lead to higher levels of nitrogen fixation. Phenotypic and genotypic correlation indicated that nitrogen fixation could contribute to yield but to a lesser extent than to vegetative growth. Shoot dry weight and total dry weight may be useful as selection criteria for high nitrogen fixation indicated by root nodule mass (Pimratch *et al.,* 2004a; 2004b).

The studies carried out by Arunachalam *et al.,* (1980) have also brought out the strong positive association between kernel yield and root nodule mass in both alternate and sequential branching genotypes of groundnut. The root nodule mass has also exhibited positive association with other stable attributes of canopy development as described by Nagbhushanam and Prasad (1992c) and Nagabhushanam (1989).

Stable characters to form selection criteria: Groundnut breeders have been using pods per plant, pod and kernel weight per plant as the selection criteria for selecting the high yielding segregates (Norden,1982; Yadava *et al.,* 1984). Since it is fairly well established that the characters such as pod number and pod weight per plant in groundnut are subjected to larger degree of interaction with environment due to subterranean pod development mechanism (Wynne and Coffelt, 1982; Chen and Wan, 1968; Coffelt, 1983, Madhavi, 1988, Sandhu and Kehra, 1977 and Sangha and Jaswal, 1975), selection based on such unstable character combination may not be fruitful. This is amply reflected in the yield performance of the available groundnut varieties, which have not exhibited any significant superiority in yield over seasons (Rao, 1976; Duncan *etal.,* 1978; Gregory and Gregory1976)since their selection was based on the above characters. Therefore, it would be more appropriate to develop selection criteria based on stable characters, which are positively correlated to kernel yield in addition to pod number and pod weight. This is a more valid approach since in a homeostatic genotype the component characters shift in a compensating manner in a changing environment in order to give a consistent performance of the final character (Grafius, 1956). Based on an earlier investigation carried out by Habibuddin (1987) in wheat on assessment of stability of diverse characters an exercise was carried out by Nagabhushanam and Prasad (1992a;1992b) to work out the stability of a wide range of characters viz., seedling dry weight, days to flower, canopy circumference, canopy diameter and leaf area at 60 days of emergence matching with peak flowering, nodule number at 0-5 cm, 5-10 cm, and 10-15 cm of root depth at 60 days of emergence, plant height, number of primaries and number of secondary branches, number of immature and mature pods, weight of mature and immature pods, number of mature and immature kernel and their weight, number of aerial pegs and total pegs, total plant dry matter at 60 days and at harvest, 100 kernel weight, shelling percentage, oil content and harvest index. Based on the values of variances, the characters could be classified as to their stability by way of percentage of genotypes to the total number taken into consideration possessing non-significant variances for a particular character and also taking into consideration those possessing non-significant variation for the particular character with a consideration of the contribution of non-linear component (Habibuddin, 1987). It was found that characters such as plant height, number of primary branches, number of secondary branches, canopy diameter at 60 days, leaf dry weight and shoot dry weight at 60 days and total dry matter at 60 and days and oil content were less subjected to the interactions with the environment, in view of low pooled deviations as compared to pod number per plant and pod weight commonly employed as the selection criteria.

Among the above characters showing low pooled deviation thereby tending towards being more stable as compared to pod number and pod weight, the attribute of oil percentage could not be considered as a criterion for selection in view of it non-feasibility and its strong negative correlation with kernel yield (Nagabhushanam, 1989). The attribute of plant height may not be an appropriate choice considering the lack of its relevance, as it showed no strong positive relationship with kernel yield. Out of the rest of the characters showing low pooled deviation, the root dry weight, nodule number and nodule dry weight at 60 days may not be practically feasible for rapid field evaluation of the breeding materials. The remaining characters particularly number of primary branches especially in *spanish* and *valencia* genotypes, number of secondary branches in *virginia* genotypes, canopy diameter, leaf dry weight, shoot dry weight and total dry matter at 60 days have exhibited a strong positive phenotypic correlation and considerable degree of genotypic correlation with kernel yield. Findings of Madhavi (1988) that initial canopy growth and total dry matter at 60 days may be used as the selection criteria in addition to yield components are in agreement with the above observations of Nagabhushanam and Prasad (1992b).

Finally, it was observed that canopy development at peak flowering at 60 days of groundnut could be an important stable attribute in terms of groundnut productivity. (Prasad 1994d: Singh, 2004, Singh 2011). Canopy development at peak flowering indicated by canopy diameter could also serve as a selection criterion for groundnut breeding (Prasad and Nagabhushanam, 1991)

Genotypes of different canopy categories in relation to yield stability

One of the major problems faced in groundnut production and productivity has been the lack of clarity regarding yield stability of groundnut varieties recommended for cultivation more particularly under conditions of dryland agriculture (Kambiranda *et al.,* 2011; Hamidou *et al*., 2012; Patra and Mohanty, 1987).

An analysis of stability pattern carried on genotypes pertaining to various canopy categories in respect of the *Virginia, Spanish* and *valencia* genotypes (Nagabhushananm, 1989) has revealed that genotypes viz., PGN1 (Canopy type 2), TMV2NLM (Canopy type 3) and MH2BC28 (canopy type 2) exhibited greater degree of stability than the rest. Among them the genotype TMV2 NLM a *Virginia* type is an induced mutant of the age old *spanish* variety TMV2 (Canopy type 2) and MH2 BC28 is another induced mutant of sequential branching variety MH2 pertaining to canopy type 1(Prasad 1988). It may be noted that the above two mutant types representing a shift towards enhanced canopy development from sequential branching parental systems exhibited much higher degree of stability for yield than the respective parents (Nagabhushanam, 1989).

In the case of *virginia* varieties, the compact mutant of M13 and strain 32-2-5 (compact mutant of the variety MK374) (Prasad *et al.,* 1984) pertaining to canopy type 3 showed higher yield stability than the spreading M13 variety which falls in the canopy type 4 and the parental variety viz., MK374. Both the above mutant types exhibited higher yield together with low environmental standard deviation as compared to the parental varieties of the extended canopy types (Nagabhushanam, 1989)

It has been suggested that genetic manipulation of canopy attributes represented by canopy diameter at peak flowering towards canopy enhancement in sequential branching *spanish* types and towards canopy compaction in alternate branching *virginia* types could bring stability for pod yield. These are in conformity with the findings of Swarnalata *et al.,* (1984a) regarding the stability of induced mutants of *virginia* and *spanish* types. The mechanism behind this phenomenon could be that the enhancement of canopy development of sequential branching systems might bring in appropriate genetic balance between canopy and yield components ensuring the stability of canopy sustenance at the time of pod development.

In respect of *Virginia* genotypes that are characterised by unproductive canopy growth at post flowering stage and pod development which drains out assimilates depriving of the developing pods, might be eliminated through canopy compaction (Ashley,1984; Duncan *etal.,* 1978; Prasad, 1988).

Selection Criteria for Groundnut breeding

It has been stated that the canopy characters such as canopy diameter at 60 days and number of secondary branches in *virginia* genotypes have exhibited positive direct effect on kernel yield together with mature pod weight, mature kernel number and 100 kernel weight (Nagabhushanam, 1989; Prasad and Nagabhushanam, 1991) It has also been shown that canopy growth measured as canopy diameter exerts an indirect effect on kernel yield through mature pod weight , number of mature kernels, 100-kernel weight, total nodule number and number of secondary branches. Thus it looks plausible to enlarge the selection criteria for groundnut breeding including most stable attributes such as canopy diameter at 60 days, number of branches at peak flowering (60 days) which have exhibited great degree of stability in addition to strong positive correlation with pod number and pod weight.

The work carried out by Nagabhushanam and Prasad (1992a) has indicated that pod number and pod weight per plant of groundnut showed a large degree of environmental interaction as compared to canopy development per plant at peak flowering, measured by canopy diameter, which exhibited low pooled deviation. This attribute also exhibited higher degree of stability coupled with strong positive correlation with kernel yield per plant apart from the feasibility and ease with which it could be employed as a dependable selection criterion. Based on their stability, the other characters that merit inclusion in the selection criteria are (i) dry matter at peak flowering stage, (ii) number of primary branches per plant at 60 days in *Spanish* bunch forms and number of secondary branches per plant at 60days in in *Virginia* types and (iii) importantly the canopy development indicated by the canopy diameter at peak flowering.

Nagabhushanam and Prasad (1992c) also reported that among different canopy types the extreme canopy types such as the most compact and the most spreading types did not show higher stability for kernel yield. On the other hand, the intermediate canopy types exhibited the highest level of stability and positive correlation with the kernel yield per plant. It was also suggested by Prasad and Nagabhushanam (1992c) that the parental combinations involving intermediate canopy types of groundnut resulted in more heterotic F1 generation and wider variance and wider segregation patterns in the subsequent segregating generations as compared to the parental combinations involving most compact and most spreading canopy categories.

Parental choice and heterosis in groundnut breeding: Heterosis in the F1 generation expressed in terms of superiority over the better parent is of direct influence for developing hybrids in cross pollinated crops; but is also important in self-pollinated crops, as such heterotic crosses may help breeders in selecting appropriate parental combinations which might lead to the isolation of desirable transgressive segregants (Arunachalam *et al.,* 1984).

The manifestation of heterosis for different economic traits of groundnut was first observed by Stokes and Hull (1930) in eleven crosses of different parents of groundnut. Higgins (1938) working with a system of diallele crosses involving 16 cultivars observed marked heterosis for vegetative traits as well as pod yield.

It has also been reported that levels of heterosis in crosses involving different groundnut varieties was not very high (Arunachalam *et al.,* 1982; Raju 1982). According to Wynne and Gregory (1981), heterosis in peanut is most often observed in crosses between sub-specifc groups, viz., *viginia* x *Spanish* and *Virginia* x *Valencia.* However, Arunachalam *et al*., (1982) and Madhavi (1988) reported occurrence of heterosis in F1 generation involving *spanish* x *valencia* combination. In most of the investigations carried out on heterosis in groundnut, the focus was mostly on the crosses involving different sub-specific groups (Wynne and Gregory, 1981). There was no systematic attempt to select parental types based on canopy development. Such a basis for selection of parents would be important, since the canopy development could account for variations in yield components as has been suggested by Reddy *et al* (1984), Ashley (1984) and Duncan *et al.,* (1978).

Prasad (1987) reported higher levels of heterosis for pod number and pod yield in the F1 generations involving TAP 5, the aerial pod bearing genotype (Prasad, 1985) as male parent and other *spanish, valencia* and *virginia* genotypes. Madhavi (1988) found that *virginia* bunch genotype R33-1, *virginia* runner genotype M13, *virginia* bunch mutant of *sapnish* parent, a *spanish* genotype JL24 and the aerial podding *valencia* type TAP5 and another *valencia* mutant genotype MH2-BC28 exhibited higher levels of heterosis in different cross combinations for yield and yield components on one hand and plant physiological traits on the other. According to Madhavi (1988) the above genotypes are genetically divergent parents, which could be used to widen variability in groundnut breeding.

The investigations carried out by Nagabhushanam (1989) on the other hand on the genotypic and phenotypic correlations as well as path coefficient analysis clearly pointed out the direct influence of canopy growth in terms of canopy diameter, leaf area and total dry matter at 60 days on yield components and kernel yield as well as on nodule number and nodule dry weight which in turn have exhibited influence on kernel yield.

An analysis of heterosis in the F1 generations of the matings involving different combinations of parents characterised for canopy development has revealed that negative heterosis for canopy diameter in canopy combinations viz., 2 x 4, 1 x 4 and 3 x 4; while significantly high level of positive heterosis was observed in the canopy type combinations viz.,2 x 3, 3 x 2, and one cross viz., M13 x Nc Ac 2821 involving canopy combination of 4 x 4. Rest of the parental combinations have not shown any perceptible levels of heterosis for canopy diameter at 60 days. Syakudo and Kawabata (1963) found appreciable levels of heterosis for shoot weight for *virgina* x *Spanish* and *valencia* x *virginia* crosses. They also observed absence of heterosis in the crosses between the cultivars with in each sub-specific forms, but not in *spanish* x *valencia* crosses. Wynne *et al.,* (1970) reported that the F1 hybrids of *virginia* x *valencia parents* gave greater degree of heterosis than other crosses for vegetative plant characters. Nagabhushanam and Prasad (1992c) observed a similar trend since the hybrids recovered from crosses involving canopy types 2 x3 and 3x2 pertained to the crosses of *virgina* x *spanish* and *virginia* x *valencia* types. Madhavi (1988) also observed heterosis for canopy spread and leaf area index in the crosses involving *virginia* x *spanish* and *virginia* x *valencia,* with the recovery of the highest leaf area indices from the latter cross combinations. Considering the heterosis for nodule mass, it was found that F1s of the cross involving canopy types 2 viz., (GAUG-1) x 3 [ICG(C)8] showed significantly high positive heterosis (Nagabhushanam, 1989). While the canopy combinations 2 x 3 exhibited significantly positive heterosis, the other parental combinations involving 2 x 4, 3 x 1, 3 x3, 1 x 4 showed negative heterosis (Nagabhushanam, 1989).

With regard to the kernel yield, all the crosses pertaining to canopy type combinations of 2 x 3 and 3 x2 together with one canopy type parental combination of 2 x 2 (TMV2 x GAUG1) showed high degree of significant positive heterosis for kernel yield (Nagabhusham and Prasad 1992c).

A number of investigations carried out work on heterosis in groundnut (Gor *et al.,* 2012, Jayalakshmi *et al.,* 2000, Dwivedi *et al.*, 1989, Reddy *et al*, 2017) reported levels of heterosis and combining ability. A perusal of the heterotic parental combinations reported in most of the cases indicate that the parents involved pertain to intermediate canopy grades, similar to the combinations of canopy types (2 x 3 and 3 x 2) reported by Nagabhushanam (1989) and Nagabhushanam and Prasad (1992c). One key point with regard to expression of heterosis is the genetic diversity of the parents. According to Dwivedi *et al., (*1989), presence of significant heterotic crosses for several characters would be an indication of genetic diversity present among the parental lines. The kind of diversity for expression of heterosis in groundnut is elucidated by Arunachalam *et al.,* (1984) that the frequency of heterotic crosses and the magnitude of heterosis for yield and its components were found to be higher in crosses between the parents in intermediate divergence classes than extreme ones. This finding is in conformity with the results of the work carried out by Nagabhushanam (1989) and Nagabhushanam and Prasad (1992c) on parental choices of intermediate canopy types for heterotic combinations. This should enable groundnut breeders to choose appropriate parental combinations with a clear focus and understanding, instead of casting or spreading the net very wide.

An analysis of pattern of means and variances in the F2 generations of wide ranging crosses involving the different canopy types described earlier, resulted in the following observations (Nagabhushanam *et al.*, 1992a; Nagabhushanam, 1989).

- The variances and means for the above characters viz., canopy circumference and diameter, nodule number, nodule mass and total dry matter which have a bearing on kernel yield were the least in the F2 generations of the crosses involving canopy types 1 x 4 and 3 x 4.
- The parental combinations for canopy types viz., 2 x 4 and 2 x 1 have exhibited high degree of variance with high mean in F2 generations, high variance with medium mean, high variance with low mean and medium variance with high mean for the above characters only in 14.29% and 7.14% of their F2 progenies respectively.
- It was observed that a large proportion of the crosses involving the parental combinations of canopy categories 3 x 3 (21.43%) and parental combinations of canopy categories 3 x 2 and 2 x 3 (50%) exhibited high variance together with high mean, high variance with medium mean, high variance with low mean and medium variance with high mean for the above characters which are important in groundnut breeding as described earlier.

This observation is indicative of appropriate parental choices for groundnut breeding.

Inheritance of Narrow Leaf shape: Leaf shape and leaf orientation are considered to be important attributes of canopy development, which have bearing on photosynthesis and transpiration rate of groundnut plant (McCloud *et al.*,1980; Mahapatra, 1966). The narrow leaf character is supposed to be associated with low level of transpiration and as such is considered important for developing genotypes with tolerance to moisture stress (Harris *et al.*, 1988).

Gopani and Vaishnani (1970) reported a *virginia* type Gujarat Narrow Leaf Mutant (GNLM), which has been reported to be tolerant to drought. Prasad *et al.*, (1984) recovered a narrow leaf *virginia* mutant from the M2 progeny of EMS treated TMV 2 *spanish* bunch variety (Fig.8).

The earlier investigations carried out by Matlock *et al.,* (1970) indicated a partial dominance of narrow leaf mutant with monogenic inheritance.

Balaiah *et al.,* (1977) conducted studies on the inheritance of narrow leaf character using GNLM and reported the dominant nature of the narrow leaf attribute with monogenic inheritance.

Vanisree (1992) confirmed the earlier observations and reported the dominant nature of the narrow leaf character, since the F2 data indicated a typical monogenic inheritance in respect of the narrow leaf attribute of GNLM.

However, the segregation pattern of the F2 generation of the cross involving TMV2NLM (induced TMV2 mutant for narrow leaf character) on one hand and the normal leaved parents on the other hand resulted in the segregation pattern of 1 narrow leaf: 2 intermediate leaf: 1 normal (Vanisree 1992).

The F2 generations of the crosses involving GNLM and normal leaved parent exhibited a segregation pattern of 3 narrow leaf: 1 normal leaf (Vanisree, 1992) as reported by earlier workers.

The cross involving TMV2NLM and GNLM both narrow leaf types resulted in an interesting segregation pattern. While the F1 of the cross showed absolutely narrow leaf attribute with alternate branching pattern, the corresponding F2 generation segregated for 1 needle shaped narrow leaf type with sterility: 60 narrow leaf types: 3 normal leaf types (Vanisree, 1992), suggesting the trigenic model of inheritance involving a recessive allele and sterility due to complementary gene action (Vanisree, 1992). Such results were also reported by coffelt and Hammons (1971) and Tai *et al.*, (1977). The results reported by Vanisree (1992) bring out that TMV2 Narrow Leaf Mutant (TMV2NLM) of Prasad *et al.*, (1984) is genetically different from the Gujarat Narrow Leaf Mutant (GNLM) reported by Gopani and Vaishnani (1970).

It was also reported that in the case of TMV2NLM, the narrowness of the leaflet starts manifesting after first 4 to 5 normal leaves, while in the case of GNLM, the narrow leaf gets expressed after first two normal leaves and from the 3rd leaf onwards (Prasad *et al.,* 1984). Such diverse genetic variability for the same phenotypic attribute could be of immense use in overcoming genetic vulnerability (Hammons, 1976).

Inheritance of aerial podding: The aerial podding of groundnut was first reported by Prasad (1985) as a spontaneous mutation in 5 plants (TAP 1 to TAP5) of the population of Brazilian groundnut variety 'Tatu'. (Fig.5), The mutants exhibited a large proportion of well-developed aerial pods. The aerial pegs that otherwise go waste without developing into pods in traditional groundnut varieties, develop into pods with good seed filling in the aerial pod mutants reported by Prasad (1985). This phenomenon of aerial pegs developing into well-developed pods could be due to a different mechanism of calcium translocation to developing pegs (Prasad and Muralidharudu, 1991) which is absent in normal traditional varieties of groundnut. Wynne and Gregory (1981) suggested structural alterations of groundnut plant involving the enhancement of fruiting sites would be one of the possible ways of breeding groundnut genotypes possessing perceptibly higher level of productivity. The aerial podding genotype of groundnut fulfils the above requirement and offers a tremendous potentiality to enhance the productivity of groundnut plant. An aerial podding segregant was also reported by Nagabhushanam *et al* (1990) in the progeny of MH2BC28 x ICG (C) 8 (Fig.6). In view of the tremendous potentiality of aerial podding genotype, the inheritance of this attribute is of great interest to groundnut breeders.

The investigations carried out by Vanisree (1992) indicated that the pattern of inheritance of aerial podding attribute varies with the female parent employed in the cross, the male parent being TAP5 (an aerial podding genotype). The crosses involving all the sequential parents as well as a lone alternate branching variety Kadiri-3 as a female parent crossed with aerial podding genotype as male parent have resulted in F1s with aerial podding attribute thereby indicating the dominant nature of the attribute. The F2 generations of the above crosses excepting the cross Kadiri-3 xTAP5, exhibited monogenic Mendalian inheritance pattern of 3 aerial podding: 1 non-areal podding.

In the cases of all the other crosses where the female parents used were of alternate branching *virginia* types, the F1 generations did not exhibit aerial podding phenotype, while in the subsequent F2 generations of these crosses, the segregation pattern was of the order of 1 aerial podding: 3 non- aerial podding. The same kind of segregation was observed in the cross Kadiri-3 x TAP 5. From this kind of data, one could draw the inference that in the F2 generations of the crosses involving *virgina* as the female parent, the heterozygote did not express the aerial podding attribute; but got grouped with the other non-aerial podding segregants irrespective of the genetic constitution. The similar observation would hold good with regard to the F1 generations of the above crosses except that of the female parent Kadiri-3

Fig. 5: Aerial pod bearing Mutant (TAP 5) of Brazilian groundnut variety Tatu

In the cases of F2 generations of the crosses involving sequential branching types as female parents, the heterozygote expressed the phenotype of the male parent and that is of the dominant homozygote showing aerial podding. Hence the observed segregation ratio of 3 aerial podding: 1 non-areal podding.

From the foregoing it appears that the expression of the aerial podding attribute is very much conditioned by the genetic background. While the genetic backgrounds of the sequential branching *spanish* and *valencia* types are conducive for the full phenotypic expression of the attribute, the genetic background of the alternate branching *virgina* types tends to modify the expression of the aerial podding character.

Fig. 6: A recombinant exhibiting aerial pod bearing

Inheritance of canopy compaction: The cultivated groundnut offers clear-cut and perceptible levels of variability for canopy development including branching and leaf type. In view of the seemingly discontinuous variation, there have been a number of investigations to work out and understand the mode of inheritance of various components of canopy development (Hammons ,1973; Wynne and Coffelt, 1982). By and large some exceptions to monogenic type of inheritance have been attributed to the branching pattern (Shchori and Ashri, 1970, Balaiah *et al.,* 1977 and Jadhav and Shinde, 1979) and leaf type (Hammons, 1964; Balaiah *et al.,* 1977, Branch, 1987).

Vanisree (1992) used the genotypes categorised for canopy development as 1 (extremely compact), 2 (compact), 3 (less spreading tending towards compact type) and 4 (Extremely spreading) (Nagabhushanam, 1989) and studied their inheritance of canopy compaction. The results of her investigation indicate the universal dominance of canopy category 4 over the rest. In respect of combinations not involving canopy category 4, the F1 exhibited the phenotype of the canopy type 3, indicating its dominance over 1 and 2, barring a few exceptions of occurrence of F1 with canopy category 2in the cases of certain combinations of 2 x 3 viz., PGN1 x TAP5, TMV2 x TAP5 , JL24 x TAP5, J11 x TAP5 and PI350680 x TAP5. Segregation pattern in respect of the crosses studied was not uniform. Combinations involving MH2, MH2BC28, TMV2NLM, GNLM and M13 segregated trigenically in the F2 generation.

Other crosses involving all the above and TAP5 and 32-2-5as one of the parents, exhibited a digenic segregation in the F2 generation involving duplicate and complementary factors. Also the crosses which exhibited monogenic segregation in F2 generation for canopy development consisted of 32-2-5, TAP5, TMV2NLM and GNLM as one of the parents.

The trigenic inheritance was observed in respect of 10 crosses out of which seven involved combinations with an extremely spreading canopy types such as 1 x 4, 1 x 3, 1 x 2, and 2 x 4. There were also combinations involving intermediate canopy types such as 2 x 2, 3 x 2, and 3 x 3, which exhibited trigenic inheritance. In terms of proportion of the total number of combinations, the crosses that involved one of the parents of the extreme category 4 dominated (<66%)in exhibiting trigenic inheritance, while those of both the parents of intermediate canopy formed only one third of the total number of crosses that exhibited trigenic inheritance (Vanisree, 1992).

In the case of cross combinations that exhibited digenic inheritance, it was observed that around 70% of the total number of crosses formed the parental combination of intermediate categories viz.,2 and 3. Only 30% of the total crosses involving extreme canopy types such as 1 and 4 as one of the parents exhibited digenic inheritance in F2 generation.

In the case of crosses exhibiting monogenic segregation for canopy in F2 generations, it was found that all of the cross combinations involved both the parents of the intermediate canopy category. There was no case of involvement of any parent of extreme canopy types.

From the foregoing it appears that the parents of the intermediate categories of canopy development represent intermediate levels of genetic divergence as suggested by Nagabhushanam (1989).

In accordance with the finding of Coffelt and Hammons (1971), it appears probable that the combinations of extreme canopy categories such as MH2 (1) and M13 (4) bring about possible involvement of genetic modifiers including zygotic and developmental lethals for canopy development. (Vanisree, 1992)

Induced Mutations in Groundnut

Low genetic variability for key parameters of plant productivity and lack of resistance to certain biotic stresses among the varietal forms of *Arachis hypogaea* has been the bottleneck in groundnut improvement. The vast potential of wild species, reservoir of new alleles remains under-utilized. Direct utilization of the wild species of *Arachis* in groundnut improvement gets hindered since most of them being diploids with pronounced crossability barriers with cultivated groundnut. Despite extensive breeding involving wide range of hybridizations, so far there have been no spectacular cases of generation of novel genetic recombinant materials superior in terms of agronomic attributes (Rao, 1976; Duncan *et al.*, 1978; Gregory *et al.*, 1980). Wynne and Gregory (1981) however, suggested that major advances in groundnut productivity could be achieved through structural alterations conducive for enhancement of fruiting cites in peanut.

Nevertheless, in such cases of intricacies, mutation breeding is worth taking up to induce new variability not available in the genetic resources of cultivated groundnut, in view of its practical feasibility and reduced time lag. Mutagenesis is the process by which the genetic information of an organism is changed in a stable manner. Mutations do occur in nature spontaneously at a low frequency in a vast population, since mutation is an established property of gene. Mutations are the primary source of all genetic variations existing in any organism, including plants. Through mutation breeding the breeder tries to enhance the probability and frequency of occurrence of mutations for needed traits employing physical and / or chemical mutagens (Alemu, 2016). Even in the case of other species of *Arachis* too, their mutational rectification before their use in groundnut breeding is an important and productive strategy as has been done in other cases such as inducedmutants for superior grain characters of *Triticum carthlicum*, for use in breeding of *T. durum*. (Prasad, 1972b).

Many mutants for seed coat colours with consumer appeal have been identified through induced mutations in groundnut (Suvendu *et al.*, 2007). Mutants with increased pod size were reported in different varieties by Ramanathan, (1979) with the use of radiations.

Irradiation of small seeded groundnut variety resulted in, high yielding and disease resistant variety "Georgia Brown". Several large-seeded lines with high variability for disease incidence, pod yield, total sound matured kernels, pod weight, seed weight and seed size distribution were isolated (Branch, 2002).

Gamma rays have so far been the most widely used physical mutagen, since more than 80% of the mutant varieties were developed using such mutants in regular breeding programs(http://www-mvd.iaea.org). According the FAO/IAEA Mutant Variety Database (http://www-mvd.iaea.org), out of the total number of registered mutant varieties, there are 71 varieties are of *Arachis hypogaea*, L. As per the cited reports there were 65 varieties developed through the use of radiations in various mutation breeding programmes. Out of the total of 65 varieties 41 were recovered through the use of gamma rays in the mutation breeding programmes (Alemu, 2016).

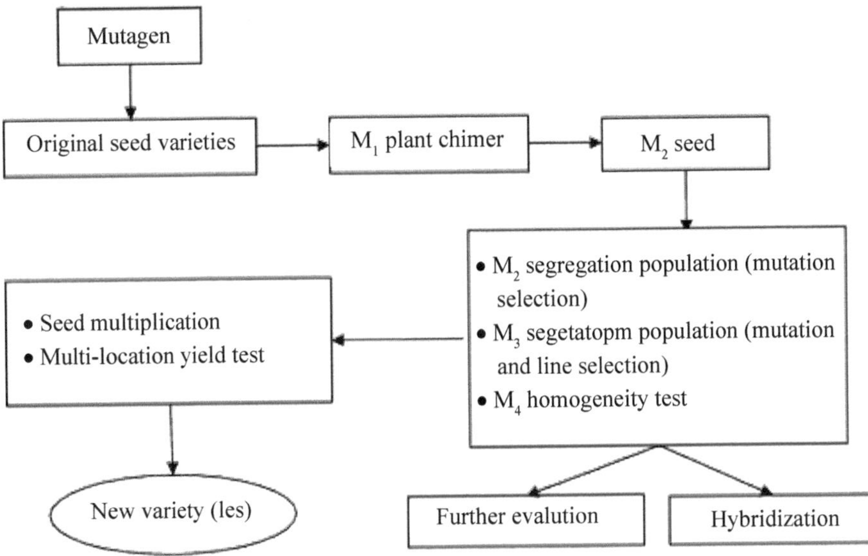

Fig.7: Schematic representation of the procedural steps involved in a regular Mutation Breeding programme.(Oladosu et al.,2016)

Mutational analysis has also resulted in unravelling the differential genetic nature of the genotypcs of groundnut despite their same level of ploidy as indicated by the studies carried out by Nagabhushanam (1989). Based on the variances for canopy attributes and yield components in different generations following mutagenic treatments, it was observed that the sequential branching *spanish* (GAUG-1, NC AC 17090, TMV2) and *valencia* (MH2)genotypes in general exhibited higher degree of variances in the M1 and M2 generations and low level of variances in the M3 generations (Nagabhushanam, 1989). In other words, the response of these genotypes was diploid like disomic nature as in the case of cereal barley and other diploid grain legumes (Brock, 1971).

Considering the response of the *virginia* genotypes viz., Robout 33-1 and M-13, it was observed (Nagabhushanam, 1989), that the variances for the above characters was low in M1 and M2 generations and high in M3 generations, which is the characteristic feature of polysomic genetic system (Swaminathan *et al.,*1962; Brock, 1971).

Although both the above groups viz., sequential branching and alternate branching genotypes of the cultivated groundnut are of the same ploidy level with 2n=4x= 40 representing tetraploid nature (Stalker and Dalmacio, 1981), it is interesting to note that the sequential branching types have attained a higher degree of genetic diploidization as compared to the alternate branching *virginia* types (Nagabhushanam, 1989).

Regarding varietal or genotypic response to mutagens, the variety GG-6 was reported to be more resistant to gamma rays and EMS as compared to GG-20. In M2 generation of the above varieties, chlorophyll and morphological mutants occurred in a low proportion, of which the chlorophyll mutations were higher in the variety GG-20 followed by GG-6 in the treatment with 0.6% EMS. Induced morphological mutations included those affecting plant height, growth habit, leaf morphology, pod and seed characteristics, testa colour and sterility. The mutation rate increased in both the varieties with increasing doses of gamma rays and EMS. For gamma irradiation, 20 KR dose had saturation effect in both the varieties. The spectrum of chlorophyll and morphological mutations varied with the kind and dose of mutagens and genotypes used. In M2 generation, significant reduction in mean values were recorded with increasing doses of mutagenic treatments, while mean values increased with lower doses of mutagenic treatments in both the varieties. Phenotypic ranges were enlarged in majority of the treatments. Similarly, increase in PCV, GCV, heritability and genetic advance was considerable under different mutagenic treatments for most of the traits studied in both the varieties. Magnitude of these estimates varied with mutagen dose, character and the genotypes. However, M2 generations indicated the potential of the higher doses of gamma rays and EMS to induce greater variability for yield contributing traits. (Dahake, 2009)

Groundnut mutation breeding programmes (including the use of induced mutants in recombination breeding) of the Nuclear Agriculture Division of Bhabha Atomic Reseach Centre (BARC), Government of India have resulted in a number of groundnut mutants and their derivatives for various attributes as given below over the years.

(www.barc.gov.in/pubaware/agri_crop_varieties.html)

The mutants and their derivatives for different attributes are as follows:

(i) bolder seed size: (1991-2009): TGS-1, TG-19A, TG-22, TG-39, TG-41, TG-45, TG-47,

(ii) Higher shelling percentage: (2006-2008): TG-38 & TG-51

(iii) Earliness of maturity, higher harvest index and higher yield: (1973-1997): TG-1, TG-3, TG-24 & TG-26

(iv) Resistance to drought and collar-rot disease and high yield: (2004): TG-37A

Some mutants showed interesting genetic behaviour, while others exhibited differential expression in different seasons leading to masking of mutant characters. In addition, mutants having economically useful characters, such as large kernel size and increased yielding potential were also isolated. Using these and other mutants new Trombay Groundnut (TG) varieties were developed which had large kernels suitable for export trade, improved oil content and increased yields. Among them TG-17 was unique for its extreme *fastigiata* character leading to flowering at all nodes and reduced number of vegetative branches. Kale *et al.*, (1997)developed a new Groundnut Variety 'TG-26' by using induced mutants in cross breeding. Demonstrations of TG-varieties for high yielding potential, on the fields of cultivators were successful.

Mutants of groundnut involving canopy changes in varieties MK 374, 32-2-5, R33-1, M13, TMV-2, MH-2 were developed which indicated that canopy compaction in *virginia* varieties and canopy enhancement in *spanish* and *Valencia* genotypes resulted in enhancement of plant yield (Prasad, *et al*., 1984). These results suggested that the sequential branching varieties viz., *spanish* and *valencia* have attained a critical genetic balance for canopy development, compaction of which would result in the reduction of their yields. The *virgina* genotypes on the other hand have not attained the critical genetic balance for canopy, manipulation of which towards compaction has resulted in higher pod and seed yield (Prasad, 1988).

The mutants developed for canopy frame in groundnut plant have paved the way for an efficient groundnut breeding (Madhavi et al., 1993).

A systemic mutant developed from the M2 progeny of EMS treated *spanish* variety TMV2 (Prasad *et al.,* 1984) resulted in alternate branching pattern, which is the characteristic feature of *virgina* types combined with narrow leaf (TMV2NLM). The narrowness of the leaflet of TMV2 NLM manifested after the 4 th to 5th normal leaves in the initial stages of plant growth (Prasad *et al.*, 1984). The further investigations (Prasad *et al.,*1984; Vanisree 1992) proved that TMV2NLM is genetically different from Gujarat Narrow Leaf Mutant (GNLM) a *virginia* type reported earlier by Gopani and Vaishnani (1970). TMV2NLM has exhibited enhanced biomass, rhizobial root nodulation in the deeper zones of root system, tolerance to budnecrosis and leaf spot diseases apart from being a better combiner in hybridization programmes as compared to the parent variety (Prasad, 1988; Nagabhushanam, 1989)

More importantly TMV2NLM has also exhibited 68.66% oleic acid as compared to 38.52% of its parent TMV2, thereby resulting in 78% enhancement in oleic acid as compared to the parental variety representing G to A mutation in AhFAD2A gene for high oleate trait (Mondal and Badigannavar, 2013).

Fig. 8: TMV2 Narrow Leaf Mutant (TMV2NLM)

Dwivedi *et al.,* (1998) treated the seeds of groundnut cultivars JL24 and ICGV88448 with 0.3% ethyl-methane-sulfonate (EMS) for 8 hrs and grew the M1, M2 and M3 generations in the field. Twenty-five M3 progenies with altered fatty acid composition were selected and advanced by 4 generations. Finally, 14 phenotypically uniform mutants were evaluated for seed quality at ICRISAT farm at Patancheru during 1995-97. Oil contents, oleic acid contents and oleic acid/ linoleic ratio were found to have increased in 6 mutants.

Prasad *et al.*, (1985) worked on the mutational improvement of the Brazilian groundnut variety 'Tatu'(*valencia*) using the chemical mutagen Sodium Azide (NaN3). The investigations culminated in the development of mutants for higher pod number. Also the defective seed filling of the parental variety was corrected through development of the mutants for better seed development and pod-filling.

Among the productive mutants selected, three mutants viz., Dwarf, Sd-HP and V3 consistently maintained their superiority in seed filling, pod number and kernel yield at significant levels over a period of three years characterised by low rainfall and drought. The mutants Dwarf and Sd-HP demonstrated significant improvement in pod number per plant, seed number per pod, shelling percentage and higher harvest index in the M4 generation and significantly higher levels of yield.

Wang C T *et al.,* (2012) adopted the technique of injection of 0.3% Ethyl Methane Sulfonate (EMS) into flowers of the groundnut variety Huayu-16 at 9:00-9:30 A.M., and subsequently selected and developed the a high-yielding peanut cultivar - Huayu 40. Through seed treatment with sodium azide (Na N3) and Diethyl sulphate (DES) three high yielding mutant lines derived from the *virginia* type variety Huayu 22, were developed. Sodium azide mutagenesis carried out by Wang C T et al., (2012) also resulted in a peanut mutant with elevated oleate (70%) content. The induced mutant of *A. duranensis* developed by Wang C T *et al.,* (2012) was found to possess thicker branches, and larger leaflets, pods and bolder seeds, which could be useful in recombination breeding of peanut. The mutants of *A. duranensis* contained more than 55% oil, with 60.33% being the highest as against the value of 52.62% of the original parent.

Among chemical mutagens employed in mutation breeding of groundnut, Ethyl Methane Sulfonate (EMS) was mostly used (Levi and Ashri1978) which resulted in isolation of six mutants for various characters in the M2 and the succeeding generations. Additionally, Ethidium bromide was also reported to be an efficient mutagen by Levy and Ashri (1978) in groundnut. Nagabhushanam *et al.,* (1992b) reported that the chemical mutagen Sodium Azide was effective and efficient with regard to the induction and recovery of groundnut mutants for canopy frame development.

Breeding for disease resistance: The yield losses in groundnut due to the incidence of diseases are considerable (Tanzubil and Yahaya 2017) as mentioned under the topic 'Biotic Stresses'. The foliar diseases causing perceptible damage to the crop are late leaf spot (*Phaeoisariopsis personata) (Mycosphaerella berkeleii),* early leaf spot (*Cercospora arachidicola) (Mycospharella arachis*) and rust (*Puccinia arachidis*). In addition to the pod yield reduction caused by the fungi, a considerable damage is also inflicted to the vegetative parts (haulmns) which form the valuable fodder to cattle. Many times, fungicidal control is only partial apart from being expensive and hence growing disease resistant varieties is the durable strategy to prevent the crop losses.

The genetic factors of the host effecting longer latent period of the pathogen, reduced sporulation, smaller but infective lesions and lower infection percentage (Waliyar,1991; Dwivedi *et al.*, 2002) contribute to the resistance of the host to the diseases.

The components of genetic resistance may involve 3 to 4 alleles. which are reported to be recessive (Nevill.,1980). In terms of gene action, non-additive and epistatic nature are reported for inheritance of components of resistance (Green and Wynne, 1986). Additive gene action for resistance against late leaf spot (Jogloy *et al.*, 1987) and low to broad sense heritability (Anderson *et al.,*1991.) as well as significant narrow sense realized heritability (Jogloy *et al.*, 1999) have been reported. Role of maternal and nuclear gene effects in the expression of late leaf spot disease is reported by Janila *et al.*, (2013b). The attribute of resistance to late leaf spot disease of groundnut is reported to exhibit high heritability and genetic advance (Ashis *et al.,* 2014).

Chaudhari *et al.*, (2019) evaluated around 340 genotypes of ICRISAT germplasm and collections from UAS Dharwar (India) for resistance to rust and late leaf spot along with their yield attributes. Out of the above,109 genotypes which had ≤3 disease severity score for rust and late leaf spot at ICRISAT and Aliyarnagar along with a susceptible check (TMV 2) were subjected to stability analysis to identify stable sources of disease resistance and pod yield. The genotypes that exhibited stability for their resistance were those with both lowest disease severity score and least vector length from Average Environmental Axis (AEA).

In respect of late leaf spot, the genotypes GPBD 4, CGV 00248, ICGV 06142, ICGV 02411, ICGV 00246, ICGV 00068, SPS 11, and ICG 11426 were found to be stable resistant genotypes based on their disease score and vector length from AEA. The genotype ICGV 86699 had lowest disease score of LLS compared to others with greater vector length from AEA. (Chaudhari *et al.,* 2019)

In the case of rust disease, the genotypes ICGV 99052, ICG 11426, ICGV 99051, ICGV 86699, GPBD 4, ICGV 06422, ICGV 07223, ICGV 07235, ICGV 05100, ICGV 06142, ICGV 02411, ICGV 05155, ICGV 00362, ICGV 00362, ICGV 00248, ICGV 01361, ICGV 99160, ICGV 02323, ICGV 87846, SPS 2, SPS 7, SPS 11, SPS 21, and ICGV 02446 may be considered as stable resistant genotypes against late leaf spot and rust based on their low disease score. (Chaudhari *et al.,* 2019a)

The genotypes of *Arachis hypogaea* L., exhibiting stability of resistance to above foliar diseases along with good yield levels could form suitable parental materials for transferring disease resistance without adversely affecting the productivity.

Among the wild species of *Arachis, A. chacoense* and *A. stenosperma* have been found to possess higher levels of resistance to early and late leaf spot diseases (Subrahmanyam *et al.,* 1980). Among the 91 accessions of wild species of *Arachis,* 54 species of A genome viz., *A. stenosperma* (24 accessions), *A. kuhlmannii* (6 accessions), *A. helodes* (4 accessions), *A. simpsonii* (3 accessions), *A. diogoi* (one accession), *A. microsperma* (2 accessions), *A. linearifolia* (one accession), and *A. cardenasii* (one accession) were reported to possess resistance to both the kinds of leaf spot diseases (Favero *et al.*, 2009).

Among the non A-genome species, *A. cruziana* (one accession). *A. hoehnei* (3 accessions), *A. magna* (3 accessions), *a. valida* (2 accessions), *A. batizocoi* (one accession) and A. *williamsii* (one accession) were found to be highly resistant to the above two leaf spot diseases. (Favero *et al.*, 2009).

Most of the wild species of *Arachis* being diploid, posing crossability barriers with the cultivated groundnut, the breeding strategies suggested by Simpson (1991and 2001) have to be adopted to effect introgressions of desirable levels of resistance into cultivated groundnut.

Breeding for resistance against insect pests:Yield losses in groundnut due to insect pest attack could result damage of the tune of more than 30% as stated earlier. Nearly 400 species of insects inflict damage on groundnut crop at various stages. (Smith Jr. and Barfield 1982) Plant resistance to insects is more complex than that of resistance to fungal pathogens since the feeding habits and adaptation of insect pests are variable. Redlinger and Davis (1982) found that more than 80 species of *Arthropde* spp. attack post-harvest groundnuts.

The levels of resistance to insect pests in cultivated groundnut germplasm are quite low (Sharma *et al.*, 2003).

The wild species of *Arachis* on the other hand offer some tangible levels of genetic variability for insect pest resistance. Accessions belonging to *Arachis cardenasii, Arachis duranensis, Arachis kempff-mercadoi, Arachis monticola, Arachis stenosperma, Arachis paraguariensis, Arachis pusilla, and Arachis triseminata* showed multiple resistance to the leaf miner *Aproaerema modicella, Helicoverpa armigera, Empoasca kerri, and to rust, Pucenia arachidis* Speg., *and late leaf spot, Cercosporidium personatum* (Berk. et Curt.). *Arachis cardenasii* (ICG 8216), *Arachis ipaensis* (ICG 8206), *A. paraguariensis* (ICG 8130), and *Arachis appressipila* (ICG 8946) showed resistance to leaf feeding and antibiosis to *Spodoptera litura* under no-choice conditions. Six lines, derived from wild relatives, showed resistance to *H. armigera* and *S. litura,* and/or leaf miner (Sharma *et al.*, 2003).

Plant morphological characteristics such as (a) main stem thickness, (b) hypanthium length, (c) leaflet shape and length, (d) leaf hairiness, (e) standard petal length and petal markings, (f) basal leaflet width, (g) main stem thickness and hairiness, (h) stipule adnation length and width, and (i) peg length showed significant correlation and/or regression coefficients with damage by *H. armigera, S. litura,* and leafhoppers. (Sharma *et al.*, 2003). Hence, these traits can possibly be used as markers to select for resistance to these insect pests. Principal component analysis placed the *Arachis* spp. accessions into five groups, and these differences can be exploited to diversify resistance to the target insect pests in groundnut.

Many runner type cultivars, exhibiting field resistance to Tomato Spotted Wilt Virus (TSWV), representing diverse pedigrees evaluated against thrips in the greenhouse revealed that thrips preferred some cultivars over others, suggesting that antixenosis "non-preference" could contribute to thrips resistance in peanut. (Srinivasan *et al.,* 2018). In other crops, morphological traits such as (i) leaf architecture, (ii) leaf waxiness and (iii) spectral reflectance have been associated with thrips non-preference. It is not clear if foliar morphological traits in peanut are associated with reduced preference or non-preference of thrips. Besides thrips non-preference, thrips larval survival to adulthood and median developmental time were negatively affected in some peanut cultivars and in a diploid peanut species *Arachis diogoi* (Hoehne) and its hybrids with a *virginia* type cultivar, indicating that antibiosis (negative effects on biology) could also be a factor influencing thrips resistance in peanut. Available field resistance to ortho-tospoviruses in peanut is not complete. It is evident that cultivars can suffer substantial yield loss under high thrips and virus pressure. Integrating thrips resistance with available virus resistance would be ideal to limit losses. (Srinivasan *et al.,* 2018)

Among different genotypes of groundnut studied, the narrow leaf mutant of TMV2 (Prasad *et al.*, 1984) exhibited least incidence of thrips, leaf miner, foliar feeding larvae including leaf miner and bud-necrosis as compared to the parent variety TMV2 (Prasad, unpublished). The reason could be enhanced thickness of leaflets associated with narrow leaf together with presence of some degree of hairiness, provided the mutant genotype with certain kind of non-preference mechanism, which was not the case with the parent variety TMV2.

Leaves with long and dense tricomes is reported to be linked with higher levels of jassid resistance (Alagirisamy, 2016). According to Dwivedi *et al.,* (1986) the host resistance to jassids due to long trichomes is governed by additive genetic variance associated with a high general combining ability.

Lynch *et al*., (1981) used host suitability index to classify the resistant and susceptible genotypes in respect of fall army worms.

Sources of genetic resistance to insect pests

A review of the available literature indicates that the wild species of *Arachis* indicated above offer dependable sources of genetic resistance to many insect pests afflicting groundnut crop. Additionally, some specific sources of resistance both from *Arachis hypogaea* L. and other species of *Arachis* are listed below.

(A) Foliar feeders

(i) *Spodoptera litura:* Nc Ac343, R9227 and Mutant 28-2 (Prasad and Gowda, 2006); ICGV8603 (Sharma *et al.,* 2002). Among the wild species of *Arachis, A. ipaensis, A. kemp-mercadoi, A. appresipila, A. major* and *A.*

parguariensis are reported to be promising (Sharma *et al.,* 2003; Stevenson *et al.,* 1993)

(ii) *Spodoptera frugiperda* (Fall army worm): Genotypes of the cultivated groundnut: IAC 886 and IAC 22 (Campos *et al.,*2010), Tifton 8 and NC 6 (Todd *et al.*, 1991). Wild species of *Arachis:A. correntina, A. villosa,* and *A. burkartii*

(iii) *Helicoverpa armigera: Arachis kempff-mecadoi, A. triseminata, A. stenosperma, A. monticola, A. durnensis, A. pusilla, A. cardenasii & A. paraguariensis* (Sharma *et al.*, 2003)

(iv) *Aproaerema modicella* (Groundnut Leaf-miner): The wild species of *Arachis* indicated above in the case of *Helicoverpa are* reported to be highly resistant to groundnut leaf-miner according to Sharma *et al.,* (2003).

Arachis hypogaea L.: NCAc 2575 (Rao, 2000) ICGV 86031, ICG (FDRS) 10, ICG 5040, NCAc 17090 and ICGS 2741 (Rao *et al.*, 1998)

(B) Sap sucking Insects:

(i) Thrips (*Caliothrips indicus, Scirtothrips dorsalis, Thrips palmi* and *Frankliniella schutlzei*) : *Arachis hypoagaea* L. : NC Ac.2230, NCAc 2240, NC Ac 2214, NC Ac 2242, NC Ac 2243, NC Ac 2232 (Amin *et al.,* 1985)

Other species of *Arachis: A. duranensi,* and *A. chacoense* (Amin 1985)

A. repens, A. paraguariensis, A. pusilla and *A. villosa* (Stalker and Campbell, 1983)

(ii) *Empoasca kerri* (Jassids): Resistant wild species of *Arachis: A. correntina, A, cardenasii, A. duranensis*, and *A. villosa (*Stalker and Campbell 1983) In addition to the above Sharma *et al.*, (2003)reported the following species of *Arachis* as resistant to jassids. *A. kempff-mercadoi, A. monticola, A. stenosperma, A. paraguariensis, A. pusilla, A. triseminata.*

(iii) *Aphis craccivora* Koch. (Aphids)*: Microtermes spp., Odontotermes spp., Pseudacanthotermes:* Amin (1985) reported that the wild species of *Arachis* viz., A. *glabrata, A. chacoense, A. correntina* and *A. villosa* are resistant to groundnut aphids

(C) Termites (*Ancistrotermes* spp., *Amitermes spp., Coptotermes spp., Macrotermes spp., Microtermes spp., Odontotermes spp., Pseudacanthotermes spp.)* Amin *etal.,*1985 reported that NC Ac 2243, NC 343 and NcAC 2240 possesses considerable resistance to termites.

(D) Certain genotypes of *Arachis hyopgaea* L indicated below have been found to possess multiple resistance to insect pestsNC6, NCAc 343, NCAc

1705, NCAc 2142, NCAc 2214, NCAc 2230, NCAc2232, NCAc 2240, NCAc 2242, NCAc2243 and NCAc 2460. In addition to the cultivated types, resistance to many insect pests has also been found in several wild *Arachis* species viz., *A. correntina, A. chacoense, A. stenosperma and A. villosulicarpa.* (Rathnakumar, 2014).

Breeding for drought tolerance: Drought is one of the most important environmental stresses affecting the productivity of most field crops. Elucidation of the complex mechanisms underlying drought resistance in crops will accelerate the development of new varieties with enhanced drought resistance. Drought incidence can be broadly of three kinds in relation to crop growth. (a)Initial or early-season drought (b) Midseason drought and (c) End season drought (Prasad, 1973; Alagirisami, 2016).

(a) *Initial phase or early drought* is the moisture stress that the crop encounters from the seedling to pre-flowering phase. Normal recommendation is that the irrigation be withheld for about 21 days soon after emergence. No doubt, the moisture stress that the crop encounters is not considered to adversely affect the growth and development of the plant or the final yield. It has been observed that the initial with-holding moisture supply for the first three weeks does help the crop in acquiring deeper root system and also promote condensation of the internodes towards the pod-bearing zone of groundnut plant (Prasad and Sudhakarbabu, 1997) resulting in profuse flowering, soon after availability of moisture following the initial drought of 20 to 23 days (Nageswara Rao, *et al.*, 1988, Puangbut *et al.*, 2009) leading to better kernel yield.

(b) *Midseason drought:* This is the kind of moisture stress that groundnut crop commonly encounters under conditions of dryland agriculture, which occurs at the stages of peak flowering, gynophore development into pod and the initial phase of pod development. The midseason drought is detrimental for recovery of pod and kernel yield. If this kind of drought is prolonged for more than 15 to 20 days, it shall have an adverse effect on pod development and results in decline in kernel yield. While the root system could tap moisture from deeper layers of the soil, the crop canopy may not be much affected. Nevertheless, the gynophores at the superficial layers of the soil devoid of moisture for a longer period fail to develop into filled pods (Prasad and Reddy, 1992). Among the sub-specific forms of *Arachis hypogaea,* the *virginia* types can withstand the midseason drought while *spanish* types in general are more susceptible to this kind of drought. The major difference between the above two types is that *virginia* types recover rapidly from the prolonged midseason drought, while *spanish* types fail to do so (Prasad and Reddy, 1992)

(c) *End season drought:* This kind of drought occurs at pod maturity stage. Normally the end season drought occurs at the final phase of the crop growth coinciding the seed filling and pod hardening stage. The *virgina* runner types or spreading type of groundnuts are more vulnerable to the end season drought due to their staggered or spread-out pod formation and late maturity, while *spanish* and *virginia* bunch types can escape this kind of drought due to their synchronous pod-development and earliness of maturity (Prasad and Reddy, 1992). The end season drought does affect the oil and protein contents and fattyacid composition to certain extent (Dwivedi *et al.*, 1996), in addition to the total yield.

Groundnut is often stated as a drought resistant crop (Wright, 2004). Groundnut is able to withstand very low relative leaf-water contents (e.g., 29%) since the deep-rooting habit and indeterminate reproductive system which gives enormous plasticity in rainfed environments typified by intermittent and end-of-season patterns of drought stress (Wright, 2004). Nevertheless, drought in the case of groundnut crop means the adverse effect of moisture stress at the peak flowering and gynophore-pod developmental stages, when the crop has to be safeguarded against this kind of drought due to the sub-terranean pod bearing nature of the crop (Prasad and Reddy, 1992), since the drought has direct effects on plant water status and assimilate supply, or through effects on peg penetration and calcium uptake in dry surface soils. Peg entry and pod development are extremely sensitive to soil hardness, which often increases with decreasing soil water content.

The depth of penetration of gynophore and subsequent pod formation and development in groundnut plant declines linearly as soil strength (as measured by penetrometer resistance), increased from 0.1 to 2.0 MPa, with cessation of peg growth above 2.0 MPa (Wright, 2004)

Table 2: Depth of peg penetration (mm) as a function of penetrometer resistance (MPa) in the surface 1.5 cm of soil during a 21day pegging period for peanut plants (cvs. Robut 33-1 and McCubbin) grown in pots

	Penetrometer resistance in surface 1.5 cm of soil (MPa)							
Cultivar	0.5	1.0	2.0	3.0	4.0	5.0	6.0	7.0
Robut 33-1	43	39	12	9	6	5	5	5
McCubbin	60	50	14	9	5	4	5	5

(Source: Wright, 2004)

Significant genetic variation in these traits exists and breeding programs aimed at incorporating these traits in combination are underway to further enhance peanut drought resistance. Significant progress has been achieved in understanding the underlying mechanism of drought tolerance in groundnut over the years which has resulted in development of efficient physiological trait-based and empirical selection approaches (Nigam *et al.,* 2005) to breed for drought tolerance in groundnut.

Empirical approach that measured yield under water limiting conditions is widely used.According to Songsri *et al*., (2008) high heritability for biomass, mature pod yield and specific leaf weight would be important for drought resistance. In other words, a combination of the attributes of profuse early synchronous flowering and pod development as reflected by high harvest index and together with high dry-matter (biomass) production are the needed combinations for a groundnut cultivar to be drought resistant. Such a combination could be attained in a *virginia* bunch varieties. The stated characters can be combined through recombination breeding, since additive gene effects and additive x additive type epistasis are involved in their expression (Nigam *et al.,* 2002).

Sources of Drought Resistance: At the International Crops Research Institute for Semi-Arid Tropics (ICRISAT) certain number of drought tolerant accessions superior to the existing standard varieties of groundnut have since been identified. The selected accessions (Upadhyaya, 2005) exhibited good pod yield (ICG 862, ICG8285, & ICG 11855), protein content (ICG 5236, & ICG 6654) and oil content (ICG 118, ICG 862, & ICG 1121). Other lines selected for specific drought conditions are ICG 3068, ICG3141, ICG 2738 & ICG 91151 for midseason drought), ICG2213, ICG 76, ICGV 92028, ICGV 9020 & ICG 93232 for end season drought (Nigam *et al.*, 2002).

Trait-based approaches measuring WUE (water use efficiency) employs SPAD (soil plant analysis development) and SLA (specific leaf area) for drought tolerance are often used in combination with empirical approach. Root traits are identified as drought adaptive traits. However their use as selection criteria for drought resistance is limited as they required elaborate phenotyping protocols. So far, studies on heat tolerance in groundnut were limited to few screening studies reporting tolerant lines for heat stress (Craufurd *et al*., 2003; Hamidou*et al.,* 2012).

The investigated morpho-physiological traits like transpiration, specific leaf area, root dry matter, root length density, and yield components decreased under intermittent water stress including at flowering phase (Falke *et al*., 2019). Significant year effect and genotypic variation were observed on most of investigated traits. The selection for Specific Leaf Area and Harvest Index can be effective in early generations in some crosses and to exploit the additive × additive type of interaction, it can be done in large populations of later generations, particularly to enhance water use efficiency (Nigam *et al.,* 2005). However, it may be observed that drought in the superficial zone of soil where pegs penetrate to swell into pods has more adverse impact on the pod yield as explained by Prasad and Reddy (1992).

Drought and heat stress can result in aflatoxin contamination (AC) of peanuts especially when this occurs during the last three to six weeks of the growing season. Identifying drought-tolerant genotypes may aid in development of peanuts that are less susceptible to aflatoxin contamination. Factors leading to AC are drought, high geocarposphere temperature, kernel/pod damage, and reduced phytoalexin synthesis by the plant. If the fungus colonises a kernel with reduced water activity, the plant cannot synthesise phytoalexin and then, the fungus synthesises aflatoxin. Breeding for resistance to AC is complicated because aflatoxin concentration is highly variable, and influenced by the environment. Since drought tolerant cultivars have resistance to AC, traits of drought tolerance have been used as indirect selection tools for reduced AC (Jeyramaraja *et al.*,2018).

Table 3: Some Drought Resistant varieties of Groundnut released in India

Variety	Pedegree	Duration (days)	Pod yield in kharif(t/ha)	Oil content (%)
Tirupati -1	EC106983/3	100	1.8-2.0	49
Tirupati -4	JL24 x Ah 316/s	105	2.0-2.5	48-50
Narayani	JL24 x Ah316/s	100	2.0-2.5	48-50
Abhaya	K134 x TAG24	105-110	2.0-2.5	52
DRG 17 (Mukta)	R33-1 x TAP5	110-115	2.4-2.7	50
ICGV 91114	ICGV 86055 x ICGV 86533	110	2.5-2.6	48
ICGS 76	TMV 10 x Chico	120	2.5-2.7	43

Fig. 9: A groundnut Variety combining (i) high harvest index, (ii) high biomass and (iii) synchrony of pod development. (https://www.icar.org.in/content/icar-directorate-groundnut-research-develops-groundnut-variety-high-oleic-acid)

Genomics in Groundnut breeding

Limited genomic resources existed for groundnut prior to 2005 (Pandey *et al.*, 2012). However, significant advances have been made in recent years in genome sequencing, development of molecular markers, construction of genetic maps and quantitative trait locus (QTL) analyses.

The genetic basis of agronomic and quality-related traits of peanut remains unclear. A genetic linkage map was constructed for cultivated peanut containing 470 Simple Sequence Repeat (SSR) loci, with a total length of 1877.3 cM and average distance of 4.0 cM between flanking markers. For 10 agronomic traits, 24 QTLs were identified and each QTL explained 1.69–18.70 % of the phenotypic variance (Huang *et al.*, 2015). For 8 quality-related traits, 12 QTLs were identified that explained 1.72–20.20 % of the phenotypic variance. Several QTLs for multiple traits were overlapped, reflecting the phenotypic correlation between these traits. The majority of QTLs exhibited obvious dominance or over-dominance effects on agronomic and quality traits, highlighting the importance of heterosis for breeding. A comparative analysis revealed genomic duplication and arrangement of peanut genome, which aids the assembly of scaffolds in genomic sequencing of *Arachis hypogaea.* The QTL analysis enables to clearly understand the genetic base of agronomic and quality traits in cultivated peanut, further accelerating the progress of map-based cloning of major QTLs and marker-assisted selection in future breeding. (Huang *et al.*, 2015)

Molecular markers are important tools for genotyping in genetic studies and molecular breeding. The SSR and SNP (Single Nucleotide Polymorphism) are two commonly used marker systems developed from genomic or transcript sequences. The objectives of such study should be to (1) assemble and annotate the publicly available ESTs (Expressed Sequence Tags) in *Arachis* and the in-house short reads, (2) develop and validate SSR and SNP markers, and (3) investigate the genetic diversity and population structure of the peanut breeding lines. Various marker systems including RFLP (restriction fragment length polymorphism), RAPD (random amplification of polymorphic DNA), AFLP (amplified fragment length polymorphism), DArT (diversity array technology), SSR (simple sequence repeat) and SNPs (single-nucleotide polymorphisms) were developed (Pandey *et al.,*2012; Varshney, 2016) and have been utilized for genetic diversity analyses, constructing genetic maps, mapping of traits of breeding interest and marker-assisted breeding. Consequently, a large number of expressed sequence tag(EST) based SSR markers ranging 26 (Hopkins *et al.,* 1999) to6455 (Peng *et al.,* 2016) have been reported. Similarly, large numbers of SNP markers have been developed including 8486 candidate SNPs from a screening of sequences of 17 genotypes assembled along with sequences from the reference '*Tifrunner*'transcriptome (Alves *et al*., 2008; GCP 2011), which was used to construct 1536-SNP *GoldenGate assay* (Nagy *et al*., 2012)

Development of linkage maps of groundnut during the last decade was followed by identification of markers and quantitative trait loci for the target traits. Consequently, the last decade has witnessed the deployment of molecular breeding approaches to complement the ongoing groundnut improvement programs in USA, China, India, and Japan. The other potential advantages of molecular breeding are the feasibility to target multiple traits for improvement and provide tools to tap new alleles from wild species.

Genetic Transformation is another novel and useful technology by which gene of interest could be transferred to the desired varietal genetic background.As a superior alternative, biotechnological intervention like transformation of foreign genes, either directly (biolistic) or via *Agrobacterium*, significantly aided in the development of advanced groundnut genotypes equipped with integral resistance against stresses and enhanced yield attributing traits. Several genes triggered by biotic and abiotic stresses, were detected and some of them were cloned and transformed as major parts of transgenic programmes (Gantait and Mondal, 2018). Application of modern molecular biological techniques, in designing biotic and abiotic stress tolerant/resistant groundnut varieties that exhibited mechanisms of resistance, relied on the expression of specific genes associated to particular stress. The genetically transformed stress tolerant groundnut varieties possess the potential to be employed as donor parents in traditional breeding programmes for developing varieties that are resilient to fungal, bacterial, and viral diseases, as well as to drought and salinity (Gantait and Mondal, 2018). Transgenic groundnut plants were produced efficiently by inoculating different explants with *Agrobacterium tumefaciens* strain LBA4404 harbouring a binary vector pBM21 containing *uidA* (GUS) and *nptll* (neomycin phosphotransferase) genes. Genetic transformation frequency was found to be high with cotyledonary nodal explants followed by 4 d co-cultivation (Venkatachalam *et al.*, 1998). This method required 3 days of pre-cultivation period before co-cultivation with*Agrobacterium*. A concentration of 75 mg/l kanamycin sulphate was added to regeneration medium in order to select transformed shoots. Shoot regeneration occurred within 4 weeks; excised shoots were rooted on MS medium containing 50 mg/I kanamycin sulphate before transferring to soil. The expression of GUS gene (*uidA* gene) in the regenerated plants was verified by histochemical and fluorimetric assays. The presence of*uidA* and*nptll* genes in the putative transgenic lines was confirmed by PCR analysis. Insertion of the *nptll* gene in the nuclear genome of transgenic plants was verified by genomic Southern hybridization analysis (Venkatachalam *et al.,* 1998).

Highly reproducible regeneration systems through proliferation of axillary buds around a cultured meristem, *de novo* shoot organogenesis and somatic embryogenesis are available today as target explants for experiments on transformation. Success has been achieved in the development of transgenic plants using both *Agrobacterium*-mediated and direct DNA transfer using particle gun bombardment. Several useful genes have already been transferred to peanut using gene transfer techniques (Eapen, 2003). Peanut production is severely limited by a number of diseases and pests and one of the most challenging needs of the day is to improve resistance to *Aspergillus* species which produce aflatoxin — which are potent carcinogenic metabolites. In addition, introduction of value added traits such as altered protein/oil composition today lies within the realms of biotechnology. Developing peanut plants with genes for abiotic stress or edible vaccines is not far away. The candidate genes which are currently available for transfer to peanut with possible implications in groundnut breeding programme need to be considered.

Turlapati *et al.*, (2006) generated putative promoter tagged transgenic lines in *Arachis hypogaea* cv JL-24 using cotyledonary node (CN) as an explant and a promoterless *gus::nptII* bifunctional fusion gene mediated by *Agrobacterium* transformation. MS medium fortified with 6-benzylaminopurine (BAP) at 4 mg/l in combination with 0.1 mg/l α-napthaleneacetic acid (NAA) was the most effective out of the various BAP and NAA combinations tested in multiple shoot bud formation. Parameters enhancing genetic transformation viz. seedling age, *Agrobacterium* genetic background and co-cultivation periods were studied by using the binary vector p35SGUSINT. Genetic transformation with CN explants from 6-day-old seedlings co-cultivated with *Agrobacterium* GV2260 strain for 3 days resulted in high kanamycin resistant shoot induction percentage (45%); approximately 31% transformation frequency was achieved with p35S GUSINT in B-glucuronidase (GUS) assays. Among the *in-vivo* GUS fusions studied with promoterless gus::nptII construct, GUS-positive sectors occupied 38% of the total transient GUS percentage. Over 141 putative T0 plants were generated by using the promoterless construct and transferred them to the field. Among these, 82 plants survived well in the green house and 5 plants corresponding to 3.54% showed stable integration of the fusion gene as evidenced by GUS, polymerase chain reaction (PCR) and Southern blot analyses. Twenty-four plants were positive for GUS showing either tissue-specific expression or blue spots in at least one plant part. The progeny of 15 T0 plants indicated Mendelian inheritance pattern of segregation for single-copy integration. The tissue-specific GUS expression patterns were more or less similar in both T0 and corresponding T1 progeny plants. The differential patterns of GUS expression were identified in the putative promoter-tagged transgenic lines.

As developments in plant biotechnology unfolds, high frequency disease resistant, drought tolerant and good tasting peanuts, which are safe to eat, will continue to be the goals of peanut biotechnologists

***Breeding for Oil Content and Quality*:** The average oil content of groundnut cultivars is around 50% with a predominance of Oleic Acid (55%) (Nagaraj,

1995) Several studies have shown that in a diverse collection of germplasm, the oil content ranges from 46 to 63% in wild species and 43 to 56% in cultivars (Murphy 1994). The constitution of groundnut oil involves monosaturated fatty acids (MUFA) as well as polyunsaturated fatty acids (PUFA).

The ICRISAT has around 8,000 germplasm lines which have been screened for protein content (16 to 34%) and oil content (31 to 55%) (Nigam *et al.,* 1991). The variability in oil content especially in wild species indicates the opportunity for breeding for higher oil content in groundnut (Wynne and Gregory 1981).A high amount of good-quality oil in seeds has an overwhelming contribution to the groundnut (*Arachis hypogaea* L.) cultivation throughout the world. However, oil content is usually associated with lower yields. Breeding efforts will be required to break this negative association. On the other hand, oil content in groundnut is found to be stable character for which genetic selection could be practised (Nagabhushanam *et al.,* 1992a). The investigations of Prasad (1984) have indicated that substantial enhancements in oil production could be achieved through development of varieties of groundnut with higher oil content, since the attribute of oil content of a variety is a stable character.

The oil content ranged from 31.7 to 57.0%, the average being 48.8%. The genotypes ACG 224, ACG 234, ACG 153, ACG 225 and ACG 173 had the highest values for this trait while the genotypes ACG 57, ACG 39, and ACG 177 had oil content lower than 35% (Yol *et al*., 2017) In the groundnut germplasm subsp. *fastigiata* exhibited higher mean for the oil content as compared with subsp. *hypogaea.* Statistically significant variation among genotypes was also observed for oil yield.

The highest amount of oil yield was recorded for ACG 140 in the germplasm, with a notable lower value in the quantity of oil yield observed in ACG 128 (Yol *et al.,* 2017)

One of the highlights of the Groundnut Breeding Program has been the rapid identification and release of high-oil lines from existing ICRISAT 'elite genetic resources'. Out of the 160 lines from 'elite genetic resources' available at ICRISAT's groundnut breeding program, 47 superior lines with high seed oil content were identified and shared with partners in India when high-oil trait was prioritized as a key target trait for markets in India (Nigam *et al*., 1991)

Oil and protein contents are significantly negatively associated. Variation of Oil content within a genotype showed a significant linear increase as the seed mass increased in the graded samples, but no such relationship was observed with protein content (Dwivedi *et al.,* 1990) Seed mass is not important in genotypes developed for oil extraction. However, due to a positive association between seed mass and pod yield (Badwal and Harbans Singh1973; Gupta *et al.,* 1982, Labana *et al.,* 1980; Layrisse *et al.,*1980) with the lack of association between seed mass and oil content it should be possible to select indirectly for high pod yield through the selection for larger seed mass in early-generation's segregating material generated for developing oil-type genotypes. The positive association between oil content and 100-seed weight of the graded seed samples among certain number of genotypes re-emphasizes the need to develop genotypes with uniform seed maturity to realize the full genotype potential in oil content (Dwivedi *et al*., 1990).

Regarding the impact of drought on oil content of groundnut, Dwivedi *et al.,* (1996) found that mid-season drought had no significant effect on the content of oil, protein and fatty acids other than eicosenic fatty acid. On the other hand, end-season drought significantly reduced total oil, and linoleic and behenic acid content, and significantly increased total protein and stearic and oleic acid content. In the case of five groundnut varieties viz., ICGVs 88369, 88371, 88381, 88382 and 88403, total oil content remained unaffected while oleic acid content increased when they were subjected to end-season drought. These were identified as desirable parents for a breeding program to develop groundnut varieties suitable for rainfed cultivation (Dwivedi et al., 1996).

General combining ability estimates were significant for oil concentration, weight of 50 sound mature kernels (50 SMK), and mean milligrams oil produced per SMK (OPS). Specific combining ability was significant for oil concentration. Reciprocal effects were detected for OPS. Narrow-sense heritability estimates were very high for oil concentration and 50 SMK and low for OPS. The low OPS heritability estimate was caused by the negative correlation between oil concentration and seed size. Consequently, oil concentration and seed mass alone can be improved through early generation selection, but large segregating populations from high oil crosses will be needed to identify progeny with elevated oil concentrations that maintain acceptable seed sizes (Wilson *et al.,* 2013; Dwivedi *et al.,* 1990))

A considerable variability for fatty acid content has also been observed in genetic resources (Hamilton 1987 and Knauft *et al.,* 1987). *Virginia* genotypes in general are reported to have higher quantities of oleic acid in their oil as compared to either *spanish* and *valencia* genotypes (Murphy 1994). Also the effect of the agro-ecological conditions under which the groundnut crop is grown in terms of location is found to have significant effect on the oleic: linoleic ratio of groundnut oil. (Salunkhe *et al.,* 1983, Hamilton1987). Crosses among pure breeding strains of these types differing in oleic and linoleic acid levels, have shown that the genotype of the developing embryo determines the oleic: linoleic ratio, in which certain additive genes are involved (Coffelt 1989; Khan *et al.,* 1974).

Moore and Knauft (1989) showed that the high oleic character in groundnut line F435 is determined by two recessive alleles (*ol1 & ol2).* The systemic mutant viz., TMV2 Narrow Leaf Mutant (TMV2 NLM) (Prasad *et al*., 1984) has been found to possess higher oleic acid content as compared to that of its parent TMV2, resulting in 78% enhancement in oleic acid as compared to the parental variety representing G to A mutation in AhFAD2A gene for high oleate trait (Mondal and Badigannavar, 2013). Hence, mutation breeding may also open genetic avenues involving new genes that may not be allelic to those found in germplasm, thereby widening genetic variability for the quality attributes.

The oil of genotypes with more than 80% oleate may exhibit superior keeping quality and frying quality, since oils with high levels of linoleate are more susceptible to oxidative rancidity (Worthington *et al.,* 1972, Salunkhe *et al.,* 1992). Significant negative correlation was observed between oleic and linoleic acid (Yol *et al.,* 2017).

Fatty acid desaturase (FAD) enzyme facilitates the conversion of oleic acid to linoleic acid by adding a double bond to oleic acid. This enzyme is coded by two homologous genes (ahFAD2A and ahFAD2B) located on A and B subgenomes, respectively. Both FAD genes have 99% sequence homology and inactivation of this enzyme results in high oleic acid in mutants due to substitution (G:C to A:T) and insertion (A:T) of one base pair in FAD genes located on A and B subgenomes, respectively (Chen *et al*., 2010). These mutations lead to the accumulation of more oleic acid and decreasing linoleic and palmitic acid ensuring oil quality improvement by increasing shelf life. With reference to the preferences of consumers and profitability to growers, groundnut breeding mainly focuses on developing improved varieties with desirable levels of oil content and other fatty acids, especially oleic, linoleic, and palmitic acids. Successful introgression of the FAD mutant alleles from Sun-Oleic 95R with increased oleic acid has already been demonstrated using genomics-assisted breeding (GAB) approaches such as marker-assisted backcrossing (MABC) and marker-assisted selection (MAS) in both high and low oil containing genotypes (Chu *et al.,* 2011; Janila *et al.,* 2016b).

The systematic speciality trial conducted in India to evaluate high-oleic groundnut lines resulted in 16 high-oleic lines developed at ICRISAT (Janila *et al.,* 2016a) being recommended by the All India Coordinated Research Project on Groundnut for advanced trials. Enhanced rates of genetic gain for high-oleic trait were achieved with the help of advanced genotyping and phenotyping tools such as near-infrared reflectance spectroscopy (NIRS) and rapid regeneration advancement. This specialty trial was conducted asper the recommendation of the workshop of the All India Coordinated Research Project on Groundnut. (ICAR-DRG 2019).

Groundnut kernels are destined for two very different markets viz., edible or table type and industrial i.e., for oil extraction. The qualities in respect of the above two types are markedly different. For the groundnut kernels used for table type the selected variety should possess the attributes of low oil, high protein and sugar combined with bold and well developed seed without any blemishes, which often might cause reduction in total yield. This type is labelled as hand-picked selections (HPS) (Narasimham *et al.,*1987) in the Indian market parlance. It has been seen that decline in yield in this context might be offset by the higher and more attractive market price. On the other hand, the table types or HPS types have to be grown under a different and intensive agronomic package where the above mentioned attributes are favoured. The acceptable testa colours in north American market are tan, pink and red. The inheritance of these important colours has been investigated (Branch, 1995).

The groundnuts destined for industry on the other hand should have higher oil content, high shelling percentage, long dormancy and good storability. Other important factors from farmers' point of view in this context are earliness and synchrony of maturity, higher seed yield, seed dormancy and resistance to biotic stresses to safeguard against any yield losses (Prasad and Reddy, 1992).

Groundnut is also used for confectionery purpose, besides high yield and large kernels, seed with lower oil and higher protein and sucrose contents are preferred. The negative correlation between oil and protein content and the lack of any association of these two traits with 100- seed mass revealed that it is possible to breed large-seeded peanut genotypes with low oil and high protein content (Dwivedi *et al.*, 1990)

Fig. 10: Pods and kernels of some Groundnut varieties displayed (www.agrocorps.com)

Biotic Stresses of Groundnut

Groundnut plant is a host for number of arthropods, fungi and viruses. Ahir *et al* (2018) reported that the insect pests caused significant reduction in mean height of plant (20.50%), primary branches (24.93%), pods per plant (25.26%) and yield per plot (35.71%); while increase in mean pod damage (50.99%) and mean kernel damage (29.61%). According to Ahir *et al* (2018) the mean yield data recorded form protected and unprotected plots indicated that insect pests caused 35.71 per cent loss in yield.

Totally neglecting pest and disease control resulted in 57.3% yield reduction in Sudan, while controlling soil pests, foliar diseases and foliar insects reduced yield losses to 27%, 32% and 37% respectively. (Tanzubil and Yahaya 2017)

Damage occurring during the bloom and vegetative stages resulted in maximum yield loss.

Insect Pests: The Important arthropod pests and the adverse symptoms caused by their incidence on groundnut are described below.

Aphids (*Aphis craccivora*)

- Infested plants have colonies of nymphs and adults of aphids.
- Nymphs are usually dark brown.
- Adults are winged or wingless with colour ranging from green or greenish brown to greenish black.
- Aphid infested plants are stunted in growth with distorted foliage and stem.
- Sooty mould on foliage and ant associations are also common

Jassids (*Empoasca kerri*, *Bachlucha* spp)

- Nymphs and adults are yellowish green.
- Whitening of veins and presence of chlorotic patches at the tips of leaflets and necrotic leaf-tips with typical 'V' shape are the symptoms of jassid damage.
- Crop presents yellow and scorched appearance known as 'hopper burn' under severe infestation by jassids.

Red hairy caterpillar (*Amsacta albistriga; A. moorei*)

- Larvae have a red head with reddish brown dense hairs on the body.
- Anterior and posterior parts of the body of the caterpillar may have black bands encircling a red band.
- Caterpillars are voracious feeders and cause defoliation of the crop.
- Much of foliage damage is done during night time.

Cotton bollworm *(Helicoverpa armigera)*

- The larvae are dark greenish to brown in colour.
- Young larvae are small and found feeding on tender leaflets.
- Grown up larvae feed on flower and foliage, and defoliate the plants.

Groundnut leaf miner

(Aproaerema modicella Deventer)

- Young larvae initially mine into the leaflets, feed on the mesophyll and form small brown blotches on the leaf.
- Later stages larvae web the leaflets together and feed on them, remaining within the folds.
- Severely attacked field looks "burnt" from a distance.

Cultural management of Groundnut leaf miner

Stray planting of cowpea or soybean as trap crop.

Crop rotation with non-leguminous crop is advised in case of severe recurring problem.

Crop rotation of groundnut with soybean and other leguminous crops should be avoided.

Use resistant/tolerant varieties.

Collect and destroy egg masses and early instars larvae.

Install pheromone trap @ 5/ha for mass trapping.

Spray neem based formulation @ 5%.

The foliar feeding larvae of groundnut may be successfully managed to avoid crop losses following the cultural control measures as detailed below

Deep summer ploughing

Early sowing to escape insect pest damage.

Irrigate once to avoid prolonged midseason drought

Planting castor or sunflower plants as trap crop for egg laying and destroying eggs or 1st stage larvae help in reducing the incidence.

The sunflower plants may be employed to act as bird perches as well.

Apart from the above mentioned biotic agents that cause yield losses in groundnut crop, white grubs, termites, millipedes, wireworms and earwigs bore holes in developing groundnut pods throughout the semi□arid tropics. White grubs and termites also damage the roots in Asia.

Aphids transmit a number of virus diseases. A species of thrips viz., *Frankliniella schultzei* (Trybom) is the vector of bud necrosis disease, which is a serious problem in Asia, southern USA and Australia. Jassids can cause much foliar damage, but their pest status is not certain, except as virus vectors.

The pest control options through employing chemical insecticides must be kept to the minimum levels due to their adverse impact on the ecosystem apart from being too expensive. Efforts towards exploiting host plant resistance, cultural management and employing management techniques have been successful in restricting the pest activity.

Other important management methods to check pest incidence in groundnut

- Seed treatment with *Trichoderma viride* @ 4 gm/kg seed.
- Apply NSKE 5% (neem seed kernel extract) to control sucking pests.
- Augment the release of *Cheilonenes sexmaculata* @ 1250/ha.
- Conserve bio-agents like flower bugs (anthocorids), lady bird beetles (coccinellids), praying mantis, hover flies (syrphids), green lace wing (chrysopids), long horned grass hoppers and spiders.
- Use of chemical insecticides once or twice only at the initial phase of the pest attack. Repeated use of chemical insecticides must be avoided.

Beneficial organisms which predate on the insect pests

Coccinellids

- Larvae (grubs) are slender and with well-developed thoracic leg.
- Pupae are fixed to plant surfaces with exposed appendages.
- Adults are brightly coloured with yellow, orange, or scarlet elytra.
- Small black spots on their fore wings are common.
- Legs, head and antennae of adults are black in colour.
- Some species have forewings that are entirely, black, grey or brown.

Spiders

- Spiders have eight legs and two body parts - a head region (cephalothorax) and an abdomen.
- They lack wings and antennae.
- Nymphs resemble adults.
- Spiders may be web spinners, jumping type and varied in colour and size.

Diseases of Groundnut Crop: Prominent diseases of groundnut, their symptoms and the corresponding casual organisms are given below

Collar rot (*Aspergillus niger*)

- Circular light brown lesions are seen on the cotyledons upon seedling emergence.
- The hypocotyl tissues of stem region upon seedling emergence show water soaked and light brown discoloration.
- Seedlings collapse and die due to rotting of the hypocotyls.
- The collar region becomes shredded due to growth of fungus and black colour becomes visible.

Stem rot (*Sclerotium rolfsii*)

- Infection takes place on the stem just above the soil surface or at the foot of the plant.
- Initially deep brown lesions appear around the main stem at the soil Level
- They occur on the stem below the soil surface under dry conditions or above the ground in wet weather.
- The lesions become covered with white radiating mycelium that encircles the affected portion of the stem.
- The sclerotia of the size and colour of mustard seeds appear on the infected area as the disease develops.
- The distinct rot occurs beneath the fungal weft, leading to wilt like symptoms characterized by yellowing and browning and drooping of the foliage
- In many cases one or two branches may be killed which may subsequently cause the death of the plant.

Dry root rot (*Macrophomina phaseolina*)

- Initially symptoms appear as water soaked lesions on the stem just above the soil surface.
- Marginal zonate and irregular spots characterize the symptoms of the leaf infection.
- Minute spots are also quite common and expand into bigger wavy spots on leaves.
- Gradually the lesions become dark and spread girdling the stem.
- The affected plants show wilting, followed by defoliation.
- The infected stem portion is shredded and becomes black and sooty in appearance with the development of sclerotia. The plant turns brown and subsequently dies.

- Usually rotting of stem is associated with rotting of roots.
- Rotting of pegs and pods is also seen.
- The dead lesions are covered with abundant minute black sclerotia giving a charcoal or ashy appearance.

Peanut bud necrosis disease (PBND)

- Young quadrifoliates have mild chlorotic spots that turn into necroticand chlorotic rings.
- Necrosis of the terminal bud is the characteristic symptom that occurs
- The virus is transmitted by the following species of thrips

  Caliothrips indicus, Scirtothrips dorsalis,Trips palmi, Frankliniella schultzei
- The secondary symptoms are stunting, auxiliary shoot proliferation and malformation of leaflets with severe leaf deformity
- Necrosis of the terminal and auxiliary buds are typical.
- If plants are infected early, they are stunted and bushy.
- If plants older than one month are infected, the symptoms may be restricted to a few branches or to the apical parts of the plants.

Fig. 11: Groundnut plant affected by the Peanut Budnecrosis disease

Management of PBND

- Early sown crop during kharif and rabi/summer seasons is less infected due to checked movement of migrant thrips.
- Optimum plant population should be maintained.

- Rogue out bud necrosis affected plants up to 6 weeks after sowing.
- Tolerant varieties like ICGS 11, ICGS 44 and R8808 should be used.
- Seed treatment with chloropyriphos @6ml per kg seed followed by mancozeb @ 3grams or carbendazim @ 1gram per kg seed.
- Sowing should be done at optimum soil moisture level.
- Intercropping of groundnut with castor or bajra or jowar in 7:1 ratio.

Spray Imidachloprid 0.33 ml or Dimidiate 1 ml and neem oil 2 ml/ litre water mixed with surfactant at 20 days after sowing

Peanut stem necrosis disease (PSND)

- Large necrotic lesions are seen on young quadrifoliates.
- These necrotic lesions coalesce and cover the entire leaflet leading to complete necrosis of young quadrifoliates.
- These symptoms will be followed by necrosis of the entire stem located below the necrosed quadrifoliates.
- If plants are affected less than one month at an age of the entire plant is often necrosed.
- In the case of older plants one or more branches will show necrosis.
- These plants are stunted but do not show any auxiliary shoot proliferation

Peanut mottle virus (PMV)

- The symptoms appear on young leaves as irregular dark green islands.
- Older leaves show mild mottle symptoms visible in transmitted light.
- Some genotypes show characteristic interveinal depression and upward rolling of leaflet margins

Peanut clump virus (PCV)

- Clump virus occurs in patches in the field.
- Young leaves show mosaic mottling and chlorotic ring symptoms.
- Older leaflets are darker green with faint mottling.
- Early infected plants are conspicuous in the field because they are severely stunted and dark green.

Early leaf spot (*Cercospora arachidicola)*

- Early leaf spot symptoms first appear on the upper surface of lower leaves of the plants about 10-18 days after emergence as pale areas.
- Spots develop to become yellow; necrosis occurs from the centre of the lesion and later on, the entire spot become necrotic.
- Spots are circular to irregular characterized by a yellow halo of variable width.
- At maturity the spots turn reddish-brown to black and lower surface of the spot is orange in colour.

Late leaf spot (*Phaeoisariopsis personata*)

- Late leaf spot symptoms appear after 28 to 35 days of emergence or at the time of harvest at the lower leaf surface.
- Shape of spot tends to remain distinctly round, and yellow halos are seen only around mature spots.
- Spots are almost black on both surfaces, but lower surface of the spot is distinctly carbon black.
- Conidiophores are always found confined to the lower leaf surface and these are usually in the plainly visible concentric circles.

Alternaria leaf spot

(*Alternaria alternata, A. arachidis & A. tenuissima)*

- Lesions are small, chlorotic, water soaked and spread over the surface of the leaf.
- Lesions are brown in colour, and irregular in shape surrounded by yellowish halos.
- Light to dark brown blighting of apical portion of leaflets are seen.
- The blighted leaves curl inward and become brittle.
- Veins and veinlets adjacent to the lesions become necrotic.

Management of Biotic Stresses of Groundnut.

Pheromone trap catches for pest management.

- Pheromone traps for two insects viz., *Helicoverpa armigera* and *Spodoptera litura* @ 2 bumber / acre fixed field have to be installed from the start of the season.

- Species specific lures have to be used with the traps. Install the traps with lures randomly in the selected field each separated by a distance of 25m from the other traps.
- Fix the traps to the supporting pole at a height of one foot above the plant canopy.
- Use a cotton swab dipped in diclorvos inside the polythene bag to kill the insects getting trapped (take care that the insecticide does not come in contact with funnel at any one time).
- If insecticide is not used, see to that the live moths are killed before counting/ emptying.
- Change of lures should be made once a month.
- Emptying of the moths from collecting container/bag should be made after counting and recording. Ensuring the presence of traps and readiness to replace in case of breakage/missing events should be followed meticulously. Therefore, after initial installation of traps with lures, surveillance team or member should carry few traps and lures during each week of surveillance for attending to missing traps immediately.
- Since both the species are polyphagous year round deployment of the traps and monitoring are recommended.

Several biotic stresses could be managed successfully by following the following practices devoid of chemicals:

- Summer ploughing
- Crop rotation with non-leguminous crops
- Foliar application of aqueous neem leaf extract (2-5%) or 5% neem seed kernel extract at 2 weeks' interval 3 times starting from 4 weeks after planting is good.
- Inter-cropping trap crops and those that attract biological controlling mechanisms.
- Enhancing the populations of natural enemies of pests by restricting the use of chemical pesticides to the minimum.

Aflatoxin contamination in groundnut: Aflatoxins are the secondary metabolites produced by *Aspergillus* species in several agricultural products. These aflatoxins are found to have carcinogenic, teratogenic and hepatotoxic effects in the consumers (Foster and Rosche, 2001). Of various agricultural commodities, groundnut is a major oilseed crop that is often contaminated by these aflatoxins (Bhat *et al.*, 1996) The permitted level of aflatoxin in groundnut for human consumption, according to international standards, is 4 parts per billion (ppb) in the European Union and 20 ppb in the US (Haque *et al.,* 2018)

Drought-stressed plants lose moisture from pods and seeds and physiological activity of plant is greatly reduced. Both factors increase susceptibility to fungal invasion. Low-altitude with warmer ecologies with low precipitation support high occurrence and distribution of aflatoxigenic *Aspergilli* in soil and high aflatoxin B1 contamination in groundnut (Monyo *et al*., 2012).

Poor harvesting and storage conditions can lead to rapid development of the fungi and thus high toxin can be produced. Aflatoxin is one of the major problems in groundnut, which hinders not only the domestic consumption but also export of groundnut since the international regulation for minimum standards for aflatoxin contamination is becoming stringent. Aflatoxin contamination is a qualitative problem induced by *Aspergillus flavus* and *A. parasiticus*. Aflatoxins are stable small molecules and cannot be destroyed by heat treatment or during processing. It has also been reported that *A. flavus* spores invade the peanut flowers and then travel down the pegs and become established in the developing seeds (Styer *et al*., 1983). It is an ascomycetous fungus that can infect plants, animals, insects and human.In addition, a study, conducted in Nairobi, Kenya in August 2018, stated that aflatoxin contamination had severe health impacts on milk drinkers, causing stunting in children under the age of five years. "The exposure to AFM1 from milk is 46 nanograms per kilogram (ng/day) on average, but children bear higher exposure of 3.5 ng/kg body weight per day (BW/day) compared to adults, at 0.8 ng/kg BW/day. This causes stunting among children (Chetia, 2020).

Six peanut genotypes that have exhibited lower aflatoxin and/or drought tolerance in previous researches (Tifguard, Tifrunner, Florida-07, PI 158839, NC 3033, C76-16) were compared to an aflatoxin-susceptible genotype, A72. The phenotyping methods included visual ratings, chlorophyll fluorescence (PIABS, φEO, and Fv/Fm), SPAD chlorophyll meter reading (SCMR), normalized difference vegetation index (NDVI), canopy temperature (CT), canopy temperature depression (CTD), and pod yield. Based on these traits, Tifguard and Tifrunner exhibited greater drought tolerance mechanisms than the other genotypes and may be good candidates to be incorporated in future drought tolerance studies. After the aflatoxin content of the different genotypes was measured, aflatoxin contamination showed high correlations with visual ratings (0.85), CTD (0.81), NDVI (0.79), and CT (0.73), and moderate correlations with Fv/Fm (0.62) and SCMR (0.57) ($P \geq 0.05$). These easily measurable, rapid and cost-effective phenotyping methods may be used as alternative to more tedious and costly methods of identifying genotypes that are less susceptible to aflatoxin contamination. Using a combination of these methods is beneficial but not always practical. The combined use of visual ratings, CTD and NDVI is advised for initial evaluation of drought tolerance in peanut genotypes.

As reported by some authors, approximately 80% reduction in total aflatoxin content could be achieved after treatment with *ajowan* or *ajwain* (*Trachyspermum ammi)* seed extracts over the control used. Aqueous extract of *ajowan* seeds was found to contain an aflatoxin inactivation factor Hajare *et al.*, (2005).

Mere adoption of good varieties and good agricultural practices can help reduce the level of aflatoxin to 20 ppb. Breeding efforts have focused on reducing groundnut maturity periods to escape end-of-season drought, and the emphasis has been on the identification of short-duration farmer preferred lines with resistance to or tolerance of *Aspergillus.*

Among the recommended practices at production stage, applications of lime alone can reduce aflatoxin contamination by 72 per cent, while application of farm yard manure reduces aflatoxins by 42 per cent under field conditions. When combined, aflatoxin contamination can be reduced by up to 84 per cent.

Harvesting at appropriate stage, adoption of proper drying method, reducing kernel moisture to the level of 8 per cent, etc., prevents the accumulation of aflatoxin significantly (Haque *et al.,* 2018)

The Directorate of Groundnut Research (Junagarh) recommends certain measures such as crop rotation with onion or garlic, selection of short- and medium-term varieties, advanced sowing, supplemental irrigation during end of the season, etc., to reduce aflatoxin.

Three resistance mechanisms have been focuses of aflatoxin resistance breeding: (a) pre-harvest natural seed infection, (b) aflatoxin production and (c) *in vitro* seed colonization (IVSC). Nigam *et al.* (2009) described a large number of groundnut lines that showed IVSC resistance (15% or fewer seeds colonized) and seed infection resistance

(<2% seed infection) including five elite lines recommended for cultivation. In West and Cental Africa (WCA), three varieties were reported for resistance to aflatoxin (Mayeux *et al.,* 2003)

The following genotypes of groundnut were identified to be resistant to pre-harvest contamination of Aflatoxin as reported by Desmae *et al.,* (2017)

Table 4: Genetic Sources of Resistance to Aflatoxin.

Genetic Sources of Resistance to Aflatoxin of Groundnut	Reference
73□33, ICGV 89063, ICGV 89112	Mayeux et al., 2003
ICGV 88145, ICGV 89104, ICGV 91278, ICGV 91283, ICGV 91284	Nigam et al., 2009
ICG 13603, ICG 1415, ICG 14630, ICG 3584, ICG 5195, ICG 6703, ICG 6888	Waliar et al., 2016.

Peanut allergy is a type of food allergy to peanuts. It is different from tree nut allergies, with peanuts being legumes and not true nuts. Physical symptoms of allergic reaction can include itchiness, hives, swelling, eczema, sneezing, asthma attack, abdominal pain, drop in blood pressure, diarrhoea, and cardiac arrest (Loza and Brostoff 1995). Anaphylaxis is also reported to have occurred due to peanut allergy in Western countries (Loza and Brostoff, 1995). Those with a history of asthma are more likely to be severely affected (Loza, and Brostoff1995)

The allergy is recognized as 'one of the most severe food allergies' due to its prevalence, persistency, and potential severity of allergic reaction. (Loza, and Brostoff 1995)

Prevention may be partly achieved through early introduction of peanuts to the diets of pregnant women and babies. It is recommended that babies at high risk be given peanut products in areas where medical care is available as early as 4 months of age (Greer *et al.*,2019) The principal treatment for anaphylaxis is the injection of epinephrine (Al-Muhsen *et al*.,2003). In the United States, peanut allergy is present in 0.6% of the population (Boyce *et al.*, 2010). Among children in the Western world, rates are between 1.5% and 3% and have increased over time (Chen *et al.,* 2018). It is a common cause of food-related fatal and near-fatal allergic reactions.(Boyce *et al.,* 2010)

Peanuts harbour 12 allergens and multiple isoforms recognized by the Allergen Nomenclature Sub-Committee of the International Union of Immunological Societies, 70 % of which fall into these families. These 12 allergens, can be categorized into the four most common food allergen families: The Cupin superfamily (Ara h 1, 3), the Prolamin superfamily (Ara h 2, 6, 7, 9), the Profilin family (Ara h 5), and Bet v-1-related proteins (Ara h 8), as well as two additional families, Oleosin (Ara h 10,11) and Defensin (Ara h 12, 13). Currently, structural data exist for Ara h 1, 2, 3, 5, 6, and 8 (Geoffrey *et al*., 2014).

According to Knoll *et al.,* (2011) peanut allergy can cause severe symptoms and in some cases can be fatal, but avoidance is difficult due to the prevalence of peanut-derived products in processed foods. One strategy of reducing the allergenicity of peanuts according to them is to alter or eliminate the allergenic proteins through mutagenesis. Other seed quality traits could be improved by altering biosynthetic enzyme activities. Targeting Induced Local Lesions in Genomes (TILLING), a reverse-genetics approach, was used to identify mutations traits in peanut (Knoll *et al*., 2011)

Food allergy is the consequence of maladaptive immune responses to common and otherwise innocuous food antigens. India, with a diverse and unique food culture, has acquired its own portfolio of allergens too.In many countries in the developing world and in some emerging economies (e.g., China, India and Brazil) there continues to be a paucity of information regarding the prevalence and incidence of food allergy and other food hypersensitivities. Current assumptions are that prevalence rates are much lower and rare (Boye, 2012).

Most of the reactions in India tend not to be life-threatening, yet the most common symptoms, according to Professor Mahesh's work, appear to be headache, itching, joint pain and vomiting (Boye, 2012). The above observations in developing countries are mostly for general food allergies. There are no specific instances of peanut allergy in the above countries.

Groundnut Production Practices

In the case of groundnut production, the crop management assumes very significant role. Groundnut was termed 'unpredictable legume' due to its obvious lack of response to several crop management inputs and its sub-terranean pod-bearing habit. (Gregory *et al* 1951).

Nevertheless, the recent advances in the management technology of groundnut have helped in managing successfully the unpredictable nature of groundnut plant.

Groundnut is grown in two main seasons in India viz., (1) *kharif* (June to October / November in the monsoon season) when groundnut crop is normally grown on drylands without any extraneous irrigation and (2) *rabi* which is the post monsoon season (October/November to April) when the crop is grown under irrigated conditions. Groundnut crop is also grown in summer season (January /Februay to May) under irrigation, employing short duration varieties which mature in 90 to 95 days.

Fig. 12: Groundnut crop grown on broad bed-furrow system

Table 5: New generation situation specific released Groundnut varieties through AICRP-Groundnut (AICRIPDA-2004)

Varieties	Salient Features	Zones where recommended
BAU 13, B 95, Somnath, TKG, 19A, GG 20	Large seeded (HPS Type) for Export	Gujarat, Maharashtra, Jharkhand
ALR 1, ALR 2, ALR 3	Multiple foliar disease resistant	Pollachi tract of Tamil Nadu for rainfed kharif
ICGV 86590	Resistant to Rust and LLS	Andhra Pradesh, Tamil Nadu, Karnataka

Varieties	Salient Features	Zones where recommended
ICGS 11, ICGS 44, DRG 12	Field tolerance to PBND under Rabi/Summer situation	Maharashtra, Tamil Nadu, Karnataka, Andhra Pradesh
K 134, BSR 1, VRI 2, VRI 3, VRI 4, Co 4, AK 159, GG 7	High yielding Spanish bunch	Andhra Pradesh, Tamil Nadu,
CSMG 884, CSMG 84-1, , HNG 10, LGN 2, DRG 17	High yielding Virginia varieties with medium maturity	Maharashtra, Madhya Pradesh Uttar Pradesh, Rajasthan, Punjab
TG 17, TG 26, VRI 1, BSR 1, Dh 40, TAG 24, DRG12	Fresh Seed dormancy for 2-3 weeks, Spanish bunch types	Maharashtra, Tamil Nadu, Karnataka, Orissa, Gujarat
TG 3, TG 22, TG 26 Gujarat,	Tolerant to acid soils suitable for rabi/summer season.	Orissa, Maharastra, Kerala, Goa and N.E. States
ICGS 37, ICGS 76, CSMG 84-1, CSMG 884, TAG 24, K 134, DRG 17	Tolerant to drought	Chattishgadh, Rajasthan, Punjab, Maharashtra, Andhra Pradesh
RSHY 1, OG 52-1, Dh 40, VRI 3, /, ICGS 44, TAG 24	Suitable for Paddy fallow/ residual moisture situations	Orissa, Coastal Andhra Pradesh, TamilNadu, Karnataka, West Bengal for post –rainy situation.
ICGS 1, SG 84	For spring situation in potato/ Toria fallow	Punjab, Uttar Pradesh

- **Summer ploughing** after the harvest of the previous crop to ward off several biotic stresses.
- **Nutrient management** should be based on soil test values
- Incorporation of organic manure in the form of FYM (4tons/ha) or poultry manure (3 tons/ha) in adequate quantities at the time of first ploughing
- Incorporation of phosphatic fertilisers to provide @ 60 kg P2O5/ha and potassic fertilizers @ 30 kg/ha at the initial ploughing.
- Good soil preparation and moisture conservation in the field, following the Broad Bed-Furrow (BBF) system (Fig. 12).
- Groundnut seed is extremely susceptible to physical damage and should be handled with care at all times. Damaged or split kernels will not germinate and grow. It is also important to plant high quality seed. Better seed will produce healthier plants.
- **Seed treatment** with fungicides viz., thiram or mancozeb or bavestien @ 3 g/kg of seedApplication of Rhizobium strains NC 92, IGR 6 and IGR 40 TAL 1000, PGPR 1, 2 as seed treatment @ 500 g/ha. Seed treatment with Phosphorous Solubilizing Micro-organism @ 500 g/ha to mitigate soil acidity and low phosphorous availability in North Eastern Hill regions.

- **Seed rate** @125 kg/ha to maintain good plant density with inter row spacing of 30 to 40 cm and intra row spacing of 10 cm between the plants within each row.
- **An optimum planting depth** is crucial for optimum plant population density and uniformity of plant stand.It has been observed that a seeding depth of 5 cm ensures rapid seed germination, good seedling emerge and uniform plant stand. Seeds placed at depth of more than 7 cm take longer time to emerge accompanied by low seedling vigour. In deeper sowing the hypocotyl and surrounding flowering and pod bearing zone remains buried in the soil causing reduction in yield. Ensure optimum moisture level and temperature of more than 19°C at plating and germination.
- **Planting** may be carried out on broad bed-furrow (BBF) system which facilitates moisture conservation under rainfed dry-land conditions and proper irrigation in the post rainy season when the crop is grown under irrigated conditions.
- **Plant population**: Maintain an optimum plant population density of 3,30,000/ha under irrigated conditions to be raised in the post rainy season in India
- Maintain a plant population density of 150,000/ha with inter-row spacing of 45 cm and with intra row spacing of 10 cm from plant to plant with in a row for rain grown dryland production in *kharif* season without any artificial irrigation.
- **Earthing-up not beneficial**: The practice of pushing-up the soil around the base of the plant at the main stem known as 'earthing-up' is followed in some cases of groundnut production. This is a drastic yield limiting practice since pod formation is not promoted but adversely affected at the lower highly productive nodes and causes disease incidence. This also has another drawback in the form of distorted plant development, when done at the earlier stages of plant growth, causing poor flowering at the productive nodes. It has been observed that yields were invariably low with the practice of earthing-up.
- **Effective weed control** implies good control of weed throughout the growing season particularly during the first 60 to 70 days following seedling emergence. The choice of the mechanisms for weed control, however, depends upon the species of weeds in question and the level of infestation. Summer ploughing before the planting season and initial intercultural operations during the first 40 to 45 days of seedling emergence are more effective in checking weed proliferation. Weed control may be achieved employing certain chemical herbicides as follows.

(i) Pre Planting Incorporation of Fluchloralin at 0.9 to 1.2 kg/ha controls some broad leaf weeds.

(ii) Two hand weedings along with two intercultural operations followed by application of Kolpa+ Diuron 0.75 or Pendimethalin 1.5 kg/ha.

(iii) An application of 750 ml Diuron 80 WP/ha at the emergence stage in 700 - 800 l water/ha.

(iv) An application of Pendimethalin as Pre-emergence application @1.5 kg/ha is also found effective in controlling the weed population.

(v) Pre-emergence application of Alachlor @ 2.0 to 2.5 kg/ha.

- **Water Management** of groundnut crop: Total water requirement of groundnut crop is estimated as 500 to 550 mm. During the initial phase of crop, which is 21 days following sowing, there is no need for any irrigation. The crucial stages at which drought should be avoided are (i) peak flowering (28 to 45 days from emergence) and (ii) pod filling stage (65 to 95 days from emergence).

  The first irrigation following seedling emergence should be given only at 20 -21 days and not before that. In the case of irrigated crop of groundnut, scheduling of irrigation should be based on IW/CPE, i.e., the ratio between fixed irrigation water (IW) and net cumulative pan evaporation (CPE). This facilitates irrigation at 0.4 and 1.0 IW/CPE ratio at the vegetative and reproductive phases respectively, which is found to be optimum.

- **Cropping Systems:** Cropping system refers to growing a combination of crops in space and time. An ideal cropping system is the one that uses natural resources efficiently,provides stable and high returns and does not damage the environment. Commonly practised cropping systems are, crop rotation practices and intercropping systems.

  Mixed cropping is practised in traditional subsisting farming to meet the domestic needs of the farmer's family. Thus, the number of crops grown mixed in one system of cropping varies depending on the family needs.

  Intercropping is the refinement of traditional mixed cropping towards achieving sustainable income levels at the same time providing tangible insurance against crop failure due to drought one hand and biotic stresses on the other

- Generally, as a rain fed crop, groundnut is grown year after year in a mono-cropping system as done in Sourashtra zone of Gujarat. This practice may lead to the crop failure due to adverse weather conditions during crop growth period. The groundnut in general should be rotated with cereals like maize, wheat, sorghum (jowar), pearl millet (bajra) or minor millets. It was also found that groundnut does well after wheat, maize, sorghum (jowar), pearl millet (bajra) and tobacco.

Crop rotations: Based on several years' data the following crop rotations are suggested

Table 6: Groundnut based crop rotations

State	Rain fed- Two-Year Rotation	Residual Moisture Double Cropping
Andhra Pradesh	Groundnut-Sorghum Groundnut-Millet Groundnut-Tobacco	Groundnut-Bengal gram(Chickpea) Groundnut-Saflower Groundnut-Sesame
Gujarat	Groundnut-Sesame	Groundnut-Sorghum (Fodder) Groundnut-Mustard
Karnataka	Groundnut-Sorghum Groundnut-Ragi	Groundnut-Saflower
Maharastra	Groundnut-Sorghum	Groundnut-(Saflower) Groundnut-Maize(Fodder)
Madhya Pradesh	Groundnut-sorghum	Groundnut(Saflower)

Additionally, the following Crop sequences are recommended for other regions as per the data generated over a number of years

Table 7: Groundnut based crop sequences for some states of India.

State	Rain fed- Two-Year Rotation	Residual Moisture Double Cropping
Orissa	Groundnut-Sorghum/s Pearlmillet-bajra	Groundnut-Bengal gram Groundnut-Sesame Groundnut-Sorghum(Fodder) Groundnut-Mustard
Punjab	Groundnut-Sorghum/Pearlmillet-(bajra)	Groundnut-Mustard
Rajasthan	Groundnut-Pearlmillet (bajra)	Groundnut-Barley
Tamil Nadu	Groundnut-Sesame Groundnut-Cotton	Groundnut-Sesame
Uttar Pradesh	Groundnut-Sorghum/Pearlmillet (Bajra)	Groundnut-Mustard

Intercropping

An ideal intercropping system should aim to:

- Produce higher yields per unit area through better use of natural resources;
- Offer greater stability in production under adverse weather conditions and with disease and insect infestation;

- Meet the domestic needs of the farmer;
- Provide an equitable distribution of farm resources.

  The following intercropping practices were found to be remunerative than sole crop of groundnut even under drought or excessive rainfall in AP.

  Groundnut + Red gram (pigeon pea) 7:1 ratio

  Groundnut + Castor 7:1 ratio

  Groundnut + Sorghum 6:2 ratio

  Groundnut + Pearl millet 6:2 ratio

  (7:1 ratio indicates 7 rows of groundnut and one row of the other crop)

  Groundnut + Sorghum inter cropping system is recommended to farmers to meet the fodder needs of cattle and milch animals.

The following (Table 8) intercropping practices were found to be remunerative in other groundnut growing States in India.

Table 8: Groundnut based intercropping systems

State	Crop combination
Maharastra	Groundnut + Red gram (6:1/4:1) Groundnut + Soybean (6:2) **Groundnut + Sunflower(6:2/3:1)**
Gujarat	Groundnut + Castor (9:2/3:1) **Groundnut + Sunflower (3:1/2:1)** Groundnut + Red gram(4:1)
Tamil Nadu	Groundnut + Bengal gram (4:1) Groundnut + Pearlmillet (4:1) Groundnut + Greengram(2:1)
Karnataka	Groundnut + Chilli (2:2) **Groundnut + Sunflower (3:1)** Groundnut + Red gram(4:1)

Fig.12: Groundnut +Maize (Corn) intercropping system

- **Drought management**: Drought is most damaging to groundnut crop, since it has a broad physiological spectrum affecting many metabolic processes including photosynthesis. Drought affects the groundnut plants at all the stages, with varied sensitivity and crop cultivars that show differential responses (Singh AL, 2004, 2011; Nautiyal *et al*., 2002)The photosynthetic rate of leaves decreases as the relative water content and water potential decreases (Singh AL, 2004).

  In the case of groundnut crop grown on drylands totally depending on rainfall, the initial moisture conservation through (i) summer ploughing in addition to bunding of the field to accumulate soil moisture and (ii) adoption of broad bed-furrow system of planting are the important practices to avoid the adverse effect of drought at the above mentioned crucial developmental stages. The furrows act as the seats of moisture conservation, which help the crop in overcoming the adverse effects of drought. In the event of drought of more than 20 days, a foliar spray of Urea @2% shall enable the crop to withstand well.

  As described earlier, in the event of unusual drought prevailing for more than 50 days, efforts should be made to give one or two crop lifesaving / protective irrigations through external source or from the rain water harvested in farm ponds and /or dug wells. The pod bearing zone should be promote to facilitate proper peg penetration and pod development (Prasad and Reddy 1992). The groundnut varieties recommended specifically for rainy season *(kharif)* cultivation described earlier may be preferred, since they possess adequate levels of drought tolerance.

- **Nutritional requirement of groundnut at the time of peak flowering and gynophore development into pods:** The most important developmental stage of groundnut in relation to its production is the gynophore or pegging phase in which the fertilized ovary of groundnut flower is carried by a peg to penetrate into the soil where pod development takes place. The groundnut plant is especially fascinating because, while flowering is areal, pods develop below ground level and the process is known as geocarpy. It is evident that gynophore acts a root also and absorbs plant nutrients particularly Calcium, needed for pod development. It has been observed that the translocation of Ca to the developing pod through the root system or shoot system doesn't seem to happen. Hence it becomes imperative to ensure that adequate Calcium is provided to the developing gynophores. At the stage of gynophore development, the top soil should have Ca and moisture, which the developing pegs can absorb. Therefore, it is necessary that gypsum ($CaSO_4$) be applied @500 kg/ha at peak flowering that follows peg development. Application of Gypsum as recommended has always resulted in higher pod yields under dryland conditions (around 1.5 to 2.0 tons/ ha) in rainy season as well as under irrigated conditions (around 3.0 to 4.0 tons/ha). In order to sustain the canopy at peak flowering, a foliar spray of 1to 2% urea solution once a week would be needed. Adequate soil moisture must be ensured in the pod bearing zone at peak flowering and pegging stages (Prasad and Reddy, 1992).

Needless to emphasize that the needful plant protection schedule should be in place for effective control of pest and diseases.

- **Harvesting of groundnuts** may be carried out based on pod maturity empirically shown by yellowing of leaves of mature plants. The right stage of harvest, however, is when approximately 75% of the pods have reached maturity. The inner wall of the pod usually has a dark discolouration at maturity. The pods should make rattling sound when shaken, which is the indication of the right stage of harvest. Pods of sampled plants must be looked into for the above attribute. Colour of seed in the pod can also be used as indicators. The colour of the young immature seed is usually white, which gradually turns pink as the seed reaches maturity. The approximate duration from sowing to harvesting of groundnuts is around 95 -110 days for *spanish* and *valencia* types and 120 -125 days for *virginia* varieties (Prasad and Reddy, 1992).
- It is recommended that the soil should have optimum moisture, in case the harvesting has to be carried out manually or using a bullock drawn diggers to uproot the plants. **Groundnut diggers** drawn by a pair of bullocks or by tractor are available in market. The bullock-drawn groundnut digger can harvest groundnut crops over an area of 0.75 ha in 8 hours. Some farmers use conventional 76 inches' blades attached to cultivator frame to dig groundnuts. Care should be taken to avoid pod losses during harvesting.
- **Yield:** With the adoption of the appropriate management practices including a recommended improved variety, it is always possible to achieve average pod yields of 1.0 to 1.5 t/ha under dryland rain-fed conditions and around 2.5 to 3.0 to/ha under irrigated conditions in the post rainy season.The pod yield however, varies with the groundnut variety employed and prevailing agro-ecological conditions of the locality.

The crop cutting experiments conducted by ICRISAT at Yarlagadda Village (red sandy soils) of Anantapur district of Andhra Pradesh in India in the year 2018 showed that the crop productivity in these fields varied from 1.9 to 2.4 tons per ha for the variety K9 and from 2 to 3.2 tons per ha for K6 variety, which is nearly twice the average productivity of Anantapur district (ICRISAT 2018)

3

Rappeseed - Mustard (*Brassica* sp.)

Introduction

Mustard and rapeseed crops were grown since 3000 BC in the Indus valley and their use has been mentioned in the *Ayurveda Simhata.* Use of mustard oil is recorded in the last few centuries of the pre Christian era.

The global production of rapeseed-mustard is at 72 million tons (USDA), while that of the oil is 24 million tons. Currently China is the largest producer of mustard in the world. China, India and Pakistan grow about 90% of the world production of mustard. Rapeseed and mustard form the third important group of oilseed crops in the world after soybean (*Glycine max*) and palm (*Elaeis guineensis* Jacq).

The word 'rape' and 'mustard' have been derived from the word *rapum* meaning turnip and European practice of mixing the sweet '*must*' of old wine with crushed seeds of black mustard (*Brassica nigra* L. Koch) to form a hot paste, respectively (Hemingway 1976). Among the seven edible oilseeds cultivated in India, rapeseed-mustard (*Brassica* spp.) contributes 26.2% in the total production of oilseeds. In India, the rapeseed-mustard group is the third most important edible oilseed after soybean and groundnut, sharing 27.8% in the India's oilseed economy. In India rapeseed and mustard are grown during the post-rainy winter season over an area of 7.49 million ha. Around 20% of the total area under rapeseed-mustard in India is totally under dryland cultivation subject to vagaries associated with dryland agriculture.

These consist of traditionally grown indigenous species, namely *toria, brown sarson*, yellow *sarson*, Indian mustard, black mustard and taramira along with non-traditional species like *gobhi sarson*, white mustard and Ethiopian mustard or *karan rai.*

Gobhi sarson and *karan rai* are the new emerging oilseed crops having limited area of cultivation. *Gobhi sarson* is a long duration crop confined to Haryana, Himachal Pradesh and Punjab.

This group of crops need about 18°C - 25°C temperature for their better growth and expression. Incidence of rains, high humidity and cloudy weather during winter cause damage to the crop together with the attack of aphids.Mustard crop may be produced on diverse kinds of well drained soils, with pH of higher than 6.0.

The moisture requirement of rapeseed-mustard complex is around 350 to 400 mm.

The seed oil content of this group of crops varies from 37- 45%. The oil is used as the main medium for cooking throughout northern India and also used as condiment in preparation of pickles and for flavouring curries and vegetables throughout India.

In commerce, the mustard oil is used for preparation of hair oils and medicines in addition to making of soaps and in manufacture of lubricants. Mustard oil is used for softening leather in tanning industry.

Oilseed cake is used as cattle-feed and organic manure. Green stems and leaves are a good source of fodder for cattle. Leaves of young plants are used as green vegetable as they supply enough sulphur and minerals in the diet.

The rapeseed and mustard group of oilseed crops consist of different oleiferous plant species collectively termed as *Brassicae*, in the family *Crucifereae*, which include

1. Indian mustard (*Rai)* (*Brassica juncea*),
2. Mustard (*Karan Rai*) (*Brassica carinata* Braun)
3. Brown *sarson* (*Brassica campestris* var. brown *sarson*),
4. Yellow *sarson* (*Brassica campestris* var. yellow *sarson*),
5. *Toria* (*Brassica campestris* var. *toria*),
6. Rocket salad orTaramira (*(Eruca vesicara* L.Cav. subsp. *Sativa* Mill) and
7. *Gobhisarson (Brassica napus).*

Among the above *Brassica campestris, Brassica napus* and *Eruca sativa* are termed rapeseed and *Brassica juncea* and *Brassica carinata* are known as mustard or *Rai.*In India there are two different ecotypes of brown *sarson* viz., *lotni* (self-incompatible)and *tora* (self-compatible). The *lotni* predominantly is cultivated in colder regions of the country, particularly in Kashmir and Himachal valley. The *tora*, on the other hand is cultivated in limited areas of eastern Uttar Pradesh (Chauhan *et al.*,2011)

It is reported that *Brassica* group of crops including certain vegetables too in addition to oilseeds were among the crops domesticated by early man (Murphy, 1994).

In India and China rapeseed and mustard oils have been used for edible purposes. Rapeseed oil, which is the largest domestic source of edible vegetable oil in China, has played decisive role in country's edible oil market, and currently accounts for 57.2% of the oil production (Yin *et al* 2010). In western countries, the rapeseed oil was used initially only for non-edible purposes due to the presence of erucic acid and glucosinolates which are considered anti-nutritional factors that could be detrimental to human and animal health (Appelqvist and Ohlson 1972). Later with the development of low erucic acid and low glucosinolate genotypes of rapeseed (Kramer *et al.,* 1983; Stefansson 1983), a new version known as 'canola' came to the fore. Canola oil is widely used edible oil in the west and other parts of the world.

All these crops are grown during the period from October to April which falls in the winter season. The rapeseed-mustard production trends represent fluctuating scenario in India, with all time high production of 8.13 million tonnes from 7.28 million ha during 2005–06. The yield levels also have been variable ranging from 854 in 2002–03 to 1,159 kg/ha in 2009–10.

This group of crops have imparted stability to crop production in India, both under the intensive crop management as well as the harsh conditions of dryland agriculture with low input management. Additionally, this group of crops can fit in diverse intercropping and sequential cropping situations, added to their suitability for solid pure cropping. The rapeseed variety *toria* has found a niche as an important catch crop. The role of *Brassicae* in enhancing the total oilseed production leading to 'yellow revolution' in India has been very well documented (Prasad 1994a; Sharma, 2001).

Ecology and Adaptability

Brassica group of oil bearing crops demonstrate a wider degree of adaptability since they can grow well under temperate, semi temperate and relatively less cool winter conditions of Indian subcontinent. Their yield is fairly high touching even 2 to 3 tons/ha under conditions of optimum supply of moisture and nutrient inputs. At the same time this group of crops impart stability to dry-land agriculture with a reasonable degree of seed yield ranging from 1.5 to 2 tons/ha with the adoption of appropriate and recommended crop management technologies. There are varieties of *Brassica campestris* var. *toria* very well suited for even pre-winter conditions between September and October as ideal catch crop. *Taramira (Eruca sativa)* can be grown successfully even under extreme drought conditions.

The mustard growing areas in India exhibit diversity in the agro-ecological conditions, with the cultivation of diverse species of rapeseed-mustard. There is a wide disparity in yields obtained under conditions of assured water availability on one hand and dryland agriculture depending on soil moisture conserved from the monsoon rains on the other. Under conditions of dryland production characterised by marginal resource situation, production of rapeseed-mustard often becomes less remunerative and hence the urgent need to ensure the adoption of recommended production technologies depending upon the site specific needs to reap reasonable yield levels.

Plant Characteristis of *Brassica* Spp.

Cultivated rapeseed-mustard group consists of herbaceous annual plants. Plants of members of the two groups have a lot of common features and similarities despite some differences. It is somewhat artificial and uncommon to present the descriptions of rapeseed and mustard separately. Hence the morphological features of the both are discussed together, which would help distinguish their principal features as well understand the commonalities.

The rapeseed (*Brassica campestris* or *rapa*), brown sarson /yellow sarson/toria , plant is shorter in height (45-150 cm) than that of mustard. The leaves are broad and flat veined and can be toothed and lobed (meaning the leaves of the mustard plant may not be smooth on the sides). It was observed that leaf angle was prostrate type, leaf blade's shape was spatulate, leaf division dentate lyrate type, leaf apex and leaf blistering was intermediate, leaf tip and lamina attitude was straight, sparse hairs were present. Petiole length, width and thickness were intermediate types. They're usually dark green, but mustard plant's leaves can also have purple streaks and appear curly or smooth. The roots of rapeseed plants are more or less confined to surface layers with an extensive lateral spread. The stem is usually covered with a waxy deposit. In rape, leaves are borne sessile. Fruits (siliquae) of *B. campestris* are thicker than those of mustard, and are laterally compressed, with a beak one-third to half their length. Seeds are yellow if the plant is yellow sarson otherwise brown with a smooth seed coat and are glabrous and hairy. The flowers are cross pollinated and self-incompatible in brown sarson and surprisingly self-pollinated in yellow sarson.

Mustard (*B. juncea*) is known as *rai, raya* or *laha* in India. The plants are tall (90-200 cm), erect and more branched. The plant bears normally long and tapering roots. The leaves are not dilated at the base and clasping as in the case of rape, but are stalked or petiolated, broad and pinnatified. The fruits (pods or siliquae) are slender, about 2.0-6.5 cm long, strongly ascending or erect with short and stout beaks. The colour of seed is dark-brown or black. There are few yellow seeded varieties of mustard too. Seed coat is rough. Mustard is self-pollinated, but cross-pollination also takes place to some extent mediated by pollinating agents of which honey bees are more important. The entire mustard plant can be used as a food source. Mustard leaves (and shoots) can be eaten as a vegetable dish, much like spinach leaves.

Flowers of both the species have 4 sepals and 4 petals of deep yellow to pale yellow colour. Each flower has 6 stamens; 4 with long and 2 with short filaments. The pistil is compound, which is separated by a false septum, thus providing 2 chambers. Mustard plants will have around 20 seeds in each silique.

Fig. 1: Diagrammatic representation of a plant of *Brassica juncea* and its parts

Physiology of seed yield in *Brassica*: Rapeseed-mustard follows C_3 pathway for carbon assimilation. Therefore, it has efficient photosynthetic response at 15–20°C temperature. At this temperature the plant achieves maximum CO_2 exchange range which declines thereafter. Apart from the above, the expression of seed yield is influenced by many interacting processes and their dependence on diverse environmental factors. During the period from sowing until the onset of flowering, seed yield potential is set by the number of flower buds and leaf area index at onset of flowering (Mendham *et al.,* 1981, Leterme 1985). Leaf area index peaks around flowering and then declines, and stems and silique (pods) take over assimilate production (Chapman *et al.,* 1984). Conditions during flowering are particularly important for pod formation and seed set (Daniels *et al.,* 1986; Leterme,1985; Keiller and Morgan, 1988). Seed filling is determined either by the sink capacity of the crop represented by number of silique (pods) and seeds and sink capacity of individual seeds or by the source which is the assimilate availability during seed filling (Mendham *et al.,* 1981). Remobilization of reserve carbohydrates accumulated during earlier growth phases in roots, stems, leaves and pod hulls may also contribute to seed filling (Quilleré and Triboi 1987).

Insight into the relative importance of yield-determining processes and crop characters requires quantitative analyses of components of yield formation in relation to crop physiological, phenological and morphological characters and the environmental factors, radiation, temperature and day length, followed by integrated analysis at crop level. This could result in a model, which would include the key processes of yield formation. Under conditions of moisture stress certain genotypes exhibit a compensating mechanisms invoving some special attributes to safe-guard satisfactory levels of productivity. Singh and Chaudary (2003) suggested that genotypes with thicker leaves had greater water-use efficiency and thus the negative effects of drought during seed development may be mitigated by remobilization of assimilates that is by increasing harvest index.

The Yield Determining Processes

The development of the ideotype concept, (Donald 1968; Thompson1983; Sedgley,1991 Rasmusson 1991) has led to integrating the researches in the areas of breeding, plant and crop physiology, thereby focusing on identification of certain clear-cut plant attributes, which influence the physiological processes that would influence final seed yield (Thompson, 1983; Daniels *et al.*, 1986; Thurling, 1974; Thurling, 1991). For example, crop biomass at flowering has been suggested as a criterion of yield potential (Mendham *et al.,* 1981, 1984;), smaller petals or apetalous flowers may improve light absorption and dry matter production by the green canopy (Mendham *et al.,* 1981; Yates and Steven, 1987)

Increase in yield components viz., pod number per plant and seeds per pod may increase the sink capacity and the partitioning of assimilates to the seeds (Mendham *et al.,* 1981; 1984).

Improvement in these aspects may enhance the effectiveness of breeding programmes and thus lead to increases in seed yields of rapeseed-mustard complex.

Plant physiology research has improved understanding of different aspects of yield formation and provides starting points for yield increase. However, crop models are needed to analyse the effects of various processes and crop traits on seed yield since yield is the outcome of the interaction between these processes and their dependence on the environment. Crop growth modelling also provides the opportunity to design high-yielding crop types (Duncan *et al.*, 1978).

Cytogenetics and Evolution of the *Brassicae*

Among the annual oil bearing plant species, *Brassica* group is unique with regard to its evolution of cultivated rapeseed and mustard starting from a three diploid ancestors each with varying chromosome numbers and concomitant cytogenetic background, forming a famous *Brassica* Triangle (Fig.2). The three diploid progenitors of the current cultivated rapeseed-mustard group are

1. *Brassica nigra* L. Koch (BB 2n=16)
2. *Brassica oleracea* L (CC 2n=18) and
3. *Brassica campestris* L (AA 2n=20)

- The cultivated mustard *Brassica juncea* L Czern & Coss (AABB 2n=4x= 36) is the result of hybridization followed by chromosome doubling involving *B. campestris* and *B. nigra.*
- *Brassica napus* (AACC 2n=4x= 38)has evolved due to the hybridization and amphidiploidy involving *B. campestris* and *B. oleracea.* In India this goes goes by the popular name *Ghobi sarson.*

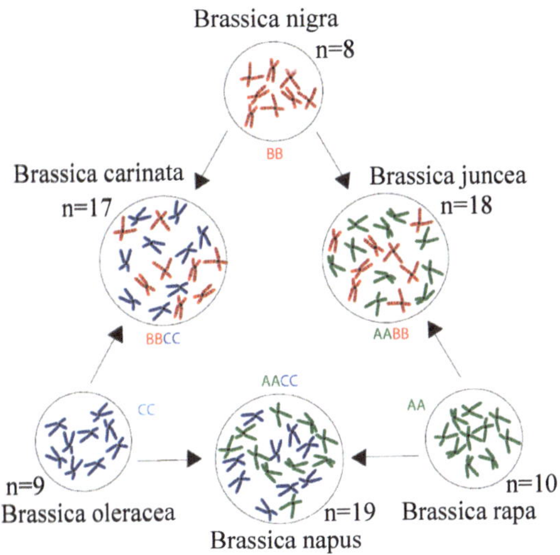

Fig. 2. The famous Brassica Triangle (*Source*: Adenosine at English Wikipedia based on work by Nashville Monkey at English Wikipedia Created: 17 March 2006)

- Another kind of mustard known as *Karan Rai* is *Brassica carinata* Braun (BBCC 2n=4x=34) has resulted due the hybridization followed by chromosome doubling involving *B. oleracea* and *B. nigra*

The origin of the above three diploid parental species is somewhat complex; but it is suggested these are secondary polyploids that have undergone diplodization originating from two common ancestors with chromosome numbers n=6 and n=7 with two basic evolutionary pathways (Singh.H and Singh 1992).

It is suggested that rapeseed *B. napus* originated in Mediterranean region (Murphy 1994), while the centres of diversity of *B. juncea* could be middle Asia, India and west of China. Northwest of China could be the centre of diversity of *B. campestris*.

Song *et al.,* (1990) proposed the geographical distribution and possible evolutionary path way for *B. rapa (B. campestris).* Despite their occurrence since long, *Brassica carinata* and *B. napus* are gaining more cultivable area in recent years and are attracting the attention of the farmers.

In the context of enhancement of breeding of *B. napus,* it would be necessary to develop hybrids easily between *B. campestris* and *B. oleracea*. Development of such hybrids, however was very difficult through the procedures of artificial pollination and embryo culture (Hosoda *et al*., 1963). Development of these hybrids, facilitates development of new cultivated plants and introduction of the genes of *B. campestris* and *B. oleracea* into *B. napus* plants, and also result in exchange of genes between *B. campestris* and *B. oleracea* (Hosoda *et al.,* 1963)

Hakansson (1956) made embryological studies on seed development in reciprocal crosses between *B. campestris* and *B. oleracea*; usually no viable seeds were obtained in the reciprocal crosses, because of the failure of their normal seed development. These phenomena were similar to the failure of seed development in the reciprocal crosses between diploid and autotetraploid *B. campestris* (Nishiyama and Inomata 1966).

Instead of embryo culture technique, ovule and ovary culture can be done *in vitro* for producing the new plants. When excised ovules and ovaries are cultured *in-vitro* on the medium, the normal development of embryo, endosperm, and seed coat are observed and viable seeds may be obtained. On the basis of this idea, excised ovules and ovaries were cultured and new triploid hybrids were obtained in ovary culture in reciprocal crosses between diploid and autotetraploid *B. campestris* (Inomata 1976).

In recent studies, the production of hybrids between *B. campestris* and *B. oleracea* was carried out, and many interspecific hybrids between them were obtained comparatively easily by embryo rescue and ovary culture techniques , and the progenies of hybrids were realized (Xiaohui Zhang *et al*., 2016).

A high frequency regeneration protocol for *Brassica juncea* cv. NRCDR-2 has been standardised using cotyledonary petiole explants. The best result was obtained on MS medium supplemented with 1.0 mg/l BAP and 0.10 mg/l NAA, where it gave an average no. of 4.56 shoots per explants and 86.66% explants regeneration. The highest percentage of root regeneration (100%) was obtained on MS medium supplemented with 0.30 mg/l of IAA. (ICAR-DRMR, 2019)

Twenty-one different wild species belonging to ten different genera of *Brassicaceae* are being maintained and genomic STMS markers suitable for their molecular characterization have been identified (ICAR-DRMR, 2019)

Haploidy for improvement of *Brassica* Spp.: Production of androgenic haploids and their subsequent use by doubling their chromosome numbers is one of the novelties of genetic improvement of *Brassica* Spp. (Gupta, 2016; Ferrie and Keller, 2002). The major advantage of the 'double haploid' technology eliminates several generations needed in order to attain uniformity and thus genetic stability in breeding lines. This also facilitates reduction of the population size that is needed to come up with desirable genetic segregants. Double haploid breeding is possible through microspore culture in the species of *Brassica* (Maluszynski *et al.*, 2003; Xu *et al.*, 2007). The double haploid technology is also useful in mutation breeding programmes (Zhu *et al.*, 1993), *in-vitro* screening of traits like drought resistance, salinity tolerance and in the context of forming mapping populations for linkage maps using molecular markers (Pratap *et al.*, 2007). Among several techniques used for double haploid development, microspore culture, anther culture and ovary or ovule culture are largely adopted (Gupta, 2016).

The advantage with the microspore culture is that it provides the best material for induction of mutations in the haploid background (Szarejko and Forster, 2007).

According to Zhou *et al.*, (2002) breeders now possess new tools of developing novel varieties of *Brassica* following the possibilities of producing haploids of *Brassica napus* from anther culture (Keller and Armstrong 1978) and availability of the microspore culture (Lichter, 1982).

After chromosome doubling, the surviving plants could be raised *in-vitro* and later screened for the specific traits such as drought resistance, cold tolerance, resistance to salinity etc.

In *Brassica napus*, the isolated microspores have been treated with EMS (Ethyl methane sulfonate) by Beversdorf and Knott, (1987) and other chemicals such as Sodium azide (Na N3) by Polsoni *et al.,* (1988). The microspores have also been irradiated with gamma rays (Beversdorf and Knott, 1987, McDonald *et al.,* 1991) and x-rays (McDonald *et al.*, 1991).

Nevertheless, the microspore embryogenesis is affected by factors such as genotype of the donor plant, pre-treatment, growth stage of the anther and microspore to be cultured, culture media and environment and diplodization process (Dunwell, 1996; Zhang *et al.*,2006; Pratap *et al.*, 2007)

Genetic Diversity: The genus *Brassica* is perhaps one of the richest in terms of the genetic diversity consisting of over 3,200 species representing wide array of genetic systems. In the task of genetic improvement of *Brassica* spp., the availability of appropriate genetic resources with desirable attributes becomes very important. Interestingly, a rich diversity for morphological traits of *Brassica* spp., is available in Indian sub-continent. Himalayan region is reported to be the primary centre of diversity for *B. campestris* (Hegde 1976). The germplasm of rapeseed from the north-eastern states of India represents dwarf and early maturing types, while on the other hand the collections made from north-western Himalayas are late maturing, tall and small seeded.

The collections obtained from the hills of Himachal Pradesh and Uttar Pradesh represent mostly leafy types with extended vegetative growth.

Rich diversity for *Toria* types was observed in Assam, Tripura, hills of Meghalaya, Bihar and Odisha and *tarai* region of Uttar Pradesh.

The Eastern U.P, Bihar and West Bengal regions have wide germplasm of yellow *sarson* with the characteristics of tall habit, deep green and dissected leaves and high degree of wax deposition at flowering and maturity. The wide diversity of *B. juncea* for plant type, seed size, pod length and maturity could be observed from north-western states of India (Rana and Singh 1992).

Genetic Resources of *Brassica* spp. and Breeding

Despite their close genetic inter-relationships, a wide variation exists in the mating systems of *Brassica* group of plant species. While the species viz., *Brassica juncea, B. napus, B. carinata* and *B. campestris* var. yellow *sarson* are self-pollinating systems, the others exhibit self-incompatibility and as such are cross pollinated.

Although *B. juncea* is considered self-fertile and self-pollinated, there exists outcrossing to the tune of 20 to 30% in this species. The main vector of pollination is bee that visits the flowers for nectar and thus carry the sticky pollen on their legs (Rakow and Woods,1987).

The floral biology of *Brassica* species consisting of self-incompatibility and thereby total out-crossing leading to allogamy in one group and autogamy coupled with significant proportion of out-crossing in the other with no crossability barriers among them assumes significance for breeding strategies to be adopted in the improvement of this group of oilseeds. The mating complex which is characteristic of *Brassicae* together with impressive genetic variability that exists therein,offers wide range of opportunities and scope for the genetic enhancement of this important group of oilseed crops.

Brassica group being the second most important oilseed crop complex in India, the genetic improvement of rapeseed-mustard can create a positive impact on the oilseed economy of the country.

The major breeding objectives for the improvement of rapeseed-mustard in India are as follows.

(i) Increasing the oil yield and enhancing the oil quality by (a) breeding for high seed oil content (b) breeding for higher seed yield and (c) breeding for low to zero erucic acid and high linoleic and oleic acid contents in the oil.

    The oil content varies from 37 to 49% in the current varietal complex. The highest obtainable seed yield is 12 to 20 quintals per hectare in the case of rapeseed and 20 to 25q/ha in mustard.

    Looking at the above ground dry matter production potential of the order of 9700 to 10000 kg/ha (Benga and Angadi, 2016) in *Brassica*, there is immense scope for scaling new heights in seed yield if dry matter production and partioning coefficient could be used as indirect selection ceiteria in addition to seed yield which is being currently employed in the breeding programs.

(ii) Incorporation of genetic resistance to biotic stresses consisting of wide range of pests and diseases which cause enormous yield losses as described earlier.

(iii) Rapeseed-mustard being grown on drylands over an area of 1.6million ha in India depending upon the natural rainfall, which often becomes erratic causing drought, it is necessary to develop crop varieties with good degree of drought resistance.

Genetic Resources: The genetic variability and diversity in *Brassica* group is very wide and impressive, providing access to several usable genetic attributes which have been used extensively in breeding for wide range of characters. The status of working germplasm of *Brassica* sp. in India is presented in Table 1., and the rapeseed-mustard germplasm registered for diverse attributes at the National Research Centre for Rapeseed-Mustard is presented in Table 2.

Table 1: Status of working germplasm at the Directorate of Rapeseed-Mustard Research (DRMR) and other centres in India (Chauhan et al., 2011)

Crop of Brassica group	DRMR	Centres
Indian Mustard (B. juncea)	3641	6079
Karan rai (B. carinata)	115	128
Toria (B. campestris var. toria	966	952
Yellow srason (B. campestris var. yellow sarson)	533	529
Brown sarson (B. campestris var. brown sarson)	231	247
Ghobi sarson (B. napus)	272	323

Crop of Brassica group	DRMR	Centres
Taramira (Eruca stiva)	381	197
Others	60	68
Total	6201	8521

Table 2: Rapeseed -Mustard Germplasm registered in India for diverse attributes

National Identity	Donor Identity	Special attributes
Brassica juncea		
IC 296501	Heera	Low glucosinolate content (16.96 μ moles/g of seed) and low erucic acid in oil (0.1%)
IC 296507	NUDHYJ 5	Low glucosinolate content (9.3 μ moles/g of seed) and low erucic acid (0.1%)
IC 296682	CMS Trachy (Restorer)	Restorer gene for CMS Trachy
IC 296683	CMS Trachy	Cytoplasmic male sterility
IC 296684	Swarna [TERI (0E)M21]	Zero erucic acid, yellow-seeded, early maturing
IC 296689	NDYR 8	Yellow seeded mustard with high oil content (45.7%)
IC 296701	CMS	Cytoplasmic male sterile carrying Cytoplasm of Brassica oxyrrhina
IC296702	CMS	Cytoplasmic male sterility
IC296703	CMS	Fertility restorer for CMS (Moricandia arvensis) B.juncea
IC 296705	CMS	Cytoplasmic male sterile with Raphanus/ ogu-B. juncea
IC296706	CMS	Chlorosis corrected, green, cytoplasmic male sterile mustard
IC 296741	RB1 (Raya Bawal 1)	Tetralocular siliqua
IC 296746	Diplotaxis erucoide	Cytoplasmic male sterile
IC 296829	TM (YSM2)	Yellow seeded Indian Mustard
IC 401570	RH 8814	Resistance to salinity reaction (up to ECE 10 dS/m
IC 405233	TERIGZ 05	High oleic and linoleic acid, yellow seed and double low material
IC 443623	BIOYSR	Yellow seeded, high oil, white rust resistant
IC 546946	PAB 9511	Tolerant to blight (Alternariabrassicae and A. brassicola
IC546947	PRQ2005-1	Low erucic acid with yellow seed
IC 553910	Divya (PPMS 1)	Dwarf, early maturity, distinct whorl around the axis
IC 555891	NRCDR 515	White rust resistant
IC 573439	PRB 2006-5	Bold seed
IC 0583386	BPR 541-4	Thermo tolerance at terminal stage and salinity tolerance at juvenile stage and high water use efficiency
IC 0583448	BPR 543-2	Thermo tolerance at juvenile stage and high water-use efficiency.
IC 296684	Swarna [TERI (0E) M21]	Zero erucic acid, yellow seeded, early maturing

National Identity	Donor Identity	Special attributes
Brassica napus		
IC 296685	Phaguni [TERI(0E)R 03]	Zero erucic acid and early maturing
IC 296688	Shyamali [TERI(0E)R09]	Zero erucic acid, high oleic acid (70.1%) and high oil content
IC 296704	CMS (B. oxyrrhina)	Cytoplasmic male sterile
IC 296707	CMS	Cytoplasmic male sterile carrying cytoplasm of Brassica oxyrrhina
IC 296731	Teri Gaurav [Teri (00) R986]	Zero erucic acid, low glucosinolate (15.3 µm/g oil free mean), early maturing (125 days)
IC 296732	Teri Garima [Teri(00)R985]	Zero erucic acid, low glucosinolate (12.2 µm/g oil free mean), high oleic acid (57%)
IC 296747	Erucastrum gallicum	Earlyflowering variant (31 – 33 days) carrying the cytoplasm of Erucastrum gallicum
IC 296827	NUDB 38	Early maturing double low gobhi sarson
IC 405232	TERI (00) R 9903	Double low with early maturity (130-135 days)
IC 0582897	NRC GS-1	Early flowering and dwarf plant
Brassica carinata		
IC 296748	Erucastrum gallicum	Useful variant of Ethiopian mustard (Brassica carinata) showing B. juncea type branching
IC 0582898	NRCKR 304	Early maturity, long main shoot and bold seed
Brassica rapa var. toria		
IC 296597	Amphidiploid	Amphidiploid Brassica
IC 552726	PHOP 2-2	High oleic acid (70.1%), low erucic acid (0.2%) and brown seed
IC 552727	PHOT8 2-11	Low linolenic acid (3.63%) and brown seeded

From the diverse and divergent germplasm available the following characterised genotypes have been identified for various attributes needed in breeding programmes (Singh and Singh 1992)

1. Genotypes resistant to diseases of *Brassica* sp.

(a) ***Tolerance to Alternaria blight*** : RH 781,KRV-tall,

CSR, 142-4PHR-1,*Jatai sarson*, CSR-43,CSR142, CSR-142-2, IC 546946

CSR-343,35,72,106,CSR-622,CSR-741, Gulivar,RC-781, Tower,YRT-3,

TMV-2, *Brassica alba, B.carinata, B.napus* .IC 546946 (PAB 9511)

(b) ***Resistance to white rust (Albugo candida) and Downy mildew (Peronospora parasitica)***

(i) ***B.juncea***: RC 781,PYSR 8, PYSR 13 and PR 10, Newton, Blaze, Metapolka and Purbiara. Exotic lines: EC 126743, EC 126745, EC 126746, Zem 1, Lethbridge 22, Stock and Domo, Blaze, Domo-4, Gulivar

(ii) ***B. campestris*var. yellow *sarson*:** PYS 6,

(iii) ***B.campestris* var.*toria*.**: PT 77 , *Bhabhari, B.carinta, B.alba, and B.napus* are highly resistant to white rust and downy mildew

(c) ***Resistance to Downy mildew:***Gulivar, Candle, Rc-781, Tower, YRT-3,45,72, Domo, Midas, Metapolka, Zem,EC-126743, *Brassica alba, B.napus*

(d) ***Resistance to Powdery mildew:****Brassica alba, B.carinata*

2. **Genotypes tolerant to mustard aphid (*Lipaphis erysimi* Kalt):** NRCM 120, NRCM 353, Rayad 9602 Vardan (Chaudary and Patel, 2016)

3. Genotypes resistant to abiotic stresses

(a) **Drought tolerance:** RH7361, RH819, RH785, *Pusa barani*, RH 781, *Durga-mani,* RH7513, *Vardan, B. tournifortii* cv local, <u>Eruca sativa</u>

(b) **Frost tolerance:** RH 781 and RW 32-2of *Brassica juncea,Bale, Candle, Span* and *FR 80* of *B. campestris*

(c) **Salt tolerance:** PR 52, PR 2030, RH 851, RH 781, DIRA343, NDR8604

4. Genotypes for quality characters

(a) **High Oil:** PR 36, EC 126743-1, EC 126743-2 and EC-126745 of *Brassica juncea,* TK 101, TK 8503, PT5, PT25, and PT 48 of *B.campestris*

(b) **Low erucic acid:** Zem-1and Zem-2 of *B. juncea,* and Tobia of *B.campestris*

(c) **Low glucosinolate:** RC 247, RC 270, RS 79 and RS 81 of *B .juncea*; Tobin, Bele, Torch and Span of *B.campestris*\

(d) **Yellow coloured seed coat:** TM2, TM4, TM17, TM21, YRT1, RIK1-1, Domo and UU22.

5. Genotypes for other desirable characters

(a) **Earliness of maturity:** IB 1755, IB 1886, Seeta, BR 46.

(b) **Non-shattering habit:** RH 30, RIK 8, P 26-31, PR10 Pusa Bold

(c) **Bold seeded character:** RH30, Pusa Bold, R74-3, PUsa 1002, KR 5610. NDR 8501

(d) **Suitability for Inter-cropping:** PP 43, RH30, Vardan, RH 785, RH 781, Laha 101, Sanjucta, Asesh

6. The ICAR-Directorate of Rapeseed-Mustard Research have registered a number of novel genetic stocks of rapeseed-mustard for diverse attributes viz., CMS, restorer, low erucic acid, low erucic acid & low glucosinolates,

high oil content, high oleic acid and low linolenic acid, dwarf, earliness, long main shoot, bold seed, yellow seed, tetra-locular siliquae, white rust resistance, tolerance to high temperature and salinity during juvenile stage, high temperature tolerance during terminal stage and high water use efficiency including eleven from DRMR with National Bureau of Plant Genetic Resources, New Delhi till 2013. The Directorate also maintains 2,452 accessions of genetic stocks as indicated in the parentheses. (Indian mustard-1868, Toria-245, Yellow sarson-150, Gobhi sarson-105, Taramira-48, Karan rai-17 and Others-19) (Table1)

Some Promising Varieties of Mustard (*B.juncea*) and *B.carinata* developed at IARI NewDelhi as exhibited by ZTM & BPD Unit, ICAR-IARI (http:// ztmbpd.iari.res.in/technologies/varietieshybrids/oil-seeds/mustard/)

(1) Pusa Double Zero Mustard 31 (PDZ-1): This Variety has been released for timely sown irrigated conditions of National Capital Region of Delhi including Delhi and adjoining areas in the states of Haryana, Rajasthan and UP. It is a yellow seeded variety with 40.56% oil content. Improved oil and seed meal quality (Canola quality) makes this variety beneficial for farmers, traders and consumers. (Fig. 3)

Fig. 3: Pusa double zero mastard 31 (PDZ-1)

Average seed yield of this variety is 2379 kg/ha. It is the first double zero (Erucic acid <2% and glucosinolates <30ppm) Indian mustard variety in the country which matures in about 144 days.

Pusa Mustard 24 (LET-18) is suitable for cultivation in Rajasthan, Punjab, Haryana, Delhi, plains of J&K and Western parts of U.P. It is a low in erucic acid (<2.0%) variety of Indian mustard, suitable for timely sown irrigated conditions. (Fig.4)

Fig. 4: Pusa Mustard 24 (LET-18)

It is a small seeded variety (4.0 g/ 1000 seeds) with an average seed yield of 20.25 q/ha and 36.55% oil content. It matures in about 140 days.

Pusa Mustard 21 (LES-1-27) is suitable for cultivation in Punjab, Haryana, Delhi, Rajasthan, UP, Uttrakhand, plains of J&K, MP, Chhatishgarh. It is a low erucic acid (single zero) variety of Indian mustard with a wider adaptability. (Fig.5)

Average seed yield in north western plain zone is 21.1 q/ha and in central zone is 18.6 q/ha, Seeds of this variety possesses 36% oil. It matures in about 142 days in NWPZ and in 133 days in CZ.

Fig. 5: Pusa Mustard 21 (LES-1-27)

Pusa Karishma (LES-39) It is first single zero (< 2 % erucic acid) variety of Indian mustard and have tolerance to white rust.It is highly suitable for timely sown irrigated conditions. (Fig.6)First single zero (< 2 % erucic acid) variety of Indian mustard. Average seed yield is 22.0 q/ha with 38.0 % oil content.

Fig. 6: Pusa Karishma (LES-39)

Pusa Swarnim (IGC-01)

This is a variety of *Brassica carinata* (Karan rai/Ethiopian mustard) recommended for cultivation in Rajasthan, Punjab, Haryana, Delhi, Himachal Pradesh, Jammu & Kashmir and parts of UP. The variety is known for high degree of drought tolerance and immunity against white rust disease. Severity of *Alternaria* blight disease is reported to be low in this variety. (Fig.7)Average seed yield under irrigated conditions is about 16-17 q/ha. Whereas, under rainfed conditions it is about 14-15 q/ha. Its seeds are yellow in colour with 40-43% oil content. It matures in about 165 days.

Fig. 7: Pusa Swarnim (IGC-01)

Pusa Mustard 30 (LES-43) is recommended for cultivation in Uttar Pradesh, Uttrakhand, Madhya Pradesh and eastern Rajasthan. This variety has low erucic acid content and as such termed single zero variety of Indian mustard. (Fig.8)

Average seed yield of this variety is 18.24 q/ha with 37.7% oil content. It holds bold seeds and matures in about 137 days. High productivity per unit area under the specified agriecological situation in the designated zone with minimal production cost.

Fig. 8: Pusa Mustard 30(LES-43)

Table 3: Varieties of Rapeseed and Mustard recommended for specific situations (Chauhan et al., 2011)

Specific trait / situation	Varieties
Brassica juncea (Indian Mustard)	
Earliness	Kanti, Narendera Ageti Rai 4, Pusa Agrani, Pusa Mahak, Pusa Vijay, Pusa mustard 25 (NPJ 112), Pusa mustard, 27 (EJ 17)
High oil content	**Narendra, Swarna Rai 8, NRCDR 02, Rohini**
Tolerance to higher temperatures	Kanti, Pusa Agrani, RGN 13, Urvashi, NRCDR 02, Pusa mustard 25 (NPJ 112), Pusa mustard 27 (EJ 17)
Suitability for Inter-cropping	RH 30, RH 781, Vardan
Suitability for late sown conditions	**Ashirwad, RLM 619, Swarnjyoti,** Vardan, Navgold (YRN 6), RGN 145, NRCHB 101, CS 56 (CS 234-2), Pusa mustard 26 (NPJ 113
Recommended for non-traditional zones	Pusa Agrani, Pusa Jaikisan, Rajat, Gujarat Mustard 2, TPM 1, Shatabdi (ACN 9)
Recommended for rain-fed situations	Aravali, Geeta, GM 1, PBR 97, Pusa Bahar, Pusa Bold, RH 781, RH 819, RGN 48, Shivani, TM 2, TM 4, Vaibhav, RB 50
Tolerance to saline conditions	S 52, CS 54, Narendra Rai (NDR 8501
Tolerance to Frost	**RGN 13, RH 819, Swaranjyoti, RH 781, RGN 48**

Specific trait / situation	Varieties
Low Erucic Acid types	Pusa Karishma, Pusa Mustard 21, Pusa Mustard 22, Pusa Mustard 24, ELM 079 (RLC 1)
Brassica napus (Ghobi sarson)	
Low erucic acid and glucosinolate	GSC 5, Hyola 401, NUDB 26-11, OCN 3 (GSC 6) TERI, Uttam Jawahar

Breeding for Oil Quality through Genetic rectification of anti-nutritional principles in *Brassica* group

In *Brassica* group of oilseed crops one specific character which forms a distinguishing feature from other oil seed crops is the presence of erucic acid and glucosinolates, which are considered anti-nutritional factors.

Erucic acid is a monounsaturated omega-9 fatty acid (CH3(CH2) 7CH=CH (CH2)11COOH) present particularly in rapeseed and mustard. There is a variation for erucic acid content among the members of the above group. The high erucic acid types contain as much as 25 to 50% of erucic acid.

Investigations carried out on animals revealed that ingesting oils containing erucic acid over time can lead to a heart condition called myocardial lipidosis, which is temporary and reversible (https://stylesatlife.com/articles/mustard-oil-side-effects/). Other potential effects observed in animals are the changes in the weight of the liver, kidney and skeletal muscle at higher doses.

The canola varieties developed with low erucic acid in *B. napus* (Stefanson *et al.*, 1961), *B. campestris* (Downey, 1964) and *B. juncea* (Kirk and Oram 1981) paved the way for human consumption of the rapeseed and mustard oils in the western countries. In the case of low erucic acid varieties, the carbon chain elongation system, which adds two carbon fragments and end of oleic and eicosenoic acids to form erucic acid is genetically blocked, resulting in new a vegetable oil with high oleic acid.

Some genotypes identified for low erucic acid content in the All India Coordinated Research Project (AICORPO) are, Zem-1 & Zem-2 of *B. juncea* and Tobin of *B. campestris* (Singh and Singh 1992). However, the rapeseed and mustard oil has been used commonly in India as cooking oil for human consumption without any deleterious effects.

Nevertheless, oil with high erucic acid is needed by the industry. In this case, erucic acid fraction of the oil is converted into amide, which is sprayed on plastic products and used in extruding equipment as a slip agent.

Glucosinolates occur as secondary metabolites of almost all plants of *Brassica* species. These are organic compounds that contain sulphur and nitrogen, which are derived from glucose and an amino acid and are water-soluble anions belonging to the glucosides. The tissues of *Brassica* species contain varying number of glucosinolates, some times as many as 90 (Fenwick *et al* 1983). They tend to give rise to thiocynates, isothiocynates or nitriles upon hydrolysis in the presence of the enzyme myorsinase (Ratzka *et al.,* 2002)

Glucosinolates are reported to have toxic effects mainly as goitrogens and anti-thyroid agents in both humans and animals at high doses with varying level of tolerance (Felker *et al.,* 2016) It was also found that the glucosinolates consumed through *Brassica* species are likely to cause colorectal neoplasms in human beings (Tse and Eslick 2014). They reduce the palatability of the feed in non-ruminant animals (Mawson *et al.,* 1993).

Glucosinolates and their products have a negative effect on many insects, resulting from a combination of deterrence and toxicity (Ratzka *et al* 2002 and Bridges *et al* 2002). The insects such as sawflies and aphids sequester glucosinolates (Ratzka *et al.,* 2002).

The task of developing *Brassica* varieties with low glucosinolates is rendered arduous since the specific genes for such attributes are not specifically come across in the current germplasm collections (Kumar 1994)

Nevertheless, genetically stable low glucosinolate characteristic of selection 1058 has been developed by Love *et al.,* (1990). The progeny of plant 1058 had plant morphology and seed coat reticulation of *B. juncea*, but was poor in fertility @< 5 seeds per pod (Love *et al.,* 1990).

The genotypes identified for low levels of erucic acid and glucosinolates including doule zero types are presented in Table2.

Induced Mutations in *Brassica* Spp

The *Brassica* group is characterised by enormous degree of variability for wide range of characters including productivity, quality and resistance to biotic stresses, which apparently precludes any use of induced mutations, which are resorted to, in the cases of paucity of genetic variability for certain agronomic attributes.

Nevertheless, in the cases of certain intractable negative linkages, which are not resolvable through recombination breeding, such as oil quality and content with seed yield and resistance to mustard aphid, it becomes necessary to resort to induced mutations to generate new variability which can be recombined at convenience. Secondly, well planned and executed mutation breeding results in cutting short of prolonged time that involves a number of generations to secure appropriate recombination. Certain wild or genetically distant species posing crossability barriers with the cultivated crop species could be genetically ameliorated through induced mutations to render them amenable and useful in recombination breeding with the cultivated crop species.

Mutants for oil content in Rapeseed-Mustard: Increasing the oil content might be one approach towards increasing the oil yield. In other words, if a high yielding variety which has moderate oil content of less than 37% needs to be upgraded with higher oil content, the recombination breeding might take considerable time and effort considering the fact of a negative correlation between yield and oil content (Singh and Chowdhury, 1983). Under such conditions, it would be appropriate to resort to mutation breeding, since the objective is to retain the high yielding genetic background of the variety and add mutant gene(s) that might augment the oil content.

Khatri *et al.,* (2005) isolated mutants with 3% higher oil content of the mustard variety S-9, following treatment with gamma rays and EMS. Verma and Rai, (1980) also recovered induced mutants with 1 to 5% higher oil content of the parent variety '*Pant Rai-5*' following irradiation of its seeds with gamma rays.

Mutations for Pest Resistance: Among the insect pests that inflict rapeseed-mustard complex, aphid is the most devastating biotic stress. Success of time consuming and extended programme of plant breeding for aphid resistance is limited to a considerable extent due to the paucity of usable stable genetic resources for this trait (Choudhary and Jambhulkar, 2016). Genotypes with waxy leaves were found to be susceptible to the incidence of aphids. Srinivasachar and Malik (1972) used the mutagens viz., Gamma rays, ethylene imine and hydrazine to treat the seeds of *B. juncea* following which, isolated non-shattering mutants that also exhibited resistance to aphids.

Mutations for Seed Yield: Seed yield being a complex character, selection for seed yield should aim at enhancement of the yield component characters and /or plant physiological attributes that influence yield indirectly. Mutations that brought about positive changes in attributes such as plant height, branching, number of silique per plant, number of seeds per silique, seed weight and oil content have been isolated (Rehman *et al.,* 1987; Chauhan and Kumar,1986; Rehman, 1996; Shah *et al.*,1999; Robbelen, 1990).

It may be mentioned that the most important yield component that has direct bearing on yield is branch number consisting of primary and secondary branches. Additionally, silique per plant, silique on the main fruiting branch and seed per siliqua too are important. Mutants with higher number of branches and silique per plant have also been isolated by Chauhan and Kumar, (1986); Shah *et al.*, (1990) and Naz and Islam, (1979) as reported by Choudhary and Jambhulkar (2016).

Out of seventeen mutants of cv. S-9 developed for the attributes earliness, dwarf stature and higher seed yield through the use of gamma rays and EMS, three were found to be significantly superior in terms of seed yield (Khatri *et al.,* 2005).

Mutants exhibiting higher yield levels were developed from *Brassica napus* by Hao-jie *et al*., (2005), Raza *et al*., (2009), Malek *et al.,* (2012): Udin *et al*., (2012) and Ali and Shah (2013)

It should be noted that the number of primary and secondary branches, which are correlated with seed yield are likely to suffer a serious decline in plants with dwarf stature due to which there could be an adverse effect on seed yield. Hence it is necessary to note that mutants with dwarf stature may not result in higher seed yield despite some unusual exceptions with tremendously higher branch number in a dwarf plant type. In order to breed for higher seed yield, the breeder should look for higher branch number, higher number of silique instead of dwarf plant habit.

Allogamy in *Brassica* spp.: Allogamy or cross-pollination and fertilization leading to the expression and recovery of diverse and variable progenies is an important factor to be considered in heterosis breeding. As explained earlier, this phenomenon offers immense scope and feasibility for the development of productive hybrid progenies.

In *Brassica* Group *B. juncea, B. napus* and *B. carinata* and *B. campestris* var. yellow *sarson* are autogamous and self-fertile with certain level of vector mediated cross pollination (8 to 20%) (Labana and Banga1984, Dhillon and Labana 1988 Abraham1994). On the other hand, *Brassica campestris* with the exception of var. yellow *sarson* are cross pollinated and self-incompatible.

This kind of diversity in the mating systems of the same botanical group could be of an added advantage in heterosis breeding since this offers adequate heterotic parental combinations.

Self-incompatibility in *Brassicaceae*

Self-incompatibility in *Brassica* species is an interesting mechanism that allows stigmas to recognize and discriminate against "self" pollen, thus preventing self-fertilization and inbreeding. In the *Brassiceae*, the self-incompatibility phenotype of both stigma and pollen is determined sporophytically by the diploid *S*-locus genotype of the parent plant and dominance relationships between *S*-locus variants are observed in both stigma and pollen (Thompson and Taylor 1966).

Pollen inhibition at the stigma surface occurs when the pollen and stigma express the same *S*-locus variant. Self-incompatibility in this family is controlled by a single *S* locus containing two multi-allelic genes that encode the stigma-expressed *S*-locus receptor kinase and its pollen coat-localized lignan (*an ion or a molecule that binds to a central metal atom to form a coordination complex.*) , the *S*-locus cysteine-rich protein. Physical interaction between receptor and lignin encoded in the same *S* locus activates the receptor and triggers a signalling cascade that results in inhibition of self-pollen(Kitashiba and Nasrallah 2014).

Genetic studies carried out in the 1950s revealed that this trait is controlled by variants of a single locus, the *S* locus, (Bateman 1955). In the 1960s, immunochemical analysis of stigmas led to the identification of proteins encoded by genes at the *S* locus (Nasrallah and Wallace 1967). In the decade of 1980s, molecular genetic analyses identified the *S*-locus genes that are responsible for the recognition of "self" pollen by the stigma. Many alleles of these genes have now been sequenced, and the sequence information has contributed to biological, physiological, biochemical, genomic, and genetic studies of self-incompatibility as well as to practical breeding utilizing this trait.

Sporophytic incompatibility has been used in breeding of *Brassica* based cruciferous vegetables like cabbage and cauliflower (Kucera *et al.,* 2006) Single-cross hybrids are more uniform and easier to produce. The top cross is commonly used. A single self-incompatible parent is used as female, and is open pollinated in a crossing block by a desirable cultivar as the pollen source.

In the case of *Brassica napus*, doubled haploid and self-incompatible segregates with improved seed quality were derived following two crossing cycles using different donors for oil quality for "0" erucic acid and "0" glucosinolates. The original genetic resource of recessive sporophytic SI was characterised by a high glucosinolate and erucic acid content. SI segregates of satisfactory seed quality parameters were obtained from the second cycle of crossing (Koprna *et al.,* 2005).

Heterosis breeding: Till recent years, rapeseed-mustard breeding has been revolving around development of standard varieties for various attributes. The differences in productivity among different varieties have not been very spectacular. The best yield levels were in the rage of 2.0 to 2.5 tons/ha. It could be observed that yield levels have not increased significantly due to a narrow operative genetic base in the case of *Brassica juncea.*

However, the availability of exhaustive reservoir of gene pool, existence of different mating systems and occurrence of significant levels of heterosis in *Brassica* group warrants exploitation of heterotic background to achieve a major yield leap.

The basic prerequisites for heterosis breeding are (1) expression of adequate exploitable levels of heterosis for yield components and yield over the best varietal check and (2) availability or development of supporting systems of genetic / cytoplasmic male sterility and fertility restoration.

Singh and Mehta (1954) in India initially reported heterosis in brown *sarson.*

Later investigations, however, reported significant level of heterosis of the order of 13 to 91% in *B. juncea* (Banga and Labana 1984, Kumar *et al.,* 1990, Rai 1995, Verma *et al.* 1998) 25 to 110% in *B. campestris* (Yadav *et al.*, 1998) and 10 to

72% in *B. napus* (Rai 1995, Dhillon *et al.*, 1996, Thakur and Sagwal 1997). The realisable heterotic levels were much higher when genetically divergent parents with good combing ability were used.

The extent of success of commercial hybrid production would depend on (1) the yield advantage the hybrid offers over the best variety, (2) the cost/benefit ratio of using hybrid seeds as compared to seeds of straight commercial variety, and (3) the efficiency of hybrid seed production,

Combining ability and gene action: Analysis of combining ability in relation to heterosis in *Brassica* species have been carried out by several workers. (Amrithadevarathinam *et al.* 1976, Sachan and Singh 1986, Sachan and Singh 1988, Wani and Srivastava 1989, Diwakar and Singh 1993, Thakur and Bhateria 1993, Kumar *et al.* 1997).

The studies indicated significant GCA and SCA effects for yield and component characters suggesting that both additive and non-additive gene action were important in the inheritance of these traits. The yield components of *Brassica* spp., usually showed a preponderance of GCA variances compared to SCA variances (Labana *et al.* 1978, Dhillon *et al.* 1990)

Gametocide induced male-sterility for hybrid production in *Brassica*: The chemical male gametocides such as MG No.01 and MG4 were used to induce male sterility to obtain a suitable female parent for hybrid development (Singh and Singh 1992). This kind of research was carried out in China with perceptible success. A hybrid viz., (*Shuza 2*) developed through the use of male sterile line induced by above chemical gametocides has been selected for cultivation in Sichuan province of China (Singh and Singh 1992). The hybrid exhibited superiority of the tune of 24% over the check variety as stated by Singh and Singh (1992).

In India, the work was carried out by Banga and Labana (1983b) using ethrel induced male sterility to produce *Brassica juncea* hybrids. Male sterility was achieved by spraying 0·25% (v/v) ethrel twice before the emergence of the first flowering shoots. The proportion of hybrid seed set was maximal in the 2:4 combinations in a circular design. Outcrossing varied from 19 to 79%. Heterosis was evident in yield trials, but F_1 hybrids showed carry-over effects due to ethrel treatment. It is stated by Banga and Labana (1983b) that ethrel induced male sterility has only a limited role in screening parental combinations for their yield assessment.

Spontaneous hybrids of *Brassica:* Through the exploitation of natural self-incompatibility system in the population of *Brassica campestris*, it should be possible to plan for natural cross pollination among parental types of good combining ability planted in alternate rows leading to recovery of superior hybrid strains. The resulting hybrids have recorded impressive levels of yield superiority of the order of 30% as compared to the parents used (Singh and Singh 1992).

One step ahead in this direction would be development of homozygous self-incompatible lines of *B. campestris* exploiting the twin properties of *(i) slow reaction of self-fertilization and (ii) low seed set under selfing*. Such homozygous lines of *B. campestris,* which are of a low frequency, are used as female parents in a crossing block of the design mentioned above. In this context of development of spontaneous hybrids, some identifiable markers such as leaf colour (green *vs* yellow) and certain seedling markers may be employed to identify hybrids and discarding non-hybrids as has been carried out by Huang (1980) in China.

Alternatively, to develop homozygous incompatible population of *B. campestris*, one could resort to spraying 5% aqueous salt solution at flowering to supress incompatibility and promote self-pollination as has been accomplished by Fu (1981) and Siring (1986).

Natural intraspecific hybrids between *Brassica rapa* var. *chinensis* and *B. rapa* var. *oleifera* are documented and analysed from a single wild population near Ashburton, Canterbury, New Zealand. Principal Component Analysis of 13 leaf morphological characters identified putative F_1 hybrids that are intermediate between the parents. A second group of plants that are highly variable probably represent F_2 and/or backcross hybrids. The putative parents and the hybrid plants all had high percentages of pollen and seed fertility. Flow cytometry data are consistent with the hybrid plants being derived from intraspecific *B. rapa* crosses. This study by Heenan and Dawson (2005) has shown that closely related intraspecific taxa in *Brassica* can form natural hybrid populations thereby not only augmenting the genetic variability; but also facilitating recovery of productive hybrids.

Frequencies of spontaneous hybridization between oilseed rape (*Brassica napus*) and weedy *B. campestris (B. rapa*) were measured in agricultural fields by Jorgensen and Andersen (1994). The interspecific hybridization was from 9% to 93% depending on the experimental design. Hybrid seed germinated in the field and adult hybrids were observed in the next growing season. The pollen fertility of the hybrids was reduced to 35% on an average, and they produced 102 seeds/ hybrid in spontaneous backcrosses with the parental species. The inheritance of 17 oilseed rape specific markers was analysed in selected backcross plants. The production of spontaneous hybrids was confirmed in field experiments with mixtures of transgenic, herbicide tolerant oilseed rape and *B. campestris*. The interspecific hybrids were herbicide tolerant and PCR (Polymerase Chain Reaction) confirmed the presence of the transgene construct. The spontaneous backcrossing to *B. campestris* was likewise confirmed in field experiments with mixed populations of *B. campestris* and interspecific hybrids with the transgene construct.

In the interspecific combination *B. juncea* x *B. napus* a frequency of 3% spontaneous hybridization was observed in the field. According to the authors, transgenes could be dispersed from oilseed rape to related *Brassica* species.

***Brassica* hybrids produced using Self-Incompatibility System:** As mentioned earlier good progress has been achieved in producing hybrids of col crops viz., cabbage and cauliflower connected to the *Brassica* group exploiting the self-incompatibility system. The work pertaining to exploitation of self-incompatibility system for production of hybrids has progressed well leading to the production of some hybrids; but such investigations are not extensive in India in oilseed *Brassicae.*

Perhaps, the first successful report of production of such hybrid was by Thompson (1964). He produced a triple cross hybrid *Kale*; which is a single cross between two three way crosses viz., (AxB)xCx(DxE). In this context the sporophytic incompatibility has been successfully exploited to ensure the recovery of 100% hybrid seed. Such an exercise wouldn't be possible through gametophytic incompatibility.

Further, Liu (1975) succeeded in coming up with the hybrids of self-incompatible lines viz., 211 and 271 of *Brassica napus.* Fu (1981a) researching on the maintainer and fertility restorers for self-incompatible lines, succeeded in producing hybrids using the said three lines.

In order to overcome the incompatibility whenever it is not needed, Fu (1981) and later Siring (1986) suggested use of 5% salt water to be sprayed at flowering time, to supress self-incompatibility in *Brassica campestris* (also known as *B.rapa*)

Male Sterile Systems

The female parent used for the development of hybrids should be male sterile i.e., the male component of the flowers viz., stamens should be sterile without production of any fertile pollen to avoid self-pollination and facilitate cross pollination with the pollen from the chosen male parental strain.

This kind of male sterility of the female parent could be either genetic i.e., controlled by nuclear genes or cytoplasmic i.e., caused by the cytoplasmic factors particularly of mitochondria.

(a) **Genetic male sterility (GMS) in *Brassicae* for hybrid development:** The researches carried out in the decades of n1980s and 1990s have resulted in a number of sources of genetic male sterile strains as indicated below in each category along with the authors contributing to the identification / development of genetic male-sterile lines.

(1) *B. juncea*: Badwal and Labana (1983), Banga and Labana (1983a), Banga and Labana (1985),

(2) *B. campestris* var. brown *sarson*: Das and Pandey (1961), Chowdhury and Das (1966, 1967a, 1967b,1968), Katiyar (1983), Liu (1975)

(3) *B. campestris* var. yellow *sarson*: Chowdhury and Das (1966), Singh *et al.* (1987), Ram Bhajan *et al.* (1993), Gupta (2016).

(4) *B.campestris* var. toria: Singh *et al.* (1987), Anonymous (1997).

The vast genetic variability available in *Brassica* group facilitated the identification of the above indicated lines, which are of spontaneous occurrence. The pattern of inheritance of almost all of the identified genetic male sterile lines is of monogenic nature. A dominant genetic male sterility trait obtained through transformation in rapeseed (*Brassica napus*) was studied in the progenies of 11 transformed plants. The gene conferring the male sterility consists of a ribonuclease gene under the control of a tapetum-specific promoter (Dennis *et al* 1993) Li *et al* (1988) reported the discovery of Genetic male sterility (GMS) in the rape (*Brassica napus* L.) cultivar, No.23, which is conditioned by genes at two loci. The genes were designated as Ms_1 and Ms_2. Plants homozygous or heterozygous for the dominant Ms_1 allele (Ms_1Ms_1 or Ms_1ms_1) and homozygous for the recessive ms_2 allele (ms_2ms_2) were sterile while plants homozygous or heterozygous for the dominant Ms_2 allele were fertile regardless of the alleles present at the Ms_1 locus. The double recessive was also fertile. Pollination of the homozygous male sterile, $Ms_1Ms_1ms_2ms_2$ with the pollen from the double recessive fertile, $ms_1ms_1ms_2$ ms_2 will produce a generation of sterile plants. These sterile plants ($Ms_1ms_1ms_2ms_2$) when pollinated with pollen from genotypes homozygous for the Ms_2 allele (Ms_2Ms_2), will produce a generation of fertile plants. The various genotypes which can be isolated from this GMS system are being used to develop hybrid rape at the Research Institute of the Shanghai Academy of Sciences. This indicates dominance epistasis at the Ms_2 locus over the Ms_1 locus.

Liu (1975) after discovering genetic male sterile line of *Brassica rapa 87A* in 1965, he used it in 1972 for the development of another genetic male sterile line Yi-3A of *Brassica napus.* With this discovery it was possible to come up with a double hybrid '*11-hybrid-03*', with significant degree of earliness of maturity and definite yield superiority of more than 11 % over the best available check variety. With further discovery of another genetic male sterile line viz., 117A involving two pairs of recessive alleles, the low erucic acid hybrid of *Brassica napus* viz., "*Shuza-1*" was developed in Sichuan Province of China in the calendar year 1998-90. This hybrid exhibited very significant superiority with seed yield of the order of more than 20% over the best check variety.

Despite discovery of genetic male sterility systems in *B. campestris* var. brown *sarson* and var. yellow *sarson* and more importantly in *B. juncea* as cited above in India, there have not been any reports of hybrids based on genetic male sterility.

According to Li *et al* (1988) it would be feasible to employ GMS system to produce first generation hybrid rape, if the required parental genotypes possessing good combining ability could be isolated. In China around 40,000 hectares of area is covered by *B. napus* GMSbased hybrids (Fu, 1981).

It could be perceived that the use of the genetic male-sterile lines for the production of hybrids is beset with certain problems with regard to their maintenance, since rouging of the fertile segregates every time before anthesis involves significant extra expenditure in this activity.

In the case of dominant allele involved in genetic male-sterility, the problems associated with segregation occurring in the resulting hybrid cannot be ruled out. Also absence of linked seedling markers and/ or pleiotropic effect of the gene for male sterility also render the task of maintenance of the genetic male sterile lines difficult.

(b) Cytoplasmic male sterility (CMS): CMS is a maternally inherited trait encoded in the mitochondrial genome, and the male sterile phenotype arises as a result of interaction of a mitochondrial CMS gene and a nuclear fertility restoring (*Rf*) gene (Yamagishi and Bhat 2014). The cytoplasmic male sterility has its own specific advantages as compared to genetic male-sterility and also self-incompatibility for hybrid development, provided an effective fertility restoration system is in place.

As back as five decades ago from now, Ogura (1968)) reported cytoplasmic male sterility in in *Raphanus sativus.* A number of authors too reported cytoplasmic male sterility as follows. (i)Pearson (1972) in *Brassica nigra*, (ii) Aanad and Rawat (1978) in *B. juncea*, (iii) Hinata and Konno (1976) in *Diplotaxis muralis*, (iv) Shiga and Baba (1973) in *Brassica napus* (v) Fu (1981) in *B. napus*, (vi)Prakash and Chopra (1990) in *B. oxoyrrhina* and (v) Prakash *et al.*, (1998) in *Moricandia arvensis.*

Before proceeding further in the subject it is necessary to know about different kinds and sources of cytoplasmic male sterility. The oilseed rape plant, *B. napus*, possesses two endogenous male sterile cytoplasms, ***pol*** and ***nap*** (another example of a natural mitochondrial variant within B. *napus*)

***Ogura (0gu)* type** of cytoplasmic male sterility was first discovered in Japanese wild radish (Ogura 1968) and other male-sterile *Brassicae* (Ogura bearing cytoplasm) derived from interspecific crosses. Ogura CMS was introduced into *B. napus* by intergeneric hybridization and repeated back-crossing (Bannerot *et al.* 1974; Heyn 1976). The resultant alloplasmic lines of *B. napus* exhibited male sterility; but possessed chlorotic leaves, yellowing at low temperatures of less than15°C. (Pelletier *et al.* 1983). This kind of chlorophyll deficiency is stated to result from functional incompatibility between the *B. napus* nucleus and *R. sativus* chloroplasts.

While ***ogu***cytoplasm (Ogura 1968) induces sterility in almost all cultivars, ***pol***kind of cytoplasm (Fu 1981) is found to induce sterility in some varieties. The *pol* CMS system can be used in both a three-line system (*pol* CMS sterile line, maintainer line and restorer line) and a two-line system (pol temperature-sensitive CMS sterile line and restorer line) in commercial hybrid seed production (Fu *et al.,* 1990).

The ***nap***cytoplasm (Shiga and Baba 1973) is not very versatile in inducing male sterility but is confined to only very few cultivars that respond to it in terms of induced male sterility.

Initially the interest of the breeders was confined to *pol* system towards the production of hybrids in the background of *Brassica napus*

Nevertheless, the main disadvantage of pol CMS is its unstable male sterility. The male sterile lines could become partially fertile at relative lower or higher temperature situations (Yang and Fu, 1991). But it is possible to select out a relatively stable *pol* CMS line.

The *pol* CMS system has been employed extensively in China to develop hybrids involving *Brassica napus.* One classical example of successful use of *pol* CMS system is the production of the hybrid '*hyola*' which has been tested in Canada (Witt *et al.,* 1991). Despite the extensive use of pol CMS system in *B. napus* background, it has been observed this system is of little use in the case of *Brassica juncea* since the pol cytoplasm has originated from *B. campestris,* which has also contributed to the cytoplasm of *B. juncea* (Wan *et al.,* 2008; Yamagishi and Bhat, 2014).

The *ogu* CMS is the most stable in all the *Brassica* systems since it results in complete male sterility and the corresponding restorer genes have been introduced in rapeseed (Trail *et al*., 1989). Efforts of transferring *ogu* CMS to *B. oleracea* and *B. napus* (Jarl *et al.,* 1988, Bannerot *et al.,* 1974) resulted in chlorophyll deficient plants with restorer genes causing female sterility (Sharma and Anand 1994). The interaction between *Raphanus* chloroplasts and *Brassica* nucleus has resulted in this case chlorophyll deficiency, which is more expressive under the conditions of low temperature (Jarl *et al*., 1988) These problems arising out of the use of *ogu* cytoplasm were rectified by adopting protoplast fusion technology (Pelletier, 1988). Subsequently, two lines of *Brassica napus* which are devoid of the above problems viz., FU 27 and F58 have been secured under the enhanced system viz., *Ogu*-INRA (French Patent No. 90-1670). However, another problem for *ogu* CMS is that its fertility restoring gene genetically linked tightly to the genes with high content of glucosinonates (Renard *et al*.,1997).

A mention must be made about another CMS system viz., *Oxyrrhina* arising out of the hybridization involving *Brassica oxyrrhina* and *B. campestirs* (Prakash and Chopra, 1990).The male sterile cytoplasm of *oxyrrhina* system could be transferred to *B. juncea* through regular hybridization process. The intricacies arising out of this CMS system such as 'chlorosis' were rectified at the Indian Agricultural Research Institute (Sharma and Anand 1994). In India the success in terms of hybrid development could be achieved using "*ogu-INRA*" system.

Moricandia system of cytoplasmic male sterility recovered from ***Moricandia arvensis*** (known as *purple mistress* or *violet cabbage*) also a member of *Brassicaceae* (Praksah *et al.,*1995 and 1998) has been successfully used in developing productive mustard hybrids.

Further advances in terms of the exploitation of recombinant DNA technology to come up with novel male sterility systems have been pursued by Mariani *et al*., (1990), employing RNA *(Barnase)* to abort normal pollen functioning leading to male sterility. In this context, a gene construct made of *tapetum specific promoter* and bacterial RNA were used. It was found that the construct retains its specificity in the transformed tobacco-rape plant. Anthers of such one are **not** found to produce functional pollen. Despite the success of novel recombinant DNA technology in producing above indicated new CMS system, some more problems such as the dominant nature of the male sterility attribute and difficulties in fertility restoration system are yet to be resolved. The dominant attribute, however, could be utilized provided it is linked to the gene conferring herbicide resistance *Bar* (Sharma and Anand 1994). *Barstar* technology (protein against *Barnase*) is considered for the development of dependable male fertility restoration system (Mariani *et al.,* 1992)

A new cytoplasmic male sterility (CMS) in *Brassica napus* (*inap* CMS) was selected from the somatic hybrid with *Isatis indigotica* (Chinese Woad, a flowering plant of the family *Cruciferae* related to *Brassicae*) by recurrent backcrossing (Kang *et al.,* 2017). The male sterility was caused by the conversion of tetradynamous stamens into carpelloid structures with stigmatoid tissues at their tips and ovule-like tissues in the margins, and the two shorter stamens into filaments without anthers. The stamens didn't have pollen grains. This attribute was observed to be stable in the following years under diverse environmental conditions. The gynoecium of *inap* CMS exhibited normal morphology and good seed-set following pollination by the pollen of *B. napus*. AFLP analysis of the nuclear genomic composition with 23 pairs of selective primers detected no DNA of Woad bands in *inap* CMS. Twenty out of twenty-five mitochondrial genes have originated from *I. indigotica*, except for *cox2-2*, the recombinant between *cox2* from woad and *cox2-2* from rapeseed. The novel *cox2-2* was transcribed in flower buds of *inap* CMS weakly and comparatively with the fertile addition line of *B. napus,* with one particular woad chromosome. The other restorers of autoplasmic and alloplasmic CMS systems of rapeseed failed to restore the fertility of *inap* CMS. Screening of *B. napus* wide resources, however, found no fertility restoration variety indicating its distinct origin.

The cytoplasmic male sterile systems available in *Brassica* species in India are presented below in Table (Chauhan *et al.*, 2011).

CMS system	Discovered by
tour	Rawat and Anand (1979)
oxyrrhina	Prakash and Chopra (1988)
siifolia	Rao et al., (1994)
trachystoma	Kirti et al., (1995a)
moricandia	Prakash et al., (1995 & 1998)
catholica	Kirti et al., (1995b)
alba	Prakash et al., (1995)
lyratus	Banga and Banga (1997)
canariense	Prakash et al., (2001)
erucoides	Bhat et al. (2006)
126-1	Sodhi et al., (2006)
barthauti	Bhat et al., (2006)

Among the above, the fertility restoration in respect of *trachystoma* and *moricandia* systems is reported to occur through single dominant genes (Chauhan *et al.*, 2011)

Fertility Restoration systems for hybrid development in *Brassica* Sp

Cytoplasmic male sterility (CMS) is a maternally inherited trait causing failure of production of functional pollen. It is associated with chimeric mitochondrial open reading frames. In a number of cases, transcripts originating from these altered open reading frames are translated into specific proteins that interact with mitochondrial function and pollen development. Nuclear restorer genes function in suppressing the negative effects of CMS-associated mitochondrial abnormalities involving certain mechanisms (Schnable and Wise 1998)

Environment-sensitive GMS (EGMS) mutants may involve epigenetic control by noncoding RNAs and can revert to fertility under different growth conditions, making them useful breeding materials in the hybrid seed industry.

The dominant restorer gene of *pol* CMS (Rfp) is found in the nucleus and is a key component of hybrid production by achieving F1 progeny with complete fertility restoration. To identify the molecular markers associated with the *Rfp* gene, a near isogenic line (NIL) population of 2,000 individuals segregating for the Rfp locus was generated by crossing and backcrossing for 12 times. This NIL population was used to screen Rfp markers by AFLP technique. Based on the sequence information of AFLP markers that have been identified in previous research, a homologous region of Rfp locus has been identified in chromosome 1 of Arabidopsis. Then, six sequenced *Brassica rapa* (*B. campestris*) BAC clones corresponding to this target region were chosen to design microsatellite (SSR) primer pairs. Twenty-two SSR markers were designed and one of them, *KBrDP1*, was verified in the 2, 000 NILs population and proved to be strongly linked to *Rfp locus*. The genetic distance between *KBrDP*1 and *Rfp* was 0.2 cM. KBrDP1 marker was found located on linkage group N9 of a published DH mapping population. This SSR marker was useful in marker assisted selection and breeding of the elite pol CMS restorer lines in rapeseed (Li *et al* 2011).

The nuclear restoration of *pol* cytoplasmic male sterility (CMS) is conditioned by a gene, *Rfp*, that is also involved in modifying transcripts of the *pol* CMS-associated *orf224/atp6* mtDNA region. It has been observed that the *nap* nuclear restorer gene *Rfn* apparently is identical to *Mmt*, a gene that conditions the modification of transcripts from several different mtDNA regions, including one that is associated with *nap* CMS and contains *orf222*, a chimeric gene related to *orf224*. *Mmt*, in turn, is found to be allelic to *Rfp*, suggesting that restorer genes for the two cytoplasms represent different alleles or haplotypes of a single nuclear locus. This view is supported by restriction fragment length polymorphism mapping studies that indicate that *Rfn* and *Rfp* map to the same chromosomal position. Thus, in contrast to CMS in other species, different forms of *Brassica* CMS are restored by alleles of a single nuclear locus, and the restoration properties of these alleles reflect their involvement in the modification of transcripts of corresponding CMS-associated mtDNA regions. A survey of 51 varieties from 8 *Brassica* and *Sinapis* species failed to find evidence of *Rfn(Mmt)* in other than fertility-restored, *nap* cytoplasm *B. napus*. This suggests that *Rfn(Mmt)* arose in *Brassica* with *nap* cytoplasm and that the necessity for fertility restoration may have provided the selective pressure for its origin and maintenance (Li *et al* 1998).

Hybrids of *Brassica* released for cultivation in India

- The ICAR-Directorate of Rapeseed-Mustard Research have released **the first CMS based hybrid NRCHB 506 through the use of *moricandia* cytoplasmic genetic male sterility system.** The hybrid mustard has shown superiority for oil yield by a margin of 20 per cent over the existing popular varieties of *Maya* and *Varuna* during 11 trials across 5 states.
- The ICAR have also identified for release, another hybrid namely **DMH 1** developed by Dhara Vegetable Oil and Food Company for cultivation in Punjab, Haryana and parts of Rajasthan.

• Pioneer ® Brand Mustard Hybrids

- **45S42** – It is a high yielding hybrid medium maturity hybrid (125 -130 days) with bold grains and high pod density. It performs consistently across all soil condition.
- **45S35** – High yielding early maturity hybrid with attractive grain color.

- **45S46** – A high yielding medium maturity hybrid with bold grain and better oil percentage.
- Today, approximately half a million farmers are expected to grow the DuPont™ Pioneer® mustard hybrid seed in 850000 acres. As per the projections, by 2017 DuPont™ Pioneer® F1 mustard hybrid seeds will be planted in approximately 3 million acres.
- **Genetically modified Mustard Hybrid** viz., **Dhara Mustard Hybrid-11**, otherwise known as **DMH - 11,** is a genetically modified hybrid variety of the mustard species *Brassica juncea.* The transgenic mustard DMH - 11 was developed in 2002 using genetic material isolated from non-pathogenic soil bacteria (Srinivas and Krishna Ravi, 2017)and techniques in transgenic systems for pollination control, which primarily involved the *Barnase-Barstar* system. (Siriguri *et al.,* 2015). Three genes *Bar, Barnase and Barstar*, were extracted from *Bacillus amyloliquefaciens* to produce the hybrid seed (Kumar-Ravi 2018)
- It was developed by Prof. Deepak Pental of the University of Delhi, with the aim of reducing India's demand for edible oil imports (Jayaraman 2017). DMH - 11 was created through transgenic technology, primarily involving the *Bar, Barnase and Barstar* gene system (Hartley 1989). The *Barnase* gene confers male sterility, while the *Barstar* gene restores DMH - 11's ability to produce fertile seeds. The insertion of the third gene *Bar*, enables DMH - 11 to produce phosphinothricin-N- acetyl-transferase, the enzyme responsible for Glufosinate resistance as a distinguishing marker.
- This hybrid mustard variety has come under intense public scrutiny, mainly due to concerns regarding DMH - 11's potential to adversely affect the environment as well as consumer health (Jayan 2018).
- However, DMH - 11 was found not to pose any food allergy risks, and has demonstrated increased yields over existing mustard varieties (Prabhu 2017). Conflicting details and results regarding the field trials and safety evaluations conducted on DMH - 11 have delayed its approval for commercial cropping.
- DMH - 11 demonstrated yield heterosis in the range of 19%–40% over some of the best Indian varieties. Another multi-site trial conducted under field conditions indicated that the DMH - 11 to produce 30% more yield than existing strains of traditional mustard varieties (NAAS 2017).

Biotechnology encompasses tissue culture techniques to create somaclonal variation, anther culture to produce haploids and homozygous lines, protoplast culture and somatic hybridization, marker-assisted selection, development of transgenics for biotic and abiotic stresses. The research on biotechnology including genetic engineering of *Brassica* spp. being rather too vast to cover in this section, only certain relevant and representative investigations are presented here. For an extensive study of the role of biotechnology in rapeseed-mustard, readers are advised to access the references cited and the cross references there-of.

Regeneration protocols have been optimized for most of the *Brassica* species using different explants such as cotyledons (Narasimhulu and Chopra 1987), hypocotyls, leaf segments and protoplasts (Kirti and Chopra 1989), cotyledon petiole and shoot apex (Verma and Singh 2007). In vitro culture techniques offer a unique opportunity to explore somaclonal variation for desirable attributes.

BIO-902, a somaclone of Varuna variety which possesses shattering resistance along with high yield was selected and commercially released (Katiyar and Chopra, 1995)

Currently Plant breeders are in a position to address new problems by incorporating biotechnological tools and techniques in their breeding programmes. In India, new methodologies based on biotech tools have been added in breeding programmes.

Special attention was paid to the investigations on the standardization of conditions conducive to regeneration of plantlets from microspores. Prakash *et al.* (2004) analysed the research efforts made towards anther/ microspore culture in different *Brassica* species and concluded that normal plants can also be regenerated by culturing anthers of CMS line of *B. juncea* carrying *Diplotaxis erucoides* cytoplasm. Transfer of various desirable traits from parents to hybrids has since been achieved through somatic cell fusion between sexually incompatible species. Fusion of mesophyll protoplasts of *B. juncea* and *B. spinescens* has also been found to be an important step in the development of inter-specific hybrids (Kirti *et al.*, 1991). Scientists at the National Research Centre on Plant Biotechnology, New Delhi developed many cytoplasmic male sterile lines, through protoplast fusion technology. Prakash *et al.* (1998) applied this technology in *B. juncea* by protoplast fusion of *Moricandia arvensis* with the fertility restoration function of the male sterile *B. juncea* by introgression. However, such CMS lines were of chlorotic nature. In this context, protoplast fusion of chlorotic male sterile *B. juncea* with green male fertile *B. juncea* resulted in green male sterile plants (Kirti *et al.*, 1998). Thus, the *moricandia* based cytoplasmic genetic male sterility system has been successfully utilized in mustard hybrid development. Various markers such as RAPD, RFLP, AFLP and SSR have been used in marker-assisted selection of plant genotypes with the desired combinations of traits. Among them markers linked resistance to white rust (Prabhu *et.al.,*1997), fatty acids, oil content, yellow seed colour and fertility restoration have been reported. Transgenic approaches have been followed to develop the transgenic for aphid resistance, male sterility, herbicide resistance and drought tolerance. *Bar, Barnase* and *Barstar* based herbicide resistance and genetic male sterility have been used in the development of experimental hybrids (Jagannath *et al.*, 2002). Transgenics expressing Cod A (from *A. glabiformis*) gene for tolerance against abiotic stresses (salt and drought) have also been produced at the IARI, New Delhi (Singh 2003). At present efforts are in progress on the development of transgenics using lectin gene for aphid resistance and DREB gene construct for drought tolerance under Network Project on Transgenics in Crops(ICAR). The other transgenes being used in rapeseed-mustard include osmotin (from tobacco) for drought and salt tolerance annexin gene for stress tolerance, Chitinase, Glucanase (from *Arabidopsis*) for tolerance to *Alternaria* blight disease (Hegde 2000) and FAE1 gene for low erucic acid mustard cultivars.

Apart from conventional methods of selection and hybridization, several *in vitro* techniques have been utilized to employ genes to incorporate resistance/tolerance against important diseases in *Brassica* species. Gupta *et al.,* (2004) reported the use of biotechnological tools such as embryo rescue, somatic hybridization, somaclonal variation and molecular techniques for incorporation of resistance against white rust disease (*Albugo candida)* in the varieties of cultivated *Brassica* species.

With a view to enhancing the efficiency of predictive plant breeding, Bancroft *et al.*, (2011) sequenced leaf transcriptomes across a mapping population of the polyploid crop oilseed rape (*Brassica napus*) and representative ancestors of the parents of the population. Analysis of sequence variation1 and transcript abundance enabled to construct twin single nucleotide polymorphism linkage maps of *B. napus*, comprising 23,037 markers. They used these to align the *B. napus* genome with that of a related species, *Arabidopsis thaliana*, and to genome sequence assemblies of its progenitor species, *Brassica rapa* and *Brassica oleracea* and developed methods to detect genome rearrangements and track inheritance of genomic segments, including the outcome of an interspecific cross. By revealing the genetic consequences of breeding, cost-effective, high-resolution dissection of crop genomes by transcriptome sequencing will increase the efficiency of predictive breeding even in the absence of a complete genome sequence.

With regard to herbicide resistance, Monsanto company genetically engineered new cultivars of rapeseed resistant to the effects of its herbicide, *Roundup* in 1997 (Hartley, 2008).

Biotic Stresses

Diseases: A number of pathogens inflict damage on *Brassica* group of oilseed crops. Importantly the following fungal diseases are the major yield drainers. It is estimated that average yield losses range from 30 to 50% if the disease complex does not assume epidemic proportions. In the event of disease epidemic, the yield losses could be as high as 90% (Petraitiene and Brazauskiene, 2011). It is estimated that *Alternaria* blight disease alone causes yield loss up to 35-38% (Mishra *et al.*, 2017)

- ***Alternaria* blight** caused by *A. brssicola& A. brssicae*: The damage caused by this disease is to the tune of 18 % in *B. juncea,* 25 to 27% in brown *sarson* and 38 to 45% in yellow sarson. The symptoms are characterised by very pronounced and deep lesions on succulent leaves and pods, causing short and shrivelled pods and lower number of seeds per pod. The disease becomes preponderant with low light intensity and high relative humidity of the order of 90%. Under the above stated conditions leaves get wet for more than 4 hours and combined with a temperature of over 25°C the disease proliferates (Saharan 1992). Although the seeds get infected with the fungus, the fungus cannot survive the succeeding severe summer temperatures and hence seed transmission of the disease is a remote possibility (Chahal, 1981).

In addition to *Brassica* group of crops, *Alternaria* is also found to infect *Anagallis arvensis* and *Convolvulusarvensis,* under natural conditions and hence these plants may be alternate hosts for the disease proliferation.

Three physiological races of *Alternaria brassicae* have been reported by Saharan and Kadian (1984), which are reported to be virulent on all the cultivated species of *Brassica.* To these races viz., RM1, RM2 and V3, almost all types of Brassica were susceptible, with the exception of *Brassica oleracea* var. *capitata* that showed resistant reaction to RM1 and RM2; but was susceptible to V3.

Brassica oleracea var. *botrytis* exhibited resistance to RM2 while it was susceptible to RM1 and V3.

Gupta *et al* (1984) reported that the varieties such as RC-781 resistant to *Alternaria* blight showed more of phenol and less of sugars and nitrogen as compared to the varieties showing susceptible reaction.

Among cultivated species, *B. carinata* and *B. napus* are tolerant, followed by *B. nigra, B.juncea* and *B. campestris* in that order (Skoropad and Tewari 1977). Sharma and Singh (1992) found *B. hirta* as most resistant. Among the related species, *Camelina sativa,Capsella bursa-pastoris* and *Neslia paniculata* showed no disease symptoms mainly because of their inhibition capacity of fungal growth on leaf surfaces by producing phytoalexin shortly after infection. *Eruca sativa* (Miller) Thell showed localized flacks, while *Armoracia rusticana* Gairn., showed slow necrosis and chlorosis.

According to Saharan (1992) the following varieties and species of rapeseed-mustard group were found to be resistant to *Alternaria* blight: CSR-43, CSR-242, CSR-142-2, CSR-343, CSR-448, CSR-622,CSR-741, KRV-Tall, Gulivar, Midas, PHR-1, RC-781, Tower, YRT-3, TMV-2, *Brassica alba, B. carinata, & B.napus.*

The disease management using efficient chemical fungicidal sprays viz., Difolatan (0.2%), Dithane M45 (0.2%), Blitox (0.3%), Miltox (0.3%) Baycor (0.2%), Dconil (0.2%), Duter (0.2%), Brestan(0.1%) and Blitane(0.2%) have resulted in control of the disease and increased seed yield (Saharan 1992). Four sprays of each of the fungicide recommended are necessary to achieve efficient disease control.

An integrated management of the *Alternaria* blight disease and aphid pest population may be achieved employing three fungicidal mixtures involving Dithane-M45, Difolatan and Blitox and three insecticides viz. Dimethoate, Metasystox and Dimecrone at lower cost leading to significant profits. A maximum net profit was reported with the sprays of Difolatan +Metasystox followed by Difolatan +Dimecron (Tripathi *et al* 1987).

Biological control of *Alternaria* blight was reported using antagonistic fungi like *Streptomyces hygroscopicus* and *S. rochei*. (Jayant and Sinha 1981; Sharma and Gupta 1979; Sharma and Gupta 1980)

White rust caused by *Albugo candida, Scleretonia* rot caused by *S. scelrotiorum*, Downy mildew caused by *Peronospora parsitica*) and Powdery mildew caused by *Erysiphe polygoni* are the major ones. Additionally, the viral disease viz., Phyllody caused by Spot Virus, Turnip Virus -1and Mycoplasma too are important.

• **White rust (*Albugo candida*) and Downy Mildew (*Peronospora parasitica*)** go together due to the mixed infection of both the diseases observed particularly at flowering stage causing decline in the pod number to the tune of 37 to 47% and loss of seed yield of the order of more than 30% (Bains and Jhooty 1979). White rust incidence alone at flowering stage is reported to cause yield reduction ranging from 23 to 54.5% (Saharan and Verma1989).

Additionally, downy mildew could be noticed earliest even on cotyledonary leaves soon after germination. The severe infection of downy mildew at seedling stage is common, which causes pronounced seedling mortality causing considerable reduction in the plant stand. The severity of downy mildew is much more pronounced with the incidence of mustard mosaic virus, leading to extensive mycelium with bulbous *haustoria* and intense sporulation (Bains and Jhooty 1978; Bains and Jhooty 1980)

Under Indian conditions, the downy mildew and white rust proliferate intensely when crop is sown after the last week of October to the first week of November. With the progressive change in the date of sowing from the normal date in the first week of October towards the end of November, one could notice the increased intensity of both the diseases from initial 4.5% to gradual enhancement towards 69%. The late sown crop is subjected to the built up fungal inoculum under more favourable weather conditions (Saharan,1984).

In terms of physiological races of *Albugo candida,* five distinct races were identified which perhaps exhibited morphological variations for sporangial size. The sources of identification of these races are given below (Lakra and Saharan 1988)

S.No	Source	Race identified
1	Raphanus sativus	Race 1
2	Brassica juncea	Race 2 & Race 3
3	Brassica chinensis	Race 4
4	Brassica pekinensis	Race 5

Race 2 was found to be the most virulent attacking three hosts viz., *B. nigra, B. juncea* and *B. campestris* var. *brown sarson,* while Race 1 was the least virulent attacking only one host viz., *Raphanus sativus* (Lakra and Saharan 1988).

While a wide range variation from total susceptibility with well-developed pustules to no disease was observed for the white rust disease reaction among different cultivars, the varieties exhibiting resistance such as RC-781 were found to have a higher concentration of phenols than the susceptible cultivar viz., Prakash, which contained more sugars than the resistant variety (Dhawan *et al* 1982).

Considering the host resistance to *Albugo candida* at the genotypic level, a large number of varieties were reported to be resistant at diverse locations in India. Nevertheless, following multi-location testing, the genotypes viz., RC-781, YRT-3, CSR-142, CSR-741, EC-126743, Gulivar, Candle, Tower, *Brassica alba, B. carinata* and *B. napus* were found to be stable for resistance against white rust (Saharan 1984 and Saharan and Verma 1989). Additionally, genotypes Bio-YSR, EC-399296, EC 399299, NPJ-127, NRCDR-515, JM-1, EC-399313, JMY-11 were found to be resistant to white rust (ICAR-DRMR 2019). The most promising varieties recommended by ICAR-DRMR in this context are Basanti, JM 1 and JM 2.

• **Powdery Mildew** caused by *Erysiphe polygoni* hither-to was not a serious disease; but is assuming alarming proportions of late, assuming epidemic form in rapeseed and mustard group of crops. The disease causes serious losses when it infects the early stage of the crop growth and later at pod development stage, when the pods get loaded with fungal spores (Saharan and Kaushik 1981)

Moderate temperatures (18°C to 27°C) coupled with low humidity (less than 60%) and relatively dry season with minimum rainfall coinciding with the periods of February and March months are congenial for the proliferation of the disease. The disease is reported to have caused yield losses to the tune of 17 to 18% of seed yield at Hissar in India. The corresponding decline in the oil content is of the order of 6 to 7% (Saharan and Kaushik 1981) *Brassica alba* and *B. carinata* were free from the powdery mildew and hence considered resistant to the disease.

Powdery mildew may be controlled by spraying Karathane (0.1%), Calixin (0.1%) and Sulfex (0.2%) at intervals of 10 days. While the damage caused by powdery mildew is of the order of 18 to 20% in terms of yield reduction, the rest of the above are more serious causing crop damage of the order of 40 to 45%.

Yield losses in Rapeseed-Mustard Complex due to incidence of Insect Pests

The crop losses due to pest infestation in these crops has been mounting since the decade of 1970s, when the number of pests inflicting damage were limited to 24 (Rai 1976). In the later years there have been several changes in Indian agriculture in terms of increasing cropping intensity, altered cropping patterns in diverse agro-ecological situations leading to the changes in the pest dynamics, which have resulted in more than 38 insect pests damaging the rapeseed-mustard crops (Bakhetia and Sekhon 1984).

Yield losses due to the insect pest complex in rapeseed-mustard complex may range from 25 to 95% in diverse locations and cropping situations as revealed by the analysis carried out by various workers (Bhaketia and Singh 1992). With the increasing popularity of *B. carinata* and *B. napus,* it is likely that new pest dynamics may emerge. For example, the aphid species *Brevicoryne brassicae* hitherto less serious pest, currently is infesting *B. napus* at several locations.

Among the diverse insect complex affecting rapeseed-mustard complex, the pest viz., *Lipapis erysimi* (Kalt) known as mustard aphid is the most important in terms of crop losses inflicted by the pest.

Other insect pests in the order are *Athalia lugens proxima (*Klug) the mustard saw fly, *Bagrada hilaris(*Burm) the painted bug, *Chromotomyia horticola* (Goureau), leaf-miner and the *Diacrisia obliqua* Bihar hairy caterpillar are important.

1. Bihar hairy caterpillar

Eggs are laid in clusters of 50-100, on the lower side of leaves. The larvae are covered with long yellowish to black hairs and are up to 5 cm long. Pupation takes place in the dey soil. The adult is a brown moth with a 40-50 mm wing span and a red abdomen.

Young larvae feed gregariously mostly on the under surface of the leaves.

Caterpillars feed on leaves and in severe infestation the whole crop is defoliated.

Fig. 9: Early Instar Bihar hairy Caterpillars scrapping the leaf

2. Mustard Aphid (*Lipaphis erysimi* Kalt)

Aphid eggs are white in colour and laid along the veins of leaves. There are four nymphal stages (instars). The general appearance of each stage is similar except for increase in size during subsequent instars. The first, second, third and fourth nymphal stages last 1-2, 2, 2, and 3 days respectfully. Adult aphids are small, soft-bodied, pear-shaped insects that have a pair of cornicles (wax-secreting tubes) projecting out from the fifth or sixth abdominal segment. Wingless, female, aphids are yellowish green, gray green or olive green with a white waxy bloom covering the body. The winged, female, adult aphids have a dusky green abdomen with dark lateral stripes separating the body segments and dusky wing veins. Male aphids are olive green to brown in colour. The aphid attacks generally during 2nd and 3rd week of December and continues till March.

Fig.10: Aphids feeding on Mustard siliquae

- Both nymph and adults suck the sap from leaves, buds and pods. Curling may occur for infested leaves and at advanced stage plants may wither and die. Plants remain stunted and sooty moulds grow on the honey dew excreted by the insects. Mustard aphid reduced seed yield and oil content up to 86.1% and 3.6%, respectively in *B. juncea.* The corresponding losses were up to 92.1% and 4.6% in *B. napus*. Total phenol, protein, surface wax, glucosinolates, proline and lectins negatively and total sugars were positively correlated with mustard aphid infestation. Sprays of Neem oil (2%) and neem seed kernel extract (5%) reduced mustard aphid infestation by 80%.*Verticillium lecanii* @ 108 CS/ml reduced aphid population by 80% under humid conditions (ICAR-DRMR 2019).
- Regarding chemical control, Spray of dimethoate @ 0.1% and release of *Coccinella septempunctata* @ 5000 beetles/ha 10 days after release reduced mustard aphid population more than 90%.

3. Painted Bug (*Bagrada hilaris*)

Painted bug lays its eggs in clusters on leaves or on the soil underneath host plants. Eggs are barrel shaped, initially white and turn orange with age. A single female can lay as many as 100 eggs within 2 to 3 weeks. The incubation period is 5 to 8 days. Nymph pass through five stages changing colour from bright orange to red with dark markings, gradually acquiring the colouration of the adult. Initially they do not have wings; wings gradually develop, as the nymphs grow. Wing pads are visible in the last instar nymph. The adult bug is typically shield-shaped, 5 to 7 mm long and 3 to 4 mm broad at its widest area. The upper surface has a mixture of black, white and orange markings, which gives the insect its common names viz., harlequin bug or painted bug. The life cycle lasts 3 to 4 weeks and several generations may occur in a year. Period of activity of painted bug starts from September.

Adults and nymphs suck sap from all parts of the plant. Young plants wilt and wither as a result of the attack. Adult bugs excrete a resinous substance which spoils the pods. Quality and quantity (31% losses) of yield is affected when grown up plants are infected. Harvested crop in threshing floor is also infested.

- Seed treatment with imidacloprid 70 WS (5g/kg) provided good control to painted bug up to 30-35 days after sowing.

Fig. 11: Painted bug feeding on Mustard flowers and stem

4.Mustard Sawfly (*Athalia lugens)*

Fig.12: and 13. Larva and adult of Mustard Sawfly

Eggs are spherical, about 0.5 mm in diameter, light bluish green in colour.

Larva is greenish black with wrinkled body and eight pairs of pro-legs. On slightest touch the larva falls to round and feigns death. The larvae had 6 instars, fed on the leaves, and pupated after 14-16 days.

Pupae look like sand particles and have salivary secretions; the pupal stage lasted 11-12 days.

Adults are orange bodied with smoky transparent wings. The pest is active during seedling stage of the crop i.e. October - November. (https://www.agristudyinfo.com/mustard-sawfly/)

Initially the larva nibbles leaves, later it feeds from the margins towards the midrib.

The grubs cause numerous shot holes and even riddle the entire leaves by voracious feeding.

They devour the epidermis of the shoot, resulting in drying up of seedlings and failure to bear seeds in older plants.

The yield losses up to 5 to 18 %. In severe case at the seedling stage, the crops have to be re sown.

5. Mustard Leaf-miner [*Chromatomyia horticola (Phytomyza atricornis)]*

Fig. 14: and 15. Pictures of leaf miner attack

Mines on leaves

Mine linear, whitish, both upper and lower surface. Pupation internal, at the end of the mine with the anterior spiracles projecting through the epidermis. Upper-surface, less often lower-surface corridor, Pupation within the mine and soil; pupal chamber in a, usually lower-surface.

Leaves show mines; the attacked leaves wither; vigour of the plant gets reduced. Its damage is often more prominent on the older leaves

6. Diamond Back Moth (*Plutella xylostella L)*

Minute yellow coloured eggs, laid singly or in groups on upper surface of leaves. Egg hatches in about 7 days. Pale yellowish green caterpillar. Larval period is 14 days. Pupation takes place on the foliage in a transparent cocoon. Pupal period is about 7 days. The adult is small greyish brown moth with three white triangular spots along the inner-margin which give diamond shape at rest position. Hind wings with fringe of long fine hairs Caterpillars feed on the foliage.

Pod damage symptoms

Fig.16: Damage to pods (siliquae) by diamond black moth

The leaves give a withered appearance but in later stages larvae bore holes in the leaves which may be eaten up completely. It also bores into pods and feeds developing seed.

Broom rape *(Orabanche* spp): Broomrape *(Orobanche)* is a major devastating parasitic weed of mustard. Broomrape weed infestation caused 28.2% average reduction in Indian mustard yield. Among *Orobanche* spp., O. *aegyptiaca* is the most important parasitic weed causing severe yield and quality reducing factor in rapeseed-mustard. It is endemic in semiarid region and may reach epidemic proportions depending upon soil moisture and temperature. Preceding crops of cowpea, black gram, moth bean, sunnhemp, cluster bean, and sesame significantly reduced *Orobanche* menace in succeeding mustard crop while sorghum, pearl millet, chilies, and green gram did not influence broomrape infestation in mustard (Kumar 2002). At Bharatpur, S. K. Nagar and Bawal directed spray of glyphosate (0.25–1.0%) and 2 drops of soybean oil per young shoot of *Orobanche* showed effective control and recorded 91.9% higher seed yield over infected sick plot (Shekhavat *et al* 2012)

Strategies for Efficient Pest / Disease Management in Rapeseed-Mustard complex

Many times chemical control of pests and diseases may not be effective; but might add to the total cost of crop production,

On the other hand, the following recommended strategies avoiding excessive use of chemicals, proved to be effective.

(i) Choosing the sowing time is very important as it affects the disease incidence significantly. Deep summer ploughing to be carried out to control pests and diseases.

(ii) Use of disease / pest resistant or tolerant or early maturing pest/ disease escaping cultivars. Use of good quality seeds.

(iii) Seed treatment with biocontrol agents viz., *T. viride*, *G. virens* or botanicals like *Allium sativum* bulb extract (1 % w/v) or carbendazim @ 0.1% active ingredient. or mixture of carbendazim with Apron 35 SD (6 g/ kg).

(iv) Seed treatment with *Imidaclorpid* 70ws @ 5.0 g per kilogram of seed for controlling painted bug and other insect pests.

(v) Pluck and destroy the infected twigs or dip them in kerosene.

(vi) To control aphids, resort to foliar spray of Metasystox 25 EC or Dimethoate 30 EC @ 1ml/litre of water at the low pest concentration of 25 to 30 aphids / plant

(vii) In case chemical disease control becomes inevitable, mixture of fungicides may be employed to avoiding development of pathogen-resistance to fungicides.

(viii) Non-chemical pesticide methods like spraying 5% neem seed kernel extract (NSK) and 2% neem oil spray give good results provided done when the pest load is low.

(ix) Aphids can be managed by sprays of *Verticillium lecani* in combination with 5% NSk and 2% Neem oil

(x) Use of biocontrol agents is advantageous as they are often effective against a wide range of soil-borne pathogens. Moreover, they are eco-friendly, cost effective and their use avoids the risk of development of resistance in the pathogen towards the control agent.

(xi) Application of recommended doses of N, P and K fertilizers with split application of N

(xii) Maintaining optimum plant population with recommended spacing and

(xiii) Proper drainage to avoid water stagnation which promotes the incidence of pests and diseases.

Production Technologies for Rapeseed & Mustard

- Growing the crop on well drained sandy-loam soil is the basic requirement for higher seed yields. The seedbed should be firm, moist, and uniform which allows good seed-to-soil contact, appropriatc planting depth and quick moisture absorption leading to a uniform germination.
- The rain-fed / dry land crop is to be grown under the conditions of conserved soil moisture from rainy season. About 25 % of the total rapeseed-mustard area is rain-fed.

Therefore, special care has to be taken to ensure good moisture conservation by carrying out light harrowing after every shower during the rainy season where the field is left fallow to allow rainwater to sink into the field by appropriate methods of soil conservation viz., contour bunding and laying out the ridges and furrows to ensure good moisture conservation. Zero tillage is preferred in mustard as it conserves more moisture in the soil profile during early growth period. Adoption of reduced tillage practices (3-4 plough/harrow) improves soil health without any loss in mustard seed yield in comparison to conventional tillage (6-10 plough/harrow) (ICAR-DRMR 2019).

Subsequent release of conserved soil moisture regulates proper plant water status, soil temperature, lower soil mechanical resistance, leading to better root growth and higher grain yield of mustard (Rathore *et al.,* 1999). In the case of minimum tillage, care needs to be taken to ensure proper distribution of crop residues adopting well-designed crop rotation and evenly distributing residues so as to create a firm, moist and uniform seedbed during end of September to the mid October, which is the proper time of quick land preparation and planting. Soak the seeds in water overnight or covering the seeds with wet gunny bag over night before sowing results in good germination.

A spacing of 45cm between the rows and 15 cm from plant to plant with a row is recommended adopting a ridge-furrow method. The seeds must be planted on ridges. The varieties recommended for dry-land conditions must be chosen.

- **Under irrigated conditions**, preparation of soil following a pre-sowing irrigation of over 60 mm to achieve a good pulverised seed bed helps in good germination and adequate plant stand.

It is very much necessary to formulate the crop nutrient schedule based on the soil test values.

The following recommendations are tentative and are valid for normal soils. The suggested values may be adjusted accordingly based on soil test values.

- **Before planting the soil has to be refurbished** by incorporating Farm Yard Manure (FYM) @ 6-7 tons/ha if the soil is reasonably fertile. Otherwise, in thecase of poor soils around 20 tons of FYM must be incorporated with adequate tillage. It is advisable to add 50 to 60 kg of P205 and 40 to 45 kg of K2O/ ha at the time of application of FYM. Basal application of 40 kg S/ ha through gypsum or elemental sulphur to mustard crop enhances its seed oil content up to 2% and seed yield up to 25% (ICAR-DRMR 2019).
- Seed dressing with nitrogen fixing bacteria *Azotobacter* saves up to 30 kg N/ha and increases mustard seed yield up to 21%. Seed dressing with phosphorus solubilizing bacteria (PSB) enhances phosphorus use efficiency from 5.2 to 19.8 kg seeds/kg P2O5 applied and increases mustard seed yield up to 37% over application of phosphorus only (ICAR-DRMR 2019)
- **The dose of total Nitrogen** that the crop needs under irrigated conditions is around 100 to 120 kg. It is advisable to manage the Nitrogen nutrition of the crop by applying in split doses. The time of nitrogen application in rapeseed-mustard is also important for getting better response to applied nitrogen. In multi-location trials of AICORPO split application of nitrogen @ 50% as basal and 50% as top dressing after first irrigation gave better utilization of nitrogen and higher seed yield under irrigated condition. Methods of application of nitrogen play an important role to increase the efficiency of applied fertilizer as well as the seed yield of mustard.

- **The nutrient management of the crop under rain-grown dryland** conditions depends on the total rainfall of the zone and its distribution during the crop cycle. In such cases a standard application of 40 Kg of P2O5 and 30 kg of K2O may be applied initially. The recommended dose of Nitrogen under dryland conditions should be around 40 Kg/ha which may be applied in depending upon the extent of conserved soil moisture.
- **The foliar application** of plant nutrients, more particularly Nitrogen is normally recommended under conditions of dryland agriculture where soil moisture may be the limiting factor. The foliar feeding of Nitrogen in the case of mustard under dryland conditions is reported to have resulted in cost saving to the tune of 25% of Nitrogen. (http://gcirc.org/fileadmin/documents/Bulletins/B25/B25_06Rapeseed.pdf)
- **Sulphur** is an important plant nutrient particularly for rapeseed-mustard group of crops, which have exhibited perceptible response to this nutrient to the tune of 30%. It is suggested that at least 12 kg of Sulphur is needed to produce one ton of oilseed. An optimum dose of 40 kg S/ha should be applied for wide range of soils to enhance the crop yield to the tune of 12 to 48% depending upon the varying soil situations under irrigated conditions. The most appropriate source of sulphur for all oilseed crops is reported to be Gypsum, which was found more effective as compared to single super phosphate (SSP).
- **Micronutrients:** Among other nutrients that have good bearing on productivity of rapeseed-mustard, Boron and Zinc are very important. These nutrients become very effective in enhancing yield particularly in combination with Sulphur. It has been reported (Kumar 1992) that application of 1 kg of Boron /ha together with Sulphur @ 20 kg/ha has resulted in increase in seed yields of the order of 31.9%. Similarly, application of Sulphur @ 20 kg/ha + Zinc @10kg/ha too has increased the seed yield of the same order. Basal application of 5 kg Zn/ha to mustard crop enhances number of silique/plant, nitrogen use efficiency and thereby seed yield up to 12%.
- The response of Indian mustard varieties, viz. 'Pusa Bold' and 'Vardan' to applied zinc was found higher (AICRP-RM, 2000) as compared to the varieties Varuna, RH- 30 and Aravali. (Fig.17)

On the other hand, the seed yield also increased significantly (16–47%) with the application of boron. The mean response to boron application in terms of seed yield ranged from 21 to 31% (AICRP-RM 2004 and AICRP-RM 2005)

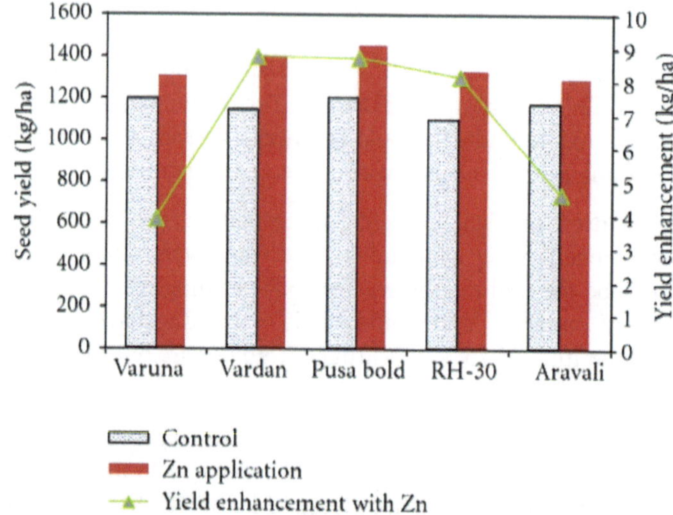

Fig. 17: Yield enhancements with Zinc application on Improved varieties of Mustard (Shekhavat et al., 2012)

Influence of zinc application on seed yield of different cultivars of mustard. (from Rathore *et al.*, 1999)

• Weed Control

(a) One hoeing and weeding mechanically at 20-25 days following sowing is needed to conserve moisture and eliminate weed competition to the crop in growth phase.

(b) The chemicals viz., Flucholralin @ 1 kg of a.i./ha as pre-planting incorporation, Pendemethalin @ 1kg of active ingredient as pre-emergence spray and application of Isoproturon @ 0.75 kg of active ingredient/ha are the important practices in this context.

- **Time of planting** varies with the crop type. (i) *Toria* (*B. campestris* var. *toria*) may be sown in any time during the third week and end of September depending upon the day temperature, which should not exceed 28*C. (ii) Mustard (*B. juncea*) may be planted after the first fort night to end of October depending upon the day temperatures.

- **The recommended inter-row spacing under irrigated conditions** is around 30 cm for rapeseed and 45 cm for mustard, adopting a seed rate of 4 to 6 kg / acre depending upon the seed size. Seed should not be planted deep. The optimum depth of sowing is around 3 to 4 cm.

- **Water Management**: Appropriate water management includes avoidance of drought at crucial developmental phases of the crop. Two irrigations, given one at flowering stage and at siliqua formation stage increased seed yield by 28% over the dryland conditions (Ghosh *et al* 1994). Furrow irrigated raised bed (FIRB) system increases the mustard seed yield by 10% and saves 33% water over conventional practices. (ICAR-DRMR 2019). Alternatively, Use of sprinkler and drip irrigation system improves seed yield of mustard by 25-30% over check basin. It also reduces total requirement of irrigation water by 50% and nitrogen fertilizer by 25% (ICAR-DRMR 2019).
- **Yield:** The productivity in terms of seed yield of rapeseed-mustard depends on the factors viz., the crop variety grown and the management practices adopted during the growth of the crop under the prevailing agro-climatic factors as well as the time of harvesting.

Nevertheless, it is seen that under optimum conditions of crop and water management and with adoption of an improved recommended variety, it would be possible to obtain seed yield of the order of 2.0 to 2.5 tons /ha (Kumar 1992)

Under rain-fed dryland conditions, with the recommended improved crop management seed yield of the order of 1.0 to 1.5 ton/ha could be obtained.

Mustard Based Phytoremediation

Phytoremediation of soils with heavy metals using *Brassica juncea* along with the addition of organic matter, organic chelates, soil amendments, adoption of suitable cropping systems, intercrops and fertilizer selection is reported to augment the ability of mustard plant to absorb, uptake and concentrate heavy metal under contaminating soils (Rathore *et al.*, 2017)

Taramira (Rocket Salad) (*Eruca vesicara L.* Cav.subsp. *sativa* Mill). (Syn *Eruca sativa* Mill)

In the group of *Brassicae,* taramira has carved its own special niche in terms of its growth habit, floral biology, ecological specificity and physiological and biochemical constitution. Taramira is widely popular as a salad vegetable.

The taxonomic classification of taramira was rather nebulous. Initially, it was included in the genus *Brassica* under the name *Brassica erucoides* (Roxburgh, 1832). Later, Hooker (1872) re-named it as *Brassica eroca*. Plaxton (Hereman, 1868) described it as *Eruca tournefort* and listed 3 species such as *Eruca hispida, Eruca sativa, and Eruca vesicaria.* Basically all the above are hardy annuals with white to yellow flowers, with shorter silique, each with a pointed beak and seeds slightly smaller than those of *Brassica rapa.*

The species of *Eruca* is native to the Mediterranean region, from Morocco and Portugal in the west to Syria, Lebanon, and Turkey in the east (Hereman, 1868).

Eruca sativa Mill., grows 20–100 centimetres (8–39 inches) in height. The pinnate leaves have four to ten small, deep, lateral lobes and a large terminal lobe. The flowers are 2–4 cm (0.8–1.6 in) in diameter, arranged in a corymb in typical *Brassica* fashion, with creamywhite petals veined in purple, and having yellow stamens; the sepals are shed soon after the flower opens. The fruit is a 'siliqua' (pod) 12–35 millimetres (0.5–1.4 in) long with an apical beak, containing several edible oilseeds. The beak is seedless. The seeds and leaves offer a pungent odour and taste; much stronger than that of rapeseed-mustard. Seed oil of taramira is often adulterated with mustard oil to enhance the pungency

The species has a chromosome number $2n = 22$.

Fig. 18: Taramira plants**By Dave Whitinger - http://davesgarden.com/pf/showimage/74928/, CC BY-SA 3.0, https://commons.wikimedia.org/w/index.php?curid=1723529

The crop of taramira (*Eruca sativa* Mill.) is normally grown in the post-rainy season with sowing taking place during late September to middle of November. The crop is grown scattered in patches on soils of poor fertility, in regions of low rainfall in north and north-western India without any irrigation facility.

The species is allogamous with 100% cross pollination due to its self-incompatibility mechanism. Partial homology between genomes of *Eruca* and *Brassica* has been reported by Mizushima (1950) and a fertile hybrid between *Eruca sativa* and *Brassica campestris* with 5 bivalents has been recovered indicating a distant evolutionary relationship between *Eruca* and *Brassica* (Matsuzawa and Sarashima 1986).

Regarding the gene action governing the seed yield and its components, dominance gene action has been reported by Kumar and Yadav (1986 a & b) except for silique length which is stated to be governed by the additive gene effects. Kumar *et al.* (1988) have reported preponderance of non-additive gene effects for primary branches/ plant under normal and saline environments.

The main advantages of taramira crop are (i) the ease of its cultivation, (ii) its hardiness to withstand adverse weather and its resistance to frost (Yadava, 1976)

and salinity (Kumar and Ydav,1992), (iii) its suitability to dryland agriculture with soil moisture conserved from showers of rainy season and (iv) its least susceptibility to the attack of insect pests and diseases. (v). One more added advantage of taramira is that its root system is robust and deep facilitating accessing moisture from deeper layers of soil. This robust root system of taramira is reported to be advantageous under frost and on saline soils too.

The seeds contain oil (31 to 33 %); but the oil is not considered fit to be used as edible oil (Kumar and Yadav 1992), although it is being used for cooking in several communities in North-western India. The oil of taramira is highly pungent and upon extraction, acrid. The pungency differs from that of mustard oil, although taramira oil can be used to make a sort of mustard preparation (Sindhukanaya and Kantaraj-Urs, 1989). In India, the oil is used for pickling, after aging to reduce the acridity, as a salad or cooking oil. The seed cake, a by-product of oil production, is also used as animal feed (Das *et al.*, 2004). Taramira seed oil also contains thiofunctionalized glucosinolates along with erucic acid (Bennett *et al.,* 2002; Lazzeri *et al.,* 2004) and hence it may not be a preferable cattle feed, though it is being used so.

The oil is used mainly in industry as a lubricant, for soap making, as an illuminating agent, in massage and in medicine (Khalil *et al.,* 2002; Miyazawa *et al.,* 2002). The oil is rich in erucic acid, an important industrial compound, which could turn *Eruca sativa* into potential industrial crop plant. It could be seen that there is an increasing global demand for the amide of erucic acid, namely *erucamide* used in the manufacture of cosmetics, detergents, and polymers. The biodegradable property of taramira crude oil, promotes it as an alternative to mineral oil in industry (Khalil *et al.,* 2002).

Fig. 19: Taramira Flowers & Silique *

*By Dave Whitinger - http://davesgarden.com/pf/showimage/74928/, CC BY-SA 3.0, https://commons.wikimedia.org/w/index.php?curid=1723529

The average yield levels of taramira are lower as compared to those of rapeseed-mustard complex and lie in the order of 0.4 to 0.7 tons/ha. Nevertheless, the crop responds to improved agronomic management. Hence with high yielding varieties in place together with improved crop husbandry it is possible to harvest seed yield levels of the order of 1.0 to 1.4 tons/ha (Kumar and Yadav 1992). With the oil content of taramira ranging from 39 to 40% with improved management, the expected oil yield will be 720 to 830 kg/ha (Maliwal, 1984)

Major problems of taramira that need improvement are as follows:

1. Long duration of the crop spanning over a period of more than 150 days; particularly under moisture stress conditions, leading to instability in its production/ productivity.
2. Physiologically plant type is stated to be least efficient (Kumar and Yadav 1992) perhaps in terms of its harvest index.
3. The agro-ecological situations under which the crop is grown are of very low soil fertility and subject to drought.

The above issues could be overcome by

(1) Breeding crop varieties with early and faster rate of growth and maturity with good yield potential, which can complete the reproductive cycle before the soil moisture gets exhausted:

(2) Focus on moisture conservation from rains in monsoon season, adopting the standard procedures of moisture conservation and water harvesting

(3) Efficient Nutrient management schedule and

(4) Enhancing the niches in which Taramira could find a suitable role in cropping systems of dryland agriculture.

Suitable Varieties

The work carried out in India (Kumar and Yadav 1992) has resulted in identification of the following varieties of Taramira for the attributes indicated.

(a) Earliness of maturity: TC4, TC40, TC52 and TC59. These varieties are reported to mature in 125 to 150 days with better mean yield performance of 450 to 550 kg/ha with a harvest index of around 32%.

(b) Productivity: RTM 522, RTM 521, JOB TC 2, RTM 523, TMH 24, Narendra, RTM 2002 (Tara), Vallabh Tara-1 (PUT 93-11) and Vallabh Tara -2 (PUT 93-1)

The reported seed yields of these varieties are in the range of 550 to 700 kg/ha under the conditions of receding soil moisture in the post rainy season. Their yield levels may go up with better soil moisture conditions.

(c) Suitability for arid conditions characterised by extreme soil moisture deficit: TMH24 and RTM 52 have exhibited average productivity levels of the order of 500 to 600 kg/ha under arid conditions

(d) Among one hundred genotypes of taramira evaluated for oil content and fatty acid composition, it was observed (Yadava *et al.*,1998) that oil content of the genotypes ranged from 31.86 to 41.32%. Large variations were observed for erucic acid (26.1 to 52.4%), oleic acid (14.1 to 23.4%), linoleic acid (6.9 to15.7%), linolenic acid (8.3to 15.3%) and eicosenoic acid (9.3-18.3%). It is suggested that the genotypes possessing low erucic acid (26.7%), could help in improving the oil quality of taramira for human consumption.

Crop Production

Tillage and Moisture conservation: Since taramira crop is grown totally under unirrigated conditions, depending upon the moisture conserved in the monsoon season, the following moisture conservation procedures need to be adopted to avoid the adverse impact of drought during the rainless post-monsoon season.

Before the onset of rains, in summer months the field should be ploughed deep breaking the soil hard pan if any, followed by an appropriate field bunding in consideration of slope of the field to ensure run-off collection in the field. With the initial soaking rains, FYM @ 2 tons /ha or one ton of vermi compost /ha may be incorporated and ploughed so as to enhance the organic matter content of the soil before sowing (Yadav *et al.,* 2013).

After every shower, light ploughing may be carried out across the slope to facilitate conservation of rainwater in soil profile. Subsequently, it is recommended to prepare the field in ridges and furrows across the slope spaced at 45 cm to allow the rainwater to sink into the furrows (Prasad and SudhakarBabu 1997).

Sowing time and Sowing: The optimum sowing time for obtaining higher yields is reported to be the period from the end of September to middle of the second week of October in the post rainy season under the conditions of conserved soil moisture (Maliwal *et al.*, 1985). It is recommended that the field be laid out into ridges and furrows with inter-ridge spacing of 45 cm as stated above and sowing done on the ridge adopting15 cm spacing with in the row (Prasad and SudhakarBabu1997). A seed rate of 3kg/ha is reported to be appropriate to maintain the needed plant population (Kumar and Yadav 1992).

Spacing: The inter and intra row spacing depends upon the moisture level of the soil. Though the spacing of 30 x 15 cm is indicated to maintain good population density for taramira production, it has been observed it would be safer to adopt spacing of 45 x 15 cm to avoid the adverse impact of drought, considering the moisture deficit conditions under which the crop is generally grown (Anonymous1989).

Crop Nutrient Management: Application of 2 tons of FYM or one ton of vermi compost / ha with preparatory tillage as described above should enhance the organic matter of the soil and facilitate better moisture conservation and higher yield levels (Yadav *et al* 2013).

Taramira is reported to have shown very good response in terms of seed yield to the application of Nitrogen @ 45 kg/ha (Kumar and Yadav 1992) and Sulphur @ 90 kg/ha (Singh, 1983).

Basal application of 45 kg of Nitrogen, 30 kg of P2O5, 20 kg of K2O and 60 kg of Sulphur (preferably as Gypsum) per hectare at sowing may cater to the needed crop nutrient requirement. Foliar spray of Urea @ 2% at flowering stage may help good capsule development.

Although Fazli *et al.,* (2005) reported that application of Sulphur and Nitrogen in three split doses has resulted in increased oleic acid (18:1) content and decline in erucic acid (22:1) content, the stated observation, however, is relevant to taramira grown under irrigation where it is possible to apply split application. It may not be possible to resort to split application of nutrients under total dryland conditions with the absence of rains during growing season and without irrigation.

Weed Control: It is important to ensure good weed control in the dry-land crop of taramira to avoid depletion of precious soil moisture due to weed infestation. Studies on the utility of employing herbicides for weed control in taramira crop indicated that pre-planting incorporation of Fluchoralin @ 0.75 kg/ha with one hand weeding was effective in controlling weeds (Jat *et al.,* 1987). Nevertheless, it was found that one good hand weeding at 20-25 days following seedling emergence was effective and more economical than any chemical weed control measures (Kumar and Yadav 1992).

Inter cropping with taramira: Among various inter-cropping systems involving taramira and other dryland crops of the post-rainy season, inter-cropping of chickpea + taramira in 2:3 proportion has reported to have yielded highest economic returns (Yadava, 1976).

Yield: Yield of taramira depends to a large extent on the soil moisture available to the crop in addition to the above mentioned factors. Absence of rains during crop growth precludes any top dressing of crop nutrients to boost the crop yield. Given favourable level of soil moisture conducive to crop growth, with the implementation of the recommended production technology with an appropriate crop variety, yield levels of the order of 1.0 to 1.5 tons/ha could be achieved (Kumar and Yadav 1992).

4

Soybean (*Glycine max* L. Merril)

Introduction

Soybean has earned a place of prominence in world agriculture as one of the most important food crops contributing to protein and oil production. Pertaining to the family Fabaceae, soybean is a legume with over 42% protein and 20% oil in its seeds. The crop fits into the production system of monsoon rainy season of India during the period from June to October and can be grown on wide variety of soils; but performs best on fertile heavier soils of clay and silt.

Soybeans originated in Southeast Asia and were first domesticated by Chinese farmers around 1100 BC. By the first century AD, soybeans were grown in Japan and many other countries. Soybean seed from China was planted by a colonist in the British colony of Georgia in 1765.

Introduction of soybean to Indian subcontinent dates back to 1000 AD through silk route from north-eastern India and Himalayan mountains. The black-seeded soybean has been traditionally grown in North and north-eastern regions of India and further spread to different parts of the country. This black-seeded soybean, *Kalitur*, was the vehicle for soybean revolution in India. The major initiative on soybean cultivation was undertaken during 1963-64 by G.B. Pant University of Agriculture and Technology (GBPUAT), Pantnagar and JNKVV Agricultural University, Jabalpur in collaboration with University of Illinois, USA. The cultivation has further picked up after the researchers took advantage of yellow-seeded material to develop high-yielding varieties that suited Indian conditions. Soybean cultivation got momentum during the 1970s as the vast monsoon fallow lands of Madhya Pradesh provided appropriate niche for its cultivation. Presently soybean is cultivated on nearly 10 million ha land with likely production of above 10 million tonnes.

The crop now occupies 5th position in the world and has acquired premier position in oilseed production in the country. The first survey carried out by Soybean Processors Association of India (SOPA) crop for the season of 2018-19, has estimated the total area under soybean for 2018 at 108.396 lakh hectare.The area has increased by 6.83 lakh hectare (6.7%) as compared to previous year.

The crop also has a prominent place as the world's most important seed legume, which contributes 25% to the global vegetable oil production, about two thirds of the world's protein concentrate for livestock feeding and is a valuable ingredient in formulated feeds for poultry and fish.About 85% of the world's soybeans are processed annually into soybean meal and oil. Approximately 98% of the soybean meal is crushed and further processed into animal feed with the balance used to make soy flour and proteins. Of the oil fraction, 95% is consumed as edible oil; the rest is used for industrial products such as fatty acids, soaps and biodiesel

It plays a significant role in Indian economy by contributing 20% of the edible oil produced in the country and fetches around Rs. 40,000 million by way of export of soya-meal. This striking growth rate is yet to be experienced in India in the case of any other crop. On account of its resilience to withstand aberrant climatic conditions and the amount of good done to transform socioeconomic conditions of Indian farmers, it appears that the crop has come to stay in India. Based on the analysis of continued unabated growth, it is projected that the area under the crop will stabilize around 13 million ha in near future with a production of over 8.6 to 9.0 million tonnes (Fig.1)

At present, India exports 55% of its soya meal. However, the Soya industry of the country is crippled by low yield together with limited domestic demand.

(https://ncsoy.org/media-resources/history-of-soybeans/

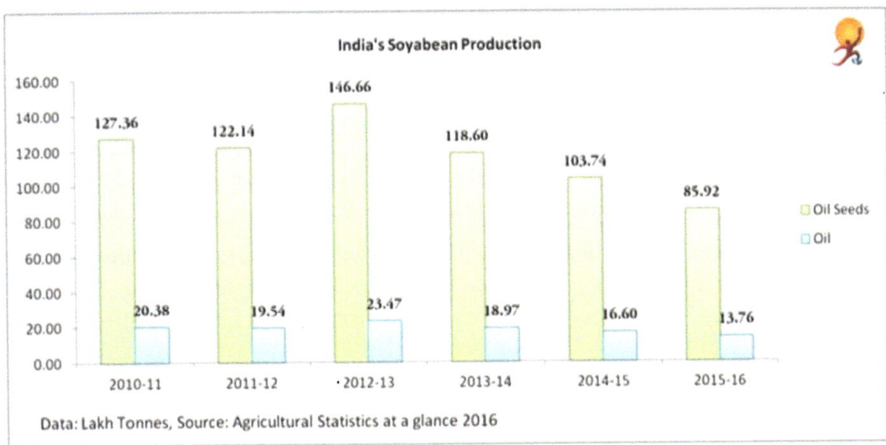

Fig. 1: Area (blue) and Production (yellow) of Soybean in India during 2010-2016. (Ministry of Agriculture, Govt. of India)

Ecology of Soybean

A fast growing herbaceous annual, soybean is native to Asia but currently grown worldwide. Its tap-root can extend to 2 m deep in good soil conditions, with secondary roots exploring the upper 15-20 cm of the soil. Roots bear nodules resulting from *Bradyrhizobium japonicum* infection (in most cases). Leaves are trifoliate and leaflets are oval to lanceolate, mostly broad in commercial cultivars (Ecoport, 2010). Soybean is reported to be warm temperate, short day plant (Weiss 2000).It was domesticated in North China 3000 years ago, and is now produced in almost all continents between 53°N to 53°S, and from sea level up to an altitude of 2000 m. The main producing countries are the USA, Brazil, Argentina, China and India.Optimal growth conditions for soybean are average day-temperatures of around 30°C, 850 mm of annual rainfall, and not less than 500 mm of water during the growing season, and soil pH ranging from 5.5 to 7.5 with good drainage. Soya is sensitive to soil acidity and aluminium toxicity. It can withstand short periods of waterlogging and short droughts (Ecoport, 2010).

With intensification of breeding programmes worldwide, soya has come to be grown in wider ecological situations throughout the world. The yield potential of soya in tropical zones is similar to that obtained in temperate zones, with varietal component being variable. Soya is highly photoperiod sensitive and even a slight variation in photoperiod could affect the flowering of the variety in question. In other words, the short season varieties require longer days to flower and mature as compared to long season varieties. The sensitivity of soya to photoperiod has been the major constraint for its wider commercial production (Weiss, 2000) and hence the varieties have to be specifically tailored for each site specific situation. In USA twelve maturity groups were recognised in the decade of nineteen seventies to assist the growers in different regions of the crop's adaptation.

Day temperatures of less than 25°C or less delays flowering and prolonged cloudy weather results in extended vegetative growth cutting into reproductive efficiency. Low night temperatures at seedling emergence also lead to slow growth rate. High temperatures in the fag-end period of seed development and maturation can adversely affect viability and vigour, resulting in lower emergence, poor growth and reduced yield (Spears *et al.*, 1997). Soybean can't be considered drought resistant as compared to other oilseed crops like sesame, since the crop needs a minimum rainfall of around 500 to 1000 mm (Nimje, 2015) depending upon the varying soil conditions to produce optimum levels of seed yield. The crop suffers when there is drought during the periods from flower bud formation to seed development.

The data from the International Soya Variety Experimental Trials of 1973 to 1982 reveal that the (i) protein and oil content were the highest between latitudes 0 to 20°C, (ii) protein content was the highest between 500 and 1000 m; and (iii) oil content was the highest at zero and 500 m, subject to the seasonal variations (He,1996*)* and confirmed by later trials (Wan *et al., 1998)*

Root system: The root system of soya plant consists of a tap root of 1.5 m along with a network of secondary roots in the soil depth of 0 to 30 cm. There are variations among the cultivated varieties for the root characteristics too. At physiological maturity, 50% of dry weight of rain-fed soya is in 0 to 15 cm soil horizon and 30% in the 15 to 90 cm horizon, while the values of root dry weight for irrigated soya are 70% in the top soil zone and 20% in the deeper soil zone. The phase of flowering and pod formation of the plant is related to root growth and development to 75 cm of soil depth.

Thus, the crop varieties to be grown on drylands depending on rains, should be faster growing with deeply penetrating tap root system.

The root nodules of soybean are due to the symbiotic association of a species specific bacterial strain *Rhizobium japonicum.* The root nodules of soybean are small, lobed to spherical and smooth. If soybean is to be cultivated on soils devoid of the above strain of rhizobium, it needs to be inoculated with the above rhizobial strain before planting. The root nodules appear in about 8 to 10 days following planting. The primary nodule that appears soon following germination contribute to the rapid growth of the seedling and the nodules on the lateral roots are known to support flowering and seed development. The nodule development and number could be affected by light intensity, which can inhibit or enhance the nitrogenase activity (Vidal *et al.,* 1996) It is estimated that the root nodulation contributes around 20 kg of residual Nitrogen per hectare; but hyper nodulation that could enhance the nitrogen addition could be achieved through discovery of a gene that regulates nodulation (Weiss, 2000). The gene is controlled by a chemical signal from young leaves in the first 4 weeks of plant growth (Weiss2000).

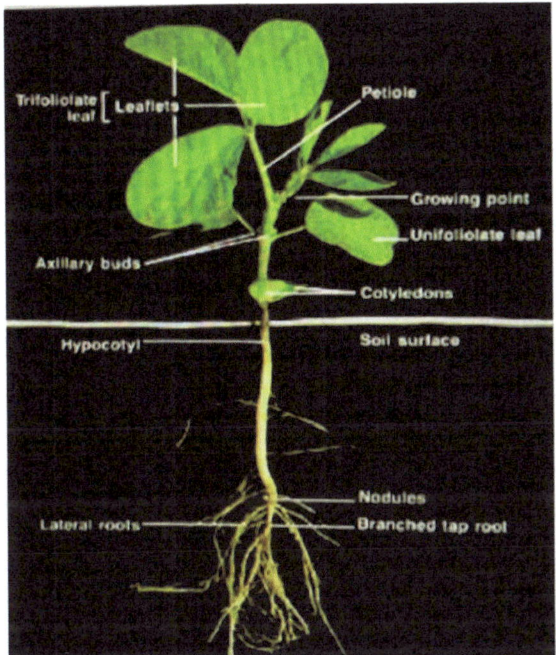

Fig. 2: Soybean seedling with all its parts labelled

Shoot system: The aerial part of soybean attains a height of 70 to 75cm. Though soybean plants are branched, the currently cultivated varieties produce less than 6 to 7 branches. The aerial branching pattern influences the growth and seed yield (Asanome and Ikeda, 1998), which can form selection criteria for varietal development (Ambuel *et al.,* 1997). The plants achieve maximum height at flowering and the lower internodes tend to become woody with age. Fasciated types exhibit shorter internodes, more number of smaller leaves per plant and longer inflorescence at the terminal portion with a larger number of flowers.

Leaves are alternate, trifoliate and dark green. The maximum leaf area index of 5 to 8 is discernible at flowering; but goes down to 4 to 6 at physiological maturity of the plan. At flowering and pod development stages, the upper leaves are reported to exhibit higher rate of photosynthesis. The plants shed leaves as the seeds ripen in the pods.

Flowers are borne in leaf axils. The inflorescence bearing short racemes have around 20 small paplinaceous white flowers. The duration of flowering is around 3 to 5 weeks. As in the case of other legumes, self-pollination is the rule in the case of soya, notwithstanding a very low level of outcrossing to the tune of 0.5 to 1.0%. Despite the minor role of insects as pollinators in soybean, honey bees have been used in soya hybrid development, as the studies indicated an increase in seed yield to the tune of 20% (Weiss,2000).

The flowering pattern varies in indeterminate and indeterminate types. The indeterminate and determinate types exhibit flowering starting at the fourth to the eighth node and then progressing upwards. The flowering on the side branches occurs later than that of the main stem.

The determinate types on the other hand exhibit flowering at the 8th and 9th nodes. Then, flowering proceeding rapidly, ensures flowering at every node and on the side branches as well. The proportion of flowers to pods is reported to be 20 - 25%.

Pods and Seeds: The pods are normally short, hairy, usually brown or back with occasional green, red or purple tinge. A single pod on an average is of 2 to 10cm x 2 to 4 cm dimensions. Pod shattering is seen in soybean; but the extent of shattering varies with the genotype in question. Pod number per plant is an important yield component, which varies from 30 to 55 and even more depending upon the variety, climatic conditions during the growth and development, particularly flowering. Each pod contains 3 or more small hard seeds. The seeds could be round or oval in shape, measuring 5 to 10 mm in diameter.

Seed coat or testa is smooth and shiny with a conspicuous hilum. Seed colour also varies as yellow or black or brown or red or green or slightly mottled. Seed weight varies from 5 to 40 g/100 seeds depending upon the variety. The general preference appears to be for yellow seeded varieties.

Oil content of seed varies from 15 to 22%, which is basically specific to each variety; but influenced by the environmental conditions. Soybean oil is basically richer in oleic acid the proportion of which is found to vary from 50 to 57%. The other fatty acids of the seed oil are linoleic (30%) and linolenic (10%). The relatively higher proportion of linolenic acid as compared to maize oil (1%) specifically is the cause for its poor flavour or flavour instability.

The Protein content of soya seed is of more commercial importance since the seed contains as much as 40 to 50% protein with all complex of amino acids including those essential ones for human diet. Protein content of soya seed is negatively correlated with the oil content and high erprotein content is associated with lower seed yield. Higher temperatures during crop growth favour higher oil content and relatively lower protein content (Weiss 2000).

It is stated that the more rapid rate of oil accumulation perhaps reduces the assimilates needed for protein conversion (Weiss,2000).

Apart from containing the protein and oil, soya seed is also rich in nutrients viz., high calcium, iron and vitamin content, which are of great value in human diet.

Taxonomy, Cytogenetics and Evolution of Soybean: The taxonomy of the species within the genus *Glycine* is somewhat nebulous (Weiss 2000). *Glycine max (*L.) Merill (syn. *G. hispidaSoya max, Dolichos soja etc.)* is placed in the subgenus *Soya (*Moench) F.J. Herm., together with *G. soja* Sieb & Zucc. (syn. *G.*

ussuriensis). Both have the chromosome number 2n=40 along with most of the subgenus *Glycine*. Verdc.

Soybean is reported to be an ancient polyploid (2n=40), which has got stabilized or diplodised over its long period of evolution (Weiss 2000).

As per the available cytogenetic evidence, the current soybean and its putative ancestor *G. soja (G. ussuriensis)* are of the same species, but have been maintained separately as a practical aid to plant breeders (Weiss 2000).

While the currently grown cultivars of soybean are erect annuals, the supposed ancestor cited above is a wild twining wine. According to Li, (1995) the cultivated soya has evolved from the wild ancestor through the inter- mediate types; the observation being based on seed coat characteristics.

Shou *et al.*, (1998*)* observed that the wild soya has thicker seed coat which makes it resistant to disease and deterioration in storage. Through long domestication, the currently cultivated soya has lost much of its original diversity through continuous selection for high yield and uniform maturity (Weiss 2000). The narrowness of genetic diversity reflected among the existing American varieties of soya is due to the fact that 80 ancestral genotypes contributed to all the foundation germplasm giving rise to around 250 commercial varieties released between 1947 and 1988, of which only 50% contributed by only six (Burton 1997). The cytogenetic knowledge of soybean lags far behind that of other important crops viz. rice, maize, wheat, tomato, because its somatic chromosomes are symmetrical and only one pair of satellite chromosomes can be identified. Molecular linkage maps have been associated with specific chromosomes, and soybean genome has been sequenced (SinghRJ, 2017)

Glycine max its wild progenitor, *G. soja*, have originated in eastern Asia and are the only two species in the annual subgenus *Soja.* Wild soybean germplasms exhibit high genomic diversity and hence may be an important source of novel genes. The perennial species of *Glycine* available in Australia contain several useful characters, especially resistance to disease causing organisms such as cyst nematode and leaf rust. The perennial species are being investigated with a focus on genomic regions of genes that could be used in breeding. The homoeologous regions in the genus are due to the role of polyploidy in the original ancestor *Glycine* species. Various processes related to the evolution of polyploidy are being investigated. (Broyles *et al*., 2014)

Singh and Nelson (2015) crossed several soybean cultivars with six accessions of *G. tomentella* with 78-chromosome as well as one accession each of 40-chromosome *G. tomentella*, *G. argyrea* and *G. latifolia*. They were selected due to their resistance to soybean rust. A fertile soybean cv. 'Dwight' *G. tomentella* accession PI 441001 with and 78-chromosomes were produced. The rest of the hybrids were discontinued either at F_1 or amphidiploid stage. Having faced with the abortion of F_1 seeds, the developing seeds from 19 to 21 day old pods were cultured aseptically in various culture media formulations leading to the successful seed maturation and multiple embryo generation. F_1 plants with shoots and roots ($2n = 59$) were transplanted to pots in greenhouse. Amphidiploid ($2n = 118$) plants were backcrossed to 'Dwight'. BC_1 ($2n = 79$) plants and were propagated in *vitro* and 43 mature BC_2F_1 seeds were harvested. Fifteen surviving BC_2F_1 plants were morphologically distinct, sterile, and had chromosome numbers ranging $2n = 56–59$. Chromosome numbers of the BC_3F_1 plants ranged $2n = 40$ to 49. The fertile soybean derivatives were taken to field in 2008 (Singh and Nelson,2015).

Soybean Genetic Resources and Breeding: The rapid growth in production and importance of soybeans as a crop is unequalled in U.S. agriculture. Production of soybeans is centred in the Corn Belt and the lower Mississippi Delta. The rapid increase in acreage could not have been accomplished without the progress that was made during the same period in the breeding of highly productive, disease-resistant varieties, adapted for specific ecological areas and for the mechanization of production and harvest (Poehlman, 1987). Soybean genetics, cytogenetics, molecular genetics, and genomics have made important contributions to soybean breeding and soybean production. As the sequencing of the soybean genome is completed, new opportunities will arise in each of these areas that will impact soybean breeding and ultimately soybean production. Soybean breeding methods for development of commercial cultivars continue to evolve. Although soybean breeders are successful in producing cultivars using traditional conventional breeding methods, the most widely used methods have shifted in recent decades to wide use of single seed descent and, with the advent of transformation, renewed use of backcrossing. Many breeders are using molecular-based plant breeding methods and techniques, such as marker-assisted selection, as a part of their cultivar development programme (Orf, 2008).

Despite the availability of modern tools of genetic analysis and plant breeding, the need for enriching the genetic diversity among cultivated varieties of soya is felt due to the proliferation of a wide gamut of biotic stresses that inflict soya as the consequence of narrowness of genetic diversity among the cultivated varieties. Plant breeders and geneticists are concerned with losing many traditional local land races in China and long standing traditional varieties in the USA (Hymowitz, 1976). The soybean breeders, worldwide, are confined to crossing within the primary gene pool; thus, genetic base of soybean is very narrow. Wild perennial *Glycine* species of the tertiary gene pool have been recently exploited to broaden the genetic base of modern soybean cultivars. The United States possesses a vast and major collection of soya genotypes numbering over 2,000 grain types at Urbana and around 1000 forage types at Stoneville. China has around 1500 types of soya at Sichuan crop Institute in addition to other collections of soya maintained elsewhere (Weiss, 2000).

The ICAR-Indian Institute of Soybean Research has an active germplasm repository consisting of over 4,591 accessions (https://iisrindore.icar.gov.in/) out of which some accessions possessing unique attributes have been identified for use in breeding, which are presented in Table 1.

Table 1: Soybean Germplasm Accessions identified for various attributes.

Character	Accessions identified
Photoperiod insensitivity	MACS 330, EC 325097, EC 333897, EC 34101, EC 325118, EC 390977 and EC 483049
Prolonged juvenile phase	AGS 25
Drought Tolerance	EC 538828, PI 416937, Young, EC 602288
Tolerance to Rust	EC 241778 and 241780
Tolerance to Yellow Mosaic Virus	UPSM 534, PLSO 84, EC24660, B463
Tolerance to Rhizoctonia root rot	AGS-48, EC 325099, EC 34117
Tolerance to Girdle beetle	TGX 863-26E, SREC 56A, TGX 302-2A, TGX 849-D
Tolerance to defoliator insects	EC 333902, VP 1165
High Oil Content	AGS 191, NRC 7, G76 (22-23%)
High oleic acid	IC 210 & NRC 106 (42%) low linolenic acid: VLS 59 (4%)
Null trypsin Inhibitor	NRC 101, NRC 102
High Protein content	G 288, G 688 (44%)

The currently cultivated types of soya for seed yield are of bushy, erect annuals of 60 to 75 cm height with good degree of branching and well developed root system. The pods of cultivated soya are small with yellow or black seeds. The various growth stages of soya are very well defined by FAO through published international standards (Fig.3).

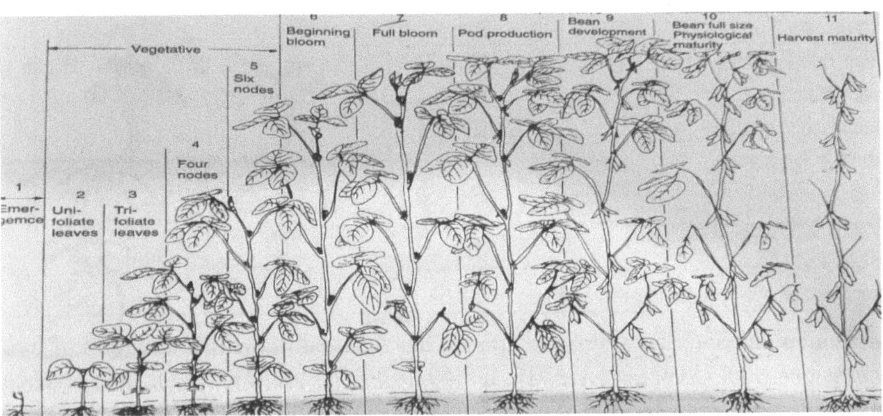

Fig. 3: Growth stages of Soybean plant (Weiss, 2000)

Canopy development as a selection criterion in soybean breeding for yield: Certain genotypes of soybean exhibited the potential to attain full canopy coverage (100%) by 50 days following planting. However, others did not exhibit full canopy coverage even at 64 days after planting, thereby indicating their incapability of reaching the higher levels of canopy development in a production system employing a row spacing of 76 cm (Xavier *et al.*,2017). The investigations carried out earlier on soybean (Weaver and Wilcox 1982; De Bruin and Pedersen 2008) indicate that yield and canopy traits are under similar genetic control when grown using narrower row spacing, with slight changes in heritability.

As a selection parameter, the character of canopy development presents attractive genetic properties, such as high heritability and strong genetic correlation with grain yield that allows for indirect selection or prediction of grain yield potential (Xavier *et al.*, 2017). In order for the indirect selection to be functional and effective, the response of grain yield potential when selecting for average canopy coverage must be greater than the response of selecting for yield directly. Certain cases of successful indirect selection have already been reported (Bernardo 2010). Now it is evident that efficient indirect selection for yield at the same selection intensity is possible with average canopy coverage because the correlated response to selection for grain yield is greater than the response through direct selection (Xavier *et al.,* 2017). Employing canopy development at peak flowering as a selection criterion has also been suggested for efficient groundnut breeding, since the attribute exhibited higher stability and direct positive correlation with kernel yield (Nagabhushanam and Prasad, 1992).

The Indian Institute of Soybean Research has reported the breeding work resulting in several soya varieties developed for specific agronomic attributes as detailed in Table 2.

Table 2: Soybean Varieties developed at ICAR-IISR India for specific attributes

Characters	Specific varieties
High Yield Potential	JS 97-52, JS 95-60, NRC 37, JS 93-05, NRC 7, JS 335
Early Maturity	JS 95-60, JS 93-05, LSb 1, NRC 7, JS 20-34
Tolerance to drought	NRC 7, Hardee, JS 71-05
Multiple Resistance	VLS 63, Pusa 9814, PS 1347
Resistance toYMV	JS 97-52, SL 688, SL Pusa 9814, SL 525, PS 1347
Tolerance to Insect Pests	TAMS 978-21, Pratap Soya 1 and 2
High Seed Longivity	JS 335, MAUS 61-2, JS 93-05, NRC 37, JS 97-52
Improved oil quality (Low linolenic acid)	VLS 59

Molecular characterization of germplasm has resulted in the identification of five accessions (PI 175192, EC 333915, CF 492, EPS 71 and DT 21) for the presence of unique SSR alleles. An advanced backcross mapping population (BC2F4) has been developed from interspecific cross JS 335 x Glycine soja and QTLs identified for 100 seed weight, number of pods and seed yield per plant. The varieties viz., JS 97-52, DS 97-12, JS 93-05 and MACS 57 have been converted to their null-KTI (null *KunitzTripsinInhibitor*) version.

Fig. 4: Soybean Variety NRC 7 developed for high yield at ICAR-IISR, India

Resolving biochemical constraints that affect the nutritional value of Soybean through plant breeding

Notwithstanding its nutritional value, soybean contains certain bio-chemical components that limit its nutritional value. Two of such constraints are the presence of trypsin inhibitors viz., Kunitz Trypsin Inhibitor (KTI) (20kDa) and Bowman-Birk factor (8kDa). These inhibitors forming 25% of soya protein are responsible for growth inhibition, pancreatic hypertrophy and hyperplasia (Rawlings *et al.*, 2004; DiPietro and Liener, 1989)

KTI constitutes 80% of trypsin inhibitors. The absence of KTI protein is inherited as a recessive allele. Hence the development of KTI free soya bean through introgression of the allele into the high yielding varieties should be an important mandate. Molecular marker-assisted backcross breeding (MABB) approach was adopted for genetic elimination of KTI from two popular soybean genotypes, DS9712 and DS9814. PI542044, an exotic germplasm line was used as donor of the kti allele which inhibits functional KTI peptide production (Shivakumar *et al.*, 2016)

The fatty acid profile of the soybeans has been enhanced through mutation breeding and this approach has also resulted in the production of soybeans tolerant to herbicides (SinghRJ, 2011). By using recombinant DNA technology, Monsanto has produced stable glyphosate tolerant soybean lines known as *'Round Up*

ReadySoybean' and DuPont is producing transgenic soybean lines with improved fatty acids content. However, the feasibility of developing hybrid soybeans is still an open question (SinghRJ, 2011).

Mutation breeding: Soybean seeds of the germplasm accession JTN-5203 were treated with with 60 mM EMS (Ethyl Methane Sulfonate) and population of 1,820 M1 individuals were produced from 15,000 treated seeds. The resulting M2 population was screened for seed traits including total oil, protein, starch, moisture content, fatty acid and amino acid compositions measured by NIR. Phenotypic variations observed in this population include changes in leaf morphology, plant architecture, seed compositions, and yield. Of most interest, were the plants with increased amounts of total protein (50% vs. 41% for control) and plants with higher amounts of total oil (25% vs. 21.2% control). Also identified were the plants with increased oleic acid content and decreased linoleic acid and linolenic acid contents. The EMS mutant populations were used for further studies including screening for various traits such as amino acid pathways, allergens, phytic acids, and other important soybean agronomic traits. In addition, these mutant individuals were evaluated in the next generation to assess the heritability. Beneficial traits from these mutants can be exploited for future soybean breeding programs (Espina *et al.,*2018)

Efficient targeted mutagenesis in soybean through the use of TALENs and CRISPR/Cas9 has been carried out by Du *et al.,* (2016), who presented the comparative analysis of these two technologies for two soybean genome editing targets, *GmPDS11* and *GmPDS18*. They found GT in soybean hairy roots with a single targeting efficiency range of 17.5–21.1% by TALENs, 11.7–18.1% by CRISPR/Cas9 using the *AtU6-26* promoter, and 43.4–48.1% by CRISPR/Cas9 using the *GmU6-16g-1* promoter, suggesting that the CRISPR/Cas9 using the *GmU6-16g-1* promoter is probably a much more efficient tool compared to the other technologies. Similarly, the double mutation GT efficiency experiment with these three technologies displayed a targeting efficiency of 6.25% by TALENs, 12.5% by CRISPR/Cas9 using the *AtU6-26* promoter, and 43.4–48.1% by CRISPR/Cas9 using the *GmU6-16g-1* promoter, suggesting that CRISPR/Cas9 is still a better choice for simultaneous editing of multiple homoeo-alleles. Furthermore, Du *et al.,* (2016) observed albino and dwarf buds (*PDS* knock-out) by soybean transformation in cotyledonary nodes. The results demonstrated that both TALENs and CRISPR/Cas9 systems are powerful tools for soybean genome editing.

***Genome Sequence of soybean**:* Schmutz *et al., (*2010) sequenced the 1.1-gigabase genome of soybean by a whole-genome shotgun approach and integrated it with physical and high-density genetic maps to create a chromosome-scale draft sequence assembly. Prediction was made of 46,430 protein-coding genes, 70% more than *Arabidopsis* and similar to the poplar genome which, like soybean, is an ancient polyploid, which might be called "*palaeopolyploid*". It was observed that around 78% of the predicted genes occurred in chromosome ends, which comprised less than one-half of the genome but accounted for nearly all of the genetic recombination. Genome duplications occurred at approximately 59 and 13 million years ago, resulting in a highly duplicated genome with nearly 75% of the genes present in multiple copies. The two duplication events were followed by gene diversification and loss, and numerous chromosome rearrangements. An accurate soybean genome sequence will facilitate the identification of the genetic basis of many soybean traits, and accelerate the creation of improved soybean varieties

Soils: Soya has exhibited its adaptability under wide range of soil conditions; but the highest seed yields are obtained on well-drained fertile loamy soil (Weiss, 2000). The soil pH levels favourable to the crop are 6.0 to 6.5; but the crop's performance was good even under pH 5.8 to 7.0 (Weiss, 2000). Higher levels of soil acidity shall bring down the root nodulation. Availability of magnesium and calcium doesn't promote root nodulation. Even very low levels of Boron (even 1ppm) in the soil is reported to be toxic to soya. Soya also doesn't tolerate soil salinity. Wild perennial types of soya native to Australian coast possess genes for salt tolerance (Weiss, 2000), which can be availed of to enhance the salt tolerance of the otherwise adaptable varieties.

Crop nutrition: The data generated so far indicates that soybean is a heavy feeder, since the crop is reported to have removed more phosphorous, potassium, magnesium and calcium as compared to maize crop (Weiss 2000). The total uptake of three major nutrients by soybean as per the data generated at the University of Illinois USA is given below in the Table 3. (Weiss, 2000).

Table 3: Total uptake of NPK at two yield levels –USA (Weiss, 2000)

	Seed Yield (kg/ha)	
	3500	6500
Nitrogen (kg)	340	620
Phosphorous (kg)	30	56
Potassium (kg)	170	225

Out of the total quantum of nutrients utilised, 70 to 80% of nitrogen and phosphate and 60% of potassium are accumulated in the seed at harvest. These values may not be constant; at the time of leaf senescence, 80% of the total nutrients are found in seeds (Weiss, 2000). Among the nutrients, Phosphorous is mostly needed by the plant, since the nutrient is widely dispersed throughout the root zone. Application of lime to the soil where soya is grown, results in increased seed yield. Hence, it is recommended that lime be applied to soil where seed production is undertaken (Weiss,2000). Variable varietal response nutrients has been observed.

Nitrogen (N): The uptake of N by the soybean crop is at the rate of 5 kg/day after 90 days, after which the uptake of N is observed to decline (Weiss, 2000). This daily need of N by the soya plant is adequately provided by the root nodules containing *Rhizobium* bacteria. According to commercial growers of soybean in USA, well nodulated soya root system can provide with N adequate to obtain seed yield of 3500 kg/ha. The plants possessing hyper nodulating root system could produce double the above yield. Young soya plants may exhibit sever N deficiency when irrigation is given in excess or following a heavy rain at the time of seedling emergence (Weiss, 2000) and in such cases top dressing of nitrogen is recommended, before the nodules start supplying adequate quantity.

Nitrogen application at planting may not be needed since that would supress nodulation.

Phosphorous (P): Though soya plant needs more phosphorous when planted in rotation with other heavy feeder crops, it is advisable to resort to soil test values regarding the availability of Phosphorous, since yield response is unlikely if soil tests show more than 30 kg of available P / ha (Weiss, 2000).

The effect of P on root nodulation is erratic since both deficit and excess can adversely affect nodule numbers. It was observed that potassium level is an important factor in this context. Inoculation of seed with *pseudomonas striata* increased the uptake of P, resulting in higher seed yield (Dubey, 1996). Adequate levels of P are absolutely necessary to ensure optimum seed protein and amino acids in the protein component of seed. On the otherhand excess P level may bring down seed protein and oil content or induce zinc and iron deficiencies (Weiss, 2000).

Potassium (K): Soya plant requires large amount of K during vegetative growth, since the highest concentration occurs in the stem and lower leaves. Pods that develop later and seeds at maturity also show higher levels of K. Uptake of K slows down as seeds begin to form (Weiss, 2000). Potassium deficiency retards nodulation; but the reasons behind this phenomenon is yet to be investigated. Application of right quantum of K produces healthier plants with very good nodulation. Although the quantum of Potassium needed equals that of P, the soil test values should be the guide. Adequate level of K reduces the incidence of fungal diseases (Weiss, 2000).

Sulphur (S): The sulphur requirement of soya plant may be higher as compared to that of legumes. In India application of sulphur @ 50 kg /ha through gypsum appears to be necessary to get satisfactory yield of soya (Weiss, 2000). An interaction of sulphur uptake with other micro nutrients was observed to affect the plant growth and seed yield (Fontes and Cox, 1998). If the sulphur deficiency is small, it can be corrected by the application of single super phosphate as the source of P.

Calcium (Ca): Calcium may be necessary to stabilize the growth of soya by correcting higher levels of soil acidity. Apart from alleviating the calcium deficiency, the element also maintains the balance between calcium and magnesium – potassium availability. Calcium is also necessary for the development of nodule bacteria and effective nodulation. Spehar and Galwey (1997) developed a method of screening cultivars for tolerance to low levels of calcium.

Manganese (Mn): Differential genotypic response of soya in terms of accumulation and distribution with in the plant in respect of Mn has been observed (Abreu *et al.,*1996 and Oliveria *et al.,* 1997). Soya is very sensitive to Mn deficiency in soils of high pH. The deficient plants remain stunted with yellowish leaves. The deficiency may be corrected either through soil application or by foliar spray. The manganese: iron ratio is important since deficiency of one will reflect in terms of the toxicity of the other. A desirable ratio could be manganese 2.5: 1.5 iron. Manganese toxicity usually appears on soils of below pH 5.0. Liming corrects the toxicity in majority of cases.

Other nutrients

- Boron (Bo) content is reported to have a positive influence on germination, since low Bo content in seed is known to reduce its viability and seedling vigour (Rerkasem *et al.,* 1997; Rodrigues *et al.,*1997). Wang *et al.,* (*1*992) reported that urea with 4% borate not only resulted in higher yields; but also enhanced the soya crop response to Phosphorous and nitrogen.
- Seed treatment with cobalt, copper or molybdenum can improve the yield, where ever such nutrients are deficient. Additionally, it may cut down the *Septoria* infection.
- Zinc deficiency which is very common and widespread in India can be rectified by application of 10kg of zinc/ha as zinc ammonia chloride which is reported to be more effective than zinc sulphate (Raghuvanshi *et al.,* 1997).
- Selenium is found to increase resistance to spider mite in China (Liu *et al.,* 1998)

Soybean based cropping systems

(i) Crop sequences

- The researches carried out at the Indian Institute of Soyabean Research over 9 years (2001-10) revealed that the inclusion of maize in crop rotation either as sole or as intercrop in soybean significantly enhanced the soybean as well as wheat productivity, enzyme activities (dehydrogenases, fluoresce in diacetate, ß-glucosidase, acid and alkaline phosphatases, aryl sulphatase and urease enzymes), soil microbial respiration, biomass net returns, AM Fungi spore density, B:C ratio, net energy output, energy use efficiency and sustainability yield index (Anonymous, 2010) .

- Among the cropping systems, soybean- wheat was found to be most productive, remunerative, energy efficient and productive than soybean-chickpea and soybean – mustard cropping system (Anonymous, 2010).
- The soybean-potato-mung bean system recorded the highest system productivity in terms of soybean seed-equivalent yield (7.68 t/ha). However, soybean-chick pea-fodder sorghum system showed the highest net energy (333.3 x 10x10x10 MJ /ha), energy efficiency (9.56), energy productivity (179g MJ) and energy profitability. This system also maintained highest carbon out-put and carbon efficiency (Prajapat *et al*., 2018)
- Soybean-chickpea system had an edge over soybean-wheat in case of renewable energy productivity and intensiveness (Billore *et al*., 2010)
- Mandal *et al*., (2002) reported that energy consumption was the highest in soybean–wheat (23705 ± 29.48 MJha−1) and the lowest in soybean–chickpea (15672 ± 26.29 MJha −1). The total bioenergy output of the crop production systems followed the order: soybean–wheat (131277 ± 29.26 MJha−1)> soybean–mustard (101661±28.91MJha−1)> soybean–chickpea (92658±28.87MJha−1). But this order was reversed for energy-use efficiency (EUE): soybean–chickpea (5.91)> soybean–mustard (5.86)> soybean–wheat (5.54). Specific energy was highest in soybean (9173MJt−1 grain) followed by mustard (8912MJt−1 seed), chickpea (7190MJt−1 grain) and wheat (6646MJt−1grain) indicating that soybean is the most energy-investment intensive crop. Though the net return from soybean–wheat was marginally higher than other systems, the soybean–chickpea system is more suitable in the central ecological niche of India due to its low requirement for non-renewable resources, higher EUE and benefit–cost ratio.
- Cereal crops grown in rotation with soybean, get benefitted from the residual nitrogen particularly when the soya crop residues are incorporated into the soil following its harvest (Vance and Graham, 1995). This will result in a significant reduction of nitrogen needed for the following crop (Bloem and Barnard 1995, Kessavalou and Walters 1997)
- Jacques *et al., (*1997) reported double cropping of soybean was the most profitable as compared to single and relay cropping. However, the damage inflicted by the diseases increases after two successive crops of soybean (Weiss, 2000). Soya root vestiges left in the soil after harvest might have an

adverse effect on the growth of the succeeding crop of soybean (Han and Xu, 1988).

(ii) Intercropping

- Kamara *et al.*, (2017) reported that it was advantageous to grow maize and soybean in association in an inter cropping system than in pure stands.
- ZhangY *et al.*, (2015)found that maize intercropping with soybean (maize - soybean) was the optimal cropping system in summer. Compared to conventional monoculture of maize, maize - soybean had significant advantage in yield, economy, land utilization ratio and reducing soil nitrate nitrogen (N) accumulation, as well as better residual effect on the subsequent wheat crop. Also, the intercropping systems reduced use of N fertilizer per unit land area and increased relative biomass of intercropped maize, due to promoted photosynthetic efficiency of border rows and N utilization during symbiotic period. Intercropping advantage began to emerge at tasseling stage after N topdressing for maize. Among all treatments with different row ratios, alternating four maize rows with six soybean rows (4M:6S) had the largest land equivalent ratio (1.30), total N accumulation in crops (258 kg ha^{-1}), and economic benefit (3,408 USD ha^{-1}). Compared to maize monoculture, 4M:6S had significantly lower Nitrate-N accumulation in soil both after harvest of maize and after harvest of the subsequent wheat, but it did not decrease yield of wheat. The most important advantage of 4M:6S was to increase biomass of intercropped maize and soybean, which further led to the increase of total N accumulation by crops as well as economic benefit.
- Raza, (2019) revealed that strip relay-intercropping increases the N, P, and K uptake and distribution across plant organs (root, straw, and seed) of maize and soybean, accelerates the dry-matter production of intercrop-species, and compensates the slight maize yield loss by considerably increasing the soybean yield. In a two-year experiment, soybean was planted with maize in different planting patterns (SI, 50:50 cm and SII, 40:160 cm) of relay-intercropping, both planting patterns were compared with sole cropping of maize (SM) and soybean (SS). As compared to SI, SII increased the N, P, and K accumulation in each organ of soybean by 20, 32, and 18 (root) %, 71, 61, and 76 (straw) %, and 68, 65, and 62 (seed) %, respectively, whereas decreased the N, P, and K accumulation in each organ of maize by 1, 4, and 8 % (root) %, 1, 10, and 3 %(straw) , and 5, 10, and 8% (seed) , respectively. Overall, in SII, relay-cropped soybean accumulated 91% of total nutrient uptake (TNU) of sole soybean plants, and relay-cropped maize accumulated 94% of TNU of sole maize plants.

- An investigation was carried out on relay inter cropping of soybean with winter wheat, in which soybean was planted earlier than the winter wheat by Goldmon (1991). Soybeans grown in narrow row spacing and those with indeterminate growth habits produced the greatest yields in both sole and intercrop systems. In 1990, the land equivalent ratio was 1.18 with the wheat component comprising over 80% of the total. Sole soybeans utilized more soil water and intercepted more light than intercropped soybeans after wheat was harvested. Among the intercropped treatments, soybeans grown in narrow row spacing and those with an indeterminate growth habit were also advantageous in light interception.
- Maize intercropped with soybean in 2:6 paired row (45/180 cm) recorded the least light transmission ratio and the highest light absorption. Though sole maize recorded significantly higher yield (70.92 q ha-1) than intercropped maize under varying geometry and row proportion, it was on par with maize intercropped with soybean in 1:1 row proportion with 60 x 20 cm (70.03 q ha-1). Highest net return (` 57,926 ha-1) and B:C ratio (3.57) were obtained with maize intercropped with soybean in 2:6 paired row. Similarly, higher land equivalent ratio (1.54) and maize equivalent yield (94.70 q ha-1) were also noticed with Maize+Soybean intercropping in 2:6 paired row than other treatments (Yogesh *et al.,*2014).
- In intercropping soybean and sorghum cultivars, results in the first season showed that, seed yields of early and medium maturing cultivars of soybeans (TGX536-02D and SAMSOY-2) were at par, however, significantly higher than the late maturing ones (TGM 344 and Malayan). Soybean seed yields in the second and third seasons of intercropping differed significantly with TGX536-02D cultivar producing higher yields, when cropped with a semi-dwarf sorghum (SAMSORG-17) variety. The land equivalent ratios (LERs) based on the sole crop yields of individual crop and legume components, provides a quantitative evaluation of the yield advantage due to intercropping. Total LER observed, ranged from 1.11 to 2.60; indicating a greater advantage in legume based-intercropping (Dantata 2014).
- Relay intercropping takes advantage of the relatively high value of soybeans compared to wheat, but there is a trade-off of reduced wheat yield with later soybean seeding which results in higher soybean yields (Jefers *et al.,* 1977)
- soya intercropping in India enhances sugar cane yield by a minimum of 7 to 10 per cent as compared to sole cropping of sugar cane. In addition, soybean crop residue can provide cattle fodder to sugar cane farmers who face immense problem in raising enough cattle fodder till cane-harvest. In view of the indirect benefits that soyabean intercropping with sugar cane provides, agriculturists are convinced that the pattern of intercropping

with sugar cane is "agronomically advantageous and economically more remunerative to, the needy farmers" (Down To Earth, 2015).

- Incidence of biotic stresses having adverse effect on soybean is found to be much less on soybean intercropped with other cereals, legumes and other oilseed crops. According to Gangwar *et al.*, (1994), there was a significant reduction in the incidence of insect pests and the extent of damage to soybean. This, however, varied with the crops involved (Shukla, 1995).
- A review of intercropping of soybean crop has since been published by Mehtre *et al.*, (1997)
- Sanbagavalli *et al.,* (2016) reviewing the work on a wide range of crop species grown in association with soybean reported that soybean can be advantageously intercropped with most of the traditional crops grown in different agro-climatic regions of India. In addition to higher combined yield, crop diversification, fertilizer economy, salutary soil environment for plant growth, smothering effect on weeds, natural check on pests and diseases and risk coverage, adoption of the system is likely to provide sustainability.

Insect pests of Soybean

A wide range of insects are known to damage soybean crop out of which the following are the most important.

1. **Stem fly (*Melanagromyza sojae* Zehntner) (Diptera: Agromyzidae) :** A major seedling pest of soybean crop in India, which causes serious damage at the seedling stage. The adult stem fly lays its eggs often in the first pair of leaves leaf tissue of 10-day old soybean seedlings. The emerging maggots mine the leaves or bore into the leaf petiole or tender stem resulting in tunnels leading to death of plant.

Fig. 5: Maggot of stem-fly boring into stem tissue of soybean plant

Fig. 6: Adult stem-fly *Melanagromyza sojae* Zehntner

2. **Tobacco caterpillar** ***(Spodoptera litura*** Fabricius) (**Lepidoptera: Noctuidae)** : Female moth lays egg mass on the lower side of young leaves. The eggs hatch out as the gregarious caterpillars feed on leaves. The larvae also feed on stems, buds, flowers and pods

Fig. 7: Larva of Spodoptera litura Fabricius

3. **Green semiloopers (*Chrysodeixis acuta*** Walker, ***Gesonia gemma*** and ***Diachrysia orichalcea Fabriciussensu*** Hübner) The semilooper larva is green in colour with several thin light lines running the length of the body. When crawling, it forms a characteristic loop or hump so is known as the semilooper. The full grown larva feeds on foliage, flowers and pods. In severe infestation, it defoliates the plant leaving behind only midribs. The female moth lays tiny white creamy eggs on the underside of the leaves or young stems. The eggs hatch 3-4 days later. The newly emerged caterpillars feed on the lower surface of the upper canopy of soybean leaves in a characteristic manner, leaving the top epidermis intact background

The semilooper feeds on soybean crops and causes yield losses up to 30-40%. It affects the crop at the vegetative stage and if the infestation is at the podding or flowering stage, it can cause considerable yield losses. Usually the infestation occurs before late August and early September.

Management

- Practise deep summer ploughing
- Install bird perches at 20/ ha in the field
- Spray *Beauveria bassiana* 4 ml/litre of water during high humidity
- Spray 5 % Neem Seed Kernel Extract (NSKE) at flowering stage as a preventive measure
- Spray quinalphos 25% EC at 30 ml or chlorpyriphos 20% EC @ 30 ml in 10 litres of water

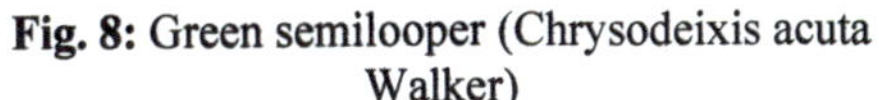

Fig. 8: Green semilooper (Chrysodeixis acuta Walker)

Fig. 9: Adult moth of Chrysodeixis acuta Walker

4. Girdle beetle (*Obereopsis brevis* Gahan) (Coleoptera)

The incidence of girdle beetles is observed at the seedling stage. The presence of 2 circular cuts on the branch or stem is a characteristic symptom. Larvae bore into the stem of soybean. The inside of the stem is eaten by the larvae, forming tunnels within the stem. The infected portion above the circular cut is unable to get any nutrition so it dries up. In the later stages of infestation, the plant is severed at about 15 to 25 cm above the ground.

The main damage is caused by the larvae of the insect. The attack of the insect initially begins in the last week of July to the first fortnight of August. The insect remains active from July to October, damaging the crop most severely during August and September. Heavy incidence may reduce the yield by up to 40%.

Management

- Practice deep summer ploughing
- Sow the crop at the proper time i.e. at the onset of monsoon, with the recommended seed rate of 75 kg/ha. The crop should not be overcrowded
- Sow tolerant varieties such as NRC-12 or NRC-7
- Avoid excess nitrogenous fertilizers
- Destroy crop residues
- Collect and destroy infested plant parts at least once every 10 days
- Spray NSKE 5% or azadirachtin 10,000 ppm @1 ml per 1 litre of water to avoid the pest laying eggs
- Apply pesticides if pest population crosses the Economic Threshold Level (5 per cent damaged plants)

- Apply Cartap hydrochloride 4 G @ 4 kg/acre at the time of sowing
- Spray lambda cyhalothrin 5% EC @ 10 ml or dimethoate 25% EC @ 2 ml per liter of water 30-35 days after sowing and repeat 15-20 days after the first spraying if infestation is observed
- Always use a hollow cone spray nozzle for spraying

5. Pod borer (*Helicoverpa armigera* Hubner) (Lepidoptera: Noctuidae)

The green coloured caterpillars feed initially on the tender leaflets by scrapping green tissue and then move on to flower buds and tender shoots. Later the pest turns into a pod borer, feeding on the developing seeds with a portion of larvae burrowing the pod. Each cater pillar can feed on 20 to 25 pods.

Fig. 10: Lava of Heliothes armigera Hubner

6. White fly (*Bemisia tabaci* Gennadius)

Whiteflies are small, approximately 1/25 inch or 1 mm in length, with a pale yellow body and two pairs of white wings and covered with a white waxy powder. At rest, wings are held in an inverted V position. Their compound eyes are red.

Adults usually emerge from their pupal cases in the morning hours and may copulate a few hours later. Oviposition occurs from 1 to 8 days after mating. Adult life span ranges from 6-55 days' dependent on temperature. Females live only 10-15 days under southern continental U.S. summer conditions, but can live several months during the winter.

In this species, reproduction can occur with or without copulation. Unmated females can reproduce by parthenogenesis in which the females produce only male progeny. Females lay 80 to more than 300 eggs in their lifetime. The plant host reportedly plays an important role in female fecundity.

Adult females oviposit preferentially on young foliage and the immature stages tend to be distributed vertically on the plant with older stages found on progressively older leaves.

The whitefly has a preference for hairy-leafed varieties of cotton and less preference for glabrous-leafed varieties (Butler and Wilson, 1984).

Whiteflies have two different flight patterns: short-distance and long-distance flights (Berlinger, 1986). Short-distance flights remain within the plant canopy and travel from plant to plant within a field. Short-range movements within and between cultivated and weed host plants are known to take place regularly. There is also evidence of long-range whitefly migration. However, the extent to which migration results in colonization of a new habitat by this whitefly has not been satisfactorily established. The direction of whitefly flight is primarily dictated by the wind.

Direct damage due to whitefly occurs when the stylet of whitefly pierces the leaves and suck the sap, thus causing chlorosis in plants. While, the indirect damage occurs due to the accumulation of honeydew that catalyses the growth of sooty mould on the entire surface of the leaf and disrupts the process of photosynthesis. In addition, whiteflies act as vectors of Yellow Mosaic Virus (YMV) resulting in 15-17% of yield loss

Management of Whitefly

High reproductive rate and multiple host sequences provide optimal conditions for whitefly population development. The varied habitats, seasonal population development and intra and inter-crop and wild host movement present an extremely complex and difficult challenge requiring new and innovative approached for formulating control and suppression methodology.

There is really no easy way of controlling the whitefly. Egg mortality is usually minimal. Weather and predation may cause high mortality rates during the crawler and first nymphal stages, but has only moderate effects on the later nymphal stages. In the past adults were easily killed with insecticides but pesticide resistance in whitefly populations is a common problem faced by many growers today. whitefly has become resistant to chemical insecticides quite rapidly in other parts of the world, and the wisdom of relying only on chemical insecticides is questioned. Moreover, regular insecticide applications can result in resurgence of other pests.

A combination of cultural practices and chemical application would provide growers the best chance of controlling this pest. The use of sound cultural practices that may avoid, delay, or lessen the severity of the whitefly infestation is a good foundation to begin with. Careful selection of insecticides can help regulate whitefly populations to reduce losses. Little can be done to reduce losses due to virus diseases.

Fig. 11: White flies of leaf tissue.

Fig.12: An adult White fly.

7. Bihar hairy caterpillar (*Spilosoma(Diacrisia obliqua* (Walk) (Lepidoptera: Arctidae)

Soon after their emergence, the young caterpillars feed gregariously on the lower (ventral) surface of the leaves and cause defoliation. Leaves give an appearance of net or web with holes all over the leaves. The larvae feed on the pods too additionally. Well-developed caterpillars enter the pupal stage in the soil under dry foliage and debris,

Fig. 13: Various stages of Bihar Hairy Cater pillar.

8. Semilooper (*Thysanoplusia orichalcea*)(Lepidoptera: Arctiidae)

The green coloured larva is a semilooper in its mobility and feeds on leaves causing skeletalization of foliage. This leads to defoliation and severe yield losses.

Fig.14 : Green coloured Semilooper (Thysanoplusia orichalcea)

9. Soybean stem borer (*Dectes texanus texanus*) (Coleoptera: Cerambycidae)

The grubs (larvae) of the beetle make tunnels through leaf petiole and enter the stem.

The leaf tissue above this point wilt and die. Due to the tunnelling plant mortality results

The yield decline due to the pest is estimated to be of the order of around 15%.

Fig. 15: Soybean stem borer (Dectes texanus texanus)

Fig. 16: Adult weevil of Soybean stem borer

10. Ash weevil (*Myllocerus undecimpustulatus)* (Coleoptera : Curculionidae)

The adult weevils feed inward from the leaf margins and veins, causing the typical leaf notching. This leads to total defoliation of the plant. The adult stage of the pest is migratory and moves on to fresh vegetation.

Fig. 17: Ash weevil

Integrated Pest Management in Soybean

1. **Mechanical management of the pests**: Collection and destruction of girdle beetle infested plant parts, egg masses and larvae of hairy caterpillar and tobacco caterpillar
2. **Erection of bird perches @ 10-12/ha**.
3. **Installation of pheromone traps for monitoring incidence of *S. litura* and *H. armigera.***
4. **Use of Castor as trap crop for tobacco caterpillar and Dhaincha for girdle beetle. 5.Biological Control**:
5. **Biological Control**

(a) Conserve spiders, coccinellid beetles, tachinid fly, praying mantis, dragon fly, damsel fly, *Chrysoperla* and meadow grass hoppers through minimum use of broad spectrum pesticides, so as to exploit maximum potential of bio-control fauna.

(b) Release *Telenomus remus* @ 50000/ha against *Spodoptera*

(c) Spray *Bacillus thuringiensis* var. kurstaki, Serotype H-39, 3b, Strain Z-52 @ 0.75 to 1.0 kg/ha for the management of semilooper complex and other defoliators).

(d) Spray SlNPV @ 250 LE/ha. Spray NSKE @ 5% for management of early larval phases and the sucking pest complex.

According to the research carried out at the Indian Institute of Soybean Research, following are the parasites and predators of several insect pests of soybean. (https://iisrindore.icar.gov.in/)

1. **Parasites: *Brachymeria* spp., *Appenteles* spp. and an unidentified dipterous larval parasite**

2 **Predators:** Rhinocoris fuscipes, Cantheconidia furcellata, Chrysopa carnea, Dragon fly and Spiders

3. **Insect pathogens:** *Beauveria bassiana*, *Noumeria* (*Spicaria*) *rileyi*, Bacteria and Virus Maximum activity of bio-control agents, especially parasites was observed during 2nd week of August causing up to 16-20 % parasitisation. High humidity conditions were congenial for entomo-pathogenic fungi (*Beauveria bassiana* and *Nomuria rileyi*) infection in green semilooper and tobacco caterpillar larvae. 30 to 40 % larval mortality was recorded due to *B. bassiana* and *N. riley*.

4. **Cultural methods to manage whitefly:** Barriers such as row covers, and repellent mulches that effect phototactic responses of whiteflies, have shown some promise in delaying or reducing the incidence, but are not useful when whitefly populations and virus innoculum levels are high. In a study with tomatoes in Israel soil mulching with yellow polyethylene sheets delayed the spread of tomato yellow leaf curl virus for 20 days against a control crop without mulching (Cohen and Melamed-Madjar, 1978).

Control of weeds adjacent to cultivated fields, the use of trap crops, and implementation of crop free periods may be effective in reducing vector populations in certain cropping systems.

Intercropping can be an alternative method for the reduction of pests in certain situations. Beneficial insects are often increased and their activity enhanced on intercrops. Recently, intercropping of tomatoes with soybean has been shown to be beneficial. Various intercropping systems have shown to contribute to the reduction of whitefly populations. Cucumber planted in alternating rows 30 days before tomato delayed infection of the tomato with the whitefly-vectored tomato yellow leaf curl virus. (Mau and Martin 2007)

Genetic Sources of Pest Resistance: Potential donors for insect resistance have been identified by the Indian Institute of Soybean Research through large-scale field screening of germplasm and advanced breeding lines, e.g. TGX 855-53D, TGX 1073-55E, DS 396, EC 34500, EC 39739, EC 109545 (against defoliators), and L 129, L 592 (against girdle beetle). Some of the lines showing multiple insect resistance were also identified.

Wild soybean (*Glycine soja*) was found to be highly resistant to stem fly and girdle beetle. (https://iisrindore.icar.gov.in/)

Pest Population in relation to season: Kalyan and Ameta (2015) studied the incidence and development of the major insect pests of soybean such as white fly, *Bemisia tabaci* (Genn.), semilooper, *Chrysodeixis acuta* (Wlk.), girdle beetle, *Obereopsis brevis* (Gahan), gram pod borer, *Helicoverpa armigera* (Hubn.) and tobacco caterpillar, *Spodoptera litura* (F.), during the rainy and post rainy seasons of 2012 and 2013.

The peak population of whitefly was observed in the last week of August (35th SMW) and 2nd week of September (37th SMW) during the Southwest monsoon period. The occurrence of the population of whitefly exhibited a significant positive correlation with the maximum temperature and sunshine hours, but significant negative correlation with rain fall and morning and evening relative humidity only during the rainy season of 2013.

The maximum population of semilooper was recorded in the 2nd week of September (36th SMW) and last week of September (39th SMW), respectively. The semilooper exhibited a significant and positive correlation with minimum temperature, morning evening humidity and rain fall during both the years.

The highest girdle beetle damage was recorded in the last week of August (35th SMW) and 2nd week of September (37th SMW), respectively; whereas, the maximum incidence of gram pod borer was recorded in the 2nd week of September (36th SMW) and last week of September (39th SMW), respectively. The larval population of gram pod borer exhibited a significant negative correlation with sunshine and minimum temperature during experimental period; whereas, it exhibited a significant positive correlation with rainfall and relative humidity during both the years.

The maximum incidence of tobacco caterpillar in soybean crop was recorded during 2nd week of October (41st SMW) and 3rd week of October (42nd SMW), respectively. The maximum temperature and sunshine hours showed a significant positive correlation with the larval population of tobacco caterpillar, while; significant negative correlation with rainfall during both the years.

Twenty-two insect pests were recorded on soybean during the experimental period. During the year 2012, the mean density (%) value was the maximum 79.26 per cent for white fly; whereas, during the year 2013, the mean density (%) value was the maximum 73.33 per cent for semilooper.

During kharif-2012, the whitefly was relatively more 35.87 per cent while, during kharif-2013, maximum 28.87 per cent relative density (%) was recorded for semilooper. The Shannon Diversity Index was 1.45 and 1.55 during the year 2012 and 2013, respectively. During kharif -2012 and 2013 the reduction in plant height and number of pods per plant was 8.58 and 11.68 per cent and 12.23 and 24.31 per cent, respectively.

Likewise, the loss in mean number of seeds per pod and weight of 100 seeds was 11.06 and 14.86 per cent and 7.28 and 8.00 per cent, respectively. The loss in mean yield was 23.79 and 37.16 per cent; during 2012 and 2013, respectively.

During the year 2012 and 2013 the crop sown during 1st week of July had significantly higher incidence of whiteflies and girdle beetle as compare to crop sown in the 3rd week of July; whereas, the incidence of semilooper, gram pod borer and tobacco caterpillar was significantly lower in the crop sown in the 1st week of July as compared to crop sown in the 3rd week of July.

During the kharif- 2012 and 2013 significant lowest incidence of whitefly and tobacco caterpillar was recorded in the soybean variety of RKS-24 while, the significant lowest incidence of semilooper, gram pod borer and girdle beetle was observed in the JS-95-60 variety of soybean.

The significantly highest yield with mean of 1564 and 1650 kg ha-1 was recorded in case of soybean crop sown during the first week of July in the year 2012 and 2013, respectively.

Among the varieties, the highest seed yield of 1840 kg ha-1 and 1812 kg ha-1 was recorded in RKS-24 during the year 2012 and 2013, respectively; which gave higher yield than other varieties except, JS-335. Two sprays were given in soybean crop from different insecticidal spray schedules, of which first spray was given against whiteflies, semilooper and girdle beetle at 35 days after germination (DAG) and second spray was given at 55 DAG against gram pod borer and tobacco caterpillar. During both the years in first spray, the significant maximum per cent reduction in white fly population was recorded in case of thiamethoxam 25 WG at 3, 5 and 7 days after spray. While, significant highest per cent reduction in the larval population of semilooper and girdle beetle was recorded in case of profenophos 50 EC at 3, 5 and 7 days after spray. In the second spray, indoxacarb 14.5 EC @ 345 ml ha-1 caused significantly maximum reduction of gram pod borer and tobacco caterpillar larvae at 3, 5 and 7 days after spray.

The highest seed yield, net profit and C : B ratio were obtained in the case of spray schedule comprising first spray of triazophos 40 EC @1.25 lit ha-1 at 35 DAG followed by second spray of flubendiamide 480 SC @ 100 g ha-1 at 55 DAG with the corresponding respective values 1925 kg ha-1, Rs. 15,008 ha-1 and 1:8.52.

Diseases of Soybean

Soybean crop is inflicted by a number of diseases of diverse kind, which cause severe yield decline every year. The following diseases are considered more important in terms of the yield losses caused.

1. Myrothecium Leaf Spot: The disease is caused by *Myrothecium roridum* Tode and is seed borne. It used to be minor disease earlier.Recently it has acquired a major proporition. The disease spreads rapidly under rainy conditions, it is more severe during warm and wet weather. The developmental stages involving Flowering to pod filling are affected causing decline in the yield of the range of 40 to 50%

Symptoms

- Small round or oval, brown, spots develop with dark brown or purple margin on leaves in the infected plants.
- Translucent areas develop around these spots which appear as concentric rings.
- Later on dark green sporodochia develop in these areas.
- The symptoms may appear on the other aerial parts, such as stem, petiole, pods etc, also.
- The spots may merge to form irregular shape and leaves become dry.
- The pathogen survives in crop derbris and seed.

Fig. 18: Myrothecium Leaf Spot of Soybean

2. Alternaria leaf spot: The disease is caused by *Alternaria alternata* (Fr.) Keiss & *Alternaria tenuissima* (Kunze) Wiltshire.

The symptoms include distinct foliar lesions as brown spots surrounded by concentric rings. Several lesions coalesce leading to defoliation. Leaves may turn reddish or yellowish. In the severe cases of pod infection, damage may be noticed together with seed decay in the cases of delayed harvest. In turn this may lead to pod and seed infestation by insects. The disease which is seed borne damages the post flowering stage leading to 50% yield losses.

Fig. 19: Alternaria leaf spot lesions of Soybean. Picture courtesy of Robert Mulrooney, University of Delaware. https://www.dekalbasgrowdeltapine.com/en-us/agronomy/a-guide-to-common-soybean-diseases-in-the-midwest-.html

3. Rust: Soybean rust is caused by*Phakopsora pachyrhizi* PrH. Sydow & Sydow

Primary source of rust inoculum for south India is lying on the bank of Krishna River and its tributaries in the districts ofKolhapur, Sangli, Satara and Belgaum. Self-sown and winter-sown sole or intercrop soybean in irrigated areas might be harbouring rust pathogen in off-season and acting as a source of primary inoculum for rainy season soybean crop. There may be little or no role of collateral hosts in the initiation of rust. Four hot spot areas in Maharashtra and Karnataka have been identified where rust appears first and from here rust spread to other areas. Study clearly indicated that it is not the amount of rainfall but high relative humidity and congenial temperature, which are the main guiding factors for the onset and spread of rust. Study involving differential hosts and morphological parameters of urediniospores indicated presence of different races of rust.

Fig. 20: Leaf underside showing symptoms of soybean rust.
Image: D. Mueller (https://cropprotectionnetwork.org/resources/articles/diseases/soybean-rust-of-soybean)

Symptoms

- In the beginning, brown spots are seen on the leaves.
- Afterwards, they cover the whole leaf. Brown coloured powder is found on leaves and after sometime whole leaf turns brown.

The stages of flowering to pod development are very crucial in this regard. Yield losses of the tune of 50 to 70% are caused by rust disease of soybean,

4. Collar rot: The causal organism for this disease is *Sclerotium rolfsii* Curzi, which is a soil borne fungus. Sclerotia overwinter in soil and can remain viable for 3 to 4 years.

The crop phases during the first four weeks after planting and mid to late flowering are affected.

Fig. 21: Collar rot of Soybean. Picture courtesy of Clemson University - USDA Cooperative Extension Slide Series, Bugwood.org. (https://www.dekalbasgrowdeltapine.com/en-us/agronomy/a-guide-to-common-soybean-diseases-in-the-midwest-.html)

Symptoms

- The fungus infects plants when conditions are wet and hot (77 to 95°F).
- Seedlings are subject to damping-off.
- Brown spots develop and expand on leaves. Leaves finally turn brown and remain attached.
- Lesions can develop at the soil line and extend up the stem several inches with a white fungal mass on or above the lesion.
- Infected plants may have fungal growth.
- Small, yellow/red/brown fruiting structures (sclerotia) can be observed on the stem (Figure 21).
- Soybean plants are susceptible from emergence through pod fill. The greatest concern is when infection occurs during vegetative growth stages.

The disease is found to cause 30 to 40% yield losses.

5. Pod and stem blight: Seed borne disease. Pod and stem blight is caused by the fungus *Diaporthe sojae*.

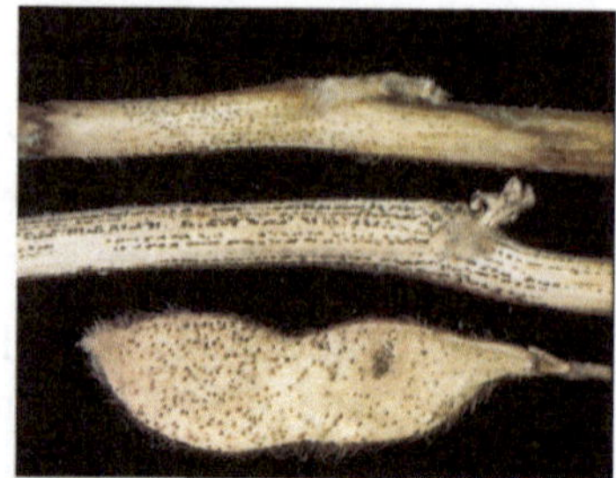

Fig. 22: Pod and Stem blight of Soybean.(Picture / Image by C.Grau: https://cropprotectionnetwork.org/resources/articles/diseases/pod-and-stem-blight-of-soybean)

Symptoms

Pod and stem blight is characterized by black, raised specks that appear in linear rows on mature soybean stems. These specks are fungal reproductive structures known as pycnidia. Pycnidia can also cover pods, but they may not follow the linear pattern seen on stems.

The fungal pathogens survive in infested residue and seed. Infection occurs when spores are splashed by rain onto plants in early vegetative growth stages. Pod and stem blight is favored by warm, humid weather when soybean plants are maturing. Also, disease is more severe if harvest is delayed.

Yield losses of 60 to 65% are observed due to the disease.

Fig. 23: Anthracnose lesions on a soybean stem. Picture courtesy of Daren Mueller, Iowa State University, Bugwood.org. https://www.dekalbasgrowdeltapine.com/en-us/agronomy/a-guide-to-common-soybean-diseases-in-the-midwest-.html

6. Anthracnose: The disease which is seed borne, arises due to the infection of *Colletotrichum dematium f. sp. truncatum* (Pers.) and /or *Colletotrichum truncatum* (Schwein.) Andrus & W.D. Moore. The disease is prevalent all through the season.

Symptoms

Disease is characterised by appearance of irregularly shaped red to dark brown blotches on stems and petioles can during reproductive stages. Petiole infection can result in a shepherd's crook (curling or "hooking" of plant tissue). Near maturity, black fungal bodies that produce small, black spines and spores are evident on infected stems, petioles, and pods.

Warm, wet weather favors infection and disease development.

The yield losses due to this disease are of the order of 16 to 25%

7. Bacterial pustule: The disease is caused by *Xanthomonas campestris* pv. *glycines* Pammel. The disease is seed borne and appears after35 days following sowing and continues to harbour the crop till its maturity. The mean yield losses caused by this disease are around 15%.

Fig. 24: Bacterial Pustules of Soybean

Symptoms

- Small yellowish green spots with reddish brown centres are formed on the upper surface of leaves.
- The central portion of the individual spot appears slightly raised.
- Small pustules are found on the lower side of the leaves.

Similar pustules also occur on pods.

- The pods may merge to produce large irregular dead areas.
- The dead tissue sometime may rupture and tear away. Heavily infested leaves turn yellow and fall and in susceptible plants complete defoliation can occur.
- The bacterium survives in crop debris and seed.

Fig. 25: Yellow Mosaic Virus of Soybean

9. Yellow Mosaic Virus: The early vegetative stage to the physiological maturity stage of the crop are adversely affected by this viral disease caused by *Mung Bean Yellow Mosaic Virus* Nariani.

Symptoms

- Characteristic symptom is conspicuous systemic bright yellow mottling of leaves.
- The yellow area are scattered or occur in indefinite bands along the major veins.
- This virus has a wide host range and is not sap or seed transmissible.
- Insect vector for Mung Bean Yellow Mosaic Virus(MBYMV) is whitefly. (*Bemisia tabaci*).
- If the plant is infected wthin 75 days after planting, the yield reduction are significant but infection after 75 days results non-significant yield losses.

The soybean yield losses caused due to this disease are of the order of 15 to 75%.

8. No podding syndrome: The malady consists of phyllody and bud proliferation leading to the total failure of pod development. The etiology of the disease still remains obscure. The developmental stages adversely affected are from flowering to podding. At the field level the disease is observed to cause30 to 40% yield losses.

Integrated Disease Management

Management of diseases of soybean involves a multipronged strategy in view of the large number of plant pathogens involved and complexities of disease development.

While development of resistant varieties has been advocated periodically as the sustainable approach, in the case of soybean due to complexities of disease syndrome, it is absolutely necessary to come up with crop varieties with multiple resistance, in addition to other practices of disease management.

The Indian Institute of Soybean Research has come up with following multipronged disease management strategy in respect of soybean crop (Anonymous 2010).

- Multiple disease resistant lines/varieties like PK 262, PK 327, PK 471, PK 695, PK 1169, PK 1243, PK 1251, SL 432, SL 459, SL 517, SL 528, TS 99-128, JS 71-05, JS 72-280, JS 75-46, JS 76-206, Bragg, Punjab 1, MACS 58, MAUS 52-1, VLS 2, Monetta, JS (SH) 91-33, JS(SH) 93-01, Himso 1569, NRC 35, NRC 41, NRC 44, RAUS-3, RSC 1, RSC 3, AMS 243, AMS 358, AMS 56, JS 20-29, SL-958, MACS 1336, DS 2614, DS 12-13, PS 1042, SL 688, JS 20-69, JS 20-89, SL 955, SL 983, MACS 1410, MACS 1407 and RVS 2002-4 have been identified.
- Rust resistant/tolerant varieties/lines like, PK 1024, PK 1029, JS 80-21, C3P27, JS 90-225, PK 1197, RSC 2, EC 389170, EC 389178, EC 241778,

EC 241780 and and DSb 21, DSb 23-2, Phule Kalyani and KDS 726 have been identified.

- Varieties/lines resistant to bacterial pustule including germplasm lines viz. EC 389150, EC 389164, EC 390981, EC 390989, EC 390975, EC 390976, EC 390977, EC 391152, EC 391172, EC 391181, EC 393222, EC 393223, EC 393225 and EC 393237 have been identified. Similarly varieties/lines resistant to myrothecium leaf spot, collar rot, soybean mosaic, Indian bud blight have been identified.
- On the basis of multi-year and multi-location screening AMS 243, AMS 358, AMS 56, JS 20-29, SL-958, MACS 1336 were identified as potential sources for resistance to Charcoal rot and DS 2614, DS 12-13, PS 1042, SL 688 for YMV.
- Intercropping with maize and sorghum and pearl millet in 4:2 ratio, sowing up to 10th July, spacing of 30 to 45 cm and plant population of 4 lakh plants/ha were found promising with less seedling mortality and higher yields. However, with pigeon pea incidence of collar rot increased.
- Seed treatment with carboxin + thiram @ 2 g or thiram and carbendazim in the ratio of 2:1 @ 3g/kg seed was found very effective for the control of seed and seedling diseases. Result of seed treatment was also encouraging on seeds of poor grade and ungraded seeds. Seed treatment 50 days prior to sowing was found much more effective than at the time of sowing.
- Seed treatment with biocontrol agents viz. *Trichoderma viride* and *Pseudomonas fluorescens* increased seedling emergence, plant population and reduced pre and post emergence mortality. Consortium of three strains of *T. harzianum* was also quite effective for the management of charcoal rot and collar rot.
- Seed treatment plus soil application of zinc and boron plus irrigation at the time of pod formation was quite effective for the management of charcoal rot
- For the management of foliar diseases two sprays of carbendazim or thiophanate methyl and for rust hexaconazole, propiconazle, triadimefon and oxycarboxin (0.1%) were found effective.
- Foliar sprays of carbendazim and mancozeb at R2 and R6 stage increased seed germination and reduced seed borne fungi.
- A method was developed to use agrowastes for mass multiplication of bio-control agents *Trichoderma.*
- *Trichoderma* spp was found to be compatible with seed dressing fungicides like vitavax and thiram

Microbiology and Root nodulation

ICAR-Indian Institute of Soybean Research (IIISR) Indore (Anonymous 2010) reported a new Soybean rhizobia *Bradyrhizobium liaoningense* from Central India (Accession no. MTCC 10753/NCBI No. JF 792426). The bacterium is capable of surviving at 36*C, enhancing N-fixation by 19% over control and possesses plant growth promoting traits.

- The IISR Identified three new potential novel bacterial strains e.g., *Paenibacillus mucilaginosus* and *Bradyrhizobium daqingense* and *B. liaoningense* were isolated and identified from root nodules of three drought tolerant lines e.g., EC 538805, PK472 and EC 538828 respectively
- Soybean rhizobia *Bradyrhizobium japonicum* (MTCC 10751) recovered from soybean cultivar JS 93-05 has been found to be highly effective under field conditions and found to be compatible with AM fungi.

New generation microbial bio-fertilizers for mitigating zinc deficiency of soil and crops

- In *Malwa* region of central India, zinc-solubilizing *Pseudomonas* isolates viz., UP1, UP4, UP8, SP4, SP8 have been isolated from rhizosphere soils of soybean. These isolates have been found to increase zinc concentration in soybean seeds significantly (Anonymous 2010).
- Further, *Pseudomonas*isolates KHBD-6 (57.34 μg/g), KHBAR-1 (55.67 μg/g), and strain ATCC 13061 (53.10 μg/g) were found useful for solubilisation of Zinc salts, by increasing Zn concentration in soybean seeds significantly (Anonymous 2010).
- The investigations at ICAR-IISR yielded Zinc-phosphate solubilizing fungi *Aspergillus flavus, Fusarium solani* and *Scytalidium* sp. from soybean rhizosphere (Anonymous 2010)

Arbuscular Mycorrhizal Fungi (AMF)

- Arbuscular Mycorrhizal Fungi mass needed for one hectare area has been produced from 0.3 IP/g soil to 22.24 IP/g soil in three years involving seven consecutive multiplicative cycles utilizing marigold, sorghum, barseem, maize and fenugreek as potential trap plants (Anonymous 2010).
- The work carried out also resulted in the recovery of one mycorrhizal helper bacteria *Burkholderia arboris* (MTCC 10752) from soybean root nodules (Anonymous 2010)

Crucial Components of Production Technology

Soybean is the most widely grown oil bearing legume in the world since the plant has exhibited a remarkable adaptation to wide variety of soils and climatic conditions as compared to any other oilseed crop. In the previous couple of decades, the scientists and technologist have come with a large gamut of products from the crop as well as more streamlined and focussed soybean production technologies.

Soya crop varieties: There are a large number of varieties of soybean out of which more popular ones grown in India are Bragg, JS2, PK71-21 (Alankar), Ankur, UPSM19 (Silajit), DS73-16 (Pusa soybean), PK327, PK271, PK262,PK 308,Moti and Kalitur (native to Northeast India mostly popular for fodder).

Plant types of soya for drylands: Soybean is also being grown as a rain-fed crop in India, where availability of soil moisture at different growth stages of crop is a major limitation to the crop productivity. While some of the varieties indicated above are being grown on drylands, the soya varieties for rain-fed conditions should have a growth and developmental rhythm matching with the pattern of moisture availability under such conditions. Therefore, optimum phenology of soybean varieties for achieving maximum yields under conditions of dryland agriculture was worked out. The results of the investigations indicated that the grain yield exhibited a strong curvilinear relationship with days to flowering (R2=0.51**), days to maturity (R2=0.47**) and seed fill duration (R2=0.51**). It was concluded that varieties with ~36 days to flowering, ~88 days to physiological maturity and ~34 days of seed fill duration would give the maximum yield and would be best suited for dryland situations of central India. Hence for wide range of conditions of dryland agriculture the soya varieties with the above attributes may be chosen to attain sustainable grain yield levels (Anonymous 2010).

In addition, certain more promising varieties suited for specific situations of crop growth are listed under the section Genetic Resources and Breeding.

Seasons

Soya crop is grown in two major seasons in large parts of India viz., Kharif (rainy season from June to October) and Rabi (Post rainy season from October-November to April-May).

Additionally, the crop is also grown as Summer Crop from March to June with irrigation.

In the hilly zones of Northeast the crop is also grown during the months of May to April.

Tillage

The conventional tillage consists of well pulverised seed bed before sowing. However, the conservation tillage followed by the conventional tillage in the succeeding season resulted in better microbial activity favouring crop growth leading to higher productivity levels. The rhizosphere under conventional followed by reduced tillage system in soybean based rotations involving maize showed higher AM Fungi spore density and colonization. Reduced tillage encouraged population of siderophore-producing bacteria (Berenguer and Joseph, 2019) in rhizosphere soils ensuring better availability of Iron.

Seed Bed and Water Management

Broad-bed furrow (BBF) and ridge tillage in conjunction with application of FYM @ 10 t/ha + recommended doses of NPK fertilizers were found to help in better moisture conservation, better soil health in terms of soil enzymes activity and ultimately in improving the productivity of soybean and soybean based cropping systems. The crop needs a minimum water quantum of 500mm to 800mm depending upon the soil and conditions of crop growth. Ridge tillage and broad bed furrow significantly increased soil microbial biomass, soil enzyme activities, nutrient uptake and soil particulate organic carbon and seed yield as compared to minimum tillage and flatbed planting under soybean-wheat and soybean-chick pea systems. The recommended seed rate is 60 kg/ha to be planted at an inter-row spacing of 30 to 40 cm depending upon the soil type, plant type and growing conditions. In the case of varieties bushy growth and branching and under the situations of dryland agriculture an inter-row spacing of 40 cm is to be adopted. The intra-row spacing of 10 cm is recommended. Before planting, the seed treatment with Thiram @ 3 g/kg of seed is to be adopted, following which the rhizobial inoculation is to be carried out.

Application of manures and fertilizes

Basal application of 10 t of FYM per ha at the time of initial tillage facilitates its proper incorporation. Subsequently at sowing application of nutrients per hectare @ 20 kg N:80kg P2O5:20 kg K2O: 20kg S is recommended. Before planting, the seeds must be inoculated with *Bradyrhizobium japonicum*. The micro-nutrient schedule as described in the preceding section of Crop Nutrients must be adopted inevitably, depending upon the site specific requirements. There is a widespread Zinc deficiency in Indian soils and hence recommendation of nutrient schedule should take into consideration this important micronutrient as already explained earlier.

It is recommended that a 3% solution of Muriate of Potash be sprayed between 30 to 35 days following sowing to facilitate good grain development.

Weed management

The Indian Institute of Soybean Research recommends the following procedures to check the weed competition to the crop (Anonymous 2010).

- Flumioxazin 50% WP @ 45 g/ha and Diclosulam 80% WDG @ 22 g/ha as pre-emergence were found to be promising herbicides for the management of weeds in soybean.
- Haloxyfop 10% @ 100 g/ha, Clethodim 24% EC @ 48 g/ha + NIS + AMS and Quizalofop ethyl 10% EC @ 37.5 g/ha as post emergence could effectively contain the grassy weeds in soybean.
- The tank mix combination of Chlorimuron ethyl 25% WP @ 9 g/ha + quizalofop tefuryl 4% EC @ 40 g/ha could effectively control the monocot as well as dicot weeds in soybean.

Appropriate Plant Protection technologies have already been outlined in the preceding sections under insect pests and diseases.

Fig. 26: Soybean reaching the stage of harvest with leaves turning yellow (https://www.farms.com/ag-industry-news/u-s-soybeans-dropping-leaves-464.aspx)

Harvesting: At the maturity of crop the leaves turn yellow and drop-off, when the crop is to be harvested either mechanically or manually. Soybean crop is amenable to mechanical harvesting.

Seed Storage: The harvested grain produce must be dried well to ensure that the moisture content doesn't exceed 10% at storage. Appropriate storage containers along with the suitable storage structures must be employed to cut down the damage by storage pests and to ensure low moisture content of the produce.

If the seeds are to be used for sowing in the subsequent season, seed treatment with fungicide Thiram @ 3 g/kg is absolutely necessary.

Soybean Products

Soybean is the richest and cheapest sources of protein known so far and is a staple part in the diets of people in many parts of the world since times immemorial. The Soybean seed contains around 18% percent oil and 40 percent protein. (Chapin,

2008). Being an affordable and good source of protein and energy, defatted Soy in meal form is also widely used globally as a part of livestock, poultry and aqua feeds.

According to industry experts, soy consumption, as a part of human diet is growing rapidly in the backdrop of its high protein density and protein quality as well as its extensively researched and well established health benefits. Soy protein contains all the essential amino acids and meets or exceeds the essential amino acid requirements as prescribed by FAO/WHO for different age groups. Soy protein has a protein quality score- PDCAAS (Protein Digestibility Corrected Amino Acid Score) of 1.0 on par with milk and egg proteins. Soybean Oil is a healthy edible vegetable oil with a healthy fatty acid profile and is a good source of Omega 3 fatty acid. Soybean and Soy protein have been the subject of extensive clinical research during the last 50 years and Soy protein's health benefits in relation to reducing heart disease and preventing cancers, particularly hormonal cancers like breast cancer and prostate cancer are well established. High in protein, low in fat and credited with aiding heart health, soy is turning up as a healthy ingredient for food and nutrition products. Soybeans are a low glycaemic food and is ideal for diabetics. In East and South East Asia the bean is extensively consumed in the forms of soy milk, a whitish liquid suspension, and tofu, a curd somewhat resembling cottage cheese. Soybeans are also sprouted for use as a salad ingredient or as a vegetable and may be eaten roasted as a snack food. Young soybeans, known as Edamame, are commonly steamed or boiled and eaten directly from the pod. Soy sauce, a salty brown liquid, is produced from crushed soybeans and wheat that undergo fermentation in salt water for about six months to a year or more; it is a ubiquitous ingredient in Asian cooking. Other fermented soy foods largely consumed in East and South East Asia include tempeh, miso, and fermented bean paste.

Fig. 27: Tofu making process in China (*Source*: wikipedia)

Modern research has led to a remarkable variety of uses for the soybean. Its oil can be processed into margarine and shortening type products. Industrially, the oil is used as bio-diesel, apart from other minor uses. But, large part of Soybean Oil is used globally as a healthy edible oil. Defatted Soy Flour is widely used in production of texturized meat analogues as well as high moisture meat analogues. Higher value ingredients like Soy Protein Concentrates and Soy Protein Isolates are extensively used as food protein ingredients for functionality as well as fortification applications in food and nutrition products. (https://www.britannica.com/plant/soybean)

Chapin (2008) reports that a recent study by Packaged Facts, a New York-based research firm, found that the soy food and beverage market grew 6 percent between 2006 and 2007 for a total of more than $2 billion. In the report, "Soy Foods and Beverages in the U.S.," Packaged Facts predicts that the soy market will continue to rise through 2012 when it is expected to reach $3 billion. Further, the United Soybean Board reported in its annual survey, "Consumer Attitudes about Nutrition - Insights into Nutrition, Health &Soy foods," that 33 percent of Americans now consume soy-based foods or beverages once a month or more.

In fact, 85 percent of consumers now identify soy products as healthy, reports the United Soybean Board, St. Louis. This accounts for a 3 percent rise over 2006 and an 18 percent increase during the last nine years. Specifically, the Soybean Board's report indicates that consumers recognize that soy foods are: low in fat (18 percent), high in protein (17 percent), help heart health (16 percent), lower cholesterol (11 percent) and potentially relieve menopause symptoms (10 percent).

Innovative applications (Chapin, 2008)

There are several new products capitalizing on soy's health benefits, while also delivering on innovation. Launched i recently Kellogg's Morningstar Veggie Cakes combine vegetables, rice, soy protein and spices to create a low fat, high protein entrée (when combined with salad), snack or appetizer, says the Kellogg Co., Battle Creek, Mich. Garden burger, also owned by Kellogg, added to its vegetarian offerings with Garden burger Gourmet Garden Steaks last year. Kellogg says that these 156-gram "steaks" come in four exotic spice varieties: Baja (chilies and pepper jack), Fire Dragon (jasmine rice, Asian vegetables), Hula (pineapple and ginger) and Tuscany (Gorgonzola and couscous).

For its part, Solae Company planned to showcase Mini Barbeque Vegetarian Beef Sandwiches, in addition to a smoothie and protein bar, at the Institute of Food Technologists' Annual Meeting & Food Expo in New Orleans this summer. Featuring shredded vegetarian "beef" and tangy barbeque sauce, the product is made with Solae's Supro Max - a form of protein that "emulates whole-muscle meat texture - chewiness, fibrosity and appearance," Gallegos says.

Recent innovative development by Wenger Manufacturing Inc., a US based Food extruder manufacturing company is Soy Protein and cereal based 'Dal / Lentil Analogue. This extruded product – 'Dal Analogue' or 'Dal Maker' contains higher protein content and higher protein quality compared to traditional dals and is an excellent substitute for dal in many traditional food products. This product was promoted as an ideal and affordable protein food in India and South Asia by the US Soybean Export Council as part of its promotional campaign of Soy for health and wellness.

5

Sesame (*Sesamum indicum* L.)

Introduction

Cultivated sesame (*Sesamum indicum* L.), has originated in India (Bedigian,1984; Bedigian *et al*., 1985; Bedigian and Harlan,1986) and is considered to be the oldest oilseed crop known to man (Ram *et al* 1990). Archaeological records indicate that it was known and used in India for more than 5,000 years and is recorded as a crop in Babylon and Assyria some 4,000 years ago (Borchani *et al.,* 2010). Sesame crop is being grown in many parts of the world. Sesame seed is an important source of edible oil and is also widely used as a spice and condiment. The seed contains 50-60% oil which has excellent stability due to the presence of natural antioxidants such as sesamolin, sesamin and sesamol (Brar and Ahuja 1979).

Sesamum indicum L., belongs to the family Pedaliaceae, which contains sixteen genera and sixty species under the generic name *Sesamum.* The other species related to sesame viz., *Sesamum angustifolium, Sesamum radiatum* and *Certotheca sesamoides* (false sesame) are cultivated in Africa. The fact that the crop is cultivated in India from times immemorial is testified by the excavations of Harappa and Mohenjo-Daro of undivided India (Watt,1966). The crop was introduced into Brazil by the Portuguese (Weiss, 2000).

World production of sesame seed didn't show any marked increase for the past three decades, although sesame oil is used largely in Asia and Africa. The current world production of sesame stands at 6.1 million tons as in the year 2016. The sesame production declined world over since the crop had to face a tough competition against other oilseed crops (Weiss, 2000). Most of the world trade on sesame revolves around the seed. The shipments of seed cake and oil are negligible. The availability of sesame seed in

Sesamum indicum

the world for export is of the order of 200,000 tons per annum. Nevertheless, most of the commerce pertaining to Sesame seed is among the neighbouring countries of Southeast Asia. All the same, the area under sesame has been found to increase in Myanmar and China perhaps to cater to the local demand (Weiss 2000). The world demand for sesame seed is greater than its production, which is indicative of the urgent need to increase the productivity and production of sesame.

The chemical composition of sesame shows that the seed is an important source of oil (44-58%), protein (18-25%), carbohydrate (~13.5%) and ash (~5%) (Borchani *et al.,* 2010). Sesame seed is approximately 50 percent oil (out of which 35% is monounsaturated fatty acids and 44% polyunsaturated fatty acids) and 45 percent meal (out of which 20% is protein) (Gandi, 2009; Hansen, 2011)The seeds also contain significant amount of important minerals with the Potassium concentration being the highest, followed by Phosphorus, Magnesium, Calcium and Sodium (Loumouamou *et al*., 2010).

The impressively wider degree of diversity of the types of sesame and the crop's adaptation almost to all the diverse environmental conditions and considerable variability for seed oil content create an exceptionally impressive gene pool for sesame breeding.

Ecology: Sesame is cultivated in tropical and sub-tropical regions of world. It grows well on planes and also hills up to 1200 m elevation. Sesame is photo-sensitive and also sensitive to lower temperatures, heavy rain and prolonged drought. In the planes of north India, the crop is grown in the rainy (*kharif*) season. In Peninsular India, sesame is grown throughout the year. In recent years the crop is coming-up as a catch crop in the short summer season under irrigation in India. Under conditions of dryland agriculture, the crop can succeed even with a rainfall of 500 mm, which offers the minimum needed quantity of moisture. The crop comes up on diverse soil types; but sesame yields well on well drained soils with better water holding capacity. The heavy clay soils with poor drainage, saline and alkali soils and shallow soils with poor depth are not suitable. Despite the fact that the crops perform well on neutral soils, it can be grown on moderately acidic and slightly alkaline soils in the pH range of 5.5 to 8.0. Sesame can't tolerate water-logging. Heavy continuous rains during peak vegetative and reproductive phases are not conducive to the growth and development of sesame (Naidu *et al.,*2006)

Sesame Plant: Sesame is a branched plant type of annual habit, but occasionally perennial growing to a height of 0.5 to 2.0 meters, possessing a well-developed root system that can withstand drought of moderate intensity. The shoot is multi-flowered and the fruit is a capsule with number of small oilseeds neatly piled-up in rows inside.

While most of the cultivated types of sesame are annuals with varying degrees of duration, one could come across sesame types of longer duration tending towards biennial to perennial habit. The short duration types possess less extensive root system unlike the long duration types that have more penetrating and spreading roots. The growth and maturity patterns of sesame land races depend upon the climatic parameters including rainfall, soil type and temperature of the areas. It has been observed that the varieties of sesame exhibit a varied expression with the variation in the weather patterns over the years of the region. For example, the varieties grown in monsoon season (*kharif*) exhibit a different kind of expression when grown in the post-rainy season i.e., *rabi* season (Weiss, 2000).

The single stemmed short duration varieties exhibit a rapid root elongation than the varieties of longer duration, which have a faster rate of root spread (Weiss, 2000). Yoopraset and Sooksathan (1992) found that sesame roots may exert allelopathic effect on some crops, perhaps grown as inter crops with sesame. The sesame roots are sensitive to soil salinity which may affect the crop performance adversely (Kumar 1998). Stem strength could be useful attribute particularly in the case of single stemmed varieties in promoting higher number of capsules per node and incorporating resistance to lodging (Murthy *et al.,* 1995A)

Sesame being photo-sensitive, it responds to long days in general and the yield of the crop is affected with the reduced photo-period which occurs generally under late sown conditions in the monsoon (*kharif*) season (Prasad 2019b). The light interception is related to growth of sesame, since it exhibits exponential functional relationship between the light intensity in the canopy and leaf area index (Kang, 1994).

Phyllotaxy or the arrangement of leaves on the stem of sesame plant is an important attribute, which determines its flowering and capsule bearing habit. Leaf morphology of sesame is variable depending upon the plant type. Also leaves vary much with in the same plant. The lower leaves are broader and lobed with serrated leaf margin. Leaves on the intermediate part of the plant are entire, lanceolate and sometimes slightly serrated, while the upper leaves are narrower and lanceolate (Weiss, 2000). Length of the leaf varies from 3.0 to 17.5 cm with a width of 1.0 to 1.7 cm (Weiss, 2000). Loss of leaves due to either biotic stresses or any other calamity to the tune of 30% may not adversely affect the yield. Any defoliation over and above 35% may have an adverse effect on the seed yield, the degree of which varies with the plant type.

The flowers are formed at the leaf axils, the number of flowers at each node varying with the plant type. The carpel consists of bicarpellary, bilocular superior ovary with a number of anatropous ovules. Flower at the leaf axil is borne on very short pedicel with five lobed corollas, the lower lobe being the longest forming the 'lip' of the flower. Rincon and Salazar (1997) found that low temperatures at the time of flowering might result in sterile pollen and / or premature flower drop. On the other hand, higher temperatures of the order of 40°C or above would adversely affect fertilization resulting in the reduction in the number of capsules.

Fig.1: *Sesamum indicum* (2n = 2x= 26), shoot, root, floral parts, capsules and seeds.

Anthesis of sesame

The statement that sesame is self-pollinated with only 10% out-crossing due to insect pollination (Weiss, 2000) needs to be reconsidered and revised. Sesame has exhibited varying levels of cross pollination up to 68% and more (Richaria,1939; Yermanos ,1980; Thangavelu,1994; Said *et al.,* 2013). Sesame is stated to be cross pollinated (Ashri, 1988) depending upon the degree of insect activity. The perceptible levels of out-crossing in sesame are due to the visits of several insect pollinators (Fig.2) including species of Hymenoptera viz., *Apis dorsata, Apis millefera. Megachile* sp, *Vespa cincta and Componotus sericius,* Diptera viz., *Sarcohaga* sp and *Musca domestica* and Lepidoptera viz. *Danauschrycippus, Amata bicinta* and *Pieris* sp. (Das and Jha, 2019) . According to Kamel *et al.*, (2013), insect pollinators of sesame consisted of four groups out of which18 belonged to Hymenoptera, 7 to Diptera, 3 to Lepidoptera and one to Coleoptera. Honeybee, *Apis mellifera* was the most dominant species in the 2011 season and the second one in the 2012 season. The small carpenter bee, *Ceratina tarsata* was the most dominant species in the 2012 season and the second one in the 2011 season. The percentage of Hymenoptera was higher in the two studied seasons by 90.94% and 89.59%, followed by Diptera by 3.93% and 5.38%, then Lepidoptera by 3.58% and 3.62, and in the last Coleoptera by 1.53% and 1.39%, respectively (Kamel *et al.,* 2013).

Sajjnar and Eswarappa (2015) observed that14 insect species were found visiting the sesame flowers out of which, ten species belonged to Hymenoptera and four belonged to Diptera.

Sesame flowers are not cleistogamous as in the case of leguminous crops like groundnut.

The anthers open longitudinally after flower opening (Weiss, 2000) ensuring pollen dispersal providing for cross pollination. Secondly stigma being receptive after flower opening too (Weiss, 2000) opens the door for further cross pollination. A minimum isolation distance suggested for avoiding out-crossing and ensuring self-pollination in sesame is from 150 to 360 meters (Prasad 2019b) depending upon the frequency of occurrence of insect pollinators.

The functional receptivity of stigma of sesame is also testified by the report that pollen from non-members of the family Pedaliaceae also will germinate on the stigma of *Sesamum* species but may not result in viable seed (Weiss, 2000.) According to Weiss (2000) the pollen of the American weed *Martynia dianalta* introduced into India germinated and grew into the styles of *S. indicum* and in Venezuela too. The receptivity of stigma of sesame is factor that adds to the promotion of cross pollination in sesame.

Fig. 2: Sesame Flower being foraged by honey bee the pollinator

Sliqua and Seed Development

By the third day after open flower, the capsules are visible and will grow to their final size about 4 days later. The speed of growth will vary by variety. (Fig.3). The seed will be mature in each capsule from 40 days at the bottom of the plant to 25 days at the top. However, the bottom capsules will not dry until after all the capsules ripen.

Fig. 3: Sesame capsules with nectar gland at the base. (Picture Courtesy WP Amstrong 2009)

The first capsule is formed at 1-2 ft. from ground. Physiological maturity (PM) normally occurs 85-110 days after planting. PM is when 75% of the capsules on the main stem have mature seed. The seed will also have a dark seed line on one side. PM is important because after that point, the crop is less vulnerable to yield loss due to frost. It is also an indicator that the time to use harvest aids is approaching. The capsule size is highly variable with in the shapes of flat to cylindrical depending upon the genotype. The length of capsule is variable from 2.5 to 8.0 cm with diameter ranging from 0.5 to 2.0 cm. The number of locules per capsule varies from 4 to 12. When the capsule loses moisture and turn dry the splitting of the capsule happens from top toward bottom all along the septa of locule. Dehiscence of the capsules results in shattering causing considerable yield losses, particularly in mechanical harvesting. To obviate the yield losses and to encourage commercial mechanised operations, a need was felt to come up with a genotype with non-dehiscent capsules. Langham (1946) developed indehiscent capsule mutant of sesame; but there was a problem of the varieties with difficulty in threshing. However, the introduction of thin-shell capsuled varieties resolved the problem of threshing (Weiss, 2000). The finding that the indehiscence is controlled by a recessive allele of the single locus for shattering, which has no correlation with yield has facilitated improving yield combining with capsule indehiscence (Delgado *et al.*, 1994). The indehiscent capsuled plants are short, erect-branching, lodging resistant and precocious. The indehiscent capsules are bicarpellate, medium long thin-shell and one per leaf axil (Weiss, 2000). Despite significant progress in sesame breeding, harvest losses due to dehiscence continue to limit domestic US production (Oplinger *et al.,* 1990)

The seeds are tan in colour, medium sized and of uniform shape and size. The number of capsules per plant are directly related to the number of flowers, expression of which is subject to interaction with environmental factors. Seed yield has positive correlation with the number of branches as much as the total number of capsules.

Kang (1994) reported that oil content of seed increased up to 35th day of flowering. On the other hand, the protein content increased at a faster rate up to the 20th day and then increased gradually up to the 45th day following flowering. Interestingly the oleic acid reached a level of 60% on the 10th day following flowering and then reduced rapidly till the 20th day of flowering.

Regarding the fatty acids (Kang 1994), the unsaturated ones (Linoleic +Oleic) are observed to increase till the physiological maturity reaching the highest level on the 45th day of flowering. On the 10th day of flowering, oleic acid reached a level of 60% and declined fast up to the 20th day following flowering. Plamitic and Stearic acids together with arachidic, erucic and linolenic acids registered faster decline up to 25th day of flowering. It appears that the above phase of flowering and seed development is an important period with regard to the sesame oil quality.

It is not easy to dry small and flat seeds of sesame, since the circulation of air in the micro-environment of seed is observed to be difficult. Hence it is recommended to harvest the capsules with seeds as dry as possible and store at 6% moisture or less since moist seed would get quickly heated up and become rancid (Hansen,2011).

Crop Nutrients: There is a misconception even among farmers that sesame doesn't respond adequately to nutrient application. This is because of the vast area of the crop both in Asia and Africa is managed by a large number of small and marginal farmers who don't have enough financial support to take to application of crop nutrients (AICORPO 1992). However, the front line demonstrations conducted by the ICAR-Directorate of Oilseeds Research in 1980s and 1990s on the farmers' fields amply demonstrated the good response of sesame to the improved agronomic management including recommended doses of crop nutrients even under conditions of dryland agriculture (AICORPO 1993). Another important factor in this context is that the local land races grown by small and marginal farmers don't exhibit perceptible response in terms of seed yield to a very inadequate quantum and time of crop nutrient application. Hence It becomes necessary to select varieties that exhibit higher response to nutrient management with better adaptability to local situations (Weiss, 2000)

In the case of sesame crop grown on lighter and sandy soils, incorporation of organic manures and plant residues shall increase the humic content and moisture holding capacity of soil(Weiss, 2000), at the same time exhibiting better response to applied crop nutrients (AICORPO 1993).

Xu *et al*., (1992) working in Korea has shown that to produce 1755 kg of seed yield / ha, the crop required 6.75 kg of N, 1.0 kg of P and 5.25 kg of K. The nutrient content in different plant parts declined during growth. N and P in leaves were higher than in stems, but stems had higher K content. K content in above ground parts was in decreasing order in capsules, shell, stem, leaf and seeds in that order.

According to Weiss (2000), in the cases of commercial production of sesame, NPK mixtures of 5:10:5; 12:12:6 and 10:14:10 @ 500-700kg/ha are generally applied at planting. In the case of dryland grown sesame in Israel, fertilizers are applied preferably to the preceding crop of wheat, which benefits the succeeding crop of sesame. However, the irrigated crop of sesame received additional quantum of fertilizes (Weiss, 2000).

The nutrient schedule may vary depending upon the variety and plant population recommended and the soil moisture at planting in the case of un-irrigated rain-grown crop.

Nitrogen

Application of Nitrogen must be related to the available P, so as to ensure adequate vigour, seed yield and seed oil content. It is also known that the increased doses of Nitrogenous fertilizers result in the incidence of pests and diseases. The interaction between Nitrogen applied and total water given as irrigations at different intervals is an important factor, in which case an optimum ratio must be established, since the excess of either component might adversely affect the seed yield.

The top dressing of Nitrogen is advisable with the availability of moisture when the plants are of 20cm height much before the appearance of first flower buds. Foliar sprays of urea solution may be beneficial under condition of dryland agriculture if moisture is limiting for top dressing of Nitrogen.

Phosphorous

Phosphorous as a nutrient is very important particularly for the development of capsules. When the sesame plants are 60 days old, the proportion of dry matter registers a decline along with the uptake of N and K. The decline of K is to a lesser degree since it is related to the capsule number. Despite the decline of uptake of Phosphorous, the uptake of P continues at level higher than N and K as the number of capsules increases (Weiss, 2000).

Normally a basal application of 40 kg P/ha may be optimum unless the special conditions of soil etc., warrant application of higher dose. In case the crop preceding sesame was fertilised heavily, the corresponding adjustment of dose of P, say @ 10-20 kg/ ha might be adequate (Weiss, 2000). In the case of irrigated commercial crops, a higher dose of P, up to 60 kg/ha is being applied; but there is no economic data to justify this practice (Weiss, 2000). The recommendation of P should take into consideration, the availability of residual P in the soil.

Among the phosphatic fertilizers, single superphosphate may result in better crop response due to its sulphur content which is often deficient in soils of tropical zones.

Potassium: The studies carried out in Korea (Li *et al.,*1996) indicated that for every 100 kg of seed produced, 6.7 kg of potassium (K) was needed. Uptake of K during the 46 days from emergence of the sixth leaf to cessation of apical growth was 95% of total K usage by plants (Li *et al.,* 1996). Potassium (K) content is usually the highest in stems and capsules of mature plants. When substantial amounts of N and P are applied to irrigated crop of sesame, it goes without saying that potassium becomes necessary to maintain the nutritional balance.

Jat *et al* (2017a) found that progressive increase in level of potassium up to 50 kg K2O /ha and sulphur 40 kg S/ha significantly increased the quality of sesame. Jat *et al.*, (2017b) reported that as per the economic and marginal analysis, increased level of potassium up to 50 kg K2O /ha and sulphur 40 kg S /ha resulted in higher net returns, benefit-cost ratio and optimum dose of potassium and sulphur for sesame crop were computed as 63.94 kg K2O/ ha and 53.93 kg S /ha, respectively.

Sulphur: As the soils of tropical zones are deficient in Sulphur, application of sulphur @40 kg/ha is beneficial to the crop. However, this deficiency can be overcome by choosing fertilizers like single superphosphate which also provide the crop with sulphur in addition to Phosphorous (Weiss, 2000).

Micronutrients

According to Weiss (2000) sesame is sensitive of boron, which causes typical toxicity symptoms such as firing of leaf margins and tips.

Shehu (2014) working in Nigeria on Lithosols, reported that stem height, number of branches, leaves, and capsules were not significantly influenced by Mn and Zn. Seed yield was depressed at 1 kg Mn/ ha, 1 kg Mn + 1 kg Zn / ha and 0.5 kg Mn + 0.5 Zn /ha by 17.57 and 41% respectively. Number of seeds / capsule was depressed by 9% at 1 kg Mn /ha. Combined Mn and Zn had dry matter yield disadvantage by 16 and 18% at 0.5 and 1 kg/ha of Mn and Zn rates, respectively. Shoot N, P, K, Mn, and Zn contents of sesame, N and P uptake were not significantly influenced. Combined Mn and Zn depressed K, Mn and Zn uptake while Mn uptake increased by 57%.

On the other hand, in India perceptible positive response in terms of increased seed and oil yield of sesame are reported following the application of micronutrients such as Zn and Mn and also the micronutrient mixtures either as soil application or as a foliar spray. Yadav *et al.*, (2009) working on agronomic management of sesame in Bundelkhand tract of India reported that the treatments with 100% of the recommended NPKS fertilizer along with micronutrients and organic manure recorded significantly higher growth and yield attributing characters, seed yield and net returns over half of the recommended dose of fertilizer (20:10:10:15 kg NPKS/ha) coupled with micronutrients and organic manure. Combined use of zinc sulphate and ferrous sulphate along with farmyard manure mixed with 100% recommended dose of fertilizer exhibited a pronounced effect on the growth characters viz., number of branches/plant and dry weight/plant, and yield attributes viz., number of capsules/plant, seed weight/plant and test weight, The seed yield and net returns were significantly influenced by the same treatment, which resulted 29.7 and 37.2% increase, respectively, over 100% recommended NPKS fertilizer alone. The same treatment also registered the highest net return/ rupee of investment (Rs1.85). The individual application of zinc and iron along with farmyard manure ranked next places in all respects.

Seervi *et al.,* (2018) reported that foliar spray application of micronutrient mixture consisting of Bo, Zn, Mo, Cu, Fe, Mn and Co @ 5 ml/ 2 litres of water resulted in significantly higher seed yield and oil content of sesame.

Khan *et al.,* (2017) found that Micronutrients micronutrient fertilizers applied @ 10kg ZnSO4/ ha, 10kg Borax /ha and 5kg MnSO4/ ha enhanced yield and yield related traits provided proper cultivar is selected.Zn, B and Mn increased capsule length, capsule weight, capsules/ plant, seeds /capsule, 1000 seed weight, seed yield and oil yield of sesame.

Shirazi *et al.,* (2017), found that foliar application of different concentrations of micronutrients improved the morphological characters and seed yield of sesame. Based on the results it is suggested, use of 60 kg N/ ha+ 150 ppm of micronutrients have produced highest seed yield of sesame by adjusting in plant morphological characters and yield

Application of 125%of the recommended NPK + composted coir pith@12.5 t/ha +Zn SO4 @25 kg/ha + Mn SO4 @ 5kg/ha found to be effective in bringing about the positive effect of the micro-nutrients viz., Zn and Mn in increasing significantly yield of sesame on coastal soils of Tamil nadu in South India (Elayaraja, 2018). This also resulted in increased soil nutrient availability.

There are clear recommendations from the Indian Agricultural Research Systems suggesting the application of the deficient micronutrients based on the symptoms of sesame crop and soil test values (http://agritech.tnau.ac.in/agriculture/agri_nutrientmgt_sesame.html)

The contrasting results reported on the effect of micronutrients from different regions of the world clearly brings out the role of soil in the expression of the positive or negative effects of micronutrients on sesame crop.

On one hand, in India where there is a wide spread deficiency of soil micronutrients, their positive effect on the yield and yield attributes of sesame has been established.

On the other hand, in the case of Litho soils of Nigeria and elsewhere the effect of micronutrients yielded negative results, due to the peculiarities of interacting factors of soil. It is known that the higher levels of P in the soil renders Zn unavailable to the plants. Similar such mechanisms must be operating in the areas, where negative effects on the crop growth were observed with the application of some of the micro-nutrients.

Sesame based Nanoparticles: Zinc oxide nanoparticles(ZnO-NP) are extensively exploited in pharmaceutical and cosmetic industries due to their astonishing properties such as compatibility with human skin, non-toxicity, self-cleansing property, UV-blocking property for sunscreens, efficient resisting capacity against microorganisms etc. and are found to enhance growth in several food crops

(Padmavathi and Vijayaraghavan,2008; Prasad, TNVKV *et al.,* 2012; Zhao *et al.*, 2014).

Green synthesis of Zinc oxide nanoparticles (ZnO-NPs) is a novel and non-toxic method as compared to the hazardous conventional physical and chemical methods. Manokari *et al.*, (2019) report production of ZnO-NPs for the first time using whole vegetative parts of *Sesamum indicum* L. The aqueous extracts of various parts of *S. indicum* were used to synthesize nanoparticles in this study. The synthesized nanoparticles were evaluated using UV-visible spectroscopy for confirmation and characterization. The maximum UV-visible spectral absorption peaks were observed from 293 to 296 nm wavelengths. Leaf and stem reaction mixtures exhibited the sharpest absorption peaks of all the variations at 293 nm and root at 296 nm. This study leads to the development of cost-effective ZnO-NPs synthesis with possible further exploration to serve mankind.

Cytogenetics and Species Relationship

The cytogenetics of the species of the family Pedaliaceae with specific reference to the genus *Sesamum* is nebulous despite the ancient history of sesame. The family Peadliaceae consists of around sixty species falling in 16 genera. The genus *Sesamum* consists of about 38 species a majority of which are not domesticated. Most of these wild species are distributed in tropical Africa and Asia. The important species of *Sesamum* domesticated and cultivated over a long period of more than three to four thousand years is *S. indicum* with a somatic chromosome number 2n=26 (Morinaga *et al.*, 1929). The other species such as *S. capense,* Burn (aka *S. alatum*, Thonn), *S. malabarium* Nar., and *S. schenkei* Aschers., possess the same chromosome number as that of cultivated sesame.

It may be observed that the chromosome numbers of only three genera are known, namely *Pedalium* (x = 8), *Sesamum* (x = 8, 13) and *Ceratotheca* (x = 8). The chromosome numbers of several wild species of *Sesamum* of Africa are unknown even today (Kobayashi, 1991).

The other species of *Sesamum* such as *S. laciniatum* (2n = 28), *S. prostratum* (Fig. 6)and *S. angolense (*2n=32); *S. occidentale* and *S. radiatum* (2n=64) have different chromosome numbers. Another genus *Ceratotheca sesamoides* related to Sesame possesses 2n=32 chromosomes. Another species of *Sesamum* found in India viz. *S. mulayanam* possesses many similarities with sesame and could be a donor of valuable genes for resistance against phyllody and wilt (Weiss,2000) The zones of spread of the wild varieties are not adequately documented (Ashri, 2007). The actual number of species of *Sesamum* is not clear. Totally around 40 species have been described and 36 have been included in the *Index Kewensis.* (Weiss, 2000).

From all the above cases of cytogenetic diversity of the species *Sesamum*, it appears that the species has two basic numbers i.e., x= 8 and 13. It appears that polyploidy /amphidiploidy have played significant role in the evolution of the genus *Sesamum.* Many wild species numbering over 18 are endemic to Africa. Eight species are reported to occur in India- Ceylon(Srilanka) region, out of which, two wild species viz., *S. lacinatum and S. prostratum* are found in India and Srilanka; but *S. malabaricum* (Fig.4) is available in India alone (Ashri,2007).

Fig 4: Sesamum malabricum (2n = 2x =26)

According to He *et al., (*1994), the sesame evolved from symmetric to asymmetric types and from *S. capense* to *S. indicum* in the sequence black, brown, yellow and white seeded types. Based on the investigations conducted on the fatty acid compositions of the seed oils of *Sesamum indicum, S. alatum, S. radiatum* and S. *angustifolium* (Fig. 5),it was observed that *S.radiatum* and *S. angustifolium* are more closely related to each other than to other two species (Kamal-Eldin and Appleqvist, 1994)

Fig. 5: Sesamuam angustifolium (2n=4x=32)

Sesamum prostratum (Fig. 6) documented for its high degree of disease resistance, appears to be apart genetically from *S. indicum,* in terms of crossability. However, Ramanujam (1942) has reported the development of interspecific hybrid involving *S. orientale (S. indicum)* x *S. prostratum* and the derivatives of the resulting amphidiploid realised.

Table 1: Some known species of Sesamum and their 2n chromosome numbers (Ashri, A. 2007).

2n=26	2n=32	2n=64	Unknown
S.alatum Thon	**S. angolense Welw.**	S.radiatum Shum & Thonn	S. abbreviatum Merxm
S.capense Burn.f.	S.angustifolium, Eng.		**S. calycinum Welw.**
S. indicum L	S.lacinatum Wild.		S. marlothi Engl.
S.malabaricum Burm.	S. latifolium Gillet		**S. parviflorum Garbow-Seindist**
	S.prostratum,Retz.		S.peadlioidesHeirn
			S.rigidumPeyr
			S. shinzianum Aschers ex Shinz
			S. trphyllumWelw.

According to Weiss, (2000) the induced polypoloids of *Sesamum* tend to produce low seed yields even with enhancement of pollen fertility levels.

The species of *Sesamum* available in Australia are stated to be due to imports by Chinese immigrants in the mid-nineteenth century (Bennett,1996).

Fig. 6: Sesamum prostratum (2n=4x=32)

The crossability barriers among the species of *Sesamum* are practically very rare. Interspecific hybridization has been carried out with good seed set (Prabhakaran and Rangasamy, 1995). With discovery of male sterility in *Sesamum* (Osman 1981), the door for production of hybrid seed is open (Tu, 1993; Ganesan, 1995; Wang *et al.*, 1995).

According to the ICAR-Indian Institute of Oilseeds Research (Anonymous 2014-15), interspecific hybridization of TKG-22 and GT-10 (male parents) was carried out with sesame wild species *(S. malabaricum, S. mulayanum S. laciniatum S. alatum and S. radiatum*) of which the crosses with only *S. malabaricum* were successful. The pollen sterility ranged from 81.0% *(S. malabaricum* x GT-10) to 84.5% (*S. malabaricum* x TKG-22) in the F1 plants.

In India *Sesamum indicum* has a large number of land races with varying characteristics of branching, leaf type, flower colour, capsule bearing per node, seed colour and seed size (Ashri, 2007). Each country possesses variable varietal materials represented by distinct ecotypes specific to the region (Hildebrandt, 1931, Vavilov, 1951)

The ecotypes that occur in Russia are of uniform plant type with single stem, broad and deeply divided leaves possessing only one flower in the axil being resistant to powdery mildew (*Odium)* and wilt (*Fusarium).* The Central Asian Region has ecotypes adapted to diverse environmental situations. The ecotypes from Eastern Mediterranean region have highest seed oil content. Tadzhik zone has local strains resistant to *Fusarium.* Oil content is the highest in eastern Mediterranean types, particularly Syria and Palestine. On the other hand the oil content was reported to be low in Japanese types The numbers of types found in Pakistan were 62, in Myanmar 49 and 54 in Sudan (Weiss, 2000). South Korean plant breeders claim to have developed 'The Ideal Sesame Plant' for their own environmental conditions (Kang *et al.,* 1995)

Genetic Resources and Sesame Breeding

There exists an impression that *Sesamum* has not received adequate attention of the research and developmental agencies in relation to other oilseed crops; notwithstanding its several desirable genetic attributes (Bathkal, 1994; Thangavelu, 1994a; Ashri, 1994). It is also stated that the introduction of several genotypes of cultivated sesame has resulted in the rapid replacement close to the extinction of primitive cultivars and land races of sesame and also the wild species of the *Sesamum* from the centres of diversity like India causing depletion of valuable genes for resistance to pests and diseases and other desirable developmental attributes (Bathkal, 1994). The wild species are poorly represented in the current germplasm collections (Ashri, 1994). Hence, it becomes necessary to augment the genetic resources of sesame on priority basis.

Genetic diversity within *Sesamum indicum* L. The history of germplasm collections of *Sesamum* got initiated during 1920s in India and the first collection was established in 1925 (Brar and Ahuja, 1979; Joshi, 1961) Later, a world collection of sesame was established in Soviet Union (Weiss, 2000). Subsequently, large collections of *Sesamum* were assembled in Venezuela and the USA (Ashri, 1994). Cytogenetic work on Sesamum has been reviewed by Joshi (1961) and most of it is confined to investigation of species hybrids.

With the support of International Bureau of Plant Genetic Resources (IBPGR) and German-Israel Fund for Research and International Development (GIFRID) large germplasm nurseries of *Sesamum* consisting of 3130 accessions were grown at Rehovot in Israel (Ashri,1994) The collection, however, had only a limited number of wild species. The gene banks were reported to have faced problems in rejuvenation of the accessions since sesame may be cross pollinated (Ashri, 1988) depending upon the degree of insect activity.

It was observed that geographical isolation of the sesame types played an important role in the differentiation /diversification of sesame varietal forms and land races

(Pham *et al*., 2009) which could be termed of polytopic origin (Vavilov, 1951) since India and Abyssinia, were basically the centres of primary diversity, the central Asia being the other centre. China was considered the secondary centre of origin for dwarf genotypes. African continent, which is has rich diversity for different species of *Sesamum* is poor in different forms of cultivated *Sesamum indicum* L and India on the other hand possesses very rich diversity for the cultivated sesame, but restricted diversity in respect of the wild species of *Sesamaum* (Rheenen, 1981; Hilderbrandt, 1931). Africa was suggested as the primary centre of origin of sesame and India the secondary centre of origin of sesame (Thangavelu, 1994a). Nevertheless, Bedigian *et al.,* (1985) and Bedigian *et al., (*1986) found that phytochemical evidence as well as other data including the archaeological findings supported the suggestion that the progenitor of sesame originated in India and that domestication of *Sesamum* occurred in India.

This kind of phenomenon explains the current state of genetic diversity. Apart from the recognised and recommended varieties of cultivated sesame, there exists a wide spectrum of diversity among several land races of the crop which have been in cultivation since times immemorial, since India is the home of origin and domestication of cultivated sesame (Bedigian *et al*., 1986). A comparative account of two known wild land races of cultivated sesame are presented in the Table 2 (Devarathinam and Sundaresan, 1990) Therefore, it is important to augment the genetic resources with collections from as many diverse geographical entities as possible.

A phenomic approach was undertaken by Prasad and Gangopadhyay (2011) to characterize seventy-one accessions of sesame (*Sesamum indicum* L.) procured from India, Venezuela and USDA following the descriptor of NBPGR and IPGRI (*Phenogram is a diagram depicting taxonomic relationships among organisms based on overall similarity of many characteristics without regard to evolutionary history or assumed significance of specific characters: usually generated by computer).*

The phenotypic traits varied extensively and the degree of polymorphism was the highest in case of seed coat colour and corolla characters (Fig. 7), while few traits exhibited only binary type character state. Morphological observations revealed Venezuelan genotypes as the most derived ones while the Indian accessions possess some of the wild character state of the traits. The hierarchical axial representation of phenogram indicated that one Indian accession under USDA repository (USIN06) was unique while the Venezuelan genotypes fell in a common cluster in neighbour-joining WPGMA tree.

Table 2: Comparative description of some wild land races of sesame(Devarathinam and Sundaresan, 1990)

Character	Sesamum indicumL. (regularly cultivated)	S. indicum var. yanamalai ADR & MS	S. indicum var. sencottai ADR &NS
Branching	Branching and non-branching/ single stemmed and bushy types	Open type, yet with more number of primary and secondary branches.	Open type with mostly primary branches and very few secondary branches
Leaves	Dimorphic, mostly serrate or dentate margin in basal leaves and entire margin in top leaves	Highly heteromorphic with serrate margin in basal leaves and entire leaf margin in top leaves	Dimorphic, with dentate margin in basal leaves and entire leaf margin in top leaves
Days to First flowering	25 - 40	55 – 60	35 - 40
Flowers	Mostly white & occasionally with purple tinge	Purple on the exposed side with deep purple lower lip	Pink coloured with purple coloured lower lip
Filament	White with no streaks	Cream yellow with prominent purple streak at the line of dehiscence.	Cream yellow with ight purple streak at the line of dehiscence.
Capsule	1.5 to 3.2 cm long & 0.6 to 0.8 cm broad	1.8 to 2.0 cm long & 0.3 to 0.4 cm broad	1.8 to 2.2 cm long & 0.6 to 0.8 cm broad
Seeds	Black, brown, white. with smooth surface, elongate and orbicular	Jet black reticulately rugose, ovoid.	Brown, rugose, orbicular

Principal coordinate analysis further resolved the relative distances between genotypes within each cluster/sub cluster. The traits related to trichomes of different plant parts showed highest correlation coefficients. A simple homology based algorithm; considering the character-states as different colour codes and comparison of the codes of each genotype with that of a postulated 'target sesame' resulted in few Indian and exotic accessions of sesame on the basis of high 'scores', which will probably lead to a more precise means of selection of genotypes for future marker assisted breeding program.

The cluster analysis of 22 accessions of sesame revealed (Pham *et al.,*2009) that they fall into four diverse groups with a clear division based on their geographical origin. Nevertheless, some geographically distinct accessions got clustered in a different group, which might indicate the play of human factor in spreading the sesame varieties.

Genetic diversity among 128 sesame genotypes representing 10 geographically distinct populations in Ethiopia was assessed by Abate and Mekbib (2015) at DNA level using RAPD analysis. Eleven RAPD primers used amplified a total of 149 bands, of which 142 (95. 45%) were polymorphic. Each primer generated 7 to 23 amplified fragments with an average of 13.5 bands per primer. Percent of polymorphic loci (P%), number of different (Na) and effective (Ne) alleles along with Shannon information index (I) and Nei's gene diversity (He) values suggested that the population of Oromia was the most diverse of all populations, while populations from Afar (cultivars) and AM-NSh were found to be the least diverse. Based on average dissimilarity values obtained with RAPD primers, AM-NG-25, SNNP-7 and SNNP-8 were the most distinct of all genotypes, while genotypes ORO-20 and TIGR-5 showed maximum similarity with others. The UPGMA (*UPGMA is unweighted pair group method with arithmetic mean or a simple agglomerative bottom-up hierarchical clustering method by Sokal and Michener, 1958*) clustering based on the dissimilarity matrix clustered the genotypes into 3 major groups and 11 subgroups, while three genotypes viz., BENSH-6, ORO-14 and SNNP-5 were found to be out of all the groups from the rest without falling in any of the clusters, indicating that they are the most divergent genotypes. Generally, both clustering and PCoA patterns revealed that most genotypes located geographically far apart were found to cluster in the same group, while those genotypes from the same origin dispersed. Overall results indicated that RAPD technique revealed a high level of genetic variation among sesame genotypes collected from diverse ecologies of Ethiopia.

Bedigian *et al.,* (1986) studied cultivars of sesame from 20 countries for morphological variability. Taximetric methods, including factor, cluster, discriminant, and principal components analyses, established patterns of similarities and were used to generate groupings among the taxa. The complementary results of the analyses indicate that 8 major groups can be discerned. Plants with tetracarpellate capsules have a distinctive form and comprise the initial separation from the entire collection on the dendrogram. The second branch is a group of purple-tinged plants from India that also includes the proposed progenitor. A short, bushy, early maturing genotypes consists of predominantly Turkish cultivars. Another group, composed primarily of Korean accessions, consists of unbranched plants with strap-shaped leaves. Other groups are less easily typified. This characterization of the genetic variation in sesame can be used to identify sources of genetic materials for crop improvement, as well as to provide information about the evolution and genetic differentiation of the crop. In this context, in order to augment the utilization of the needed germplasm from the vast and diverse gene banks, it would be necessary to develop 'core collections' (Frankel 1984). In other words, the core collection represents a condensed representative of the gene pool.

Fig. 7: Polymorphism for seed and Corolla characters in Sesame (Prasad &Gangopadhyay, 2011)

Wild species of *Sesamum* and their utility in Sesame breeding: The genus *Sesamum* consists of 38 species (Kobayashi, 1991) out of which 18 species are reported to be endemic to Africa (Rheenen, 1981) including those listed below.

Sesamum alatum, S. angolense, S. angustifolium, S. Prostratum, S. indicum, S. laciniatum, S. latifolium, S. Occidentale, S. radiatum, and *S. schenckii*

According to Kobayashi (1991, out of the above 38 wild species, around 30 species are distributed in the Savannah of tropical Africa., that is from West

(Upper Guinea) to East Africa and Lower Guinea (Congo and Angola etc.,) to South Africa. Ten of the above wild species occur in the Indian sub-continent including India and SriLanka and five in East Indies (Kobayashi, 1991). Bedigian (1984) reported the presence of *Sesamum malabaricum* (2n=26) in India.

Considering the breeding value of the wild species of *Sesamum,* limited studies are found in literature on the desirable attributes and their role in the genetic improvement of sesame (Thangavelu, 1994). Nevertheless, often statements are made that the wild species of *Sesamum* possess resistance to various insect pests and diseases, since they are believed to have undergone natural selection under rigorous wild conditions with diverse stresses.

Sesamum angustifolium is reported to possess good drought resistance (Uzo and Adedzwa, 1985a) in view of the high degree of fruiting of the order of 89% to 90% during the dry season.

The special adaptive attributes of *S. angustifolium* are reported to be (i) fleshy root system facilitating water conservation in the plant system, (ii) small linear leaves that offer resistance to water loss due to transpiration, hairiness, location of stomata only at the adaxial surface and high fruit set in the dry season, when other annuals around dry-up due to harsh dry conditions. The species possesses grey seeds.Based on the available reports in the literature, the special attributes of some wild species of *Sesamuam* are listed in the Table 3 below.

Table 3: Special attributes of some wild species of Sesamuam

S.No.	Wild species	Special characters	Reference
1	Sesamum angustifolium	Drought resistance & good fruit set; grey seed	Uzo and Adedzwa, (1985a)
2	S. radiatum	Resistance to shoot- webber; freedom from foliar diseases	Uzo and Adedzwa, (1985a)
3	S. alatum	Resistance to shoot webber, Immunity to Phyllody, Resistance to Antigastra and Powdery mildew	Srinivas (1991) as reported by Thangavelu (1994a)
4	S. malabaricum	Resistance to Antigastra & Powdery mildew ; Resistance to waterlogging	Thangavelu (1994a) John et al.,(1950)
5	S. occidentale	Resistance to Antigastra and Powdery mildew	Thangavelu (1994a)
6	S. laciniatum	Drought resistance and resistance to Antigastra	Thangavelu (1994a)
7	S. prostratum	Practically unaffected by pests and diseases, drought resistance and suitability for sandy tracts	Ramanujam (1942a)

It is necessary that the wild species are tested rigorously in diverse environments to confirm their characters useful in sesame breeding. However, some of the wild species have been crossed with cultivated sesame by different workers. The studies on interspecific hybridization have resulted in valuable information with regard to the species relationship, the compatibility or otherwise and genetic affinity. According to Thangavelu (1994), close affinity was observed between *S. malabaricum* (2n=26) and *S. indicum* (2n=26); *S. prostratum* (2n=32) and *S. laciniatum* (2n=32); *S. indicatum* (2n=52) and *S. radiatum* (2n=64) and *S. occidentale.* (2n=64)

Prabhakaran (1992) found that among the diverse interspecific crosses, only three species, viz., *S. malabaricum* (2n=26), *S. laciniatum* (2n=32) and *S. prostratum* (2n=32) were cross compatible with *S. indicum* (2n=26). However, the combination of *S. alatum* x *S. indicum* var. CO-1 set viable seeds. The investigations carried out by Prabhakaran (1992) have revealed for the first time that cytoplasm of *S. malabaricum,* in interaction with the genome of *S. indicum,* resulted in male sterility with the basis of cytoplasmic and genic interaction. This could be of immense importance with regard to the development of hybrid sesame.

The desirable characteristics associated with the wild species of *Sesamum* that could be useful in sesame breeding are as follows (Thangavelu, 1994).

(1). Erect branching, (2). Linear /deeply palmate / lanceolate leaves, (3). Long capsules, (4). Fleshy roots, (5). Larger number of seeds per capsule, (6). Resistance to *Antigastra,* (7). Resistance to diseases viz., Phyllody, powdery mildew, wilt complex, bacterial blight etc., (8). Drought resistance and (9). Tolerance to heavy rainfall

Though the wild species are potential sources for the major genes for pest and disease resistance, transfer of these into cultivated sesame has not been accomplished over the past 50 years due to several reasons including post fertilization barriers including embryo abortion (Ramesh *et al.,* 2003). Nevertheless, with present day molecular tools (Tiwari *et al.,* 2011), identification and transfer of desirable genes from wild to cultivated *S. indicum* would be much faster and less cumbersome. The availability of a transformation system for sesame can hasten this process.

Sesame breeding: The objectives of any crop breeding programme are dictated by the needs or requirements of the specific agro-ecological system in which the crop is grown. Nevertheless, the over-all major objectives in sesame breeding are as follows

1. Enhancement of productivity levels under the conditions of dry land agriculture (from current average yield of 500 kg/ha to progress towards a minimum yield level of 1,000 kg/ha)
2. Incorporation of durable levels of pest and disease resistance
3. Developing (a) plant types with 2 to 4 branches and higher number of capsules per branch for irrigated agriculture and, (b) moderately branched type with robust root system for dryland situations.

4. Developing photo-insensitive and slow senescent types for adaptability to diverse weather conditions
5. Developing suitable plant types for inter cropping situations on drylands
6. Increasing the oil and protein content of sesame seed and
7. Developing non-shattering varieties to avoid seed losses at harvest.

The emphasis on the objectives might vary depending upon the needs of each agro-ecological conditions.

Some types, especially bicarpels, monocapsule, branch (BMB) and bicarpels, tricapsules, branch (BTB), were considered as ideal plant types in breeding for high-yielding varieties (Baydar, 2005)

The wide genetic diversity of plant types and the ease of making crosses among them, together with the remarkable adaptability of sesame plant as exemplified by the diverse environments in which it has established (Weiss, 2000) are the assets which need to be capitalised for sesame breeding. According to Weiss, (2000) great scope exists for improvement of currently grown varieties, and breeding new cultivars. Sesame exhibits a very wide range of adaptation since the crop is grown throughout the Indian sub-continent and other south-east Asian countries under diverse agro-ecological conditions. More than 55 recognised and improved varieties of sesame are being grown in India in addition to many land races representing a wide spectrum of genetic diversity. Spectrum of the diversity among the types grown over different zones of India is reflected in the sesame varieties given Table 4 (Rana *et al.*,1994; MNOOP 2018).

Table 4: Sesame varietal diversity in different zones of India
(Rana et al.,1994; MNOOP 2018)

Zone	States	Diverse varieties found in the region.
East	Bihar	B-67. Kanke, White, Krishna, OMT-11-63,
	Orissa	Vinayak, Kanak, OMT-11-63, Nirmala, Shubra, Prachi, Amrit, Smarak. Uma, Usha
	West Bengal	Savitri, B-67, Punjab Til-1, Improved Sesame-5 (Rama)
West	Gujarat	Gujrat Til 1,2, 3&10 Mrug-1,Purva-1,
	Maharashtra	Phule Til-1008, Tapi, AKT-64& 101, TC-25, JLT-26 & 408, PKVNT-11
	Rajasthan	Pratap (C-50), TC-25, T-13, 351, 346 RT-46,54 103, 125 &127
North	Haryana	Haryana Til, RT-46 HT-1, HT-2
	Punjab	Punjab Til-1, TC-289, TC-25, RT-46
	Uttar Pradesh	T-4, T-12 & 78,RT-46, Sekhar
Northeast	Assam	Madhavi, Gauri, ST-1683, RS-1, RT-1, OMT 11-6-3
South	Andhra Pradesh+ Telangana	Gauri, Madhavi, T-85, Sweta, Rajeshwari, Varaha, Gautama, Chandana, Hima, YLM 66(Sarada)
	Karnataka	E-8, TMV-3, Co-1, DS-1, DSS-9
	Kerala	Soma-1, Surya, Kayam, Kutu-1,Tilothama,
	Tamil Nadu	TSS-6, Payur-1, VRI-1&2, TMV-3,4,5,6, &7 Co-1
	Andaman & Nicobar	B-67
Central	Madhya Pradesh	TKG-21, 22, 55,306,308, JTS-8,PKDS-8, 11&12, N-32, JT-7, RT-46,Improved Selection-5, OMT-11-63

In order to improve the plant type of sesame, pedigree selection was applied (Baydar 2005) to segregating generations of crosses of genotypes with contrasting characters. In the segregating F2 population, the progenies were classified into eight types according to the combinations of carpel number per capsule, capsule number per axil and branching habit. Progeny of individual F2 plants including a particular type were advanced to the F6 generation. It was possible to improve the lines with eight types at the end of the selection process. Some types, especially bicarpels, monocapsule, branch (BMB) and bicarpels, tricapsules, branch (BTB), were considered as ideal plant types in breeding for high-yielding varieties. While the low-yielding type quadricarpels, tricapsules, non-branch (QTN) was the highest in oil content (49.3%), the high-yielding type BMB was the lowest (43.2%), Although the QTN-type had the lowest content of oleic acid (41.3%) and the highest content of linoleic acid (43.1 %), the bicarpels, monocapsule, non-branch (BMN) type had the highest content of oleic acid (48.4%) and the lowest content of linoleic acid (36.6%). Total tocopherol varied between 175.6 and 368.9 mg/kg in the seed oil of the sesame types. The best high-yielding type BMB was one of the types containing less tocopherol.

Ideal Plant-type of Sesame: A plant type within 150–160 cm in height, with a limited branching and with capsules inserted not closed to the ground and not more than 70 cm, is considered ideal to non-lodging, high-yielding and mechanical harvesting (Baydar, 2005). For the latter aspect, however, the major difficulty remains the uneven ripening of the capsules due to the indeterminate growth habit, which is typical of the most currently available cultivars, including those considered in the present study (Baydar 2005). The main component involved in yielding ability of sesame seems to be the number of capsules per plant as reported by Baydar (2005). The height of sesame plant suggested may not be applicable to the situations of arid and semi-arid ecosystem, where excessive vegetative growth results in late maturity not matching with their rainfall patterns. While trying to arrive at an ideal plant type of sesame, the important plant physiological attributes related to the growth, development and productivity of sesame must be considered. For example, significant point emerging from the investigation by Ruchi (2008) is that in sesame the yield is related to harvest index. Higher yield in sesame is related to higher rate of Relative Growth Rate (RGR), Net Assimilation Rate (NAR) and dry matter accumulation. The longer duration from flowering to harvest is an important yield determinant. The continued late dry matter accumulation and Crop Growth Rate (CGR) resulted in low harvest index and grain yield. Higher harvest index with higher grain yield was recorded in Phule Til-1 followed by LTK-4. Saha and Bhargava (1980) also noted that maximum harvest index gave higher seed yields due to higher number of fruits and better retention of capsules in sesame after fertilizing crop. In this context it must be appreciated that higher harvest index is a function of efficient portioning of the assimilates to the sink thus achieving an appropriate balance of source – sink relationship (Duncan *et al*., 1978). At the same time, it is also important that higher degrees of harvest index should always be combined with higher dry matter production. Otherwise the harvest index without higher dry-matter can't result in higher seed yield.

In sesame, the crop growth rate (CGR) showed a gradually increasing trend till maturity. This is because the crop growth rate increases with the increase in crop age (Chauhan *et al.,* 1996), which also results in higher dry matter. Dry matter is positively and significantly correlated with CGR at all stages and hence both the characters are found to increase with crop age. Interestingly, the highest crop growth rate in Phule Til-1 also coincided with maximum dry matter production (Ruchi, 2008). The lower leaves at the bottom of thc canopy under shaded conditions have low CO2 assimilation rates and contribute less to other plant parts. Similar results were obtained earlier by Durga *et al.,* (1997), who reported early senescence of the crop due to the above cited situation. One of the important limitations of sesame plant is its photosensitivity (Langham, 2007), which hinders the crop's performance under conditions of low photoperiod. In India, the crop is normally sown with the onset of monsoon showers in June, when the yield of sesame is of normal degree. On the other hand, if the monsoon is delayed, sesame crop often fails to yield. In summer season from March to May or June, sesame is grown as a catch crop in certain areas under irrigated conditions, when the productivity of the crop is good due to the prevalence of favourable photoperiod promoting good crop growth and development. In order to enhance the adaptability of sesame, it is necessary to incorporate photo-insensitivity in the cultivars (Prasad, 2019b). Growing variable genetic materials under different situations of diverse photoperiod might give an opportunity to select for photo-insensitivity.

Selection Criteria: Correlation analysis illustrates the association among different yield related traits exist in plant population under study. Genotypic and phenotypic coefficients of variability reveal the extent of differences among the accessions, due to the genetic factors and their response to environmental condition of the experiment. Path coefficient analysis reveals the relationship between variables in multivariable system. Previous studies of various sesame breeders proved that plant height, number of branches per plant, capsules per plant, seeds per capsules, 1000-seed weight are the traits which have significant and positive correlations with yield per plant at both genotypic and phenotypic levels. These characters also have high path analysis values especially for capsules per plant had highest direct effect on seed yield followed by 1000-seed weight. So, these traits may be used as selection criteria in breeding programs for the improvement of seed yield of sesame (Mustafa, *et al.*, 2015). In addition to the above, the physiological attributes such as harvest Index and crop growth rate (CGR) which exhibit positive relationship with seed yield (Chauhan *et al.,* 1996, Ruchi, 2008) should be included in the selection criteria for breeding productive varieties of sesame.

Development of Hybrid Sesame

First report on the hybrid vigour in sesame in India was by Pal (1945). Since then several workers reported the phenomenon of hybrid vigour in the crosses of sesame varieties. Hybrids manifesting levels of heterosis up to 328% were reported (Singh VK *et al.,* 1983;Krishnaswamy *et al.*, 1985).

Availability of wider genetic diversity in sesame (Bedigian *et al.*, 1986, Pham *et al.*, 2009) is one of the important factors in favour of identification of heterotic parental lines.

According to Tu (1993) seed produced from one hectare of crossing blocks can provide for the adequate F1 hybrid seeds for 60 to 80 ha of area. According to Ashri (2010), availability of productive hybrids can boost sesame productivity levels at least by 50% as compared to the existing varieties of sesame.

Due to the paucity of ready availability of cytoplasmic male sterile, maintainer and fertility restoration systems for hybrid seed production, the other option could be to take advantage of the inherent floral attributes of sesame to effect mechanical hybridization among the best heterotic combinations, despite the cost of manual labour involved for the operations (Prasad 1994b, Manivannan and Ganesan,1995, Sharma1994). The cost of the hybrid seed produced through mechanical emasculation and pollination was estimated as more than IRs.300 /kg (Barwale and Panchabhaye, 1994).

Sesame has the advantage of epipetalous flowers, and thus, emasculation and pollination are easy. A large number of seeds (40–50) are produced per flower. In addition, low seed rate (2.0–2.5 kg/ha) and high seed multiplication ratio (1:300) facilitate ease of hybridization and recombination breeding programs. Insects favoured cross-pollination in sesame can also facilitate natural hybridization (Duhoon 2004).

The following steps have been suggested for the production of sesame F1 hybrid seeds adopting hand emasculation and pollination (Barwale and Panchabhaye, 1994).

- The female and male parental materials must be planted in an isolated crossing block to avoid any contamination by extraneous pollen. The minimum isolation distance suggested is 100 meters. The parental lines may be planted in a ratio of 4 females to 1male (4:1) ratio at 2m spacing.
- Emasculation of the flowers of the female parent may be carried out in the afternoons between 15.00 hrs to 18.00 hrs and the corresponding pollinations using the pollen of the male parent may be carried out in the subsequent day at 7.00 hrs to 12.00hrs.
- Care should be taken to eliminate all flowers left without emasculation (un-emasculated flowers) before they open and compulsorily before pollination
- The plants of the male parent must be uprooted after seed set in the pollinated flower buds to avoid any mechanical admixture while harvesting the female parental lines for recovery of crossed seed.
- The estimated production of hybrid seed was 2.5 to 3.0 kg/40 square meters' area.

Bavbiskar (1994) produced hybrid seeds of sesame adopting hand emasculation and pollination. Male and female parents were sown on the same day in a row proportion of 1 male to 2 females (1:2), with boarder row of male parent. The inter-row spacing of 60cm was adopted between the male and female lines. The inter-row spacing in the case of female parental lines was 30 cm, while a uniform intra-row spacing of 15 cm was adopted from plant to plant. The emasculations were carried out in the evening hours between 16.00 hrs and 18.00 hrs and the emasculated flowers were covered. The pollinations were effected the in the morning hours between 6.00 AM and 8.00 AM and pollinated flowers were covered. The details of the area covered and the quantities of hybrid seed produced are given the Table.5.

Table 5: Production of sesame hybrid seed through hand emasculation and pollination (Bavbiskar 1994)

Hybrid parental combination	Area of hybrid seed production (meter squares)	Quantity of hybrid seed produced (g)
Tapi x HT-24	180	555
ATPT 85-6 x Tapi	135	365
Tapi x CST 785	135	295
Total	450	1,215

The hybrids produced were evaluated for their yield performance against the varieties viz., Tapi, Padma. the local variety Hawari and national check TC25 (Table 6.)

Table 6: Yield Performance of the Sesame Hybrids produced over check varieties. (Bavbiskar, 1994)

Hybrid/Check	Seed yield (kg/ha)	%increase over Tapi	%increase over Padma	%increase over Hawari	% increase over TC25
Tapi x HT-24	1200	24.86	90.17	119.37	139.04
1994)ATPT 85-6 X API	1161	20.81	83.99	112.24	131.27
Tapi x CST-785	1077	12.07	70.68	96.86	114.54
Tapi	961				
Padma	631				
Hawari	547				
TC-25	502				
Mean	867				
SEm+	36				
C.D @5%	106				

One of the means of cutting down the cost of labour involved in the manual production of hybrid seed of sesame could be through installing honey bee colonies in the hybrid seed production plots.

Under the national crossing programme of the ICAR-Indian Institute of Oilseeds Research (IIOR) (Anonymous 2014-15) sixty experimental hybrids were evaluated in which the hybrids of DS-5 x JLS-9707-2 and DS-5 x TKG-22 performed better with 20% and 17% standard heterosis, respectively over the best check, GT-10. In the studies to estimate the extent of heterosis in sesame, four good combiners were selected from different Oilseeds Research Centres based on the available data. Twenty-four common male parents were distributed to all the six centres identified, to initiate a crossing programme with the best female parental lines available at each centre to develop 72 new hybrids (Anonymous 2014-15). Also, *Sesamum* wild species including *S. lacinatum* collected from Kerala facilitated in developing 80 crosses.

At the ICAR-Indian Institute of Oilseeds Research, 69 new crosses were generated by crossing a set of 24 common male parents with three females (DS-1, DS-5 and DSS-9) during kharif 2013. Among 69 experimental hybrids evaluated for heterosis and combining ability in summer 2014, four early experimental hybrids viz., DSS-9 x MT-10-81, DSS-9 x Prachi, DS-5 x JLS-9707-2 and DSS-9 x DS-30 were identified. A similar trial is being conducted at Dharwad in summer.

The available male sterile lines are of nuclear / genic type (Osman, 1981; Osman and Yermanos, 1982). Maintenance of the genic male sterile system to be used as female parent in hybrid seed production shall be through rouging of the male fertile segregates in the F2 generation based on a marker character. Crosses between the male-sterile plants and male-fertile cultivars produced a proportion of F_2 segregates that were also male-sterile. Seed set on male-sterile plants following hand pollinations with normal pollen was high, indicating that these plants were female-fertile. Greenish anther colour at pollen dehiscence distinguished male-sterile from male-fertile plants, the latter having whitish anthers. The male-sterility trait reported is stable and may be maintained under greenhouse and open field conditions. In the segregating F2 generation, male sterile plants are recognizable 3 to 4 days before anthesis. The male-sterility described here can be used for commercial hybrid seed production.

In the context of developing genic-cytoplasmic male sterile –maintainer –fertility restore systems, the repository of genetic wealth available in the wild species of *Sesamum* must be tapped (SethupathyRamalingam *et al.*, 1994). Incorporation of desirable genes from wild species to cultivars was attempted by several researchers, using controlled crossing. Close compatibility or incompatibility among the interspecific crosses is well established, which should form the basis for their exploitation. Among the various wild species, *Sesamum malabaricum* has exhibited promise for developing cytoplasmic male-sterile system in crosses with sesame varieties (Prabhakaran,1992).

Mutation Breeding in Sesame

Despite the wider degree of variability available in sesame germplasm, certain highly desirable attributes could not be accessed such as good seed retention and resistance to certain diseases. Hence in accordance with the recommendations of FAO Expert Consultation (Ashri 2001) that induced mutants be used to strengthen sesame breeding, work was carried out to enhance genetic variability in sesame in respect of easily identifiable characters such as seed retention (non-shattering habit), modified plant architecture, modified growing period and resistance to diseases and pests. The project has resulted in fourteen officially released cultivars developed through mutation breeding and the derivatives of induced mutants. In this regard the variety 'Ahnsan' of South Korea for increased disease resistance released in 1985 continues to be the major cultivar grown in 1996 over 30% of the sesame area in the country. Other two prominent varieties of sesame released through induced mutations are 'Uma' and 'Usha' of Odisha state of India. (Ashri, 2010). Mutagens employed for seed treatment in this project were radiations (mostly gamma rays), Ethyl Methane Sulfonate (EMS) and Sodium azide (Na N3).

Since the varieties differ in their response to mutagenic treatments depending on their genetic background, it is always advisable to use a mutagenic dose around LD 50 with an adequately large population size, so as to have a basis for comparison. The choice of the mutagens should be based on their relative effectiveness and efficiency (Prasad, 1972a). The details of sesame varieties derived through mutation breeding and released by FAO/IAEA are given in Table.7.

Table 7: Varieties of Sesame derived through Mutation Breeding and officially released as per the FAO/ IAEA Division, Vienna. (Ashri, 2010)

Country	Variety	Year	Mutagen employed	Main Characters
Egypt	Cairo White 8	1992	Gamma rays	Non-branching
	Sinai White 48	1992	Gamma rays	Seed colour
India	Kalika	1980	EMS	Short stature
	Uma	1990	Chemical mutagen (not specified)	Uniform maturity
	Usha	1990	Chemical mutagen (not specified)	Higher yield
Iraq	Babil	1992	Gamma rays	Earliness
	Rafiden	1992	Gamma rays	Earliness
	Eshtar	1992	Gamma rays	Capsule size
South Korea	Ahnsan	1985	x-rays	Disease resistance
	Suweon	1991	Cross of Induced Mutant	Lodging & Disease resistance
	Yang baek	1995	Sodium azide	Higher oil content
	Pungsan	1996	Cross of induced mutant	Determinate habit and seed retention
	Seodun	1997	Sodium azide	Higher oleic acid, Phytophthora bight tolerance
Sri lanka	ANK 2	1995	Gamma rays	Disease resistance

Cagirran (2006) developed three true botanical determinate mutants (dt-1, dt-2 and dt-3) of *Muganlı-57* variety which is of indeterminate growth habit and three short flowering period leafy determinates (dt-4, dt-5 and dt-6) of the variety *Çamdibi* of indeterminate type and grown in Antalya, Turkey, in 1995 and 1996, respectively.

Mutant varieties resistant to *Phytophthora* blight in sesame have been developed in the Republic of Korea (Kang *et al.,* 1995) and Sri Lanka (Pathirana, 1992). In India (Kumari *et al.,* 2018) 416 gamma-rays induced M3 mutant progenies were raised at Palampur, 164 mutants at Kangra and 314 at Akrot. The mutants were screened for *Phytophthora* blight resistance both under natural field conditions and also under the laboratory conditions through 'detached leaf technique' at Palampur. Out of these, 186 and 62 M4 mutant progenies were screened again under natural field conditions at Palampur and Kangra, respectively during rainy season of 2016. Based on the screening for 2 years through 'detached leaf technique' as well as under natural field conditions, twelve mutants having moderate to high resistance against *Phytophthora* blight could be isolated. One mutant P97-1 (300 Gy) exhibited highly resistant reaction against *Phytophthora* blight during both seasons. The isolated mutant can further be utilized in hybridization programme for the introgression of *Phytophthora* blight resistance in sesame.

A monogenic recessive mutant with determinate growth habit and unique plant architecture characterised by clustered capsules was recovered by Ashri in the progeny of Gamma ray (500Gy) treated sesame cultivar No. 45 (Ashri 1995, 2010). The mutant possessed 5 to 7 capsules arranged on the tip of the main stem and its branches and internodes were telescoped, giving smaller plants. The apical flowers of the mutant were often 6-parted instead of the standard 5-parted condition in corolla lobes and the anthers were bell shaped. The mutant termed *dt45* possessed large seeds with oil characters as in the parental cultivar (Ashri, 2010).

Mutagen induced variability for oil content and oil quality in sesame have been reported by Begum and Dasgupta (2015). Thirty mutant lines isolated for higher yield from the progenies of mutagen (gamma rays and EMS) treated genotypes of sesame viz. Rama, SI 1666 and IC 21706 were evaluated for oil content and quality against their respective control genotypes. Ten of the thirty high yielding mutants, possessed high oil percentage with a better oil profile having relatively more polyunsaturated fatty acid content especially linoleic acid, than the control, indicating potentiality of mutation breeding to restructure plants with high yield, improved oil percentage and quality.

The oil content of the mutants reported by Begum and Dasgupta (2015) varied from 46.63 to 53.25% indicating 0.09 to 29.42% improvement over respective controls. Highest oil content was recorded in selection no. 8, followed by selection no. 9 which were developed from the population of IC 21706 isolated from 0.5% EMS treatment. The mutants selected from induced population of SI 1666 exhibited higher percentage increase in oil content ranging from 19.46 to 29.42% over control, while the increase was 0.09 to 7.32% and 0.12 to 10.71% in selections from mutant populations of Rama and IC 21706, respectively.

Substantial genetic variation was also recorded in fatty acid profile among the selected mutants reported by Begum and Dasgupta (2015). The results revealed that linoleic acid was the component which got enhanced in the range of 42.21 to 46.29% of the total fatty acids followed by oleic (35.35 to 38.56%), palmitic (7.05 to 10.27%) and stearic (2.98 to 3.44%) acid. Four mutants namely selection no. 1 (selected from mutant line no. 5 of Gamma irradiated by 400 Gy γ-rays), selection no. 5 (derived from mutant line no. 4 of SI 1666 treated by 200 Gy dose of γ-rays), selection no. 7 (selected from mutant line no. 10 of SI 1666 induced by 0.5% EMS) and selection no. 8 (developed from mutant line no. 7 of IC 21706 treated by 0.5% dose of EMS), were characterized by improvement in both oleic and linoleic acid contents over their respective controls. In general, positive enhancement of linoleic acid content was more pronounced than oleic acid in all mutants.

Some of the genetic problems of cultivated plants, which are intractable for amelioration through standard breeding methods, could be resolved through mutation breeding (Prasad, et al., 1984; Ashri,2001). Couple of such problems associated with sesame are (1) photosensitivity (2) early senescence (Prasad 2019b) and (3) enhancement of dry-matter production (Prasad *et al.,* 1984). Additionally, seed size of sesame needs enhancement. With the present size of seeds, it is becoming difficult to attain the needed spacing and population density at the level of small farmers, who are the major producers of sesame. Through mutation breeding it should be possible to enhance the seed size of sesame as has been done in other crops including *durum* wheat (Prasad, 1972b)

Higher levels of heterosis was observed when induced mutants of sesame were used as parents instead of the original parental variety (Murthy, 1979).

The mutation breeding programs in sesame should prioritise the work pertaining to developing mutants with photo-insensitivity and slow senescence, which would enhance the adaptability of sesame crop (Prasad, 2019b).

Biotechnology of Sesame: Progress pertaining to the genetic improvement of sesame has not been rapid as compared to that of other oilseed crops. Major constraint in this context has been the lack of efficient genotypes of sesame with spectacular productivity levels of more than 50% as compared to traditional varieties (Batkal, 1994, Thangavelu, 1994, Dossa *et al.,* 2017). Additionally, a wide range of biotic stresses inflicting the crop need to be resolved by developing sustainable levels of resistance. In this back drop of slow pace of growth of this important and ancient crop, it becomes absolutely necessary to take recourse to modern biotechnological tools which might strengthen the existing research efforts to achieve rapid genetic advance of the crop.

While plant breeding programmes release genetic variation leading to selection, mating systems and population improvement, molecular breeding techniques ranging from cell and tissue culture, vegetative propagation, genetic mapping and gene transfer help to enhance the efficiency of the crop breeding resulting in sustainable productivity. The genetic and molecular biology study of sesame began very late with only one genetic map published and no report on quantitative trait loci (QTL) mapping before 2013. However, over the last few years, some significant progress has been made in the development of large-scale genomic resources including informative molecular markers, ultra-dense genetic maps, transcriptome assemblies, multi-omics online platforms etc. In addition, the release of the draft genome of sesame (Wang *et al.,* 2014) triggered functional analyses of candidate genes related to key agronomic traits. Tiwari *et al.,* (2011) has made a comprehensive review of the important researches carried out in the area of biotechnology of sesame.

(a). Plant tissue culture: Despite large volume of work done in the area plant tissue culture and regeneration, there has been practically little progress in the area of *in vitro* culture in respect of *Sesamum indicum.* The work carried out by Lee *et al.,* (1985) on shoot tip cultures and George *et al.,* (1987) on the use of different explants need a mention in this context.

The investigations on callus induction as influenced by the explants and hormones carried out by Kim *et al.,* (1987) was in relation to developing herbicide tolerant sesame through *in vitro* selection. The successful effort in this regard was carried out by Chae *et al.,* (1987) though achieving the regeneration from herbicide tolerant callus. Nevertheless, such advances have not been taken to field of sesame production. The research pertaining to the organ culture as influenced by growth regulators (KimYH, 2001) and the combination of growth regulators on anther-culture with cold pre-treatment carried out by Lee *et al.*, (1985) brought out the genotypic effects.

In sesame, micro-propagation is achieved from shoot tip (Rao and Vaidyanath, 1997), nodal explants (Gangopadhyay *et al.*, 1998) and leaf (Sharma and Pareek, 1998) cultures. Somatic embryos have been obtained from zygotic embryos (Ram *et al.*,1990) and seedling derived callus (Mary and Jayabalan 1997; Xu *et al.,*1997) Adventitious shoot regeneration in low frequencies from hypocotyl and cotyledonary explants has been reported by Rao and Vaidyanath, (1997) Taskin and Turgut (1997) and Younghee (2001).

High-frequency plant regeneration through direct adventitious shoot formation from deembryonated cotyledon segments of sesame was achieved by Seo *et al* (2007). Chattopadhyaya *et al.,* (2010) established efficient protocol for shoot regeneration from internodes using transverse thin cell layer culture method. Abdellatheef *et al.,* (2010) evaluated *in vitro* regeneration capacity of sesame varieties exposed to culture media containing ethelene inhibitors such as cobalt chloride and silver nitrate. Growth promoting effects were observed due to reduction of ethylene concentration followed by inhibition of ethelene action.

(b) Genetic transformation: The wild species of *Sesamum* possess a wide range of useful genes which can create a positive impact on the genetic enhancement of cultivated sesame.

Nevertheless, the efforts to achieve gene transfer through conventional methods from wild species to cultivated sesame have not succeeded due to post-fertilization barriers (Ramesh *et al.,* 2003). In order to overcome the problem, genetic transformation could be a viable option to effect the gene transfer from some of the wild species of *Sesamum* possessing desirable genes and cultivated sesame. The obstacle, however, in this context could be the recalcitrant nature of cultivated *Sesamum indicum* due to the excess production of secondary metabolites (Bhaskaran and Jayabalan, 2006). However, Rao and Vaidyanath, (1997) Taskin and Turgut, (1997)and Younghee, (2001) have been able to effect shoot regeneration. On the other hand, Ogaswara *et al., (*1993) and Jin *et al., (*2005) have successfully carried out hairy root cultures using *Agrobacterium rhizogenes*.

Notwithstanding the fact that sesame is vulnerable to inflectional manipulation by *Agrobacterium tumefacience,* no transformed plants have been recovered till the past decade (Taskin *et al*., 1999).

However, the development of an efficient method of plant regeneration through direct multiple shoot organogenesis from cotyledonary explants, leading to the establishment of an optimal selection system, facilitated generation of successful *A. tumefaciens* mediated transformation protocol for generation of fertile transgenic sesame plants (Yadav *et al*., 2010). Using hypocotyl and cotyledon explants from sesame seedlings, Chun *et al.,* (2009) established hairy root cultures and a peroxidase gene was cloned from the hairy roots.

It was found that hypocotyl explants were better for gaining the higher frequency of hairy root formations as compared to cotyledonary explants (Chun *et al.,* 2009). It was also observed that the expression of peroxidase gene was different in different tissues of sesame.

C. Molecular techniques: Molecular marker technologies have significantly speeded up modern plant breeding in enhancing the genetic gain and reducing the breeding cycles in many crop species. Different types of molecular marker systems have been developed and applied to sesame genotyping and breeding efforts. The first class of molecular markers including Random Amplified Polymorphic DNA (RAPD) (Bhat *et al.,* 1999) and Amplified Fragment Length Polymorphism (AFLP) (Laurentin and Karlovsky, 2006) were designed and employed mainly for genetic diversity studies. The second class of markers involved basically Simple Sequence Repeat (SSR) types such as Inter-Simple Sequence Repeats (ISSR) (Kim *et al.*, 2002), Expressed Sequence Tags-SSR (EST-SSR) (Wei *et al*., 2008 ; Badri *et al*., 2014; Sehr *et al*., 2016), cDNA-SSR (Spandana *et al*., 2012

Surapaneni *et al.*, 2014; Wu *et al.,* 2014), Genome sequence-SSR (gSSR) (Dixit *et al*., 2005; Wei X. *et al*., 2014; Uncu *et al*., 2015; Dossa *et al.,* 2016; Yu *et al.,* 2017), Chloroplast SSR (cpSSR), (Sehr *et al.*, 2016). By compiling all developed SSR marker resources, there are in total more than 7,000 validated and 100,000 non-validated SSR markers available for sesame research. Interestingly, a new study is underway to set up an online database gathering all SSR information and providing an integrated platform for functional analyses in sesame. Many of these markers were used for genetic and association mapping, molecular breeding and genetic diversity studies in sesame (Wei *et al.*, 2013; Li *et al.*, 2014; Liu *et al.*, 2015; Uncu *et al.*, 2015; Dossa *et al.*, 2016). Finally, with the next generation sequencing (NGS) technology, the third class of molecular markers came into existence, in recent years. SNPs are more useful as genetic markers than many conventional markers because they are the most abundant and stable form of genetic variation in most genomes. Therefore, the available high-throughput methods for SNP discovery and genotyping have been employed in sesame research including Restriction Site-Associated DNA sequencing (RAD-seq); Wu *et al.*, 2014; Wang *et al.,* 2014), Specific Length Amplified Fragment Sequencing (SLAF-seq) (Zhang *et al.*, 2016), RNA-Seq (Wei *et al.,* 2014), Whole-Genome Sequencing (WGS) (Wang *et al.,* 2014; Wei *et al.*, 2015; Zhang *et al*., 2016), Genotyping by sequencing (GBS) (Uncu *et al.*, 2016). Another important marker system referred as insertion/deletions (Indels) has also been reported in sesame (Wei L.B. *et al.*, 2014; Wu *et al.*, 2014).Molecular techniques are of great value in assessing the genetic diversity and also in identification of a cultivar. In this regard the most popular molecular technique that has been employed in diverse crops is Random Amplified Polymorphic DNA (RAPD). Abdellatef *et al., (*2008*)* used RAPD to study the genetic diversity in 10 selected germplasm accessions of *Sesamum.* in Sudan. Diversity measures applied to crops usually have been limited to the assessment of genome polymorphism at the DNA level.

Occasionally, selected morphological features are recorded and the content of key chemical constituents is determined, but unbiased and comprehensive chemical phenotypes have not been included systematically in diversity surveys (Tiwari *et al.,* 2011)

Laurentin *et al.,* (2007) assessed metabolic diversity in sesame by non-targeted metabolic profiling and elucidated it between metabolic and genome diversity. They also observed di□erent patterns of diversity at the genomic and metabolic levels, which indicated that selection plays a significant role in the evolution of metabolic diversity in sesame. Another important molecular tool is Amplified Fragment Length Polymorphism(AFLP) which can be employed with success to distinguish cultivars of sesame and to elucidate the genetic relationship among genotypes (Laurentin and Karlovsky, 2007; Ali *et al.,* 2007). Distinguishing the genetic differences among cultivars has since been becoming increasingly important particularly with regard to the process of patenting the genetic discoveries and protection of Plant Breeders' Rights (PBR) (Tiwari *et al.,* 2011).

In this context, Sharma *et al*., (2009) characterized sixteen sesame genotypes using RAPD and ISSR markers, thereby finding that putative variety of specific RAPD and ISSR markers could be converted to co-dominant Sequence Characterized Amplified Region/Sequence Tagged Site (SCAR/STS) markers to develop *variety-specific markers.*

(d) Marker assisted selection: The first case of marker assisted selection was in respect of 'closed capsuel'mutant (appressed mutant) identified through AFLP marker, using Bulk Segregant Analysis (BSA) (Uzun *et al.,* 2003) applied to the segregating population of the cross involving parents viz., 'closed capsule mutant' (*CC3*) and a Turkish variety 'Mugalni-57'. This marker is stated to be of immense use in accelerating breeding programmes aimed at modifying unwanted side-effects of the 'closed capsule' mutation through marker-assisted selection.

The first report of molecular tagging of the *dt* gene which regulates determinate growth habit in sesame came from Uzun and Cagirgan (2009). Breeding of determinate sesame cultivars has become a high priority objective in sesame breeding programmes. RAPD and Inter Simple Sequence Repeat (ISSR) techniques were investigated for the development of molecular markers for the above induced mutant characteristic.

Using the F2 segregating population and bulked segregant approach, two ISSR marker loci originating from a (CT) AGC primer were detected. This marker might be of considerable use in assisting breeding programmes through marker assisted selection. The marker might also facilitate the integration of determinate growth habit into the varieties of new genetic backgrounds.

(e). Genomics: Despite a long history of sesame production and wider genetic diversity together with breeding work carried out, the crop is still plagued with certain serious and intractable limitations such as photo-sensitivity and early senescence restricting the adaptability of the crop (Langham, 2007; Prasad, 2019b). There are other problems pertaining to the seed quality. The sesame meal that is left out after oil extraction, corresponding to 50% of seed dry weight is mostly used for feed. Identification of novel genes involved in the biosynthesis of sesame-specific flavour or lignans, and a better understanding of the metabolic pathways involved in the pathways from photosynthates towards oil, are needed to improve the quality and quantity of oil in sesame cultivars.

Transcriptome or EST sequencing is the first step to access the gene contents of a species and has emerged to be an efficient way to generate functional genomic data for non-model organisms. Like genome sequence resources, sesame holds several transcriptome data generated from various organs of the plan (Dossa *et al.*, 2017).

Over the last few years, some progress has been achieved in the development of genomic resources including some molecular markers, ultra-dense genetic maps, transcriptome assemblies, multi-omics online platforms etc. The release of the draft genome of sesame (Wang *et al.*, 2014) triggered functional analyses of candidate genes related to key agronomic and developmental traits. With these invaluable efforts, sesame holds some important genomic resources and platforms for its improvement which for the time being are inexistent even in some important oilseed crops such as groundnut.

Expressed Sequence Tags (ESTs) generated by large-scale single-pass cDNA sequencing have proven valuable for the identification of novel genes in specific metabolic pathways (Tiwari *et al.*, 2011). Suh *et al.*, (2003) obtained 3,328 ESTs from a CDNA library of 5-25 days old immature sesame seeds. ESTs were clustered and analysed by the BLASTX or FASTAX pro-gram against the Gen-Bank NR and *Arabidopsis* proteome databases.

Comparative analysis was carried out between developing sesame and *Arabidopsis* seed ESTs for gene expression profiles during development of green and non-green seeds. Analyses of these two seed EST sets helped to identify similar and different gene expression profiles during seed development, and to identify a large number of sesame seed-specific genes. EST candidates for genes possibly involved in biosynthesis of sesame lignans viz., sesamin and sesamolin were identified, following which a possible metabolic pathway was suggested for the generation of cofactors required for synthesis of storage lipid in non-green oilseeds (Suh *et al.,* 2003).

Sesame genotypes with better mineral contents: Evaluation of germplasm for mineral content and selecting varieties with high quantities of essential minerals and incorporating those varieties in breeding program can assist in developing mineral-efficient crops with higher yield which can accumulate minerals from marginal soil. Pandey *et al., (*2017*)* studied 60 sesame genotypes of diverse origin from Bangladesh, Bulgaria, India and USA and examined their acid-digested samples by atomic absorption spectrophotometer for Fe, Zn, Cu, Mn, Cr and Co contents. All elements except Cr were found to be highly variable among genotypes. A significant discrimination showed that content of the elements in the sesame seed coat colour was a specific character. High-yielding varieties of India contained high Zn but low Fe concentration in seed. The concentration of mineral elements in black-seeded genotypes was significantly higher than those in white seeded. The indigenous collections were found to be a good reservoirs of mineral elements. Correlation study among trace elements and yield attributes indicated that though Fe and Zn were not correlated significantly with yield and its components, the two elements were interrelated. Phenotypic and genotypic coefficient of variability and heritability were high for Fe and Zn. It was observed that large genetic variability for element concentrations in the genotypes provides good prospects to breed improved sesame cultivars with elevated levels of micronutrients to mitigate mineral deficiency (Pandey *et al.,*2017)

Biotic Stresses

(i). Insect Pests: Sesame is infested by several insect pests that are varied in their feeding habits and the crop losses caused in different sesame growing tracts. For example, in East Africa 30 insect species were found to feed on sesame; but only six of them are of some economic importance (Weiss, 2000). In India sesame is infested by as many as 29 insect-pests belonging to 8 species which are potential pests of the crop, out of which the leaf roller/capsule borer *(Antigastra catalaunalis*) is the key pest of the crop (Gopinath *et al.,* 2011) followed by sesame gall fly (*Asphondylia sesame*) (Hegde, 2012). The extent of sesame yield losses due to insect pest attack is estimated as 25% (Weiss, 2000). The direct attack of insects by boring into the capsules and feeding on the seed caused 15.63 and 21.43 percent seed yield reduction during 1997 and 1998, respectively. The major insect pest causing such loss is the sesame shoot webber capsule borer *Antigastra catalaunalis* (Hegde, 2012)

1. Sesame leaf roller and capsule borer (*Antigastra catalaunalis)*

The female *Antigastra* lays minute and conical shaped eggs singly on the lower surface of leaves, branches and capsules. The eggs which are white in colour hatch into cylindrical larvae. The caterpillars which are greenish in colour with black head, pupate after undergoing 5 instars during which period they cause damage to the crop. (Figs. 8 and 10) The moth is medium sized with brownish forewings. In the initial stages the larvae roll-up the top leaves and start feeding from inside the roll and later move to more leaves. At flowering, caterpillars penetrate the flower and the fertilized ovary and feed on the developing capsules and tender seeds. (Fig.8).

In the initial stages, the pest kills the plant totally. In the later stages of the host plant, the infected shoots can't grow further.

The yield losses due to the infestation of the pest ranges from 20.34 % to 64.65% in Haryana state of northern India (Rohilla *et al*., 2003). The crop sown on 1 July in rainy season, 1 November in cold season and 1 March in summer season in peninsula India recorded the lowestincidence of shoot- webber, ie 11.2, 3.2 and 3.5% respectively (Bhaskaran and Mahadevan, 1991)

https://eurekamag.com/research/007/260/007260490.php (for further information on non-pesticidal management of Antigastra)

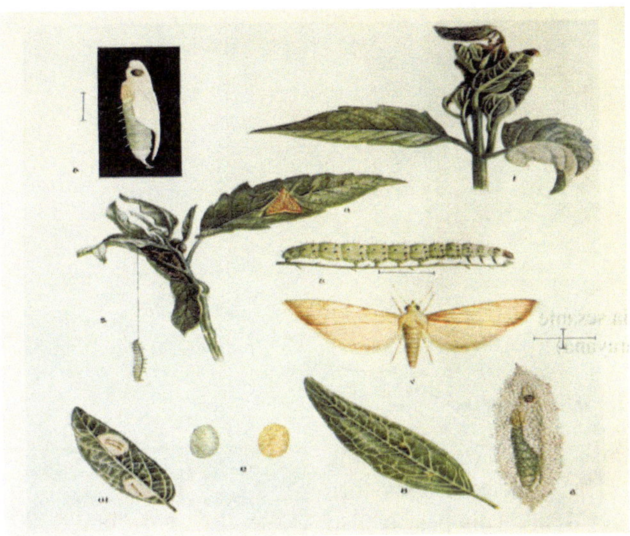

Fig. 8: Life cycle of Antigastra catalaunalis (Source: Wikipedia)

Natural Enemies, which check the proliferation of the pest are (i)Parasitoids: *Pteromalus fasciatus* and (ii) Predators: Spider, ladybird beetle, lacewing (*Chrysopa perla*) (Fig.9)

Fig. 9: Lacewing: Predator of Antigastra and other pests of sesame

Fig. 10: Sesame crop as damaged Antigastra catalaunalis

2. Gall fly (*Asphondylia sesame* Folt.): The female lays eggs by ovipositor inside flower buds. (Fig. 11). Following an incubation period of 2 to 4 days, maggots of the Depteran pest remaining inside the flower-bud, keep on feed on the floral parts especially pistil portion, leading to failure of flowering and capsule formation.

(Fig.12) The pest is active during flower formation stage of the crop. Capsule damage due to gall midge of up to 29-34.3% has been reported in Uganda (Egonyu *et al.*,2005)

Fig. 11: Adult of Asphondylia sesame
(*Courtesy*: Dr M. Lakshminarayana)

3. Leaf hopper: *Orosius albicinctus*

The pest is a sap sucking insect which causes curling of leaves on the edges. The affected leaves turn reddish brown and later get dried. This pest is also considered vector of phyllody virus of sesame.The adult leaf hopper is light brown in colour (Fig.13)

Fig.12: *Orosius albicinctus*

Natural enemies of leaf hopper are Predators viz., Spider, ladybird beetle, lacewing etc.

Spinx Moth's caterpillar: *Acherontia styx*

The adult moth lays globular shaped eggs on the under surface of leaves. The duration the egg stage is 2 to 5 days. Larva is stout caterpillar with green and yellowish oblique stripes and a curved anal horn. The larval phase lasts for approximately 60 days. The fully grown caterpillar pupates in soil in an earthern cocoon. After 14 to 21 days during summer season the adult moth emerges out of the cocoon. The pupation lasts for a longer period of around 7months in winter. The adult is a giant hawk moth brownish in colour with a characteristic skull marking on the thorax and violet yellow bands on the abdomen. Hind wings are yellow with black markings.

Fig. 13: Larva Pupa and Moth (Acherotnia styx)

The younger caterpillars roll out the top leaves and start feeding on them. Later it turns a voracious feeder of vegetative parts and flowers too. (Fig. 13)

5. Other insect pests which inflict sesame are *Spilosoma obliqua* (Bihar hairy caterpillar) and sap-sucking pests viz., *Aphis gossypi, Aphis craccivora, Deltocephalus* spp. and *Nazara viridula,* occurring in Americas through Africa and Asia to Australia (Weiss, 2000).

6. Integrated pest management in sesame

- Select a variety resistant/tolerant to major pests
- Select healthy seed material
- Treat the seeds with recommended pesticides especially bio-pesticides
- Follow proper spacing (Avoid higher plant population density leading to over-crowding that alters the crop micro-climate encouraging pest proliferation)
- Soil health improvement (mulching and green manuring wherever applicable)
- Nutrient management especially organic manures and biofertilizers based on the soil test results. If the dosage of nitrogenous fertilizers is too high the crop becomes too succulent and therefore susceptible to insects and diseases. If the dosage is too low, the crop growth is retarded. So, the farmers should apply an adequate amount for best results. The phosphatic fertilizers need not be applied each and every season as the residual phosphate of the previous season will be available for the current season also. (Weiss, 2000). Hence the application of Phosphatic fertilizers should be guided by soil test values.
- Proper irrigation: Irrigation should be need based.
- Cropping patterns: (i)Avoid growing sesame after sesame. In the case of dryland agriculture where only one crop is possible in the monsoon season, adopt the rotations viz., sesame followed by pearl millet / Pigeon pea / groundnut / cotton / castor / long duration legume fodder crop. (ii) Adopt appropriate inter-cropping systems in addition to common cultural, mechanical and biological practices viz., Intercropping of Sesame+redgram (6:1) and / or Sesame + pearl millet (6:1) or sesame + castor (6:1)and
- Spraying of neem oil @50ml/l of water for control of leaf hopper, which is helpful in managing the phyllody also.
- The tolerant varieties are TKG 21, RT-125 and RT-103. Intercropping of Sesame+Pigeonpea (1:1) is helpful for the management of phyllody (Gupta *et al.,* 2018).

In the case of irrigated crop of sesame, the crop rotations viz., sesame followed by safflower / wheat /rice /vegetables like okra / cotton /soybeans/ groundnut are appropriate for adoption (Gupta *et al.,* 2018)

Researchers at Texas A&M and Auburn University have found that sesame reduces root-knot nematode populations that attack peanuts and cotton. (http://baylor.agrilife.org/files/2011/05/sesamegrowerguide2008.pdf)

The relevant pest management practices *in-situ* in respect of sesame are as follows:

- If the insect pest is noticed in the initial stages, spray neem oil emulsion@50 ml/litre of water.
- If the larvae pests are over-grown, practice manual picking of the caterpillars, since
- the insecticide can't control the larvae of the later instars.
- Avoid spraying chemical insecticides and encourage the insect predators described previously.
- In the case of gall fly, carry out clipping of galls, picking and burning the fallen buds as prophylactic measures.
- Monitor the field situations at least once a week (soil, water, plants, pests, natural enemies, weather factors etc.)
- Make decisions based on the field situation and Pest: Defender ratio (P: D ratio)

***Pest: Defender ratio (P: D ratio*)**: Identifying the number of pests and beneficial insects helps the farmers to make appropriate pest management decisions. Sweep net, visual counts etc. can be adopted to arrive at the numbers of pests and defenders. The P: D ratio can vary depending on the feeding potential of natural enemy as well as the type of pest. (https://farmer.gov.in/imagedefault/ipm/sesame.pdf)

The natural enemies of Sesame pests can be divided into 3 categories 1. parasitoids; 2. predators; and 3. pathogens.

- Take direct action when needed (e.g. collect egg masses, remove infested plants etc.)

(iii) Diseases of sesame

Among the diseases of sesame crop, Phyllody is the most critical and damaging not only in India, but in several other Asian and African countries too. Leaf curl is also reported to be causing heavy losses at the initial phase of the crop. Bacterial leaf spot by *Pseudomonas sesami* too causes considerable crop damage. Bacterial blight by *Xanthomonas sesami* is also serious in the rainy season (Jume to October) particularly in the young crop.

Among the diseases caused by fungi, leaf blight caused by *Alternaria sesame*, leafspot by *Cercospora* sp., and leaf blight caused by *Phytophthora parasitica* are important. Charcoal rot caused by *Macromophomina phaseolina* too is widespread and destructive.

Powdery mildew (*Oidium erysiphoides*) is also observed but may not result incrop losses. Yield losses ranged from 5 to 50% by phyllody, from 10 to 75% by charcoal rot (root and stem rot), from 5 to 50% by *Cercospora* leaf spot, and 5% by bacterial leaf spot. Other abnormal symptoms such as discolouring of root, seedling death and leaf yellowing were also observed and the yield losses ranged from 5 to 50%. (Min and Toyota, 2019)

Gupta *et al., (*2018) have given an account of the diseases of sesame in India along with their integrated management highlighting the indigenous practices adopted by farmers.

1. **Phyllody of sesame (*Phytoplasma*)**: As the name indicates it is the disease that affects the total shoot system (Fig. 10). The infected plant is characterized by transformation of all floral parts in to green leafy structures followed by abundant vein clearing in different flower parts. In severe infection, the entire inflorescence is replaced by short twisted leaves closely arranged on a stem with short internodes and abundant abnormal branches bend down. Finally, plants tend to get stunted with floral plants transformed into green leafy structures and give a look of witches' broom (Abraham *et al*., 1977; Vamshi *et al.,* 2018). It is a serious and wide spread disease of sesame, caused by a pleomorphic mycoplasma-like organism (MLOs), which is now called as *phytoplasma* (Choopanya, 1973). In the earlier decades the cause of the disease was attributed to virus. Phyllody transmitted by the insect vector *Orosisius albicinctus,* which is also a pest of sesame (Gupta e*t al.,* 2018). The disease has been recorded in India, Iran, Israel, Burma, Sudan, Nigeria, Tanzania, Pakistan, Ethiopia, Thailand, Turkey, Uganda, and Mexico (Akhtar *et al.,* 2008). At times, Phyllody can causes total crop losses. The crop losses of the order of 90% have been reported by Gopal *et al.,* 2005. The crop sown on 1 June, 1 December and 16 January of rainy, cold and summer seasons recorded 15.9, 3.5 and 1.7% incidence.

Fig. 14 and 15: Grades of Phyllody disease of sesame

Among 20 genotypes tested by Vamshi et al., (2018), ES-62 and 12-JUN ranked as highly resistant.

2.Alternaria leaf spot: The disease caused by (*Alternaria sesami*) is one of the most common and economically important foliar diseases of sesame. The disease is reported in different sesame growing parts of the country by many workers (Mehta and Prasad,1976; Dolle, 1984). All stages of the plant are affected with symptoms of small dark brown water soaked, round to irregular lesions with concentric rings varying from 1-8 mm in diameter (Mohanti and Behera,1958). Brown to black, round to irregular and often zonate lesions measuring up to 1 cm diameter are produced on the leaves and in severe attacks the leaves dry out and fall off. Stem lesions are either in the form of dark brown spots or streaks. Dark brown, circular lesions are produced on the capsules (Verma *et al.*, 2005) which cause the capsule to drop The most visible symptoms are the leaf spots which are dark, irregular patches mostly on the edges and tips of the leaves, but the stem rots can be more significant. Kushi and Khare (1979) observed that *A. sesame* caused seed rot, pre- and post-emergence losses as well as stem rot and leaf spots. In severe infections several spots involving major portions of leaf blade and later drop off from the plants. The plants were observed to be most susceptible at 8-10 week's age (Ojiambo *et al.,* 1999). The pathogen leaves circular or irregular brown spots on leaves, and brown stripes on stem. (Fig.15) Correlation-coefficient studies revealed significantly and positively correlation between temperature (Maximum), relative humidity (RH - I) and rainfall with *Alternaria* blight disease intensity (Pawar 2017).

Maximum mean blight disease intensity was observed in Central Maharashtra Plateau Zone (37.99 %), followed by the zones viz., Western Maharashtra Plain Zone (36.21 %), Central Vidarbha Zone (34.27 %) and Western Maharashtra Scarcity Zone (31.06 %); whereas, it was minimum in the zones of Western Ghat

(20.24 %) and North Konkan Coastal Zone (21.82 %) (Pawar 2017).

Although considered an important fungal disease of sesame, there is little information about actual economic impact of *A. sesami.* Kumar and Mishra (1992) state that most of the common diseases of sesame cause yield losses of 20-40%. This yield loss is caused by the premature defoliation of the plants leading to smaller capsules and loss of capsules due to infection.

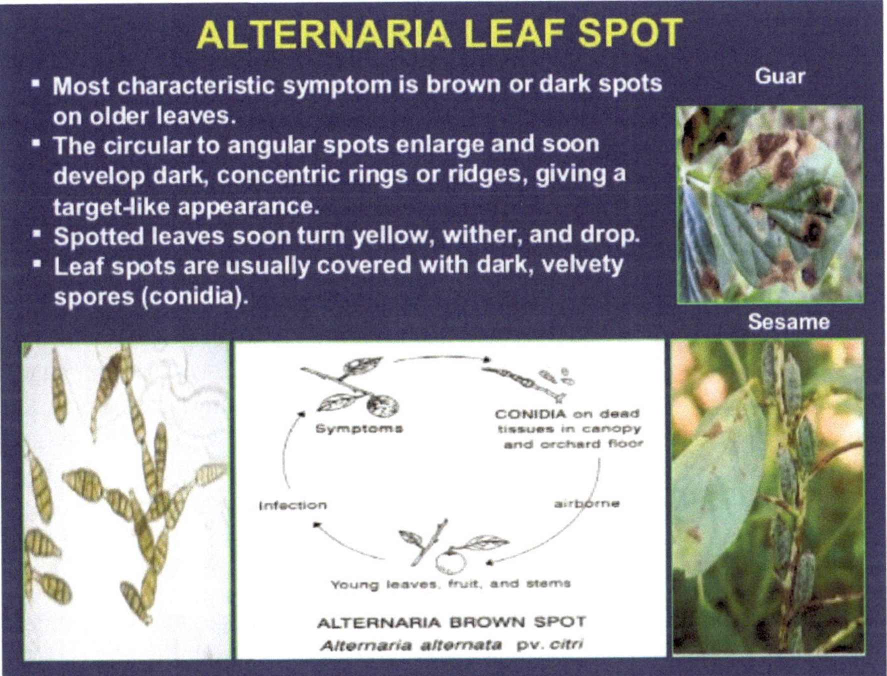

Fig. 16: Alternaria Leaf spot. Source: Arun Kumar: Farmers' Training-Insect Pests and Crop Diseases, Jodhpur (India). https://www.slideshare.net/Arpur/farmers-training-insect-pest-and-crop-diseases

3.Stem and Root rot

Stem and Root rot are caused by *Macrophomina phaseolina* (Tassi) Goid (Mihail and Taylor 1995). The fungus showed black discoloration on PDA media and mycelium grown up in right direction. The fungus produced sclerotia, pycnidia, pycnidiospores (Savaliya, 2015). *Macrophomina phaseolina* is a diverse, omnipresent soil borne pathogen. The pathogen survives as sclerotia in the soil and in host tissue for varying periods. It showed typical symptoms like black discoloration of stem and rotting of secondary roots. The flowering stage is most susceptible stage and the seedling stage is also attacked (Savaliya 2015). The symptoms include rupturing of the stem and appearance of black dots on the infected stem. The roots of the diseased plant also become brittle. The diseased plants exhibit black capsules. which open prematurely exposing shrivelled seed (Gupta *et al.*, 2018). The diversity of host species and the geographic range have suggested that *M. phaseolina* is quite heterogeneous. Its variability has been confirmed by reports demonstrating differences in pathogenicity of isolates obtained from single plant and a single host species (Babu *et al.,* 2010: Jana *et al.*, 2005; Saleh *et al.,* 2010). Despite having a wide host range, only one species is recognized within *Macrophomina.* The disease also causes severe losses right from seedling to maturity of the crop. High temperature and water stress during growing season favour the pathogen's incidence (Adhikari *et al.,* 2019).

Vyas (1981) reported *M. phaseolina* as very serious and destructive pathogen in all sesame growing areas and causes 5-100% yield loss while Maiti *et al*., (1988) estimated yield loss of around 57% at about 40% of disease incidence. The tolerant varieties are RT-46, RT0125, MT-75, TKG-22 and Nirmala. Intercropping of Sesame + Mothbean (1:1 or 2:1) is helpful for controlling stem and root rot.

4. Phytophthora blight: Phytophthora blight (*Phytophthora parasitica* var. *sesami*) produces initial symptoms of water soaked spots on leaves and stem. The spots are brown in the beginning which later turns to black. Disease can attack at all the stages of the crop (Roy *et al.*, 2007). High soil moisture favours the development of the pathogen. The disease is severe in the areas of heavy soil with high rainfall.The pathogen is seed borne. Temperature of 20-30° C and high humid conditions favour the disease. The inoculum density and seedling stage were adjusted for large-scale screening. The optimal stage for inoculation was the seedlings grown to emerged-first-true leaves.

The screening method consisting of the following steps is suggested. (1) Seedlings be grown two about V1 stage; (2) Inoculum density is to be adjusted to 1 × 104 to 1 × 105 zoospore per mL and 5mL of zoospore suspension is to be used for soil-drenching; (3) After the inoculation, soil water should be maintained to be saturated; and (4) Disease severity may be rated seven days after inoculation (Oh *et al.,* 2018)

Phytopthora tolerant varieties are MT-75, TKG-22and TKG055. Mutant varieties resistant to *Phytophthora* blight have been developed in the Republic of Korea and Sri Lanka. Based on the screening for 2 years through 'detached leaf technique' as well as under natural field conditions, twelve mutants having moderate to high resistance against *Phytophthora* blight could be isolated. One mutant P97-1 (300 Gy) exhibited highly resistant reaction against *Phytophthora* blight during both seasons (Kumari *et al.,* 2018). Intercropping of sesame + Pearl millet (3:1) also helpful for controlling *Phythopthora* disease. Gujarat Til 10 (GT 10) found to be resistant against Phytophthora blight disease (Kapadiy *et al.,* 2015).

5. *Cercospora* leaf spot

The disease is caused by (*Cercospora sesami* Zimm) and is one of the economically important diseases of sesame in almost all the production area. The crop is affected by the pathogens at all stages of the growth and causes heavy economic losses. The released varieties are highly susceptible to *Cerospora* leaf spot. It appears as small, angular brown leaf spot 5-15 mm in diameter on both leaf surfaces. Under favourable conditions, the disease spreads to capsules, leaf petiole, stem and capsules producing linear dark coloured lesions. The damage to plant growth and seed yield depend on the severity of infection on the stem and the stage at which the infection takes place. The fungus is seed borne, both internally and externally, but can also survive in the plant debris Thus, primary infection in the field may be from seed and infested plant debris and secondary spread may be through wind borne conidia. Extensive infection of foliage and capsule by *Cercospora* leads to defoliation and damage of sesame capsule and yield losses may range from 22 to 53% (Enikuomehin and Peters, 2002). Intercropping of Sesame+Pearl millet (3:1) is helpful for controlling *Cercospora* disease.

6. Powdery mildew

Powdery mildew is caused by *Erysiphae orantii* Cast (Rajpurohit, 1993) / *Erisiphae cichoracearum* (Reddy and Haripriya, 1993) / *Odium erysiphoides* (Mehta, 1951 and Roy, 1965) It is an important disease of sesame, infesting the crop throughout India and causing yield losses to the tune of around 25 to 50% (Rao *et al.,* 2012). The initial symptoms of the disease consist of appearance of greyish-white powder on the upper surface of leaves. Later, the several spots coalesce and the entire leaf surface may be covered with a powdery coating. (Fig.17).

In severe cases the disease symptoms do appear on flowers and young capsules causing premature shedding. The affected leaves may be twisted and malformed. In the advanced stages of infection, the mycelial growth changes to dark or black colour due to the development of cleistothecia. Defoliation of severely infected plant occurs before maturity.

The mycelium of the pathogen sends haustoria into the host epidermis and feed on the host tissue. Dry humid weather with low relative humidity favour the disease. Since the pathogen is an obligate parasite, the disease sustains through cleistothecia, on the infected plant's debris. The infected and dead plants must be removed to check the proliferation of the disease.

Fig. 17: Powdery mildew disease of sesame

Sesamum mulayanum the wild relative of cultivated sesame is reported to be resistant to Powdery mildew (Mehetre *et al.,* 1994) According to Hiremath (1976) two genotypes of sesame viz., Si-1926 and KRR 2 were found to be immune to powdery mildew. Sadakshari *et al.*, (1989) reported that ES 227 and IS-401to be resistant to powdery mildew disease of sesame. Some other cases of moderate susceptibility and moderate resistance have also been reported; but the only four genotypes indicated above are reported to be resistant.

7. Bacterial Leaf Blight

Sesame leaves showing purple water soaked brown specks on the under surface turning into large angular spots subsequently leading to defoliation indicate the incidence of Bacterial blight (*Xanthomonas Compestris* pv. *sesami*) (Rao, 1962). Bacterial leaf blight of sesame adversely affects growth and metabolism of the plant at all ages. The bacterium infects the seed too, due to which the disease gets dispersed easily (Kottle, 1985). The primary infection could be through the infected seed and infested plant debris. The secondary spread may be through wind borne conidia.

8. Bacterial Leaf Spot

The symptoms of the disease include light brown angular spot with dark purple margin appears in the leaf veins and petioles. The infected area becomes necrotic and disintegrate, giving a tattered appearance. The disease incidence is favoured by high rainfall and high humidity. The disease of Bacterial Leaf Spot is caused by *Pseudomonas syringe* pv *sesame* (Sabet and Dowson,1960; Sutic and Dowson 1962) *Pseudomonas* and other Gram negative bacterial genera infect plants through natural opening such as stomata and wounds. The mode of infection of causal bacterium was studied in a susceptible genotype to elucidate the process of infection within host tissues (Firdous *et al.*, 2014). Bacterium was identified as dark blue masses in the infected tissues. Results showed that pathogen colonized sub-stomatal and intercellular spaces of the spongy parenchyma cells when initial water soaking symptoms developed at 2-3 days after inoculation. Upon water soaking progressed, disruption of mesophyll cells occurred, mesophyll tissues were surrounded by bacterium followed by thinning and disruption of the cell walls. Later, when bacterial cells increased in space previously occupied by mesophyll cells, there were empty spaces without any differentiation of tissues. Bacterium did not occur in vascular bundles of leaf and stem, but some phloem tissues of stem sections were found infected. It was concluded that damage to chloroplast might be due to chlorosis or necrosis inducing toxins and toxins, which played a crucial role in pathogenesis of *Pseudomonas syringae pv.sesami* in sesame plant.

The disease is also seed borne and transmitted (60%) through seeds. Hence seed treatment of the seed material is necessary to check the proliferation of the disease.

9. Integrated management of diseases of sesame

- Carry out seed treatment before sowing with Apron 35 SD(0.3%) or Thiram (0.3%) or *Trichoderma harzianum* or *T. viride* or *Bacillus subtilis* (0.4%).
- Do one-year crop rotation as indicated under cropping systems of sesame production practices.
- Destroy all the crop debris to check the proliferation of source inoculum of sesame diseases.
- Rogue out disease infected plants and destroy since the affected plants carry fungal / bacterial inoculum
- Carry out Intercropping practices particularly sesame +pearl millet (3:1) and /or other intercropping systems described under cropping systems of sesame production practices.
- In addition to common cultural, mechanical and biological practices, Intercropping of Sesame+redgram (6:1) and / or sesame + pearlmillet (3:1) and spraying of neem oil @50ml/l for vector (leaf hopper) control is helpful in managing the Phyllody. The tolerant varieties are TKG 21, RT-125 and RT-103. Intercropping of Sesame+Pigeonpea (1:1) is helpful for the management of phyllody and other diseases like *Phytopthora* (Gupta *et al* 2018).
- Sesame varieties resistant to diseases as indicated under Sesame production practices are the best option for managing the respective diseases.
- In vitro studies on fungicides revealed Hexaconazole (100 %), followed by Propiconazole (98.02 %), Penconazole (95.19 %), Difenconazole (87.59 %), Carbendazim 12 WP + Mancozeb 63 WP (82.84 %) and Mancozeb as most effective fungicides with significantly higher inhibition of mycelial growth of *Alternaria sesami*.
- Among all bio-agents to control fungal diseases, *Trichoderma viride* (87.96 %), followed by *T. harzianum* (80.74 %) and *T. hamutum* (79.44 %) are the most effective (Pawar 2017)

Sesame Crop Production

Sesame is grown in areas with annual rainfall of 500-1100 mm and temperature of higher than 27°C. The crop is reasonably tolerant to drought, but not to water logging and excessive rainfall. Sesame is adapted to a wider range of soils, with good depth and drainage. Sandy loams of good soil fertility are well suited for sesame production

Normally the optimum temperature required during sesame's life cycle is between 25 – 35 °C. If the temperature is more than 40 °C with hot winds the oil content was found to decline. If the temperature goes beyond 45 °C or less than 15 °C there is a severe reduction in yield of sesame. The crop is normally grown in the rainy season i.e., *kharif* in arid and semi-arid tropics and in summer season in the cooler zones. In semi-arid tropics too the crop may be grown in the post rainy season under irrigation.

The reasons for the low productivity levels of sesame are as follows (Prasad 2019b):

(i) The crop is being grown under rain-fed conditions on drylands by small farmers who have no access to the updated knowledge of right production practices including water management, sustenance of adequate plant stand and plant protection.

(ii) The crop is subject to the vagaries of weather including inadequate or excess of rainfall to which sesame is sensitive

(iii) Sesame is photo-sensitive and hence any delay in planting due to late rains resulting in reduced photo-period, causes yield reduction

(iv) Sesame also suffers from early senesce causing yellowing of older leaves at crucial stages of post flowering and capsule formation leading to lack adequate 'source support'.

(v) Lack of control over harvest losses due to shattering of capsules and

(vi) Unlike other dryland grown crops including castor bean and safflower sesame suffers from the disadvantage of lack of productive hybrid varieties at the farmers' level.

• Promising sesame varieties

Table. 8: Some Indian varieties of sesame possessing multiple resistance or tolerance to biotic stresses with seed yield over 500 kg/ha in monsoon (Kharif) season (under rain-grown conditions) (Source: http://www.mpkrishi.org/EngDocs/AgriTop/e-agrotech/Kharif/Sesamum.aspx)

Variety	Yield (kg/ha)	Duration (Days)	Oil content (%)	Special features
N-32	770	90-100	53.0	White seeded, Moderately resistant to leaf spots, resistant to gall fly and capsule borer
TKG-21	500 -690	75-78	55.9	White seeded, tolerant to Antigastra, Bacterial spot and Cercospora leaf spot
TKG-55	630	76-78	52.6	White seeded, tolerant to Phytopthera, resistant to Macrophomina stem-root rot
RAMA	1000- 1500	85-90	45.0	Reddish brown seeded, resistant to Macrophomina stem-root rot, Phyllody and Bihar hairy caterpillar

Variety	Yield (kg/ha)	Duration (Days)	Oil content (%)	Special features
RT103	680-800	75-88	48.0	White seeded, tolerant to Macrophomina, bacterial leaf blight and Phyllody and Resistant to insect pests
TC-25	700	80-85	48.4	White seeded, moderately resistant to diseases and insect pests
RT-46	600-800	76-85	Not given	White seeded, resistant to oozing complex, tolerant to Macrophomina, capsule borer and gall fly
RT-125	600-900	71-83	49.0	Synchronous maturity, White seeded, tolerant to Macrophomina, Alternaria, Cercospora, bacterial leaf spot and phyllody
RT-103	600-800	75-88	48.0	White seeded, tolerant to Macrophomina, Bacterial blight and Phyllody and resistant to insect pests
RT-127	600-900	85	50.6	Synchronous maturity ,Bold white seeded, drought hardy, tolerant to Macrophomina, Phyllody, Bacterial leaf spot, powdery mildew and gall fly
Tilottama	900 – 1000	75-80	40.0	Blackish brown seeded, Resistant to Macrophomina, Phyllody and Bihar hairy caterpillar
Haryana Til-1	500	80	52.0	White seeded, early maturing, Resistant to major diseases, tolerant to Phyllody and Leaf curl
Krishna	750	85-90	46	Black seeded, tolerant to Alternaria and Capsule borer
E-8	500	100	53	White bold seeded,moderately resistant to powdery mildew, resistant to Bacterial leaf spot.
Swetha Til	600 -700	80 -85	45.0	White seeded determinate plant type, tolerant to gall fly, capsule borer, Phyllody, leaf curl and powdery mildew
JTS-8	700-800	83-87	48.0	White seeded, tolerant to Alternaria leaf spot and Bacterial blight

(*Source*: http://www.mpkrishi.org/EngDocs/AgriTop/e-agrotech/Kharif/Sesamum.aspx)

- **Preparatory tillage**: This is an important practice to attain good seed bed that ensures emergence and sustenance of seedling growth. It is necessary to resort to deep ploughing in summer with the receipt of pre-monsoon showers to achieve better moisture conservation and keep the soil ready for seed bed preparation with the receipt of regular monsoon showers in June-July in India.

Incorporation of organic manure in the form of FYM or poultry manure in adequate quantities @ 3 to 4 tons /ha at the time of first ploughing towards providing better organic matter content.

- **Nutrient schedule**: Since most of the soils used for sesame cultivation in India are deficient in crop nutrients, it is always advisable to get the soil test values based on which the doses of the crop nutrients should be worked out.

  As a general recommendation, incorporation of phosphatic fertilisers to provide @ 40 to50 kg P2O5/ha and potassic fertilizers @ 20 kg/ha at the initial ploughing is recommended. In areas of uncertain rainfall post sowing Nitrogen @ 40 kg/ha may be applied at the time of sowing.

- **Planting:** Seed should be treated with fungicides like Thiram or Mancozeb or Carbendazim @ 3g/kg of seed to avoid seed borne diseases.

  The recommended spacing for sowing varies with the rainfall and soil. In the areas of light soils with low rainfall and higher degree of drought, 40cm of inter-row spacing and 10 cm of intra-row spacing may be adopted.

  Normal recommended spacing is 30 cmx 10cm. The seed being small a seed rate of 5kg/ha is recommended. It is always better to mix the seed with sand or dry soil or well sieved farm yard manure @ 1:20 proportion at sowing to get even distribution of seed in the row.

- **Seed treatment:** Treat the seed with *Trichoderma* @ 4g/kg. This can be done just before sowing. ***Trichoderma treated seed should not be treated with fungicides like Thiram or Mancozeb etc.***

- Alternatively pre-treat the seed with Thiram 4 g or Carbendazim at 2 g/ kg of seeds initially. After 4 to 5 hours the same seed can be treated with *Trichoderma*.

• Application of *Azospirillum*

Saving of Nitrogen to the tune of 25% o may be achieved by the use of *Azospirillum* @600 g/ha and *Phosphobacteria* @ 600g/ha or *Azophos* @1200 g/ha by seed treatment.

Alternatively, soil-application of *Azospirillum* @2000 g/ha and *Phosphobacteria* @ 2000g/ha or Azophos @4000 g/ha may be resorted to.

Manganese deficiency: Leaves develop interveinal chlorosis, chlorotic tissue, later develop light brown or husk coloured necrotic lesions. Mix $MnSO_4$@5 kg / ha with 45 kg of soil and broad-cost evenly in the beds after sowing.

Zinc deficiency: Middle leaves develop chlorosis in the interveinal areas and necrosis along the apical leaf margins. Mix 5 kg/ha of Zinc sulphate with 45 kg of soil and broadcast evenly in the beds after sowing.

Weed management: Sesame crop is very vulnerable to competition from weeds, more particularly in the initial pre-flowering phase. Hence, the first weeding must be carried out within 15 days of crop emergence. The second weeding which is also important must be effected before 30 to 35 days following seedling emergence. It is recommended to keep the field weed free during the first 40 days to ensure better moisture retention in the soil and encourage good crop growth.

Dryland grown sesame

A large part sesame area is under dryland agriculture which is totally dependent on rainfall without any facility of extraneous irrigation. Under such conditions, it is absolutely necessary to conserve rainwater before sowing through summer ploughing as indicated under preparatory tillage and accumulate harvested rainwater in farm-ponds and dug wells in the vicinity of the sesame farm.

(i) At sowing the soil moisture should adequate i.e., at least 50 to 60 mm.

(ii) With every intercultural or weeding operation the soil should be stirred to create a soil mulch to check loss of soil moisture and achieve weed free situation on the field.

(iii) At critical stages of flowering and pod development crop should not be subjected drought. If possible one or two crop-life saving irrigations may be given from farm ponds or dug-wells in case rainfall fails. In the case of drought persisting for more than 15 days, the intercultural operations should be carried out at weekly intervals to create soil mulch and to check loss of soil moisture through evaporation.

Irrigated crop of sesame

At sowing the soil moisture should adequate i.e., at least 50 to 60 mm.

The first irrigation may be given 10 days after emergence, depending on the soil and climatic conditions.

Give one pre-flowering irrigation at 25 days, one irrigation at peak flowering and another one or two at the stages of capsule-setting. An irrigation during the flowering period is critical. Each irrigation should be of 40 to 50 mm of water.

Management of Biotic stresses of Sesame is presented in the preceding pages

- **Harvesting and Threshing:** The appropriate time of harvesting would be at the physiological maturity of sesame crop, when leaves turn yellow and lose their turgidity and turn flaccid. The capsules in the lower portion of the plant too turn yellow, which will gradually spread to the capsules in the middle zone of sesame plant.

In the case of irrigated commercial plantations, mechanical harvesting may be carried out which is very effective.

In case of small farms, the harvesting may be carried by manual labour, by cutting the whole plant at the base below the first formed capsule.

Any delay in harvesting may result in shattering. Therefore, farmers should not wait for the drying of crop in the field, which would result in shattering causing yield losses.

After harvesting the crop manually, the harvested plants may be bundled and placed in the erect position with the top facing the sky to avoid any loss of seed.

Later the harvested crop bundles may be threshed carefully on a clean threshing yard to recover the sesame seeds.Usually the crop is threshed by gentle beating of well dried plants with sticks.

• Cropping patterns

Dryland agriculture: Sesame followed by Groundnut /Rice / Horsegram / Finger millet (Ragi)/Sorghum/Horsegram /Upland Rice.

Irrigated situation: Early Rice - Potato-Summer Sesame/Green gram, Kharif Sesame-Maize/Pigeonpea/Chickpea, Wheat-Summer Sesame/Green gram, Sesame – Mustard

Intercropping systems for drylands: Sesame + castor (6:1), Pearlmillet+Sesame (6:1), Sesame + Groundnut (4:1), Sesame + Pearl millet (6:1) and Sesame + Cotton (6:1)

- Sesame is an excellent soil builder. Roots have as much mass as the visible plant. Stalks get incorporated into soil easily and break down quickly. Soil thus turns very mellow and requires little work for next crop. Many farmers have done one light disking and planted wheat after sesame. Tilth and moisture retention is improved. Farmers report that after sesame, the soil retains moisture better for planting the next crop. (http://baylor.agrilife.org/files/2011/05/sesamegrowerguide2008.pdf)
- Many farmers have incorporated sesame into their wheat rotation because it increases yield and provides a second cash crop. Sesame ahead of wheat will use resources - moisture and fertility. In dryland conditions in a dry year, there may not be enough moisture for both crops. No additional total fertilizer is necessary for wheat, but wheat will need more up-front nitrogen (10-20 units/ac), since the breaking down of the sesame stalks will

tie up a bit of the nitrogen early. (http://baylor.agrilife.org/files/2011/05/sesamegrowerguide2008.pdf)

Sesame Oil: The sesame accessions varied widely in their oil content and fatty acid compositions. The oil content varied between 44.6 and 53.1% with an average value of 48.15% (Kurt, 2018)

The content of oleic acid, linoleic acid, linolenic acid, palmitic acid, and stearic acid in sesame oil varied between 36.13–43.63%, 39.13–46.38%, 0.28–0.4%, 8.19–10.26%, and 4.63–6.35%, respectively (Kurt, 2018). On other-hand, Uzun *et al.*, (2008) reported the variation of oil content of sesame ranging from 41.3 to 62.7%, the average being 53.3%. The percentage content of linoleic, oleic, palmitic and stearic acids in the seed oil ranged between 40.7–49.3, 29.3–41.4, 8.0–10.3 and 2.1–4.8%, respectively according to Uzun *et al.*, (2008).

Oil and minor components of sesamin and sesamolin were studied in 42 strains of *Sesamum indicum* L. The oil contents of the seed ranged from 43.4 to 58.8% and varied inversely with the percentage of hull (r = --0.804, significant at the 1% level) (Tashiro *et al.,* 1990). The hull percentage was used as a criterion to predict oil content. The percentage of sesamin in the oil ranged from 0.07 to 0.61% and that of sesamolin from 0.02 to 0.48%. There was a significant positive correlation between the oil content of the seed and the sesamin content of the oil (r = 0.608, significant at the 1% level); no correlation was found between the oil and sesamolin contents. The average oil content found for the white-seeded strains was 55.0% and for the black-seeded strains 47.8%, the difference of 7.2% being significant at the 1% level. The white- and black-seed strains also differed significantly in sesamin content, but not in sesamolin content.

Cheng *et al.,* (2006) studied the compounds sesamin and sesamolin, the abundant lignans found in sesame oil. According to them, these lignans have been demonstrated to possess several bioactivities beneficial for human health. Excess generation of nitric oxide in lipopolysaccharide-stimulated rat primary microglia cells was significantly attenuated when they were pre-treated with sesamin or sesamolin. The neuroprotective effect of sesamin and sesamolin was also observed *in vivo* using gerbils subjected to a focal cerebral ischemia induced by occlusion of the right common carotid artery and the right middle cerebral artery.

Wei *et al.*, (2015) sequenced 705 diverse sesame varieties to construct a haplotype map of the sesame genome and *de novo* assemble two representative varieties to identify sequence variations. Fifty-six agronomic traits were investigated in four environments and identified 549 associated loci. Examination of the major loci identified 46 candidate causative genes, including genes related to oil content, fatty acid biosynthesis and yield. Several of the candidate genes for oil content encoded enzymes involved in oil metabolism. Two major genes associated with lignification and black pigmentation in the seed coat are also associated with large variation in oil content.

At the ICAR-Indian Institute of Oilseeds Research, Hyderabad (India) set of 320 germplasm accessions were evaluated for agronomic traits. Among them, the oil content ranged from 28- 53%. The accession IC-526435 showed highest oil content (53%) as well as high test weight (4.0 g). Six accessions viz., IC-526435, IS-104, IS344, IC-264692, KMR-85 and IS-349 were found promising for yield and related traits and oil content (Anonymous, 2018-2019)

6

Sunflower (*Helianthus annuus* L)

Introduction

The Heliantheae are the third-largest tribe in the sunflower family (Asteraceae). With some 190 genera and nearly 2500 recognized species, only the tribes Senecioneae and Astereae are larger. The name is derived from the genus *Helianthus*, which is Greek for sunflower.

This is one of the large and predominantly flourishing flowering botanical groups occurring in almost all parts of the world. Except for three species in South America, all other *Helianthus* species are native to North America and Central America i.e., Southwest United States-Mexico area (Heiser, 1976). The common name, "sunflower", typically refers to the popular annual species *Helianthus annuus*, or the common sunflower, whose round flower heads in combination with the ligules look like the sun. This and other species, notably Jerusalem artichoke (*H. tuberosus*), are cultivated in temperate regions and some tropical regions as food crops for humans, cattle, and poultry, and as ornamental plants.

The species *H. annuus* typically grows during the summer, post summer rainy season and later in post-rainy season too with the peak growth during pre-summer season with temperatures ranging from 28°C to 35°C in India.

Domestication of sunflower started around 4,000 years ago in Eastern North America by the native tribes of American continent (Blackman *et al.*, 2011). The current annual oilseed yielding sunflower was cultivated originally by local inhabitant tribes (Weiss, 2000), who used the seed as a source of food. The adaptations in annual sunflower to human cultivation include a dramatic increase in apical dominance, an increase in seed size, the loss of natural seed dispersal and seed dormancy, and the loss of self-incompatibility (Wills and Burke, 2007).

Perennial sunflower species are not as common in garden use due to their tendency to spread rapidly and become invasive. Notwithstanding the rapid spread of wild *Helianthus* spp, the whorled sunflower, *H. verticillatus*, was listed as an endangered species in 2014 when the U.S. Fish and Wildlife Service issued a final rule protecting it under the Endangered Species Act. The species grows to 1.8 m (6 ft) and is primarily found in woodlands, adjacent to creeks and moist, prairie-like areas.

Sunflower cultivation started in North America with the re-introduction of cultivated varieties from Europe in the late 19th century. Sunflower, however, was introduced into Europe in the 16th Century from Mexico via Spain (Weiss, 2000). The crop was imported from Holland into Eastern Europe and later into Russia in the 18th century, where the oil was extracted for the first time during the decade from 1830 to 1840. This paved the way for further expansion of the area under sunflower to 6 million ha in Europe.

Sunflower is also the source of various chemical constituents which are used for the treatment of many fatal or life threatening diseases. It is one of the best plants for the antimicrobial activity. The leaves of this plant are source of maximum photochemical than other parts like seeds, stem, flower etc., (Dwivedi and Sharma, 2014).

Sunflower as oilseed crop: In 1960s the Russian varieties viz., *Peredovik, Mennonite* and *Sunrise* gave a fillip to the development of sunflower as a commercial crop. The first commercial hybrid of sunflower was developed in 1972 which exhibited yield superiority by 25% and catapulted sunflower into the international arena as world's important oilseed crop. The year 1979-1980 is important in the history of sunflower production in the sense that there was a record global oil production of the order of 15 million tons much of which was attributed to sunflower production.

The success of sunflower as a major world oilseed crop could be justifiably attributed to the introduction of short-stemmed, high yielding hybrids, which apart from being high yielding, facilitated mechanised cropping. The health benefits of sunflower oil which is of poly-unsaturated nature also widened the use of sunflower oil.

The global sunflower production was dominated by the erstwhile USSR, which accounted for 50% of the commercial crops of sunflower leading to the harvest of 6 to 7.4 million tons annually till 1974. Since the mid-1990s Argentina has become the largest exporter of sunflower seeds (Weiss, 2000).

Sunflower was introduced into India in 1969 and in 1985-86 the crop production was 0.3 million tons; but by 1997-98 a production figure of 3 million tons was achieved, largely due to the cultivation of the crop in Karnataka, Maharashtra, erstwhile combined Andhra Pradesh, Haryana & Uttar Pradesh.

Ecology: Looking at the area of its origin, it is obvious that sunflower is of warm-temperate zone. However, the breeding and selection over a large number of years resulted in cultivars and hybrids with wider degree of adaptability over diverse environments. Sunflower is being cultivated 20 and 50*N and 20 and 40*S. Sunflower can be grown from sea level to 2500 meters from MSL. Nevertheless, the highest levels of productivity in terms of oil yield/ha are obtained in the zone below 1500 MSL. Sunflower is comparable to sesame (Weiss, 2000) in its adaptability and the crops ecological belt is expected to expand. In this context, mention must be made of early maturing dwarf varieties adapted to higher latitudes in Canada like the variety 'Sanola' commercially planted in 1995 (Weiss, 20000). Bange *et al*., (1997) developed a method of determining the suitability of varieties for sub-tropical environments. In Russia tall and strong stemmed cultivars of sunflower are grown in the northern limits as snow breaks, in order to protect the spring wheat crop (Weiss, 2000). Apart from protecting spring wheat, sunflower growing enhanced the soil moisture due to its accumulation of snow. In Canada too sunflower is being grown as snow trap crop. Sunflower may be considered 'day neutral' plant and the effect of photoperiod on sunflower is negligible as compared to the effect of other environmental factors such as high temperature, which can reduce the time to crop maturity. While sunflower is day neutral, high levels of ultra violet radiation shall have an adverse effect on the crop (Rajendiran,1996).

As sunflower plant becomes more resistant to low temperatures at the adult stage, it is adversely affected by hard frost at eight leaf stage when floral initiation takes place. The optimum temperature for good growth of sunflower is in the range of 27-28*C; but the crop grows satisfactorily at 20 to 30*C. It can also tolerate much the temperatures in the wider range of 8 -34*C (Weiss,2000) without any significant yield reduction. Temperatures higher than 30*C and those lower than 16*C adversely affect the seed development, consequently seed and oil yield (Weiss,2000). Higher temperatures result in higher linoleic acid and lower temperatures are found to increase the oleic acid content. Hence there is an option for the date of sowing in order to obtain the kind of fatty acid composition needed.

Sunflower crop is normally grown under rain-fed conditions in India, since the plant is considered drought resistant. The drought occurring at capitulum initiation stage coinciding with eight leaf stage can adversely affect the seed filling and consequently oil yield (Anonymous 1976).

Sunflower crop yields satisfactorily resulting in a minimum seed yield of over 1t/ha with a well distributed rainfall of 350 to 500 mm (Anonymous 1976). In

Argentina the average yields of sunflower have risen to 1.5 t/ha in 1995 due to the increase in summer rainfall (Pascale and Damario, 1997).

Sunflower crop is susceptible to waterlogging like other oilseed crops grown under rain-fed conditions, the degree of damage being proportional to the duration of waterlogging period (Weiss, 2000).

Weiss, (2000) has given a useful formula relating rainfall and soil type to potential yield developed in South Africa, which might suit wide raging conditions of crop growth. The basic formula is yield (kg/ha) = rainfall (mm) x soil depth (mm) x F, a constant related to the cultivar selected (Weiss, 2000).

Sunflower grows well on wide range of soils that are well drained with good depth. However, soil acidity is an impediment to sunflower growth, particularly when soil pH is less than 5 and saline (Kumar,D.1998). Salinity adversely affects the plant growth, development and some seed characteristics leading to the decline in the oil content. Soil salinity can also decrease the crop's resistance to diseases such as *Macrophomina* and charcoal rot (Weiss, 2000). Growth of the crop is satisfactory on neutral soil as well as those that are mildly alkaline with pH less than 8 (Kumar,D. 1998).

Cytogenetics and species relationship of sunflower

As stated earlier, the genus *Helianthus* has over 70 species. In the gamut of the cultivated sunflower *H. annuus* L., there are three main subspecies viz., *H. annuus* subsp. *annuus* Dougl., *H. annuus* subsp. *lenticularis* Dougl and *H. annuus* var. *microcarpus*. Dougl.

The subspecies *lenticularis* is closer to the cultivated oil seed sunflower.

The Russian Geneticists have included *H. petiolaris* and further divided subsp. *annuus* Into *annuus, pustovojtii, armeniaca* and *australis*. (Weiss, 2000)

The chromosome number of *H. annuus* L is 2n = 34, as is the case with majority of species of the genus *Helianthus.* But, there are also the species with chromosome numbers commonly 2n = 68 and 102. Nevertheless, there are certain exceptions like the ones with 2n = 14 and 28 or 2n = 32 (Weiss, 2000). Most of the species are said to be diploids. However, autotetraploid sunflower has been produced in breeding programs (Weiss, 2000).

Srivastava and Srivastava (2002) working on the induction of polyploidy in sunflower variety Morden, found that the tetraploids showed increase in seed weight as compared to diploids in C_1 and then to C_2 generations. The octoploids on the other hand showed increase in seed weight in C_1, but decrease in C_2 generation. Low seed setting in the C_1 generation of tetraploids improved in the C_2 generation. Meiotic behaviour of the new polyploids exhibited various associations of chromosomes such as univalents, bivalents and quadrivalents and abnormalities such as laggards, unequal separation of chromosomes, bridges, stickiness in C_1 generation which got reduced in the C_2 generation of tetraploids. However, in the case of octoploids, there was increase in the meiotic abnormalities in the C2 generation. Higher number of bivalents and lower frequency of abnormalities at metaphase of tetraploids in C_2 generation led to an improvement in pollen and seed fertility of tetraploids.

The cultivated sunflower *(H. annuus*) can be crossed with other wild species without any barriers. The wild species of *Helianthus* possess many useful genes which can be of importance in sunflower breeding. The classical example in this context is that of use of *H. tuberosus* to introduce disease resistance into sunflower in the erstwhile USSR (Weiss, 2000).

There exist several wild species of *Helianthus* that are not of direct economic value but might possess certain valuable genes for sunflower improvement. Many of them might present sterility in hybrid combination when crossed with cultivated sunflower. Nevertheless, with the availability of modern biotechnological tools it might be possible to overcome the barriers to their use in sunflower breeding. One such is the wild sunflower species *H. laevigatus* a perennial species which contains desirable characters such as resistance to *Sclerotininia* stem rot and high contents of protein and linoleic acid in seed. Despite the fact that it has not been extensively studied, it may be considered for sunflower breeding as a potential source of some desirable genes indicated above.

A set of six *H. laevigatus* populations was crossed to cultivated sunflower lines (Atlagic and Korić, 2008) and produced nine F_1 (2-14 plants) and 66 BC_1F_1 hybrid combinations (1-13 plants). Male sterility occurred in F_1 and BC_1F_1 hybrid combinations and pollen viability was lower in the progenies than in the parents (51.6-77.2%in F_1 and in F_1 and 4.8-34.0% in BC_1F_1). Meiosis was normal in the *H. laevigatus* populations. Numerous irregularities were observed in the meiosis of the F_1 interspecific hybrids. During diakinesis, quadrivalents and hexavalents were recorded in addition to bivalents. Dislocated chromosomes and chromosome bridges were present in the other phases The chromosome number in F_1 was 68 (tetraploid). Irregularities in chromosome pairing were observed in the interspecific hybrids at BC_1F_1. There were many univalents, and additionally trivalents, quadrivalents and hexavalents were also present The chromosome number in the BC_1F_1 generation ranged from 34 to 60. The occurrence of meiotic irregularities in the F_1 and BC_1F_1 interspecific hybrids indicates that *H. laevigatus* and the cultivated sunflower differ in genome constitution. The recovery of rare interspecific hybrids by overcoming the constraints is an important task. Palmer and Keller (1994) have described the various techniques that can be employed in this context.

Sunflower Plant

Stem: Sunflower plant is unique in the sense that it possesses a tall and strong and robust stem of 3 to 6 cm in diameter with small short stem hairs which are rough to touch. Sunflower stem at times has long longitudinal ridges. Sunflower stem can reach a height of 8 to 10 m with green to greenish-yellow colour. The height of the stem varies depending upon the kind of variety. The dwarf varieties grow to a height of 1 to 2 m, while most other giant varieties reach a height of 5 m. During the active growth phase the stem that looks solid and stiff with white pith, later as the plant reaches seed maturity turns hollow losing the green colour (Weiss, 2000).

The major part of the main stem consists of cellulose (50-55%), lignin (17-20%), pentosane (15-20%), crude protein (5%) and ash (8-10%) (Weiss, 2000). The strength and sturdiness of the stem depends on the genotype and soil and water quality in which it is grown. Generally, hybrid varieties of sunflower possess stronger stem. Branching of the main stem is not uncommon and can be seen in the genotypes other than the cultivated varieties and hybrids in which branching is supressed by breeding.

Leaves: Leaves are alternate on the upper portion of the stem and opposite in the lower part. Leaves are dark green, large, ovate to cordate with a long petiole. A sunflower plant has around 20-40 leaves, which are heliotropic before anthesis, which diminishes with maturity of the plant. Sunflower plant produces leaves continuously till flowering. With advancing of flowering, the number of leaves decline. Leaf production and stem elongation will continue till the opening of the capitulum. Defoliation of the upper part of the stem adversely affects the seed yield (Weiss, 2000), since the leaves in the upper part of the plant actively participate in photosynthesis. The leaf area index which is indicative of the total leaf area may be correlated with the growth of sunflower plant, its transpiration, light interception and disease incidence (Debaeke and Raffailac, 1996). The role of leaf area index and its sustenance at higher levels particularly in relation to seed yield expression is an important aspect which needs detailed investigation.

Capitulum (Head) and Anthesis: The head or capitulum is borne terminally on the main stem size of which varies depending upon the genotype, soil type and other seasonal factors. Soil organic matter content and atmospheric temperature are some important factors involved in the above (Weiss, 2000). Normal size of head observed in this case ranges from10 to 30 cm. However, depending upon the genotype, the size may go up to as much as 80 cm. diameter (Heiser, 1976). Once bud initiation starts, more leaves will not develop, but the leaves will continue to unfold. The plant will begin developing the bud and during this time each of the disc florets in the bud will be fully differentiated. This process will be completed by around 2 weeks prior to the start of flowering. During this time of bud formation, any stress can cause floret abortion, which is often visible in large numbers in the centre of the head as missing seeds (Anonymous, 2012).

Head or capitulum size in terms of diameter is one of the most important factors which has significant bearing on sunflower seed yield. The most critical phase of sunflower plant is from the capitulum formation and flowering. The proportion of well-developed and sound seeds per head is the key factor component of sunflower head determining the seed yield (Weiss, 2000), which forms the important criterion for selection of productive genotype of sunflower. According to Lou, (1996) capitulum of sunflower plant is composed of glucose and various pectins. The receptacle of sunflower capitulum is pithy and succulent. The heliotropism of capitulum continues till the flowers of the capitulum are fertilised. The capitulum consists of outer row of showy and brightly coloured ray florets which are sterile and ligulae and the fertile disc florets at the inner whorls of the capitulum (Fig.1). The number of florets of a capitulum varies from 1000 to 4000 arranged in spiral whorls. The maturity of the florets starts from the peripheral outer whorls and proceeds towards the centre. The opening of all the florets of a capitulum lasts for 5 to 10 days(Weiss,2000); the individual florets, however, can remain receptive up to 14 days (Weiss, 2000).

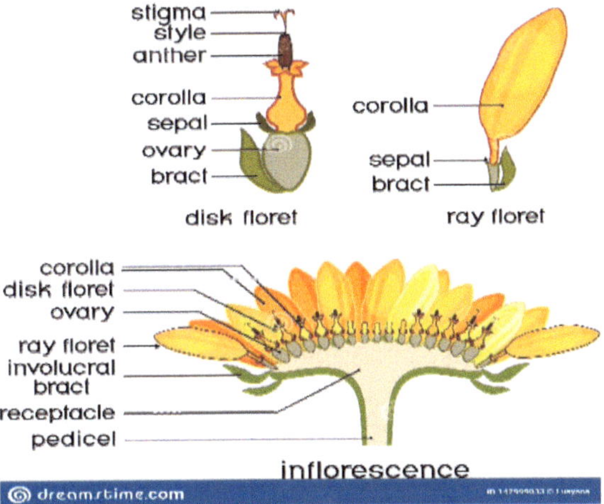

Fig. 1: Section of Sunflower capitulum with single disc and ray florets shown separately (Picture from dreamsite.com)

The duration of reproductive phase of a capitulum is stated to be influenced by the day length in addition to various other factors influencing the complex process. (Weiss, 2000). Over all observation, however, has been that the floral initiation of the head occurs at the eight leaf stage of the plant. The florets of sunflower head are protandrous and cross fertilization is the rule. The cross pollination is effected by insects, principally honey bees. wind pollination is rare (Low *et al.,*1975). One of the main causes for poor seed in farmers' fields could be lack of bee pollinators in adequate numbers resulting in inadequate pollination. Since sunflower exhibits sporophytic self-incompatibility (Wills and Burke 2007), the role of pollinators is very essential for proper seed set.Sunflower pollen is fairly large, yellow and oval in shape, with a prickly appearance which allows it to be transported by insects. Sunflower pollen is generally not moved about by wind (Anonymous, 2012).

If the disc floret is fertilised, then a seed which will comprise the hull and the kernel, starts to develop and if the floret is not fertilised, then a hull may begin to form but the kernel will be absent, resulting in empty hulls at harvest. (Anonymous, 2012). Nectar production in a sunflower capitulum, which is the major attractant to honey bees, also varies with the date of sowing that influences the time of flowering (Sharma and Sharma, 1997).The problems encountered in older cultivars due to the above factors have been nullified to a great extent in the modern hybrids which are virtually self-fertile.

Sunflower Seed: What is commonly known as sunflower seed is the fruit, which is achene. Wild *H. annuus* achene is of wide obovate shape, measuring 0.29 mm to 3.3 mm in width, and 0.41 mm to 6.7 mm in length. Its colour is somewhat brown, with two or three dark stripes that are variable in width. However, cultivated sunflower achene is significantly larger, measuring 7 mm to 25 mm in length, and 4 mm to 13 mm in width. During the extensive breeding process of sunflower, hull content has decreased from 40–50% to 20–25% (Radanovic *et al.*, 2018). The seed development can be split into two stages: 1. Hull (pericarp) development – which typically stops developing 14 days after anthesis. 2. Kernel development – which starts around 8 days after anthesis and is approximately 1/3 of its final weight when the hull development ceases. The formation of oil in the embryo begins several days after the start of the embryo development. As a result, very little oil is deposited during the first third day of the seed filling period (Anonymous, 2012).

Some growth regulators such as TIBA sprayed on capitulum in flowering stage particularly when ray florets open-up might increase the seed yield and 1000 seed weight and reduce the unfilled seeds, while gibberellic acid or ethefon applied when heads are forming may produce complete sterility (Weiss, 2000).

The colour rating system normally used in sunflower breeding is black in India. Also the colours of black with grey stipe, black with white or light grey stripe are also employed (Weiss, 2000). Although there exists an association between dark hull and high average oil content, many recently bred hybrids have light coloured seed with high oil content. The seeds recovered from the outer whorl or rim of sunflower head are of larger size, with higher weight and oil content than those recovered from the centre of the head. The variation in oil content and fatty acid composition of oil may also be observed in seeds from heads of different sizes (Weiss, 2000). At physiological maturity, the seeds from hybrid varieties of sunflower will be of 50% dry weight of the total head. The kernel to husk ratio is of considerable importance for oil production and sunflower seed meal. Black and striped seeds normally are of lower kernel: husk ratio and total oil content as compared to white types. A negative relationship exists between oil content and hull thickness (Weiss, 2000).

The studies on oil biosynthesis in developing seeds (Mazhar *et al.*, 1988) indicated the total seed oil content is related to the rate of colyledonary cell formation. This process is found to be higher in seed with high oil content than in seed with lower oil content, which may continue for longer spell. This finding emphasizes the important role of good nutrition together with adequate soil moisture at the stage of kernel / cotyledonary development.

The seed oil content varies from 25 to 48% normally, but there exist cases of seed oil reaching higher values of 50 to 65% (Weiss,2000) particularly when the crop is grown at lower temperatures, favouring good seed development. Higher temperatures during the seed development tend to reduce the oil content. Considering the fatty acid composition of oil, it is observed that while the linoleic acid content increases gradually starting from 3 days (49.4%) following pollination to the 51st day (72.4%), the oleic acid content on the other hand gets enhanced up to the 15th day after pollination (32.8%); but decreases by the 51st day (14.7%) of pollination (Weiss, 2000). The crop produced in the temperate regions exhibits an oil content of 55 to 60%, linoleic acid of 25-30% and oleic acid content of 20%. The seed produced under the conditions of high temperature may contain 65% oleic acid, 20% linoleic acid and a protein content of 15-20%. Similarly, the oil, its fatty acid composition and protein content are subject to the influence of location and soil (Weiss, 1966). If the sunflower has to be grown on acid soils, application of lime would help to counteract the adverse effects of soil acidity.

Modelling of Sunflower: This is challenging task since the crop species combines high yield potential with great adaptability. Simulations of sunflower using two closely related models, *epic* and *almanac*, have been tried. Phenology was predicted with Growing Degree Days with a 6°C base temperature (GDD_6) summed from sowing to maturity, assuming that anthesis occurred when 0.62 of the total GDD_6 had accumulated. Growth simulation involved Leaf Area Index (LAI) development, light interception and Radiation-Use Efficiency (RUE). Inclusion of a Vapour Pressure Deficit (VPD) effect appeared to make RUE more general. A modified harvest index approach was used to simulate seed yields. The *epic* and *almanac* models gave reasonable yield simulations over a wide range of environments and management options. The models should be valuable both for assessing the impacts of different management schemes and for identifying the relevant areas of investigation where additional basic research is needed (Kiniry *et al.,* 1992).

Nutrients for Sunflower Crop

Sunflower nutrient requirements are considered similar to those of hybrid maize or cotton for NPK (Weiss, 2000), with dose of N reduced by around 50%. Balanced crop nutrition can have a positive influence on seed weight. It may be remembered that the seasonal conditions can alter the effect of crop nutrients on its growth. Hence the nutrient interactions with the conditions of crop growth should be analysed to get the actual picture of crop performance.

Nitrogen: Crop response to nitrogen depends upon the availability of phosphorous and potash. Nitrogenous fertilizers applied without proper dose of P and K shall cause decline in seed yield and oil content. Nitrogen doses for sunflower crop should be based on well conducted local nutrient management trials. In the case of very fertile soils being used for sunflower production, the normal dose of nitrogen may be reduced depending upon the soil test data. Normally sunflower crop's nitrogen requirement does not exceed 50 to 80 kg N/ha depending upon the soil and seasonal conditions including moisture availability (Weiss, 2000).

Under Indian conditions normally 60 kg of N is recommended for irrigated sunflower and 40 kg N/ha for unirrigated dryland grown crop (AICORPO 1993). This however, might vary based on site specific conditions of diverse nature. The yield increases due to N application are due to enhanced head size and not so much due to increase in seed weight. The increase in seed yield thus obtained might lead to higher oil yield /ha and may not be due to increase in the seed oil content. Dividing total nitrogen application into two or three treatments (viz., basal, at 30 days and at capitulum initiation stages) can help growers enhance nutrient efficiency, promote optimum yields and mitigate the loss of nutrients (AICORPO 1993). The increase in nitrogen dose results in enhancement of oleic acid content of the oil and causes decline in linoleic acid content.Increasing the N rate increases seed protein and decreases oil concentration (Darby *et al.*, 2013). In contrast to the implied N response formula, recent N rate experiments in many sunflower growing states, have indicated less yield response to N than expected (Darby *et al.*, 2013; Scheiner *et al.*, 2002).

Protein accumulation in seeds is stated (Weiss, 2000) to be related directly to soil nitrogen availability to sunflower crop and this is stated to affect viability and seedling vigour (Weiss, 2000). On the other hand, inadequate nitrogen is perhaps responsible for proportion of unfilled seeds in the centre of sunflower head.

Phosphorous: Phosphorus deficiency is considered the second (after N) deficiency to occur within sunflower crops. Phosphate and potash may be applied before a tillage operation. If the previous crop had received good quantum of phosphorous, the P requirement of succeeding crop of sunflower should be calculated on that basis of availability of residual P in the soil. These nutrients are not readily lost from the soil since they attach to the soil forming only slightly soluble compounds. Phosphorus can be applied pre-plant-broadcast, pre-plant banded, or banded at seeding. Band applied applications are most efficient, especially when small amounts are applied in fields low in available phosphorus. Having applied the basic requirement of P, application of any additional quantity of P does not result in extra yield. Increasing levels of needed P often results in augmentation of seed weight. On the other hand, deficiency of P results in fall in oil content (Weiss, 2000). In India the recommendations of P for sunflower production vary as per the zonal requirements. On an average 60 to 80 kg of P per hectare is recommended, the lower value is in respect of rain grown crop of sunflower (AICORPO, 1995). The higher yield reaped following the application of Phosphorous has invariably due to the presence of potash. The fertilizer viz., single super phosphate often used as a source of P can result in higher yields due to the presence of sulphur too in addition to phosphorous.

Potassium: The needs of P and K of sunflower are almost equal. Since sunflower shows response to calcium, the potassic fertilizers such as potassium chloride and potassium sulphate containing 14-18% calcium, give better yield response than those without calcium.

Potassium (K) is required by the crop for stalk and tissue strength. Sunflower is a high user of potassium, removing 30 kg K/ha, with every 1t/ha yield, Potassium deficiencies normally only occur in sandy soils. Potassium that is band placed is about twice as efficient as broadcast applications.

General recommendations with regard to potassium application for sunflower in India ranges from 30 to 60 kg K2O/ha the lower values being applicable to unirrigated dryland production systems and the higher one to irrigated sunflower, depending upon the site specific requirements.

Around the world, K rate studies indicate that about 150 parts per million (ppm) from the standard K test in the region is sufficient for maximum sunflower yield (https://www.ag.ndsu.edu/publications/crops/fertilizing-sunflower).

Sulphur: Sulphur deficiency is very common in most of the tropical regions of the world. When single super phosphate is used as the source of phosphorous, the sulphur component of the fertilizer helps in ameliorating the sulphur deficiency to certain extent.

In India use of gypsum, ammonium sulphate and single super phosphate fertilizers have helped to correct the sulphur deficiency to great extent and contributed to increased seed yield (Srimannarayana and Raju, 1993). If the above mentioned fertilizers are used in sunflower production programs, the corresponding levels of Sulphur made available to the crop through them have to be estimated before any decision for sulphur application is to be taken.

The recommended dose of Sulphur to overcome S deficiency is around 25 kg/ha. In alkaline-clay soils having 12ppm available S, an application of 10 kg S/ha more than doubled seed yield and increased oil yield.

In the case of irrigated sunflower crop on lighter to medium lighter soils of India, application of Nitrogen @ 80kg/ha together with application of Sulphur

@ 25kg/ha produces largest heads, heaviest grains and more seeds/head and increased oil yield. Application of elemental S application is not recommended. Application of any thiosulfate fertilizer with the seed is not advised. Also, application of Sulphur in rainy season might face the likelihood of getting it leached out.

Boron: Boron is a crucial micronutrient, deficiency of which causes thickening of stem apex with brown lesions appear in younger plants. This causes death of upper leaves. Boron (B) deficiency in mature plants leads to deformity of stem part below the capitulum before and during flowering. The deficiency can be alleviated by spraying or adding Borax @ 12 kg/ha (Weiss, 2000). Asad *et al.*, (2003) reported that the highest rate of Boron foliar fertilization resulted in leaf-burn but had no other evident detrimental effect on plant growth. Under B-deficient conditions, Boron foliar application increased the vegetative and reproductive dry mass of plants. Foliar application of 28-120 μm Boron increased the total dry mass of most B-deficient plants by more than three-fold and that of plants grown initially with 1.72 μm Boron in solution by 37-40 per cent. B foliar spray increased the B concentration in various parts of the plant tops, including those that developed after the sprays, but the B concentration in the roots was not observed following B foliar application. The B concentration in the capitulum of the plants sprayed at the highest rate was between 37 and 93 per cent of that in the control plants.

Ayad and Saad (2011) reported that the maximum seed yields due to addition of boron given as foliar spray @ of 3 mg/ 1lit of water. The treatment resulted in increase in the number of seeds /head and 1000 seed weight. Spraying boron fertilizer on the leaves at budding stage significantly increased leaf area, head diameter, number of seeds/head, 1000 seed weight and increased seed yield by 13 to 22 % in spring season and 15.5 to 21.43% in autumn.

Al-Amery *et al.*, (2011) Reported that Boron @ 200 and 250 mg/litre tended to decrease the empty seed percentage, and consequently seed yield increased in linearly.

Zinc: Zinc deficiency is very common in lands that are levelled causing the turning of top soil. Most of the Indian soils are deficient in Zinc resulting in characteristic symptoms on leaves and decrease in seed yield. Zinc deficiency can be corrected by applying super phosphate enriched with zinc @ 1-2 kg/ha or foliar application of zinc sulphate heptahydrate and urea @ 1 kg/ha each in 100 litres of water (Weiss, 2000).

Saad *et al.*, (2011) reported that foliar application of zinc with the concentration of 10 mg. mg/litre lead to a significant increase in plant height, significant increase in number of leaves, leaf area index, capitulum diameter, number of seeds head-1, 1000 seed weight seed yield (2.72 t ha-1) and oil yield (1.24 t ha-1) of sunflower.

Ayad and Saad (2012) reported that foliar application of Zinc on plant leaves with concentration 8 mg /litre lead to a significant increase in plant height, stem diameter, leaf area, head diameter, number of seeds per head, 1000 seed weight, seed yield, oil percentage, oil yield, protein percentage and protein yield.

Iron: Iron is a constituent of several enzymes and some pigments, and assists in nitrate and sulphate reduction and energy production within the plant. Although iron is not used in the synthesis of chlorophyll, it is essential for its formation. In sunflower, iron deficiency decreased leaf dry weight, area, iron concentration, chlorophyll a and b concentrations, and absorptance, increased reflectance and transmittance, and red edge position of reflectance curves towards shorter wavelengths. Leaf iron concentration was well correlated with leaf chlorophyll *a* (r = 0.92) and *b* (r = 0.93) concentrations across crop species. Reflectance was a nonlinear inverse function, and absorptance was a nonlinear increasing function of leaf iron concentration and leaf chlorophyll *a* concentration (Mariotti and Masoni, 1996). High levels of nitrogen, zinc, manganese and molybdenum in the soil can cause iron deficiency as well. Iron is immobile in plants and therefore, deficiency symptoms appear first on the youngest leaves. The most notable symptom of iron deficiency is chlorosis or yellowing between the veins of the youngest leaves.Iron deficient plants may over accumulate heavy metals such as cadmium, which could be toxic (Korshunova *et al*., 1999). Younger leaves have pale yellow interveinal chlorosis, plant stunting, spindly stems. Severe cases of iron deficiency develop brown necrosis in alkaline soils as iron is less available at high pH, and also in waterlogged soils, acid soils with excess Mn, Zn, Cu, Ni.

Ebrahimian *et al.,* (2010) evaluated the effect of foliar spray of iron sulphate on yield and some agronomic traits of two sunflower cultivars under drought stress conditions. The results showed that there was no significant difference between two cultivars. In general, drought stress led to decrease of vegetative growth, seed and oil yield, while iron foliar spray significantly improved dry matter production and finally increased seed and oil yield. So iron sulphate foliar application (4 ppt) is recommended in sunflower farms under normal and drought stress conditions.

Molybdenum: Molybdenum deficiency turns older leaves pale yellow; leaves may become cupped with necrotic leaf margins. Leaves develop interveinal yellowing, margins die back and leaves assume cup-shaped appearance. Reduced establishment, seedling death on acid soils especially under irrigation. Molybdenum deficiency can occur on acid soils of pH 4 to5. Application of lime or a spray of sodium molybdate @ 100 to 400 g /ha would be useful to overcome the problem.

Combination of Micro-nutrients: Elnaz *et al., (*2010) revealed that the highest seed yield, oil yield, oil percentage, 1000 seed weight, seed weight to capitulum weight ratio and protein percentage were obtained from the soil and foliar application of iron + zinc treatments. Also, the highest amounts of nitrogen, phosphorus and potassium concentration in leaves were achieved from control treatment which was an indication of non-efficiency of iron and zinc on the absorption rate of these substances in the leaves. In general, foliar and soil application of zinc and iron had the highest efficiency in aspect of seed production. The comparison of the various 17 methods of fertilization showed that foliar application was more effective than soil application. Also, micronutrient foliar application increased concentration of elements, especially zinc and iron. Antioxidant enzymes activity was different in response to treatments, also the highest palmitoleic, oleic and myristic acid were observed in Zinc + Iron and merely Zinc foliar application treatments.Accumulation of micronutrients (Zn, Cu, Mn and Fe) in the dry matter of vegetative organs (leaf opposite to the treated leaves) of sunflower are generally higher in variants treated with foliar sprays of the said nutrients, as compared with the control (treated with water) and increases are generally statistically significant (Maria *et al.*,2009)

Genetic Resources and Sunflower Breeding

Sunflower Genetic Resources: Sunflower (*Helianthus annuus* L.) is one of the few crop species that originated in North America. It was probably a "camp follower" of several of the western native American tribes who domesticated the crop (possibly 1000 BC) and then carried it eastward and southward of North America. (Putnam *et al.*, 1990). The first Europeans observed sunflower cultivated in many places from southern Canada to Mexico.

The rapid spread of new sunflower varieties and more productive hybrids that occurred since 1960s, has caused the erosion of genetic diversity of landraces even in the heartland of sunflower diversity (Putnam *et al.*, 1990). Hence, the collection and characterization of the diverse genetic resources has become very essential. Genetic resources are the biological basis of global food security. Preservation of cultivars, landraces, and wild relatives of sunflower provides the basic foundation to promote and sustain agriculture. The extensive genetic resources of sunflower are maintained by USDA –ARS (www.ars-grin.gov)

Among diverse *Helianthus* species, there are 14 diploid annuals (2n=2x=34), 27 diploid perennials (2n=2x=34), 4 tetraploid perennials (2n=4x=68), 7 hexaploid perennials (2n=6x=102), 2 mixaploid perennials-1 (2n=2x=34, 4x=68) and 2 mixaploid perennials-2 (2n=4x=68, 6x=102) possessing valuable genes for sunflower breeding.

Interestingly there is co-evolution of sunflower crop, ancestors, and pests as well. The estimated genetic diversity across cultivated sunflower was 0.47, as compared with 0.70 in the wild species, indicating a higher level of genetic diversity in groups of wild populations based on SSR markers (Mandel *et al.*, 2011)

There are 543,203 accessions of *Helianthus annuus* and 13,852 accessions of wild species (www.ars-grin.gov). Among, the wild species, *H. debilis* is maintained in Florida and *H. niveus* in California. Seed increase of the species is being carried out in Ames. IA.

The use of crop relatives in past 20 and odd years in respect of sunflower breeding has been in the direction of incorporating resistance against diseases and insects for which three genetic sources of wild species were exploited. In the context of developing resistance against abiotic stresses, only one source of wild relative was used. Only one source was exploited for incorporating male sterility (Hajjar & Hodgkin, 2007)

The Value of wild species in monetary terms is estimated at US $ 269.5 million (Phillips and Meilleur, 1998)

Wild *Helianthus* species have been and continue to be an invaluable source of new genes for the improvement of cultivated sunflower. Over the past several decades, genes for resistance to several diseases, such as rust, downy mildew, powdery mildew, broomrape, *Sclerotinia* head rot, *Sclerotinia* stalk rot, and resistance to insects such as sunflower moth have been identified from the wild *Helianthus* species (Table 1) and successfully transferred into cultivated sunflower in some cases. Other useful traits found in wild *Helianthus* species include cytoplasmic male sterility, altered fatty acid composition, salt tolerance and herbicide resistance. The majority of these gene-transfers have been from the annual *Helianthus* species, which comprise only a third of the species of the genus. Exploiting the genetic diversity of the wild perennial *Helianthus* species will present more of a challenge than the annual species for a number of reasons. Firstly, many perennial *Helianthus* species are poorly represented in the seed collections of different nations including USA, thus necessitating further exploration to make seeds available for research. Secondly, many perennial species have different polyploidy levels, which complicate the production of fertile seed in crosses with diploid cultivated sunflower. Lastly, since perennials usually do not flower the first year, this delays the speed with which interspecific crosses can be made. Despite these challenges, the perennial members of *Helianthus* represent an untapped potential for new genes for the improvement of cultivated sunflower. Their collection and preservation will ensure the availability of seed for future researchers and will also aid in conservation efforts to ensure the preservation of the rarer *Helianthus* species in threatened habitats within the country's borders (Seiler and Gulya, 2004). Maintenance of germplasm bank in a cross pollinated plant system like sunflower with varying levels of incompatibility is not an easy task. Nevertheless, sib-mating could be used to conserve the germplasm where selfing is not possible. The technique of sib-mating could be used even in the case of wild species of *Helianthus* as suggested by Weiss (2000). The seeds of many wild species of *Helianthus* have strong dormancy which require special treatments to germinate them (Weiss, 2000).

A procedure developed by UCLA California (Weiss, 2000) appears to be useful to maintain the entries. The method consists of (i) soaking the dry seeds in 10% aqueous solution of ethylene at 50*C for one hour, (ii) bottled dry and stored for 10 days at 5 °C before planting.

Table 1. (A): Sources of resistance in Wild *Helianthus* species (www.ars-grin.gov).

Disease	Annual wild species	Perennial wild species
Rust	3	5
Downy mildew	10	15
Scelerotinia	7	18
Phomopsis	7	18
Alternaria	3	9
Powdery mildew	3	9
Rhizopus	0	4
Phoma	2	8
Charcoal rot	0	5
Broom rape	5	25
Verticillium	4	3

The USDA-ARS System has identified the following genetic resources for the attributes indicated (Tables 2, 3 and 4)

Table 2. (B): Resistance against Downy mildew (New Virulent Races)

Species	Germplasm
Helianthus argophyllus-Tx	RHA 340
H. argophyllus –FL	HA 419/420
H. annuus NM	HA 428
H. annuus ID	HA 458
H. annuus TX	TX-16

(C). Rust resistance (Races 336 and 777)

Population	Germplasm
PI 413038 (annuus-SD)	PH3
PI 413048 (annuus-CA)	PH4
PI 413118 (annuus-CA)	PH5
PI 642072 (annuus-TX)	TX16R
PI 413047 (annuus-CA)	RHA 464
PI 596746 (annuus-OK)	Rf-ANN-1742

Table 3. (D): Resistance against Broom-rape (Orabanche)

Species	Source	Germplasm
RACE -F		
H. grosseserratus	PI 617026	BR1
H. maximiliani	PI 617027	BR2
H. divaricatus	PI 617028	BR3
H. divaricatus	PI 617029	BR4
RACE -G		
H. debilisssp.tardiflorus PI 468691 ------		

Table 4. (E): Salt Tolerance

Wild Population	Germplasm
H. paradoxus-1671 (NM) (PI468801)	HA 429 (PI 632388)
H. paradoxus-1673 (TX) (PI 468802)	HA 430 (PI 468802)(PI 632339)

(F). The following Genetic Sources of Cytoplasmic Male Sterility have been identified in the USDA-ARS System:

PET1-French cytoplasm; 62 sources from wild species;15 different species; 56 annual populations; 6 perennial populations

In India at the ICAR-Indian Institute of Oilseeds Research Hyderabad, around 20 exotic accessions possessing attributes such as high yield, resistance or tolerance to downy mildew, powdery mildew &*Alternaria* blight, lodging resistance, high oil content earliness of maturity, self-compatibility, semi-dwarfness and resistance to bird damage were introduced by the year 2006. (Anonymous, 2006A.)

Initially there was emphasis on attributes such as high yield, earliness of maturity and dwarfness. Subsequently, some quality germplasm lines such as R-265 (for rust resistance), SG-1 (for high yield), Composite -6 (for high oil content), *Cernianka -66* (for earliness of maturity) and BCZ -111 (for fertility restoration) were used in the breeding programmes.

Some of the introductions utilised in the development of varieties and hybrids of sunflower are listed below (Anonymous, 2006A)

Table 5: Sunflower Introductions used as straight varieties and in the plant breeding programmes in India

Character / feature	Accessions used
Straight introductions for Field Production	EC 68414, EC 68415, Surya, Morden, EC-689099, NDOL 87
Cytoplasmic Male Sterility (CMS)	CMS 234A,CMS 7-1A, CMS 851,CMS 21, CMS 207, CMS 338, CMS 852
Fertility Restoration	RHA 274, RHA 265, RHA 298, RHA 271, MRHA-1, MRHA-2, RHA 6D-1
CMS Sources	PET-1,PET-2,GIG-1,1, PF, ARG.

The Institute has also introduced 570 accessions of 43 species of *Helianthus* from Yugoslavia, France, Bulgaria, USA and the Institute of Field and vegetable crops, Novisad (Serbia), which are found to possess resistance to wide range of diseases. The species are being maintained (Fig. 2)at a 'species garden', specially developed for the purpose. The wild species with problems of seed set are being propagated by *in vitro* micro-propagation through rapid standardization of suitable protocols (Fig. 3) (Anonymous, 2006A).

Efforts are underway to collect diverse local germplasm of ornamental sunflowers for certain important characters including resistance to diseases viz., *Alternaria* blight and Bud-necrosis.

Fig. 2: Some of the variable species of Helianthus at IIOR Hyderabad (Anonymous 2006A)

Fig. 3: Micro-propagation of wild species of *Helianthus* at IIOR Hyderabad (Anonymous 2006A)

Sunflower Breeding: The native tribal Americans should be considered the first 'plant Breeders' of sunflower. They improved sunflower by selecting the types that varied widely for the length of the growing season, degree of branching and size and colour of achenes (Fick and Miller,1997). They used the crop primarily for making flour or bread and also as medicine and dye. They also used it as a source of oil for seasoning food etc. (Fick and Miller, 1997). With the introduction of sunflower into Europe including Russia, systematic sunflower breeding started at Kharkhov Research Station in 1910 and at the Kruglik and Saratov stations in in 1912 and 1913 respectively (Pustovoit, 1964; Fick and Miller, 1997). This early phase of work included studies on the biology of sunflower plant, genetic and biochemical investigations and studies on breeding, methods of selection and seed production. Limited genetic studies and investigations on improvement of sunflower for silage were also conducted in USA and Canada at about the same period. (Fick ad Miller, 1997)

The most important contributions to the development of sunflower were made by Pustovoit and Jdanov in the Soviet Union, who increased the oil concentration in sunflower seed above 500 g/ kg (Skoric,1992). Leclerc (1969) of France discovered the first usable source of cytoplasmic male sterility in a cross *H. petiolaris* Nutt. × *H. annuus* L.

Kinman (1970) of the United States developed fertility restorer lines RHA 265 and RHA 266 which allowed practical development of sunflower hybrids . These discoveries paved a way for the heterosis breeding in sunflower crop, leading to the development of productive hybrids.

History of sunflower breeding can be divided into three basic periods.

(i) The first period is the use of varieties created by mass selection,

(ii) The second is the use of varieties created by the method of individual selection, and

(iii) The third, which is still present, is the introduction of hybrids in the production of sunflower.

The phenotypic changes that sunflower has undergone during the above phases largely follow the domestication syndrome. These adaptations to human cultivation include a dramatic increase in apical dominance, an increase in seed size, the loss of natural seed dispersal and seed dormancy, and the loss of self-incompatibility (Radanovic *et al.,* 2018). Sunflower has also undergone both selection and genetic drift during domestication and improvement, which has reduced its genetic diversity (Tang and Knapp, 2003; Liu and Burke, 2006), with modern cultivars retaining 50–67% of the diversity that is present in wild *H. annuus* populations (Kolkman *et al*., 2007; Mandel e*t al.,* 2011.)

The important task before plant breeder is to enhance the seed yield, oil content, reduce the plant height and introduce resistance to biotic stresses. Now, sunflower breeding revolves around the development of breeding lines suitable for hybrid breeding. Nevertheless, it is necessary that the inbred lines in the pipeline should contain the genes for the above attributes, which can be ensured following a rigorous evaluation of the material generated.

In this context, one could visualise the basic considerations for sunflower breeding should be as follows

(1) Components of seed and oil yield, which consists of the number of seeds per plant, 1000-seed weight, low husk content and high oil concentration in the seed;

(2) Architecture of the sunflower plant: plant height, head size and shape, angle of the head, leaf area and leaf canopy;

(3) Increased harvest index through earliness of maturity and short stem and uniform height

(4) Drought tolerance and resistance to biotic stresses, which could be attained by incorporating the specific morphological and bio-chemical attributes related to them such as specific leaf weight (SLW), proline content, small leaves and high cuticular wax content (Tyagi *et al.,* 2018; Hussain *et al.,*2017) and

(5) Oil content and quality and seed protein content and quality.

The wild species of *Helianthus* could be donors of the genes of some of the above characters, since the current intra specific variability of annual sunflower is said to be limited. Wide crosses were made to transfer cytoplasmic male sterility, disease, abiotic and *orobanche* (Broom-rape) resistance (Kaya *et al*.,2014). Moreover, induced mutations were used to induce new genetic variability for earliness of flowering, increased seed weight and high oil content (Anonymous, 2006A). Marker- assisted selection has been validated for rust resistance, downy mildew resistance, oleic acid content, and fertility restorer genes (Liu and Jan, 2012; Skoric, 2012). Sunflower breeding will be greatly facilitated by the genomic tools such as CRISPR/Cas and whole genome association mapping (Du *et al*., 2016; Vear, 2011)

A strong positive correlation exists between the seed yield, plant height, head diameter and seed weight. However, oil content is negatively correlated to hull thickness. Simple recurrent selection programmes could be employed to overcome the constraints mentioned above. The classical work in this direction was carried out in erstwhile USSR at Kharkov (Radanovic *et al*., 2018). There exists considerable variability for several agronomic attributes including seed oil content in the genetic resources of sunflower and its related wild species. It is widely believed that seed oil content of 55% might be the upper limit and hull content of 20% the lower limit in commercial types (Weiss, 2000). Understanding the domestication process and assessing the current situation concerning available genotypic variations are essential in order for breeders to face future challenges.

Varietal Development: Basic foundation for varietal breeding was laid by Dr V.S. Pustovoit (Pustovoit, 1964) in Russia where several promising varieties of sunflower were developed during late 1950s to 1970s. The cultivated sunflower as we know it today, was most likely created by mass selection from the wild *H. annuus*, which has small seeds and a branched stem (Radanovic *et al*., 2018). The Russian varieties were introduced into several sunflower growing countries including India in late 1960s. These varieties represented tall genotypes with extended duration of maturity. The problems of low seed set under selfing was observed to be causing low seed yields. The situation called for varietal renovation, as per the Pustovoit scheme of breeding and recurrent selection, resulting in improved versions of Russian introductions viz., EC 68413, EC 68414 and EC 68415. Among open pollinated varieties, cultivar Morden was grown over a sizeable area due to its earliness of flowering and maturity. The variety readily found place in cropping sequences and inter-cropping systems in India. Through the process of recurrent selection, the variety Morden was renovated at Raichur Centre (India) of oilseeds research in 1991, which resulted in reduction in plant height from 1.5 m to 1.0 m and maturity duration from 90 days to 75 to 80 days. Consequently, the realized seed yield was 1000 to 1200 kg/ha with an oil content of 38.9%.

Similarly, the varietal populations of EC 68415 and EC68413 had undergone renovation at Bangalore and Coimbatore centres of the All India Coordinated Oilseeds Research Project.

In the decade of 1995 to 2005 some more promising varieties of sunflower were developed at the oilseeds research centres which are presented in the following Table 6.

The focus of the breeding programmes was towards developing varieties possessing the following attributes suited for cultivation under rain-fed conditions of dryland agriculture.

- Earliness of flowering and maturity (less than 90 days) with no adverse pleiotropic effect on roots system
- Deeper root system to withstand the intermittent drought conditions of dryland agriculture
- Synchrony of maturity in the varietal population
- Uniform seed development without unfilled seeds
- High seed yield of not less than 800 kg/ha under dryland conditions
- Seed oil content of not less than 38%
- Resistance to diseases with special reference to downy mildew and rust.

Most of the varieties developed in the decade of 1995 to 2005 confirmed to the above standards.

Table 6: Sunflower varieties released under All India coordinated Oilseeds Research Project

Variety	Pedigree	Year of Release	Seed yield Kg/ha (Rainfed)	Duration (Days)	Oil content (%)
EC 68414	Introduction	1972	800-1000	100-110	40-42
EC 8415	Introduction	1972	800-1000	100-110	40-42
Morden	Introduction	1980	600-800	80-85	35-38
Rumsun Record	Introduction	1982	2000(irrigated)	105-110	41
Surya	Selection	1983 (Akola)	800-1000	90-95	32-35
CO-1	Selection	1983 (Coimbatore)	500-700	65-70	38-39
CO-2	Selection	1986 (Coimbatore)	800-1000	85-90	38-40
SS 56	Selection	1994(Solapur)	600-800	85-90	40-42
GAU SUF-15	Mutant of Sun Sel. 10-84	1994 (Amreli)	800-1200	90-95	40-42
CO-3 (TNAU SUF-10)	Mutant of CO2	1995(Coimbatore)	800-1200	90-92	40-42
CO-4 (TNAU SUF-7)	**Dwarf derivative x Surya**	1996 (Coimbatore)	800-1200	90-95	40-42
AKSF-9 (PKVSF-9)	Selection	1996(Akola)	800-1200	85-90	38-42
LS-11	Selection from EC -689099	1999 (Latur)	800-1200	82-85	38-42
DRSF108	Selection from gene pool	2003 (DOR)	800-1200	90-95	40-42
TAS-82	Mutant of Surya	2005 (Akola)	800-1200	90-95	40-42
CO-5	Selection	2006 (Coimbatore)	800-1200	90-95	40-42

Population Development: Taking advantage of the cross pollinated nature of sunflower crop, population improvement programmes were launched under the All India Coordinated Oilseeds Research Project, with the objectives of (1) cutting down the seed cost at the farmers' level and (2) provide genetic buffering or plasticity to avoid crop failures which might arise due to genetic uniformity at the field level due to various biotic and abiotic stresses. With regard to disease resistance, at Latur Centre the breeding emphasis was on resistance to downy mildew, while at Nandyal and Coimbatore resistance against *Alternaria* and rust were mandates respectively. The research programmes undertaken thus, resulted in the development of the following populations of sunflower which were recommended for production.

Table 7: Sunflower Populations developed in India.

S.No.	Oilseed Research Centre	Sunflower Population	Chief attributes
1	DOR Hyderabad	DORF -108, 113	High yield, high oil content and harvest index
2	Nandyal	NDSV-6	Tolerance to Alternaria and Rust
3	Raichur	RSFV 901-1 RSFV 902	Derivatives of the cross H. annuus x H. argophyllus Tolerance to Alternaria and drought. High oil content
4	Akola	TAS 82-8-1/3 TAS 81-2-84/6	Drought resistance and high yield
5	Latur	LDMRM10-12	Resistance to Downy mildew and high seed yield
6	Coimbatore	TNHSF-239	Higher level of autogamy and high yiled

Mass selection was adopted (Gowda and Seetharam,2008) for developing two distinct populations regarding maturity, one early (80-85 days) and one medium (95-100 days). The derived populations MSE -2 (early) and MSM - 2 (medium) retained substantial variability for most of the characters even after two cycles of mass selection. As compared with the base population, these derived populations showed substantial improvements in seed yield, to an extent of 40 to 65%, oil content (38%) and oil yield (83 to 97%).

Breeding for drought tolerance

Sunflower is sensitive to drought and its hybrids have a limited cytoplasmic diversity. The wild cytoplasmic sources of sunflower are not well exploited to their potential for drought tolerance and hybrid development. Nevertheless, the hybrids of sunflower have registered higher level of average yields of the order of 1200-1600 kg/ha under rain-fed conditions as against the varieties which gave on average yield levels ranging from 800 to 1,000 kg/ha. Drought tolerance is conferred by a constellation of characters governed by many genes. The constellation of characters strongly associated with drought resistance in plants are higher root density and greater root depth, smaller sized leaves and denser or thicker stem to cut down transpiration losses, higher specific leaf weight (SLW), low phot-respiration, higher rate of photosynthesis, greater sink capacity and faster rate of seed development / grain-filling, earliness of flowering and maturity not linked to shallow root system, higher plant dry weight and higher moisture use efficiency (Prasad, 1973 ; Fick and Miller, 1997). As per the stated considerations, the accession CMS-234A was the best general combiner for biological yield and photosynthetic efficiency under both the conditions. Tyagi *et al*., (2018) developed the cross combinations CMS-ARG-2A × RCR-8297, CMS-234A × P124R, and CMS-38A × P124R, which were found significant for biological yield, seed yield and oil content under diverse environments. Similarly, ARG-6A x P124R recorded maximum significant SCA effects for the number of leaves/plants. The hybrids CMS-ARG-2A x RCR-8297 and CMS-234A x P100R recorded highly significant SCA effects for leaf area, specific leaf weight and leaf area index under both the environments. For proline content, two hybrids i.e. CMS-XA x P69R and CMS-ARG-2A x P124R shown the higher SCA effects under both the environments. The hybrids PRUN-29A x P69R and 234A x P69R recorded high SCA effects for biological yield/plant under both the environments. The parents (female and male) generally had high GCA (good combiners) for both the environments. According to Tyagi *et al.,* (2018) the above are suited to be exploited in the hybrid development program for drought tolerance, high yield level, and physiological efficiency under a diverse cytoplasmic background. The information about different cytoplasmic sources and their effect on traits under water-limited environment should be obtained to strengthen the programme of development of new hybrids adapted to the challenges imposed by the drought.

The investigations conducted at the Directorate of Oilseeds Research (DOR), Hyderabad (Anonymous 2006A) under different levels of moisture stress in field indicated that *H. argophyllus* among the wild species exhibited the highest ability to withstand moisture stress. Among the different strains of *H. annuus,* EC 68414, CGP-1, hybrids BSH-1, KBSH-1 and KBSH-4, Acc 465, Acc1573, Acc 1582, Acc 1616, Acc 931, CMS 353B were found to be tolerant to moisture stress.

A new screening technique viz., 'Temperature Induction Response (TIR)' has been developed (Anonymous 2006A) to screen the breeding material at a faster rate for drought tolerance. In this technique, the germinated seedlings were exposed to cyclic induction temperature (35-40-45*C) followed by lethal temperatures of 55*C for 1 to 3 hours to select induction responsive types. The surviving seedlings were further subjected to lethal stress following cyclic induction. Such plants exhibited better recovery of growth and membrane integrity as compared to the original population. As per the results obtained, it is evident that the above technique could be used for in screening segregating populations. For drought tolerant types.

Another method of screening for drought tolerance developed at the DOR (Anonymous 2006A) is based on male gametophytic and selective fertilization. This method is standardised for screening the breeding material at gametic level. In the selective fertilization, the ovule fertilization was carried out in the presence of an osmotic barrier in the style. The progeny developed from selective fertilization performed much better under moisture stress than the control. The resistance thus observed was found heritable. Thus, by serially conducting gametophytic screening, it is possible to obtain progenies intrinsically tolerant to moisture stress. The pollen grains from the plants of selective fertilization exhibited significantly higher germination and pollen-tube growth indicating the genetic tolerance even at the gametophytic stage.

With regard to breeding for better drought resistance in sunflower, breeders might use some of the USDA inbred lines viz., HA-48, HA-22, CMS 1-50, PH-BC-291. PR-ST-3, RHA-SES and RHA-583 which are reported to possess higher photosynthetic efficiency (**www.ars-grin.gov**; Kaya, 2016). Also the recent availability of molecular markers might be of help in determining drought tolerance in sunflowers (Kaya, 2016; Skoric, 2009; Fick and Miller, 1997).

Development of Hybrids: The evolution of sunflower towards cultivated type involved a larger number of genes, with the majority of them showing small or moderate phenotypic effect (Burke *et al.,* 2002; Wills and Burke, 2007), unlike in other crops like rice (Cai and Morishma,2002). While the development of productive varietal populations paved the way for stabilizing sunflower production to a certain extent, breeding of sturdy and high yielding self-fertile hybrids is another historic land mark in sunflower breeding, which has taken sunflower productivity to new heights. Despite branching was considered undesirable attribute, it was needed in the male parents of hybrids to provide continuous supply of pollen to the female in the crossing blocks. Hybrids have been highly productive as they are self -fertile and with uniformity of populations.

In India, a number of exotic hybrids were introduced for evaluation; but valid conclusions on their performance could not be drawn due to inclusion of different hybrids in different years sequentially at different centres. More over non-availability of the corresponding parental lines was another problem. The importance of indigenous hybrids in heterosis breeding was recognised with the inclusion of sunflower research under the All India Coordinated Research Project on Oilseeds for testing of experimental hybrids in 1975. With the inclusion of the cytoplasmic male-sterile lines such as CMS-2, CMS-124, MS-204 and CMS-234 in the ongoing sunflower breeding programmes, the following first set of hybrids were developed during the decade of 1980 -1990.

Table 8: Sunflower hybrids released during 1980-90 in India

S.No.	Hybrid	Parentage	Seed Yield (kg/ha) (dryland situation)	Oil content %
1	BSH-1	CMS-234A x RHA274	1000-1500	40-42
2	APSH-11	CMS 7-1Ax RHA 274	1000-1500	40-44
3	LSH-1	CMS338Ax MRHA2	900-1200	37-39
4	LSH-3	CMS207A x MRHA-1	1000-1500	38-40
5	KBSH-1	CMS 234A xRHA6D1	1200-1500	42-44
6	PSFH 67	234A X P35R	1000-1500	38-42

The yield levels of the above mentioned hybrids range from 1000 to 1500 kg/ha under un-irrigated dryland situation.

Table 9:.Sunflower hybrids released during the period from 1995 to 2006 in India

S.No	Hybrid	Parentage	Seed Yield (Kg/ha) (dryland situation)	Oil content (%)
1	PKVSH -27	CMS2A x AK-1R	1200-1500	40-43
2	DSH-1	DSF-15Ax RHA857	1200-1400	41-43
3	TCSH-1	CMS234A xRHA 272	100-1500	40-44
4	KBSH -44	CMS 17A xRHA95C-1	1300-1600	38-40
5	NDSH-1	CMS234Ax859R	1200-1600	40-42
6	LSFH-35	CMS35XRHA-1	1200-1600	40-42
7	RSFH-1	CMS103x R64NB	1300-1600	40-43
8	DRSH-1	ARM234Ax6D-1	1300-1650	42-44

The subsequent generation of hybrids that were developed in the second phase of sunflower breeding programmes have exhibited an improvement in yield and also in oil content. All the hybrids developed in both the phases possess resistance to diseases particularly to downy mildew. The yield levels presented pertain to the conditions of dryland agriculture. Under irrigated conditions the mean yield levels are reported to be in the order of 2500 to 3000 kg/ha on an average. The latest hybrids viz., RSFH-1 and DRSH-1 are reported to have given yield levels of the order of 3,500 kg/ha under irrigated conditions.

The cultivation of hybrids on a very wider scale leading to progressive replacement of varietal populations has resulted in the liquidation of the local varieties and wild species of *Helianthus* causing genetic drift and loss of some useful genes particularly in the original centres of diversity in north and south America. This warrants strengthening of the genetic resources *in situ* with specific emphasis on genetic diversity.

Three-way cross hybrids (TWCH): While the single cross hybrids have added significantly to the productivity and production of sunflower, in recent years the yields of the single cross hybrids are getting stagnant, reaching a plateau (Anonymous 2006A). To overcome this problem, three-way crosses (TWCH) are being developed (Anonymous 2006A). TWCH provide broad genetic base at the reduced seed cost too. The broad genetic base of the hybrid is of advantage in facing the climatic fluctuations and resisting the biotic stresses, more particularly diseases. TWCH are produced by initially crossing the male-sterile female with a maintainer line of selected different genetic background to produce the single cross, which is again crossed with a male fertility restorer to recover the TWCH. This is in variance with the regular straight hybrid development in the sense that the maintainer line used in the single cross hybrid is identical to the CMS female line genetically and hence the operation of maintenance of the CMS female line is like selfing. In the production of the TWCH the maintainer is of different genetic constitution from the regular CMS line and hence in this process the maintenance of the CMS itself results in a male-sterile hybrid which is a single cross. When this is crossed with R (restorer) line the way is paved for the TWCH.

Evaluation of the TWCHs with single cross hybrids carried out (Anonymous 2006A) indicated higher levels of heterosis and productivity of the TWCHs over single cross hybrids, which has been attributed to the favourable epistatic interaction of genes from the three parents and also to the buffering action against adverse environmental conditions due to the higher level of heterozygosity and heterogeneity of the TWCH population. The two three way cross hybirds viz., TWCH 245 possessing earliness of maturity and D-1 with higher seed and oil yield developed in the coordinated oilseed research project in India have proved to be very promising.

Diversification of Cytoplasmic Male Sterility (CMS) sources: One can visualise that domesticated sunflowers could be traced back to a single centre of domestication (Smith 2014), which indicates the narrow genetic diversity and a fragile genetic base on which modern sunflowers have been built. Hybrid sunflower cultivars have been derived almost exclusively from a single cytoplasmic male sterile (CMS) source, discovered by Leclerc (1969) in the progeny of a cross between *Helianthus petiolaris* and *H. annuus* and transferred to sunflower germplasm. Due to the exclusive use of female lines with this cytoplasm for hybrid seed production, all hybrids cultivated world-wide are closely related - at least with regard to their cytoplasm (Serieys, 1996; Horn *et al.*, 2003). The efforts carried out in the direction of the genetic diversification at the Directorate of Oilseeds Research (Currently Indian Institute of Oilseeds Research) Hyderabad (Anonymous 2006A) have led to the identification of nuclear-cytoplasmic interaction in the combination where, *H. argophyllus* (designated as *Arg* cytoplasm.) was used as the ovule parent. Several lines with this CMS background and additionally with diverse phenotypic effects were developed (Anonymous 2006A) and are being tested for fertility restoration and combining ability. The main constraint faced in this connection is the very restricted availability of fertility restorer lines. Nevertheless, in the case of *Arg* cytoplasm, 46 restorer lines have been identified (Anonymous 2006A).

Diversification of fertility restorer base: In the task of diversifying the fertility restorers based on the inclusion of new genotypes, the male sterile of 3CR 37 of French derivative, was pollinated with a pollen mixture of RHA 273, 274, RHA 266 and Krasnodarets in 1977 in Bangalore, India. This poly-cross had given rise to a number of new restorer lines with high oil content (Anonymous 2006A), which are being used in Indian sunflower breeding programmes. The restorer line 6D-1 used as the male parent of the popular sunflower hybrid KBSH-1 is a derivative from the above cross. Some of the new restorer lines identified from the germplasm collection of both cultivated and wild species are GMUS-602, GMUS-414, *H. petiolaris* ssp *petiolaris* and *H. praecox* ssp *praecox.* Another method in this context is to develop restorers from the crosses of R-lines and cytoplasmic male sterile line. These crosses were utilised to select fertility restorers possessing other important agronomic attributes. Widely used restorers of this kind are RHA 271, RHA 273 and RHA 274. The new CMS and R lines developed in the All India Coordinated Research Project of the DOR are given in the Table 10 below.

Table 10: New CMS and Restorer (R) Lines developed in All India Coordinated Sunflower Research Project (Directorate of Oilseeds Research, Hyderabad)

AICRIP Centre	CMS Lines	R Lines
Akola	CMS-2A, CMS-21A, ID145-2/5	PKV-101 to PKV-110
Coimbatore	COSF -1A	
Raichur	CMS CFA 1to 4	CFR 1 to 6
Latur	CMS-207, CMS-71106, CMS-338423 (Downy mildew resistant) DSF-2, KSP-11	RNS341,RLW5,RLP1,RLY1, RNDL2095, LRR3, LRR6, LLR14, LRR15,LRR73,LRR80, PSM20
Bangalore	CMS-17A, CMS-1A	RHA95-C-1, RHA95-C-2, GVK-1 GVK-2
Ludhiana	CMS44A, CMS 10A,CMS 304A,CMS395A,CMS18A, CMS 31A,CMS 32A & CMS 13A	40
DOR, Hyderabad	DCM 11 to DCM 45	DRS 1 to 14

Mutation Breeding: In order to develop novel sources of CMS and fertility restoration, it would be necessary to take up work on induced mutants, which would result in new genes of much different functional nature due to neomorphic or antimorphic gene effects (Prasad *et al.*, 1984; Prasad, 1988).

At the Bangalore entre of AICRIP 8 mutant lines of RHA-234A (female parent) and 15 mutants of RHA-274 (male parent) possessing earliness of flowering and 100 seed weight were isolated. At Akola centre mutation breeding has resulted in the development of *Improved Surya* variety with black seed coat, increased vigour and enhanced 100 seed weight to replace the old parental variety Surya with a striped seed coat (Anonymous, 2006A)

Interspecific hybridization to diversify the genetic base of sunflower Breeding: The main objectives of the interspecific hybridization is to transfer genetic resistance against the foliar disease of *Alternaria* and to initiate pre-breeding for introgression of agronomically desirable genes into the cultivated sunflower.

The three wild species of *Helianthus* viz., *H. argophyllus, H. petiolaris* and *H. debilis.* were used in the hybridization programmes. In addition to the above a wild strain of *H. annuuas* was also included in the hybridization programme. From the crosses around 350 promising pre-bred lines were recovered (Anonymous, 2006A). The pre-bred lines expressed a wider degree of variability for attributes such as stem thickness, leaf angle, leaf arrangement, leaf colour, shape, texture and hairiness of leaves and number of leaves per plant. Also perceptible variation was observed for pigmentation of petiole and stem, disc colour, stigma colour, and ray-floret morphology. Additionally, significant variation was recorded for growth and yield characters, leading to the identification of promising lines for the agronomically desirable attributes including seed weight and seed yield. Apart from their use in regular breeding programmes of sunflower, the pre-bred lines are being used as sources of maintainers and restorers in hybrid development programs (Anonymous, 2006A).

Breeding for Confectionery purposes: Seed of confectionery sunflowers has a lower oil percentage and test weight than oil-type sunflower seeds. The seeds are available in a number of ways: in shell or dehulled, raw, salted and/or roasted. Sunflower seeds are eaten as a snack or incorporated into baked foods, salads, candies, and main dishes. The smaller seeds are used as bird food. Sunflower seeds are high in iron containing twice as much as raisins and 3-4 times as much as peanuts.

Confectionery sunflower seeds are used in bakery products, but also in a number of natural health products and healthy snacks, as well as for direct consumption in the hull, dehulled, raw or backed form. Confectionery sunflower breeding is characterized by the fact that different markets have different demands when it comes to the seed and kernel size, hull colour and other traits (Fig. 4), which make this process more difficult and costlier. Market demands and production trends of confectionery sunflower, however, show a steady increase in confectionery sunflowers due to its nutritional value and role in the snack industry.

Fig. 4: Sunflower kernel and achene

Screening of germplasm for confectionery types has led to the identification of accessions with higher seed weight (6 to 10 g) and higher protein content (16-24%) and low oil content (21-33%). The cited case is that of an inbred genotype CFBB-1, which has a seed weight of 9.5 g, high kernel protein and low oil content of 22.5%. Superior inbred genotypes of this kind (IMS, PET 2 & PET 1) are being put into diverse cytoplasmic backgrounds through back cross breeding. Such experimental hybrids are being evaluated for their utility for confectionery purposes (Anonymous 2006A).

Sunflower Biotechnology: Biotechnology deals with the use of molecular and cell cultural tools for genetic characterization and manipulation of organisms. The core of biotechnology is the ability to identify, isolate, amplify and modify genes or sequences of DNA from any source and to deliver and express the novel modified sequences in the same or different species.

- Sunflower has already been successfully manipulated with various *in vitro* techniques, e.g., embryo culture ("embryo rescue"), meristem culture and protoplast culture. In particular, interspecific hybrids, produced mainly by embryo rescue, have provided a unique source of "new" traits, like cytoplasmic male sterility (CMS) or resistance to devastating diseases, like *Sclerotinia* rot. The improvement and application of anther and microspore culture would allow the rapid production of homozygous sunflower lines. Isozyme markers and molecular techniques, such as restriction fragment length polymorphism (RFLPs), DNA fingerprinting and polymerase chain reaction (peR), can be implemented to characterize and identify the regenerated plant material (Friedt, 1992). With the advent of genomic resources, it is possible to genetically improve the crop through marker assisted breeding and gene editing approaches.
- Applications of a technique that produces haploids or spontaneously doubled haploids is to accelerate the breeding process. In addition, the technique may facilitate the selection of quantitative traits, characters controlled by individual recessive genes, or genes incorporated from alien species. Although the production of microspore-derived embryos from cultured anthers has been well-established for many crop plants, in sunflower there is still a need to improve haploid techniques before its application in practical breeding so far, success in anther culture of cultivated sunflower has been poor. Several publications have described extensive callusing induced from anthers of various interspecific sunflower hybrids cultured *in vitro* (Mezzarobba and Jonard, 1986; Vasiljevic *et al.*, 1991).
- The regeneration rate of whole sunflower plant is variable depending upon the genotype used, type of explant employed and its developmental stage and culture media conditions (Bidney and Scelonge, 1997). Genetic variation is closely associated with regeneration ability. The successful regeneration of entire plants from cultured somatic tissue or even single cells, i.e. protoplasts, has been demonstrated for specific genotypes. Successful regeneration of shoots was also reported by Mezzarobba and

Jonard (1986). Furthermore, the recovery of androgenetic haploid and doubled haploid plants is basically feasible in sunflower, although the rate of regeneration strongly depends on genotype. Particular progress has been made in the field of "genome characterization", by using both biochemical and molecular markers. In particular, the chloroplast and mitochondrial genomes have been analysed and described in detail. Such analyses built the foundation for future "marker-based selection" and for the identification and isolation of specific genes as candidates for "genetic engineering (Friedt, 1992)".

- Jambulkar (1995) developed a rapid embryo raised plant system for sunflower production from immature embryos, which facilitated five cycles in 316 days. A high frequency regeneration method via somatic embryogenesis is available now for several sunflower genotypes (Fiore *et al.*, 1997). Most of the somatic embryos develop into normal plants and produce viable seeds. Embryo meristem and primordial leaf tissues can be used in the routine regeneration of 17 different genotypes of sunflower (Shin *et al.*, 2000). Two high-oleic genotypes of sunflower exhibit rapid and efficient regeneration (Mohamed *et al.*, 2003). *Agrobacterium rhizogenes* can also be used to induce rooting of *in vitro* regenerated shoots to improve the rooting rate (Devi and Rani, 2002)

- One of the constraints faced in sunflower biotechnological research is its low transformation efficiency. Nevertheless, there are several methods of transformation that can generate transgenic sunflower plants. One such technology is based on *Agrobacterium tumefaciens* mediated transformation, which is considered one of the most efficient methods of transforming sunflower. Transformation rates ranged from 0 (zero) to 82.7% GUS positive plants among the tested genotypes with hybrids showing higher rates than the inbreds (Gurel and Kazan, 1999). Hewezi *et al.,* (2003) reported a method of genetically transforming recalcitrant split embryo axes of immature sunflower (*H. annuus*, var. RHA 266). Rao and Rohini (1999) reported a simple and reproducible method of *A. tumefaciens* mediated transformation system reported to be producing about 2% transgenic plants.

- Wounding explants with particle bombardment before inoculating meristems with *A. tumefaciens* reported to increase the transformation frequency (Bidney and Scelonge, 1997) The particle bombardment method includes coating DNA on to macro-particles, such as tungsten and gold, before delivering the particles into target cells with acceleration provided by gun-power or high pressure gas (Hansen and Wright 1999).

- Chloroplast genetic engineering offers several unique advantages, including high-level transgene expression, multi-gene engineering in a single transformation event and transgene containment by maternal inheritance, as well as a lack of gene silencing, position and pleiotropic effects and undesirable foreign DNA. More than 40 transgenes have been stably integrated and expressed using the tobacco chloroplast genome to confer desired agronomic traits or express high levels of vaccine antigens and biopharmaceuticals. Despite such significant progress, this technology has not been extended to major crops. However, highly efficient soybean, carrot and cotton plastid transformation has recently been accomplished through somatic embryogenesis using species-specific chloroplast vectors (Daniell *et al.*, 2005). Also, plastid operons could be created for expression of the multiple genes needed to introduce new pathways or enzymes to enhance photosynthetic rates or reduce photorespiration (Hanson *et al.*, 2013)
- Regarding sunflower genomics, around 40,000 sunflower ESTs from multiple libraries have been assembled using CAP3 Programme and organised into composite Genome Project data base (Kozik *et al.,* 2003) About 12, 886 EST sequences were released into the public domain (db EST) in 2003. These are available through *GnpSeq* module sequence databases (genoplate-info.infobiogen.fr/). A total of 1502 unigenes were obtained from four sunflower protoplast cDNA libraries to identify embryo-genesis related genes (Tamborindeguy *et al.*, 2004). Fernandez *et al.,* (2003) reported organ specific ESTs that were identified using subtractive hybridization methods, which were made from root, stem, leaf and flower bud tissues at R1 and R4 developmental stages. About 200 novel genes were discovered from such cDNA libraries. As on date, 2 genome assemblies, 481,318 nucleotide sequences, 10,541 sequence read archives (SRA), information of 157 bio-projects are available for *Helianthus* in NCBI (https://www.ncbi.nlm.nih.gov/search/all/?term=helianthus). These EST sequences and new functional genomics will facilitate cloning of sunflower trait genes and promote the molecular breeding and genetic engineering of sunflower.
- Despite domestication and improvement bottlenecks, the cultivated sunflower remains highly variable genetically. To characterize genetic diversity in the sunflower and to quantify contributions from wild relatives, 287 cultivated lines were sequenced along with 17 Native American landraces and 189 wild accessions representing 11 compatible wild species (Hubner *et al*, 2019). Cultivar sequences failing to map to the sunflower reference were assembled *de novo* for each genotype to determine the gene repertoire, or 'pan-genome', of the cultivated sunflower. Assembled genes were then compared to the wild species to estimate origins. Results indicated that the cultivated sunflower pan-genome comprises 61,205 genes,

of which 27% vary across genotypes. Approximately 10% of the cultivated sunflower pan-genome is derived through introgression from wild species of *Helianthus*, and 1.5% of genes originated solely through introgression. Gene ontology functional analyses further indicated that genes associated with biotic resistance are over-represented among introgressed regions, which is consistent with breeding records. Analyses of allelic variation associated with downy mildew resistance provide with an example, in which such introgressions have contributed to resistance to the disease (Skoric, 1988).

Breeding for higher oil yield and improved oil quality: The higher seed yield realised through development of superior varieties and hybrids should reflect in higher oil yield which is the main objective of sunflower breeding. While the higher oil yield is achievable, it is necessary to appreciate that oil yield is subject to the impact of the yield attributes and environmental factors (Kaya, 2016). The temperatures and moisture levels at seed filling stage of the crop play a decisive role in this context. Favourable conditions for improved oil yield are the mean atmospheric temperatures of less than 25*C, good soil moisture and freedom from biotic stresses during the phase of seed development (Skoric, 1988). However, the major contributing factor for oil content is recognized as the genetic component which is amenable for manipulation (Pustovoit, 1964, Putnam *et al.,* 1990). The first major contribution in breeding for higher oil content was by Pustovoit (1964) at Krasnodor in Russia. It has been widely reported that the GCA effects are more preponderant than SCA effects with regard to the oil content. Recurrent selection, which has been successfully employed in many cases of genetic improvement is one certain method to achieve enhanced oil content of sunflower (Kaya, 2016). Skoric (2012) suggested that Pustovoit's Reserve Method employed in breeding for high oil content could be used effectively in selection for high oil content of more than 50%.

The sunflower oil is widely known for its higher proportion of linoleic acid (C18:2) and as such the oil is widely used in the food industry (Kaya, 2016). In addition to linoleic acid, sunflower seeds also possess oleic acid (C18:1) content (OAC), which is considered a very desirable component of oil from health point of view. The oil rich in oleic acid is ideally suited for frying (Skoric, 2012). Fatty acid can be expressed as the inverse rate of biosynthesis of oleic and linoleic acids during the seed development stage. Several oil properties have been characterized as quantitative traits, however, some traits such as oleic acid content (OAC), which could, to a certain extent, be considered a semi-qualitative trait since OAC is dependent not only on the environment, but also on the genetic background of the receiver line (Ferfuia *et al.,* 2015; Regitano Neto *et al.,* 2016). The inheritance of oil quality in terms of fatty acid composition is determined by joint action of the embryo and the genome of the female parent. It is primarily determined by the genotype of the embryo rather than that of the female parent, which calls for better genetic analysis of the specific origin of the seeds than that of the average seed samples in order to obviate the interaction of the environmental factors with the residual genetic variation present in the sample due to the above stated phenomenon (Skoric,2012). The fact that some inbred lines are least affected by the environmental factors in respect of the fatty acid composition (Kaya, 2016) calls for analysis of the gamut of inbred lines from different genetic sources for fatty acid composition (Skoric, 2012). The major contribution in terms of developing a genotype for high oleic acid was elucidated by Soldatov (1976) who created the sunflower variety "Pervenets" with elevated oleic acid (80%) through an induced mutant, which has become the main source of elevated OAC in sunflower breeding programs worldwide due to the beneficial properties of high oleic sunflower oil (Vannozzi, 2006). Medoleic hybrids (NuSun) with high oleic acid content of 65-75% were developed and used in the food industry in USA (Vic *et al.*, 2002) Many molecular studies especially those utilizing molecular markers were carried out for selection of developing appropriate selection criteria for the development of inbred lines and hybrids of sunflower (Pérez-Vich *et al.* 2002). Fernandez *et al.*, (2007) indicated about the feasibility of developing mutant lines of sunflower to make new kind of oils, such as low saturation oil (17%), high palmitic oil (>25%), high stearic oil (>25%), high oleic oil (>85%), high linoleic oil (>75%), as well as other combinations differing in fatty acid composition, in addition to the standard type of sunflower oil. Induced mutants could be potent tools to create wider genetic variability for fatty acid composition of sunflower oil with specific reference of oleic acid (Kaya, 2016). A partial duplication of the FAD2-1 allele caused by genic mutation leads to an increase in oleic acid content by silencing the FAD2-1 gene encoding FAD2 (oleoyl-phosphatidylcholine desaturase) (Lacombe *et al.*, 2002; Schuppert *et al.*, 2006). This enzyme catalyzes the synthesis of linoleic acid from oleic acid and by silencing its activity oleic acid gets accumulated. Inheritance of the oleic acid content (OAC) trait has been a subject of numerous studies and different results were reported from a single dominant gene to several genes influencing OAC (Lacombe *et al.*, 2009; Premnath *et al.*, 2016). Gene/genes involved in inheritance of OAC have been denoted as *Ol* genes. Different markers were employed in mapping and detecting the mutation (*Ol* mutation) in sunflower. The two RAPD markers, F15-690 and AC10-765, were linked with 7.0 and 7.2 cM to Ol_1 gene, respectively (Dehmer and Friedt, 1998). Later on, the Ol_1-FAD2-1 locus was placed onto LG14 (Schuppert *et al.*, 2006). One major QTL identified by Pérez-Vich *et al.* (2002) explained 84.5% of the variation in the OAC. Schuppert *et al.* (2006) provided dominant INDEL markers for tracking the *Ol* mutation in addition to identifying 49 SNPs and five INDELs in the 3′-region of FAD2-1. Three years later, a co-dominant SSR marker tightly linked to the *Ol* mutation and dominant markers specific for the mutation were published (Lacombe *et al.*, 2009). Recently, Premnath *et al.*, (2016) identified in addition to the QTL on LG14, two additional QTL for OAC on

LG8 and LG9. The two markers HO_Fsp_b for the QTL on LG14 (Schuppert *et al.*, 2006) and ORS762 for the QTL on LG8 explained about 60% of the phenotypic variation in OAC. Several of the markers have been used for validation across numerous sunflower lines (Nagarathna *et al.*, 2011; Bilgen, 2016; Dimitrijevic *et al.*, 2016). Dimitrijevic *et al.*, 2017 reported marker F4-R1 created by Schuppert *et al.*, (2006) as the most efficient in MAS for OAC.

Genetic and biochemical studies indicate that the quality of sunflower oil is not only determined by fatty acids, but also by the tocopherol content of the seed (Kaya, 2016), which is an important indicator of anti-oxidative imbalance of sunflower. In view of this observation, there is a need to study the inheritance of tocopherols to enhance the oxidative stability of sunflower oil. Standard sunflower oil consists of α-tocopherol (95%), β-tocopherol (3%) and γ-tocopherol (2%), which accumulate in sunflower seeds through a bio-synthetic pathway determining the build-up of triacyglycerols, that are the natural anti-oxidants involved in formation of Vitamin E (Demurin *et al.,* 1996; Skoric,2009 and 2012). Thus the incorporation of new genes for high oleic acid and various tocopherols in breeding lines, facilitates development of novel hybrids with diverse oil quality (Vera-Ruiz and Fernandez-Martinez, 2006). New developments in sunflower production with reference diverse novel qualities of oil, have led to new uses in pharmaceutical, cosmetics and other industries including that of energy involved in the production of bio-diesel, in addition to their use in edible oil and food industry. Therefore, it is suggested that sunflower breeders and food specialists should define new parameters and novel uses of sunflower oil for high quality final products (Kaya, 2016).

Sunflower breeding programmes for fatty acids, could be set up in three directions viz., (i) high oleic acid combination with low saturated fatty acids, (ii) high linoleic combination with low saturated fatty acid (<6%) combination and (iii) high saturated genotypes (Skoric, 2012).

Biotic Stresses of Sunflower: When sunflower was introduced into India in 1969, the problems of biotic stresses were not very perceptible and an impression was perpetrated that sunflower was relatively a safer crop in terms of incidence of pests and diseases (CAZRI, 1976). The crop situation changed rapidly in the decade of 1980 when sunflower acreage registered a steep rise following to its cultivation under different seasons continuously due to its wider adaptability and photo-insensitivity following which, the problems of biotic stresses got accentuated, particularly with the destructive proliferation of downy mildew disease and *Helicoverpa* pest. Now it is estimated that the defoliating insects alone cause more than 20% yield losses under varying conditions of crop growth (Jayewar *et al.*, 2017). More than 30 diseases have been reported on sunflowers worldwide (Gulya *et al.,* 1997). In India too, there is a wide ranging disease complex inflicting sunflower. While the sunflower yield losses due to diseases vary depending upon the locations, weather conditions and other factors, it is estimated that the disease of bud necrosis of sunflower alone causes drastic yield reductions ranging from 30 to 100 % (Bhargav and Meena, 2014) indicating the gravity of the situation.

Insect Pests of Sunflower: Sunflower crop is reported to be inflicted by around 50 insect pests at different stages of crop growth in India and over 251 insect pests are recorded at global level (Rajamohan 1976).

Defoliator Pests: Among defoliators, the polyphagous larval pests such as *Spodoptera litura* Fab, and *Helicoverpa armigera* Hubner, are more damaging since they bore through the head at post anthesis and seed development stage apart from being devastating foliar feeders.

***Spodoptera litura* Fab.:** The extent of average yield losses caused by *Spodoptera* is estimated to be more than that of all other pests. The early instar-larvae scrape the softer and more digestible tissue from the lower surface of the foliage leaving the upper epidermis intact causing a condition called 'windowing'. The later instar larvae can digest the leaf lamina but tend to avoid the leaf mid rib and larger leaf veins, eating the tissue from between the veins causing a condition caused 'skeletonising'. Mature larvae eat the whole leaf except for the toughest parts and large populations can completely strip the host plants of almost all foliage. The larvae also eat into buds and frequently eat into the capitulum (Smith *et al.,* 1997). As a consequence, large amounts of frass may be visible. These symptoms are not specific to *S. litura* alone, but generic for many foliage feeding Lepidoptera species (EPPO, 2015). All developmental stages of *S. litura* can be detected visually in the field. The larvae are mainly external feeders but will also bore into plant parts 30% (Ferry *et al,* 2004).

Ranga Rao *et al*., (1993) reported 71 species of parasitoids, 36 predatory insects and 12 species of spiders which preyed upon *S. litura*. The mass release of egg and larval parasitoids to control *S. litura* has been a partial success in India (Ranga Rao *et al.,* 1993). *S. litura* is preyed on by a number of predators like ants, earwigs, pentatomid bugs and predaceous beetles. *Telenomus spodopterae* Dodd and *Telenomus remus* Nixon can parasitize the eggs and reduce its population. A species of entomogenous fungus, *Nomuraea rileyi* infects some larvae in the field. Caterpillars are susceptible to the biopesticide Bt which at severe infestation could be used for spot application (Ranga Rao *et al.,* 1993)

Planting a border crop of castor 10 days earlier to the sowing of sunflower around the field served as the pest attractant. The infesting caterpillars may be collected from castor leaves and destroyed and thus ensure minimum attack of *Spodoptera* on sunflower crop.

Fig. 5: Defoliation caused by larvae of *Spodoptera* in sunflower

***Helicoverpa armigera* Hubner.:** The defoliator of next importance is *Helicoverpa armigera,* which is known as the capitulum (head)-borer of sunflower crop since the pest eats into the developing seeds of sunflower head. The multiple regression equation worked between the weather parameters and the pest (Anonymous, 2006A) indicated that an increase in minimum temperature by 1°C would reduce the larval population by 0.68 per head and increase in morning RH (relative humidity) by 1% would increase the larval population by 0.12% per head. The total effect of weather parameters on larvae of *Helicoverpa* was estimated as 51.2% (Anonymous 2006A).

The pest may be managed through adopting measures to enhance the populations of the predators and parasitoids of the pest. The predators viz., *chrysoperla carnea* and *Minochilus sexmaculatus* and parasitoids such as *Bracon* spp, *Campoletis* spp and *Trichogramma* spp have been found effective (Anonymous, 2006A).

Since capitulum initiation stage of sunflower is the most vulnerable to *Helicoverpa,* one to two prophylactic sprays of Quinalphos 3ml/l of water or Propenophos 2ml/l of water may be given to check the initial stages of the pest at 40 to 45 days after sowing. Also the pest density could be brought down successfully with the spray of NPV@250 LE/ha. The spray combined with UV protectants gave better results at Bangalore and Coimbatore. Three sprays of NPV@ 500LE/ha brought down the pest population completely and resulted in very high seed yield. Coverage of sunflower capitulum and upper plant parts thoroughly with NPV +Bt would also give excellent results of controlling the pest. Also, spray of Ethion 50 EC (0.1%) has given good results of controlling the larval pest. (Anonymous 2006A).

Monitoring of foliar feeding larvae through pheromone traps: The activity of both *Spodoptera litura* and *Helocverpa armigera* could be monitored by setting up different pheromone traps for each, since the male adult moths get attracted to their respective sex pheromones. The catches obtained from the pheromone traps help in estimating the peak activity of adult moths leading us to arrive at the phases of larval pest incidence. This information would be useful in supplementing or modifying the crop management activity in relation to pest incidence. The population dynamics of the adult moth thus gathered will enable us to relate the pest incidence in relation to weather data including the rainfall incidence.

The data available so far in this context indicated that the peak activity od *Spodoptera litura* occurred during September-November and the emergence of the adult moth was triggered by 23 to 29°C coupled with high humidity of 75-90%.

On the other hand, the peak activity of *Helicoverpa armigera* normally happened during the months of September –October as per the available records. However, the data from pheromone and light traps indicated that there were two peaks of the pest activity one in August-September and the other in October - November

Sap-sucking pests of sunflower: Sucking pests like leafhoppers, thrips, whiteflies contribute to a considerable extent of loss of crop yields. Whiteflies under favourable conditions also pose threat to the crop. Several species of thrips are associated with the crop, which do not cause direct damage but cause enormous loss indirectly as vectors of viral diseases, especially sunflower necrosis virus disease (Anonymous 2006A). Sucking pests like leafhoppers, thrips, whiteflies bring down the crop yields to a considerable extent. Leafhopper alone can cause damage up to 46 per cent (AICORPO, 1997).

Ananthakrishnan (1971) reported the occurrence of five species of thrips on sunflower viz., *Thrips flavus, Scirtothrips dorsalis,Franklinella schultzei, Thrips tabaci* and *Caliothrips indicus.* Basappa and Prasad (2005) reported that several species of thrips were associated with sunflower crop at different phenological stages, but none of the thrips species caused direct damage as a pest, but they caused enormous loss indirectly, as vectors of necrosis virus disease. *Frankliniella schultzei (*Tryb*),Scirtothrips dorsalis* (Hood), *Thrips palmi* (Karny) and *Megalurothrips usitatus* (Bagnall) were found to be associated with sunflower crop and were considered vectors of certain viral disease. *Microcephalothrips abdominalis* (Crawford), *Frankliniella damfi* (Priesn), *Thrips hawaiiensis* (Morgan) and *Thrips tabaci* (Lind.) were the other species recorded in sunflower ecosystem.

Leaf hopper population exhibited a significant negative correlation with minimum temperature. Low relative humidity and high temperature were found to have significant positive correlation with the population of leaf hoppers (Anonymous 2006A).

Correlation studies revealed that maximum temperature exhibited a significant and positive correlation. Maximum temperature of 30.1° C was found to be the most favourable.

The population of thrips (Fig.6) and necrosis disease incidence on the crop sown at different dates varied with different months and the disease was the highest in late spring and summer which indicated that the incidence was very lowin the crops sown during rainy season (September to October) and *rabi* season but started increasing in the crop sown from December onwards to summer season (Anonymous 2006A).

The thrip population (numbers/leaf) varied from 0.09 to 2.49 without much variation among seasons except on rainy season.

Fig. 6: *Thrips tabacii*

Among several pesticides studied imidaclorpid 200 SL@ 0.25 ml/litre was found to offer good control at the same time being cost effective in the management of thrips (Anonymous 2006A). Among the plant based oils, Neem oil @ 4% followed by Mahua and Pongamia oil @ 4% resulted in better control of the above pest (Anonymous 2006A).

Diseases of Sunflower: The disease dynamics in respect of sunflower crop has been changing rapidly since the introduction of the crop into India in 1969. During the initial decade of the crop's performance, the disease situation was at a very low ebb. Subsequently with the rapid spread of the crop and with the cultivation of the crop throughout the year under varying cropping systems, the disease scenario of sunflower expanded posing serious challenges. In the initial phase of crop cultivation in 1980s, the diseases of *Cercospora* leafspot, *Alternaria* leafspot, *Curvularia* leaf blight, *Phoma* leaf blight, powdery mildew and root-rot were observed. In the next decade, apart from the above, the bacterial leafspot, *Rhizopus* head rot, charcoal rot, *Helminthosporium* leaf spot, *Fusarium* wilt, *Scelrotium* root rot and *Phyllody*made their presence. Since then *Alternaria* disease has been the most devastating one, causing considerable yield losses. The disease of downy mildew caused by *Plasmopara halstedi* (Farl) Berl appeared in the form of an epidemic in in the command area of Mrathwada region of Maharashtra resulting in a serious setback to sunflower production in the region and the disease spread to the neighbouring Karnataka, Andhra Pradesh, Telangana and later to Tamil Nadu. *Cladosoprium* blight caused yield losses in sunflower grown in summer months. In the subsequent phases of the crop in late 1990s to 2009, sunflower necrosis disease(SND) assumed alarming proportions. The diseases of the next importance are *Alternaria* leaf spot, downy mildew, rust, *Sclerotium* wilt/rot in the sunflower growing areas (Anonymous 2006A). Since the year 2007, powdery mildew incited by *Golovinomyces orontii* has become a serious problem on sunflower.

Additionally, some plant parasitic nematodes viz., *Aphelenchus avenae, Aphelenchoides* spp, *Haplolaimus indicus, Meloidogyne incognita, Rotylenchus reniformis, Thrichodorus* spp, *Tylenchus* spp, *Xiphinema* spp and *Helicotylenchus* spp. have found a niche in the with sunflower crop.

***Alternaria* Leaf Spot / Blight (*Alternaria helianthi* Schwein):** This disease is found in almost all sunflower growing zones of India with varying intensities. *Alternaria* is more severe in *kharif* (monsoon season)and less in *rabi* (post monsoon season) and summer. The disease intensity is found to range from 8.0 to 36.5%, causing reduction of 7% yield with every 1% disease incidence (Anonymous 2006A). The seed yield losses due *Alternaria* was found range from 30 to 43% (Anonymous 2006A). The disease adversely affects plant height, stem girth, head diameter, seed weight and finally seed yield.

Under conditions of severe infestation, the infected plants dry prematurely as a result of rapid drying of leaves and blackening of stem due to heavily coalesced lesions. The disease appears late in *kharif* season. The early sown *kharif* crop exhibits symptoms of the disease late in the season and the late sown crop gets more adversely affected exhibiting symptoms early in its growth phase. The disease intensity is found to increase with the age. The disease intensity was found to be low at higher temperatures of 30 to 36*C and low at the lower temperatures of25.3 to 31.9*C at Coimbatore. The severity leafspot was found to increase with relative humidity, which also resulted in low seed filling.

Varieties / hybrids tolerant to *Alternaria* leaf spot are PRARUN-1329, TNAUSF-242, KBSH-44, MSFH-17, TNAU-2, KBSH-4, GAUSUF-15, NSH-27, MLSFH-47, EC68414 and EC68415

Disease management: *Alternaria* disease severity was always low in sunflower crop grown under conditions of recommended nutrient and other package of practices as compared to that of the crops grown with inadequate nutrient application and farmers' practices. Seed treatment of Captan (@3g/one kg) or Carbendazim (@ 2 g/one kg) was found to be effective against *Alternaria* disease. Four sprays of Mancozeb (0.2%) each given at 10 days' interval from 35days after sowing decreased the intensity of the *Alternaria* leafspot and rust disease. Fungicides like *Triadimefon* (0.05%) was found to be more effective than Mancozeb in reducing the *Alternaria* leafspot disease. Sprays of Propiconazol (@ 0.1%) three times each at 30, 45 and 60days after sowing effectively reduced the incidence of *Alternaria* leaf spot with higher B.C. ratio. Neem product AFF 3EC sprayed with 0.6% concentration or neem leaf extract (2%) reduced the incidence of the disease.

Sunflower crop grown under conditions of organic nutrition showed reduced incidence of *Alternaria* as compared to a crop grown with inorganic nutrients.

Downy Mildew [*Plasmopara halststedii* (Farl.) Berl. and deToni]: Downy mildew of sunflower is caused by the fungus- *Plasmopara halstedii*. Small yellow spots develop on the upper sides of the leaf, while white to bluish white fluffy growth forms on the underside of the leaf. White cottony masses of pathogen mycelium are apparent on the lower surface of leaves opposite the yellow areas.

As plants continue to grow, leaves become wrinkled and distorted. As the leaf spot disappears, the fluffy growth darkens to grey in colour.

The disease is initiated by soil-borne dormant structures called oospores of the fungus or infected seed. Oospores germinate in the spring in wet soils and produce swimming structures called zoospores, which migrate to roots, encyst, and germinate. Plants tend to become resistant to infection with age. Infection being systemic, occurs over a short period (two to three weeks) after germination. The pathogen infects plants through roots and becomes systemic within the plant. The pathogen produces zoosporangia profusely on roots and aboveground plant parts; zoosporangia are readily disseminated by wind and splashing water to other plants, where infection can occur through apical buds or leaves. The pathogen produces oospores in diseased tissues that are returned to the soil during tillage operations and serve as primary inoculum for future sunflower crops. Oospores can survive in the soil for five to 10 years. *P. halstedii* can also attack many weeds. The disease being of systemic nature, the seed yield is drastically reduced even up to 80% in a susceptible variety like Morden. The disease of downy mildew is more severe under ill drained and water-logged conditions. The investigations conducted in the Oilseeds Research Centre at Latur (Maharashtra), indicated that the disease more is intense on the crop sown during the rainy days as compared to that of sown during non-rainy days.

In India the disease of downy mildew of sunflower was recorded for the first time at Latur and Beed districts of Marathwada region of Maharashtra in 1984. Initially the disease intensity ranged from 5 to 60% (Appaji *et al*., 1996). Later the disease spread to other areas of Maharashtra and also to the neighbouring districts in states of Karnataka and erstwhile Andhra Pradesh including Telangana.

In order to screen the sunflower materials for resistance against the disease, Downy Mildew Sick Plot measuring 600sq. m. was developed at the Oilseeds Research Centre at Latur. For the effective screening of the germplasm, a radicle inoculation technique was developed at Latur. (Anonymous 2006A).

The race picture of the pathogen fungus *Plasmopara hulstedii* has since been established, through the host differentials. The sunflower strains RHA 272, RHA 273, RHA 274, RHA 265 and RHA 801the host differentials were resistant to race present in India, while the varieties Progress and Morden were susceptible. The strain RHA 265 is resistant to races 1, but susceptible to races 2 and 3. This indicates that under Indian conditions the race 1 of the pathogen is preponderant; but not the races 2 and 3.

Disease Management:Seed treatment with Metalaxyl @ 4 to 6 g/one kg of seeds was found to be effective in reducing the disease considerably. The seed treated with Metalaxyl powder could be stored safely for more than 4 months without any adverse effect on the seed quality including its germinability and the seedling vigour.

The new liquid version of seed dressing compound Metalaxyl @ 3ml/ kg of seeds was found to be highly effective in reducing the incidence of the disease. In case disease incidence is noted in the standing crop, foliar spray of Ridomil @ 4ml or 4g/ kg of seed was found to check the disease.

Sunflower Necrosis Disease (SND): Among the viral diseases affecting the crop, sunflower necrosis disease caused by Tobacco Streak Virus (TSV) belongs to the genus *Ilarvirus* of the *Bromoviridae f*amily of the order +ssRNA virus (Prasada Rao *et al.,* 2000, Bhat, 2011) and spread by thrips was observed for the first time at Bagepally near Bangalore in Karnataka during 1997 in seed production plots causing panic among farmers and seed growers (Nagaraju and Hanumantha Rao, 1999; Harvir Singh, 2005). In Andhra Pradesh, sunflower necrosis disease was observed first time in kharif 2000 in Anantapur district. Later SND was observed in the districts of Jalna, Aurangabad, Latur and Akola of Maharashtra and because of its fast spreading nature, this necrosis virus is considered as one of the deadly virus diseases on the crop in India (Nagaraju and Hanumantha Rao 1999). Epidemiological studies revealed that the per cent necrosis due to virus disease and average number of thrips per five plants was the highest during June 2004 (7.48 and 1.51 respectively), followed by July 2004 (6.86 and 0.86, respectively). Whereas, it was the least during September 2004 (1.48 and 0.38, respectively). This revealed that the disease incidence and thrip population was the highest in summer-sown crop followed by *kharif* (monsoon season) and the least in *rabi* (Post-monsoon season). The studies revealed that the positive correlation existed between thrips population and maximum and minimum temperature, relative humidity at morning and bright sunshine hours. Similarly, the SND had positive correlation with maximum and minimum temperature, relative humidity at morning and bright sunshine hours as reported by Lokesh *et al.,* (2008). In other words, the highest incidence of disease and maximum population of thrips were found during the prolonged dry spell after rainy season (Anonymous 2006A).

The disease has been noticed in an epidemic form consecutively for all the year years since its appearance in Karnataka. The incidence is ranging from 5-90 per cent both in open pollinated varieties and hybrids (Bhargav and Meena, 2014). Necrosis virus (TSV)can cause infection at any stage of the plant growth causing necrosis of a part of the leaf lamina and making the leaf to twist, followed by varied type of necrotic and mosaic symptoms (Nagaraju and Hanumantha Rao, 1999). In some genotypes, the leaves developed typical mosaic, rings and line pattern. The disease incidence is high in early sown crop resulting in failure of total seed formation. In the case of late infected plants, the disease caused malformation of capituala with poor seed setting, in which the mean unfilled and chaffy seeds ranged from 85 to 98% (Anonymous 2006A). The disease incidence was found to be very high in unfertilised plots of sunflower and the plots where no thinning was carried out and those lacking plant protection measures (Anonymous 2006A).

The disease incidence is a function of various aspects of host, vector, weather and time. On an average, yield losses due to this disease range from 30 to 100 percent. The general symptoms of the disease are as under:

1. Mosaic and chlorotic ring spot on younger leaves,
2. Marginal necrosis of leaf and necrosis on stem and stalk,
3. Twisting of stem into S shape and
4. Defoliation emaciation and drying of the head inflorescence.

The symptoms of the disease at different stages of the crop are as follows:

Vegetative stage:

- Necrosis (death of the tissue) on part of the leaf lamina near the midrib resulting in twisting of the leaf.
- Necrosis (black streak) extends through one side of the lamina to the petiole and stem and finally terminate at shoot of the plant leading to partial paralytic symptom.
- The tip of the growing plant becomes necrotic giving typical bud necrosis symptoms, the plants fail to produce flowers and finally die.
- Few affected plants continue to show necrosis on young leaves, whereas few other plants after infection show mosaic mottling symptoms with rough and narrow leaves.
- Few affected plants which showed stunting after infection recovered and later become normal.

Flowering stage:

- Necrosis at bud formation stage lends to partial twisting of the capitulum.
- Necrosis on bracts and back of the capitulum are common.
- Opening of the flower head, expansion of capitulum and seed filling are affected depending up on the stage at which the flower bud/capitulum got infected by necrosis.
- Similar symptoms were observed on weed hosts like, *Parthenium hysterephorus, Xanthium stromarium, Agrimonemexicana* and watermelon, surrounding sunflower crop.

The virus is reported to be transmitted through thrips which carry the infected pollen grains. Several species of thrips serve as vectors of TSV, including *Microcephalothrips abdominalis* and *Thrips tabacii* (Rabedeaux *et al., 2005).* Pollen grains are the real 'vectors' of the disease.

Ajith Prasad (2004) investigated the disease transmission through sap, seed, insect vector and pollen. The disease was successfully transmitted through sap and the transmission ranged from 19 to 22 per cent. The insect vector, *T. palmi* also successfully transmitted the disease and the transmission ranged from 10 to 12 per cent.Anjula and Nagaraju (2002) conducted seed transmission studies through seeds collected from natural infected plants in the field from different genotypes viz., Morden, KBSH1. MSFH-17, PAC-1091, Sungene-85, CMS-234 B and RHA-6D-1 which recorded good percentage of germination but did not show any disease symptoms even after 60 days, indicating the non-transmissible nature of the disease through seeds. By employing DAC-ELISA, the virus was detected in infected sunflower pollen, leaves and developing seeds and thrips from infected sunflower plants under field conditions. However, the virus was not detected in matured seeds of infected sunflower plants.

Alternate hosts : Among 44 of the other plant species tested, the virus infected only *Cucumis sativus* cv. Green Long producing chlorotic spots followed by necrosis of the leaves along the margin, narrowing of leaves and stunting of plants under laboratory conditions upon mechanical sap inoculation (Anil Kumar,1999) Among the different hosts used as test plants for mechanical transmission studies, the virus infected *Vigna unguiculata* cv. C-152 and *Lycopersicon esculentum*, apart from the main host sunflower (Halakeri, 1999). Studies carried out by Ramaiah *et al.* (2001), Sharman *et al.*, (2008) and Vemana and Jain (2020) revealed that this virus could infect members of plants belonging to manyfamilies viz., Amaranthaceae, Chenopodiaceae, Malvaceae, Cucurbitaceaeand Fabaceae. Out of 48 weed species tested by mechanical sap inoculation, the virus infected the plant species viz., *Acanthospermum hispidum, Euphorbia geniculata, Ipomoea aquatica, Stachytarpeta jamaikensis, Galinsoga parviflora, Amaranthus spinosa and Commelina benghalensis* (Aravind, 2002).

Fig. 7. Symptoms of sunflower bud-necrosis

Fig. 8. Final symptoms of sunflower bud necrosis

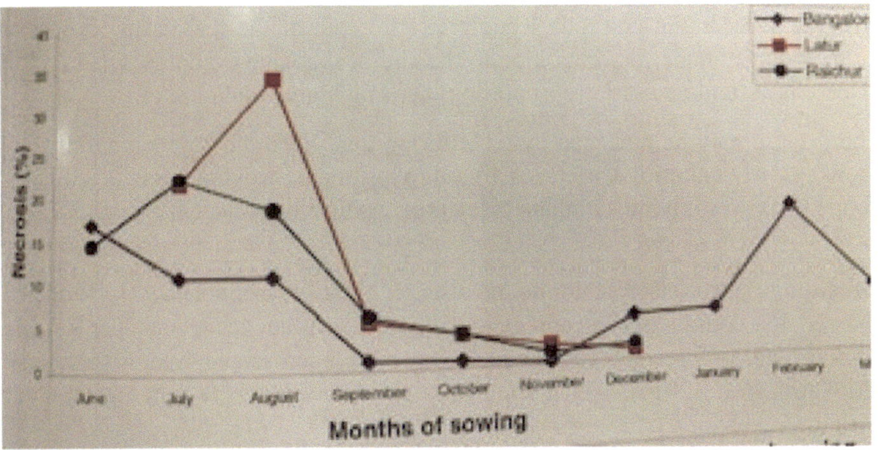

Fig. 9: Incidence of Bud-necrosis of Sunflower in different months of sowing (Anonymous 2006A)

Management of Sunflower Necrosis Disease (SND): The investigations carried out at the ICAR-Directorate of Oilseeds Research (Anonymous 2006A) revealed that the following steps taken would result in disease reduction:

1. Clean cultivation with proper thinning and removal of alternate hosts including *Parthinium* in the vicinity

2. Planting 3 to 5 rows of border crop of pearl millet (*bajra*) or sorghum (*jowar*) around sunflower field

3. Rouging virus infected plants in early stage of disease infection.
4. Seedtreatmentwithimidacholrprid @ 5g / kg of seed followed by spraying of imidachlorprid of 0.01% each at 15, 30 and 45 days following sowing. Imidacholprid residues were not detected in the harvested seeds.
5. Before sowing too remove all the alternate host plants around the field and destroy them

Fig. 10: Powdery mildew of sunflower

Powdery Mildew [*Glovinomyces (Erysiphe) cichoracearum*] : The disease produces white powdery growth on the leaves. White to grey mildew on the upper surface of older leaves could be seen. As plant matures, black pin head sized structures are visible in white mildew areas. The affected leaves are more lustrous with curls becoming chlorotic leading to the death of the leaf tissue. Powdery mildew occurs under dry condition to the end of the winter months. Off-late the disease is assuming a serious proportions causing higher yield losses.

The disease may be managed by (i) ensuring total field and crop sanitation, (ii) growing early maturing varieties, (iii) removing and destroying the infected plant debris and (iv) applying karathane or calixin one litre/ha or wettable sulphur 2 kg/ha.

Regarding the varietal reaction to the disease, highly susceptible check 'Morden' showed 468 conidia/microscopic field, while the resistant accessions like R-GM-41 and R-GM-49 recorded 52.2 and 56.8 conidia/microscopic field, respectively (Kulkarni *et al.,* 2015), which can act as good source for developing resistant varieties and hybrids.

Reliable sources of resistance to the pathogen were identified in four annual wild species *Helianthus argophyllus, H. argestris, H. debilis and H. praecox), six perennials (H. angustifolius, H. atrorubens, H. rigidus, H. salicifolius, H. pauciflorus and H. resinosus*, two interspecific derivatives viz., HIR-1734-2, RES-834-3 and two exotic lines viz., PI642072, EC537925 (Supriya *et al.,*2016)

Rust Disease of Sunflower (*Puccinia helianthi* Schwein)

Symptoms of the rust disease are small, reddish brown pustules (Fig. 11) consisting of uredia covered with rusty dust appear on the lower surface of bottom leaves. Infection later spreads to other leaves and even to the green parts of the head. In severe infection, when numerous pustules appear on leaves, they become yellow and dry. The black coloured telia are also seen among uredia on the lower surface. The disease is autoecious rust. The pycnial and aecial stages occur on volunteer crops grown during off-season. The uredospores are round or elliptical, dark cinnamon-brown in colour and minutely echinulated with 2 equatorial germ pores. Teliospores are elliptical or oblong, two celled, smooth walled and cheshnut brown in colour with a long, colourless pedicel.

The pathogen survives in the volunteer sunflower plants and in infected plant debris in the soil appear as teliospores. The disease spreads by wind-borne uredospores from infected crop.

The favourable conditions for incidence and spread of the disease are (i)day temperature of 25.5° to 30.5°C with (ii) relative humidity of 86 to 92 %. The rust disease incidence was observed to be low in the seedling stage; but increased with age and maximum intensity was observed in the crop of 75 days' age.

Rust disease was recorded in all sunflower growing zones of the country in low to moderate form. It was observed that the disease was more intense in the post rainy (*rabi*) season; but starts appearing in the late monsoon (*kharif*) season. The data recorded show that the rust intensity varied from 2.1 to 21.8% (Anonymous 2006A). The disease was at its maximum peak during the period from August to January; but gets abated during February to May when the disease was low to practically nil. The crop sown during the months of August to November becomes more vulnerable to disease exhibiting severe symptoms.

Screening of germplasm lines resulted in identification of certain rust resistant types which were used in the breeding programmes and particularly in the development of hybrids like BSH-1, KBSH-1 and LSH-3, possessing resistance to rust.

Disease Management: As in the case of *Alternaria l*eaf spot, rust disease can be controlled by spraying Mancozeb of 0.25%three times at intervals of 15 days, starting the first spray with the initial appearance of leaf spots / pustules of the disease.

Fig. 11: Rust disease on sunflower leaves

Damping off / Root-rot /Charcoal rot (*Rhizoctonia bataticola* Pycnidial stage: *Macrophomina phaseolina*) : The pathogen is seed-borne and primarily causes seedling blight and collar rot in the initial stages. The grown up plants also show symptoms after flowering stage. The infected plants show drooping of leaves and death occurs in patches. The bark of the lower stem and roots shreds and are associated with a large number of sclerotia. Dark coloured, minute pycnidia also develop on the lower portion of the stem. In addition, different soil-borne and seed-borne fungi viz., *Aspergillus niger, Rhizopous stolonifera, Sclerotium rolfsii* and *Pithium aphanidermatum* might attack sunflower seeds sown in the field. The affected seeds might rot germinate or emerging seedlings may be killed. Such seedlings exhibit damping-off, root rot and stem rot symptoms leading to patchy plant stand in the field. There is a lot of variation among the isolates of *Rhizoctonia* with regard to sensitivity to fungicides tested such as Brassicol, Captan, and Thiram. The isolates also differed in their virulence on sunflower seedlings.

Disease Management

- Closer planting of the seedling should be avoided.
- Optimum nutrition should be provided to maintain the plant vigour.
- Whenever the soil becomes dry and the soil temperature rises then irrigation should be provided.
- Seed treatment with *Trichoderma viride* formulation at 4 g/1kg seed.
- In endemic areas long crop rotation should be followed.
- Soil application of Carbendazim (400 ppm), formaldehyde (0.1%) and Sodium azide (200ppm) reduced severity of the disease.
- Summer ploughing before the onset of planting season to expose the soil to sun also helps in combating the soil borne pathogens apart from recovering other benefits that facilitate good crop growth.

Head rot (*Rhizopus oryzae*):The affected heads show water soaked lesions on the lower surface of the head, which later turn brown. The discoloration may extend to stalk from head. The affected portions of the head become soft and pulpy and insects are also seen associated with the putrefied tissues. The superficial white mycelial growth of the fungus will be apparent. The larvae and insects which attack the head pave way for the entry of the fungus which attacks the inner part of the head and the developing seeds. The seeds are converted into a black powdery mass. The head finally withers and droops down with heavy fungal mycelial nets. Pathogen produces dark brown or black coloured, non-septate hyphae. It produces many aerial stolens and rhizoids. Sproangia are globose and black in colour with a central columella. The sporangiospores are dark coloured and ovoid. The disease appears as the sunflower plant reaches the peak flowering stage.

Prolonged rainy weather at flowering and damages caused by insects and caterpillars are the predisposing factors for the proliferation of head-rot disease.

The disease can be successfully managed by (i) treating the seeds with Thiram or Carbendazim at 2g/kg, (ii) controlling the caterpillars feeding on the heads and (iii) spraying the head with Mancozeb at 2kg/ha during intermittent rainy season and repeat after 10 days, if the humid weather persists.

Nematode Pests of Sunflower

Sunflower crop is infected by certain species of plant parasitic nematodes such as *Meloidogyne incognita, Rotylenchus reniformis, Aphelenchus avenae, Xiphinema* spp., *Haplolaimus indicus, Trichodorus* spp.and *Helicotylenchus* spp.

Plant parasitic nematodes are tiny worm like animals, usually 0.25 mm to 3 mm long (1 / 100 " to 1 / 8 ") and cylindrical, tapering toward the head and tail. Females of a few species lose their worm shape as they mature, becoming pear, lemon- or kidney- shaped. Plant parasitic nematodes possess all of the major organ systems of higher animals except respiratory and circulatory systems. The body is covered by a transparent cuticle, which bears surface marks helpful for identifying nematode species. These nematodes feed on living plant tissues, using an oral stylet, a spearing device somewhat like a hypodermic needle, to puncture host cells. The plant parasitic nematodes inject enzymes into a host cell before feeding to partially digest the cell contents before they are sucked into the gut. Most of the injury that nematodes cause to plants is related in some way to the feeding process.

It is rare that above-ground symptoms of nematode infestation give sufficient evidence to diagnose a nematode problem in the roots. Since most plant nematodes affect root functions, most symptoms associated with them applied to be of inadequate water supply or mineral nutrition to the tops: chlorosis (yellowing) or other abnormal coloration of foliage, stunted aerial growth, failure to respond to nutrient applications, small or sparse foliage, a tendency to wilt more readily than healthy plants, and slower recovery from wilting. Woody plants in advanced stages of arfestation nematode may exhibit dieback of progressively larger branches(https://mrec.ifas.ufl.edu/lso/SCOUT/Nematodes.htm)

Presence of plant parasitic nematodes in the root zone of sunflower crop (Fig. 12) leads to certain decline in seed yields.

A significant reduction in seeds yield such as 35. 4%, 42. 0% and 43. 4% was observed at 3, 4 and 5 root gall indices, respectively in the case of *Meloidogyne* infestation. Protein, N, P, Mn and Zn content s of seed decreased with increasing nematode damage, in opposition to carbohydrates and Fe content which increased with increasing nematode damage, while oil and K contents were not affected.

Infested sunflower seed oil consisted of six fatty acids, four of them were saturated (myristic, palmitic, stearic and arachidic) and two unsaturated (oleic and linoleic) (Korayem *et al.*, 2016)

Regarding varietal response to nematode attack, variety France-Lever appeared to be the least affected by the diseases and consequently exhibited superiority over others in most growth and yield parameters (Fabiyi *et al.,* 2013)

Management of plant parasitic root knot nematodes

- The use of cultural control methods to manage root-knot nematodes is the most environmentally sustainable and potentially most successful method for limiting root-knot nematode damage. However, because root-knot nematodes have very large host ranges, cultural control methods require careful planning. Cover crops such as sudan grass and marigolds actually produce chemicals that are toxic to nematodes. Cover crops have the added benefits of stabilizing topsoil and improving soil quality. As with crop rotation, however, specialized equipment may be required to deal with different cover crops. Other techniques, including flooding and solarization of fields, have controlled nematodes, but only in warm climates and when a particular field can be removed from cultivation for long periods during the treatment. While cultural control methods are extremely valuable tools, they require extensive consideration, planning and economic investment before successful implementation can be achieved.
- Root-knot nematodes are very difficult to manage because they are soil borne pathogens with a wide host range. Because root-knot nematodes live in the soil, chemical control requires applications of large amounts of chemicals with specialized equipment. Fumigants (such as 1,3-dichloropropene, methyl bromide and dazomet) are commonly applied as pre-plant treatments to reduce nematode numbers, but they must thoroughly penetrate large soil volumes to be effective. Some fumigants volatilize very quickly, so treated soil must have a cover or tarp to maintain the fumigant in the soil long enough to be effective. The phasing out of methyl bromide (an efficient fumigant) has intensified the search for alternatives that can be used by farmers. Currently, the chemicals that are supposed to be the most economically feasible for control are carbamates and organophosphates. Because they are not toxic to plants, they are the only chemical options for established plants. However, they are extremely toxic to humans and other non-target organisms.
- Control measures employing biological systems antagonistic to root-knot nematodes have been attempted by many researchers. The most commonly used biological control agents are fungi and bacteria. There are many kinds of nematophagous (nematode-feeding) fungi. Some fungi use mycelial

traps or sticky spores to capture nematodes for example, *Arthrobotrys* spp. and *Monacrosporium* spp. Other fungi parasitize eggs and root-knot nematode females, e.g., *Pochonia chlamydosporia* and *Paecilomyces lilacinus*. The major bacterial antagonists are *Pasteuria penetrans* and species of *Bacillus*. Endospores of *P. penetrans* attach to the cuticle of a juvenile nematode, produce penetration structures that enter the nematode, and slowly consume it

- The most successful approaches to nematode control rely on integrated pest management strategies (IPM). IPM combines management options to maintain nematode densities below economic threshold levels. IPM techniques can still be difficult to implement against a pathogen as aggressive and resilient as root-knot nematodes. Nevertheless, a combination of management tactics/tools, including cultural practices (rotations with non-host crops and cover crops that favour the build-up of nematode antagonists), resistant cultivars, and chemical soil treatments, if necessary, generally provide acceptable control of root-knot nematodes. The extent of this success, however, is dependent upon having accurate damage threshold densities and available and readily acceptable resistant cultivars (Mitkowski and Abawi. 2003). https://www.apsnet.org/edcenter/disandpath/nematode/pdlessons/Pages/RootknotNematode.aspx

Fig. 12: Sunflower plant's roots infested with Meloidogyne incog*nita*

Integrated Disease Management: Sunflower disease management involves the integration of appropriate cultural practices with cultivation of resistant and tolerant varieties and hybrids and judicious use of chemical fungicides, without exclusive dependence on them.

Cultural Practices: Pre-sowing summer ploughing, removal and burning of infected crop residues, early and timely sowing, clean cultivation, growing border crops such as pearl millet so as to check the vectors of bud-necrosis disease, sorghum and Marigold, inter-cropping with those crops that would evade the sunflower diseases, rouging of infected plants, avoiding water stagnation in the crop-stand and its vicinities, avoiding moisture stress, crop rotation, proper tillage. Avoiding injury to the capitulum during intercultural operations

Resistant / Tolerant Varieties of sunflower: LSH-1 & LSH-2 against sunflower downy mildew, hybrids like BSH-1, KBSH-1 and LSH-3 against rust disease,

PRARVN-1329, TNAUSF-242, KBSH-44, MSFH-17, TNU-2, KBSH-4, GAUSF-15, NSH-27, MLSFH-47, EC68414 & EC68415 against *Alternaria.*

Judicious use of fungicides: Seed treatment with Captan, Thiram or Mancozeb @2 -3 g/ one kg

Sunflower Crop Production: Sunflower in relation to other traditional oilseed crops is relatively a new crop to India, since it was introduced in the country during the decade of 1969-1970. The crop has attracted the attention of researchers, extension workers and farmers as a promising oilseed crop of impressive potential, in view of its wider adaptability and photo-insensitivity. Very soon sunflower spread into wide ranging cropping systems throughout the length and breadth of India. Thus the area which was only 15thousand hectares in 1972, reached to 2.7 million hectares by the year 1993-94. However, the mean productivity of the crop remained at 566 kg/ha, due to sunflower production being carried out under sub-optimal conditions including highly drought prone dryland agriculture. In India sunflower has found a niche in Peninsular India particularly in Maharashtra, Karnataka, Andhra Pradesh and Telangana states. Other states of the next category for sunflower crop production are Tamil Nadu, Panjab, Haryana and Uttar Pradesh. Sunflower has shown excellent potential in the *rabi*-summer or spring seasons in Punjab, Haryana, Western Uttar Pradesh, Bihar, Madhya Pradesh, West Bengal and Chhattisgarh.

Season and Rainfall: Nationwide, Sunflower crop is grown in *kharif* season to the tune of 40% of its area. The rest of the crop area is confined to the seasons of *rabi*, spring and summer. In northern India the crop is grown during spring and summer (*zaid*) seasons. The rainfall need of the crop is around 500 mm; but the crop performs reasonably well even with well distributed rainfall of 400 -450 mm. The crop needs cool weather during the germination and seedling growth and warm weather during subsequent phases up to flowering and seed filling. The clear, non-cloudy weather with good sunshine is needed at flowering and seed filling stages of the crop.

Soils: Sunflower crop performs satisfactorily on well drained soils of neutral pH, with a moisture supply of around 500 to 600 mm. The ideal pH for the crop is around 6.5 to 8.0. The crop tolerates slight alkalinity of soil but not acidity. Nevertheless, the crop is being grown on wide range of soils viz., black soils, sandy loams and alluviums. A large chunk of 60% of black soil in the states of Maharashtra, Karnataka, Andhra Pradesh, Telangana and Tamil Nadu is covered by sunflower crop. In the northern region consisting of Indo-Gangetic belt confined to the states of Punjab, Haryana, Western Uttar Pradesh, Bihar and West Bengal, sunflower crop is being grown on alluvial soils. The crop's performance suffers adversely under water-logging.

Contingent Crop: Sunflower has become popular as a crop for midseason correction, in case sowings of traditional crops like cotton or sorghum get delayed or failed at seedling stage as occurs in the region of Akola of Maharashtra. Also, in the event of monsoon getting delayed for sowing regular crops like groundnut and finger millet in Karnataka and castor and cotton in Andhra Pradesh and Telangana, sunflower is the preferred choice. The crop being photo-insensitive, sunflower can be grown during any part of the year as a substitute catch-crop in order to impart stability and sustainability to cropping systems.

Inter-cropping: Inter-cropping of sunflower with other *kharif* crops like groundnut, pigeon pea, , finger millet , green gram, soybean and castor is popular. The inter-cropping of sunflower has added extra income due to its success, without offering much competition to the main traditional companion crop. With major crops such as groundnut, pigeon pea, sorghum etc., with a duration of more than 120 days, sunflower with a duration of less than 100 days fits in very well in inter-cropping systems in *kharif* season. The row proportion, however, has to be fixed depending upon the growth characteristics of each of the crop components in the system and the soil type. The efficient sunflower based *kharif* inter-cropping systems identified in different agro-ecological situations are given below in the Table 11 (Anonymous 2006A)

Table 11: Sunflower based Intercropping Systems in India (Anonymous, 2006A)

Region	Soil type	Intercropping System	Row Proportion
Karnataka	Alfisols	Groundnut + Sunflower	4:2 , 3:1
	Alfisols	Pigeon-pea + Sunflower	1:2 , 1:1
	Vertisols	Pigeon-pea + Sunflower	1:3
	Alfisols	Finger-millet + Sunflower	4:2
Maharashtra	Vertisols	Pigeon-pea + Sunflower	3:3
	Vertisols	Soybean + Sunflower	2:1
	Vertisols	Groundnut + Sunflower	6:2
Andhra Pradesh + Telangana	Alfisols	Groundnut + Sunflower	4:2
	Alfisols	Pigeon-pea + Sunflower	1:2
	Alfisols	Castor + Sunflower	1:1
Tamil Nadu	Alfisols	Groundnut + Sunflower	3:1
	Alfisols	Castor + Sunflower	1:1
Gujarat	Vertisols	Groundnut + Sunflower	3:1 , 6:1
Non-traditional Areas	Inceptisols	Black gram (Urad)/ Green gram (Mung) + Sunflower	4:2 , 3:1

Crop Sequences: Sunflower being of shorter duration and not season bound unlike other crops, it fits in well in a wide range of crop sequences. Some of the sunflower crop sequences adopted in different agro-ecological zones are given below (Anonymous, 2006A).

(i) Non-traditional areas for kharif sunflower such as Punjab, Haryana, Western Uttar Pradesh*: Wheat-Sunflower; Potato – Sunflower; Toria (Brassica campestris var. toria) – Sunflower; Cotton – Sunflower and Rice – Sunflower.*

(ii) Vertisols of Maharashtra, Madhya Pradesh & Chhattisgarh*: kharif legumes viz., Mung bean/ Urad bean – sunflower; Sunflower – rabi sorghum; Groundnut – Sunflower& Sunflower – Soybean*

(iii) Peninsular zones of Karnataka, Andhra Pradesh, Telangana &Tamil Nadu: *Groundnut – Sunflower; kharif Sunflower – Groundnut in alternate years (Red loams of Hyderabad region); Mung bean – Sunflower; Sunflower – Castor (Telangana zone) Setaria (foxtail millet) – Sunflower & Soybean-Sunflower.*

Among most of the crop sequences, Groundnut – Sunflower has been rated as the most remunerative as it recorded the highest productivity and net returns in the peninsular zone (Anonymous 2006A)

In general, rice fallows offer immense potential to grow sunflower with least inputs in terms of crop nutrients and irrigation. While this could be easily adopted in light soils of coastal rice fallows, in heavier soils and those with high water table, more precautions have to be taken for crop establishment and crop protection (Anonymous 2006A).

It needs to be emphasized that crop rotation is very important factor in sunflower production. Continuous cultivation of sunflower after sunflower due to its photo-insensitivity not only depletes soil fertility but also leads to disease build-up and progressive decline in seed yields. It is recommended to follow 2 to 3-year rotation with other traditional crops such as cereals, millets and grain legumes of the area, to improve soil productivity.

Pre-sowing land preparation: In case sunflower to be grown in the *kharif* season as rain-grown crop, appropriate land preparation is necessary. With the receipt of summer showers, the deep ploughing is recommended to be carried out to expose the soil borne pathogenic biotic agents. Application of 5 to 6 tons ha of well decomposed Farm Yard Manure (FYM) is to be applied and incorporated at ploughing.

Seed bed preparation and planting: With receipt of regular monsoon showers, the soil may be prepared in a manner amenable for planting. Adequate plant population per unit area is necessary to ensure good expression of the plant attributes including capitulum size in terms of its diameter and seed development leading to higher yield. Firstly, the crop may be raised on the ridges spaced at a minimum distance of 60 cm and maintaining the intra-row spacing (spacing from pant to plant with a row) of 30 cm. Ridge-furrow system is ideal both for rain-fed dryland system as well as for irrigated condition. Under dry land conditions the furrows act as the seats of moisture conservation and the system is also conducive for lesser humidity in the crop micro-climate avoiding water-logging. Under irrigated conditions, the timely and judicious irrigation may be effected through furrows there by avoiding flood irrigation. The water consumption of the crop sown on ridges is lower and the process imparts resistance against crop lodging. Also the crop sown on ridges escapes basal stem rot and attack of cut-worm (Anonymous 2006A).

Seed rate: Optimum seed rate for sunflower crop depends upon the seed size and its quality. The recommended seed rate is @ 5 kg/ha with planting of 2 seeds per hill. Under precision planting with high quality seed, the seed rate may be reduced to 3 to 4 kg/ha. While planting, adequate moisture status of the soil in the seed zone must be ensured. It is advised to go in for higher seed rate and later following the germination, the extra seedlings may be thinned-out. Seed size should be medium to bold with a weight of 5 to 6g/100 seeds. The germination should be more than 80%. The under-sized seeds result in low germination and the resulting seedlings will be stunted with very low seedling vigour that are more prone to abiotic and biotic stresses leading to poor crop establishment and reduced seed yield. Use of good quality seed alone is reported to result in increased seed yield of the order of 6 to 9.4%. A sieve size of 2.4 to 2.8mm was found optimum for higher recovery of good quality seed (Anonymous 2006A).

Seed hardening: Seed hardening is reported to overcome the problems of delayed germination, lack of uniformity in emergence due to soil crust in shallow alfisols. The standardised procedure for seed hardening consists of soaking the seeds in fresh water (1:1 weight / volume) for about 14 hours followed by shade drying. The shade dried seeds that have gone through the process of seed hardening may be used for sowing at a later date. This process imparts drought resistance too, to the emerging crop as a result of favourable physiological modifications that have been in place following seed hardening (2006A).

Seed treatment before sowing:Treating seeds with Captan or Thirum or Carbendazim @ 2 to 3 g/kg should be a part of the crop production schedule to ward-off incidence of wide range of diseases as explained previously

Crop nutrition: Management of crop nutrition should be taken care of with appropriate doses of fertilizers depending upon the rainfall pattern of the locality. The crop performs the best, when both organic and inorganic fertilizers are applied to it. The fertilizer schedule consisting of nitrogen, phosphorus and potash (NPK) @ 20N:30P2O5:20K2O per hectare is recommended in the case of rain-fed low rainfall dry-land conditions. In respect of high rainfall zones and irrigated conditions nutrient schedule of 40:60:40 kg (NPK) of per hectare is appropriate. Half of the N and total quantity of the P2O5 and K2O are to be given as basal dose at the sowing. Rest of the quantity of Nitrogen (50%) may be applied as top dressing at 35 days after sowing in medium and heavy soils. In the case of lighter soils 25% of the remaining quantum may be applied as top-dressing at 30 to 35 days after sowing and rest of the 25% may be given as the second top-dressing at 50-55 days after sowing. Single Super Phosphate (SSP) may be preferred as the source of phosphorous since it provides the crop with sulphur (12%) too, in addition to P2O5 (16%). A quantum of 30 to 45 kg/ha of sulphur is recommended for sunflower crop for obtaining higher yields and oil content. Additionally, application of calcium (Ca) at the rate of 20 kg / ha and Magnesium (Mg) at the rate of 10 kg/ha has resulted in increased yield by 20% (Anonymous 2006A). While the above suggested recommendations are of general nature, it is necessary that the nutrient recommendations for sunflower crop for site specific situations should be based on the local soil test values.

Micronutrients: Among wide range of micro nutrients, boron and zinc are found to be more important for sunflower nutrition in that order. Since most of the Indian soils are deficient in zinc, it is recommended that soil application of zinc sulphate @ 5 kg /ha would replenish the deficiency. Boron (B) is found to be very important in sunflower nutrition since it enhances the pollen viability and the receptivity of stigma that would result in increased seed setting. It has also been found that Boron increases the translocation of photosynthates to sink, thereby enhancing the seed yield (Anonymous 2006A). Spray of 0.2% borax (Disodium tetraborate) at ray floret opening or dusting 2 kg of Borax /ha would result in increased seed yield. Since borax is not readily soluble in cold water, the needed quantity of borax is to be dissolved in boiling water and cooled before the same is used for spraying. In the cases of acute soil deficiency of boron, soil application of 1 kg of B/ha is recommended.

Bio-fertilizers: Sunflower seed treatment with dual inoculation of *Azospirillum* and *Azotobacter* has brought in 50% reduction in recommended fertilizer Nitrogen, thereby resulting in reduction in cost of production.

In vertisols of Andhra Pradesh (Nandyal) and Karnataka (Raichur) with the crop sequence of sunflower – chick pea, it has been possible to substitute 50% P by PSB (Phosphorous Solubilizing Bacteria) with or without FYM (@5t/ha) for *rabi* grown chick pea, when the preceding sunflower crop was fertilized with recommended P(Anonymous 2006A).

Water management: The optimum water requirement of sunflower crop being 500 to 600 mm, the crop stages of seedling, star-bud (capitulum initiation), peak flowering and seed-filling are critical in terms of moisture requirement. Moisture stress at the above stages would have a strong negative impact on seed yield. In the case of irrigated crop of sunflower, irrigation may be given at intervals of 20 to 25 days on black soils and at intervals 8 to 10 days on lighter red soils.

In the case of unirrigated crop grown on drylands, moisture conservation has to be strengthened so as to avoid drought at the above stages. Alternatively, provision for extending protective irrigations should be in place to save the crop from the adverse impact of drought at the above crucial stages, either through run-off water collected in ponds or by alternate means.

Harvesting and Threshing: Sunflower crop should be harvested at physiological maturity, which is indicated when back of the capitulum turns lemon yellow in colour as the bottom leaves start withering and drying. Further delay in harvesting the crop might cause lodging, termite attack and invasion of rodents leading to yield losses. After separation of the capitula, they may be dried for 2 to 3 days to facilitate easy threshing and separation of seeds. Threshing of the dried capitula may be done manually by beating with sticks or rubbing manually by the workers engaged for the purpose. The seeds many be dried to bring down the moisture content to around 9 to 10%.

Sunflower Seed Production: Seed production in sunflower crop consists of (i) Seed production of varieties, (ii) Seed production of parental materials of hybrids viz., Female parent (A – line), Maintainer line (B – Line), and Restorer Line (Male Parent or R-Line) and (iii) Sunflower hybrid production

(i). Breeder Seed production of sunflower varieties: Sunflower being a cross pollinated crop, all the procedures to avoid genetic deterioration of sunflower varieties should be practised in its varietal seed production. Firstly, the minimum isolation distance of 1.5 km for breeder seed production, 1.2 km for foundation seed production and 1000 meters for certified seed production must be maintained for the sunflower varietal seed production plots.The isolation distance suggested for breeder, foundation and certified seed production plots are much higher than those currently prescribed (600 meters) under seed certification. In view of the role of honey bees in pollination of sunflower crop, their flight range, problems encountered so far in maintaining the genetic purity of parental stocks of commercial varieties and hybrids and the standards prevailing in other important sunflower growing countries of the world, the above isolation distances are needed to be observed,

- Avoid low lying areas and the fields where previous sunflower crops were grown. The field should be thoroughly prepared to ensure weed free seed bed. Incorporate 4 to 5 tons of Farm Yard Manure per hectare at the time of initial land preparation. Make necessary provision for surface drainage through adoption of recommended soil and water conservation measures particularly in soils which are ill drained. Two to three ploughings followed by harrowing / planking the in case of black soils should be adequate to achieve the required seed bed conditions. Avoid flood irrigation. Plant the crop on ridges and irrigate in furrows. Avoid moisture stress at all stages of the crop. Adopt the recommended nutrient schedule @ 60 kgN, 80 Kg P2O5 and 40 kg of K2O /ha. Apply 50% of N and the entire dose of P2O5 and K2O at the time of planting preferably as band placement by the side of seed rows. The remaining 30 kg N/ha may be applied as top dressing in two equal splits of 15 kg / ha each, first at 30 days following planting and the second dose, a fortnight later. Maintain the recommended plant population by dibbling the seeds in rows, maintaining the inter-row and intra-row spacing. The spacing would vary depending upon the soil type and moisture availability.

- Rouging of the off-types must be carried out rigorously two times before flowering; later at weekly intervals from capitulum initiation to peak flowering and another 3 times during the seed development and maturity phases. All the plants not conforming to the standard genotypic characters should be pulled out and destroyed.
- In order to ensure better seed set from seed production plots, hand pollination must be practised. The duration of flowering extends for about 15 to 18 days. During the flowering phase of the crop, practise hand pollination by gently rubbing the capitulum with palm covered by gloves every day between 8 and 11 AM.
- Planting time: In the case of traditional sunflower growing areas, the best planting time is late *rabi* to summer seasons. This is recommended to avoid pollen contamination from standing *kharif* crop of sunflower, which will be wide-spread in the traditional sunflower growing areas. In respect of new areas where sunflower is not being grown, the seed production could be taken up in late *kharif* to *rabi* seasons.
- Plant Protection: All the needed steps must be taken to ensure total plant protection of the seed crop. Any damage to the crop due to biotic factors would be disastrous and affects adversely the seed yield. There is additional danger of some of the fungal, bacterial and viral diseases that could contaminate the seed quality and would get transmitted through the seed from the infected seed crop. Seed treatment with recognised fungicides at the recommended doses is obligatory for the seed crop. Before sowing, it is recommended that the seed is thoroughly checked for the seed borne diseases despite the seed dressing with the recommended fungicides.
- Harvesting of the seed crop must be done with utmost care. The plants with deviant capitulum characters must be harvested separately not mixing them with the selected bulk. Harvesting must be carried-out at physiological maturity of the crop to strictly avoid shattering of seeds that would happen at total drying of the standing crop.
- The harvested capitula should be dried in a separate and secluded drying yard without any chance of admixture from other kinds of seeds. Similarly threshing must be carried out on clean threshing floor without any other

foreign sunflower seed materials. In case mechanical threshers are to be used, it must be ensured that the thresher is totally clean and free from other kinds of seeds including sunflower seeds of other categories.

- The threshed seed should be dried to the standard moisture content and stored in bags and containers of the prescribed standards with a prominent label of identification

(ii) Seed production of parental materials of hybrids

(a) ***Breeder Seed Production of Female Parent (A-line)***: The seed plot should be raised under isolation maintaining the isolation distance prescribed above. All the package of practices described above should be followed. Avoid the plots, where sunflower was grown in the previous season. Grow the A-line (female parent, which is male sterile) and B-line (Maintainer line of the same genetic constitution of A line except for its male fertility) in a proportion of 3A:1B lines and the field borders may be planted with two rows of B (Maintainer) line. Manual transfer of pollen from B to A line should be ensured 4 to 6 times during flowering period in order to ensure good seed set. Removal of B line is recommended after pollination/ seed development in A line, to avoid any mechanical mixture. The seed from Aline is harvested carefully and separately as per the procedures suggested above.

Note:*There have been some cases of occurrence of "pollen-shedders" (male fertile plants) in A-line population. This is due to residual variability of the maintainer B- line. Hence, in order to check this problem, the B-line should be grown in diverse environments for rouging out the variable plants not conforming to the typical characters of the B-line. Secondly, the "pollen-shedders' can be identified by their characteristic feature of "Dark Anthers" containing pollen, which should be removed before the dehiscence of such kind of anthers.*

(b) ***Breeder Seed Production of B- Line (maintainer line)***: The seeds of B-line harvested from the nucleus B-line seed production plots should be used as the base material for B-line seed production. This production system is as per the strict norms outlined for varietal seed production, with particular regard to isolation distance, rouging etc. The seed thus produced should be harvested carefully maintaining the genetic purity and guarded safely without any kind of contamination

(c) ***Breeder Seed Production of [R]Restorer (male parent) Line:***In the case of the seed production of Restorer Line [R] which will be the male parent in the certified hybrid seed production program, the guidelines explained at item (b) above and at the item (i) above should be adopted without fail with regard to strict isolation distance, rouging and other production parameters. The R-lines have been designed as branched types for continuous supply of pollen in hybrid seed production plots. The R-line should be 100% genetically pure so as to produce quality hybrid seed.

(iii) Certified Hybrid Seed Production

The ideal period for hybrid seed production is *rabi* to summer seasons in tropical zones. Sowing for the *rabi* season must be in the months of mid-October to early November. For raising the summer crop of hybrid seed production, the ideal time is during December to January. During these periods, the incidence of biotic stresses would be at a very low ebb, at the same time the sunshine hours during the period would be ideal for good crop growth.

The proportion of planting shall one R - line after every three lines of A- line (3A: IR) with isolation distance prescribed above.

Rouging operations have to be strictly carried out at all the three stages as suggested at item (i) to maintain 100% genetic purity of the hybrid seed. All the off-types not conforming to the genetic attributes of A and R lines should be removed. The R- line is designed as branching type so as to provide continuous supply of pollen to the A-line. Hence any non-branching off-type occurring in the population of R-line must be removed and destroyed before anthesis. The branches other than main top two branches may be nipped during flowering of R line to ensure good pollen production.

Recently non-branched R lines possessing the attribute of copious pollen production are also being used. The practice of employing both branched and no-branched R lines is in vogue. In such cases rigorous rouging should become part and parcel of the seed production programmes of R lines.

Pollination being the most important component of hybrid seed production, transfer of pollen from R line to A line should be ensured. Wind pollination in sunflower is negligible in sunflower and hence hand pollination is to be resorted to ensure good seed development. This is carried out by gently rubbing the heads of R line over the capitula of A-line. Also gently passing the palms covered with muslin cloth of workers from R line to A line.

All the subsequent stages of the seed production crop should be managed strictly as per the guidelines described.

7

Safflower (*Carthamus tinctorius* L.)

Introduction

Safflower belongs to the family Asteraceae, which is also called Compositae, the aster, daisy, or composite family of the flowering-plant order Asterales.tribe Cardueae (thistles) and subtribe Centaureinae . With more than 1,911 genera and 32,913 accepted species of herbs, shrubs, and trees distributed throughout the world, Asteraceae is one of the largest plant families. The species of the family are annuals and perennials of worldwide distribution, consisting of ornamentals, vegetables and spice or medicinal plants.

The genus *Carthamus* L., contains about 25 valid species distributed from Spain, North America and West Asia to India. Many are indigenous to the Mediterranean region, but only one viz., *Carthamus tinctorius* L. (safflower) is cultivated for oil. Safflower is a diploid species (2*n*=24). The probable origin of cultivated safflower is the area bounded by the Eastern Mediterranean and the Persian Gulf.

Safflower is grown as edible oil crop in pots, as strip crop, on field borders or around home-steads in the tropical Asia, Africa, Russia and China. Safflower is more drought resistant than other oilseed crops and can yield well on drylands, exploiting sub-soil moisture. (Lovelli *et al.,* 2007). Safflower is also salt tolerant and can be grown on soils affected by some degree of salinity (Bassil and Kaffka, 2002).

The safflower plant has a history of long domestication, initially for orange red dye obtained from florets of the plant. The plant was identified as growing in Egypt in 2000 BC, introduced from Euphrates (Weiss, 2000). Safflower oil has unique variation in fatty acid composition (Knowels, 1969). In general, linoleic acid is the dominant fatty acid ranging from 70 to 78% in local types. In improved types oleic acid ranges from 70 to 75%.

The procedures of harvesting and preparation of dye from safflower flowers was described by Hasselquist (1762). Until recently, the local eye cosmetic *kohl* was made of shoot from charred safflower plants.

The Arabs believed safflower to have alexipharmic and diaphoretic properties and used it to promote sweating to alleviate fever. Mesua (C. AD. 1000) an expert on contemporary Arab & Greek medicine called safflower plant *cincus,* and that the large white-seeded variety was the most valuable. From a dye giver, safflower graduated to an oil yielder in Russia, after it was introduced there from Irano-Afghanistan area.

Old Sanskrit literature mentions medicinal properties of safflower including the oil used as a purgative. Archaeological safflower was found in Rajasthan which was dated to around 2000 B.C. (Pokharia, 2008). Safflower cultivation was mentioned in a Ptolemaic (325–30 B.C.) papyrus (Monson 2007). In Afghanistan, Pakistan and India, safflower has been in use to produce edible oil. Safflower was introduced into China around AD 200-300, initially to produce dye to be used as cosmetic and for medicinal value.Transgenic safflower has also been developed to produce human insulin from seeds, which provides a cheaper option for meeting the global demand for human insulin (Boothe *et al*., 2010)

Safflower was introduced into Central and South America by Spaniards and Portuguese. Safflower became a major oil seed crop in USA after world war II, when private breeders produced high yielding, high oleic acid containing varieties with greater resistance to diseases (Weiss, 2000). However, safflower suffered a severe competition from rape seed-mustard complex. Since 1975, world production of safflower has declined due to competition by increased planting of sunflower and soybean. Despite stiff competition from soybean and sunflower, individual growers in Australia and elsewhere have found niche markets producing organically grown safflower oil, which sells at a premium locally and for export.

Safflower Plant is highly branched, herbaceous, thistle like annual, varying in height from 30 to 150 cm, generally with yellow flowers. Seeds contain 25-45% oil.

Li, *et al.,* (1995) defined five growth stages of safflower based on organ development, plant growth, dry matter accumulation, photosynthate distribution and metabolism. The plant after emergence remains in rosette stage, wherein the internodes are condensed and corresponding palatable fleshy leaves are produced very closely. These vegetative leaves in early growth were used as leafy vegetable in olden days. The period of rosette stage varies depending on the genotype and the prevailing temperatures. The short duration genotypes take around 30 to 45 days for elongation. On the other hand, long rosette stage ends by 70 to 90 days in late maturing genotypes especially those developed and evolved under temperate climate. Irrespective of early or late maturity, all genotypes take 10 to 20 days more under severe winter conditions as experienced in northern India when compared to traditional southern planes, wherein the winter is short.

In India normally the sown safflower seed takes a week to nine days to emerge and the elongation of the young plant takes around 30 days and 45 to 55 days to full branching. The flowering takes place after 80 days from sowing. The crop will be ready to be harvested from 125 to 135 days depending upon the genotype and the ambiental conditions including temperature and soil type prevailing at the location.

Safflower has tap root system which goes down deep as 2.2 meters (Dajue and Mundel 1996); but most of the laterals spread in the zone of first 30 cm. The root growth gets retarded if there is an impediment or hard layer with pebbles and stony material, which again affects adversely the plant growth. The deep rooted character facilitates safflower plant to draw moisture from the deeper layers of soil, which renders safflower more drought resistant.

Flemmer *et al.*, (2014) presented description of the phenological growth stages of safflower plants from seed germination to harvest maturity according to the codes of the extended BBCH scale, a scale widely adopted throughout the world for the description of growth developmental stages of weeds and crops. This scale uses a two-digit code, where the same code is applied to the same growth stages of different plant species. For each code, a description based on easily observable external morphological characteristics is given.

The stem is cylindrical and stiff with a thick base. Stem and branches are encompassed with leaves having numerous spines. As the plant grows towards branching, the stem becomes thinner, smooth, glabrous of light grey or light green or white in colour with fine longitudinal grooves. The stem turns brittle as the plant matures. Stem anthocyanin pigment has been identified as a marker for breeding experiments. Stem girth of mature plant (ranging from 3 to12mm) is one important character indicating productivity (Weiss, 2000). After the emergence, the plant enters a short rosette stage with a bunch of leaves on the soil surface. In the case of other species of *Carthamus*, the rosette stage is prolonged unlike the case of cultivated safflower.However, the rosette phase of safflower gets prolonged in the event of occurrence of very low temperatures as in the case of northern hemisphere following autumn season. Safflower plant registers a rapid growth when once the stem elongation starts.

Li *et al.* (1995) reported the dry matter accumulation in different plant parts under the conditions prevailing in China. The leaves exhibited 87% dry matter accumulation at rosette stage and. 50% in stem at elongation stage as the dry matter gets translocated to reproductive organs from the flowering stage onwards. Use of growth regulating hormones like gibberellic acid at seedling stage could be counter-productive resulting in lodging and reduced reproductive growth

(Potter *et al.,* 1993). Lodging is one serious problem in safflower, which may occur in thin-hulled genotypes which have weak stem due to reduced secondary wall-thickening of stem. Use of TIBA and chlormequat increased seed yield of safflower in India as reported by Deotale *et al.,* (1996).

Height of plant varies (50-200cm) as per the genotype and also influenced by the environmental conditions prevailing. The tallest genotypes are from Turko-Afghanstan area and the shortest varietal forms are from India. The main stem at 15-20 cm height to give rise to secondary stems. The secondary stems in-turn produces branches, each branch terminating in a flower head. Since the attribute of number of effective capitula which is contributed by branch number is strongly correlated with seed yield (Jadhav *et al.,* 2018), it is needed to enhance the effective branching of safflower plant to secure higher yield. It is reported (Weiss, 2000) that removing shoots just before flowering may increase the branching, the number of effective heads (seed heads) per plant and total seed yield. The excised tips are used as vegetable (Weiss,2000).

There exists a considerable variation for leaf size and shape among varieties of safflower. The leaves are long, large and deeply serrated in the zone of lower stem; but they become short, stiff and ovate to obovate at the capitulum. At the inflorescence, the leaves get superimposed as the involucral bracts. According to Zope *et al.,* (1994), a relationship may exist between the total area of outer involucral bracts, seed yield and moisture utilization. The lower leaves may not have spines, which are present on the upper leaves. The extent of spiny nature of leaves varies with the genotype. Often spiny attribute may offer problems at harvesting when the process is not mechanised, particularly in the case of small and marginal farmers. Nevertheless, with the availability of mechanical harvesters on hire in most of the safflower growing regions, even groups of small farmers tend to opt for mechanised harvesting.

The degree of spiny nature expressed as spine index is measured by the mean number of spines and the average spine length on the outer involucral bracts (OLB) (Claassen *et. al.*,1950). In most non-spiny genotypes, the size of OLB is markedly reduced and the number of staminate flowers is increased (Ramachandram and Goud, 1982). However, the weak association of spine index with either seed yield or oil content (Claassen *et al*., 1950; Ashri, 1975; Rao and Ramachandran,1977 and Ramachandram and Goud, 1983) suggests the possibility of developing high yielding and/high oil spineless varieties. Two non-spiny varieties: CO 1 from Tamilnadu and NARI-NH-1 (PH-6) in 2001 have been released for cultivation in India, These have not become very popular for their high susceptibility to safflower aphids and high proportion of staminate flowers resulting in low seed number per capitulum as reported in CO 1 by Ramachandram and Goud (1982).

Non-spiny varieties of safflower namely, JSF-97, JSI-73, JSI-7, NARI-6, PBNS-40, have been developed and released under All India Coordinated Project on Safflower for cultivation especially to spread the crop in non-traditional areas. However, non-spiny varieties yield around 25-30% less as compared to the spiny varieties.

The inflorescence of safflower is head or capitulum of the family Asteraceae or Compositae. Ramachandram and Goud (1982A) observed two types of florets: staminate florets located in the outer one or two whorls and perfect florets in all other inner whorls. Total number of florets per flower head varied from 60 to 224 in ten select genotypes representing considerable variation in flower head size. Staminate florets constituted 21.5 to 43.2% of the total florets. The floral structure of staminate floret is similar to that of perfect floret except empty ovary sac in the former. The florets of the inflorescence are bound closely together on a receptacle, with varying floret numbers, ranging from 20 to 180. The base of the inflorescence forming the receptacle is framed by several layers of involucral bracts, with the spined outer whorl. The individual florets possess small hairy bracts. The entire structure of involucre is closed with an opening through which the corolla tubes of the florets project out.

The corolla colour varies from whitish yellow to red-orange, deep yellow being the most common colour. The corolla yields two kinds of materials viz., (a) water soluble yellow pigment, Carthamidin and (b) an orange red dye Carthamin, the colouring material, which is soluble in an alkaline solution. Fukushima *et al*., (1997) have developed method of identifying changes in floret colour and red compounds employing a spot scanning technique.

Flowering of safflower begins at the head margin and proceeds centripetally, which takes around 3 to 5 days to complete. The total flowering of all the heads may take 10-40 days, which is specific to each variety. Incidence of higher temperatures during the early growth phases accelerates flowering. The delayed planting also reduces the duration of flowering by 25 to 40 days. Among other environmental factors, soil salinity reduces the number of flowering capitula and consequently reduced seed yield. The temperatures of the order of 24-32°C are favourable for safflower growth and development.

Although safflower is considered self-pollinated, bee activity results in enhanced seed yield. Before flowering the stigma is enclosed by five fused anthers, which are attached by a short filament to the tip of the corolla tube. Florets elongate by sunrise before flowering during the day. The anther dehiscence takes place soon after the style emerges through anther tube and the pollen form a column along the style on the top of the anther tube up to the base of stigma. In that process, usually most of the florets get fertilized by the pollen of the same floret or the other florets of the same capitulum. The extent and the time of anther dehiscence are genotype dependent. Ramachandram and Rao (1984) have observed considerable variation in seed setting on covering the flower heads in butter paper bags before anthesis in a set of cross combinations. Detailed studies in the subsequent crop season have led to the identification of partial anther dehiscence being the major reason for poor seed setting as low as 7 to 9% in certain genotypes on bagging the flower heads ahead of anthesis. There was little or no difference between bagged and open pollinated flower heads in case of genotypes, which had complete anther dehiscence. In most of the genotypes, the anther dehiscence takes place soon after almost simultaneously with stigma emergence from the anther tube. In some genotypes, the anther dehiscence takes place by releasing a small quantity of pollen at the tip of the anther tube quite some time up to an hour or more after stigma emerges. In some other genotypes like thin hull mutants, anther dehiscence is almost negligible and the fertile pollen remains in the anther tubes.

Dehiscence of anthers occurs soon after sunrise, when the tip of the anther column releases the pollen, as the stigma emerges through the anther tube. The elongation of the style pushes the stigma, which is like a brush, through the anther tube beyond the tips of anthers. Consequently, the stigma gets covered by the pollen of the same flower. Based on this process, a technique of mass emasculation is developed at NARI, India (Singh *et al.,* 1995).

The major pollinator of safflower is bee species *Apis millifera,* which is attracted by the abundant pollen and nectar of safflower plants. Higher temperatures, however, reduce the hours of bee visits which is reflected in the reduced out-crossing (Ahmadi and Omidi, 1997) and consequently lesser crop yields.Classen (1950) examined outcrossing rates for safflower plants grown either with or without insect exclusion cages. Depending on the cultivar, uncaged plants had outcrossing rates averaging 8.2–35% (range 6.3–58%).

Anthesis of disc florets takes place between 0530-0800 hrs. The number of pollen grains produced per anther and per flower is 236+74 and 5906+372, respectively. The stigma remains receptive up to 32-56 hours after anthesis (Kumari 2008). The unpollinated, elongated stigma may remain receptive for several days. Bees, bumblebees and other insects seek out safflower blossoms for both pollen and nectar and can increase levels of outcrossing.

Wind-pollination does not contribute to safflower seed set. Absence of pollinators results in self-pollination. Developed capitula contain 15-30 or more achenes which mature from 4 to 5 weeks after flowering where about 85.99% seed set achieved in open pollinated heads followed by 38.15% in muslin cloth and 35.54% in butter paper bagged flowers (Kumari, 2008).

A small study in Germany found the level of outcrossing between plots of safflower ranged from 0–33%, with averages of 6.5–18% depending on the location of the sampled plants (Rudolphi *et al.,* 2008).

According to Nabloussi *et al.,* (2013) the observed average outcrossing rate in safflower under Moroccan conditions was 26.6%, ranging from 8.3 to 53.0% at the plant level, and from 0 to 79.2% at single-head level.

The fruit is achene, which is called seed. It is more fibrous due to hull than sunflower. The testa is generally cream coloured off-white. Sometimes grey mottled types too occur. A black or brown testa is an indication of out-crossing with wild species. The seed-coat consists of two brown epidermal layers, with the layer of parenchyma. The endosperm is normally of only few cells in thickness. In the case of thin hulled type, the outer cells of sclerecnchyma are very thin and the melanin layer is visible (Weiss., 20000) producing a grey or brownish colour of seed. The period to anthesis and days to maturity have a bearing on seed filling. In general, seeds reach maximum dry weight, oil content, germination percentage, iodine value and minimum hull percentage four weeks after flowering (Zope *et al.,*1994; Kumar, 1996)

Seed size is specific to each variety. The general proportions of seed components are seen as 25-45% of hull, 55-65% of kernel and 35-60% of oil (Weiss, 2000).

Origin, Species Relationship and Cytogenetics

Based on the genetic diversity in respect of *Carthamus* species prevailing in different geographical zones of Asia and Africa, Vavilov (1951) proposed six centres of origin, which are presented below.

Centre 1: China, Japan, and Korea

Centre 2: India, Pakistan & East Pakistan (Bangladesh)

Centres 3 and 4: Afghanistan to Turkey, southern USSR to the Indian Ocean

Centre 5: Egypt's zone of northern Aswan High Dam. and Sudan bordering the Nile in northern Sudan and southern Egypt

Centre 6: Ethiopia and

Western Part of the Centre 5: Spain, Portugal, France, Italy, Romania, Morocco, and Algeria.

Considering the common features among the safflower variability of the above centres Knowles (1969A) termed them as 'centres of similarities'.

Classification of the genus *Carthamus* L. remains incomplete. However, there are comprehensive reviews by Ashri and Knowles (1960), Hanelt (1961), Weiss (1971) and Estilai and Knowles (1978), describing many species of the genus in detail. Cultivated safflower, *Carthamus tinctorius* L., has 12 pairs of chromosomes (2n=24), together with *C. oxycantha* M.B., *C. palaestinus* Eig., and *C. flavescens* Spreng. The first three can be readily crossed, which yield fertile hybrids. Isoenzyme patterns have confirmed close genetic relationships between

C. tincotrius and *C. oxycantha.* Other species have 2n=20, 22, 44 and 64. Some could be crossed with safflower successfully and used to produce interspecific and intergeneric hybrids. Tetraploids of *C. tinctorius* have been induced using certain chemical treatments and irradiation.

As per the evolutionary history of *C. oxycantha* and *C. flavescens*, it appears that their origin and domestication coincided with the adoption of settled agriculture by nomadic man (Weiss, 2000). High rate of cross transferability of SSR markers in *C. oxyacantha* (97%) followed by *C. palaestinus* (87%) has supported homology between these species and their possible contribution to the safflower genome (Ambreen *et al.,* 2015). Sehgal *et al.,* (2008) suggested on the basis of chloroplast diversity that either of *C. oxyacantha* or C. *palaestinus* or both were to be progenitors of cultivated species. The recent studies on species-genomic relations indicated that *C. palaestinus* was genetically the closest species to the cultivated species (Chapman and Burke 2007; Sasanuma *et al.,* 2008; Sehgal *et al,* 2009; Agrawal *et al.*, 2013).

Carthamus palestinus, on the other hand is a wild desert species, ancestral to rest of the species of *Carthamus*. (Weiss, 2000). Cytogenetic studies of Imrie and Knowles (1970) and Khidir and Knowles (1970) to suggest that *Carthamus palaestinus* Eig, a self-compatible wild species restricted to the deserts of southern Israel and western Iraq (Zeven and Zhukovsky 1975), with white and yellow flowered forms, is the progenitor from which derived the weedy species *C. oxyacantha* Bieb., a mixture of self-compatible and self-incompatible types, and *C. persicus* Willd., a self-incompatible species. These in turn are considered the parental species of the cultivated species *C. tinctorius* L. (Ashri and Knowles 1960).

A gene bank of *Carthamus* is maintained at the Washington State University. At Solapur in India a World Germplasm Collection under the aegis of the Indian Institute of Oilseeds Research of ICAR is also being maintained.

Ashri and Knowles (1960) attempted a ***classification of the species*** of the genus *Carthamus* on the basis of their chromosome numbers among species of *Carthamus.* Among the twenty-five (25) species of *Carthamus* so far identified in different geographical regions, it is only safflower (*Carthamus tinctorius* L.) that is widely cultivated and possesses chromosome number of 24, which form regular 12 bivalents in meiotic metaphase (Richharia and Kotval, 1940; Ashri and Knowles, 1960; Kumar *et al.*, 1981). On the basis of 4 different chromosome numbers viz., 2n = 20, 24, 44 & 64, the genus was categorized (Ashri and Knowles, 1960) into four sections with the differing basic number of 10 and 12. This is different from the work done by Darlington and Wylie (1956) who proposed the basic chromosome numbers of 8 and 12.

The classification as per Ashri and Knowles, (1960) as presented by Dajue and Mundel (1996) is as follows:

Section 1 (2n = 24): This section includes the annual species *C. tinctorius, C. palaestinus* and *C. oxyacantha.* All three species cross readily and produce fertile hybrids, show high homology between genomes and hence are considered related (Ashri and Knowles, 1960). The possibility of natural gene flow between *C. tinctorius* and its wild relatives *C. palaestinus* and C. *oxyacantha* appear to be very remote since they are geographically isolated inhabiting diverse ecological zones of globe and growing in different seasons. *Carthamus oxyacantha* occurs in the zones of north-western India to Iraq, while *C. palaestinus* is found in southern Israel alone.

Carthamus tinctorius is cultivated in diverse locations of India, in parts of Pakistan, northern and central Iran and also in pockets of Jordan, Syria, Turkey, and Israel (Ashri and Knowles, 1960). *C. oxyacantha* is suggested to be the wild ancestor of the cultivated safflower (Bamber, 1916; Deshpande, 1952; Ashri and Knowles, 1960).

Ashri and Knowles, (1960) described the species of section-1 *as pubescence minor or none; outer involucral bracts green, ovate to linear; inner bracts entire at apex; florets not saccate; corollas yellow, orange, red, or white; pollen grains yellow; pappus none or chaffy.*

Section 2 (2n = 20): Section 2 consists of C. *alexandrines C. glaucus, C. syriacus* and *C. tenuis*, inhabiting the eastern side of the Mediterranean Sea. They possess blue or pink flowers, and there exists morphological similarity among *C. alexandrines, C. glaucus* and *C.syriacus.C. glaucus* are distinct morphologically from the others due its larger head and bracts that are ovate rather than linear.

Other species in this group are *C. boissier* Halacsy, *C. dentatus* Vahl. (representing genome, A 1 A 1), *Carthmus leucocaulos* Sibth. and Sm. (possessing genome, A A2), *Carthamus glaucus* subsp. *anatolicus* (Boiss.) Sam. subsp *glandulosus* Han., *C. ambiguus*

Ashri and Knowles (1960) indicated that the species of sections 1 and 2 are not closely related. Hybrids have been produces without much constraints by crossing the species of the sections 1 and 2; but were sterile due to lack of homology between their chromosomes. Consequently, there cannot be any gene flow between the said species.

Section 3 (2n = 44) There is a solitary species viz., *C. lanatus,* (2n = 22) in this category, with natural habitat in Portugal, Spain, Morocco, Greece, and Turkey. As observed in respect of the previous section, *C. lanatus* when crossed with the members of Section 1, exhibited least level of chromosome homology between the chromosomes of the parents of different sections. This observation ruled out the earlier supposition that *C. lanauts* evolved through hybridization between members of Sections 1 and 2, followed by chromosome doubling. On the other hand, C. *lanatus*, when crossed with members of Section 2, the hybrids exhibited s good pairing with the chromosomes of the members of section 2, thereby indicating that some species of Section 2 would have contributed 10 pairs of chromosomes to the genome of *C. lanatus* (Ashri and Knowles, 1960).

Section 4 (2n = 64): There are two species in this group viz., C. *baeticus* (Boiss and Reuter) Nyman and C. *turkestanicus* M. Popov. One of the two viz., *C. baeticus* is allohexaploid possessing (i)one genome of 12-chromosome and (ii) two different non-homologous genomes of 10 chromosomes. The hybrids of *C. baeticus* × *C. lanatus* demonstrate pairing of 22 chromosomes, indicating that *C. lanatus* is a progenitor of *C. baeticus* as per Ashri and Knowles, (1960). *C. glaucus* with 2n = 20 chromosomes is considered the progenitor of *C. turkestanicus*. The zonation of *C. baeticus* is confined to eastern Mediterranean, North Africa, and Spain. On the other-hand, the habitat of *C. turkestanicus* is around west Asia, east of Kashmir, and Ethiopia. The complement of 22 pairs of chromosomes of the species is as in the case of *C. baeticus* with white coloured pollen. The species similarity to *C. lanatus* in its phenotype indicates gene exchange between the two species as reported by Khidir and Knowles, (1970).

***Carthamus* species with 2n = 22**: *C. divaricatus* (Beg and Vace.) Pamp., is the lone species with 11 bivalents, inhabiting a very restricted zone in Libya (Knowles, 1988). The species is recognizable with yellow, purple, or white flowers with yellow pollen. The species is reported to be self-incompatible (Knowles, 1988), but reported to be cross-compatible with species of 10 pairs of chromosomes, giving raise to partially fertile hybrids. Though it is reported to hybridize with *C. tinctorius*, the resulting hybrids are sterile.

Other species of *Carthamus* with 2n=24: In this category there are *C. arborescens*, *C. caerules*, *C. rhipaeus* and *C. nitidus* each with 12 pairs of chromosomes, having distinct set of morphological attributes.

The above species are not cross-compatible with any otherspecies of *Carthamus* as reported by Ashri and Knowles (1960). This is the basic reason for these species not being clubbed with the other species of *Carthamus* of the other four sections.

(i) The habitat of *C. arborescens* is reported to be southern Spain and its adjoining areas of North Africa.

(ii) *C. caeruleus* is inhabiting Iberian Peninsula which includes Spain and Portugal and also North Africa. The species is much distinct from *C. arborescens* and other species of *Carthamus*.

(iii) *C. rhiphaeus* Font Quer, the third species with12 pairs of chromosomes, is morphologically closely related to *C. arborescens* (Ashri and Knowles, 1960). It is found in a small area of northern Morocco.

(iv) *C. nitidus* Boiss, which was considered to possess 20 chromosomes by Ashri and Knowles (1960), has been reclassified subsequently as having 12 pairs of chromosomes (Lopez-Gonzalez, 1990). This species has been grouped into a separate section due to its isolation from other species.

Reclassification of the genus, *Carthamus*

Reclassification of the genus *Carthamus* has been carried out by Lopez-Gonzalez (1990), who proposed a new classification system for genus based on anatomical characteristics, biogeographic distribution, and bio-systematic information.

The proposed classification consists of two sections viz., Section *Carthamus* with 24 chromosomes and Section *Odontagnathius* (DC.) Hanelt (including section *Lepidopappus* Hanelt) having 20 or 22 chromosomes.

In the new classification system, genera *Carthamus* and *Carduncellus* are grouped along with two new genera, *Phonus* and *Lamottea.* The species of *Phonus, Lamottea,* and *Carduncellus* are perennials with 24 chromosomes. The genus *Carthamus*, on the other hand, is an annual including species with 2n = 20, 22, 24, 44, and 64 chromosomes (Singh and Nimbkar, 2007).

The molecular classification of the genus *Carthamus* species with two sections has been proposed by Vilatersana *et al.,* (2005) consisting of two sections viz., (a)Section *Carthamus*: similar to the one proposed by Lopez-Gonzalez (1990) and (b) Section *Atractylis*, which includes sections *Atractylis, Lepidopappus*, and *Odontagnathius* as indicated in the classification suggested by Hanelt (1961). The controversy in the above system is that in the redefined section of *Atractylis*, the inclusion of *C. nitidus* having chromosome number n = 12, as the rest of the members of the section have n = 11 and n = 10 (Singh and Nimbkar, 2007).

Cytogenetics of *Carthamus* species: The cytological investigations in *Carthamus* that were carried out during the decades of 1960s revolved around identification and establishing the chromosome numbers in different species of the genus and producing the inter-specific hybrids to study the chromosomal homologies ((Ashri and Knowles, 1960; Hanelt, 1961; Schank and Knowles, 1964; Harvey and Knowles, 1965; Khidir and Knowles, 1970; Estilai and Knowles, 1978; Estilai, 1977). These investigations did not throw much light on the location of specific genes on chromosomes and genetics pertaining to certain characters of importance.

The first step in this direction was perhaps was taken by Estilai and Knowles (1980) who tried to relate certain genes located on specific chromosomes through the aid of aneuploids, that occurred following interspecific crosses. Thus, they identified one primary and one secondary trisomic in the progeny of an open pollinated triploid plant and used them for assigning genes to chromosomes in safflower. The morphological dissimilarity of the aneuploids from diploids could not be related to the attributes of the missing or extra chromosomes of the aneuploids in question mainly because progenies from open-pollinated plants had diverse genetic background (Knowles and Schank, 1964).

Kumar *et al.* (1981) reported three satellite chromosomes in *C. tinctorius*, compared to one reported by Knowles and Schank (1964). Karyotype analysis in depth including measurements of chromosome length and arm ratios were carried out (Kumar *etal.,*1981*)* in order to identify translocation homozygotes in safflower. This study resulted the identification of 10 translocation homozygotes out of 42 translation heterozygotes isolated in the M1 generation of gamma-irradiated populations. The chromosomes involved in interchange homozygotes were 6-8 (line 58), 1-3 (line 66), 3-12 (line 123), 3-10 (line 131), 3-6 (line 153), 4-6 (line 186), 4-8 (line 197), 3-8 (lines 208 and 279), and 5-9 (line 290). Chromosome 3 (sat-chromosome) was involved in 6 of the 10 translocations, and thus the segment interchanges were not random (Pillai *et al.*,1981b).

The chiasma frequency per bivalent was 1.48 for section I (2n = 24) (Ashri and Knowles, 1960), 1.59 for section II (2n = 20) (Schank and Knowles, 1964), 1.95 for section III (2n = 44) (Ashri and Knowles, 1960), and 1.78 for section IV (2n = 64) (Khidir, 1969).

The genetic data together with the linkage maps are not available in safflower due to the non- availability of karyological characterization of individual chromosomes (Singh and Nimbkar 2007). Singh V., *et al*., (2005) reported existence of polyembryony in safflower strain D-149. The genotype possessed the tendency to produce twin seedlings at the rate 0.6 to 10%, of which 30% were male sterile assumed to be due to changed ploidy level.

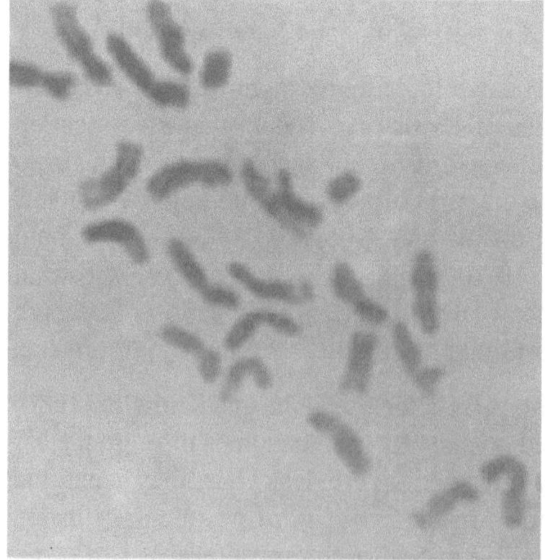

Fig. 1: Somatic chromosomes of safflower
(Sujatha and Dutta Gupta, 2013www.researchgate.net)

Biotechnology of *Carthamus* spp.

(a). Molecular Cytogenetics of *Carthamus* spp.: Molecular Cytogenetics involves the combination of molecular biology and cytogenetics; using various reagents, molecular cytogenetics seeks to study cytological phenomena. It encompasses all aspects of chromosome biology and the application of molecular cytogenetic techniques in all areas of biology including structural and functional organization of the chromosome and nucleus, genome variation, expression and evolution, chromosome abnormalities and genomic variations. Role of Molecular Cytogenetics in resolution of phylogenetic problems in safflower is demonstrated by Raina *et al.,* (2005) who isolated two novel repetitive DNA sequences, pCtkpn1-1 and pCtkpn 1-2 of 343-345 bp and 367 bp, respectively with 37% sequence similarity between the two. Florescence *in situ* hybridization (FISH) on metaphase chromosomes of safflower using the two sequences indicated exclusive localization of pCtkpn1-1at telomeric regions of 22 and 24 chromosomes, while pCtkpn-2 clone is distributed on nucleolar and non-nucleolar chromosome pairs. The variability in the number, size and location of the two repeated sequences provide identity to most of the chromosomes. Raina *et al.,* (2005) were the first to apply FISH technique to *Carthamus* spp. This technique would be particularly useful to study phylogeny and evolutionary pathways in *Carthamus* spp.

In order to clarify the phylogenetic relationships among species of *Carthamus*, Sasanuma *et al.,* (2008) performed sequence analysis of the nuclear stearoyl acyl carrier protein desaturase (SACPD) gene and the chloroplast intergenic spacer region between leucine and phenylalanine tRNA genes (*trn*L–*trn*F IGS) in 13 taxa of *Carthamus*. According to them, the previous division of the genus into 4 taxonomic sections and allocation of particular genomes to various taxa on the basis of morphological, cytogenetic, and biosystematic analyses does not draw support by the data presented in their study. Sasanuma *et al.*, (2008) provide evidence of the occurrence of 5 nuclear genomes (A, B, C, X, and Y) and 3 cytoplasm types (A, B, and C) in the genus *Carthamus*. The cultivated safflower, *C. tinctorius* ($2n$ = 24), has the B genome and type B cytoplasm. Both of these are not present in the polyploid taxa, which contradicts the earlier view that one of the genomes involved in the origin of the polyploid taxa of *Carthamus* is the B genome. Comparison with an outgroup species (*Cirsium japonicum*) indicated that *C. arborescens* is the most primitive species in the genus. As per their results, *Carthamus palaestinus* is suggested to be genetically the closest to the cultivated safflower.

(b) Tissue Culture: Safflower tissues are amenable for manipulation through *in-vitro* cultures. In the initial phase of the decades of 1970s and 1980s research efforts were in the direction of developing suitable culture conditions for whole plant regeneration. With refinements of the techniques, the regeneration frequencies were high, and regeneration was effected through embryogenesis and organogenesis pathways. (George and Rao, 1982; Tejovathi and Anwar,

1987, 1993; Sujatha and Suganya, 1996; Nikam and Shitole, 1999; Walia *et al.*, 2005). Direct as well as callus and shoot mediated root regeneration systems through organogenic and embryogenic pathways are reported for spiny and non-spiny genotypes of safflower (Sujatha, 2007). Among the various basal media tried the Murashige and Skoog salt medium (Murashige and Skoog, 1962) was found promising for induction of morphogenic response in somatic as well as gametic tissues (Prasad, B.R. *et al.*, 1991, Chatterjee and Singh, 1993) Anthers of safflower too exhibited morphogenic potential on medium supplemented with 2.0 mg/l of BA alone or in combination with 0.5 mg/l of NAA, through which haploids were recovered at a frequency of 64% (Prasad *et al.,* 1991). The Use of a combination comprising TDZ+NAA has yielded the best results of regeneration in terms of responding explants, response from all tissue types, tested genotypes and number of shoots per responding explants in the case of both American and Indian genotypes (Orlikowska and Dyer, 1993; Radhika *et al.*, 2006).

Rooting of the regenerated shoots, which has been the problem in safflower, could be resolved by increasing the sucrose to a level of 9% in the rooting medium (Tejovathi and Anwar 1984, Nikam and Shitole, 1993) and 1 mg/l each of NAA and riboflavin (Orlikowska and Dyer, 1993). Nevertheless, rhizogenesis depends to great extent on the genotype, shoot quality, medium to which shoots are habituated and period of culture.

Tejovathi and Anwar (1987) and Radhika *et al.*, (2006)were able to induce flowering in tissue culture medium characterised by the production of capitulum, which depends upon the kind of growth regulators, their concentration employed and the genotype used.

The progress achieved in safflower tissue / cell culture has resulted in the following, which could be of considerable advantage in safflower breeding and production (Sujatha, 2007):

- Protocols for *in vitro* multiplication using shoot apices are available, which could be extended for the multiplication and conservation of wild safflower species and rare and quality germplasm lines.
- The phenomenon of *in vitro* flowering could be exploited towards recovering interspecific hybrids of *Carthamus* through *in vitro* fertilization and in overcoming asynchronous flowering.
- Protocol for somatic embryo genesis is developed (Mandal *et al.*, 1995), which will be helpful in enhancing the frequency of plantlet conversion and also for use in genetic transformation.
- Cell cultures of safflower have been used in the production of Kinobeon A (Wakayama *et al.*, 1994), which is useful in protecting biological systems from oxidative stress (Kanehira *et al.,*2003).

- Plant cell cultures can also be used for the production of red and yellow pigments from safflower flowers, which are of commercial value.

(c)Transformation studies in safflower were limited to constructs harbouring commonly used reporter (uidA, GFP) and selectable marker genes (nptII) and such studies indicated influence of co-cultivation conditions, *Agrobacterium tumefaciens* strain, type of explant, genotype, etc. Transgene expression was confirmed with transient assays by GUS expression and molecular analysis through PCR and Southern hybridization assays of the primary transformants (Sujatha and Dautta Gupta, 2013).

Notwithstanding the high frequency of shoot regeneration from transformed tissues (15–34.3 %), rooting of transformed shoots is yet to be resolved. Till recently safflower has been used to produce valuable plant made pharmaceuticals through agro-infection. However, the tissue culture based transformation protocols are yet to be exploited for the development of transgenics with agronomically desirable traits (Sujatha and DuttaGupta, 2013).

(d) Genomics of Safflower:*Sem-Biosys* has started a programme on safflower genomics with the objective of generating safflower BAC genomic library and seed EST library, isolation and characterization of olesin and other seed storage protein genes and also genes involved in lipid metabolism and identification of high expressing seed specific promoters (Sujatha, 2007). Garnatje *et al.*, (2006) assessed the nuclear DNA content of 16 *Carthamus* species with objective of extending knowledge base on interspecific variations and its implications for infrageneric classification of genus *Carthamus,* in addition to assessing the variation in plant genome size in relation to allopolyploidization. The 2C values ranged from 26 pg for *C. leucocaulos* to 7.46 pg for *C. turkestanicus,* while the values for monoploid genome ranged from 1.13pg for *C. leucocaulos* to 1.53 for *C. alexandrines.* The 2C values for cultivated safflower 2.70pg. In the case of the allopolyoloid species, the nuclear DNA content was as per the exact sum of that of the putative parental genomes in the case of *C. creticus.* On the other-hand in the case of C. *tercastnicus* there was a slight reduction as compared to that of *C. creticus.* The allopolyploid species, *C. creticus* and *C. turkestanicus* have originated from the hybridization of a tetraploid ancestor, *C. lanatus* with a diploid progenitor lineage, *C. leucocaulos* and *C. glaucous*, respectively (Vilatersana *et al.* 2007). Mayerhofer *et al.* (2010) have developed genetic linkage maps of *C. tinctorius* and *C. oxyacantha* based on SSR and RFLP markers.

High genetic diversity was reported in the cultivated safflower using several molecular markers *viz.,* RADP, AFLP, SSR, ISSR and SNPs (Johnson *et al.*, 2007; Yang *et al.*, 2007; Khan *et al.*, 2009, Sehgal *et al.*, 2009; Chapman *et al.*, 2009; Ambreen *et al.*, 2015). Identification of marker-trait association is necessary for successful marker-assisted breeding. Ambreen *et al.* (2018) have developed a large set of novel microsatellite markers for safflower using next generation sequencing. These can be used for diversity analysis, synteny studies, construction of linkage maps and marker-assisted selection. Markers associated with seed oil content, oleic acid content, linoleic acid content, 100-seed weight, plant height, number of primary branches, days to 50% flowering and number of capitula per plant have been identified. García-Moreno *et al.* (2011)could tag candidate gene of the high gamma-tocopherol in safflower, and suggested the role of the gamma-TMT gene in determining high gamma-Tocopherol content in safflower.

(e) Safflower's role in Commercial Biotechnology: The technological merits of safflower in this context are as follows (Sujatha, M., 2007).

- Safflower is easily transformed with *Agrobacterium.*
- High levels of recombinant proteins in safflower seeds
- Safflower is amenable for large scale production and purification
- Safflower is non-allergic and
- Possesses a high degree of self-pollination.

A Canadian Plant Biotechnology Company viz., ***Sem-Biosys*** – Genetic Incorporation (Sujatha, 2007) has formed a strategic partnership with several companies to utilize safflower as production vehicle for various pharmaceutical, nutraceutical and cosmaceutical applications, towards generating products such as human insulin, apolipoprotein A1, growth hormones, healthy fatty acids, antibody production and safflower oil-bodies for use in cosmetic industry(http://www.isb.vt.edu/articles/oct0605.html).

Sem-Biosys has commercialized oil-body –oleosin technology. Safflower is found to be emendable for molecular farming (Ganapathy, 2016). Nykiforuk *et al.*, (2011) have established that high levels of biologically functional ApoAI(Milano), which can be produced using plant-based expression system in safflower.

Safflower Genetic Resources

The Collection, evaluation, categorization and preservation of diversity of safflower genetic resources assume tremendous importance in the present situation so as to meet future challenges associated with climate change, disease evolution, and the increasing needs of a growing population.

Safflower with its numerous and varied uses (Dajuae and Mündel,1996), has benefited from the diversity of genetic resources conserved and distributed by gene-banks. A germplasm directory for safflower was compiled by Zhang and Johnson (1999) which documented 18 different collections in 14 countries. (http://safflower.wsu. edu/). India reported the largest collections with nearly 10,000 total accessions held at both the National Bureau of Plant Genetic Resources in

New Delhi (2,393 accessions) and ICAR-Indian Institute of Oilseeds Research, Hyderabad (7,525 accessions). Other significant collections are in China, Mexico, and the USA. The US safflower collection was developed starting in the late 1940s and is at Pullman, Washington,(http://www.ars.usda.gov/main/site_main.htm?modecode=53481500). It now includes more than 2,400 *C. tinctorius* accessions. The US collection is represented by germplasm from more than 50 countries, and accessions are available to scientists worldwide. Data on many descriptors have been gathered on a large number of accessions in the collection and have been entered into the Germplasm Resources Information Network (GRIN), and this information is easily available to Internet users (Bradley and Johnson, 2001). The core collection was evaluated for seven quantitative factors, which indicated considerable diversity within the core collection (Johnson *et al*,.2007) Correlation analysis showed the strongest association between plant height and flowering and also the association between outer involucral bract width and length. The evaluation of accessions as per their places of origin, viz., the Americas, Australia, China, East Africa, Europe, Japan, the Mediterranean, South-Central Asia, Southwest Asia, and Thailand, revealed significant differences among regions for all factors except outer involucar bract length and yield per plant (Bradley and Johnson, 2001)

The National Active Germplasm Site for safflower (*Carthamus tinctorius* L.) of India is located at the ICAR-Institute of Oilseeds Research, Hyderabad, India. The germplasm collection includes indigenous as well as exotic accessions representing most of the safflower growing regions of the world. The collection has been characterized and evaluated for 30 to 59 descriptors and wide variability has been recorded for some of the important traits viz. days to 50% flowering, plant height, number of primary branches and capitula, capitula size, seed weight and oil content. Multilocation testing and evaluation of the promising accessions for different cropping situations is in progress. These include early sowing, late sowing in rice fallows, minimal irrigation and salt affected areas. Accessions possessing resistance/tolerance to important biotic stresses viz. aphids, *Alternaria* leaf spot, *Fusarium* wilt have been identified. The collection is being submitted for conservation in the Long Term Storage Facility (-20°C) at the National Gene Bank, National Bureau of Plant Genetic Resources, New Delhi, India in a phased manner (Mukta, 2008).

Although wide diversity exists in safflower germplasm, the divergence pattern among the regions was not always distinctly distinguishable. No clear relationship was observed between diversity pattern and geographical origin. (Ashri 1975; Khan *et al*., 2009; Elfadl *et al.,* 2010).

Genetic diversity and relationships among 48 safflower accessions were evaluated by Yang *et al.,* (2007) using 22 Inter-Simple Sequence Repeats (ISSR) primers. The results showed that the polymorphism of the safflower germplasm was higher at the DNA level. All the 48 accessions could be distinguished by ISSR markers and were divided into 9 groups based on ISSR GS by using UPGMA method. The genetic relationships among the accessions from different continents were closer. Comparatively, the genetic diversity of the accessions originated from Asia was higher than those assembled in Europe. The results also showed that the genetic variation of accessions from Indian and Middle Eastern safflower diversity centres were relatively higher. ISSR is an effective and promising marker system for detecting genetic diversity among safflower accessions and give some useful information on its phylogenetic relationships.

Kiran *et al.*, (2017) studied a core subset of 148 safflower accessions representing 15 countries, predominantly of Indian origin, for agronomic traits and characterized for genetic diversity, population structure and linkage disequilibrium (LD) using 44 simple sequence repeat (SSR) loci across 11 linkage groups to enable its utilization in breeding and genetic mapping purposes. The collection had substantial variation for seed yield-related traits. SSR allelic variation was low as indicated by average number of alleles (3.6) per locus, gene diversity (0.314) and polymorphism information content (0.284). Cluster analysis (neighbour-joining tree) revealed five major genotypic groups with very low bootstrap support. STRUCTURE analysis showed recognizable population structure; based on membership coefficients ($\geq$ 0.75), 52% accessions were classified into four populations (K= 4) and the remaining 48% accessions into admixture group. High F_{st} values (0.30–0.48) suggested that the populations were substantially differentiated. Analysis of molecular variance results showed that maximum of genetic variation (85%) was explained between individuals within the population suggesting that the population structure was only weak. About 1.9% of SSR locus pairs were in LD, which appeared to be low. High phenotypic variation, mild population structure and low level of LD among unlinked loci suggested that the core subset can be explored for association mapping of seed yield components in safflower.

Genetic variation in world germplasm collections of safflower and its species was evaluated using sequence-related amplified polymorphism (SRAP) and Start codon targeted (SCoT) molecular markers (Peng *et al.*, 2008; Golkar and Mokhtari , 2018). Analysis of molecular variance showed a significant difference across cultivated safflower genotypes possessing a high intra-population variation.

Wild species of *Carthamus* constitute a rich repertoire of genes for biotic and abiotic stresses besides oil quality traits. Despite this, the exchange of genetic material between cultivated and allied species is lacking due to the sterile nature of the hybrids resulting from diverse interspecific crosses (Ashri and Knowles, 1960). This indicates that there has been no exchange of genetic material between cultivated species and the species possessing other than those with 2n = 24 chromosomes. Natural hybrids of *C. tinctorius* and *C. oxyacantha* have been documented in India, Pakistan and Iran (Knowles 1989). However, the possibility of such crossing is very limited, particularly in India and Pakistan, as harvesting of *C. tinctorius* is competed as *C. oxyacantha* starts flowering (Ashri and Knowles, 1960). No instances of natural crossing between *C. tinctorius* and *C. palaestinus* or with the other wild relative with 2n = 24 chromosomes, has been reported, due to their geographical isolation. The introgression of genes from wild relatives to *C. tinctorius* has not received due attention. However, introgression of *phytophthora* root rot resistance gene from a wild composite of 12 species into cultivated safflower has also been recorded (Rubis,2001).

Pre-breeding through intraspecific and interspecific genetic enhancement coupled with marker aided selection will accelerate the breeding programmes. The wild species, *C. oxyacantha, C. palaestinus, C. lanatus, C. creticus* and *C.turkesthanicus* are resistant to *Fusarium* wilt (Pallavi *et al.*, 2007).Anjani *et al.* (2018) have introgressed resistance to *Fusarium* wilt from *C. oxyacantha* and *C. palaestinus* into the cultivated species. Species-specific SSR markers co-segregating with *Fusarium* wilt résistance have been identified, markers marker-assisted selection was practiced using these markers and developed stable wilt resistant interpspecific inbred lines (Anjani *et al.*, 2018).

Alternaria leaf spot (ALS) is one of the devastating diseases in safflower. No reliable resistance source is available in cultivated species repository. Wild species are potential sources of resistance to ALS. Prasad and Anjani (2008) reported high resistance to ALS in safflower wild species, *viz. C. palaestinus*, *C. lanatus*, *C. creticus* and *C. turkestanicus* and moderate resistance in *C. oxyacantha* and *C. glaucus.* Heaton and Klisiewicz (1981) reported resistance toALS in *C. lanatus* and an alloploid derived from (*C. tinctorius* × *C. lanatus*). Rcently Anjani *et al.*, (2019) have successfully introgressed resistance to ALS from *C. palaeastinus* and *C. lanatus* into *C. tinctorius.*

An effective cytoplasmic-genic male sterility system was developed in India through wide hybridization, in which C. *oxyacantha* was donor of sterile cytoplasm (Anjani 2005a)

As a result of utilization of well-characterized germplasm resources, safflower has become an important oil seed crop, bred for specialty niches through the development of healthier or more heat stable oil constituents, winter hardiness, and disease resistance. Molecular methodology has facilitated characterization of the world-wide diversity of safflower and identified geographical regions of similarity to assist breeders in the exploitation of available diversity. The development of molecular markers from expressed sequences should aid researchers in mapping genes of importance and reducing population size and generations required for the development of new varieties by using marker-assisted selection. Sequencing technology has established relationships among species of *Carthamus*, further aiding in the exploitation of diversity within the secondary gene pool. A coordinated, collaborative effort among safflower researchers in the development of marker-assisted characterization of global diversity would further increase the utility of available germplasm resources. (Kisha and Johnson, 2012).

Safflower Breeding

The breeding objectives in safflower might vary depending upon the regional needs dictated by the prevailing agro-ecological conditions. Nevertheless, there are certain common objectives such as breeding for high yield and oil content. The oil quality in terms of fatty acid composition: selection for high oleic acid or high linoleic acid, minimizing or elimination of undesirable odour etc, are important for oil trade.

Earliness of maturity could be another objective for most of the safflower growing zones, the degree which would vary as per the regional requirements including cropping patterns. While breeding for resistance or tolerance against biotic stresses is an important objective, diseases such as *Alternaria* leafspot, *Fusarium* wilt and *pyhtopthera* root rot could be common for many situations.

Safflower germplasm collections have a wide diversity and variability for many agronomic attributes, which provides choice for the selection of suitable parents for recombination breeding. The collections represent reasonably wide diversity for many agronomic attributes including the quantity and quality of oil, resistance to biotic and abiotic stresses, plant type etc, which in fact provided the basic genetic stocks for the crop improvement programs in USA, India, Australia and elsewhere during the last six decades or so. Quite a bit of literature has been compiled on genetic basis of seed and oil yield and their components. Further, the crop has undergone certain degree of genetic improvement in respect of seed yield, oil content, oil type, resistance to diseases etc.

Globally safflower growing area falls under 14 degrees and 22 degrees N in India, 73 degrees to 79 degrees E in USA, between 35 degrees and 45 degrees N in erstwhile USSR and 15 degrees and 35 degrees S in Australia. Safflower is traditionally grown as a post-rainy season crop starting from October to following year February in India mostly by utilizing receding moisture conditions. A number of soil related feature such as soil moisture status, nutrient levels, salinity etc., and climatic factors especially day length, temperature and relative humidity directly or indirectly influence the growth and development of the crop. Safflower can be grown on a wide range of soils varying in depth, available moisture and nutrients. Unless an ideal genotype is identified for a particular crop growing situation, either the traditional or new crop suffers very badly. As an example, safflower has been grown as post rainy season dry land crop in medium deep to deep black soil region of Peninsular India. Over the years their depth is considerably reduced, thereby making them medium (< 90 cm depth) to shallow (< 30 cm depth) and consequently reduced water holding capacity. As a result, the productivity of traditional safflower varieties maturing in 120 to 135 days declined, and the crop has become less competitive when compared to short duration sunflower varieties and hybrids maturing in 85 to 100 days. In North-West, where the soils are deep, slightly saline and the winter is severe, safflower takes over with 150 days' duration with high yields and also 3 to 4% high seed oil content. Almost all tested varieties suffered from frost. The crop holds promise in such situations provided suitable genotypes are developed. Drought tolerance at different crop growth stages, response to mineral nutrition, salt tolerance are the important soil related stresses, which require the attention of safflower breeders.

Although safflower is considered as day neutral crop (Weisss,2000), it has the prolonged rosette stage and consequently the crop duration. It is thermo-sensitive and the ideal temperature for normal growth and development is 15 to 20 degrees in the early crop growth stages and 24 to 32 degreess at and after flowering. Relative humidity is another critical factor for safflower. Prevalence of high relative humidity at grand crop growth period and flowering for a week or more encourages the diseases like leaf spots and wilts. These are some of the climatic factors, which have to be considered while formulating the crop improvement strategies

A mass emasculation technique has been developed for use in breeding programmes (Singh *et al.,* 1995). *In vitro* techniques have also been developed that enable large scale production of selected strains and methods have been reviewed (Palmer and Keller, 1994).

Using principal component analysis, Ali, *et al.*, (2020) found that days to flower initiation, days to 50% flowering, days to flower completion, seed yield per plant, capitula per plant, branches per plant, seeds per capitulum, and capitulum diameter were the major contributors to the observed genetic variability. Seed yield per plant reflected a significant and positive correlation with capitula per plant, branches per plant, and capitulum diameter, and these traits could be considered for inclusion in the selection criteria in safflower breeding

The basic considerations for safflower parental selection could be (1) selection of the parents on the basis of their per se performance for seed yield and other desired traits, (2) consideration of extent of expression of the above mentioned yield components, (3) consideration of usable genetic diversity of the parents to bring desired genes of diverse origin together, and (4) identification of the best general combining parents as well as the best cross combination by following the diallel cross approach (Golkar *et al.,* 2017; Fasahat *et al.,* 2016).

While the above guidelines are indeed useful in parental selection, it is observed that the selection of parents is often based on the seed yield per se performance. It is also known that yield is not only a polygenically controlled attribute; but also an unstable character. Hence, it is needed to analyse the stability of various attributes associated closely with seed / grain yield based on which the selection of the suitable character should be made for parental choice. (Nagabhushanam and Prasad, 1992b &1992c; Prasad, 1992). Ineffectiveness of direct selection for yield in safflower is also reported by Nie *et al.* (1988).

Considerable work has been carried out on gene action for various traits in safflower. While the attributes such as seed yield and the number of capitula per plant are stated to be controlled by non-additive gene action, (Makne and Choudhari,1980; Ramachandram and Goud, 1981 and 1982; Ranga Rao, 1983; Parameshwarappa *et al.*, 1995; Ghorpade and Wandhare, 2001; Gadekar and Jambhale, 2003), the other attributes viz., oil content, seed weight, and seed number are stated to be under additive gene action (Makne *et al.*, 1979; Deokar and Patil, 1980; Ranga Rao, 1982; Mandal, 1990). The importance of both additive and non-additive gene actions for the above traits was revealed by Prakash and Prakash (1993) and Singh V., (2004).

Flower yield in addition to seed yield exhibited the importance of both additive and non-additive gene actions, with non-additive predominant for flower yield in safflower (Singh V., 2004). The major yield contributing traits in safflower are number of capitula per plant, number of seeds per capitulum and seed weight and all these traits are negatively correlated with each other; but each has direct effect on seed yield (Anjani, 2000a). Hence, increase in one yield component would reduce the other components. Ashri (1971) and Narkhede and Patil (1987) have reported that capitula per plant was the most important yield component in safflower. According to Ranga Rao *et al., (*1977) capsule number, capsule weight and hull per cent were the most important components in breeding for higher yield and oil content in safflower. Guan, (2007) reported that the number of effective cones per plant and number of total cones per plant had more direct effect on the seed yield per plant. Both of these attributes were the focus on the high yield breeding and cultivation of safflower varieties. The other important attributes such as seed weight and oil content are negatively associated (Ranga Rao *et al.*, 1977; Mandal, 1990).

While hull content is negatively correlated with oil content; but is positively associated with seed yield (Ranga Rao *et al.,*1977). Hull content and oil content are negatively related while seed weight and hull content are positively associated (Ranga Rao *et al.* 1977; Koutroubas and Papakosta 2010). Reduction in hull content would increase oil content but simultaneously reduce seed yield because of reduction in seed weight.

The magnitude of negative association between seed weight and seed number is approximately equal to that of positive correlation between seed number and oil content (Ramachandram and Goud, 1982A; Patil *et al.*, 1987. A review of the investigations carried out in this context brings out the importance of number of capitula per plant, yield per capitulum, and seed weight for seed yield improvement in safflower.

While giving weightage for the above mentioned characters forming the important yield components in selection of the parents, it becomes necessary to ensure adequate levels of genetic diversity in the selected parents (Joshi 1961) without which the degree of variability which forms the basis for selection will be marginal in the succeeding generations. The genetic diversity among the parental types needs to be considered while selecting the parental material.

Parental selection in respect of safflower needs to be effected based on the multi-location data at least for two years. In the event of developing hybrid varieties too, greater degree of genetic diversity in the parents will help in recovering heterotic hybrids with a greater yield advantage. Also the detailed comparative analysis of the parents and the F1 hybrids, will be useful in selecting the parental genotypes for generation advancement (Singh and Nimbkar, 2007). The population size of F2 generation should be adequately large so as to allow the expression of diverse recombinants in good frequency so as not to miss any valuable segregant.

In addition to the availability of thin or reduced hull types, enhancement of oil content from current 30% to at least 50% should be attained along with the increased seed yield, in order for safflower to compete with sunflower and soybean in oil yield per unit area (Weiss, 2000).

Desirable oil quality could be attained with high oleic acid content. Keeping with the local preferences, Australians have developed high linoleic acid type (76-78%) of safflower with 10-12% of oleic acid (Weiss, 2000). Increasing the seed protein and lysine levels and reducing the fibre content resulted in more attractive cake or meal along with the removal of matairesinol monoglucoside, a bitter principle of safflower (Palter and Lundin, 1970)

Selection for high oil can be carried out by following the thumbnail technique (Sawant, 1990). In this method, the dry seeds obtained from the main capitulum are pressed hard by using thumbnail pressure (Classen *et al.,* 1950). The seeds with high oil content in general contain thin hulls and are thus pressed easily, in contrast to seeds with thick hulls, which cannot be pressed at all with thumbnail pressure. Thus, the selection for high oil types can be done in the field as suggested by Singh and Nimbkar, (2007). The increase in oil content in the United States was attained incrementally by reducing hull content and using single genes such as those associated with striped hull, reduced hull and partial hull (Mundel, 2008). Most of the safflower varieties developed in India have oil contents ranging from 28 to 32%. The efforts in developing high oil varieties of safflower in India have resulted in the breeding of non-spiny variety of safflower NARI-6 with seed oil content of 35% (Singh and Nimbkar, 2007).

Safflower cultivar 'UC-1', released by Knowles in 1968 (Mundel and Bergaman 2009)was essentially the first commercial variety of a new type of safflower oil, a new crop. Its oil properties (high oleic acid) differed in fatty acid composition from that of traditional safflower oil (high linoleic acid) as a result of a mutation discovered in a safflower accession from India. The oil of UC-1 was chemically similar to olive oil. The identification of the novel oil type was made possible only by screening diverse collections of safflower germplasm for variation in fatty acid properties. The Indian accession was crossed in Knowles' breeding program to a Nebraska cultivar (N-10) in 1957 to initiate the incorporation of the novel oil type into an agronomically acceptable background. By 1967, after a backcrossing breeding program, certified seed of the new cultivar was available and it was the foundation for California production for several years and for commercial breeding programs thereafter (McGuire *et al.*, 2012)

Considering the high demand and imports of high oleic safflower oil, India too was successful in modifying its traditional safflower oil and has developed several high oleic varieties (Anjani and Praduman 2017). Recently the high yielding-high oleic (76%) safflower variety was released in India for commercial cultivation. A high oleic safflower containing 90% oleic acid was developed in Australia through gene silencing (Wood *et al.*, 2015).High oleic content is controlled by a recessive allele *ol* at the single locus *OL*. The *ol* allele was incorporated in breeding lines to enhance oleic acid levels in safflower (Mundel and Bergaman 2009; Anjani and Praduman 2017). The *ol* allele in high oleic genotypes was found to be defective in *FAD2-1* with a single nucleotide deletion in the coding region that leads to premature termination of translation and subsequent nonsense-mediated mRNA decay of *FAD2-1* (Guan *et al.*, 2012; Liu *et al.*, 2013). Rapson *et al.*, (2015) have suggested ancient hybridization and introgression of the *FAD2-1* gene into *C.tinctorius* from its wild relative, *C. palaestinus.*

Despite the current novel molecular techniques and transgenic approaches, thorough and systematic exploration and tapping of novel and useful genes and gene combinations from *Carthamus* species, primary and secondary gene pools and their introgression into desired genetic backgrounds through classical breeding are still the most important approaches for genetic improvement of safflower. The most common breeding methods used for improvement of seed yield, disease resistance and oil content and quality in safflower are pure line selection, mass selection, pedigree and backcross breeding, recurrent selection, single-seed-descent method and a combination of biparental mating followed by reciprocal recurrent selection and diallel selective mating. Genetic improvement will continue in safflower through these methods but may not occur at higher degree than that occurred before unless some new strategies are employed to tackle the current challenges the crop experiences under the changing climate.

Heterosis Breeding

Hitherto safflower breeding was revolving around breeding for increased seed yield as being done in many self-pollinated crops. Safflower is described as a self-pollinator; but considerable rates of out-crossing of over 50% (Dajue and Mündel,1996) have been reported depending upon the environmental conditions. Bees of several genera are attracted to safflower for pollen and nectar (Karve *et al.,*1981).

Heterosis breeding in many crops has resulted in advancing the productivity levels to very impressive heights irrespective of the mating system.

(i) ***Gene action***: The importance of both additive and non-additive gene action for the of seed yield and capitula per plants has been highlighted by Prakash and Prakash (1993) and Singh V., (2004). The high degree of non-additive gene action available in safflower for seed yield (Ramachandram and Goud, 1981, Narkhede *et al.*,1992; Patil *et al* 1992; Parameshwarappa *et al.*,1995; Ghorpade and Wandhare, 2001; Gadekar and Jambhale, 2003) is indicative of the impressive potential for exploitation of the phenomenon of heterosis for seed and oil yields in safflower (Makne and Choudhari, 1980; Ranga Rao, 1982; Narkhede *et al.,*1992). The investigations carried-out outside India too, confirm the feasibility of exploitation of heterosis as a powerful tool for safflower breeding.

(ii) ***Male-sterility Systems***: The availability of usable male -sterility system has been the limiting factor for the exploitation of heterosis in safflower. Early efforts in that direction were to make use of functional male sterility (Rubis, 1967) through delayed anther dehiscence. These attempts were not successful because of the pleiotropic effect of thin hull on weak stem and high percent of self-fertilized population in the F1 hybrid (Urie and Zimmer, 1970). The efforts to find workable male sterility system have been intensified in India and USA by public and private sector safflower breeders. In that direction, all available male sterility systems along with the progress made in case of promising male sterility systems are reviewed.

In order to put the safflower hybrid seed production on a sound footing, there is a need to evaluate and exploit the genetic tools of male sterility.

(i) *Genetic male sterility system:* Dominant and recessive type genetic male sterility (GMS) systems are reported in safflower (Heaton and Knowles, 1982; Joshi et al., 1983; Deshpande, 1940). Heaton and Knowles (1982) developed two genic-male-sterile lines viz., UC-148 (from the progeny of colchicine-treated line PI 253914 of Afghanistan) and UC-149 (from the cross of UC-148 x PI 34008), which are controlled by recessive allele. No pleiotropic effect of the allele is noted so far in its original genetic background. The genetic male sterile line segregates into sterile and fertile plants in 1: 1 ratio.

Ramachandram and Sujatha (1991) have also developed genic male sterile line based on a recessive allele. Anjani (2005a) has described the method to stabilize male sterility per cent and its maintenance in GMS lines.

The GMS sources in safflower controlled by single recessive alleles are as follows:

1. UC-148 and UC-149 GMS lines developed by Heaton and Knowles (1982)
2. Genetic Male Sterile (GMS) lines developed by Ramachandram and Sujatha (1991)
3. MSN and MSV Male-sterile lines developed by Singh V., (1996)
4. Male-sterile lines associated with dwarfness developed by Singh V., (1997)

Male-sterile and male-fertile plants are identified at flowering by the presence of a pinched capitulum opening in case of male-sterile plants and a normal opening in case of male-fertile plants.

There is also another genic male sterile system (Joshi *et al.,* 1983) expression of which controlled by a dominant allele. Identification of sterile and fertile plants is to be done at flowering. It should be possible to cull out fertile plants which occur in lower frequency as per monogenic segregation.

Due to the dominant nature of the male sterile allele the hybrids and the male-sterile line in this system segregate in a ratio of 1 MS (male-sterile) to 1 MF (male-fertile) plant. The successful performance of hybrids based upon this source is hindered due to the occurrence of 50% MS plants in the hybrid population, which adversely affects the yield of the hybrid. This however, could be countered with adequate honey bee activity to ensure 100% seed set in the male-sterile plants (Meena *et al.,*2012).

(iii) ***Marker-linked GMS System (MGMS):***In this case the male-sterility attribute is linked to a genetic marker so as to facilitate easy identification. The Directorate of Oilseeds Research (Indian Institute of Oilseeds Research) has carried out pioneering work of linking the male-sterile factor to spineless character. The marker linked male sterile plants could be distinguished from fertile plants at about 35 to 40 days after planting giving sufficient time for rouging of fertile plants prior to flowering in hybrid seed production plot.

(iv) ***Dwarf Male sterility (DMS)*****:** DMS male-sterile lines associated with dwarfness were developed by Singh (1997) at NARI, Phaltan (India). In case of DMS lines (namely DMS-H-5-1 and DMS- N-11-1), owing to a linkage between sterility and dwarfing genes, the sterile and fertile plants become obvious at 30 to 40 days after sowing. The male-sterile plants attained a height of only 5 to 10 cm at 30 to 40 days after sowing; however, the male-fertile plants attained a normal height of 20 to 25 cm. The height difference between the two types enabled the rouging out of male-fertile plants at this stage, leaving behind a 100% pure stand of dwarf male-sterile plants (Singh and Nimbkar 2007).

(v) ***Thermo-sensitive genetic male sterility (TGMS):*** This is a case of male sterile factor is associated with temperature. Singh and Nimbkar (2007) at NARI, Phaltan (India) reported the thermo sensitive genetic male-sterility (TGMS) in safflower. The male sterility expresses during the winter season of India when temperatures will be of the order of less than 13°C. Fertility restoration is effected during summer season when temperatures are fairly high ranging from more than 21°C and go up to 39 °C. Safflower being a winter season crop, this kind of male-sterility could be of practical use in the production of commercial safflower hybrid seed.

Development of agronomically superior genetic male-sterile lines in India has paved the way for the development and release of spiny safflower hybrids DSH-129 and MKH-11 in 1997, the first non-spiny hybrid safflower NARI-NH-1 in 2001 (Singh *et al.*, 2003), and the spiny hybrid NARI-H-15 in 2005. These hybrids in general show a 20 to 25% increases in seed and oil yield over the national check A-1 (Anjani and Mukta, 2008).

(vi) ***Cytoplasmic-genic male-sterility (CGMS):*** Stable cytoplasmic male-sterilitytogether with genetic mechanism to restore fertility in hybrid should be an ideal mechanism for successful hybrid seed production. The first report of cytoplasmic-genic male-sterility in safflower is by Hill (1989). The research in progress at the Directorate of Oilseeds Research (Currently Indian Institute of Oilseeds Research) Hyderabad has resulted in the development of successful and usable cytoplasmic-genic male-sterility system in safflower (Anjani, 2005b).

This system is developed using the wild species *C. oxyacantha* and 77 restorer lines (R-lines), which could restore 100% male fertility on various CMS lines through conventional breeding (Anjani, 2005). The procedure of developing CMS system involved a series of back crosses to the wild species used as female and safflower varieties that are restorers as male parents.

This resulted in the development of DSH-185, the first public sector CGMS-based safflower hybrid developed at ICAR-Indian Institute of Oilseeds Research. It has been released and notified for all-India cultivation. DSH-185 is a cross between A-133 (CGMS line) x 1705-p22 (a restorer line). Wild species, *Carthamus oxyacantha* is the source of cytoplasmic genetic male sterility in A-133.

On an average, DSH-185 gave the seed yield of 14.3 q/ha under unirrigated dryland situation, 21 q/ha under irrigated conditions. At the national level the hybrid gave a seed yield of 17.4 q/ha at national level, surpassing all the checks. It gave the oil yield of 4.12 q/ha under unirrigated dryland situations, 5.7 q/ha under irrigated conditions and 4.89q /ha at national level. On an average, DSH-185 exhibits 25-30% superiority in seed yield over the best check varieties, A1 and PNBS-12 and 15.2% superiority over the GMS-based national hybrid check, NARI-H-15. It has 28-29% oil content and recorded 25-28% superiority in oil yield over A1, PNBS-12 across test locations.

The potential of DSH-185 in comparion with variety has been demonstrated in farmers' fields under dry and irrigated conditions in Maharashtra, Madhya Pradesh, Telangana and Chhattisgarh States. DSH-185 recorded on an average, 17 q/ha seed yield under dry conditions in Chhattisgarh as against 5 q/ha seed yield of check variety, A1. In Maharashtra, it gave 21 q/ha against 16 q/ha yield of variety, A1 under irrigated conditions while in Telangana, DSH-185 gave 10-14 q/ha against 4-5 q/ha of state variety, *Manjira* under dry conditions. It is resistant to *Fusarium* wilt which is the major disease of safflower.Seed production technology of DSH-185 and A and B lines has been perfected (Anonymous 2018-19).https://www.icar.org.in/content/dsh-185-new-cgms-based-safflower-hybrid-0

(vii) ***Chemical induced male sterility***: One of the earlier reports of successful induction of male sterility was in rice through the spray of 6000-8000 ppm of ethrel one week earlier to boot leaf stage or two sprays with 4000-6000 ppm, one week earlier to boot leaf stage (Parmar *et al.*, 1979).

The researches carried out at the Directorate of Oilseeds Research in Hyderabad (Prayaga *et al.,* 2001) resulted in the enhancement of male sterility in safflower with the use of different concentrations of the growth regulators viz., gibberellic acid, ethylene and the chemicals such as NaCl and Maleic hydrazide. These growth regulators and chemicals were sprayed on the crop at uninucleate stage (85 DAS) for two consecutive days. The concentration of GA@ 300 ppm showed maximum sterility (95%), but the plants became tall and lanky. Therefore, lower concentrations ranging from 5 to 50 ppm were applied at the same stage. Among these, GA@ 40 ppm recorded the highest sterility (85%) without altering the plant form. Baydar *et al.,* (2003) reported that three successive sprays (75, 82 and 89 days after sowing) of 100 ppm of GA3to safflower buds of <0.5 cm diameter at pre-meiotic interphase stage resulted in reduced pollen viability.

(viii) ***Transgenic male-sterility:***Nuclear male sterility is common in flowering plants, but its application in hybrid breeding and seed production is limited because of the lack of feasibility to propagate a pure male sterile line for commercial hybrid seed production.

Transgenic male-sterilitysystem is fundamentally advantageous over the current cytoplasmic male sterile and photoperiod/thermo-sensitive genic male sterile systems. One of such cases in oilseed crops is that of the transgenic mustard DMH – 11, which was developed in 2002 using genetic material isolated from non-pathogenic soil bacteria (Srinvas and KrishnaRavi, 2017) and techniques in transgenic systems for pollination control, which primarily involved the *Barnase-Barstar* system. (Siriguri *et al* 2015). Three genes *Bar, Barnase* and *Barstar*, were extracted from *Bacillus amyloliquefaciens* to produce the hybrid seed (Kumar-Ravi 2018)

In the case of safflower, Dinesh Kumar and Narasimha Rao (2008) described three genes i.e., a mitochondrial, male sterility inducing gene orfH522 from PET1 CMS line of sunflower and unedited version of mitochondrial genes, atp9 (u-atp9) and nad3 (u-nad3) from safflower which have been cloned under TA29 promoter, in-frame with the coxIV pre-sequence, so that the resultant cytoplasmic proteins (ORFH522, u- ATP9 and u-NAD3) would be targeted into mitochondria. *Bar* gene that confers resistance against the herbicide *basta* (phosphinothricin) has been used as selectable marker. All cassettes have been cloned in pCAMBIA0390. To silence the sterility inducing transgenes for restoring the male fertility in three different PTGS strategies, that include (i) cloning full length antisense version, (ii) cloning inverted repeat of complete/selected part of the genes separated by a intron and (iii) cloning antisense version of the selected part of orfH522 upstream of the inverted repeat of a heterologous 3'-UTR separated by an intron have been employed. These gene cassettes have been cloned in pCAMBIA1300. All the developed binary vectors have been mobilized into *Agrobacterium* strain LBA 4404.

Fig. 2: Safflower Hybrid DSH-185

The transgenic male-sterility system is of GMS system. This system could not be exploited for developing hybrids due to the lack of reliable protocol for regeneration of trasformants. *Agrobacterium* mediated transformation has largely been unsuccessful in regeneration of whole plant trasformants in safflower due to inherent problems associated with tissue culture of safflower (Sankara Rao and Rohini 1999).

(ix) ***Self-incompatibility (SI) as a genetic tool for hybrid development*:** This system has been successfully used in the case of *Brassica campestris* by Thompson (1964) and later by Liu (1975) to produce hybrids of *Brassica.* Sp.

Imerie and Knowles (1971) have also reported self-incompatibility in *Carthamus flavescens.* Ramanamurthy (1963) however, reported self-incompatibility in C. *oxycantha.* Considcring the success achieved in *Brassica campestris* possessing sporophytic incompatibility, it should be possible to produce such hybrids of *Carthamus* using the sporophytic incompatibility.

Table 1: Safflower Hybrids released in India

Hybrid	Year of Release	Yield (kg)/ha	Oil Content %	Salient features	Institution
NARI-NH-1 (PH-6)	2002	1936	31	Non-spiny, Moderately resistant to Cercospora & Wilt.Tolerant to Alternaria	NARI Phaltan Maharashtra
NARI-H-15	2005	2200	29		NARI, Phaltan, Maharashtra
MRSA 521	2006	1000 -1500® 2200-2500 (I)	27	Resistant to Wilt and Tolerant to Alternaria	MAHYCO Jalna, Maharashtra
NARI-H-23	2013			Recommended for irrigated & assured rainfall areas	
DSH-129 **[MS9(o) xA1]**	1997	2200-3500	30-32	Tolerant to Wilt	ICAR-IIOR Hyderabad
DSH-185 **[A-133 (CGMS line) x** 1705-p22 (a restorer line).]	2017	1430® 2100(I)	29	Recommended for irrigated and Rainfed **conditions. Resistant to Fusarium wilt**	ICAR-IIOR Hyderabad

Table 2.: Safflower varieties released in India

Safflower (Carthamus tinctorius L) Varieites recommended for various regions in India						
Variety	Year of release	Releasing Centre	Yield potential (kg/ha)	Oil content (%)	Recommended states/regions situations	Salient features/ traits
PBNS-40	2007	MAU, Parbhani	1500	27	Madhya Pradesh	Non spiny
AKS-207	2007	PDKV, Akola	1200-1400	27	Maharashtra	-
NARI-38	2007	NARI, Phaltan	2038 (irrigated)	28	All India	Resistant to wilt
SSF-658	2009	MPKV, Solapur	1430	28	All India	Tolerant to wilt and aphids
SSF-708	2012	MPAU, Solapur	1300-2200	29	Western Maharashtra	Moderately tolerant to aphid
PKV-Pink (AKS-311)	2013	SRS, Oilseeds, PDKV, Akola	2200-2500	33	Vidharbha region of Maharashtra	Moderately tolerant to wilt
NARI-57	2015	AICRP (Safflower) centre, Nimbkar Agricultural Research Institute, Phaltan	1500	30	Irrigated areas of all safflower growing states in India (Maharashtra, Karnataka, West Bengal, Rajasthan, Uttar Pradesh., Punjab, Jharkhan)	
Manjira	1976	AICRIP Tandur Centre ANGRAU Hyderabad	1000- 1100	28	Telangana, Andhra Pradesh, Maharashtra	Spiny
PBNS-12	2001	AICRIP Parbahani MAU	1200-2000	29	All India	Moderately tolerant to Aphid
JLSF 414	2003	AICRIP Rahuri MPKV	1200-2000	28-29	All India	---
A-2	1997	AICRIP Centre Annigeri UAS Dharwar	1000-1200	31	Dryland zones of Karnataka	Aphid tolerant, Recommended for late sowing
Sharda (BSF-168-4)	1990	AICRIP Parbhani MAU	1000-1800	29	Marathwada of Maharashtra	Moderately tolerant to Aphids and Wilt
ISF- 1	2018	ICAR-Indian Institute of Oilseeds Research Hyderabad	1236 (rainfed), 1864 (irrigated)	31	All India	High oleic type (76%)
ISF-764	2018	ICAR-Indian Institute of Oilseeds Research Hyderabad	1583 (rainfed), 2274 (irrigated)		All India	Moderately resistant to wilt and Alternaria

A variety of techniques of genetic engineering, including gene mapping, recombinant DNA etc., are now available (Weiss, 2000) for achieving genetic advancement of safflower crop. When the results thus gained can be applied to commercial production, especially in respect of increased oil yield, desirable oil components and disease resistance, then the safflower's position may get enhanced as an important edible oil.

Vegetative hybridization with grafts of safflower on sunflower, sunflower on safflower and sunflower on *Xanthium strumarium*, were successful and viable seeds obtained from the first and last. Grafted safflower cultivars in the USA showed no influence of rootstock on scion.

Heterosis has been widely reported in safflower for yield, oil percentage, plant height and other characters and the discovery of a gene which caused partial male sterility was important in the study of heterosis and related processes (Carapetian, 1994).

Another aspect of safflower improvement is the use of new extra early dwarf safflower varieties that have been developed by Oilseeds Coordinated Research Project, college of agriculture, Indore (M.P) centre in India which have numerous advantages as extra earliness, higher per day yield, bald grain, minimum disease and aphid problem, easy harvesting, as per Deshpande *et al.,* (2005).

An ideal safflower plant

According to Weiss (2000) the concept of an ideal safflower plant consists of a plant of medium size of 60-80 cm height, with maturity of 130-150 days possessing appressed or semi appressed branches with 6-8 primary branches bearing 12-14 large heads with 30 to 40 seeds/head with bold seeds of 5g/100 seeds. The minimum seed oil content of such plant should be of 50% together with high protein content. Additional assets of an ideal safflower plant should be rapid seedling growth, hardiness, low seed-hull content and resistance to major pests and diseases. Knowles (1991) too suggested that the appressed plant type of safflower should be more efficient than the traditional one with expanded secondary branches in terms of population performance and harvest index.

The investigations carried out by Ramachandram and Ranga Rao (1991) on intra-plant variability for the yielding ability of capitula flowering over a period of 25-35 days, revealed that the capitula which flowered during the first 15 days which are mainly from the primary branches, contributed to a bulk (over 65%) of the yield of the safflower plant. The later order capitula borne on secondary branches which flowered later by 10-15 days could not add much to the seed yield and hull percent for seed oil content (Ramachandram, 1985). Significantly high negative correlation between total capitula per plant and seed yield per capitulum as well as its components, viz., seed number and diameter of capitulum poses serious problem to the indirect selection for seed yield through direct components. Thus, the simultaneous selection for yield per capitulum and total capitula need to be based on appropriate weights (Ramachandram and Goud, 1982). Alternatively, restricted selection of Kempthorne and Nordskog (1959) could be followed where in the improvement has to be brought in directly for yield per capitulum through seed number by keeping total capitula as a constant variable. It is possible to increase the number of capitula per unit area by manipulating the population density. Ashri (1975) emphasized the potentiality of a plant type with fewer but larger capitula having maximum number of sound seeds under limiting moisture conditions of arid and semi-arid regions. The orientation of branches on the main stem in terms of angle and length of branch in such a plant type play an important role in increasing the response to population levels. In that direction Ramachandram (1985) proposed the desirable plant type having 8 to 10 shortest primary branches (10 to 12 cm length) oriented at an angle of around 20 to 30 degrees bearing 12 to 15 high yielding capitula per plant (2.5 to 3.0 grams as against 1.0 to 1.5 gram in local varieties) and growing up to 60 to 75cm height. Ideal seed or achene characteristics include: 4.0 g to 4.5 g of 100 seed weight, 28 to 35% hull content and around 45 to 48% oil on whole seed basis. Non-spiny plant is yet another desirable character. Other characteristics like maturity duration, oil type in terms high linoleic acid or high oleic acid type, resistance to biotic and abiotic stresses are location specific requirements. Nevertheless, in recent times both high linoleic and high oleic acid varieties and hybrids are in demand.

Cloreto (2008) suggested an ideal safflower plant type for Mediterranean region should possess extra earliness characterised by short rosette stage with day neutral habit.

The range of variability in safflower germplasm being high, it should be possible to select for the characters indicated above.

In addition to the availability of thin or reduced hull types, enhancement of oil content from current 30% to at least 50% should be attainable along with the increased seed yield.

Desirable oil quality could be attained with high oleic acid content. Keeping with the local preferences, Australians have developed high linoleic acid type (76-78%) of safflower with 10-12% of oleic acid (Weiss, 2000). Increasing the seed protein and lysine levels and reducing the fibre content resulted in more attractive cake or meal along with the removal of the bitter principle *matairesinol monoglucoside.* (Polyphenol-proponoid).

Rubis (2001) produced safflower lines of reduced height that were flowering very early (selected within *Gila* dwarf type) and characterized by day-length ncutral. Improvement in earliness associated with good productivity was possible as a result of crossing with very big stiff stem plants and with tall late plants. Therefore, the first step is to produce a line characterized by short rosette stage, insensitive to day-length that will soon after start the stem elongation followed by bud initiation that represents the beginning of the reproductive plant development stage (Tanaka *et al.*, 1997).

Considering development of an efficient ideotype of safflower, it is to be appreciated that one kind of plant type may not suit diverse situation under which safflower crop is being grown. The earlier stated appressed plant type with only primary branching may be suitable for rain-fed dryland conditions. In the case of irrigated situation with assured moisture availability for safflower production, one could think of a plant type with reduced primary branching with one to two primary branches or even a single stem type with enhanced capitulum size, which would facilitate higher plant population density per unit area, to take the crop towards much higher levels of seed and oil productivity.

Breeding for Disease Resistance

Wide range of biotic agents such as 57 pathogens, including 40 fungi, 2 bacteria, 14 viruses, and 1 mycoplasma are reported to cause diseases in safflower among which Alternaria leaf spot caused by *Alternaria carthami* and wilt caused by *Fusarium oxysporum* are the most devastating ones, that are reported to cause 13 to 49% losses and wipe out the entire crop in the region under conditions conducive to their development in India (Patil *et al.*, 1993). Breeding resistant genotypes for such serious diseases should be the priority item of research.

Mundel and Huang (2003) described in detail the methods of controlling major diseases of safflower through breeding and by crop management practices. Despite the fact that the germplasm lines of safflower have been screened or major diseases, the basic data with regard to the genetics of disease resistance / susceptibility including pattern of inheritance of disease resistance and tolerance are not adequately worked out in the case of safflower (Dajuae and Mundel 1996).

Safflower is adversely affected by number of diseases, which vary in incidence and severity from one region to other. Among them the most relevant are Fusarium (*Fusarium oxysporum* f. sp. *carthami* Klisiewicz & Houston) and Vetrticillium (*Verticillium dahliae*) wilts, Rhizoctonia blight (*Thanatephorus cucumeris* (A.B. Frank) Donk), rust (*Puccinia carthami* Cda.), Pythium damping-off (*Pythium ultimum* Trow), Phtophthora root rot (*Phytophthora cryptogea*; *Phytopthera dreschsleri* Tucker), Alternaria leaf spot (*Alternaria carthami* Choudhury), Ramularia leaf spot (*Ramularia carthami* Zapromeor), Cercospora leaf spot (*Cercosporacarthami* H and P. Sydow) and powdery mildew (*Erysiphe cichorac earum*). The genetics and the mode of inheritance of disease resistance and tolerance in safflower are not well defined in respect of most of these diseases. An overview of the global occurrence and conditions favouring these diseases was presented by Mundel and Huang (2003).

Four races of *Fusarium oxysporum* f. sp. *carthami* were identified in USA and India (Klisiewicz and Thomas, 1970; Klisiewicz, 1975; Prasad and Suresh, 2012.) Studies on Fusarium wilt resistance revealed role of two additive genes (Bockelman *et al.*, 1973), monogenic dominant control, (Anjani *et al.*, 2012), inhibitory gene action (Singh *et al.,* 2001) and complementary gene action of two genes (Shivani *et al.,* 2011) in expression of wilt resistance in safflower. The disparity in genetic control of wilt may be because of presence of different isolates or use of different cultivars.

Several Fusarium wilt resistant germplsm accessions have been reported in safflower (Klisiewicz and Urie, 1982)

Wide range of biotic agents such as 57 pathogens, including 40 fungi, 2 bacteria, 14 viruses, and 1 mycoplasma are reported to cause diseases in safflower among which *Alternaria* leaf spot caused by *Alternaria carthami* and wilt caused by *Fusarium oxysporum* are the most devastating ones, that are reported to cause 13 to 49% losses and wipe out the entire crop in the region under conditions conducive to their development in India (Patil *et al.*, 1993). Breeding resistant genotypes for such serious diseases should be the priority item of research.

Resistance against wilt disease and the insect aphid together with drought resistance could be achieved by pyramiding of genes in common genetic background through classical breeding methods in safflower (Anjani *et al.*, 2005).). The multiple resistant line 96-508-2-90 with the above attributes, has been registered with Plant Germplasm Registration Committee of ICAR, New Delhi, India.

Singh *et al.*, (2003) reported breeding for wilt resistance following back cross method in safflower leading to the recovery of resistant genotypes with good seed yielding attributes.

The germplasm accession VFR-1 exhibiting multiple resistance, was derived from the breeding line Nebraska 4051; the derivative showing resistance to *Verticillium* wilt, *Fusarium* wilt and root rot, and *Rhizoctonia* root rot (Thomas, 1971).

Karve *et al.* (1981) found that resistance to each of the diseases, viz., *Alternaria carthami* Chowdhari, *Cercospora carthami* Sund and Ramak, *Ramularia carthami* Zaprom, *Fusarium oxysporum* Sehl. ex. Fries, *Rhizoctonia bataticola* Bult, and *Rhizoctonia Solani* Kuhn, is imparted by single dominant allele.

Study of inheritance of wilt (*Fusarium oxysporum*) resistance in safflower revealed the control of inhibitory gene action its expression. (Singh *et al.*, 2001). The source of resistance to wilt has been identified in the local germplasm lines (Sastry and Ramachandram, 1992).

Harrigan, (1987 and 1989) developed through back cross method, the Australian safflower cultivar *Sironaria*, showing resistance to *Alternaria* blight and moderate resistance to *Phytophthora* root rot.

Resistance to *Phytophthora* is conditioned by a single pair of recessive alleles (Thomas and Zimmer, 1970). Da Via *et al.*, (1979) have identified 15 lines resistant to *Phytophthora* root rot (UC-150 to UC-164) among world collection of safflower. The cultivar Dart (Abel and Lorance, 1975) contains genes for resistance against all races of root rot disease caused by *Phytophthora drechsleri*. Mundel *et al.*, (1985) reported the incorporation of *Sclerotinia* head rot (caused by *Sclerotinia sclerotiorum* Lib. de Bary) resistance into the first Canadian safflower cultivar *Saffire* by adopting mass selection.

Safflower rust caused by *Puccinia carthami* is distributed worldwide (Kolte, 1985). Three independently inherited dominant-factor pairs *AA*, *II*, and *MM*, which individually and collectively condition the resistance to both seedling and foliage rust, are found in three resistant sources *viz*., PI 195895-95, PI 250601-109, PI 253914-5 of safflower. Nebraska 1-1-5 carried another rust resistant gene N_1N_1 which is independent of A, I and M gene-pairs (Zimmer, 1965; Zimmer and Urie, 1968). The resistance conditioned by each gene separately or collectively is sufficient to prevent yield loss caused by either seedling or foliage rust. Five improved lines (PCA, PCM-1, PCM-2, PCN_1, PCOy) each carrying a different dominant gene for rust resistance against all known races are developed in safflower (Zimmer and Urie, 1970).

Wild species of *Carthamus* are reliable sources of resistance against several biotic stresses.

Heaton and Klisiewicz (1981) developed an alloploid from a cross between *C. tinctorius* and *C. lanatus*, which was found to be resistant to Alternaria, Fusarim wilt and bacterial blight. A high level of resistance was found in an accession of *C. oxyacantha* from Iran (Zimmer and Urie, 1968).Wild species *viz., C. oxuacantha, C. palaestinus, C. creticus, C. lanatus, C. turkestanicus* and *C. glaucus* were found to be new sources of resistance to Fusarium wilt in Carthamus genus (Pallavi *et al.*, 2007). Anjani *et al*., (2018A) could introgress resistance to Fusarium wilt from *C. oxyacantha and C. palaestinus* into cultivated species though marker-assisted selection. RAPD and SSR markers flanked to Fusarium wilt resistance were identified in cultivated and wild species (Anjani *et al.*, 2012A; Anjani *et al.*, 2018A).

Prasad and Anjani (2008) identified four wild *Carthamus* species viz., *C. palaestinus, C. lanatus, C. creticus* and *C. turkestanicus* to be immune to *Alternaria* leaf spot in laboratory as well as under field screening. Resistance to Alternaria could be introgressed from the wild species, *C. oxyacantha, C.palaestinus* and *C.*

lanatus into cultivated safflower through wide hybridization and stable *Alternaria* resistant interspecific inbred lines were developed at the ICAR-Indian Institute of Oilseeds Research, Hyderabad, India (Anjani *et al.,* 2019).

The other safflower cultivars viz., Sidwill, Hartman, Oker, Girard, and Finch developed in the U.S. (Bergman and Riveland, 1983; Bergman *et al.,* 1985, 1987, 1989a, 1989b) are resistant to *Alternaria* blight. These cultivars have been derived from crossing of the existing cultivar AC-1 with *Alternaria*-resistant line 87-42-3 in a disease nursery initiated in the early 1960s. The resistant parent line 87-42-3 was developed through mass selection. Resistance to all prevalent races of root rot caused by *Phytophthora drechsleri* was incorporated into the cultivar Dart (Abel and Lorance, 1975).

Strategies for Further Genetic Improvement

In spite of its ancient culture in India and elsewhere, the safflower breeding programs were initiated only towards the middle of last century. It was Dr. Knowels, who had laid the foundation for safflower crop improvement with the collection of germplasm including a number of wild species from almost all primary and secondary centres of origin. Besides evaluation of the germplasm across all major safflower growing areas, a number of high yielding varieties with marginal improvement in seed oil content up to 34 to 36% were identified for introducing safflower cultivation in United States of America. Further efforts to increase the seed oil content through bringing down the hull percent, thin, striped and reduced hull mutants were identified. In view of the limitations of thin hull, mostly striped hull and reduced hull to certain extent facilitated to bring in a big jump in the seed oil percent of about 45 to 48 %. Simultaneously high oleic acid type identical to olive oil was also identified. In the second phase, the varieties with high yield, high oil content and diverse fatty acid composition were available. On the other hand, the Indian safflower improvement programs had concentrated more on seed yield with 2 to 4% oil improvement. So far, the breeding methods like pure line selection, pedigree breeding, back cross and recurrent selection have been followed to bring about the improvement in seed yield, oil content and other traits. More recent has been the development of hybrid safflower. These developments have led to the simultaneous exploitation of both additive and non-additive components for seed yield, oil content and their components.

In the present context of the status of safflower crop improvement and eroded crop area, it is utmost important to examine possible approaches to bring in further improvement in oil yield per unit area. As the crop has already undergone considerable improvement in terms of seed yield, oil content and various agronomic characters, there is a need to look at the improvement more closely traditional approach. Restructuring the plant type, which responds to different agronomic practices, possesses with tolerance to biotic and abiotic stresses is foremost important. With reference to its introduction in new areas/cropping systems, one need to consider the plant type, crop duration, insect pest and disease problems etc. The development of two or more distinct oil types without undesirable odour/ component, facilitating diversified utilization of oil and thereby the oil trade forms other top priority. The male sterility system is now perfected to towards developing, hybrid safflower possessing as many desirable attributes as possible. Thus it is possible to regain the lost area and production on global level. In order to achieve all set goals within reasonable time frame, one has to take advantage of biotechnological tools in addition to traditional breeding methods.

Safflower Crop Production

History of safflower crop production in India dates back to around 4,000 years. In olden days the crop used to be cultivated not only for oil; but also for reddish dye from flowers, which were also used in *Ayurvedic* medicine (Singh and Nimbkar, 2016). In recent years the area and production of the crop has declined due to various factors such as competition from more remunerative crops viz., chick-pea, *rabi* sorghum and cotton. A large chunk of the crop is grown under dryland conditions on receding soil moisture over short duration of 5 to 6 months, due to which the productivity levels had remained stagnant (Prasad, 2019b). The mean day temperature has significantly negative correlation with the duration of mean of different growth phases, more so after bud initiation (Rao and Ramachandram, 1977). Prevalence of high temperature of more than 32°C causes early termination of rosette stage and accelerate subsequent growth and consequently reducing the overall maturity duration. As a result, the plant produces smaller capitula with reduced number of ill filled seeds. Other important environmental factor influencing safflower growth and development is the humidity. Early crop growth stages up to branch initiation tolerate relatively high humidity and even intermittent rains. There after high humidity coupled with prolonged rains pre-disposes the crop to various diseases particularly alternaria leaf blight.

Nitrogen deficiency is one the major limiting factors in safflower producing regions. There are no clear data on the needs of micro-nutrients for safflower production. The efficiency of micronutrients, especially zinc and copper, significantly depend on the availability of nitrogen (Weiss, 2000). The investigations of Fattahi *et al.*, (2018) indicated that safflower can be successfully produced in a highland region with precise medium fertilizer input. An evaluation of growth and yield parameters by Fattahi *et al.*, (2018) revealed that the most effective micronutrients were zinc and copper. The nano-chelated micronutrient fertilizer with smart delivery systems, which can react to environmental changes could be a suitable option to overcome the micronutrient deficiencies in semi-arid regions. However, the efficiency and effectiveness of nano-chelated fertilizer is strongly influenced by availability and abundance of macronutrients. The findings of the study by Fattahi *et al.*, (2018) showed that there were obvious simplifications in the concept of nutrient interactions and this was greatly influenced by other environmental factors such as water availability. The results highlighted the need for low rates of fertilizer N application along with nano-chelated zinc to optimize spring safflower yields. Samadhiya (2017) reported that the treatments of100% recommended NPK@ 50N-25P-20K/ha + Fe 0.2% + Mn 0.3% + Zn 0.5% + 1% urea foliar spray was a sound combination of nutrients to get highest oil yield (529.78 kg ha-1) followed by Fe 0.2% + Mn 0.3% + Zn 0.5% with recommended NPK (508.33kg ha-1) were superior to application of recommended NPK alone (393.15kg ha-1) which increase was oil yield by 34.75 and 29.30% percent over recommended NPK alone respectively. The results clearly indicate the yield enhancement with the foliar spray of micro-nutrients at the suggested doses.

Bonfim-Silva *et al.*, (2016) clearly brought out that among primary macronutrients, nitrogen, followed by phosphorus, caused the highest reductions in all the analysed variables in safflower plants. Among secondary macronutrients, absence of Calcium caused the highest reductions, and the need for these three nutrients did not allow plants to complete their life cycle. Macro-nutrient omissions showed most severe symptoms in relation to micronutrients. The authors brought out clear foliar symptoms of micronutrient deficiencies leading to yield reductions.

From the above investigations it appears that the role of micronutrients is closely linked to the availability of macro nutrients more particularly Nitrogen. Secondly among micronutrients, boron omission, followed by manganese omission, caused the most significant reduction in development and production of safflower plants. In general, deficiency symptoms were observed in the absence of all nutrients visual diagnosis, except for omission of molybdenum. Macronutrient omission caused the most severe symptoms compared to micronutrients.

Fig. 3: Foliar symptoms of Micro nutrient deficiencies in Safflower[From Bonfim-Silva *et al.*, (2016)]

The distinct problems associated with safflower crop: (Prasad, 2019b)

- Lack of efficient plant types with high harvest index
- Lack of varietal specificity based on diverse soil and moisture situations
- Lodging of crop leading to loss of productive plant population
- Photoperiod also influences the crop growth. Short photoperiod prolongs the rosette stage. Long days are conducive to good yields.
- Seedlings are susceptible to higher temperatures and soil salinity.
- Lack of clear recommendations on micro-nutrient requirements of safflower especially Sulphur, Boron, Copper, Manganese and Zinc which are important for oilseed production as indicated several by investigations.
- Lack of early maturing and non-spiny varieties for paddy fallows.
- Lack of adequate levels of seed production of released hybrids to meet the needs of farmers.
- The rapid spread of soil borne pathogens such as *Fusarium* and anaerobic soil conditions cause plant death very quickly as suggested by Nimbkar (2019).

Research efforts have been in progress to enhance the production and productivity of safflower supported by more productive varieties and new hybrids

Seasonal Conditions for Safflower Production

Safflower comes up better in relatively drier areas. Frequent and prolonged rains and heavy dew at flowering stage adversely affect pollination and seed development.

Optimum temperature for flowering is between 24*C and 32°C, however, adequate soil moisture reduces the adverse effect of high temperature.

High temperature at seed development also tends to decrease seed weight.

Safflower is more or less a day-neutral plant but is thermos-sensitive and the crop is tolerant to low temperatures at seedling and vegetative stages

Package of Practices for Safflower Production in India

The crop grows well on fairly deep, moisture retaining and well-drained soil.

Safflower crop is fairly tolerant to saline conditions. The following package of crop management practices would ensure recovery of sustainable seed yield levels. (www.nmoop.gov.in)

Initial land preparation by deep ploughing is carried out with the incorporation of 5 tons of farm Yard Manure or 3 tons of Poultry manure.

Sowing time: Safflower crop is sown from the second fortnight of September to the first week of November.

Seed rate: 10-15 kg/ha following seed treatment with Thiram @3g/kg or Mancozeb @ 2-3g/kg or Carbendizim @ 2 g/kg of seed.

Sowing method and Seeding depth: For better stands, follow line sowing using improved seed drills or fertilizer-cum-seed drills available in different regions. Seeds can be sown behind the plough also. Place the seed at a depth of about 5 cm in the moist zone to facilitate good germination and emergence. Cover the seed with heavy plank.

Spacing: Inter-row spacing of 45 cm and intra-row spacing of 20 cm from plant to plant is normally recommended. In rainfall scarcity zone of Maharashtra, border method of planting (skipping one row after every two rows and opening a furrow in the skipped row) to ensure some degree of moisture conservation is recommended.

Nutrient Management: Nutrient application depends on the soil type and moisture availability. Soil test values must be a guide for working out the nutrient schedule.

In general, the crop needs nutrients @40 kg/ha of N and 25-30 kg/ha of P2O5 /ha under un-irrigated dryland conditions.

In the case of irrigated crop nutrients @ 60 kg/ha N and 40-50 kg/ha P2O5 are needed.

In the case of soils with Potassium deficiency application of Potash @ 20-30 kg / ha is recommended. Application of Phosphorus Solubilizing Bacteria (PSB) along with the recommended nutrient schedule to safflower gave the significantly highest seed yields of safflower (Nimbkar, 2008).

From the available literature it is observed that Sulphur, Boron, Copper, Manganese and Zinc are important for safflower production (Bonfim-Silva *et al.*, 2016)

Murthy, (2011) working on safflower crop production on vertisols of western India, found that Sulphur application at the rate of 20-45 kg/ha to safflower alone or safflower based cropping system proved to be beneficial in improving seed yield. Sources of S vary depends on the soil type. Critical Zn and B levels in vertisol with safflower as test crop were 0.63 and 0.60 mg/kg, respectively. Among the micronutrients, deficiencies of Zn, Fe and B are reported in vertisols and associated soils. It was found that safflower responded significantly to the application of Zn (as ZnSO4) @ 20 kg/ha, Fe (as FeSO4) @ 20 kg/ha and B (as borax) @ 1.5 – 2.5 kg/ha, which resulted in increased safflower seed yield. Intra and inter nutrient combination of S, Zn, Fe and B have shown significant seed yield responses in vertisols and associated soils.

Thinning: Pull out the excess seedlings within 10-15 days after emergence and maintain the desired plant to plant spacing. This practice increases the seed yield by 15 to 30% under rainfed conditions. This practice is very important to realize the potential yield but generally neglected by the farmers. The thinned plants could be utilised as green leafy vegetable.

Moisture Management: (1) In dry land areas under scanty moisture conditions, yield can be boosted by 40% to 60% by providing the crop with ***one life saving irrigation (5 to 8 cm)*** collected in farm ponds or tanks from rains of *kharif* at critical phases of crop growth viz., early stem elongation or flowering stages or before soil moisture becomes limiting factor for seed development. (2) In the case of irrigated crop, the crop needs at least 4 to 5 irrigations each of 60 mm water.

Harvesting and threshing: Crop reaches the stage of harvest, when the leaves and most of the bracteoles become dry and brown.

If the crop has to be harvested manually, use of hand gloves to protect hands of workers against spines. Cut the plant at the base.

Multi crop thresher and combine harvester are in vogue now for harvesting and threshing, which are efficient.

Biotic Stresses of Safflower

Insect Pests of Safflower: Among the insect pests that attack safflower, the aphid *Uroleucon compositae* is considered as a major pest causing severe losses to the crop throughout the world. Yield losses due to pest attack on safflower are considerable, since aphid alone is reported to cause as much as 24.2 % (Shetgar *et al.,* 1992) to around 60% (Suryawanshi and Pawar,1980)

1. Safflower Aphid (*Uroleucon compositae*): This is a sap sucking pest. The adult is black and shining. The body is spindle shaped to elongate pyriform measuring 1.5 to 2.0 mm in length.The nymphs are reddish brown in colour. Both nymphs and adults feed on tender parts like shoot apices, veins of young leaves. With the help of syringe like proboscis both nymphs and adults suck the cell sap due to which the plant growth is stunted. In case of sever attack of the aphid, the plants start showing drying symptoms from lower leaves and progress towards top of the plant resulting in premature death of plants. In addition, they also excrete honeydew, which falls on the upper surface of the leaves on which sooty mold develops hindering the photosynthetic activity. Cool nights for 1-2 weeks in November are favorable for aphid attack on safflower.(Hanumantharaya *et al*., 2007).

Fig. 4: Safflower Aphid Uroleucon compositae stages of Egg, Nymph and Adult

In a trial in 1987-88 with safflower, spraying with 0.03% dimethoate at 10 days intervals beginning 30 days after sowing to control *Uroleucon compositae* gave seed yields of 1.29 t/ha compared with 0.90 t for the untreated control. The average aphid population was 0.80 and 50.28 in treated and untreated plots, respectively. The aphid incidence resulted in decrease in plant height, number of branches and capitula/plant, diameter of capitulum, seedsper capitulum and 100-seed weight. (Narangalkar and Shivpuje, 1990)

Aphid tolerant genotypes of Safflower are GMU-579,11-17-2, 6-9-3, PBNS-59, SSF-659, SSF-679, Bhima, A-1 and NARIH-15(Hanumantharaya *et al.*, 2007).

Aphid resistance in safflower is stated to be under both additive and non-additive type gene actions. Hence it should be possible to develop productive hybrids with durable levels of resistance against safflower aphid, which is the most important insect pest of the crop./

Alternate hosts of safflower aphid: Sunflower, niger,*Euphorbia geniculata*, *Calendula*, *Glyricidia maculate*, *Ashwaghanda*, *Lactus sp.* and *Parthenium hysterophorus* are the alternate hosts of safflower aphid. (Hanumantharaya *et al.*, 2007)

Natural predators of safflower aphid are (1) Dipteran, *Pseudendaphis* sp., which is known to cause up to 10 percent parasitization of the aphid during first week of January (Anon, 2005), (2) lady bird beetles (Coccinellids) and (3) Chrysopids. (Hanumantharaya *et al.*, 2007).

Cultural methods of aphid management: Safflower intercropped with sorghum, coriander and wheat reduces the pest population (Hanumantharaya *et al.*, 2007).

Whereas, safflower intercropped with Niger harbours the aphid population on safflower. Safflower mixed cropping with coriander also reduces the aphid population on safflower and increases the predatory population

(ii) Gram pod borer/Capsule borer (*Helicoverpa armigera)*

In the early stage of crop growth larvae feed on leaves and shoot apices. Later, the larvae shift to the developing capitula. The larva causes perforation of leaves and involucre bracts. The capitula are damaged partially or completely by the larva, in the bud stage.

The moth lays single spherical eggs of creamy white colour. The caterpillar's colour varies from greenish to brown. The pupa is brown in colour and is deposited in soil, leaf, pod and crop debris. Adult is astout moth oflight pale brownish yellow colour. Forewings of the moth are olive green to pale brown with a dark brown circular spot in the centre and hind wings are pale smoky white with a broad blackish outer margin

Fig. 5: Helicoverpa armigera stages of Larva, Pupa and adult moth.

The pest may be managed culturally by(1) intercropping of safflower with non-host crop like wheat or barley, (2) avoiding inter-cropping with chickpea and (3) cutting down the excessive application of nitrogen. The other non-chemical control options are (4) application of Ha NPV@ 250-300 larval equivalents/ha, (5) Conserving and promoting Campoletis chloridae and, Enicospilus sp, which are natural enemies of the pest (Anonymous 2018A)

(iii) Safflower caterpillar (*Perigaea capensis)*

The larva feeds on the leaves bracts and capitulum and capsule.

The pest is a stout, green and smooth larva. The anal segment of the caterpillar is humped and the body has some purple markings. The adult is a dark brown moth of medium size. Forewings of the moth are dark brown with pale wavy marks and the hind wings are light brown in colour.

Fig. 6: Perigaea capensis larva

The pest may be managed by (1) Intercropping with non-host crop like wheat, (2) avoiding excessive application of nitrogen and (3) spraying carbaryl 50 WP@ 2.5 - 3.00 kg/ha. (Anonymous, 2018A)

3. Capsule fly/Safflower bud fly: *Acanthiophilus helianthi* rossi

Newly hatched larvae / maggots feed on the soft parts of the capsules. The damaged buds have small bore holes. The infested buds become rotten emitting a foul smelling ooze. The pest is a maggot is dirty white in colour. Adult flies are ash coloured with light brown legs.

Fig. 7: Safflower bud fly: Acanthiophilus helianthi rossi

The pest may be controlled by sprays of dimethoate 30 EC @ 600-650 ml/ha or malathion 50 EC @ 1.00 litre/ha or phosphomidon 100 EC@ 150-200 ml/ha. (Anonymous, 2018A)

Natural enemies of safflower fly/capsule bud fly

- Parasitoids: *Orymurus* sp., *Eurytoma* sp, *Bracon* spp. etc.
- Predators: Lacewing, ladybug beetle, reduviid bug, spider, red ant, robber fly, black drongo (King crow), common mynah, big-eyed bug (*Geocoris* sp), earwig, ground beetle, pentatomid bug (Eocanthecona furcellata), preying mantis etc

Ten hymenopterans belonging to different families as indicated below were found parasitizing pupae of safflower bud / capsule fly in Iran: *Bracon hebetor, Bracon luteator* (Braconidae); *Isocolus tinctorious* (Cynipidae); *Pronotalia carlinarum* (Eulophidae); *Eurytoma acroptilae* (Eurytomidae); *Ormyrus orientalis* (Ormyridae); *Colotrechnus viridis* and *Pteromalus* sp. (Pteromalidae); and *Antistrophoplex conthurnatus* and *Microdontomenus annulatus* (Torymidae). The average parasitization rate was 23 % (Saeidi *et al.*, 2016)

Diseases of Safflower: Safflower yield losses due to foliar diseases such as Alternaria leaf blight alone is reported to vary from 20% (Indi *et al.*, 1988) to 50% under severe conditions of disease incidence (Indi et al., 1986).There are instances of total crop losses due to Alternaria incidence in Maharashtra (Patil *et al.*, 1993)

1. Alternaria blight (*Alternaria carthami*)

Dark necrotic lesions 2-5 mm in diameter are formed first on hypocotyls and cotyledons. Symptoms also appear on stem and severely infected plant gets blighted.

Brown discolouration appears on the stem, dark brown spots with concentric rings up to 1 cm in diameter appear on the leaves which later develop into large lesions.

Seeds also may be affected. Dark sunken lesions are produced on the testa. It may rot and damping off of seedlings occur.

The disease may be managed by avoid growing safflower in low-lying areas, and avoiding flood irrigation. The diseased plants must be destroyed.

Fig. 8: Foliar symptoms of Alternaria carthami

Seed treatment with Carbendazim @ 1.5g/kg of seed is recommended to prevent the disease.

Mancozeb (0.25%) should be sprayed immediately after disease incidence which may be repeated 15 days later depending on the intensity of disease.

2. Leaf spot (*Cercospora carthami*)

Safflower plants few weeks after planting or at flowering stage are commonly attacked.

Circular to irregular brown sunken spots of 3-10 mm diameter are formed on leaves.

Spots are surrounded by yellow halos.

Symptoms first appear on lower leaves and spread to upper leaves.

Fig.9. Symptoms of *Cercospora carthami*

Stems and nodes may also be affected.

In severe infections bracts are also affected with reddish brown spots.

Affected flower buds turn brown and die.

Removing and destroying the diseased plants helps in checking the pest.

Deep summer ploughing must be carried out to expose the dormant stages of the pest to hot sun. Seed treatment with Thiram 3 g/kg and spraying of mancozeb 2.5 g; or carbendazim 1 g per litre of water helps in checking the pest. (Anonymous 2018 A)

3. Wilt (*Fusarium oxysporum f.* sp. *carthami*)

The pathogen causes lesions at soil line initially, which extends inside and affects the vascular system. The Symptoms become more pronounced, when plants reach 6-10 leaf stage. The yellowing of leaves is observed followed by wilting and vascular browning.

The infected plants show small sized flower heads with partial blossoms. The pathogen is seed borne and gets entrenched in soil and in the plant debris. The disease is more pronounced in acidic soils with high nitrogen and under warm humid conditions.

The disease may be managed by appropriate crop rotation and avoiding drought conditions in the field.

Seed treatment with *Pseudomonas fluorescens* is helpful in cutting down the disease incidence.

Fig. 10: Wilt affected plant

The other methods of disease management are (1) soil application of *Trichoderma harzianum* @2.5 kg/ha. *T. harzianum* and (2) seed treatment with *T. viride* @ 10 g/kg seed, (3) seed treatment with Carbendazim (ST) (4) seed treatment with Captan or Carbendazim @0.1-0.2% 3 g/kg, (5) (5) seed treatment with *Trichoderma viride* mutant and (6) soil. application of potash @15 kg/ha to reduced wilt incidence.

Integrated Management of Pests and Diseases of Safflower: (NIPHM, 2014.)

Ecological engineering for pest management has recently emerged as a paradigm for considering pest management approaches that rely on the use of cultural techniques to effect habitat manipulation and to enhance biological control. The cultural practices are based on ecological knowledge rather than on high technological approaches.

Natural enemies of the pests in the ecosystem may require:

- Food in the form of pollen and nectar for adult natural enemies.
- Shelter such as overwintering sites, moderate microclimate etc.
- Alternate hosts when primary hosts are not present.

Ecological engineering for pest management – Above ground

- Raise the flowering plants / compatible cash crops along the field border by arranging shorter plants towards main crop and taller plants towards the border to attract natural enemies as well as to avoid immigrating pest population.
- Grow flowering plants on the internal bunds inside the field.
- Not to uproot weed plants those are growing naturally such as *Tridax procumbens, Ageratum* sp, *Alternanthera* sp etc. which act as nectar source for natural enemies.

- Not to apply broad spectrum chemical pesticides, when the P: D ratio is favourable. The plant compensation ability should also be considered before applying chemical pesticides.

Ecological engineering for pest management – Below ground:

- Crop rotations with leguminous plants which enhance nitrogen content.
- Keep soils covered year-round with living vegetation and/or crop residue.
- Add organic matter in the form of farm yard manure (FYM), vermicompost, crop residue which enhance below ground biodiversity.
- Reduce tillage intensity so that hibernating natural enemies can be saved.
- Apply balanced dose of nutrients using biofertilizers.
- Apply mycorrhiza and plant growth promoting rhizobacteria (PGPR).
- Apply *Trichoderma viride/harzianum* and *Pseudomonas fluorescens* as seed/seedling/planting material, nursery treatment and soil application (if commercial products are used, check for label claim. However, biopesticides produced by farmers for own consumption in their fields, registration is not required).

Due to enhancement of biodiversity by the flowering plants, parasitoids and predators (natural enemies) also will increase due to availability of nectar, pollen, fruits, insects, etc. The major predators are a wide variety of spiders, ladybird beetles, reduviids, long horned grasshoppers, lacewing, earwigs, etc.

Common cultural practices:

- Deep summer ploughing to control juveniles and adults of nematodes, and resting stages of insect pests.
- Follow crop rotation with non-host crops
- Destroy the alternate host plants
- Sow the ecological engineering plants
- Sow sorghum/maize/bajra in 4 rows all around the main crop as a guard/ barrier crop

Gram pod borer/capsule borer

Cultural Control

Growing intercrops such as cowpea, onion, maize, coriander, urd bean etc.

- Rotate the safflower crop with a non-host cereal crop such as wheat or barley, cucurbit, or cruciferous vegetable

- Grow repellent plants: Ocimum/Basil
- Plant ovipositional trap crops such as marigold for Helicoverpa
- Intercropping with non-host crop like wheat or barley
- Avoid chickpea as intercrop

Biological control

- Inundatively release *Trichogramma pretiosum* @ 40,000/acre 4-5 times from flower initiation stage at weekly intervals
- Application of NPV @ 100 LE/acre in combination with jaggery 1 Kg, sandovit 100 ml or Robin Blue 50 g thrice at 10-15 days' interval on observing the eggs or first instar larvae in the evening hours
- Apply entomopathogenic nematodes (EPNs) @ 100 crore infective juveniles of Steinernema feltiae/acre

Safflower caterpillar

Cultural control

- Intercropping with non-host crop such as wheat
- Follow common cultural, mechanical and biological practice

Safflower bud fly

- Follow common cultural, mechanical and biological practices

Safflower aphid

Cultural control

- If the attack is observed in the border rows take control measures
- Intercultural operations such as harrowing and hoeing reduce weeds such as *Parthenium hysterophorus* in safflower field which serve as alternate hosts
- Intercropping with sorghum, wheat and coriander reduces aphid infestation
- Intercropping with niger should be avoided

Biological control

- Release of Chrysoperla larva @ 2-3/plant or 70,000/acre
- Spray neem oil emulsion @ 0.25%
- Spray NSKE 4% (Neem Seed Kernel Extract 4%)

Chemical control

- Spray acephate 75% SP @ 312 ml in 200-400 l of water/acre or dimethoate 30%
- EC @ 264 ml in 200-400 l of water/acre or phenthoate 2% DP @ 8,000 g/acre or
- quinalphos 1.5% DP @ 8,000 g/acre

Integrated Management of Safflower Diseases

Alternaria leaf blight

Follow common cultural, mechanical and biological practices Cultural control:

- Avoid growing in low-lying and flooding areas
- Do not delay irrigation until the crop exhibits moisture stress symptoms

Cercospora leaf spot

Follow common cultural, mechanical and biological practices Cultural control:

- Avoid growing in low-lying and flooding areas
- Avoid continuous cropping/follow crop rotation

Safflower based Inter-cropping systems

Some of the remunerative inter-cropping systems are

Coriander + Safflower (3:1)

Linseed + Safflower (3:1)

*Chick pea + Safflower (6:3)

* Though this system is remunerative, it is being discouraged due to the incidence of a common pest *Helicoverpa*, which is becoming serious off late.

Rabi sorghum+ Safflower (1:3)

Wheat+ safflower (3:1)

Barley + Safflower (3:1)

Safflower based Crop Sequences

Region	Crop Sequences
Dharward and Balgaum of Karnatak	Green gram –Safflower Soybean-Safflower Groundnut-Safflower
Bijapur and Western Bellary of Karnataka Fi.9. Wilt affected plant	Hybrid Sorghum-Safflower Green gram-Safflower Sunflower-Safflower
RRdistt, Adilabad, Medak, Nizamabad of Telangana	Green gram-Safflower Maize –Safflower Hybrid Sorghum-Safflower Sesame - Safflower
Assured Rainfall Zones of Maharashtra	Green gram- Safflower Black gram-Safflower Hybrid Sorghum-Safflower Groundnut-Safflower Sesame-Safflower Sunflower-Safflower

Safflower Products

Safflower Oil: Safflower seeds (achenes) contain an edible oil ranging from 30 to 40%. The oil contains nearly 75% linoleic acid, which is considerably higher than corn, soybean, cottonseed, peanut or olive oils. This type of safflower is used primarily for edible oil products such as salad oils and soft margarines. Researchers disagree on whether oils high in polyunsaturated acids, like linoleic acid, help decrease blood cholesterol and the related heart and circulatory problems. Nonetheless, it is considered a "high quality" edible oil and public concern about this topic made safflower an important crop for vegetable oil.Varieties that are high in oleic acid may serve as a heat-stable, but expensive cooking oil used to fry potato chips and french fries.

As an industrial oil, it is considered a drying or semidrying oil that is used in manufacturing paints and other surface coatings. The oil is light in colour and will not yellow with aging, hence it is used in white and light-coloured paints. This oil can also be used as a diesel fuel substitute, but like most vegetable oils, is currently too expensive for this use (Oelke *et al.*, 1992)

Safflower cake: The meal left after oil extraction from the decorticated seed is used for animal feed, while that obtained from un-decorticated seeds is used for manure. Safflower cake contains about 40-45% protein. The cake contains about 7.9 per cent nitrogen, 1.9 per cent potash and 2.2 per cent phosphoric acid and its application as manure is supposed to greatly improve the physical properties of heavy soils. Although cattle apparently find safflower cake palatable, its bitter taste maker it unacceptable to humans. Protein isolates prepared from debittered meal can be used to fortify bread, pasta and nutritional drinks. Only lysine is limiting, while methionine and isoleucine are borderline.

Safflower dye: Apart from the oil, which is the primary product of safflower, the crop was grown in yester years to extract dyes of two kinds from its petals. Chemical analysis of ancient Egyptian textiles dated to the Twelfth Dynasty identified dyes made from safflower, and garlands made from safflowers were found in the tomb of the pharaoh Tutankhamun.

The petals are a most magical dyestuff. Yellows, surprisingly sharp pinks, orange-reds, and corals can be extracted from safflower. Soaking petals in water at room temperature gives a yellow which can be collected and used to dye any modanted natural fibre. Repeated soaking will exhaust the yellow at which point pinks may be obtained by "turning the bath" (drastically changing the pH to alkaline and then back to slightly acidic).

The water-soluble yellow dye, carthamidin, and a water-insoluble red dye, carthamin, which is readily soluble in alkali, can be obtained from safflower florets. Dye manufacture has virtually ceased in Asia, but dye is still prepared on a small-scale for traditional and religious occasions. Carthamin is found in the florets to the extent of 0.3-0.6 per cent and imparts a bright red colour to cotton and silk fabric. In order to get a better colouring effect from carthamin, the yellow colour first has to be separated from it. Florets may be collected, after the crop ripens, so that dye and oilseed can be obtained from the same crop. Colouring 1 kg of cotton yarn crimson requires 1 kg of dye, rose pink requires 500 gm, and light pink, 250 gm. With current problems encountered due to the use of chemicals for dying, there exists reaction against the use of chemicals for colouring of yarn, with an increased use of naturally coloured cotton. Thus, the revival of safflower dyes is in the offing.

Safflower as leafy vegetable and fodder: The tender leaves, shoots and thinnings of safflower are valued as pot-herb and salad. They are high in vitamin A, iron, phosphorus and calcium. Bundles of young seedlings are commonly sold as a green vegetable in markets in India and some neighbouring countries.

Safflower can be grazed or stored as hay or silage. The forage is palatable and its feed value and yields are similar to or better than those for oats or alfalfa. Tests have shown that fairly good seed production from a ratoon crop is possible by cutting at about 3-4 cm above ground at 30-45 days after planting in crops planted before the end of September.

"In many areas of India safflower seedlings up to 20 cm in height are used as a cooked vegetable. Sometimes seedlings are 'nipped' back to provide more plant as vegetable; it is believed by many that this practice will increase branching and yields" (P.F. Knowles In: McGuire *et al.*, 2012)

Safflower food colour: Safflower flowers are occasionally used in cooking as a cheaper substitute for saffron, sometimes referred to as "bastard saffron".

The dried safflower petals are also used as a herbal tea variety.

Addition of safflower florets to food is a widespread and ancient Chinese tradition. Health concerns regarding synthetic food colourants has increased interest in safflower-derived food colouring in recent times. Carthamin is the only chalkone-type pigment suggested for colouring foods. It finds use in colouring cakes, biscuits, butter, ice cream, rice, soup, sauces, bread and pickles, yellow to bright orange. Both safflower yellow and carthamin can be used as non-toxic food colourants. True saffron, which is probably the world's costliest spice, is quite commonly substituted or adulterated with safflower florets due to the similarity in their appearance (Nimbkar, 2019.)

8

Niger (*Guizotia abyssinica)*

Introduction and History

Niger, *Guizotia abyssinica* (L.f.) Cass. [Asteraceae] is commonly known as nigerseed [English]; noug, guizotia oléifère [French]; verbesina da Índia, abisin, negrillo, ramtilla [Spanish]; ramtil, nigersaat [Dutch]; [Amharic];[Chinese]; Ram-til[Hindi]; Gurellu [Kannada]; Verri-nuvvulu (Telugu) Karalle[Marathi] and Jhuse-til [Nepali] Niger's other synonym is *Guizotia oleifera* DC., *Polynya abyssinica* L. f.

Niger (*Guizotia abyssinica* (L. f.) Cass.) is an oilseed crop cultivated in Ethiopia and India for its edible oil. Niger plant is originated in highlands of Northern Ethiopia (Weiss, 2000). The crop is reported to have been introduced into India from Ethiopia during around2000BC. The Ancient *Agau* people of Ethiopia had a well-developed agriculture, who traded with Egypt which facilitated the exchange by barter that included Niger too (Weiss, 2000). Currently, Niger is an important oilseed crop of Ethiopia. Ground niger seeds and chillies are the main components of the traditional stew known as *wat* of Ethiopia. Niger is the most important oilseed crop of Ethiopia contributing to 50% of the country's oilseed production. Although the crop is grown throughout the country, the main production centres are Begemdir, Gojam and Welaga (Weiss, 2000). Since the Ethiopian government actively supports the oilseed industry, production was on steady increase till 1974. With the fall in the production post-civil war in 1974, most of the niger produced is consumed domestically and the exports got dwindled.

Niger crop is widely dispersed in India and is an important oilseed crop cultivated in the states of M.P., Chhattisgarh, Jharkhand, Gujarat, Maharashtra, Odisha, Karnataka, Tamil Nadu and Andhra Pradesh of India and also in Pakistan for its edible oil. The crop is also present in other African and Asian countries (Sudan, Uganda, Zaire, Tanzania, Malawi, Zimbabwe, Nepal, Bangladesh, Bhutan), and in the West Indies. European countries viz., Germany, Switzerland, France, Russia and West Indies and Canada etc., tried to produce niger; but the crop was not a success commercially. In France, it is used as a cover crop (Arvalis, 2016; Sem-Partners, 2016).

The niger harvest in India is 0.84 lakh tonnes over an area of 261000 ha of which a major part is contributed by the above mentioned Indian states. The productivity of Niger in India is 0.32 t/ha, which is far below the crop's potential. The productivity of niger is the highest (0.72t/ha) in the state of West Bengal of India. Niger should be considered primarily a valuable oilseed crop for small farmers especially in those areas where other oilseed crops are few. Although niger is considered as a minor oilseed, it is very important in terms of quality and taste of its oil and export potential. It has a great potential as a source of edible oil and could support small scale extraction plants.

In addition to its edible oil, niger seed is an important source of proteins, carbohydrates, vitamins and fibre that significantly contribute to the human diet. The crop has a wide genetic basis that is reflected in the form of a high variation in desirable traits, including seed yield, seed oil content, seed oil quality and photoperiod sensitivity (Geleta and Ortiz, 2013). However, the wealth of niger genetic diversity has so far remained largely unexploited and research efforts have not yet yielded satisfactory results in the form of new and superior cultivars. The recent molecular and nutritional quality studies coupled with pre-breeding work have opened up new opportunities for the improvement of niger. A high yielding niger cultivars with oil content of up to 60 % and/or oleic acid content of up to 70 % can easily be bred based on Ethiopian niger gene pool through the combined use of novel genomic tools, traditional breeding and farmer-participatory approaches (Geleta and Ortiz, 2013). The improvement of niger will have a significant contribution towards food security and sustainable development in general and self-sufficiency in edible oil in particular. Overall, an investment in the niger improvement programs will likely be of benefit far beyond Ethiopia's borders and could potentially lead to the expansion of the crop outside the regions where it is currently grown.

Uses: In Ethiopia, niger is a major source of edible oil and provides about 50% of the country's oilseed production. Ethiopian niger seeds contain about 40% oil (Getinet and Sharma, 1996). In India, niger oil is less than 3% of oilseed production (Bulcha, 2007). Niger oil is pale yellow, with a nutty taste and a pleasant odour. It is mostly used for cooking, as well as in paints and for the extraction of perfume from flowers (Getinet and Sharma 1996). Niger seeds are used as food in many Ethiopian dishes, condiments and snacks: "*littlie*" or "*chibito*" are niger seeds roasted and ground with salt, then mixed with roasted cereals. They are traditionally served during coffee ceremonies (Bulcha, 2007) in Ethiopia.

In Ethiopia, niger oil cake is the main protein supplement for livestock (Getinet and Sharma., 1996). In Western countries, niger seeds are important components of birdseed mixtures (Lin, 2005). The whole niger plant can be used as fodder for sheep, but it is unpalatable to cattle, to which it is only acceptable as silage (Chavan, 1961). Niger is a valuable cover crop between cereal crops and it can be turned into green manure (Arvalis, 2016). Niger seed oil can be utilized for the manufacture of soaps, paints and lubricants, and the protein rich meal obtained after oil extraction is used as animal feed or for fertilizer (Getinet and Sharma, 1996

Ecology: Niger is a short day plant adapted to temperate climate. However, the crop has adapted itself to semi-tropical environment. In Ethiopia the crop is grown between 1600 m and 2500 m zone from sea level bisected by 10*N latitude, though it is sown in small patches throughout the country. It appears that the existing variability in the niger population may facilitate extending its zone of adaptability (Getinet and Sharma 1996) to wider range of climate and soil. The highest reproductive efficiency of niger is observed in the zone of 500 m to 1000 m from sea level under semi tropical conditions with availability of adequate quantum of moisture. In India the crop grows well on warmer plains and up to 2000m from sea level representing some hilly districts. The growth and successful production of niger in Ethiopia on one hand and in India on the other, representing diverse agro-ecological situations is ample testimony of the marvellous adaptability of the crop to varying ecological situations. This also brings out the differentiation of the Indian types from the Ethiopian types in their reaction to specific climatic conditions of day-length and temperature. While the moderate temperature of 18-23*C is the normal requirement for the growth and development of niger, the Ethiopian types can't tolerate the temperatures of around 30*C, which adversely affects their growth and flowering, in contrast to the Indian types which can grow satisfactorily in such ambiental situation.

In Ethiopian high lands, the night temperature that goes down to 9*C has no any visible effect on the growth and development of niger, while the incidence of frost kills the seedlings. Thus the farmers of Ethiopia try to avoid the adverse effect of frost by selecting an appropriate sowing time. Niger is a variable species adapted to different environments: cool tropical Eastern Africa, hotter tropical and subtropical lowlands of India and temperate Europe. It can be grown from sea level up to an altitude of 2500 m where average daily temperatures range from 13°C to 23°C, and night temperatures are above 2°C. Optimal annual rainfall is about 600-1300 mm, and more than 2000 mm depresses seed yield. Niger grows well on a wide range of soils, from poor sandy soils to heavy black cotton soils, at a pH varying from 5.2 to 7.3. Niger withstands waterlogged areas where there is poor oxygen supply because of its aerenchyma and its ability to form respiratory roots. Niger plants have some tolerance to soil salinity (Bulcha, 2007).

Niger requires adequate rainfall of around 600 to 1,000 mm to grow and to yield commercially viable produce under diverse soil conditions. Nevertheless, now in India niger is pushed to drylands due to the necessity to use more fertile and irrigated lands for food crops. In this context it has been observed that niger can withstand higher temperatures in the vegetative phase.

As compared to other crops, niger withstands water-logging. In Ethiopia the cultivars that withstand water-logging are planted in high rainfall areas and in zones of impeded drainage.

There exist varietal differences among genotypes of niger regarding their adaptation to different regions in Ethiopia. The varietal form grown in rainy season on well drained soils (June –December) is known as '*abat noug'*, while '*mesno noug' is* grown during September –January in high rainfall areas where waterlogging occurs.

Soils: Niger grows on wide range of soil types; but is found to grow well on clay-loams, sandy-lay soils and sandy-loams. The crop gives its best expression on soils of moderate fertility; but the crop performance declines when grown on poor laterite and gravelly soils as is happening in India (Chavan, 1961). Similarly, niger crop performance goes down on black cotton or heavy clay soils. In Ethiopia the crop is grown on dark-brown clays of south Begemdir regions, the red-brown clays of Gojam and more loamy clays of Addis Ababa. Niger seems to grow satisfactorily on acidic soils of pH of 5.5 to 6.5. The optimum soil pH suitable for niger has not yet been worked out.

Niger Plant

The niger plant of the family Asteraceae is a stout, erect annual herb that grows up to a height of 2 m (Bulcha, 2007). The stems are soft, round, hairy, hollow with a diameter up to 2 cm, and branched moderately. The degree of branching depends on the environment in which the plant population is growing. Their colour is pale green, often stained or dotted with purple, and become yellow with age. Local varieties may have a distinct stem colour, the green being replaced to some extent by red or reddish yellow. Dwarf plants have been recovered in breeding programmes, which could be used to reduce the stem height in varieties for mechanised harvesting (Dagne, 1994). Branching of stem is an important attribute for niger since the flower-heads or capitula are borne terminally on stem branches. The number of branches and their capitula account for 50% of variation in seed yield (Weiss, 2000). Generally, the stem and leaf biomass increase until the appearance of the first capitulum. Later there is a gradual decrease of the plant biomass. Shoot biomass was found to be the greatest at 120 days after emergence on selected Indian varieties (Pandey, 1997).

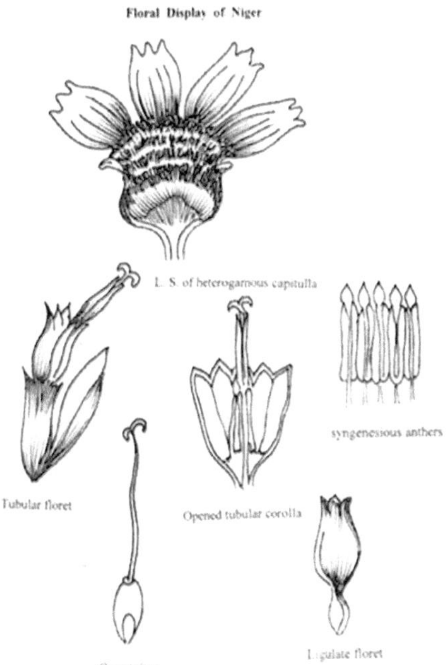

Fig. 1. L.S. of Niger capitulum and its floral parts. (Picture from Patil *et al.*, 2012)

Floral Biology and Anthesis including self-incompatibility in Niger: The inflorescences are arranged in apical or axillary cymes surrounded by leafy bracts up to 3 cm long. The flowers are in capitula, ranging from 15 to 50 mm in diameter, bright yellow becoming golden yellow as they mature. The capitula are surrounded by leafy involucar bracts arranged in numerous rows. The outer bracts are lanceolate with a length of around 2cm. The inner rows of bracts get progressively smaller and finally merge into flattened paleae of the receptacle of the capitulum, which is of 1.5 to 3.0 cm diameter. The period from emergence to peak flowering is normally around 90 days provided optimum conditions for growth of plant prevail. Nevertheless, the period often gets reduced since the crop is grown under diverse conditions of soil, moisture availability, day length and temperature. A niger plant may produce 30 to 40 flower heads, the number of which my go up occasionally depending upon the environmental factors mentioned above. The flower heads develop from leaf axel in clusters of 3 to 5. Increase in the number of flower heads is not matched with synchrony of maturity and hence at the time of harvest perceptible yield losses occur due to uneven ripening and shattering.

The flowers on the capitula are of 2 kinds viz., ray and disc florets. There may be eight to 15 ray florets which are sterile but showy with bright yellow projecting petals. The colour of the ray floret petals often turn golden yellow as the capitulum nears maturity. The disc florets situated surrounded by the ray florets are fertile, numbering around 35 to 40 or even more at times. The disc florets are arranged in three whorls and those at the periphery open first followed by the rest in the subsequent inner whorls progressing towards the centre in that order. The process of anthesis takes around a week.

While the pollen is liberated in the early hours of morning, the emergence of stigma happens in the mid-day and the stigma lobes open and curl backwards by the evening. The style emerges covered with pollen but the receptive part rarely or never comes in contact with that pollen, a phenomenon favouring cross pollination (Getinet and Sharma,1996). Hence, niger is a completely outcrossing species with certain levels of self-incompatibility mechanism (Chavan 1961; Mohanty 1964; Shrivastava and Shomwanshi 1974; Sujatha 1993).

Self and cross incompatibility in niger (*Guizotia abyssinica*) is found to be governed by multiple oppositional alleles at one locus. Here the pollen reaction is sporophytically determined. Some alleles showed dominance and some behaved as recessives in pollen or stigma. The parents contained five S alleles of which three (S_1, S_3 and S_5) were in active form and two (S_2 and S_4) in recessive condition. Two types of allelic relationship namely independence and dominance were detected in heterozygous state. Of the five alleles recovered in progeny families, three acted independently and found to be dominant over the other two in pollen and stigma (Panda and Rao, 1976).

The self-incompatibility nature of niger complicates the production of selfed seed. At Holetta, 600 accessions were tested for their ability to produce selfed seed using muslin cloth bagging (Riley and Belayneh 1989). Twenty-two out of the 600 accessions produced approximately 1 g of selfed seed per plant, indicating that niger germplasm with some level of self-compatibility exists within the Ethiopian genepool (Getinet and Sharma,1996)

Major pollinators of niger are insects, particularly honey bees. In India the experiments carried out by introducing bees in protected cages where niger plants were grown resulting 3 to 4 times seed yield as compared to those without bee populations (Weiss, 2000). Gebremedhn and Tadesse (2014) reported that the highest germination rate was found in seeds obtained from plants caged with honeybees (86.2%) followed by that of open pollinated (79%), while the lowest germination rate was found in the seeds of plants excluded from insects (60.5%)

The use of bee attractants such as Bee-Q and Fruit Boost™ in the pollination of niger was evaluated by Sivaram *et al*., (2013). Bee visitations to niger flowers were observed for two weeks and an estimation of seed yield was determined.

Different concentrations of Bee-Q and Fruit Boost™ were evaluated to understand honeybee visitation patterns on the target crop and pollination efficiency. Results indicate that spray applications of Bee-Q at 12.5 g/l and Fruit boost at 0.75 ml/l on niger plots significantly increased the number of bee foragers over control plots. In addition, plots sprayed with these bee attractants significantly enhanced the seed set, seed weight, and germination of niger. This study suggests that pheromone-based bee attractants applied to niger can increase the marginal percentage of bee visitation, seed yield, and percent germination. Niger flowers are very attractive to honey bees and for this reason niger plants are inter planted with sunflower to attract bee populations for sunflower pollinations.

Notwithstanding the cross pollinated nature of niger, it is possible to produce self-fertilized seeds, that are viable. The process involves dusting the capitulum repeatedly with pollen 3 to 4 days without emasculation after flower opens. The minimum isolation distance to protect the breeding plots from out-crossing is 250 meters. In addition to maintaining the required isolation distance, it should also be recommended to grow tall crops such as castor or sunflower or sorghum etc., around niger breeding plots to shield the plant material from out-crossing.

Atmospheric temperature affects the period of flowering, including pollen fertility. In India atmospheric temperature of more than 30*C may have adverse effect on pollen fertility. The total duration of flowering for individual heads is around 5 to16 days. The total flowering for a plant may be from two weeks to a month. Most of the immature seeds and pops arise from the central portion of head. The major problem of harvest is shattering since the well-developed and mature seeds get dislodged from the head and fall down.

Seed (Fruit) of Niger: The fruit is achene which is termed seed as in the case of sunflower pertaining to the same family. Each flower produces about 50 seeds. The seeds are small achenes, each 3-6 mm long,1.5-4 mm broad, glossy or black in colour (Bulcha, 2007). Niger seeds are lanceolate in shape, without papus. The testa is glossy, hard and black in colour. However, mottled seed also can occur at times. Each head may produce 15 to 30 mature seeds. The normal 1000 seed weight is 3 to 5 g. Niger seeds resemble sunflower seeds in shape, but are smaller in size and black. It bears a fairly thick, adherent seed coat and can be stored for up to a year without deterioration. Niger seed contains proteins, oil and soluble sugars. Niger seeds are used as bird feed worldwide.

Seed germination is epigeal, which takes 4 to 5 days to complete. The cotyledons remain adhered to seedling after completion of germination.

Roots of Niger Plant: The root system is well developed, with a taproot that has many lateral roots, particularly in the upper 5 cm of soil. Niger roots exhibit good tolerance to water-logging due to presence of aeranchyma, which varies depending upon the degree of soil saturation. It has been found that the niger roots secrete a compound that inhibit germination of monocots (Weiss, 2000).

Seed Oil of Niger: Seed oil content from open pollinated land races exhibits a wide variation from 25 to 45%. Certain improved varieties may have an oil content ranging from 45 to 50% and protein content may from 12 to 25%. Niger oil is pale yellow in colour and virtually odourless or with a slight sweet aroma. The major component of niger oil is linoleic acid measuring up to 75% and the rest of the oil contains largely oleic acid and certain recently bred varieties are reported to possess as much as 85% of linoleic acid (Dasthagiriah and Nagaraj. 1993). Oil of niger is reported to be safe for human consumption (Dagne and Jonsson, 1999)

Crop Nutrients: Since niger is grown largely by small farmers, its nutrient requirements are ignored. The crop, however, is grown on the residual fertility of the previous crops viz., cotton or a grain crop and at times the farm yard manure is also incorporated in the soil where the crop is grown (Chavan, 1961). According to Halder *et al.*, (1997) inoculating niger seed with *Azotobacter* and top dressing on nitrogen @ 30 kg/ha has enhanced crop yields.

Nitrogen: Nitrogen improves the yields of niger by enhancing the branching and consequently the number of flower heads per plant, the number of seeds per head, seed weight per plant, oil yield and oil-cake yield (Patil, 1979) as given in Table 4.

Table 1: Increased seed yield and cash returns from applied Nitrogen (Patil 1979)

Nitrogen (kg/ha)	Seed yield (kg/ha)	Oil yield (kg/ha)	Oil cake yield (kg/ha)	Total returns from oil & oil-cake (IRs/ha)
0	172	68	104	489
20	213	84	129	608
40	247	96	151	699

It is preferable to apply nitrogen as top dressing, particularly at the initiation of first capitulum buds.

Phosphorous: The soils on which niger is being grown particularly in India are deficient in Phosphorous; but application of P alone without nitrogen may not result in significant yield increases. Conversely response to nitrogen is higher with added phosphorous in niger (Weiss, 2000). According to Weiss (2000) application of triple superphosphate resulted higher yields of niger in East Africa including Ethiopia. Application of P @ 40kg/ha at the seedbed may be adequate in the soils where niger was not grown earlier (Weiss, 2000). In the case of soils where the previous crop received adequate fertilizer doses, application of P @ 10 kg/ha to the crop of niger could be adequate.

Other nutrients: Response to potassium was noted where the soils were deficient in K. There was a good response to applied sulphur in the cases of availability of major nutrients, in India (Halder *et al.*, 1997). Response to Mg, Ca, Mn, Fe, Si and Na was observed along with the content of the said nutrients in plant parts in pot culture experiments (Nalawade and Chavan, 1990; Guggari *et al.*, 1995).

Cytogenetics and Species Relationship: The nomenclature of *Guizotia* (Cassini) has been chosen to honour the French historian Guizot. The genus *Guizotia* is a small group consisting of six species including *G. abyssinica* (L.f.) Cass (syn. *G. oleifera* DC) the cultivated oilseed crop niger. Vavilov (1951) described niger as belonging to the 6th Centre of Origin of cultivated plants, which is the Abyssinian Centre. The genus has somatic chromosome number 2n=30.

Except for the species *G. zavattarii,* the rest are crossable with niger. Crosses involving. *G abyssinica* and *G. scabra* subsp. *schimperi, G. scabra* subsp. *scabra* and *G. villosa* produce well developed seeds. *Guizotia scabra* contains two subspecies, namely *scabra* and *schimperi. Guizotia scabra* subsp. *schimperi,* known as '*mech*' is a common annual weed in Ethiopia.

Guizotia abyssinica, Guizotia schimperi and *Guizotia scabra* are diploid species (2n = 2x = 30) characterized by 15 bivalents during prophase – I / metaphase – I of meiosis (Murthy *et al.,* 1993)

Interspecific hybrids between these three species were obtained and the F_1 hybrids were studied in order to get insight into the cytogenetic relationships and genetic differentiation within the genus *Guizotia.* Meiotic chromosome configurations at diakenesis / metaphase – I in pollen mother cells of hybrids averaged 0.25_{I} + 14.60_{II} + 0.15_{IV} for *Guizotia abyssinica* x *Guizotia schimperi;* 0.05_{I} + 13.6_{II} + 0.14_{III} + 0.58_{IV}*Guizotia abyssinica* x *Guizotia scabra* and 0.8_{I} + 12.7_{II} + 0.08_{III} + 0.88_{IV} for *Guizotia schimperi* x *Guizotia scabra* [I = Univalent, II = Bivalent, III = Trivalent and IV = Quadrivalent.

Based on the results, it could be concluded that the genomes of *Guizotia abyssinica* and *Guizotia schimperi* are similar and homologous, whereas the *Guizotia scabra* genome is only partially homologous to that of *Guizotia abyssinica* and *Guizotia schimperi*. It is suggested that the crop species of *Guizotia abyssinica* might have originated from *Guizotia schimperi* through selection and domestication. Chromosomal translocations have played a major role in the divergence and differentiation of the three species (Murthy *et al.,* 1993).

G. abyssinica and *G. scabra* are closely related to *G. villosa* than to *G. scabra* sub spp., *schimperi. G. zavattarii* is, however, observed to have very low affinity to rest of the species of *Guizotia* (Dagne,1994; Murthy *et al.,* 1995) Seed oil of *G. abyssinica* and that of rest of the genus are similar in their biochemical properties, but for small differences (Dagne and Johnsson, 1999). The distribution of different species of *Guizotia* and their specific characters are given in Table 2.

Table 2: Distribution and specific characteristics of species of Guizotia (Getinet and Sharma 1996)

Species of Guizotia	Distribution	Characteristics
ANNUALS		
G. Abyssinica (L.f.) Cass	Ethiopia and neighbouring countries in East Africa & Indian sub-continent including Nepal	The intensive branching account for 50% of variation in seed yield. The leaves are opposite, sometimes alternate at the apices of the stems, leaves are simple and sessile, leaf blade is lanceolate to obovate, 3-23 cm x 1-6 cm, variable in shape, with a margin that is either entire or toothed, ciliate, softly hairy on both surfaces. Leaves are usually dark green, lower ones yellowish
G. scabra (Vis.) Chiov. ssp schimperi (Schultz Bip. In Walp.)	Ethiopian highlands	Moderately branched, with outer involucar leaves, ovate leaves
G.villosa Sch. Bip. Northern and south western	Distributed in northern and south western Ethiopian highlands.	Highly branched, weedy habit, occurred as a weed
PERENNIALS		
G. scabra (Vis.) Chiov. ssp.scabra	Ethiopa to Zimbabwe in south and to the Nigerian highlands (West)	Moderately branched, scabrous, different from annual types with an outer involucar lanceolate leaves.
G. reptans. Hutch	Restricted to Mount Kenya, Aberdadares, Mt.Mount Elgon in East Africa,	Sparsely branched, mat like growth that covers the soil
G. zavattari Lanza	Endemic to South west Ethiopia, Hari hills of North Kenya.	Erect, glandulous, Predominantly a shrub
G. arborescens.I.Friis.	South west Ethiopia & Mt. Imontong on the border between Sudan and Uganda.	Rare arboreal species.

Genetic Resources and Plant Breeding

Germplasm Collections: The health and vigour of any niger breeding programme depends upon the extent of genetic diversity and usable variability instilled. Hence, the utility of effective and comprehensive genetic resources in this context can't be over emphasized. Germplasm collection and its evaluation are the key to success of breeding programme and every effort must be made to enhance the value of the germplasm by pumping in new and novel genetic resources.

The niger germplasm collections available in different countries are given below (Getinet and Sharma, 1966)

Ethiopia: 1071 accessions

India: 1,528 accessions

Germany: 4 accessions

Nepal: 36 accessions

South Africa: 1 accession

Since the genetic diversity is the basic foundation of any germplasm collection, the variability collected must be documented with passport data and evaluated for all the taxonomic and agronomic characters so as to facilitate breeders to select suitable parental materials.

Improved technique of emasculation and pollination for niger breeding: Sujatha and Angadi (1989) have developed improved crossing method in niger for obtaining F1 hybrid seeds. The different steps involved in the process are described below for the use of niger breeders.

(i) The disc floret of the female parent may be selected.

(ii) The stamens of disc floret are epipetalous and syngenesious.

(iii) Chose a cloudless, clear and sunny day for emasculation when disc floret appears swollen, turning yellow in the morning hours at around 8 AM

(iv) The anther column becomes more conspicuous with white tip, in the morning which is the appropriate time together with suitable condition of the floret for effecting emasculation.

(v) Without splitting of anther column, pull out the white tip of anther with one jerk using a fine tweezer.

(vi) The syngenesious anther tube comes off quickly without any damage to the stigmatic surface.

(vii) Pollinate immediately the emasculated floret with desired male parent's pollen.

(viii) Bag the pollinated disc floret with label where the details are entered clearly.

(ix) The above method of crossing registered more than 75 % seed set as compared to the conventional method (Sujatha and Angadi, 1989).

The reasons for enhanced seed set in this method are (i) clear exposure of stigmatic surface for convenient pollination, (ii) little or no damage to the floral parts following emasculation and (iii) adequate time and facility made available for pollination.

Among gametocides used for induction of male sterility in niger, GA_3 at a concentration of 150 ppm proved to be the most successful chemical to induce maximum male sterility in niger compared to other chemicals (Gangaprasad *et al.*,2004)

Niger Breeding: In crop breeding, the objectives of the genetic improvement of the crop in question must be specified to chalk out the needed strategies of genetic improvement: The basic problems that are faced with regard to niger plant are as follows;

- The niger plant has an extremely low harvest index which inhibits the expression of the reproductive potential.
- Some problems hampering the realization of the full potential of niger are the low-yielding capability of its cultivars and a susceptibility to diseases.
- In addition, self-incompatibility causes serious dificulties for inbred line development and maintenance which has impeded the improvement of the crop by conventional plant breeding techniques (Getinet and Sharma, 1996).

Based on the reports of Alemaw and Wold (1995), it appears that the niger land races and native strains under cultivation in Ethiopia have not been subjected to much of human selection. This draws confirmation from the biochemical investigations of Dutta *et al.,* (1994) on the pattern of variation in lipid composition of niger seed samples collected from different regions of Ethiopia.

On the other hand, in India due to the preferences of diverse local populations pertaining to different regions, the niger populations were subjected to selections pressures. The activities of varietal development to cater to the specific regional choices (Ex. Varieties like *Ootacmand* and *Alsi-1*) have also added to the fixation and restriction of genetic diversity of niger in India unlike Ethiopia where the niger populations have retained good amount of genetic diversity. This can be clearly seen in any monocrop of niger in Ethiopia.

The Indian trials conducted with Ethiopian types of niger have revealed the Indian and Ethiopian types differ in their reaction to photoperiods, temperatures and plant populations (Weiss, 2000). While Ethiopian strains out-yielded Indian varieties in Ethiopia, the reverse trend was observed in India. In other words, Indian varieties being superior to the Ethiopian strains in India was observed indicating thereby long acquired continental adaptation (Weiss, 2000).

Niger production in Ethiopia is mainly based on local landrace populations. Four improved varieties – Sendafa, Fogera-1, Esete-1 and Kuyu – were released by the Institute of Agricultural Research, Holetta Research Centre, Addis Abeba (Getinet and Sharma, 1996). The maturity duration of the above varieties varies from 136 to 148 days with seed yield ranging from 800 to 1060 kg/ha. The oil content of the varieties is of the order of 39 to 41%.

Despite breeding research being carried out on sound lines, progress of the niger breeding in India has been rather slow as compared to other oilseed crops. During the period from 2002 to 20016, 13 varieties of niger were released, out of which the varieties JNC-6, JNC-9, Utkal Niger-150 and BNS-10 are promising from the point of view of productivity. Certain important varieties recommended for different states of India are given below in the Table 3

Table 3: Varieties of Niger Recommended for different States of India

State	Recommended Varieties
Madhya Pradesh/ Chhattisgarh	JNC-1, JNC-6, JNC-9
Maharashtra	IGP-76, IGPN-2004-1 (Phule Karala-1)
Karnataka	RCR-317, RCR-18, KBN-1
Odisha	GA-10, Utkal Niger-150
Jharkhand	Birsa Niger-1, Birsa Niger-2, BNS-10
Gujarat	Gujarat Niger-1, NRS-96-1
Tamil Nadu	Paiur-1

(Source: Oilseeds Division, Department of Agriculture, Cooperation & Farmers' Welfare, Ministry of Agriculture & Farmers' Welfare, Government of India, New Delhi.)
www.nmoop.gov.in

In the above back-drop the breeding strategies to be adopted are as follows:

- Breeding for higher number of branches bearing effective capitula
- Breeding for synchrony of maturity
- Breeding for higher number of seeds per capitulum
- Breeding for bolder seed size of at least 3.5 g per 1000 seeds. (Current 1000 seed weight is 2.3 to 2.5g)

Current mean values of yield components in niger crop are as follows (Sandeep and Kubsad, 2017):

- Mean number of branches per plant- 10.5,
- Mean number of capitula per plant- 54.5,
- Mean number of seeds per capitulum - 21.4 and
- Average seed yield / per plant - 2.78 g

Normal breeding procedures that could be adopted in the case of niger are indicated below:

(1) Mass section from (a) S1 (selfed) populations, (b) Half-sib progeny selection and (c) Full-sib progeny selection. Mass selection is a powerful tool for crop improvement. In niger, this technique has been successfully employed for the development of an early to medium maturing, short plant type variety. The resulting varicty (Kuyu) was 9 days earlier maturing, 10 cm shorter in height and significantly higher yielding than standard niger varieties (Getinet and Sharma, 1996)

(2) Recurrent selection (Pandey and Gardner, 1992) is used for traits that are polygenically inherited. Polygenic inheritance occurs when many genes, each with a small effect, control the expression of a trait. It is a cyclic method of population improvement in which, the basic steps in a cycle are inter-mating, evaluation, and selection. Recurrent selection increases or decreases the frequency of alleles (an allele is one of a pair or series of forms of a gene at a specific locus on a chromosome) by selecting within a normal distribution of genotypes. Recurrent selection is a proven method in which the selected individual plants are inter-mated among themselves and the succeeding progeny is subjected to selection over certain norms. The plants thus selected were allowed to inter-mate and the steps mentioned above will be continued till the selection advance is achieved for the attributes in question. The various steps involved in the process are as follows:

- A large number of plants are to be selected based on the norms indicated from open pollinated populations. Seeds of the selected populations are analysed for oil content.
- A part of seeds of the selected plants are planted in progeny testing nursery@ one row per plant with a minimum of two replications along with appropriate check / controls.
- Based on the progeny test values the best original selections are selected and inter-mated.
- In the progeny of the inter-mated selections again selections are made based on the norms and again taken to the next cycle of inter-mating. Thus the process is repeated till the 4th generation, when the selected population is expected to get stabilized for the norms laid out for the characters chosen.

(3) In the case of paucity of variability for the genetic attributes under study as is the case of Indian niger crop, mutation breeding (Prasad, (1972a), Prasad *et al.,* (1984) may be carried out to generate new genetic variability for the attributes concerned, which widens the variability for selection process.

(4) Despite the prevalence of self-incompatibility in niger, some self-pollination is possible. Where ever self-fertilization was effected, a very high degree of inbreeding depression of the order of 91% was observed (Ramachandran and Menon, 1979), which invariably resulted in very poor seed setting. Selfing or sib-mating is normally resorted to, in order to maintain the genetic purity of the genetic resources and also to produce inbred lines for production of hybrids and synthetic varieties.

(5) Development of Synthetic varieties involves production of inbred lines which are evaluated for their general combining ability (GCA) by crossing

with an open pollinated tester. The inbred lines thus selected based on their GCA are inter-mated.

Niger Varietal Seed Production: Niger being cross-pollinated plant with incompatibility system, care must be bestowed in the seed production of the promising varieties, which in deed is technical job. Implementation of the following items would go a long way to ensure quality seed production in Niger.

- **Isolation Distance**: The two factors viz., cross-pollination and self-incompatibility influence the isolation distance which would be different from that of other cross-pollinated systems. In the case of niger the varietal material should be so isolated as to (a) prevent any outside pollen other than that of the selected variety entering the varietal population and (b) promote or force the obligate pollination with in the varietal population isolated, which at times may result in reduced seed yield, which of course should be preferred as the cost incurred for the genuine genetic purity of the variety. In consideration of the above crucial factors a minimum isolation distance of 1000 meters is prescribed to rule out intervention of any foreign pollen contaminating the genetic purity of the variety under seed production. This is specifically recommended for nucleus, breeders' and foundation seed production systems. In respect of the certified seed production, the isolation distance indicated is 600 to 800 meters depending upon the prevailing local agro-ecological factors.
- **Rouging**: This is the most crucial operation aimed at eliminating the expression of the residual variability with in the varietal population that does not contirm to the standards of character expression set for the variety. Hence it is obligatory to remove all the off-types not confirming to the standard characters emerging from the isolated varietal population. Rouging has to be done at four different stages viz., seedling, vegetative, flowering and maturity stages. The fields should be inspected soon after each rouging to confirm the implementation of removal of the off-types.
- **Seed Production:** Management of seed production can't be treated in isolation. Also, it can't be clubbed under regular agro-production programmes, since it has to go hand in hand with other items of seed production.

In the order of importance, the subjects of (1) isolation distance and (2) rouging precede, planning of seed production system. The following are the important factors to be considered for seed production agronomy.

(i) In this context season plays an important part. Normally regular niger crop is grown during the monsoon season starting from July to October in India. In the case of seed production, it is preferable to choose off-season for seed production due to two important reasons viz., (a) the seed crop will be protected in the off-season from pollen from regular crop and (b) it should be possible to implement specific seed production agronomy conducive to higher yields avoiding to a great extent the biotic stresses that plague the crop in regular monsoon season.

(ii) The nutrient management package to be implemented should aim at higher dose of Nitrogen in addition to all other components. In otherwords the recommended nutrient schedule should be 30 to 40 kg of Nitrogen + 40 kg of P2O5 + 20 kg of K2O/ha at sowing after the pre-sowing incorporation of 4 tons of FYM/ha. Seed treatment with regular fungicide viz., Thiram, Mancozeb, Bavistein @ 2 to 3 g/ 1 kg of seed and pre-treating the seed with PSB (Phosphate solubilizing bacteria) @ 10 g/ kg of seed is also recommended. Topdressing of nitrogen @ 20kg/ha after 35 to 40 days following sowing should be carried out. Spray schedules of micronutrients and application of bio-fertilizers may be carried out as per the necessity.

(iii) The seed production crop should be spaced wider enough to promote good growth and expression of varietal characters and yield components of each individual plant. While the normal recommended spacing may be 30-35 cm between the rows and 5-10 cm between the plants within a row, the seed production crop should be spaced with an inter-row spacing of 40 to 45 cm and 15 to 20 cm of intra-row (within the row from plant to plant) spacing.

(iv) Since the seed production crop is invariably irrigated, even slight moisture stress must be avoided at different stages of the crop growth and timely weeding is to be carried out.

- **Use of Bee Pollinators to augment yield in seed production plots:** It is being suggested that bee colonies be established in niger seed production plots, since it is proved beyond doubt the efficacy of bee pollination in increasing the seed yield of niger. While setting up bee colonies in regular crop production plots of niger is welcome proposition, such a step should have to be viewed with caution regarding varietal seed production. Bees are not stationary insects. It has been demonstrated that honey bees (*Apis millefera*) regularly forage several kilometres from the nest (Caron and Conor, 2013; https://en.wikipedia.org/wiki/Western_honey_bee) by which the bees may bring in foreign pollen from outside that may contaminate the seed production plots, thereby jeopardising the genetic purity. Since genetic purity can't be compromised with in varietal seed production, use of bee pollinators to augment seed production must be reviewed with caution since such a step may not be appropriate in the context of seed production, unless and otherwise a set of separate steps are taken to restrict the movement of the bees with in the ambit of seed production plots by installing barriers not to allow bees to forage outside seed production plots.

Harvesting and Post-harvest processing of seed crop

1. Harvesting of seed crop may be carried out at physiological maturity to avoid any loss due to shattering or breakage of dry plants due to delayed harvest
2. Threshing of the harvested crop must be done in *isolated clean threshing yard, which has not been used for threshing the regular non-seed crop* of niger to avoid any mechanical contamination.
3. The threshed seed should be sieved, graded and cleaned to recover bold, well developed and lustrous seed devoid of trash, colourless, shrivelled and broken seeds and foreign material.
4. Seed thus processed must be packed in bags or containers suitably designed for seed storage with clear identification labels and kept preferably under cold storage till it is dispatched for further use.

Heterosis breeding in Niger: Niger is an excellent candidate for hybrid development in view of its pollination system which is similar to that of sunflower. The basic requirement for hybrid breeding is the availability of genetically diverse and heterotic germplasm. It is anticipated that Ethiopian and Indian niger germplasm which are genetically very diverse, might express high levels of heterosis for seed yield in cross combinations. With theidentification of genetic male sterile (GMS) system by Singh and Trivedi (1993), the path is paved for studies to estimate the extent of heterosis in Niger, which is primary step towards production of heterotic hybrids. Among the six crosses studied involving diverse parents, two of them viz., GMS x RCR - 290 and GMS x Phule-1 exhibited higher levels of heterosis by virtue of the heterotic effect clearly discernible among the yield components. The hybrids exhibited 10-30% heterosis for seed yield over the better parent and 15-55% over mid-parent yields. No heterosis was observed for oil content except in one hybrid combination.

Breeding for Oil content and quality: Before discussing breeding for oil content and quality, the salient features of niger oil should be highlighted. Niger seed oil content in general in open pollinated land races exhibits a wide variation from 25 to 45%. Niger oil is pale yellow in colour and virtually odourless or has a slight sweet aroma.

The oil yield of niger crop per unit of land is very low ranging from 50 to 75 kg as compared to other oilseed crops. In view of the low oil yield of niger the priority should be to enhance the seed oil content of the crop on one hand and increasing the seed yield on the other. Breeding for higher seed yield which will result in augmented oil yield is discussed in the preceding pages.

Considering the genetic enhancement of oil content in niger, the available variability for oil content ranges from 25 to 40% in the cultivated open pollinated niger. The range of oil content among the released varieties of niger crop is within 18 to 40% (Patil *et al.*, 2012 Ranganatha *et al.,* 2016) thereby indicating the paucity of genetic variability for oil content among the varieties of niger crop. On the other hand, the analysis of 153 niger populations carried out by Geleta *et al., (*2011) revealed a twofold variation in seed oil content (27–56%) with 7% of the populations having more than 50% oil. These high oil content populations came from different regions and a wide altitudinal range (1400–2590m asl). There was also a high variation in oleic acid content between populations (3.3–31.1%). Interestingly, the populations with more than 13% oleic acid were entirely collected from elevations of less than 2000m asl. Breeding of selected genotypes by them for two generations revealed a highly significant positive correlation between the parents and their progenies both in oil and oleic acid contents. The study also suggests a significant contribution of environmental factors to the variation in both traits suggesting a moderate heritability, leading to the conclusion that there is a highly significant variation both in oil content and fatty acid composition in the niger populations that can be used for its improvement.

Baghel *et al*., (2018) reported high genetic advance in terms of a percentage of means for capitulum per plant, seed yield per plant (g), seeds per capitulum, oil content (%), harvest index (%), primary branches per plant, plant height (cm) and straw yield (g).

Based on the above reports, the population improvement strategy suggested below for genetic enhancement of niger oil content should be more relevant:

- Pool all the available diverse genotypes and varieties including the diverse Ethiopian varieties showing consistently more than 45% of oil content to be used as the parental materials
- Hybridize the selected parental lines indicated above in all possible combinations and raise F1 and F2 generations. Postpone selection for high oil content together with seed yield to the later generations. Carry-out rejection of the low yielders with low oil content in the F2 generation.
- Inter-mate the selected lines and proceed to effect selections for higher oil content in the resulting F3 and F4 generations. In this step too rejection of poor seed yielders with low oil content must be practised as done in the F2 generation.
- With the above approach it would be possible to pool diverse alleles for oil content maintaining higher yield seed in three to four cycles.

Additionally, mutation breeding could be carried out to induce new gene mutations for higher oil content and quality (Mondal and Badigannavar, 2013).

Niger seed oil has linoleic acid (C18:2) as the principal fatty acid (65.7-68.5 %, weight percent of total lipid). Oleic acid (C18:1) is the major mono-saturated fatty acid (5.4-7.5 %) (Dutta *et al.,* 1994).Niger contains two major saturated fatty acids [palmitic (9.6-10 %) and stearic (7.6-8.1 %)]. The above fatty acids represent 91-97 % of the fatty acid profile. Palmitoleic, linoleic, arachidic, eicosenoic, behenic, erucic and lignoceric acids constituted less than 1% each. The proportion of oleic acid in niger varieties of niger is stated to be 6 to 11%(Dutta *et al.*, 1994). Both the oil content and fatty acid profile would depend upon the origin of the material and maturity level of the seed (Riley and Belayneh, 1989; Stymne and Appleqvist, 1980). Ranganatha *et al.,* (2016) reported that the niger materials from Nepal would be genetically diverse from that of Ethiopia in terms of fatty acid profile since it expressed a wide variation for all the fatty acids, perhaps giving scope for selection (Table 4) .With regard to the utility of employing wild taxa of *Guizotia* in hybridization with cultivated niger varieties, Dagne and Jonsson (1999) reported that differentiation in fatty acid composition between the taxa is too small to be of taxonomic use and hence it was inferred that when gene transfer is desired, hybridisation between the wild and cultivated taxa may not affect the oil quality of the latter.

In the earlier studies, Alemaw and Wold (1995) found negative correlations between oleic and linoleic acid. Likewise, a negative correlation is expected for oleic and stearic acid since oleic acid is synthesized from stearic acid. A study by Geleta *et al*., (2011) found that seed weight and linoleic acid had a negative correlation of -0.17. Seed weight and oleic acid were correlated at 0.13, while seed weight and palmitic acid had a negative correlation of -0.15.

Benelli (2015) reported that strong negative correlations were found between oleic and linoleic acid (-0.84- -0.86). According to Benelli (2015) comparisons of palmitic and linoleic acid resulted in weak to moderate positive correlation coefficients (.07-0.55), while palmitic acid and oleic acid resulted in weak negative correlations (-0.12- -0.27) and correlations between oleic and linoleic acid as well as palmitic and linoleic were significant among all plant materials with the exception of F1 correlations involving palmitic and linoleic acid.

The quality of oil and its utility for different purposes such as human consumption and industrial use depends upon its fatty acid profile. Niger oil being rich in its linoleic acid, suffers from the disadvantage of poor shelf life despite its suitability for human consumption. On the other hand, oil with a higher proportion of oleic acid has better stability and good shelf life, in addition to its suitability for deep-frying.

The above data on fatty acid profile of niger oil from different sources should be useful to chalk out a breeding strategy towards developing niger oil for different purposes.

Table 4: Variation in Fatty Acid Profile of Niger (Ranganatha et al., 2016)

Origin	No. of lines	% Plamitic acid	% stearic acid	% Oleic acid	% Linoleic acid
Ethiopia	4	8.8 – 9.5	6.5 – 6.7	7.1 – 9.5	73.1 – 75.4
Nepal	37	8.1 – 30.7	5.1 – 24.5	3.8 – 11.3	30 -77.1
Poland	1	8.3	8.4	18.2	63.3
Portugal	1	8.8	7.6	8.6	73.4
Spain	1	8.3	6.7	14.9	68.7

Biotechnology of Niger

Plant Tissue Culture:*In vitro* technology could serve as an alternative means for genetic upgrading and its application largely depends on having a reliable plant regeneration system. Niger has been the subject of numerous cell/tissue culture studies.

Naik and Murthy (2010) obtained regeneration from suspension cultures via somatic embryogenesis. Establishment of embryogenic suspension cultures has great potential to aid crop improvement and is also suitable for *in vitro* se-lection of variants especially the selection of salt tolerant, disease resistant and cold tolerant lines in crop plants. In niger, homozygous lines obtained through anther culture (Adda *et al* 1993; Hema and Murthy, 2008) and un-pollinated ovule cultures (Bhat and Murthy,2007; 2008) can be used for crop improvement. The effects of amino acids (arginine, aspargine, cysteine, glutamine, glycine and proline) and poly-amines(putrescine and spermidine) to enhance embryogenesis and plant regeneration from cultured anthers of niger cv. Ootacamund has been reported by Hema and Murthy (2008) Jadimath *et al* (1998) and Sujatha (1997) effectively maintained male sterile plants through culture of leaves of a male sterile plant developed through tissue culture.

Disasa *et al.,* (2011) developed plant regeneration protocols of niger from leaf tissue. Efforts to develop dihaploids from ovule culture were unsuccessful. During the last 10 years modern techniques of plant tissue culture, doubled haploid technology and transformation are increasingly used by breeders for crop improvement. Protocols to regenerate plants from niger hypocotyl and cotyledon tissues and seedlings were developed in India (Sarvesh *et al.* 1993a 1994b). Plant regeneration was dependent on genotype and media composition used. If niger is susceptible to *Agrobacterium tumefaciens* infection, then it will be a good candidate for gene transfer within the compositae family. Dihaploid plants of niger have been produced by anther culture (Sarvesh *et al* 1993b, 1994a). Following *in-vitro* anther culture and induction of flowering (Sarvesh *et al.,*1996), haploids have been produced in niger. (Kishor *et al*., 1997).

Self-compatible lines, dwarfs and single-headed doubled haploid plants were obtained from anther culture of niger in India. Anther and microspore derived dihaploids can be used to develop homozygous mutant types and inbred lines in a short time (Getinet and Sharma1996). Recessive, simply inherited and easily identifiable marker traits which are important for niger seed production to ensure genetic purity of varieties could be obtained through microspore culture technology (Getinet and Sharma, 1996).

The protocol for tissue culture work for multiplication and propagation of valuable genetic material is needed in niger in view of the difficulty of procuring homozygous selfed seed from self-incompatible lines of niger. Out of the four explants used by Baghel and Bansal (2017), apical bud proved to be the best in terms of shoot regeneration and multiplication. Best shoot multiplication was obtained from apical bud, axillary bud and leaf explants on MS medium supplemented with 2.22 μM BAP + 2.85 μM IAA. Whereas supplementation of MS medium with 2.22 μM BAP + 28.55 μM IAA produced maximum number of shoots from internode explants, BAP (0.44 μM) in combination with Kn (0.46 μM) proved suitable for maximum mean shoot length. Moreover, culturing the regenerated shoots on half-strength liquid MS medium supplemented with NAA (2.68 μM) induced maximum rooting from elongated shoots (direct and indirect regeneration). The plantlets were established in plastic cups containing vermiculite, soil, sand and farm yard manure and then successfully transferred to field with 97.33% survival. Analysis of RAPD recognized 197 different amplification products and showed the presence of somaclonal variation in the plantlets arising from direct regeneration as well as from indirect regeneration. The protocol developed in this study is suitable for propagation of quality planting material for commercialization, germplasm conservation and for future genetic improvement studies.

Genetic transformation: An efficient transformation protocol will be a prerequisite for the introduction of specific and desirable genes into niger. Murthy *et al* (2003) developed an efficient protocol for *Agrobacterium* mediated genetic transformation of niger using hypocotyl and cotyledon explants fromin vitro grown seedlings. They found that cotyledon was better as explant for transformation with a frequency of 15% as compared to hypocotyls.

Molecular markers: Characterization and evaluation are important to enhance the inherent value of the conserved germplasm. The accessible collections of diverse cultivated, as well as wild, germplasm accessions have made great contributions to crop improvement. Molecular markers and their excellent attributes prove to be extremely useful for the assessment of genetic diversity as well as the identification and maintenance of germplasm collections. Molecular marker based genetic diversity analysis for Ethiopian niger was reported for the first time by Geleta *et al* (2007), who investigated the genetic diversity of 70 populations of niger collected from 11 regions of Ethiopia, using RAPD analysis to reveal the extent of its populations' genetic diversity. Ninety-seven percent of the loci studied were revealed to be polymorphic for the whole data set. UPGMA cluster analysis showed that most of the populations were clustered according to their region of origin. However, some populations were genetically distant from the majority and seem to have unique genetic properties. They concluded that the crop has a wide genetic basis that may be used for the improvement of the species through conventional breeding and/or marker assisted selection. In a similar approach, Nagella *et al.*, 2008) reported the genetic diversity of selected Indian niger germplasm accessions from different origins and pedigree backgrounds through the use of RAPD markers. It was the first attempt to estimate genetic variability among Indian niger cultivars using molecular markers. Molecular techniques are being widely used in systematic and phylogenetic studies to provide a measure of genetic relatedness based on DNA sequence variation (Soltis and Soltis 1998). The internal transcribed spacers (ITS) of the nuclear ribosomal RNA genes have been among the most widely used sequences for DNA sequence variation studies. However, in spite of the small number of species in the genus *Guizotia* there are no DNA sequences available for systematic purposes. Bekele *et al* (2007) drew phylogenetic inferences using the information from ITS sequence generated for five species of *Guizotia* and related this to the previous understanding and unresolved problems of the genus. It was found that *G. scabra ssp. scabra* , *G. scabra ssp. schimperi* And *G. villosa* have all contributed to the origin of *G. abyssinica* ,the cultivated species of the genus. Based on the findings it was suggested that the present composition of the species of genus *Guizotia* and the sub-tribe the genus is presently placed in, should be redefined.

Biotic Stresses of Niger

The problems due to biotic stresses are considered not very acute in niger as compared to other oilseed crops. Nevertheless, there are cases of considerable damage done to the crop due to biotic stresses such as *Alternaria* leaf spot in Gujarat (Kansara and Sabalpara, 2016), which has been detailed in this section. Niger being a poor farmers' crop India, scientific management of pests and diseases is rather rare at the farmers' level. There is also the parasitic plant dodder (*Cuscuta campestris*) growing on niger, which is becoming a threat to niger cultivation. Among insect pests, capsule fly or niger fly is indeed causing perceptible yield losses as described below in this section. In India, control measures for niger caterpillar, semilooper, hairy caterpillar and surface grasshopper have been developed (ICAR 1992).

Insect Pests: As per the researches carried out at the ICAR-Directorate of Oilseeds Research (Basappa and Singh, 1990) the major pests recorded are a as follows. Of the several insect pests of niger reported, niger fly (*Dioxyna sororcula* and *Eutretosoma* spp.) and black pollen beetles (*Meligethes* spp.) are the most serious insect pest of niger, both in Ethiopia and India.

The black pollen beetle is also reported from all niger-growing areas of Ethiopia. Although precise identification is lacking, five species are suspected. These insects feed on pollen grain and damage the fertilization of ovules. The adult beetles are adapted to live within the disk florets. Some of the insect pests found in Ethiopia are not yet identified.

(i) **Niger Capsule fly (*Dioxyna sororcula*):** https://en.wikipedia.org/wiki/Dioxyna This is the most devastating pest of niger causing great losses due feeding on the capitulum. The flies start mating when the flower is blooming (Getinet and Sharma, 1996). Eggs are laid within the disk florets thereby adversely affecting seed-set. The damaged flowers turn red brown and contain larvae or pupae. Maggots feed on the pulp of the developing seed residing inside the capitula. At maturity the damaged disc florets turn stony and reveal pupae when dissected. Integrated pest management needs to be adopted to manage the pest and check the yield losses.

Control measures: (1) Install the light trap one per ha. (2) Spray Quinalphos 25 EC 1.5 ml/l or profenofos 50% EC 1.5 g/l of water soon after noticing the egg laying and initial damage.

(ii) **Myrid bug (*Taylorilygus pallidulus*)** : These plant-bugs are polyphagous, mainly feeding by sucking the sap on *Asteraceae* species. The broken-backed bugs are vectors of a phytoplasma that cause phyllody in species of *Parthenium*.

(iii) **Noctuid Caterpillar *(Thysanoplusia orichalcea)*:** It is polyphagous, with an overlapping host range. It is more abundant on 7 species of plants including niger, and could become a significant defoliating pest in niger, legumes and brassicas. *Nomuraea rileyi*, a fungal pathogen of noctuid moths could be effective in checking the pest.

(iv) **Cutworm (*Spodoptera litura*)**: *S. litura* is a general herbivore pest on various plants. The lower and upper limits of habitable temperatures are 10 °C and 37 °C, respectively. Therefore, it is well suited for tropical and temperate climate regions.Eggs are spherical and slightly flattened. Each individual egg is around 0.6 mm in diameter with an orange-brown or pink colour. These eggs are laid on the surface of leaves in big batches, with each cluster usually containing several hundred eggs. As caterpillars, *S. litura* can only move short distances; but feed voraciously causing immense crop losses. Adult moths can fly up to a distance of 1.5 km for a total duration of 4 hours. This helps disperse the moths into new habitats and onto different host plants as food sources are depleted. The larval parasites such a *Nomuraea rileyi*, and NPV can bring about control of the pest.

(v) ***Heliothis armigera* [*Helicoverpa armigera*]:** The female moth can lays several hundred eggs, distributed on various parts of the plant. Under favourable conditions, the eggs can hatch into larvae within three days and the whole lifecycle can be completed in just over a month.

The eggs are spherical and 0.4 to 0.6 mm in diameter, and have a ribbed surface. They are white, later becoming greenish.

The larvae take 13 to 22 days to develop, reaching up to 40 mm long in the sixth instar. Their colour is variable, but mostly greenish and yellow to red-brown. The head is yellow with several spots. Three dark stripes extend along the dorsal side and one yellow light stripe is situated under the spiracles on the lateral side. The ventral parts of the larvae are pale. They are rather aggressive, occasionally carnivorous and may even cannibalise each other. If disturbed, they fall from the plant and curl up on the ground. The pupae develop inside a silken cocoon over 10 to 15 days in soil at a depth of 4–10 centimetres (1.6–3.9 in). It is not easy to check the pest through chemical control alone. Integrated management strategy described in this section has to be adopted.

(vi) **Cutworm-2 (*Argotis ipsilon):*** The moth hides under dried twigs during day time and lays eggs on leaves. Larvae attack the crop and plants at ground level. **Control measures**: (1) Proper weeding reduces hiding places. (2) Crop rotation is effective in reducing pest population. (3) For effective control of the pests particularly at early stage, apply phorate 10 G 10 kg /ha as basal application (4) Spray NSKE 5% or Neem based insecticide (Nimbecidin 5 ml/l water) (5) Two sprays of Chloropyriphos 20 E C 1.5 ml/l or Quinalphos 25 EC 1.5 ml/l of water if the pest persists even after the use of NSKE

(vii) **Niger caterpillar** (***Condica conducta***): The green coloured caterpillar with purple markings, feed on leaves and defoliates the plants.

(viii) **Semilooper (*Plusia orichalcea***): The semilooper feeds on the foliage and causes defoliation of the plant. **Control measures:** (1) Collection and destruction of egg masses of early instars of caterpillars (2) Spray NSKE 5% or Neem based insecticide (Nimbecidin 5 ml/l water) (3) Two spraying of any one of the following insecticides Chloropyriphos 20 EC 1.5 ml/l or Quinalphos 25 EC 1.5 ml/l or Indoxacarb 15.8 EC 0.5 ml/l

(ix) **Bihar hairy caterpillar (*Spilosoma obliqui*) :** The caterpillars remain gregarious underneath leaves in early stages and cause serious loss in yield at third and fourth instar.

The control measures are similar to what has been suggested for semilooper.

(x) **Niger flower thrips (*Haplothrips articulosus*)**: Thrips reside in the disc florets and suck sap, damaging the capitula resulting in yield losses. While the sucking pests including thrips, jassids, aphids and white fly could be effectively controlled using systemic insecticides such as Dimethoate and Metasystox, it is advisable to manage the pest by adopting integrated management strategy rather than depending upon exclusive use of chemical pesticides, which results in augmented cost and pollution of the ecosystem.

(A) The section of control measures

(1) Proper weeding reduces hiding places

(2) Crop rotation is effect ive in reducing pest population

(3) Birds readily eat the caterpillars and help to check when they are numerous, 40 -50 bird perches are sufficient for one hectare

(4) For effective control of the pests particularly at early stage, apply phorate 10G 10 kg /ha as basal application.

(5) Spray NSKE 5% or Neem based insecticide (Nimbecidin 5 ml/l water) & 376 Oilseed Crops

(6) Two sprays of Chloropyriphos 20 EC 1.5 ml/l or Quinalphos 25 EC 1.5 ml/l of water

Diseases of Niger

(i) **Ozonium wilt (*Ozonium texanum* var. *parasiticum* Thirum.):** Necrotic lesions develop on stem of well grown plants near the soil level, which gradually extends upwords. Whitish fungal mycelium grows on these necrotic areas under high humidity and the branches, leaves, inflorescence etc become soft and start rotting. The diseased stem breaks. The mycelium grows on dead plants if enough moisture is present. Root rot is also commonly observed. The fungus forms rhizomorphs and sclerotia. The sclerotia are oval, grey to dark brown. The recurrence takes place through soil borne sclerotia and mycelium.

(ii) **Collar rot (*Sclerotium rolfsii* Sacc.) :** The tissues of collar region become soft and depressed. White fungus grows on the diseased part and forms mustard seed like sclerotia. The diseased plants turn yellow and dry up.

(iii) ***Macrophomina* root and stem rot [*Macrophomina phaseolina* (Maubl.) Ashby**: It is seed and soil borne disease. Typical root rot, stem rot, charcoal rot and leaf blight symptoms are produced. *Macrophomina* infected roots are light blackish to black in colour, which are covered with black sclerotia and are brittle. The blackening extends from ground level upward on the stem giving black colour to stem. The recurrence of the disease takes place through seed and soil borne inoculum, later spread is through workers, tools and insects.

(iv) **Damping off/root rot *(Rhizoctonia solani)* :** The fungus attacks stem of the seedling at ground level, makes water soaked soft and incapable of supporting the seedling which falls over and dies. On older seedlings elongated brownish black lesions appear which increase in length and width girdling the stem often the roots are blackened due to fungal attack resulting in root rot. The disease is favoured by cold and wet weather.

(v) **Cercospora leaf spot (*Cercospora guizoticola)*, :** The disease is severe under warm and humid weather. Disease appears as small straw to brown coloured spots with grey centre on the leaves, spots may coalesce causing defoliation. 'Later the spots increase in number and size and cover the entire lamina and the leaves start dropping off. Elongated dark brown spots are produced on the stem. The capsules are also affected. The pathogen is seed and soil borne. The primary infection occurs through seed borne inoculum as well as the inoculum present on diseased plant left over in the soil. The spread of the disease is through conidia formed on infected plant parts. The fungus is active throughout the year where Niger is grown in Rabi and kharif seasons.

(vi) ***Alternaria* blight *(Alternaria* sp.)** : The disease appears as concentric rings on the leaves, which turns brown with grey centre later on. As the disease advances, the spots become oval or circular and irregular in shape. The infected leaves become dry and lead to the defoliation. Further it, spreads to other plant parts and results in to premature drying of the plant (Yirgu, 1964). The pathogen is seed and soil borne.Control measures: Seed treatment with Thiram (0.2%) & Bavistein (0.1%) and Two foliar sprays with Bavistin (0.1%) + Dithane M 45 (0.25%)

(vii) ***Curvularia* leaf spot (*Curvularia lunata*):** Small circular to irregular brown to reddish brown spots which later coalesce to form larger spots. The leaves turn yellow dry-up causing defoliation of plants. The pathogen is seed borne and survives in plant debris.

(viii) **Powdery mildew (*Sphaerotheca* sp.) :** It is caused by *Sphaerotheca* sp. in Ethiopia (Chavan, 1961), and by *Oidium* sp. (*Erysiphe cichoracearum* DC) in India. The disease occurs in *rabi* as well as *kharif* season but it is more severe in *rabi.* **The defoliation due to the disease results in significant yield** losses. Small cottony spot develops on the leaves which gradually cover the whole lamina (Vyas *et al.*, 1981). Purple rings also develop on stem. The diseased leaves turn yellow and drop off. The seeds of the diseased plants are small and shrivelled.

(ix) **Rust (*Puccinia guizotiae*)**: The disease is more severe in Ethiopia as compared to India. It is caused by *Puccinia guizotiae*. Brown pustules up to 7 mm in diameter appear on the leaves. The lesions consist of densely aggregated brown telio that measure 0.16 to 0.37 mm in diameter. The uredosori are formed on the lower leaf surface and the corresponding upper surface becomes chlorotic. Later teliospores are formed. The teliospore measures 37-49 x16-20 um in size (Kolte, 1985)

(x) **Bacterial leaf spot (*Xanthomonas campestris* pv. *guizotiae*):** This disease was first reported from Ethiopia (Yigru, 1964). Small brownish spots are formed on leaves which are surrounded by yellow halo. Most of the spots are on mainly on the leaf margin. The spots increase, coalesce and give appearance of blight. All the leaves get destroyed. Under severe cases, scabby brown needle pricks lesion produced pricks on the stem (Yigru, 1964). The bacterium X. *campestris pv. Guizotiae var. indicus* also produces similar type of the symptoms on the plants but the bacterium requires low temperature for growth and infection.

(xi) **Seed rot/ Seed borne microorganisms**: Several fungi have been found to be seed borne like *Alternaria sp., Aspergillus flavus, A. Niger, Rhizopus nigracans, Cercospora guizoticola, M. phaseolina, R. batalicola* etc. Besides pathogens like *Xanthomonas campestris* and *Actinomycetes* have also been reported to be associated with rotting of the seeds

Niger yield losses due to the infestation of *Alternaria* leaf spot disease *Alternaria alternata* (Fr.) Keissl., are a major concern of niger growers. Leaf spot is considered a most injurious disease of niger in south Gujarat, resulting in severe loss especially in the case of susceptible niger cultivars. The disease became epidemic during rainy season, caused heavy loss in grain yield. The avoidable loss of grain yield was estimated to be 34.27% due to leaf spot disease. With the proper protection measures, 71.40 per cent leaf spot diseases can be controlled. Niger crop can be protected from leaf spot disease by giving the seed treatment with mixed formulation of carbendazim (12 %)+ mancozeb (63 %) (Saaf 75 % WP) @ 2.5 g/kg seeds and two sprays of Saaf 75 WP @ 0.2%, 1st at appearance of the disease and 2nd at 15 days after 1st spray (Kansara and Sabalpara, 2016)

Based on the work carried out under the Coordinated Research Project on Sesame and Niger under the ICAR- Indian Institute of Oilseeds Research the following varieties / genotypes have been identified as tolerant to the diseases of *Cercospora* leaf spot, *Alternaria l*eafspot and Powdery mildew as indicated below.

- **Cercospora leaf spot**: JN -13, JN-106, JN-107, JN-112, JN-118, JN-128, JN130, JNS-9, PCU-45, PCU-46, N-24, N-128, AJSR48, ONS-157
- **Alternaria leaf spot:** RCR-328, JN-121, N-17, N-18, N-24, N-87, N-128, N-141, N-142, N-165, N-187, AJSR-47, AJSR-48
- **Powdery mildew:** KEC-6, RCR-238, RCR-290, JN-17, JN-19, JN-20, JN-21, JN-37, JN-60, JN-68, JN-72, JN-77, JN-78, JN-85, JN-86, JN-87

(xii) ***Cuscuta* /Dodder (*Cuscuta hyalina*):** Dodder, a parasitic leafless weedy plant infects niger crop and has assumed serious proportions in Odisha, Andhra Pradesh and parts of Madhya Pradesh. With the parasite in niger plants get stunted, turn pale yellow with a very few flowers and seeds. It is a total stem parasite drawing its nourishment from niger plant through its haustoria. Niger gets infected with dodder with its seeds getting mixed with niger seed at planting. The seeds of the parasite can be removed by soaking the seeds in 10% brine solution and sieving using 1 mm sieve before sowing. The damage due to the parasite could be controlled by (1) hand weeding i.e., manual removal of the parasite together with (2) Application of thiobencarb at 0.75 kg/ha and Anilofos at 0.50 kg/ha.

Integrated Management of Biotic Stresses involves a combination of cultural, chemical and biological methods including resistant varieties which is more effective and eco-friendly and sustainable.

The important cultural practices to avoid pests and diseases are

1. **Summer ploughing**: Before the planting the soil should be ploughed deep with the receipt of summer showers to expose the dormant phases of pests and pathogens. This will help in checking the proliferation of pests and diseases.
2. **Crop rotation**: Avoid growing niger after niger in the same field. Follow crop rotation involving cerial crops such as millets, maize, grain legumes viz., cowpeas, groundnut, pigeon pea, fodder crops etc.
3. **Scouting and surveillance:** Conduct surveys for the incidence of pest/ disease in the neighbouring zones and take precautionary actions to check proliferation of pests and diseases.
4. **Eliminate the alternate and collateral hosts of pest / pathogen**
5. **Clean cultivation** and timely weeding operations reduce the biotic stresses
6. **Seed treatment** with thiram (0.2%) + Carbendazim (0.1%), application of Trichoderma viride @ 2.5 kg/ha mixing with 50 kg FYM in the field before sowing can effectively manage the disease. Most of the diseases of niger are seed borne and as such seed treatment with the above indicated products could be very effective in managing the diseases.

7. Spray the appropriate dose of insecticide or fungicide when the pest / disease is in the initial stages for effective control. Avoid repeated sprays of chemicals.
8. Cultural methods including inter-cropping to manage whitefly and other pests and diseases:
   (i) Erecting crop barriers such as row covers, and repellent mulches that effect photo tactic responses of whiteflies, have shown some promise in delaying or reducing disease incidence. In a study with tomatoes in Israel soil mulching with yellow polyethylene sheets delayed the spread of tomato yellow leaf curl virus for 20 days against a control crop without mulching (Cohen and Melamed-Madjar, 1978).
   (ii) Control of weeds adjacent to cultivated fields, the use of trap crops, and implementation of crop free periods may be effective in reducing vector populations in certain cropping systems.
   (iii) Intercropping can be an alternative method for the reduction of pests in certain situations. Recently, inter-cropping of tomatoes has been shown to be beneficial. Various intercropping systems have shown to contribute to the reduction of whitefly populations. Cucumber planted in alternating rows 30 days before tomato delayed infection of the tomato with the whitefly-vectored tomato yellow leaf curl virus. (Mau and Martin 2007)
9. Use of resistant / tolerant varieties to pests and diseases is the least expensive and efficient method of managing the pests and diseases
10. Employing fungal pathogens viz. *Nomuraea rileyi*, and NPV to control the lepidopterous larvae could be effective (AICORPO 2003).

Niger Crop Production

In India, niger is sown as a sole or mixed /inter crop with finger millet, castor, groundnut, soybean, sorghum, green gram, chickpea and even sunflower. The crop is grown normally in monsoon / rainy (*kharif*) season; but in certain localities it is also grown in the post-monsoon season (October to March/ April) or summer (February/March to May/June) in the pockets where the temperatures do not exceed 37°C (Ranganatha *et al*., 2016). Under farmers' situation in southern India, the crop is managed without any nutrient application or plant protection and hence the yield levels are low. Niger crop responds well even to minimal inputs of nutrient application and plant protection leading to better yields. The following steps of improved niger crop production technology promises encouraging levels of productivity of the order of over 500kg/ha.

- **Sowing time:** Primarily, Niger is a *kharif* crop, but mostly sown in late July/ August.

- **Seed treatment:** Seed should be treated with Thiram or Captan 3.0 g/kg seed before sowing. Seed treatment with Phosphorus solubilizing bacteria (PSB) 10 g/kg seed gives higher yield.
- **Seed rate:** Generally, 5 kg/ha seed is required for the sole crop. Line sowing has been found beneficial with spacing of 30cm x 10cm.
- **Method of sowing:** The crop is largely sown by broadcasting. Seeds are mixed with sand/ powdered FYM/ ash to increase the bulk, 20 times to ensure even distribution of seed.
- **Nutrient Management:** The crop is mostly grown on marginal and sub-marginal land without manure or fertilizer application. However, application of recommended N @ 20kg/ha applied in split doses based on soil moisture through urea + seed treatment with PSB 10 g/kg seed enhances yield significantly. Application of Sulphur (20-30 kg/ha) increases seed yield and oil content in niger. The recommended nutrient schedule is 20N 40P_2O_5 and 20 k kg/ha

• Weed Management

First weeding is needed 15-20 days after sowing. In Orissa, *Cuscuta* (*Cuscuta hyalina/ C. chinensis*) infestation has become a major problem. Seed should be obtained from *Cuscuta* free areas. Cuscuta seeds could be separated with a 1 mm sieve.

Pre sowing soil application of Fluchloralin (1 kg a.i./ha). or Pre emergence application of Pendimethalin (1.5 kg a.i./ha).

• Water Management

It is invariably grown in the rainy season and it is seldom irrigated. There are indications that niger yields can be doubled under irrigation, it the crop suffers from moisture stress. Irrigation may be given at the seedling stage.

Intercropping of Niger

Niger can be planted as an intercrop with cereals (millet, sorghum), grain legumes (cowpea, soybean, green gram) or other plants (castor, sunflower, sesame and other annual crops grown in the season). The prevailing niger based intercropping systems in India are presented in Table 5. In Ethiopia, when cultivated as a sole crop, niger seeds is broadcasted on a well prepared seedbed, at 5-15 kg seeds/ha and covered 1-3 cm deep. In India, seed density is lower: 5-8 kg/ha. When sown as an intercrop, niger seed is allocated about 20-25% of the area. Sowing should be done at the onset of rains or when there is residual moisture in the soil (Bulcha, 2007). Establishment of niger is easy and requires only a few weeding operations since the plant competes quickly with weeds (Bulcha, 2007)

Inter-cropping could be a potent tool to check the proliferation of pests and diseases of niger. In India the widely practised inter-cropping systems are Niger + horse gram (legume) (*Macrotyloma uniflorum* or *Dolichos uniflorus)*, Niger + finger millet or ragi (*Eleucine coracana*) and Niger + little millet (*Panicum miliare*) in the niger growing tracts. In the inter cropping the row proportions and population densities of niger crop and inter-crop should be planned. It could be 1:1 alternate rows or 3 rows of niger and one row of inter-crop if the plants of inter-crop are taller and faster growing than niger. Economics of the suggested inter-cropping systems are presented in the Table 6.

Table 5: Niger crop based inter-cropping systems in different states of India (Row proportions given in parentheses)

State	Intercropping Systems	Row ratio
Madhya Pradesh & Chhattisgarh.	Niger + Kodo/ Kutki/ Pearlmillet/ Green gram (2:2)	2:2
Maharashtra	Niger + Finger millet / Horse gram/ Ricebean (2:2 / 4:2)	2:2 or 4:2
	Niger + Pearlmillet (3:3)	3:3
Orissa	Niger + Finger millet / Black gram (2:2 / 4:2)	2:2 or 4:2
	Niger + Ricebean/ Cowpea/ Frenchbean (4:2)	4:2
Bihar / Jharkhand	Niger + Finger millet / Black gram (2:2)	2:2
	Niger + Redgram (3:2)	3:2
	Niger + Ricebean (4:2)	4:2
AP and Telangana	**Niger + Cowpea** (4:2)	4:2
Karnataka and Tamil Nadu	Niger + Groundnut (3:6)	3:6
	Niger + Finger millet (1:1)	1:1

(*Source*: Oilseeds Division Ministry of Agriculture & Farmers' Welfare Government Of India) www.nmoop.gov.in

Table 6: Economics of various intercropping combinations with Niger

Centre-wise productivity potentials and economics of intercropping in niger (1997-98 to 2001-02)

Year	State	Centre	Intercoppin	No of demos.	Mean seed yield (kg / ha)		Increase in yield (kg ha)	Gross returns (Rs / ha)		Addl.net (Rs / ha)	B:C Ratio	
					IT	FP		IT	FP		IT	FP
1997-98	BR	Ranke	Niger+horsearam	5	470	242	94	4832	2488	1359	2.76	3.25
1998-99	OR	Semiliguda	Niger + little millet	6	503	166	203	6012	1985	1023	1.47	1.32
2001-02	TN	Paiyur	Niger + finger millet	5	274	200	37	5487	4081	942	1.62	1.40
2001-02	CG	Jagdalpur	Niger + finger millet	6	224	172	30	5577	3087	1296	1.97	I.SS

IT: Improved recommended crop production technology

FP: Farmers Method (Source: ICAR-Indian Institute of Oilseeds Research, Hyderabad

http://www.icar-iior.org.in/index.php/oilseeds-database)

Harvest

Harvest should be done before the crop matures, 3 weeks after half of the florets have dropped. The leaves are then dry and the heads are black. Plants are cut and stacked for a few days to complete drying. They are threshed by hand (in India) or with the help of oxen (in Ethiopia). After threshing, the seeds are winnowed and stored (Bulcha, 2007).

Yield

Seed yields are about 300-450 kg/ha when the plant is cultivated as a sole crop, and about 150-200 kg seeds/ha when intercropped (Krishna, 2013). In Ethiopia in 2009-2010, niger was cultivated on about 250,000 ha for a seed production of 150,000 t, giving an average yield of 605 kg seeds/ha (Melaku, 2013). In India, niger was cultivated on 437,000 ha for a seed production of 111,000 t, giving an average yield of 253 kg seeds/ha. Yields ranged from 150 to 645 kg/ha in the Indian plains and up to 1050 kg/ha in Bangladesh (Krishna, 2013). Since niger oil cake represents approximately 60-75% of seed weight, it can be estimated that Ethiopia and India have an annual production of 90,000 t and 66,000 t of niger oil cake, respectively.

Oil extraction

In Ethiopia, traditional oil extraction was done by grinding the dry seeds into fine powder, adding hot water and stirring it until the oil floats to the surface, and then scooping the oil off. Today, niger oil is mostly processed in small, mechanized expeller mills. In India, niger oil is extracted in traditional bullock-powered *ghanis*, in small rotary mills, or in hydraulic or screw presses. Usually, locally-extracted oil has a short storage life, but heating and storing in airtight containers can prolong the storage life of the extracted oil. (Bulcha, 2007).

Storage of Oil and cake: Niger oil cannot be stored easily due to its high content of unsaturated fatty acids. Likewise, oil-rich niger cake may have a short shelf-life. The moisture content of stored seeds must be less than 8% to prevent damage by storage pests, especially moulds (Krishna, 2013)

Nutritional attributes

Niger cake is a valuable source of protein, with a CP content varying between 22 and 42% of DM. The oil content depends on the extraction process: it is in the 7-14% DM range for mechanically extracted cake, and lower than 4% for solvent-extracted meal. Niger oil cake is rich in fibre, with a large variability: NDF ranges from 22 to 51% of DM. It is particularly rich in lignin (about 12% of DM). Niger seed oil contains 75-80% linoleic acid, 7-8% palmitic and stearic acids, and 5-8% oleic acid (Getinet and Sharma 1996). The Indian types contain 25% oleic and 55% linoleic acids (Nasirullah *et al.*, 1982). The variations in mineral, protein, fatty acid and amino acid composition depend on environmental factors, location and varieties, and on the oil extraction method (Bhagya and Samathaka Sastry, 2003; Gebremedhin *et al.*, 2009).

Potential constraints

Because niger oil cake is often stored for more than 6 months under unfavourable conditions, it is prone to *Aspergillus niger* mould infestations that produce aflatoxin B1, a carcinogenic toxin that can be subsequently found in milk (Szonyi *et al.*, 2015). In a 2014-2015 survey of feeds distributed in the Addis Ababa area, niger oil cake had the highest amount of aflatoxin B1 (290-397 μg/kg, 10 times that of wheat bran) and was the main contributor to aflatoxin contamination in milk (Gizachew *et al.*, 2016).Only one study reported values of condensed tannins, with a relatively low value of 3.65 g/kg DM (Makkar *et al.*, 1990), which if it is confirmed would not limit the intake or protein digestibility in ruminants.

Niger cake as cattle and sheep feed

Niger oil cake is a good protein supplement in ruminant diets. It is mainly used in Ethiopia in sheep and goat diets for growth or fattening, and to a lesser extent in dairy and beef cattle diets. In most cases, niger oil cake is used to supplement low quality forages (protein content < 7% DM). This supplementation has a positive effect on the digestibility of cell wall constituents which creates a positive effect in the rumen and allows better fibre degradability (Butterworth *et al.*, 1986). It also has a positive effect on diet and forage DM digestibility, provided dietary energy is sufficient as observed with dairy cows, goats, and sheep studies (Kebede *et al.*, 2009; Nuwanyakpa and Butterworth,(1987). All results presented and discussed come from studies done in Ethiopia. In Ethiopia, it was reported that some farmers prefer to feed niger cake rather than linseed cake for milk production because cows fed with the latter become fat, and milk production decreases (Gebremedhin *et al.*, 2009).

9

Linseed (*Linum usitatissimum* L)

Introduction

Flax (*Linum usitatissimum* L.), also known as common flax or linseed, is a member of the genus *Linum* in the family Linaceae. Linseed is an annual oil crop that is grown either for its fibre (fibre flax) or for its oil. The area under annual cultivation is 3 million hectares, accounting for 1% of the total world production of plant oil. The five major flax-producing countries are Canada, China, India, the United States, and the European Union.Canada with 614,000 metric tonnes of flaxseed produced in the year 2013-2014, is the world's largest producer of flax and accounts for nearly 80% of the global trade in flaxseed (Oomah 2001; Bhatty, 1995)

Flaxseed, of Mesopotamic origin, has been cultivated in Mediterranean region and India since 5000 BC, being used principally for the fabrication of cloths and papers (Soni, *et al.*, 2016).

With the adoption of cotton as a fibre crop for cloth preparation, fibre flax production declined and the major use of flax was switched from stem fibre to seed oil. It is a food and fibre crop cultivated in cooler regions of the world. Textiles made from flax are known in the Western countries as linen, and traditionally used for bed sheets, underclothes, and table linen. Its oil is known as linseed oil. In addition to referring to the plant itself, the word "flax" may refer to the unspun fibres of the flax plant. The plant species is known only as a cultivated plant and appears to have been domesticated just once from the wild species *Linum bienne*, called pale flax. (Allaby *et al.*,2005). Fibre flax is currently grown in the cool-temperate regions of China, the Russian Federation and Western Europe while linseed cultivars are grown in Canada, India, China, the United States and Argentina (Green *et al.*, 2008).

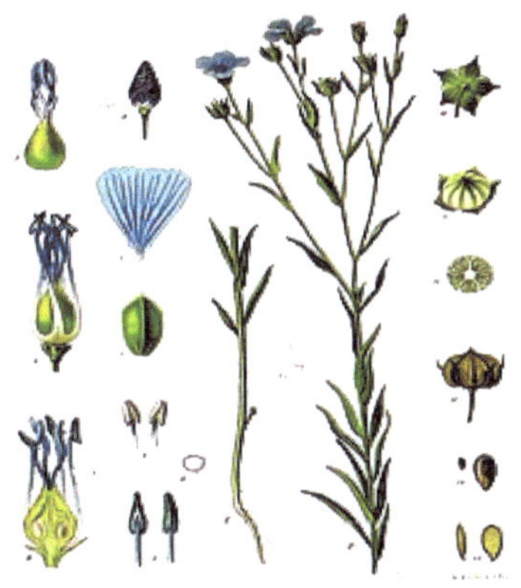

Linseed is now enjoying renewed popularity worldwide thanks to emerging market opportunities for raw material derived from seeds and stems. The dual-purpose cultivation is particularly attractive to growers for its capacity to yield both seed/oil and straw/fibre.The total world production of linseed reached approximately 2.56 million tons in the year 2014, with Canada (34 %), the Russian Federation (15 %), and China (13 %) being the main producers (FAOSTAT, 2016). India is the sixth largest producer in the world with contribution to global linseed area and production 13 % and 5.5 %, respectively. However, the average national productivity (403 kg/ha) is far below the world average (851 kg/ha) because of its cultivation in marginal and sub-marginal areas, rain-fed soils, sensitivity to fungal diseases, non-availability of improved varieties and *utera* cultivation by poor farmers. Lack of proper nutrient management at farmers' level is another lacuna bringing down the production and productivity of linseed in India.

This dual-purpose crop is believed to have originated somewhere in Central Asia, Near East, or Mediterranean region with oil flax (linseed) predating fibre use (fiber flax).

Humans first domesticated flax in the Fertile Crescent region and the evidence exists of a domesticated oilseed flax with increased seed-size by 9,000 years ago from Tell Ramad in Syria (Fu YB, 2011).In China and India, domesticated flax was cultivated at least 5,000 years ago (Cullis, 2007).Flax was cultivated extensively in ancient Egypt, where the temple walls had paintings of flowering flax, and mummies were embalmed using linen (Sekhri, 2011).

Genetic resources available in respect of flax / linseed are abundant offering substantially impressive diversity of characters to achieve desired recombinants to make the crop attractive to producers and industry. The crop is unique in offering high quality fibre and Also, flax / linseed is one of the few oilseed crops in which new biotechnological applications got integrated into the traditional crop research and improvement activities. Paradoxically, however, global crop production of flax is declining, though the trend in respect of linseed has been not so depressing; but the area and production of linseed crop too is yet to reach new heights to support the industry to come up with diverse health products. The silver ling however, is the slight increase in Canada, Egypt, and some member countries of the European Union (EU). The policy environment significantly affects linseed production in the EU and has substantial influence on the marketing and commercialization strategies of the crop.

A multipotential crop: Linseed provides raw materials for seed oil, stem fibre, paper, wax, nutraceutical and food processing industries. The flax fibre is a valuable raw material for linen clothing, the straw and short fibre for thread, rope and other packaging materials; cigarette paper, currency notes (Cullis, 2011), while the seed oil has been used in production of paints, inks, varnish and quality linoleum flooring be-cause of its unique drying properties (Kurt and Bozkurt, 2006). In addition, linseed varieties with oils suitable for culinary use are also available. Now a days, these two basic types of uses are supplemented by using of common flax as a functional food (Bassett *et al.*, 2009). Consumption of ground seeds adds nutritional benefits because flax seeds are also a rich source of lignans, compounds that have anticancer properties (Westcott and Muir, 2003; Abraham, 2011)). Its oil, characterized by a high concentration of omega-3 alpha linolenic acid (55-57 %) is used in anti-hypercholesterolemic drugs for cardiovascular diseases. Soluble fiber from flaxseed mucilage delays gastric emptying, improves glycemic control, alleviates constipation and reduces serum cholesterol. Epidemiological studies show that the intake of dietary fiber reduces colorectal cancer. Flaxseed lignans and fatty acids have been investigated in several cohort studies for their effects on breast cancer risk and there is an association between elevated serum enterolactone and decreased incidence of breast cancer. The flaxseed diet has been shown to be beneficial on prostate cancer and benign prostate hyperplasia when defined by cell proliferation index and other cancer biomarkers (Tarpila *et al.*, 2005). As a rich source of soluble fibre, linseed mucilage constitutes an excellent alternative to inulin (polysaccharide with prebiotic effects isolated from chicory root) exclusively required for the food processing industry (Diederichsen et *al.*, 2006). Recently, there has been a growing interest in the probiotic properties of linseed and its beneficial effects on coronary heart disease, neurological and hormonal disorders (Abraham, 2011), colon tumour (Dwivedi *et al.,* 2005), atherosclerosis and various kinds of health problems including cancer (Yance *et al.*, 2020) Keeping in view of increasing demand of linseed due to numerous health benefits, there is consistent need to increase genetic potential for seed yield as well as its development into a truly multi-purpose crop.

Description of flax/linseed Plant: *Linum usitatissimum* L. is a diploid species with 15 pairs of chromosomes ($2n = 30$) (Tammes, 1928). Several other species in the genus *Linum* are similar in appearance to *L. usitatissimum*, cultivated flax, including some that have similar blue flowers, and others with white, yellow, or red flowers (Liu and Zhou., 2014). Some of these are perennial plants, unlike *L. usitatissimum*, which is an annual plant.

Cultivated flax plants grow to 1.2 m (3 ft 11 in) tall, with slender stems. The leaves are glaucous green, slender lanceolate, 20–40 mm long, and 3 mm broad.

The flowers are pure pale blue, 15–25 mm in diameter, with five petals. The fruit is a round, dry capsule 5–9 mm in diameter, containing several glossy brown seeds shaped like an apple pip, 4–7 mm long.

Seeds of oil flax varieties are larger in size than fibre flax (Green & Marshall, 1981). The physical components of flaxseed include seed coat (testa, true hull, or spermoderm), embryo or the germ, thin endosperm, and two large, flattened cotyledons. The cotyledons form the bulk of the seed weight (57%) followed by the seed coat (38%) and the endosperm (5%) (Wanasundara *et al.*, 1998). The seed coat has five distinct layers. The epidermal layer or the mucilage layer and the testa, consisting of pigmented cells responsible for the colour of flaxseed, are the two important layers of the seed coat. The seeds are reddish to deep brown or smoky yellow in colour (Oplinger *et al.,* 1989). The seed of flax is small, flat, oval-shaped, and pointed at one end with a smooth glossy surface. The seed colour ranges from medium, reddish brown to a light yellow (Freeman, 1995). Of these, brown flaxseed is the most common. The seed colour is determined by the amount of tannin pigments in the pigment cells of seed coat and brown flaxseeds contain more pigments compared to yellow seeds. In yellow-seeded flax, these pigment cells are often absent or may be present without any pigments (Diederichsen and Raney, 2006, 2008). Solin (Linola) is a yellow-seeded commercial flaxseed variety released and marketed in Canada (Muir & Westcott, 2003).

Its genome is composed of an estimated 373 Mbp of nucleotides for which an assembled sequence is available (www.phytozome.net). There are 48,000 accessions of cultivated flax preserved in gene-banks around the world but only 10 000 are estimated to be unique because of the redundancy between collections. Nevertheless, these collections represent a tremendous amount of diversity for genetic enhancement (Fig. 1). Other important genetic resources include a large collection of expressed sequences (genes), a consensus genetic map, and a physical map that were used to order and annotate the genome sequence (Cloutier, 2016). A core collection of flax including more than 300 linseed types and 92 fibre types was extensively characterized in the field at two locations during four consecutive years. Agronomic traits such as yield and its components (thousand-seed weight, seeds per boll, and bolls per area), lodging, maturity, flowering time, height, and plant branching were measured. Seed quality traits such as fatty acid composition, iodine value, oil content, and protein content were also evaluated. Finally, disease ratings for rust, pasmo, and powdery mildew were recorded. The collection was genotyped with molecular markers and a genome-wide association study revealed several marker-trait associations. These important studies provided background information on the flax genome and the genetics of important traits that, combined together, will assist and accelerate improvement of linseed by plant breeders.

Production of both linen and oil is generally not compatible in the same crop. Bast fibres become more lignified and less flexible during flower and seed set, so high-quality linen fibre is harvested prior to seed ripening and harvest (Deyholos, 2006). Additionally, in many regions where flax is grown for oil, weather conditions do not permit postharvest fibre processing (retting), which requires a relatively wet fall climate. The 202,000ha of fiber flax are mainly located in China, Russia, Egypt, and near the northwestern European coast (Vromans, 2006; Green *et al.*, 2008). For oilseed production, the straw is considered an impediment during harvest and for subsequent seeding operations and until recently has been burned

(Hall *et al.*, 2016)

Fig. 1: Genetic variability in Linseed for floral and seed characters

Genetic diversity in cultivated flax, ***Linum usitatissimum.*** ***(a)*** *Flower colour varies from white to shades of pink or blue and petals' overlap, margin, and folding are also indicative of the breadth of the genetic diversity.* ***(b)*** *At maturity, some accessions have completely closed bolls (capsules) while others are slightly or completely dehiscent. The latter is a useful trait to disseminate the seeds in the wild but is undesirable and selected against by breeders for usage as a crop.* ***(c)***

The seed coat colour of linseed can vary from bright yellow to olive to dark brown and even include variegated.

Photos, reproduced with permission and courtesy of Dr. Axel Diederichson, Plant Genetic Resources of Canada Photo by Ralph Underwood, with permission Ministry of Agriculture and Food 2003 Canada.

Physiological regulation of seed oil production: Triacylglycerols (TAGs) constitute a highly efficient form of energy storage. In seeds of angiosperms, they can act as a reserve of carbon and energy allowing to fuel post-germinative seedling growth until photosynthesis becomes effective. They also constitute the economic value of seeds in many crops. In the past years, extensive tools allowing the molecular dissection of plant metabolism have been developed together with analytical and cytological procedures adapted for seed material. These tools have allowed gaining a comprehensive overview of the metabolic pathways leading to TAG synthesis. They have also unravelled factors limiting oil production such as metabolic bottlenecks and light or oxygen availability in seed tissues. Beyond these physiological aspects, accumulation of TAGs is developmentally regulated in seeds. The oil biosynthetic process is initiated at the onset of the maturation phase, once embryo morphogenesis is achieved. A wealth of recent studies has shed new lights on the intricate regulatory network controlling the seed maturation phase, including reserve deposition. This network involves a set of regulated transcription factors that crosstalk with physiological signalling. The knowledge thus acquired paves the way for the genetic engineering of oilseed crops dedicated to food applications or green chemistry (Baud and Lepiniec, 2010)

Linseed Oil

Linseed oil is very rich in omega-3 fatty acid, alpha linolenic acid (ALA) but because of its very highly unsaturated nature, it is not suitable for cooking (Hegde, 2019). Oilseed flax oil has a very high content of Linoleic Acid (LA)(40–60%) and is used as a fast-drying oil for technical applications such as linoleum flooring, paints, varnishes, and other coatings. Due to its high content of LA, oilseed flax oil has recently been increasingly used for human consumption. Its daily consumption aids in the prevention of atherosclerotic cardiovascular disease due to the presence of ALA, lignan, phytoestrogen, and soluble fibre, which makes it suitable as a source of drugs and therapeutics. The only GE flax cultivar in the world, "CDC Triffid," resistant to sulfonylurea herbicide, was considered for commercial release in Canada in 1998. Being a minor crop, flax has not fully benefited from genetic engineering approaches. Oil with >70% ALA would increase its drying quality.

Zero percent ALA cultivars of flax can be obtained by reducing the activity of Δ15-desaturase. Being the richest plant source of ω-3 FA, flax has tremendous potential for its further accumulation, which makes it a potential vegetarian alternative to fish oil. Therefore, it might be used not only as nutraceuticals but also as a valuable feedstock for fish and poultry (eggs containing high levels of ω-3 FA).

Plant and Seeds

Domesticated *Linum* plant, *L. usitatissimum* L. throughout cultivation has evolved into two ecotypes. The type that grows for seed oil extraction has a fairly short plant with many secondary branches compared to the type that is grown for the stem fibre (Gill, 1987). Today, the term flaxseed is used to describe the seeds of flax as consumed by humans, while linseed is used to describe the seed of flax when it is used for industrial and feed purposes. Flax plant grows to 12–36″ in height and has a typical life cycle of 45–60 days of vegetative period, followed by 15–25 days of flowering period, and 30–40 days of maturation period. The plant produces flowers of blue, white, pink, or violet depending on the variety. The mature fruit of the flax plant is a dry boll or capsule with five segments. Each segment of the boll contains two seeds, therefore a capsule carries up to 10 (average of 6–8) flaxseeds (Oplinger *et al.,* 1989).

Anthesis and Pollination: Linseed plant exhibits perfect flower structure and sticky pollen and as such is considered self-pollinated (Beard and Comstock, 1980). The pollen is viable for only a few hours, from the time of anther dehiscence until about the time the petals dehisce - between 4 and 7 hours (Lay and Dybing, 1989; Dillman, 1938). As the flower opens, the anthers come together and form a cap over the stigma. However, certain degree of out-crossing of the order of 1to 6% has been reported (Dillman, 1938: Masuo, 1958). According to Dillman (1938) the extent of cross pollination depends on the size of the flower, with larger flowers being disposed to higher levels of out crossing. According to Sundararaj and Thulasidas, (1993) the amount of self and cross-pollination is influenced by the position of anthers relative to the stigma. In most of the cultivars, the anthers are above and entirely surround the stigma, favouring self-pollination, whereas, in some the tip of the stigma extends above the anthers, increasing the chance of cross-pollination (Yermanos and Kostopoulos, 1970). The rate of cross-pollination increases in cultivars with large disc flowers than tubular ones. Among different pollinators, honeybees play a pivotal role in the cross-pollination of over 100 cultivated crops including flax seed plant (Bradbear and Clement, 1991) According to them, the yield attributable to honeybees' pollination, was found to be about $9.3 billion in USA during 1985. The value of bees in crop pollination is 14 times the value of honey which they produce (Rehman and Singh, 1944).

Bezdenezhoykh (1956) found that honey bees in cages increase the seed production in flax by 22.5%. Gubin (1945) found the effect of bee pollination in terms of enhanced seed yields was to the tune of 22.5 to 38.5%. Kozin (1954) also found considerable increase in seeds per capsule and seed weight when 40 colonies were placed near flax fields.

Honeybees' visits start in linseed from 07.00-15.00 hours, but the most intense visitation was recorded between 11.00-12.00 hours (Sabir *et al.*, 1999). Honeybees' visits ranged from 04.25 to 22.42/10 minutes/flower during 15-24 March, while it remained 02.85 to 16.57 visits/ 10 minutes/flower during 25 March - 03 April (Sabir *et al.*,1999) in a position to get a support of petals during foraging (Gubin, 1945).

The highest seed yield (485.98 kg/ acre), 1000 grain weight (7.085 g.) and germination percentage (96.51) were observed in the case of plots pollinated by honeybee resulting a 30.14 percent increase in yield (Sabir *et al.,* 1999) as compared to the plots pollinated by "honeybee + wind + other pollinators with 373.43 kg/acre, 5.975g and 90.16 percent respectively (Table 1) in accordance with the observations of Williams (1988).

Table 1: Efficiency of different pollinating agents with regard to the potential linseed yield (Sabir et al., 1999).

Pollinating Agent	1000 Grain weight (g)	Germination percentage	Yield (kg/acre)
Honeybees	7.085	96.51	485.98
Wind	3.870	70.19	265.65
Honeybees + Wind Pollinators	5.975	90.16	373.43

Biochemical Composition of flax seed components

Flaxseeds are rich in oil (42–46% by seed weight) with a healthy fatty acid profile. This oil consists of ~73% PUFA, ~9% saturated fatty acids, 52–57% (mean value for Canadian varieties during 2004–13. http://www.grainscanada.gc.ca/flax-lin/harvest-recolte/2014/hqf14-qrl14-en.pdf/ Accessed November 3, 2015.) α-linolenic acid (ALA) The variety *Linola* is low in ALA (5% or less) compared to brown flaxseed. Thus, the oil of *Linola* seeds is more stable for oxidation compared to that of brown flaxseed and is suitable for many culinary uses such as for cooking, salads, margarines, and shortenings (Muir & Westcott, 2003).According to Sammour (1998; 1999), the protein level of defatted flaxseed meal (DFM) may range from 35% to 40%. Seed protein level may be influenced by genetic and environmental factors. Flaxseed proteins are mainly located in the cotyledons Wanasundara *et al*., (1999) and the majority is composed of two major protein fractions; a predominantly salt-soluble fraction with high molecular weight (11–12S globulin, 18.6% nitrogen) and a water-soluble basic component with low molecular weight (1.6–2S albumin, 17.7% nitrogen) (Madhusudhan and Singh, 1983;Singh KK *et al.,* 2011). According to Madhusudhan and Singh (1983) globulins comprise 70–85% of flaxseed proteins. Total dietary fiber level of flaxseed is about 400 g/kg which is rich in pentosans. The hull or seed coat contains 2–7% mucilage that is composed of soluble polysaccharides (Venglat *et al*., 2011).

Among the minor compounds in flaxseed, cyanogenic glycosides (CG), phytates, lignans, and antipyridoxin factors are reported. Flaxseed contains the lignan secoisolariciresinol diglucoside, at levels 75–800 times greater than any other crops or vegetables currently known (Madhusudhan, 2009). The lignan level of flaxseed varies from 0.9% to 3% (Muir and Westcott, 2003) and flaxseed is the richest source of this mammalian lignan precursor. Flax lignan exhibits weak estrogenic and antiestrogenic activity and may prevent estrogen-dependent cancer growth. Hypercholestrolemic, antiartherosclerotic, antidiabetic, and antioxidative properties of flaxseed lignans have been reported (Muir and Westcott, 2003).

Currently, several flaxseed-derived products are commercially available and marketed for their functional food and nutraceutical properties. They include whole seeds, ground or milled flaxseed (flour), oil extracted (by pressing; cold-pressed or not) flax meal, seed coat, partially removed seed, flaxseed hull, cyclic peptides (orbitides) from flaxseed oil, and lignan extracts from flaxseed hull (Shim *et al.*, 2015). Hegde, (2019) reported that the severely skewed omega 6:omega 3 ratio, as one ot the maior risk tactors, responslble for the dominance of Inflammatory pathway in our populatioon, as the root cause of devastating levels of non-communlcable diseases (NCD), hurting both health of the people and the economy of our country. He presented the unique, innovative, flax bio-village concept, that for resourcing omega-3 fatty acid from linseed to enrich commonly consumed food products, including egg, milk and bread as a tool to effectively tackle the rising NCD threat in the country. He developed technologies to extract high grade omega-3 oil from linseed under non-oxidizing conditions (31), free from all the antinutrients (cyanogen glycoside, antivitamin B., phytic acid, lignan) to produce omega-3 soft gel capsules. The omega-3 oil to produce emulsiori, which is both water and oil miscible, to enrich any food products including milk chocolates and other dairy products with omega-3 fatty acid. Cake, after omega-3 oil extraction, is processed to produce poultry feed mix to produce omega-3 egg and omega-3 chicken. Cake is also processed to produce high fibre, low carbohydrate, good protein product that can be used in the production of, omega-3 cereals and bread with high dietary fibre. The technology to resource omega-3 from linseed to enrich egg and milk that reaches every home would be most powerful means tackling NCDs. ln order to get consistent quality product only high yielding high omega3 PKV. NL 260 linseed variety is used at Real World Nutrition Laboratory Foundation (RWNLF), Pune.

Origin Genetic Resources & Cytogenetics of Linseed: Linseed / Flax is an annual, self-pollinated and only cultivated species under the genus *Linum* which is believed to be originated in either the Middle East or Indian regions and spread throughout Asia and Europe, and later into the New World (Green *et al.*, 2008; Soto-Cerda *et al.,* 2013). The first recorded use of flax is in Cayonu in South-eastern Turkey about 9000 years ago. Large flax seeds, showing signs of domestication, and a piece of calcified linen wrapped around a curved bone sickle were discovered by Halet Cambel of Istanbul University and Robert Braidwood of the University of Chicago. Small flax seeds have been discovered in Syria dating from about 8000 BC, Iran about 7500 BC, and in Bulgaria in the Mesolithic era (https://ambrosiabag.com/linen/).

The genus *Linum* is a large group with around 230 species. The genus is divided into five sections, Linum, Linastrum, Cathartolinum, Dasylinum, and Syllinum, based on chromosome number, floral morphology, and interspecific compatibility (Rogers, 1982 : Gill, 1987). Cultivated flax, *Linum usitatissimum*, is placed in the section *Linum* and has 30 diploid chromosomes (Tammes, 1928).

The domestication of flax was also observed on the Indian subcontinent near the Mediterranean Sea and this region is known to have high biological diversity of genus *Linum* (Allaby *et al.*, 2005; Fu, 2011 Tutin *et al.*, 1980) along with the proposed progenitors *L. angustifolium* and *L. bienne*, which share homostylous rather than heterosylous flowers.

Cultivated species (*Linum usitatissimum*) arose from a single domestication event from *L. bienne*, involving selection for annual habit, non-shattering of capsules and more efficient self-fertilization (Fu, 2011).Molecular studies indicate that *L. bienne* is the sister species of *Linum usitatissimum* (Vromans, 2006), but some researchers consider *L. angustifolium* and *L. bienne* to be the same species (Tutin et al., 1980; Zohary and Hopf, 2000). Investigations including molecular study involving genome comparisons with molecular markers of these three species *(L. angustifolium, L. bienne, and L. usitatissimum*), have confirmed that they are very closely related genetically (Muravenko *et al.,* 2003). *L. bienne* may be considered as a subspecies of *L. usitatissimum* rather than a separate species, and *L. angustifolium* can be considered as a wild progenitor of cultivated flax (Hall *et al.*, 2016)

Interspecific variability to augment the genetic resources: It is estimated that there are around 200 species of the genus *Linum* out of which only 53 are reported to have been in the world's germplasm collection consisting of 900 accessions. (Diederichson, 2007). Among these, 279 accessions represent the variability pertaining to *Linum bienne,* which is considered closest relative of the cultivated linseed. Nevertheless, the genetic variability of *L. bienne* from the centre of origin of cultivated flax or linseed is rarely represented in the existing germplasm collections. The potential contributions of *L. bienne* as genetic resource for cultivated flax breeding are not well known. *Linum bienne* from the region of origin of cultivated flax, the Fertile Crescent, is presently rare in gene-banks and should be collected (Diederichsen, and Richards, 2003).

Since *L. bienne* is reported to be the presumed ancestor of cultivated linseed (Allaby *et al.*, 2005), it is of considerable importance to access the genetic variability of *L. bienne* for use in linseed breeding.

The genetic potentials of other species of *Linum* are yet to be assessed. Some of them exhibit promise of being ornamental plants. To ensure long term sustainability of flax production, the diverse germplasm stored in gene-banks is essential, as on-farm diversity has already mostly disappeared. Evaluation of gene-bank germplasm is critical for unlocking the potential of these genetic resources. Better access to such characterization and evaluation data would enhance the overall utility of flax gene-bank collections. The duplication within and among gene-bank collections needs closer investigation. The intra-accessions diversity in gene-bank material deserves more attention.

Modern breeding techniques are being used in limited number of selected varieties. This, however, has resulted in a loss of specific alleles and thus, reduction in total genetic diversity relevant to climate-smart agriculture. Nevertheless, well-curated collections of landraces, wild linseed accessions and other *Linum* species exist in the gene banks, which are important sources of new alleles. Availability of considerable diversity in linseed germplasm indicates a considerable potential for improving this crop for both agronomic and quality traits needed towards developing climate-resilience genotypes tailored to specific environments. Some of the species of *Linum* are ornamental and their utilization may increase in future. Blue and occasional white flowers are frequent in the different species of the *L. perenne* and *L. narbonense*. Yellow flowers characterizing the *L. flavum group* and red flowered *L. grandiflorum* are often used as ornamental plants. An account of the diverse species of *Linum* preserved in the Plant Gene Resources of Canada is presented in the Table 2.

Many related species of *Linum* are extremely recalcitrant to hybridization due to interspecific incompatibility of pollen and stigma (Friedt *et al.*, 1989). In such crosses the corresponding interspecific hybrids could be recovered by the techniques of protoplast fusion (Friedt *et al.*, 1989).Flax can be also transformed by *Agrobacterium*-mediated and biolistic methods; its regeneration system was established from protoplasts, hypocotyls, cotyledons, and leaves (Mlynarova *et al.*, 1994) Flax varieties containing single gene modifications conferring resistance to insects and herbicides (glyphosate, glufosinate, and sulfonylurea) have been commercialized (McHughen, 2002). Other important genetic resources include a large collection of expressed sequences (genes), a consensus genetic map, and a physical map that were used to order and annotate the genome sequence (Cloutier *et al.*, 2012).

Divergent selection applied over thousands of years has resulted in fibre types which are taller and less branched and linseed types that are shorter, more branched and larger seeded. Though both these types are of the same species, they differ considerably in morphology, anatomy, physiology and agronomic performance (Diederichsen and Ulrich, 2009).Germplasm from West Asia mostly belonged to the intermediate group, i.*e* *L.usitatissimum.* The few accessions representing some species of the genus *Linum* came from the Mediterranean area and adjacent West Asia and East Europe. The mean values for plant height, technical stem length and 1000 seed weight for germplasm from the different areas showed that taller flax originated from East and Southeast Asia and Europe, while flax from the Indian subcontinent was short. The distribution for 1000 seed weight in respect of flax with heavy seeds are from the Mediterranean area, the Indian Subcontinent and Australia/New Zealand. Light-seeded flax was characteristic for Africa, East and Southeast Asia, and East Europe (Diederichsen and Fu 2008)

Genetic variability for disease resistance

Linseed germplasm collections have been evaluated globally for disease resistance (Kutuzova, 1998; Zhuchenko and Rozhmina, 2000; Rashid, 2003; Singh, J. 2004) and adaptability to dry and warm regions (Diederichsen *et al.,* 2006) resulting in valuable information for use in linseed breeding. Earlier phase of linseed breeding and comprehensive agro-botanic analyses of variation (Vavilov, 1951; Dillman, 1953) revealed the importance of germplasm collections in 19th century. More information on germplasm diversity has been generated over time by studies of Vavilovian collections (Kutuzova, 1998). Further, the use of landraces for fibre flax breeding was described by Zhuchenko and Rozhmina (2000). Such studies have proven to be useful tools for efficiently preserving and using flax germplasm collections (Diederichsen, 2007); flax adaptation to stress and diversity in commercially important accessions. As result of the build-up of the changes due to climate change, the intensity as well as incidence patterns of linseed diseases is getting altered. There are diverse germplasm sources with tolerance/resistance to biotic factors that could be incorporated into the cultivars to develop multiple genetic resistance through developmental breeding or germplasm enhancement. On the other hand it may be noted that a limited number of sources of resistance against pathogens from wild germplasm systems such as wild populations of *Linum marginale* particularly against flax rust (*Melamspora lini*) have been studied (Burdon and Leather, 1990). Considerable variability for resistance has been recognized in co-evolved systems of wild germplasm and their pathogens. Knowledge of diseases and pathogens of wild germplasm allows construction of maps of geographic patterns of disease resistance that can facilitate more effective collection of resistant germplasm. Geographical zones that are rich in diversity of pathogens which are known as "hot-spots" for diseases facilitate in the natural screening of the native germplasm for resistance. Hence there is every need to map these "disease hot-spots" to survey the local germplasm to select *in situ* resistant accessions for use in breeding.

Quality Germplasm for agronomic traits: The studies on diversity and oil content in Indian linseed germplasm have led to the development of promising genotypes possessing higher oil content, early flowering and early maturity (Dikshit and Sivaraj, 2015). Investigations on genetic diversity in indigenous and exotic linseed germplasm (Abdul Nizar and Mulani, 2015) has led to the identification of trait-specific accessions for stalk height, seed yield, oil content, capsule-number per plant and earliness, thereby enhancing the use of core germplasm (5-10% 0f original base collections) and pre-breeding materials (Vikender Kaur *et al.*, 2017). The main objective behind the development of core collections and pre-breeding materials is to facilitate the breeding programmes to access more promising quality genetic materials, which are the products of initial screening of the germplasm with a focus on utilizable genes, thereby enhancing the efficiency of breeding programmes and cutting down the time lag involved in screening the germplasm for valuable genes at the level of the breeders. Before the decade of 1970s, N.I. Vavilov Institute in St. Petersburg, established "Genetic Collections", which included about 250 lines genetically defined for morphological, agronomic or disease-resistance characters (Brutch, 2002). A catalogue listing 50 linseed accessions as donors for specific genes for traits, such as resistance to the diseases of rust, wilt, anthracnose, and pasmo, fibre content and quality, lodging resistance, earliness, seed oil content, seed size and capsule indehiscence, was published by the Vavilov Institute (Kutuzova, 2000).

The Centre for Genetic Resources at Wageningen in The Netherlands has established a flax core collection of 84 accessions from 506 fibre flax accessions, 440 accessions of linseed and wild species too (vanSoest and Bas, 2002). This core subset was used in evaluations for fibre characters (Booth *et al.*, 2004). A core collection comprising of 381 accessions from the flax world collection of 3378 accessions maintained by Plant Gene Resources of Canada (PGRC) was also assembled recently (Diederichsen *et al.*, 2013).

A large part of the collection is of European origin but accessions from the USA, Canada, Australia, Turkey, Japan and several countries of N. Africa are also included in the collection. In order to conduct the characterisation and evaluation of the collection a minimal descriptor list has been developed (Soest,2001). The descriptors for *Linum* are based on the list of the International Flax Data Base (IFDB) of the FAO European Cooperative Network on Flax (Rosenberg, 2004, Pavelek, 1995). In total 16 descriptors were included on the list and after a final discussion with the flax breeders the list was divided in the following two groups:

- 10 mandatory descriptors, these traits need always to be recorded during the regeneration
- 6 optional descriptors, these traits can be recorded when sufficient time is available or, in the case of diseases for which the character can be optimal scored in the field. Some of these descriptors (quality properties) can not be screened during the maintenance of thc material as special analyses are required (e.g. fibre and oil content). (https://www.wur.nl/en/Research-Results/Statutory-research-tasks/Centre-for-Genetic-Resources-the-Netherlands-1/CGN-flax-collection.htm)

Table 2: Linum Spp. Taxa Preserved at Plant Gene Resources of Canada (Diederichsen and Fu, 2008)

Species	No. of Accessions	Chromosome number (2n)	Habit	Origin
L. altaicum	2	18	Perennial	West Siberia
L.austricum	3	18	Perennial	Europe, West Asia
L.bienne Mill	11	30	Winter annual	West Asia, Mediterranean, W. Europe
L. campanulatum	1	28	Perennial	W. Mediterranean
L. capitatum, Kitt,ex Shuttles	1	34	Perennial	Balkan
L. decumbens Desf	1	30	Annual	S. Europe
L. flavum L.	4	28 (30)	Perennial	S. Central Europe, Caucasus
L. grandiflorum	7	16	Annual	Algeria
L. hirsutum L	1	16	Perennial	C &E. Europe, West Asia
L. leonii F.W. Schultz	1	18	Perennial	France , Germany
L. lewisii Pursh	11	18	Perennial	North America
L. narbonense L	3	18 (20)	Pereniial	Mediterrranean
L. pallescens, Bunge	1	30	Perennial	W. Siberia
L. perenne	10	18 (36)	Perennial	C & E. Europe, West Asia, Siberia
L. rigidum Pursh	1	20	Perennial (?)	N.America
L. strictum L	3	18 (30,32)	Annual	S. Europe, Mediterranean, W.Asia
L.tenuifolium L.	1	16	Perennial	West Asia, Mediterranean, C.Europe.
L. trigynum L	2	20	Annual	S & Central Europe, Mediterranean, W.Asia
L. usitatissimum L	352	30	Annual	**Worldwide**

Table 3: Qualitative characters of Linseed as Genetic markers (Diederichsen and Fu 2008)

Character	Character states
Anther colour	**White, Blue, Yellow**
Filament colour	White, Blue
Style colour	White, Blue
Petal colour	White, Blue, Pink, Red-vioet
Petal colour	Yes, no
Ciliation septa	Yes, no
Seed colour	**Brown, Yellow, Olive, mottled**

Genomics role towards unravelling of utility of genetic variability:The molecular characterization of *Linum* (Fu *et al*., 2005) employed random amplified polymorphic DNA (RAPD) to examine a flax world collection of 2,727 germplasm accessions, while Cloutier *et al*.,(2012) employed simple sequence repeats (SSR) for mapping of QTLs underlying fatty acid composition traits. The study of diversity and further understanding of transcriptional level molecular mechanisms of flax adaptation to stress has been greatly accelerated by recent release of the flax genome sequence (Wang, Z. *et al.*, 2012). Coupled with this, wide range of genomic and analytical tools in public domain has furthered understanding of molecular mechanisms. However, for detailed study of the genes underlying the genetic diversity, population structure and linkage disequilibrium (LD) patterns in respect of linseed which are necessary for association mapping (AM) have not yet been assessed, because genomic resources have only recently been developed. Cerda *et al.*, (2013) characterized 407 globally distributed flax accessions using 448 microsatellite markers. The data was analyzed to assess the suitability of this core collection for AM. Genomic scans to identify candidate genes selected during the divergent breeding process of fiber flax and linseed were conducted using the whole genome shotgun sequence of flax.

Combined genetic structure analysis assigned all accessions to two major groups with six sub-groups. Population differentiation was weak between the major groups (FST=0.094) and for most of the pairwise comparisons among sub-groups. The molecular coancestry analysis indicated weak relatedness (mean=0.287) for most individual pairs. Abundant genetic diversity was observed in the total panel (5.32 alleles per locus), and some sub-groups showed a high proportion of private alleles. The average genome-wide LD (r2) was 0.036, with a relatively fast decay of 1.5 cM. Genomic scans between fibre flax and linseed identified candidate genes involved in cell-wall biogenesis/modification, xylem identity and fatty acid biosynthesis congruent with genes previously identified in flax and other plant species (Cerda *et al.*, 2013). Analysis of the population structure of crop genotypes from morphological and DNA markers will be helpful for association studies through linkage disequilibrium in populations for identifying particular alleles associated with a given phenotype.

Based on 53 variable RAPD loci observed for the 61 accessions, it was found that the landraces had a lower proportion of fixed recessive RAPD loci (0.427) (i.e., more genetic variation) than all of the flax cultivars examined (0.492). The linseed cultivars had a lower proportion of recessive loci (0.469) than the fibre flax cultivars(0.529). Canadian linseed cultivars had a lower proportion of recessive loci (0.465) than the selected world flax cultivars (0.512). A trend was also observed that the rate of loss in genetic variation in Canadian flax breeding programs over the last fifty years was approximately two variable loci per 100 lociper 10 years. Clustering analyses based on similarity estimates showed that the fiber cultivars were more related (or similar to each other) and were classified as a homogeneous group. All of the linseed cultivars were clustered in diverse groups with the nine landrace accessions (Fu *et al.*, 2002).

DNA diversity was measured using inter-retrotransposon-amplified-polymorphism and inter-simple-sequence-repeat markers in 203 Ethiopian landraces and reference varieties of linseed (*Linum usitatissimum*) and wild *Linum* species and wild *Linum* species (Mhiret and Harrison 2018). Molecular diversity was high (PIC, 0.16; GD, 0.19) compared to other reports on the species. The phylogeny analysis revealed that *L. bienne* Mill. is the progenitor of domesticated *L. usitatissimum.* Markers developed in the study will be useful for genetic mapping and selection of breeding lines. The results indicated the range of characters that can be exploited in breeding lines appropriate for smallholder and commercial farmers in Ethiopia, producing a sustainable, secure, high-value crop meeting agricultural, economic and cultural needs. Based on the study it was observed that exploitation of diversity within linseed, hybridization with *L. bienne*, and backcrossing with other wild *Linum* species, can introgress novel and useful characters to linseed for quality (oil and fibre), yield, biotic and abiotic stress resistances.

Retrotransposon segments were characterized and inter-retrotransposon amplified polymorphism (IRAP)markers developed for cultivated flax and the *Linum* genus (Smykal *et al.*, 2011). Over 75 distinct long terminal repeat retrotransposon segments were cloned, the first set for *Linum*, and specific primers designed for them. IRAP was then used to evaluate genetic diversity among 708 accessions of cultivated flax comprising 143 landraces, 387varieties, and 178 breeding lines. These included both traditional and modern, oil (86), fiber (351), and combineduse (271) accessions, originating from 36 countries, and 10 wild *Linum* species. The set of 10 most polymorphic-primers yielded 141 reproducible informative data points per accession, with 52% polymorphism and a 0.34 Shannon diversity index. The maximal genetic diversity was detected among wild *Linum* species (100% IRAP polymorphism and 0.57 Jaccard similarity), while diversity within cultivated germplasm decreased from landraces (58%, 0.63) to breeding lines (48%, 0.85) and cultivars (50%, 0.81). Application of Bayesian methods for clustering resulted in the robust identification of 20 clusters of accessions, which were unstratified according to origin or user type. This indicates an overlap in genetic diversity despite disruptive selection for fibre versus oil types. Nevertheless, eight clusters contained high proportions (70–100%) of commercial cultivars, whereas two clusters were rich (60%) in landraces. These findings provide a basis for better flax germplasm management, core collection establishment, and exploration of diversity in breeding, as well as for exploration of the role of retrotransposons in flax genome-dynamics.

Cytogenetics of *Linum* spp

Flax plant has undergone multiple whole genome duplications, one being 30 million years ago and another 5-9 million years ago (Wang *et al.*, 2012) representing genome plasticity of the *L. uititatissimum*, which is indeed useful in breeding.

According to the results of the study carried out by Samadi *et al.,* (2007), *L. austeriacum* species is diploid that its karyotype consists of 18 chromosomes (2n = 2x = 18) whose size are between 2.246-3.44 micron and chromosome No. 1 includes satellite.

All of the chromosomes were metacenteric type. *L. nervosum* species, as this study explains, is diploid and its karyotype includes of 18 chromosomes (2n = 2x = 18) with size of 2.95-4.291 micron. Also according to the results of this research, *L. usitatissimum* species is diploid, with 30 chromosomes (2n = 2x = 30), with their sizes are between 1.292-2.968 micron. Type of all the chromosomes are metacentric except chromosome No. 4 that is submetacentric and its arms ratio is 1.8. The karyotype *Linum* spp. species consists of 30 + 1 chromosome and is aneuploid, that the sizes of the chromosomes are between 1.224- 1.992 micron. Mostly chromosomes of this species are metacentric type except chromosomes No. 1 and 2. that are submetacentric. Based on investigating asymmetry, *L. austeriacum* is being estimated to be the most developed one among them.

Rachinsskaia *et al.,* (2011) studied genomic polymorphism in 15 flax (*Linum usitatissimum* L.) varieties from different geographic regions belonging to three directions of selection (oil, fiber, and intermediate flax) and in the k-37 × Viking hybrid. All individual chromosomes have been identified in the karyotypes of these varieties on the basis of the patterns of differential DAPI/C-banding and the distribution of 26S and 5S rDNA, and idiograms of the chromosomes have been generated.

Unlike the oil flax varieties, the chromosomes in the karyotypes of the fibre flax varieties have, as a rule, pericentromeric and telomeric DAPI-positive bands of smaller size, but contain larger intercalary regions. Two chromosome rearrangements (chromosome 3 inversions) were detected in the variety Luna and in the (K-37x Viking) hybrid. In both these forms, no colocalization of 26S rDNA and 5S rDNA on the satellite chromosome was detected. The SSR assay with the use of 20 polymorphic pairs of primers revealed 22 polymorphic loci. Based on the SSR data, was genetic similarity of the flax forms studied and constructed a genetic similarity dendrogram. The genotypes studied form three clusters. The oil varieties comprise an independent cluster. The genetically related fiber flax varieties Vita and Luna, as well as the landrace Lipinska XIII belonging to the intermediate type, proved to be closer to the oil varieties than the remaining fiber flax varieties. The results of the molecular chromosome analysis in the fiber and oil flax confirm their very close genetic similarity. In spite of this, the combined use of the chromosome and molecular markers has opened up unique possibilities for describing the genotypes of flax varieties and creating their genetic passports.

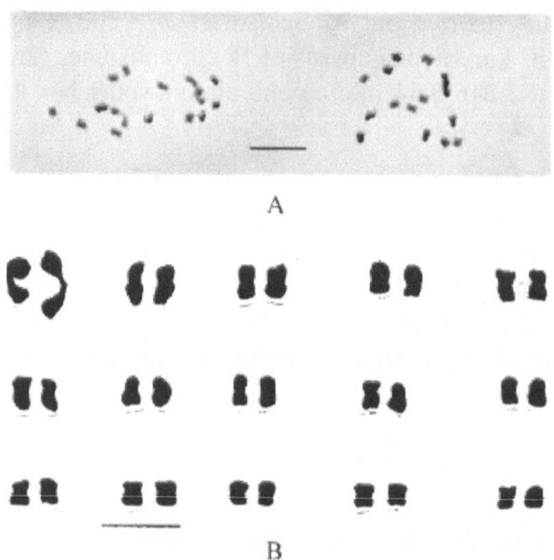

Fig. 2: A) Mitotic metaphase plate of *L. usitatissimum*, B) Karyogram , constructed from chromosomes of this plate (2n = 2x = 30), Bar = 5 µm (Samadi et al., 2007)

Fluorescence *in situ* hybridization (FISH) was employed for the first time by Muravenko *et al.,* (2004) to study the chromosomal location of the 45S (18S–5.8S–26S) and 5S ribosomal genes in the genomes of five flax species of the section *Linum* (syn. *Protolinum* and *Adenolinum*). In *L. usitatissimum* L. (2*n* = 30), *L. angustifolium* Huds. (2*n* = 30), and *L. bienne* Mill. (2*n* = 30), a major hybridization site of 45S rDNA was observed in the pericentric region of a large metacentric chromosome. A polymorphic minor locus of 45S rDNA was found on one of the small chromosomes. Sites of 5S rDNA were colocalized with those of 45S rDNA, but direct correlation between signal intensities from the 45S and 5S rDNA sites was observed only in some cases. Other 5S rDNA sites mapped to two chromosomes in these flax species. In *L. grandiflorum* Desf. (2*n* = 16) and *L. austriacum* L. (2*n* = 18), large regions of 45S and 5S rDNA were similarly located on a pair of homologous satellite-bearing chromosomes. An additional large polymorphic site of 45S and 5S rDNA was found in the proximal region of one arm of a small chromosome in the *L. usitatissimum, L. angustifolium,* and *L. bienne* karyotypes. The other arm of this chromosome contained a large 5S rDNA cluster. A similar location of the ribosomal genes in the pericentric region of the pair of satellite-bearing metacentrics confirmed the close relationships of the species examined. The difference in chromosomal location of the ribosomal genes between flax species with 2*n* = 30 and those with 2*n* = 16 or 18 testified to their assignment to different sections. The use of ribosomal genes as chromosome markers is considered important for comparative genomic studies in cultivated flax/ linseed and in its wild relatives.

Linseed Breeding: Plant breeders have a challenge before them to develop linseed cultivars adapted to changing environments and market needs.

Flax /Linseed is a diploid (2n = 30) autogamous crop plant. Improvement in flax has lagged behind other oilseeds, including rapeseed-mustard and soybean, because flax occupies a smaller niche as an oilseed and, consequently, received reduced resources for development. Genetic diversity within the crop is low, and cannot be readily supplemented by intraspecific hybridization. Finally, methods of hybrid seed production have not been developed. Inheritance of crop characteristics can range for one major gene not influenced by environment (qualitative traits) to control by many genes, much influenced by the environment (quantitative traits). Many important traits in flax including days to flowering, days to maturity, plant height, plant branching and seed yield are quantitative, and demonstrate additive gene effects that are predominant, as in other autogamous crops (Murty *et al.*, 1967; Sood *et al.*, 2007; Bhateria *et al.*, 2006; Mohammadi *et al.*, 2010). Selection of parents and in early generations is therefore critical in flax breeding programs.

Fibre and Seed yield: One important objective of linseed breeding is to increase the fibre yield per hectare, which is not easy to evaluate since it is largely influenced by the environmental factors and has low heritability (Fouilloux, 1989). Instead, breeders may use an alternate criterion of productivity, which is also termed 'fibre-wealth', which is more heritable and easier to evaluate. Fibre-wealth is defined as the ratio of the fibre weight on the dry stems without seeds (Fouilloux, 1989). However, this needs large samples of stems consequently large plots to generate the samples, which is a limitation in evaluating a large number of accessions and genotypes. In this context the length and fineness of the fibre are two important attributes to be considered for which the environmental factors and quality of retting are important factors. Fibre quality is less heritable than '*fibre-wealth*' and more difficult to evaluate. There also exists a negative correlation between fibre productivity and fibre quality. However, instead of direct selection, the possibility of indirect selection is to be explored. For early generation testing, plant height may be considered (Fouilloux, 1989). In order to safeguard the fibre quality, 'early-pulling' is to be practised to avoid the increased lignification of fibre cells. This, however, negatively affects the seed yield (Keijzer, 1989). Lignin deposition in and around the fibre bundles results in decline in fibre quality (Schilling, 1930; Crozier, 1950) Another important objective is to increase the seed yield. Since the 'early-pulling' (Keijzer, 1989) of flax fibre results in reduction in seed yield, there is a need to find a method of synchronising the maturation of the stem fibres and seed development to avoid losses of seed yield. Selection for early seed yield (seed maturity) may result in genotypes with shorter plant height, which is not desirable for fibre flax yield. It has been observed that no genotypes are available yet with seed maturity synchronising with defoliation. Stem maturation and seed maturity are independent characters (Keijzer, 1989).

Flax cultivars have been selected for production of either fibre or oil. Location of production, climatic adaptation, and morphology of these types now differ considerably. Oilseed-type plants are usually shorter, have more branches, and produce more seeds, while fibre flax types are generally taller, have few branches, and have been selected for fibre (Gill, 1987). Bast fibres from flax, derived as part of the phloem, are long (4 cm), have high tensile strength, and have a high quality of cellulose (Deyholos, 2006).

Flax types have a short tap root system with fibrous branches (Hall *et al.*, 2016) and is relatively shallow rooted, with only 4–7% of root mass deeper than 60 cm (Gan *et al.*, 2009). Production of both linen and oil is generally not compatible in the same plant type.

Bast fibres become more lignified and less flexible during flower and seed set, so high-quality linen fibre is harvested prior to seed ripening and harvest (Deyholos, 2006). Additionally, in many regions where flax is grown for oil, weather conditions do not permit postharvest fibre processing (retting), which requires a relatively wet climate (Hall *et al.*, 2016)

For oilseed production, the straw is considered an impediment during harvest and for subsequent seeding operations and until recently has been burned. However, the potential for alternative fibre uses such as industrial fibres for composites, paper, and nonwoven fibre increases the potential for whole plant utilization (Irvine *et al.*, 2010). The majority of flax in North America is grown for the oil (Hall *et al.*, 2016).

Oil seed development is a complex process and selection for seed yield in linseed breeding programmes requires study of other correlated attributes in addition to the normal yield components which are routinely used. Several important crop characteristics and environmental factors may influence seed yield, the available data on basic crop characteristics of flax that affect yield is very meagre. The plant physiological attributes such as canopy expansion and light interception, dry matter production, and partitioning are crucial attributes that need to be considered in this context. During the reproductive phase, light use efficiency and harvest index are correlated with grain production under favourable growing conditions (D'Antuono and Rossini, 1995). Selection for seed yield should be deferred to later generations along with the above cited attributes (Hall *et al.*, 2016).

With regard to the seed yield, maturity of crop is an important factor. Earliness of maturity helps the linseed crop to tide over the stresses such as disease, heat, drought, and frost. Early maturity aids harvest as immature flax remains green, which entangles machinery and reduces grain quality. The linseed crop duration from seeding to harvest can vary between 90 and 150 days (Diederichsen and Richards, 2003). Linseed plant is a long day plant, where lengthening days hasten the reproductive phase of development, but the degree of photosensitivity in flax varies greatly (Hall *et al.*, 2016). Major genes under epigenetic control contribute to the expression of early flowering. Gene methylation silences gene expression (reviewed by Suzuki and Bird, 2008 and Gehring and Henikoff, 2007) so that methlyated genes are transcriptionally inactive, while genes that are not methylated remain active. Early flowering flax genotypes from the Royal (R) cultivar and Large (L) genotroph derived from *Stormont cirrus*, a fibre cultivar (Durrant, 1971), were generated when germinating seedlings were treated with 5-azacytidine (Fieldes, 1994).

Larger seeded varieties are less prone to losses during harvest and seed cleaning. There is potential to accumulate more oil in a larger seed as the embryo is relatively larger. This is a particularly important characteristic for flax, where the seed coat constitutes nearly 40% of the seed mass, but contributes very little to the oil content. Global flax collections (ex. Plant Gene Resources Centre) (PGRC, 2011) contain flax lines with large seed size, and genetic crossing results in large seed progeny. Inheritance of seed size is quantitative and selecting from the extremes of a population distribution may result in lines with the targeted seed size (Mohammadi *et al.,* 2010).

Seed oil and protein are highly heritable traits in flax and can be selected for in early generations and seed colour could be used as a marker for selecting high-ALA acid flax associated with brown seed and low-ALA flax with yellow seed (Hall *et al.,* 2016).

Segregation analysis indicated seed color is governed by four independently inherited loci, designated as the dominant Y1 allele and recessive g, d, and b1 alleles. The variegated seed color is conditioned by a second recessive allele of the b1 locus, designated as b1vg. The recessive genes (g, d, and b1vg), when homozygous recessive, are epistatic to the other loci carrying dominant alleles (Mittapalli and Rowland, 2003).

Parental Selection: Selection of parental germplasm begins with an understanding of the goal of the breeding program (see Breeding Objectives) and the plant breeder must consider many characters for improvement and assign priorities to the traits considered for improvement. It is desirable that the parents are genetically diverse in order to facilitate selection of transgressive segregants, i.e., progeny that combine the genes of parental genotypes and with a phenotypic expression level superior to either parent. The parental lines may be from many different sources including existing cultivars, adapted elite breeding lines, or new introductions from global flax collections. The Plant Gene Resources Canada (PGRC, 2011), Saskatchewan, provides access to more than 3300 accessions of *Linum* L. germplasm, one of the larger collections of flax in the world (Diederichsen and Fu, 2008).

Breding Methods: Linseed being autogamous system, the breeding methods such a Pureline Selection and Pedigree Method have been adopted by linseed breeders in India.

- In order to obtain a pure-line cultivar that combines all the genes and meeting the breeding objectives, hybridization of homozygous lines with complementary traits is carried out, followed by pedigree selection of recombinant lines from segregating progeny (Culbertson, 1954; Kenaschuk, 1975).
- Pedigree selection method begins with the inter-crossing of homozygous lines, followed by selfing of the F1 hybrid plants to produce a large F2 seed population. F2 seed from individual crosses is sown the following growing season and single plants are harvested from F2 plots. Seed harvested from single plants planted in F3 progeny rows. Seed quality traits such as oil quality, oil content, and protein content are determined using near infrared resonance or gas chromatography from samples taken from the bulk seed lots, and data used to further scale down the population of desirable lines. Seed harvested from F3 single plants is used to establish F4 rows and the selection cycle of between sister-lines and within-family selection continues and is repeated in the F5 generation, following which the F5 bulk seed is used to establish replicated small yield plots to facilitate between line selections for seed yield, and the single plant selections are grown as progeny rows. Superior F5 lines are identified on the basis of yield and seed quality attributes in the replicated plots, and sub-lines of these are used to establish the subsequent F6 generation.
- The bulk population method of selection may be adopted in the case of small grain crops that cannot be harvested on single plant basis, which is the basic requirement of the pedigree selection method. Selection is delayed until later generations when selection for more complex characters such as seed yield is more effective. In the bulk method, a segregating population is harvested together and a representative sample of the seeds of the harvested diverse population is used to plant the next generation ie., F3. The F3 plot is similarly harvested and a plot of F4 planted. This is continued until the F6 generation. Selection begins in the F6 generation, which is largely homozygous. Large populations of F6 need to be planted for selection to be effective at this stage. Two methods of selection can be practised at this stage: either pureline selection (single plant selections, plant to row progenies, selections among homozygous lines) or mass selection (large number of individual plants selected and bulked and carried forward).
- Single Seed descent is proposed as an alternative procedure to Pedigree and Bulk methods. In this method attaining homozygosity and selection are separated in two different phases (Goulden 1939). As a modification of Pedigree method (Brim, 1966), single seed descent method allows accelerated development of recombinant inbred lines (RIL), which are derived from the single seed progenies of large F2 populations.

- Backcross method of breeding is used to incorporate simply inherited characters from one of the first donor parent by repeated back crosses of the FI and subsequent generations to the second recipient parent called recurrent parent, selecting each time for the character in question of the donor parent that is being transferred. Backcross method of breeding was first used in linseed to develop a set of rust differentials by transferring individual rust resistance alleles into the cultivar Bison (Flor, 1955)
- Recurrent selection involves selection practised within consecutive segregating progeny generations after the population has undergone selection for the major traits being transferred (Hull and Fred, 1952; Vasal *et al.*, 2010). The procedure involves reselection generation after generation with interbreeding of selects to provide for genetic recombination. Thus, it is a **cyclic selection** that is used to improve the frequency of desirable alleles for a character in a breeding population.
- Mutation breeding has been to employed in linseed to upgrade elite lines by altering one or two production-limiting or quality traits (Ahloowalia *et al.*, 2004; Rowland and Bhatty, 1990). Mutation breeding in flax led to the development of a new type of edible flaxseed oil that has nearly eliminated ALA (Green and Marshall, 1984; Rowland, 1991). The deficient ALA trait is known to be controlled by two recessive genes (ln1 and ln2) at independent loci (Green, 1986; Rowland, 1991; Ntiamoah and Rowland, 1997). Mutation breeding has a much greater potential to enhance the available traits in flax, including further modification of fatty acid profiles and the reduction of bast fibre content (McKenzie, 2011). Mutation breeding may have potential in the future to remove undesirable compounds from flaxseed, such as cyanogenic glycosides and linatine (Green *et al.,* 2008).

Added to the yield per se, the quality parameters of linseed in respect of human and animal nutrition are gaining importance. Limited success was reported in initial attempts to grow flax in saline-alkaline soil to avoid competition for land with other food crops. Using different strategies, flax breeders have made great efforts to develop a salt tolerant flax cultivar (McHughen, 1987; El-Beltagi *et al.*, 2008).

Genetics of several traits for disease resistance have been investigated in linseed, and associated genes have been assigned (Keijzer and Metz, 1992; Kutuzova, 1998), which has laid a foundation for resistance breeding. Diversity analysis based on morphology alone has a significant limitation in the fact that it is highly influenced by the environment, to overcome this problem; molecular characterization can play an important role. Hence, in-depth studies based on both morphological and molecular markers will help in better perception of the genetic diversity.DNA-based molecular markers have several advantages like abundance, environment independent early and rapid assessment and non-tissue specific characteristics. Oh *et al.*, (2000) first reported the use of DNA-based markers to study flax diversity, who compared RAPD and RFLP techniques and generated a preliminary genomic map based on these marker data. Molecular characterization of flax germplasm has been made using various molecular techniques to assess genetic diversity of cultivated flax and to examine evolutionary relationships of wild flax species.

To identify potentially novel genotypes among the flax accessions, and to assess genetic diversity for both germplasm management and core collection (Frankel, 1984) assembly, molecular markers are highly useful. A variety of marker systems, including Random Amplified Polymorphic DNA (RAPD), Inter-Simple Sequence Repeat (ISSR), Amplified Fragment Length Polymorphism (AFLP) and simple sequence repeat (SSR) have been used to analyse flax germplasm (Cloutier *et al.* 2012; Diederichsen and Fu, 2008,Fu *et al.*, 2002,) (Table 3). Taken together, these studies show that cultivated flax has low genetic diversity compared to wild relatives or some other crops (Smykal *et al.* 2008a), possibly resulting from a domestication bottleneck.

The current linseed breeding efforts in India have resulted in the development ofabout 56 high yielding, disease resistant varieties suitable for different agro-climatic zones, which has helped in improving not only the productivity but also addressing various biotic and abiotic stresses. Sources for genetic resistance have been identified for the disease like rust (*Melampsora lini* Pers. Lev.), wilt (*Fusarium oxysporum* f. lini), powdery mildew (*Oidium lini* Skoric), leaf spot (*Alternaria lini* Dey) and pests such as linseed gall midge (*Dasyneura lini* Barnes) and linseed semilooper (*Plusia orichalcea* Fabr.) The major limitation, however, has been the narrow genetic base of the breeding programs, with a predisposition to genetic vulnerability (Smykal *et al.,* 2011).

To ensure long term sustainability of flax production, the diverse germplasm stored in gene-banks is essential for search of new traits relevant to changing environmental and socio-economic needs. In this context, the potential contributions of *L. bienne* as wild progenitor of cultivated flax, presently rare in gene-banks deserves more attention as a genetic resource for flax breeding.Concerted breeding efforts involving distant and exotic germplasm, collected from main centres of diversity are required in view of narrow genetic base of the modern varieties.

In India, linseed breeding got accelerated in recent years with the development of high seed and fibre yielding double purpose varieties for the North Western Hill states particularly Himachal Pradesh (Singh and Chopra 2018). Linseed crop has gained a new interest in the emerging market of functional food due to its high content of favourable fatty acids, alpha linolenic acid (ALA), an essential Omega-3 fatty acid and lignin oligomers which constitute about 57% of total fatty acids (Singh, K.K. *et al.*, 2011).

Table 4: Linseed Varieties recommended for cultivation in India (www.nmoop.gov.in)

State	Recommended Varieties of Linseed
Chhattisgarh	RLC 92, Deepika, Kartika, Indira Alsi-32, Sharda, Js 552, Padmini
Uttar Pradesh	Sharda, Ajad Alsi, Ruchi, Padmini, Garima, Shikha
Bihar	Shival, Ruchi, Azad Alsi, Shekhar, Prvati, Shikha
Madhya Pradesh	JSL 9, Padmini, Parvati
Rajasthan	Meera
Jarkhand	Shiv, Ruchi, Azad Alsi, T397, Shweta, Shubhra, Padmini, Shekhar

New Linseed Varieties in pipe line developed in India and their characteristics (*Agropedia,2012*)(http://agropedia.iitk.ac.in/content/linseed-varieties-and-its-charectristics)

1. **Surbhi (Kl-1) A** high yielding variety resistant to lodging, rust, powdery mildew and drought. It matures within 165-170 days and contains 44% oil. Its average yield is 9-10 q/ha.

2. **Nagarkot (KL-31) :** This is a new variety, resistant to rust, drought and powdery mildew. It contains 44% oil and gives an average grain yield of 14-15 q/ha and fibre yield of 10-12 q/ha. This is a double purpose variety

3. **Jeevan (DPL-21) :** It is a dual purpose variety released for commercial cultivation. It is a tall variety with average total and technical height of 90 and 75 cm respectively. The seed is brown coloured, medium in size and flowers are blue coloured. The variety matures in 175-180 days. It is resistant to wilt, rust and powdery mildew and yields 10 q of seed and 9 q of fibre per ha. This is a double purpose variety.

4. **Janaki (KL-43):** It is a tall variety with an average total and technical height of 90 and 70 cm respectively. It is a blue flowered, having medium sized brown seeds. The variety is resistant to rust, wilt and powdery mildew. It matures in 175-180 days and contains about 43% oil. It is more suited for prepared bed conditions and yield is about 10 q/ha, and also performs satisfactorily under *uttera* system giving about 6 q/ha of seed yield. It has yield superiority of 15% over variety Himalini.

5. **Himalini :** The seed is medium and seed colour is brown. The plant is medium in height, erect in growth habit and bears white flowers. It is resistant to wilt, free from rust and moderately susceptible to powdery mildew. It matures in 180 days and yields about 40% oil. The average yield is 10 q/ha.

6. **Him Alsi-I (KL-187):** It is a new variety suitable for cultivation in the linseed growing areas of H.P. which can be grown both under prepared seed bed & utera system of cultivation. It is a semi-spreading type in early stage of growth. It has snow white, medium sized, funnel shaped flowers. It has

bold, shining brown seeds. The variety has better oil content (41.5%). It has resistance to rust, wilt & blight diseases. It has yield potential of 8-10 q/ha.

7. **Him Alsi-2 (DPL-17):** It is a new dual purpose variety suitable for cultivation in rainfed conditions in the linseed growing areas of Zone I & II under prepared seed bed conditions. The plant is medium sized with cup shaped white flowers with bold brown seeds. The oils content is 40.5%. It matures in 188 days resistant to rust & wilt and moderately resistant to powdery mildew. The average grain yield is 11.0 q/ha and that of fibre is 6-7 q/ha. This is a double purpose variety.

Fig. 3: Field view of Varietal Evaluation Trials of Linseed at Agricultural University, Nagpur in India

Regarding the role of genetic variability and the kind of gene action for agronomic attributes, highest Phenotypic coefficient of variability (PCV) and GCV (Genotypic coefficient of variability) were expressed by number of capsules per plant and seed yield per plant Kumar *et al.* (2019). High heritability with high genetic advance as percentage of mean was recorded for technical height, plant height, number of capsules per plant, seed yield per plant, 1000-seed weight and oil content (%), indicating predominance of additive gene action for these characters.

Breeding flax using haploid techniques has the potential advantages of the rapid development of completely homozygous lines within one generation and the development of efficient means of genotypic selection. Currently, anther culture is the most successful method for producing doubled haploid lines in flax (Friedt *et al.*, 1995). Given such recent development of linseed cultivars and the historical significance of flax cultivation, it is surprising to learn that very few studies have been made to assess the genetic diversity of flax germplasm with molecular techniques.

Association mapping (AM) also known as "linkage disequilibrium mapping", is a method of mapping quantitative trait loci (QTLs) that takes advantage of historic linkage disequilibrium to link phenotypes (observable characteristics) to genotypes (the genetic constitution of organisms), uncovering genetic associations.Association mapping is based on the idea that traits that have entered a population only recently will still be linked to the surrounding genetic sequence of the original evolutionary ancestor, or in other words, will more often be found within a given haplotype, than outside of it. It is most often performed by scanning the entire genome for significant associations between a panel of SNPs (which, in many cases are spotted onto glass slides to create "SNP chips") and a particular phenotype. These associations must then be independently verified in order to show that they either (a) contribute to the trait of interest directly, or (b) are linked to/ in linkage disequilibrium with a quantitative trait locus (QTL) that contributes to the trait of interest.

Association mapping (AM) takes advantage of an extensive range of germplasm including natural populations and collections of varieties and breeding lines to map traits by linkage disequilibrium (LD), which is the non-random association of alleles at different loci (Myles *et al.,* 2009).

Investigations on genetic diversity in linseed varieties carried out by Kumar *et al.*, (2012; 2019) have led to the grouping of the genotypes into six clusters as per their genetic divergence. Genotypes belonging to cluster III (A-49, L-14, A-449, A-434 and 5620-A) exhibited maximum genetic diversity within the cluster as compared to the genotypes belonging to other clusters. Hence, hybridization could be taken up among these genotypes for obtaining desirable sergeants for the yield and yield component traits. On the basis of high inter cluster divergence, hybridization programme could be taken up between the varieties of (i) cluster II (A-210, Banner, A-10-2-2 and 191 RR 9/2) and cluster IV (LC-185, A-385 and LC-2063) and also between the varieties of (ii) cluster II (A-210, Banner, A10-2-2 and 191 RR 9/2) and cluster VI (A-404, L-108, ARNY and NP-23K). Such mating between the genotypes of diverse clusters is expected to be of advantage in breeding research facilitating recovery of transgressive sergeants (Schwarzenbacher *et al.*, 2012; Soto-Cerda*et al.,* 2014a and 2014b; Chandrawati *et al.,* 2017) and also results in certain useful genetic recombinants which could be employed with profit in the further recombination breeding. Transgressive segregation creates an opportunity for genotypes with new hybridity offering higher level of fitness and productivity than the parental types, for which there is an urgent need in linseed to overcome the genetic stalemate observed in current genotypes of linseed.

Wide range of phenotypic expressions for important agro-morphological traits were observed in linseed germplasm (Kaur *et al.*, 2018). Through Principal Component Analysis (PCA) it was possible to identify (Kaur *et al.,* 2019) days to flowering, plant height, thousand seed weight, seed weight and number per boll,

Table 5: Brief description about the QTL and Linkage mapping studies in flax (Nag et al., 2015)

Marker	Population	Linkage group	QTL identified	Relevance	Reference
143 SSR markers	F2 population of 300 individuals generated from a cross between the susceptible cultivar NorMan and a resistant cultivar Linda	The 15 linkage group map covered 1241 cM	Three PM resistance QTL located on LG1, 7, and 9 were identified	Understanding the genetics of PM resistance in flax and map-based cloning of candidate genes underlying the QTL	Parvaneh Asgarinia et al. 2013
Consensus Genetic and Physical Maps of Flax					
Marker used	Population	Linkage map	feature	Relevance	Reference
SSR Primers	CDC Bethune/ Macbeth, E1747/Viking and SP2047/ UGG5-5 containing between 385 and 469 mapped markers each	Three linkage maps of flax (Linum usitatissimum L.)	The total length of the consensus genetic map is 1,551 cM with a mean marker density of 2.0 cM. A total of 670 markers were anchored to 204 of the 416 finger printed contigs of the physical map corresponding to *274 Mb or 74 % of the estimated flax genome size of 370 Mb	Resource for comparative genomics, genome organization, evolution studies and anchoring of the whole genome shotgun sequence.	Sylvie Cloutier et al. 2012

seed yield and seed size as the most important attributes responsible for variation in the Linseed Germplasm Accessions. Cluster analysis carried out has resulted in grouping the accessions under four major clusters, which indicated fair association of genetic diversity and geographical diversity. Few trait specific promising accessions such as IC0096539, IC0096496 (early flowering and maturity), IC 0 096 4 87, IC0096488 (large boll size), IC0096490 (high oil content and bold seeds), IC0054949, IC0054954 (bold seeds), EC0718827 (tall, large corolla) and EC0718835 (high seed yield/plant) were found to exhibit high estimates of heritability for the mentioned traits. SSR profiling of trait specific accessions was carried out to develop unique molecular identity. The inter-relationships between the traits suggested that accessions with short flowering and maturity duration, low plant height, large bolls and bold seeds should be given priority in breeding for enhanced yield. Donors for various traits were identified which may be used in linseed breeding to target yield enhancement and diverse geographical adaptation.

Genotypic evaluation for stability of performance: The 'cultivar × year' interaction was not significant except for seed oil content and composition. Stability analysis showed that at least half of the cultivars were unstable for oil content and α-linolenic acid fraction. Plant height was negatively associated with seed yield and oil content, but it was positively associated with straw and fibre yield. Plant density was critical for fibre yield since below 500 plants/m2 the increased plant branching makes it difficult to extract fibre. Together with plant density, plant height could be used to manipulate the seed/straw ratio according to production goals. (Fila *et al.*, 2018)

The stability analyses identified 'R12□N10D', 'Chilalo' and 'P13611'×'10314D' as more stable genotypes, while 'R11□N1266', 'R10□N27G' and 'R12□D24C' were specifically adapted to some environments. The highly significant rank correlations found among the deviations from regression, additive main effects multiplicative interaction, stability values, coefficients of determination, and stability variances indicated their close similarity and effectiveness in detecting stable genotypes over a range of Ethiopian environments. (Adugna and Labuschagne, 2008)

The degree of character association varied considerably across years and locations, due mainly to climatic factors (e.g. temperature, moisture levels, etc.) and disease incidences. However, seed yield per plot was significantly ($P<0{\cdot}01$) and positively associated with seed yield per plant, 1000 seed weight and bolls per plant across environments. These three yield attributes were also strongly and positively correlated with plant height, branches per plant, days to flowering and maturity. Oil yield was significantly and positively associated with polyunsaturated (linoleic and linolenic) fatty acids, whereas it was negatively correlated with saturated (palmitic and stearic) fatty acids. Oil yield also had a weak positive relation with monounsaturated oleic acid. The quality of linseed oil, which is dependent on the levels of these fatty acids, can be influenced by the correlated responses of these variables in reaction to different environmental factors. Thus, knowledge on the causes and effects of these correlated responses are necessary to undertake sound and effective selection programmes. (Adugna and Labuschagne, 2008)

Biotechnological Approaches in Linseed: Since linseed has a small nuclear genome, it has been used as a model system for genetic engineering techniques (Belanogova and Raldugina, 2006). Despite this, linseed biotechnology has not made much progress in comparison with that of other oilseed crops. The biotechnological mechanisms such as genetic engineering, haploidization, cellular selection, mutagenesis, embryo rescue and protoplast techniques have potential to take linseed genetic system to higher level complementing to the work on traditional linseed breeding.

The utilization of tissue cultures has played a crucial role in the application of genetic engineering approaches. The techniques for regeneration from protoplasts, hypocotyl-cotyledon and leaf-derived callus (Barakat and Cocking1983, 1985) made flax callus an attractive experimental system for somatic genetic manipulation.Soma clonal variation (O'Conner *et al*., 1991) and mutagenesis through the use of γ-rays (Green and Marshall 1981) and chemicals (Green 1986; Rowland and Bhatty1990; Rowland 1991) resulted in new genetic variation in flax seed. Following the treatment of ethyl methane sulfonate (EMS), mutant lines derived from cv. Glenelg (Green1986) and cv. McGregor (Ntimoah and Rowland 1997) exhibited very low content of linolenic acid in seeds. The same method was applied for the creation of low-linolenic linseed strain viz., Allan and medium linolenic linseed viz., Raciol (Tejklova *et al.,* 2011).Under the experimental conditions tested, 2,4-dichlorophenoxyacetic acid and zeatin were the most efficient combination of plant growth regulators for callus induction and biomass production. The induction of Somatic embryogenesis and shoot organogenesis were stated to be associated with increase in total sterols in the competent calli and increased ratio of stigmasterol to β-sitosterol in derived embryos and shoots (Cunha and Ferreira 1997). Cunha and Ferreira (1999) also reported that 4% glucose or fructose with MS media to be effective to give highly embryogenic cultures. Tejavathi *et al.,* (2000) reported initiation of callus from hypocotyl explants of three elite varieties of *Linum usitatissimum* L. on Murashige and Skoog (MS) medium supplemented with various growth regulators. They found that somatic embryos differentiated from callus grown on MS + α-naphthalene acetic acid. The origin from a single cell and subsequent developmental stages was traced. Embryos developed into normal plants upon transfer to MS + 2.69 μM α-naphthalene-acetic acid + 2.22 μM 6-benzyladenine in two varieties.

The technique for the transfer of mobile Ac/Ds elements employed in maize (*Zea mays*), which could be useful in mutagenesis, genome mapping and gene isolation, was successfully performed in flax (Finnegan *et al.,* 1993). Pretova and Obert (2003) summarized the possibilities of flax embryogenesis. The extensive research of Tejklova (1996) under the aegis of Agritec on flax seed anther culture led to testing of twenty double haploid genotypes of flax seed / linseed and their hybrids for morphogenic callus induction. Since that time, flax/linseed double-haploid technology has been routinely used in Agritec's breeding programme and has contributed to the release of several varieties (e.g. Raciol, AGT 297/12, AGT 318/08). Krause *et al.*, (2003) compared the regeneration of flax plants from anther culture and somatic tissues and established a flax regeneration system providing a basis for the creation of transgenic flax.

Obert *et al.*, (2004 **a** and 2004**b**) also described the creation of flax haploid plants through anther culture, while McHughen (2000) obtained flax haploid plants through microspore-derived culture. McHughen (2000) and Millam *et al.*, (2005) discussed these results and their possible applications in flax research and breeding. The above stated non-genetically modified (non-GM) approaches have been very helpful in flax breeding and development of new varieties. However, genetic engineering becomes indispensable when a new trait is to be conferred to plants (when genes are not naturally present in the cultured flax/linseed genome or when genes of interest are absent in accessible (compatible) gene pool. This approach enables the scientists to increase the genetic diversity and to produce new variant alleles that can be implanted from other plant species (and even from bacteria, animals etc.)

Genetically modified flax variety: Flax was among the first commodity crop species to be genetically engineered by recombinant DNA technologies (McHughen, 2002). It was also among the first plant species to be genetically modified for imparting agronomic traits such as herbicide resistance (Jordan and McHughen, 1988; McHughen, 1989) and salt tolerance (McHughen, 1987). Flax has been transformed with resistance to several herbicides including glyphosate, glufosinate, and sulfonylurea (McHughen, 2002). The only GE flax cultivar in the world, 'CDC Triffi d', resistant to sulfonylurea herbicide, was considered for commercial release in Canada in 1998 (McHughen, 2002), but after commercialization for six years, it was deregistered at the request of the flax industry because of the EU's concern with importing GE flaxseeds.

Milestones in flax transgenesis, agronomic and technological/quality traits are summarised below (Ludvikova and Griga, 2015)

1983: First report on *Agrobacterium* mediated transformation of flax cells Hepburn *et al.*,

1987 (i) Chlorsulfuron (sulfonylurea) resistant transgenic cell lines Jordan and McHughen (ii) *Agrobacterium tumefaciens(A.t.)*-mediated transformed shoots from callus Basiran *et al.*,

1988 First flax transgenic plants with glyphosate (Roundup®) tolerance Jordan and McHughen

1989 *A.t.*-mediated transfer of sulfonylurea resistance to commercial flax cultivars by McHughen

1991 First field tests with sulfonylurea-resistant transgenic flax McHughen and Holm

1991–1996 Optimization of methodology of A.t.-mediated transformation protocols mostly McHughen lab

1995 Development and field tests of glufosinate-ammonium tolerant GM flax McHughen and Holm

1996 Registration of CDC Triffid. McHughen *et al.*,

1997 Transformation of flax protoplasts Ling and Binding

1999 (i) Genetic transformation of flax by particle bombardment Wijayanto and McHughen (ii) Development of organ specific promoter from linseed for linseed transformation Jain *et al.*, (iii) First release of GM flax into environment in EU (flax lines after T-DNA insertional mutagenesis; randomly induced genetic/phenotype changes) Rakousky *et al.* (1999, 2001) since

2000 Transformation of flax with various genes of interest (GOI) and environmental risk assessment studies

2001 CDC Triffid transgenic flax was deregistrated www.cabn.ca

2004 (i) Polyhydroxybutyrate (PHB) synthesis – production of biodegradable polymers Wróbel *et al.*, (ii) Very-long polyunsaturated fatty acids biosysnthesis in linseed Abbadi *et al.*,

2005 Increased antioxidant capacity: expression of genes encoding chalcone synthase, chalcone isomerase and dihydroflavonol reductase Lorenc-Kukula *et al.*, https.//doi.org/10.1021/if047987

2007 (i)Polyhydroxybutyrate (PHB) synthesis, improved elastic properties of flax fibers Wrobel-Kwaitkowska *et al.*(2007a). (ii) Lignin deficiency, improved mechanical properties Wrobel-Kwaitkowska *et al.*,(2007b) increased flavonoid content connected with *Fusarium* resistance Lorenc-Kuluka *et al.* (2007)

2008 (i) Introduction of crtB gene into flax, enrichment of carotenoids in flax seed Fujisawa *et al.,* (ii) Reduction of pectin content, higher retting e□ciency of transgenic fibres. Musialak *et al.*,

2009 (i) Glutathion synthesis, tolerance to oxidative stress and Fusarium tolerance Czuj *et al.*(2009)

(ii) Omega-3-fatty acid (stearidonic acid) biosynthesis in flax seed,

(iii) ***delta*6**-desaturase gene from Primula vialii used Rui-Lopes *et al.*, (2009)

(ii) SsGT1 gene, higher resistance to *Fusarium* infection and significant increase of the flavonoid glycoside content Lorenc-Kukula *et al.*, (2009)

(iii) Risk assessment analysis (cross-pollination, interspecific hybridization, escape of transgenes) Tejklova *et al.*, (2009)

2010 Mitigation of adventitious presence of volunteer flax in wheat Dexter *et al.*, (2010)

2011 Risk assessment analysis: field experiments with GM and organic flax Jhala *et al.*

2012 Enhanced accumulation of heavy metals, introduction of αMT1 human metallothionein gene Vrbova *et al.*, (2013)

2013 Expression of chimeric gfp-TUA6 used for visualisation of microtubules Shysha *et al.*,2013

2014 RNAi silencing of pinoresinol lariciresinol reductase gene (LuPLR1), failed accumulation of SDG. Renouard *et al.*, (2014)

2015 High oleic flax through RNAi-mediated multiple FAD2 gene silencing Chen *et al.*,2015

Gene delivery into flax genome: Different methods to deliver genes into flax genome have been developed over the years. The most frequently used procedure has been *Agrobacterium*-mediated hypocotyl transformation – flax, like most dicotyledonous crop species, being amenable to gene transfer via Agrobacterium (Hepburn *et al.,* 1983)

Flax cells can be relatively easily transformed with *A. tumefaciens* and easily grown when the suitable inoculation/selection/regeneration procedure is applied (Jordan and McHugan 1988). Many attempts were made to enhance the efficiency of flax transformation procedures including a prolonged callus induction phase, removal of epidermis and prolonged co-cultivation. To enhance transformation efficiency, an improved procedure for flax was developed by increasing the cell transformation intensity in inoculated hypocotyls, a deliberate choice of selection agent and an optimization of the selection scheme (Dong and McHugan 1993).

Flax transformation can also be performed with *Agrobacterium rhizogenes*. The first report of the regeneration of flax transformed by *A. rhizogenes* was described by Zhan *et al*., (1988).In their work the regeneration of flax plants after transformation by either *A. tumefaciens* carrying a disarmed Ti-plasmid vector, *or A. rhizogenes* carrying an unmodified Ri plasmid, was examined. Transformed plantlets with curled leaves and short internodes were obtained from hairy roots induced by *A. rhizogenes.* Some plantlets had a more developed root system characterized by plagiotropic behaviour. These results show that the transformation by *A. rhizogenes* can be an alternative to the transformation by *A. tumefaciens.* Transformation with *A. rhizogenes* led to stable transformants for over two years, whereas transformation by *A. tumefaciens* resulted in non-regenerable transgenic calli and lasted only 6–8 weeks (Belho *et al.,* 2011)

Direct gene transfer into isolated protoplasts is a more suitable procedure for wild species such as *Linum suffruticosum,* which is easily regenerated, while cultured flax *L.usitatissimum* seemed to be more recalcitrant to protoplast technology (Ling and Binding., 1997)

Another possibility is offered by the particle gun (biolistic) method, adapted to flax in the Crop Development Centre (University of Saskatchewan, Saskatoon, Canada). Wijayanto and McHugen (1999) documented a successful biolistic process to regenerate transgenic linseed from bombarded hypocotyls cultured on a standard selection medium.

The results showed that particle bombardment could be an alternative method to flax/linseed *Agrobacterium* mediated transformation.

Bastaki and Cullis (2014) reported *Agrobacterium* mediated flax transformation via floral dip, which by virtue of its simplicity, low cost and high transformation rate, could replace the previous methods.

Linseed Crop Production

A Structural Equation Model was developed to investigate causal relationships between the productive performance and factors such as environmental variables, phenological traits, plant size and density. Rainfall was beneficial to seed yield, both before and after flowering, whereas higher post-flowering air temperature had a depressive effect. A higher oil content was favoured by pre-flowering rainfall (Fila *et al.,* 2018)

Important zones of Linseed production in India: The major linseeds growing states are Madhya Pradesh, Maharashtra, Chhattisgarh, Uttar Pradesh, Jharkhand, Bihar and Orissa. In addition to the above, the crop is also grown in Karnataka, Nagaland, Assam, Jammu & Kashmir, West Bengal , Himachal Pradesh, Rajasthan and Punjab.

Season & Soil: Linseed is grown in post rainy season *(rabi)* in Idia. It can be grown successfully on different soils ranging from light soil to heavy one. The crop , however, is often grown on marginal and sub-marginal rainfed soils.

Following crop management practices have been developed under the All India Coordinated Research Project in Linseed under the ambit of the ICAR-Indian Institute of Oilseeds Research in India (Anonymous, 2014-15)

Crop Varieties

A large number varieties of linseed have been developed which have been indicated in the earlier section. However, more promising varieties developed in recent years suited for different agro-ecological situations are listed below: .

Irrigated: *T-397, Neelum, Himalini, Chambal, LC-54, Jawahar-23, Garima, Pusa-3, Suyog (SLS-27), Binwa (KL-210), RL-914, LC-2063*

Rain-fed: *Neela, R-552, Sweta, Kiran, Sheetal, Padmini, NL-97, Sheela, Indira Alsi-32, Kartika, Sharda, SLS-67, PKV-NL-260*

Rain-fed & Irrigated: *Pusa-2, Shubhra, Laxmi-27, Triveni, JLS-9, Shekhar*

Double purpose: *Gaurav, Jeevan, Nagarkot, Shikha, Rashmi, Meera, Parvati, Pratap Alsi-1*

Land preparation

The land should be ploughed 2 to 3 times followed by 2 to 3 harrowing for fine tilth. To conserve moisture, it is advisable to create soil mulch with the help of hoe after each good shower.

Nutrient application

Linseed responds well to crop nutrients. The recommended doses vary from zone to zone. Under rainfed condition, all the fertilizers (30-40 kg N + 15-20 kg P) should be applied at the time of sowing whereas, under irrigated conditions (60-80 Kg N + 30 kg P), half dose of nitrogen and full dose of phosphorus and potash is applied as a basal dose. The remaining nitrogen is applied with the first irrigation *i.e*, about 35 days after sowing. An application of 5t FYM/ha can save 25% of inorganic fertilizer.

Sowing time: Time of sowing varies widely from October to 15th November in different zones depending upon the availability of soil moisture and water for irrigation. Early sowing helps the crop to escape attack of powdery mildew, rust, linseed bud fly in different regions.

Spacing and seed rates: Spacing and seed rates are recommended for different growing situations. Generally, row to row distance of 25 cm is followed for rain-fed as well as irrigated crop while 20 cm distant rows are sown in double purpose linseed. Broadcasting of seed is adopted for *utera* cultivation with 40-60 kg/ha seed. Seed rates of 30 kg, 25-30 kg and 45-60 kg are recommended per hectare for rain-fed, irrigated and double purpose cultivation of the crop respectively.

Seeding depth: Depending upon the soil moisture, the seed should be placed 2-3 cm below the soil. Shallow sowing in well prepared land having sufficient moisture ensures good plant stand.The sowing and the seed rate adopted should facilitate establishment population of about 400 plants/m^2.

Seed treatment: Seed should be treated with Bavistein @1.5 g/kg seed or Thiram @ 3 g/kg seed or Topsin-M @ 2.5 g/kg seed to protect the crop from seed borne diseases and to some extent from sopil borne diseases also.

Irrigation: Yield can be doubled if two irrigations are given, the first at about 35 days after sowing and the second at about 65 days after sowing. However, three irrigations (at about 35, 55 and 75 days after sowing) have proved very effective in increasing the yield.

Weed management: It is necessary to keep the crop free from weeds for the first 35 days after sowing. Isoproturon 75 WP @ 1.0 kg/ha either with or without 2, 4-D (sodium salt) @ 0.50 kg/ha as post emergence at 35 days after sowing can control weeds effectively. Post emergence application of clodinafop @ 80 g/ha may manage weeds in the areas where isoproturon is not effective.

Plant protection

- Summer ploughing and crop rotation should be followed.
- Use *Cuscuta* free seed for sowing.
- Sowing of resistant varieties against different insect-pests and diseases.
- Need based (2-3) spray of Rovral (Iprodione) @ 0.2% or mancozeb (0.25%) or calixin (0.05% to 0.1% for Alternaria blight.
- 2-3 sprays of Calixin (0.05%-0.1%) or Karathane (0.1%) or wettable sulphur (0.3%) reduces powdery mildew.
- Spray imidacloprid 17.8 SL (100 ml/ha) for Bud fly management.

Harvesting

The crop should be harvested when the leaves are dry, the capsules have turned brown and the seeds become shiny.

Storage

Initial moisture content of the harvested produce and storage conditions are the most important factors, which need to considered. A relative humidity of 70% with 80% moisture content in the seed are the best conditions for its storage.

***Utera* cropping** is the sowing of the next crop seeds (Linseed) before harvesting of standing paddy crop in order to utilize moisture efficiently under rainfed agro-ecosystem. *Utera* cropping is adopted in Rabi season, following the *kharif* (monsoon season) rice crop. This system helps farmers to use the available moisture in rain-fed areas and diversify the next cropping system by incorporating two or three pulse and cereal crops to reduce crop failure risks. In this system major disadvantage is total absence of other crop management practices other than sowing.

The linseed varieties Surabhi and Baner have been recommended for *Utera* system of cultivation.

Table 6: Intercropping systems of Linseed for various states

Region/state	Intercropping system	Row ratio (Linseed: other crop)
Bihar	Linseed + chickpea	2:1 or 3:1
Madhya Pradesh	Linseed + chickpea	1:3 or 3:1
	Linseed + wheat	1:3 or 1:4
Punjab	Linseed + chickpea/lentil	1:3 or 3:1
Uttar Pradesh excluding	Linseed + chickpea	1:3 or 1:4
Bundelkhand	Linseed + potato	row ratio of 3:3 potato at 60 cm and linseed at 20 cm. and space ratio 1:3
Bundelkhand	Linseed + lentil/chickpea	1:3 or 3:1
	Linseed + wheat	1:3 or 3:1
Maharashtra	Linseed+chickpea/safflower/durum Wheat in different ratios	-
Karnataka	Linseed+ safflower/chickpea/durum Wheat in different rations	-

(Ranganatha et al., 2011)

Double /Dual Purpose Linseed

Despite the negative correlation that exists between the fibre yield and seed production in linseed (Keizer, 1989), intensive breeding efforts carried out in linseed breeding under the All India Coordinated Research on Oilseed, it has been possible to develop dual purpose linseed varieties combining fibre ad seed yields to be grown in the northern hilly regions of India where temperatures and other weather parameters are favourable for fibre yield in addition to seed yield of linseed. Now, linseed may be grown for both seed and fibre purposes (Singh and Chopra, 2018) in northern hilly zones. The morphological frames of seed yielding linseed, the fibre yielding flax types and the dual purpose varieties differ from each other as detaild below.

- The seed type linseed is shorter with average height of 30 to 50 cm, multibranched from the base with more number of capsules.
- The fibre yielding flax plant is taller with average height of 100 to 120 cm with very few branches at the top of the plant.
- The dual purpose linseed / flax types have average height of 75 to 100 cm and technical height (height between ground to the point where first branch starts) of more than 50 cm with more branches restricted to the upper part of plant.

In general, the dual purpose linseed varieties bred in warm climates like India are inferior in fibre fineness than those in cold climates. Nevertheless, the improved production technologies including promising varietal genotypes developed in India for dual purpose linseed for the northern hill region will be helpful in changing the production status of linseed. The extraction of fibre from linseed requires cooler temperature during retting process. This becomes a major limitation in India, since after the harvest of the crop the temperatures rise and adversely affect the quality of fibre and thus Indian flax does not command good price in the market.

However, with the development of dry scutching machines, this problem has been overcome partially.

The following production package is as per the guidelines developed by ICAR All India coordinated Linseed Research Project (Singh and Chopra, 2018)

The promising varieties viz., Jeevan, Nagargot, Him Alsi 2, DPL 21, LCK 152 , 2018)and RL 993 (Meera) are recommended for double purpose linseed varieties in India.

The crop may be sown in the first fortnight of October at a row spacing of 23cm with 75 kg N/ha using a seed rate of 40 kg/ha to raise the crop for optimum fibre and seed yields.

A minimum of two irrigations are needed for proper crop growth given at 35 and 65days following sowing.

Retting is the process of treatment of stalks for fibre extraction. It is carried out for separating the embedded fibre from the linseed stem through partial rotting by immersion in water. The process is brought about by a complex enzyme action of microbes naturally present in retting water. Retting is a complex biological and biochemical process and ranks as the single most important factor governing the quality of fibre. Retting is best done in clear, slowly flowing water. Hence canal and even rivulets are best suited for retting. The depth of water should be sufficient to allow the stem bundles to float. The time taken for retting depends on a number of factors such as genotype, water quality and water temperature.When retting is done in stagnant water, the minimum ratio of plant material to irrigation water should be 1:20 by volume and pH should be around 6.0 to 7.5 to ensure good results. Retting takes 4-6 days at a temperature of around 40^{0}C at harvest. Retting process should be completed immediately after threshing to extract fibre before the onset of monsoon.

Completion of retting can be determined by a rapid test described below.

- When straw is bent around the finger, the wood in the centre should spring clearly through the sheath of the fibre.

- If a piece of the fibre is separated from the straw, it should break under strain without leaving a square broken end.

The cleaning will soften the fibre and will increase acceptability. Over retting must be avoided. It is better to get under retted straw which can be further retted again for complete retting. Among a set of micro-organisms, *Bacillus subtilis* has been identified for microbial retting for effectively decreasing the retting time without affecting the fibre quality. Same procedure as done for water retting is followed for microbial retting in which inoculum of this bacteria is added for enhancing the retting process.

Scutching :After successful retting, scutching is taken . Scutching is defined as beating and scraping of the dried stems to separate the fibres. The fibre is separated from retted stalks either by hand or by breaking and scutching with machinery.

Manual process: Take small bundle of retted and dried stalk and beat it by hand mallet (*mungri*) on plane and hard surface. Owing to this the wooden part of the stalk will split out and resultant fibre can be obtained easily. This type of scutching can be done at small-scale rightly at farmers' home.

Mechanical process: Small bundles of retted and dried stalks are placed on the feeder side of the machine and allowed to run the machine (manually and power operated). The stalks get ruptured and come out from the delivery side. Shake the ruptured stalk by hand resulting in this fibre, which will remain in hand and hard stick split out. If the rupturing of the stalk is not completed one time, repeat the process again to get the fibre out.

Hackling: The scutched fibres are combed to remove short fibres, parallelize the long fibres and also to remove any extraneous matter. Hackling is done by hand or through hackling machine.

- A wooden comb like instrument is used for dressing or combing of the scutched fibre.
- The hackling machine takes lengths of scutched flax and combs them between hackle's successively finer pins. The product of the hackling process is called line fibre.

The machine is operated manually through peddling. Short fibres and fragments of fibres separated during the scutching and hackling process are recovered and marketed

In most of the linseed growing regions in India, under rainfed conditions, lack of water availability for retting is the main cause for not extracting fibre from the crop as by product after removing seed.

The crop should be harvested when leaves are dry, the capsules have turned brown and seed have become shiny. Relative humidity of 70% with 8% moisture content in the seed is ideal for storage. The yield varies according to the varieties and growing conditions. The 800-1000 kg/ha under rainfed conditions, 1200-1500 kg/ha with heavier soils and protective irrigation and 1600-2000 kg/ha under irrigated high input management can be expected.

Biotic Stresses of Linseed Crop

Insect Pests of Linseed

***Linseed Bud Fly,* (*Dasyneura lini*)** is the most serious and destructive pest of linseed in linseed growing parts of India including, Madhya Pradesh, Bihar, Uttar Pradesh, Delhi, and Punjab. The adult is a small dipteran fly, which is a delicate, mosquito-like and orange coloured insect. The damage is caused by the maggot of the fly. The maggot feeds on the flower buds and prevents anthesis, thereby affecting seed set. Consequently, galls are produced without pod formation. The incidence of this pest is estimated as 20% (Abrol and Umashankar 2016)

Fig. 4: Damage done to the capsules of Linseed by Bud-fly [Source. Abrol and Umashankar 2016]

***Pest Management*:** According to Rabindra Prasad *et al.*, (2010) abundance of lacewings (Chrysoperla spp.). lady bird beetles (coccinellids) and rove beetle (Paedorus spp.) were found to be relatively more effective as compared to those of preying mantids, spiders and yellow wasps (Polistes spp.). Conservation, encouragement and augmentation of these natural enemies in the linseed agro-ecosystem could be a supplementary tools of IPM for realizing higher yield of linseed through application of eco-friendly approach of cultivation. Linseed bud fly is parasitized up to 50.7% in their larval stage during the month of February. Coccinellids, *Coccinella septempunctata* and *Macaria sexmaculata*, are predators of midge and predation of up to 10–15% has been reported in Kanpur (Malik and Singh, 1997). Similarly, planting of resistant varieties is a promising option in order to mitigate linseed bud fly attacks. Other options include the installation of pheromone traps for early detection of pests and therefore mass destruction of adults of *Helicoverpa* armigera. Among avian predators the bird 'Black Drongo' *(Dicrurus macrocercus*) is supposed to be more efficient in checking the proliferation of *Helicoverpa* larvae. Hence bird perches may be erected in linseed fields to harness the bird to predate upon the larvae of *Helicoverpa*. Biological control agents such as *Campoletes chloridae* are promising because they suppress *Helicoverpa* larval populations on linseed crops (Abrol and Shankar, 2011.)

Cultural methods conducive to pest avoidance are more important than chemical control. Practices such as (i) early planting of crop, (ii) adopting crop rotations involving grain legumes and cereals such as bajra, sorghum and *Setaria* in *kharif* season are known to minimise the pest incidence. (iii) Appropriate inter-cropping practices such as Linseed + chicpea, Linseed + safflower, Linseed +potato, Linseed +wheat, linseed + barley in a proportion of 2:1 are conducive for pest avoidance.

In case insecticidal control becomes inevitable, spraying of Imidachloprid (0.2 ml/ litre) or Oxydemeton methyl 600 ml/ha and repeating the spray at 15 days interval, should control the pest

The next important insect is *Helicoverpa* , a polyphagous pest. This pest has been dealt with in greater detail in earlier chapters.

Diseases of Linseed

Rust ***(Melampsora lini var. lini):*** The disease is characterised by the appearence of brown or ornge coloured pustules on the leaves and stem.Small pustules are initially surrounded by chlorotic areas. Little necrosis of the leaves is at first observed but it grows, becomes more general and the leaves prematurely die. The pustules on the leaves are uredo-pustules containing uredospores. Uredo-pustules may also appear on stems. While the uredopustules on leaves are round and small, those on stems are elongated and irregular.It is presumed that the uredospores produced on linseed at hills come down to plains to cause infection. Thus the primary inoculum, windblown, fall on the host, germinate and cause infection.

The disease generally appears in February or later, but Butler (1918) has observed the disease in central India in carly Novcmbcr. At thc timc of crop harvesting the affected plants get a fired appearance due to the presence of *telial* sori. (http:// www.biologydiscussion.com/biologyarticle)

Disease Management

- Destroy plant debris and weeds to reduce the primary sources of infection
- Use of clean seed treated with Oxycarboxin
- Grow resistant varieties like Neelam, Shubhra, Sweta, Kiran, Himalini, JLS(J)1, Jawahar 23
- Chose chemical fungicidal control depending upon the field situation. Spray Sulfex (0.05%) or Calixin (0.005%) or Calixin + Dithane M-45 twice at 15 days' interval

Powdery mildew ***(Odium lini)***:A greyish white powdery growth is seen on young, leaves (tips).

Powdery mildew of linseed is one of the economically important diseases, which caused loss up to 60 per cent particularly in Chhattisgarh.Early infections may cause severe defoliation of the flax plant and reduce seed yield and quality. The symptoms are characterized by a white powdery mass of mycelia that start as small spots and rapidly spread to cover the entire leaf surface Heavily infected leaves dry up, wither and die. Early infections may cause complete defoliation of flax plants. Some flax varieties are resistant/moderately resistant to this disease. Powdery mildew is the major problem in the linseed production during *utera* cultivation. The average yield losses due to the disease could be about 40%. (https://flaxcouncil.ca/growing-flax/chapters/diseases/)

Disease Management

- Early sowing minimises the disease incidence.
- Growing disease resistant linseed varieties such as *Kiran, Jawahar 23, Garima, Sweta, JLS (J)1, LC 54, Himalini* are recommended to check the disease.
- Hexaconazole spray with least incidence (25%), severity (14.12%), highest PDC (64.95%) gave highest yield (531.57 Kg/ha) as the best treatment followed by Tridemefon with PDC (62.91%), Propiconazole with PDC (62.59%), Dinocap with PDC (59.16%), Difenaconazole with PDC (58.63%)
- Linseed crop treated with *Pseudomonos fluorescens* (5 g/l) on 45th day followed by hexaconazole on 60th day @ 1 ml/l resulted in less disease.
- The disease can also be controlled by spray of sulfex @ 3 kg/hectare in 1000 litres water.

***Alternaria* Blight (*Alternaria lini*):** Delayed sowing from 27 Oct. to 27 Nov. decreased the incidence and severity of *Altemaria* blight of linseed. With increased seed rate, the incidence and severity of the disease also increased. Among the experimental materials. only one entry (UPN 36) showed moderately resistant reaction. Among the released varieties, Kiran and Shubra are rated as resistant.

Intercropping of linseed with mustard, wheat, safflower and gram lowered the disease pressure and significantly increased the yield over control. Seed treatment with Thiram + 2 sprays of Dithane M-45 to the standing crop resulted in good disease control. All the fungicidal treatments in vitro reduced the growth and sporulation of *Alternaria lini*. As the concentration of fungicide was increased, there was increase in inhibitive capacity. Rovral was best among all the fungicides evaluated.

Seed treatment with *Trichodrermaa. viride* (4 g/kg seed) followed by two foliar sprays (FS) of Mancozeb (0.25 %) decreased the blight intensity and bud damage with significantly lower mean blight intensity of 23.24 % and bud damage of 12.81 %, respectively. It also increased seed yield by 56.70% with higher 1,000-seed weight of 7.22 g and seed yield of 1,364.58 kg/ha (Singh RB *et al.*, 2013)

Plant extracts were found to be inhibitive for growth, sporulation and spore germination of Alternaria lini. The extract of Blackpepper, garlic, pepper, clove and Neem leaf were found to be most effective in inhibiting growth, sporulation and spore germination of *Alternaria lini.* (Tripathi, 1987)

Wilt disease of Linseed

The disease is caused by ***Fusarium oxysporum*** spp. Lini., which may appear at any time of growth stage of linseed crop. Flax wilt or *Fusarium* wilt is caused by the seed-borne and soil-borne fungus. The fungus invades plants through the roots at any growth stage during the growing season and continues infection inside the water-conducting tissue of the root. This interferes with water uptake and warm weather therefore aggravates plant symptoms from the disease. Being a systemic disease, the vascular system is plugged causing wilting indicated by drooping of leaves etc. The fungus could be isolated from seeds, stems and roots of diseased plants indicating the systemic nature of the disease. Pathogenicity experiments using sterilized soil inoculated with these isolates showed that the fungus was a virulent parasite (Wilson, 1946).

Disease Management

- Follow regular crop rotation involving the other crops not infected by the disease to minimize inoculum load. Include preferably legumes like green gram and cowpeas, and cereals like sorghum, Pearl millet, ragi and setaria in kharif season
- Carry our deep summer ploughing to expose the fungal mass in the soil to sun, since the disease is soil borne
- Treat the seed with Thiram or Topsin 2.5 g/kg seed that might prevent infection through seed
- Grow resistant/tolerant linseed varieties like *R 552, Kiran, LC 185, LC 54, JLS(J)1, Himalini, Garima*

10

Castor (*Ricinus communis* L.)

Introduction

Castor plant (*Ricinus communis* L) is an age old crop of India which has been in use in *Ayurvedic* medicine since 2000 BC, when the oil was used in lamps and also in local medicine particularly as a laxative and also to cure arthritis. The seeds of the crop yield non-edible viscous oil which has a number of applications in industry apart from its medicinal properties, which are well documented in *Ayurvedic, Siddha* and *Unani* medicine. The word *Ricinus* in the scientific name of the castor plant has cropped due to the appearance of its seed resembling dog's tick!

Castor is cultivated on commercial scale in an area of 1,296,855 ha in 30 countries with global castor seed production at 1,396,104 MT (FAO STAT, 2019). India is the world's single largest producer of castor seed accounting for more than 85% followed by China with about 7% and Brazil with about 5% of world castor seed output. There has been a consistent increase in output primarily on account of rise in yield levels in India. In recent years, however, Brazil and China have experienced stagnation in castor crop. Castor is widely grown in Ethiopia too. Among oilseed crops of India, castor crop has registered positive and high growth rate along with soybean (Mohapatra 2017).

Castor (*Ricinus communis* L.) belongs to the family *Euphorbiaceae* also known as Spurge family. The major centres of its genetic diversity are (i) East Africa (Ethiopia), (ii) Northwest and Southwest Asia and Arabian Peninsula, (iii) India and (iv) China (Anjani 2010). Castor plant is suggested to have originated in East Africa (Ethiopia) with India being its another centre of rich genetic diversity.

Ricinus communis is a monotypic genus. Certain sub-species viz., *persicus, chinensis, zanzibarinuz, sanguines, africans, mexicans* etc., were suggested (Kulkarni and Ramanamurthy 1977, Moshkin 1986 and Weiss 2000); but these entities are in no way different from *Ricinus communis* in terms the chromosome number and genetic analysis. There are no crossability barriers between them and *Ricinus communis* (Moshkin 1986 and Weiss 2000) and the genetic segregation data from the crosses of these type with regular cultivated castor is of simple monogenic inheritance (Kulkarni and Ramanamurthy, 1977). Also, there is clear lack of discretionary principles of species based on morphological characteristics, geographical demarcation and chromosome number. The variability that exists among earlier described types does not exceed the overall characteristics of *R. communis*.

ECOLOGY

Weather

Castor requires a moderately high temperature 20 to 27 °C with low humidity throughout the growing season. However, temperatures less than 18°C are not conducive to growth of castor crop. It grows best in areas where there are clear warm sunny days. Castor crop can be grown successfully in all those areas receiving 400 to 1200 mm of rainfall. The crop can yield satisfactorily even with a less rainfall of the order of 350 to 400 mm, provided the rain is well distributed. Prolonged cloudy weather with high temperature at the time of flowering resulted in poor seed setting, which is due to sex reversion. High temperature above 41 °C at flowering time even for a short period results in blasting of male flowers. Heavy rainfall at flowering reduces the yield. Castor plant is very susceptible to frost.

Soil: Castor can be successfully grown on any type of soils except clays as the castor crop is highly susceptible to water logged conditions. It is generally grown on red loams in peninsular India and on light alluvial soils in northern states. Inferior soils not fit for valuable commercial and food crops are often used for raising castor crop. The crop cannot tolerate alkalinity of soil but withstands slight to moderate acidity and salinity of soil.

CASTOR PLANT

Castor's stem colour varies from greyish-green to reddish tinge. Stems are hollow, branched, glabrous and covered by white bloom in many cases. At the maturity, castor plant stems turn greyish and brittle and somewhat woody.The leaves are alternately arranged on the stem with long petioles (10-20cm long) and large lobed leaf lamina (10-70 cm long and 15-60 cm wide). Each lobe is with a prominent vein running down its centre with serration at the border. Leaves are generally brick-red or purplish in colour when young; but turn green to bluish-green as they become older. Castor is protogynous and wind pollinated. Inflorescences are borne terminally on the main and lateral branches. The main stem ends in an inflorescence, which is the first or primary raceme. After the first raceme appears, 2 or 3 branches arise at the nodes immediately below it. Each of these branches terminates in racemes after 4 or more nodes have formed which are known as secondary racemes. Branches arise from the nodes just beneath secondary racemes, ultimately terminating in tertiary racemes. This sequence of development (indeterminate growth habit) continues.

Inflorescence and fruiting

Castor Inflorescence is raceme and monecious with clusters of staminate (male) flowers situated at the base and the pistillate (female) flowers at the toper portion of the panicle. (Fig.1&2). The female (i.e. pistillate) flowers have large tri-lobed ovaries with three bright red to yellowish stigmas and styles and are produced at the top of the flower clusters. There exists a lot of variation for sex expression in castor from monoecious nature with male flower clusters at the base and female flowers spreading toward the top, totally pistillate, totally male, to male flowers interspersed among the pistillate flowers. Castor plant flowers through-out the year and the sex expression in the panicle is influenced by the temperature, with the occurrence of more male flowers at higher temperatures. The ovoid rounded or slightly three-lobed capsules (fruits) are greenish, greenish-red or bright red in colour when young and covered in soft blunt spines or without spines. These capsules of 10-30 mm size, dry-up showing brown or grey colour as they mature and split-up into three segments, each containing a single seed. The seeds are mottled grey and brown to blakish (10-17 mm long and 6-10 mm wide), smooth in texture, oblong to ellipsoid, and have a small white cap like projection known as caruncle at its one end.

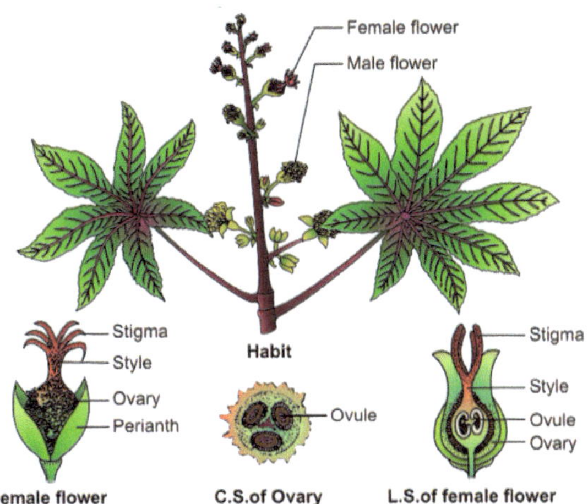

Fig. 1: Diagrams of Castor branch showing leaves and inflorescence and details of female flower

Special Features of Castor Plant

- Despite castor being a member of the family *Euphorbiaceae,* the genus *Ricinus* of castor plant is totally isolated from the rest of the members of the family, with no apparent filial or genetic relationship with any of them including the genus *Jatropha* to which castor plant has certain morphological resemblance. *Ricinus* has remained a monotypic genus with no other species. The diversity within the *Ricinus communis* is not wider genetically though one comes across perceptible morphological variability, which is a function of certain simply inherited characters. Many morphological markers exist among castor genotypes. Most of these characteristics are qualitative traits affected by a single or a few gene mutations. In other words, the castor has remained within a close gene pool. However, agro-morphological and molecular diversity studies in castor germplasm collections originated from North-Eastern Hill Province and Andaman and Nicobar Islands in India revealed presence of genetic diversity within geographically isolated castor gene pool (Ashoka Vardhana Reddy *et al.*,2002; Meena *et al,* 2014; Meena *et al.,* 2015).
- Though the castor is termed monoecious in terms of its sex type, the sex expression with in *Ricinus communis* is highly variable, with the occurrence of totally masculine, mixed monoecious nature which is variable, totally pistillate nature etc., subject to the influence of environmental temperature (Shifriss,1960; Moshkin 1986; Anjani, 2005a).
- Notwithstanding its allogamous nature characterised by 100% cross pollination, castor doesn't exhibit any inbreeding depression (de Oliveira *et al.,* 2012).
- No marked levels of heterosis, are observed in castor, since it represents a monotypic genus with diversification of parental base being restricted to intra-generic, intraspecific, or inter-varietal hybridization.(Kaul and Prasad, 1983; Lavanya *et al.*, 2018) This has certain implications in hybrid development in castor.
- The male sterility in castor is in terms of expression of the pistillate nature which is genetic subject to the interaction with temperature (Shifrisss 1960). Castor has not yet exhibited any cytoplasmic factors controlling sex expression including male sterility as in the case of several other genera.
- Castor is being grown as an annual crop; but the plant possesses the perenniating habit. Castor plant produces new shoots and panicles of later orders without any restriction provided moisture is available. In other words, castor plant is an annual with perenniating habit, which offers certain advantages to castor breeders.

- Castor oil is generally considered non-edible oil; but certain tribal populations particularly those inhabiting the banks of Indravati River is Odisha-state use castor oil as cooking oil without any adverse effects on their health (Prasad, 1989).
- Castor seed contains a poisonous substance known as 'ricin', which however doesn't get translocated into its oil; but remains in the cake left out after castor oil extraction.

Fig. 2: Castor inflorescence showing male (staminate) flowers at the base and female (pistillate) flowers towards the top.http://ecoursesonline.iasri.res.in/Courses/Principles%20of%20Seed%20Technology/GPBR112/Data%20Files/lec16.html)

Diversity in longevity of castor seed: High degree of seed longevity is a desirable trait for conservation of seeds for desirable periods. Castor seed generally maintains 100% germination when stored for one to two years under ambient conditions, this gradually decreases as the storage years increase. In Indian castor germplasm repository great variation for seed longevity was observed where some castor genotypes have ability for long longevity (11-19 years) (Anjani and Jawaharlal, 2015).

Cytogenetics: Castor has chromosome number 2n=20 and it is considered generally a diploid. However, there are reports that castor is a secondarily balanced polyploid (Richharia, 1937; Kurita, 1946; Jacob, 1957) with basic chromosome number of five (x=5), which has undergone diplodization over long time. Ten pachytene bivalents were clearly distinguished morphologically in diploid castor (Jacob, 1957;Moshkin and Dvoryadinka.1986)). Also, the meiotic phenomenon of haploid confirmed that castor had arisen through the doubling of a diploid progenitor with 2n = 10 (Narain and Singh 1968). Polyploid Castor has been produced in Israel which is reported to be as fertile as the current diploid castor plant (Avidov-Amit and Alon, 2008)). According to Avidov Amit and Alon (2008), the average size of the diploid Castor plant stomata was 150 $\mu^2 \pm 32$ and that of the tetraploid was 221$\mu^2 \pm 96$. The difference was significant at a 0.001 level. The news from Isreal's Kaiima company (https://www.kaiima.com/castormaxx-c) is that the polypolid castor is highigher yielding than the regular diploid castor. The reported polyploid castor with its stated fertility level could be of great advantage in castor breeding apart from the possibility of taking castor productivity and product development to new heights. Baghyalakshmi *et al*., (2020) treated Seeds of three castor genotypes with aqueous solutions of colchicine, at LD50 values and identified three putative tetraploids as potential polyploids based on increases in a guard cell size and reductions in the number of stomata. The putative polyploid plants were selfed and the progeny were subjected to meiotic analysis. All the progeny were found to be tetraploid. The pairing of chromosomes was abnormal with univalent to octavalent configurations during meiosis-I, but the later parts of meiosis were normal. Seasonal variations in pollen fertility indicated the possible role of temperature-sensitive male sterility in causing the sterility in tetraploid plants. The tetraploid plants were phenotypically comparable with their diploid counterparts, but produced substantially bigger seeds. Thus, these tetraploid plants are valuable resources for basic and applied research in castor

Genetic Resources and Castor Breeding

Genetic Resources: Impressively vast germplasm collections of castor are being conserved in more than 50 gene banks across the world (Anjani 2012). In India the National Bureau of Plant Genetic Resources maintains a vast castor germ plasm collection of 4,307 accessions (Anjani 2012). At the Institutional level the ICAR-Indian Institute of Oilseeds Research in Hyderabad has good collection of working germplasm of 3051indigenous accessions and 365 exotic collections introduced from 39 countries (Anjani 2010), which could support a dynamic castor breeding programme.

At the ICAR-Indian Institute of Oilseeds Research (IIOR) Hyderabad the accessions of germplasm collection were categorised and confirmed employing 23 highly heritable and distinguishable morphological descriptors (Anjani 2001) as per Shannon-Weaver Diversity Index (H') (Ortiz-Burgos, 2016), which has revealed distinct and wide morphological diversity as described and tabulated for attributes of stem colour, bloom, node number to primary spike, plant height, intermodal distance, stem nature, branching, leaf size, leaf shape, lacination, nature of raceme, raceme size, raceme shape, compactness of raceme, capsule colour, nature of capsule, capsule stalk, capsule size, capsule dehiscence, seed size, seed shape, seed colour, seed mottling and caruncle by Anjani (2010).The germplasm collections were also categorised for agronomic and economic traits, oil content, and maturity pattern. The germplasm collections of castor reflected a wide variability of spike desity and colour (Fig.3)

The variability for oil content and accessions for extra early maturity with high per day productivity in the castor germplasm collections at IIOR are indicated in the two Tables 1 and 2 below.

Table 1 : Variability for Oil Content in Castor germplasm at ICAR-IIOR (Anjani 2012)

Oil Content (%)	Number of Accessions	Oil Content (%)	Number of Accessions
28-38	31	46-47	263
38-39	12	47-48	318
39-40	15	48-49	406
40-41	27	49-50	353
41-42	64	50-51	232
42-43	78	51-52	175
43-44	111	52-53	122
44-45	158	53-54	61
45-46	211	54-55	14

Table 2 : Extra-early accessions from germplasm collection with high productivity per day (Anjani 2010)

Accession number	Productivity (g/day)
RG 26	0.82
RG 18	0.79
RG 19	0.82
RG 197	0.82
RG 210	0.94
DCS 9 (Extra early Check)	0.68
48-1 (Medium maturity check)	0.17
GCH 4 (Medium maturity check)	0.42
Sowbhagya (Late maturity check)	0.00

Though castor bean is a monotypic, it exhibits wide phenotypic diversity. In castor bean, genetic markers such as agro-morphological characters, biochemical and cytological markers were widely used in characterization of genetic variation in the germplasm from India, Nigeria, Turkey, China, Brazil, Iran, and Ethiopia which indicated a low-to-high-level diversity in the castor bean germplasm depending on the markers and the genotypes studied. Genetic diversity in castor bean was assessed by using both dominant and codominant molecular markers (random amplified polymorphic DNA, RAPD), inter-simple sequence repeats (ISSR), start codon targeted (SCoT), amplified fragment length polymorphism (AFLP), simple sequence repeat (SSR), expressed sequence tag-simple sequence repeats (EST-SSR), and random microsatellite amplified polymorphic DNA (RMAPD), and also advanced molecular markers, such as single nucleotide polymorphism (SNP), sequence-related amplification polymorphism (SRAP), target region amplification polymorphism (TRAP), and methylation-sensitiveamplification polymorphism

(MSAP). (Kallamadi and Sujatha, 2018). Molecular marker utilization in the characterization of genetic diversity in germplasm from countries including India, China, Brazil, Mexico, and worldwide collection revealed a low-modest genetic diversity in castor bean.(Kallamadi and Sujatha, 2018). The availability of castor bean genome has accelerated the molecular marker development of easy, fast, and reliable DNA markers for the assessment of genetic variation in the castor bean and further these would serve as a major tool in the characterization of genetic diversity and assist marker-assisted breeding programs for crop improvement.

Widening the genetic base of *Ricinus communis*:*Ricinus communis* pertains to a monotypic genus and hence there is a need to promote gene flow from other genera to strengthen the genetic base of castor for diverse characters. There, however, exist strong crossability barriers between castor and other genera of Euphorbiaceae. Nevertheless, with the availability of novel tools of biotechnology and genetic engineering, it is not difficult to surmount the crossability barriers and come up with novel genetic materials. Considering the remote similarity between castor and *Jatropha,* such a project is expected to yield useful results.

Fig. 3: Genetic variation for capsule colour and spike density in castor (Anjani 2012)

Fig. 4: Papaya leaf type of castor showing resistance to leaf-miner (Anjani 2012)

Genetic Improvement: Castor is a polymorphic species; normally monoecious with pistillate (female) flowers are situated on the upper part and staminate (male) flowers on the lower part of raceme. Proportion of female and male flowers on a panicle is highly influenced by environmental conditions. Despite its allogamy causing 100% cross pollination, castor breeding before 1980s was confined to varietal development through hybridization of straight varieties differing in clear-cut morphological attributes of economic value. This was largely due to the absence of inbreeding depression following self-pollination of open pollinated castor genotypes. It was possible to fix recombinants of some economic value in the second generation following crossing of two parents.

Varietal development: The earlier castor varieties (HC 6. HC8, HO, Junagadh-1, Vijapura, TMV1, TMV2, TMV3, S 20, EB-16A, T3 etc.,) developed through hybridization of selected parents, were tall, non-shattering, with yield levels of the order of 300 to 350 kg /ha possessing good oil content of the level of 48% (Kulkarni and Ramamurthy 1997). The major limitation of the older varieties was their long duration of more than 200-250 days causing late maturity, which was a great disadvantage under conditions of dry land agriculture. Their yield expression was constrained due to depletion of soil moisture before the varieties could come to maturity.

A major breakthrough in castor varietal development occurred with the development of a new plant type in the form of an early maturing castor variety Aruna (NPH 1) a mutant of HC6 induced by fast neutrons (Ankineedu *et al.*, 1968). Variety Aruna exhibited earliness of flowering (35-40 days) and maturity (100-120 days) to primary spike, relative reduction in plant height as compared to the parent and wider adaptability leading to higher levels of productivity under dryland conditions (Kulkarni and Ramanamurthy 1977).

Next important phase of castor breeding was the release of fourteen high yielding varieties out of which the most prominent are DCS-9(*Jyothi*), 48-1 (*Jwala*) and PCS-124 (*Haritha*), which are superior to NPH-1 (Aruna) by 25 -30%, in addition to their resistance to wilt decease caused by *Fusarium*. (DOR, 1987). *Jwala* (48-1) is the first high yielding variety resistant to *Fusarium* wilt (Prasad and Ankineedu, 1994) and it was also used as the male parent in the development of the first wilt resistant castor hybrid GCH4 (AICORPO 2003).

DCS 107 is another wilt resistant and high yielding variety with significant yield advantage (Anjani, 2014)

The details of some promising varieties of castor are given in the Table.3.

Table 3: Salient attributes of some Important Castor Varieties developed in All India Coordinated Research Project on Oilseeds, Hyderabad. (ICAR-IIOR Hyderabad)

Charcter	**Jwala** (48-1)	Jyothi (DCS9)	Kranti (PCS 4)	Haritha (PCA124)	TMV 6	GC 2	KIRAN (PCS136)	DCS 107
Morphological characters	Red, double bloom, Non spiny	Red, double bloom Spiny	Red double bloom spiny	Green, double bloom Spiny	Red double bloom spiny	Mahogany red, triple bloom, Spiny	Double bloom Non Spiny	Green, double bloom, spiny
Growth habit	Tall	Moderate Height	Moderate height	Medium height	Tall height	Medium height	Moderate height	Medium height
Node number	14-16	9-12	12	12-16	17.5	10-14	10-12	
100 seed weight (g)	28.5	29.0	20-24	25-28	28	28-30	22-25	30
Oil Content (%)	48.0	48.0	48 -50	48-51	52.0	47-49	48-51	48-49
Maturity (Days)	110-115	90-100	110-120	100-150	110-120	115-120	120	110-130
Disease Resistance	Resistant to **Fusarium wilt** and moderately resistant to Botrytis	Resistant to **Fusarium wilt**	Susceptible to Wilt	Resistant to **Fusarium wilt**	Moderate Resistance to Fusarium and Alternaria	Moderately tolerant to **Fusarium wilt**	Tolerant to Botrytis	Resistant to Fusarium Wilt
Insect resistance	Tolerant to Spodoptera, Heliothes and Capsule borer	Resistant to Jassids	Tolerant to Semi-looper	Less damage by Semi-looper	Tolerant to Jassids, Semi-looper & Capsule borer	Tolerant to Jassids	NIL	NIL
Yield (kg/ha)	1110 (R) 2000 (I)	1030 (R) 2460(I)	1365	1400 (R) 1600(I)	930	2165	1200(R) 1500 (I)	1375(R) 2300(I)

0I = yield under Irrigated conditions
R= Yield under dry-land conditions

Pistillate lines of Castor: Unlike other annual food crops grown on drylands like sorghum *(jowar)*, pearl-millet (*Bajra*) and maize (*Makka*), heterosis breeding leading to the development of productive hybrid varieties was very much lacking in castor. As stated earlier the major reason in this context was the non-availability of the usable systems of male sterility such as cytoplasmic and genetic male-sterility systems. The kind of male sterility we come across in castor is that of pistillate nature highly influenced by environmental conditions (Shifriss, 1960). Investigations carried out to come up with stable pistillate lines led to the identification of recessive (N) Nebraska and dominant (S) Weisman Institute types (Shifriss 1960).

The concomitant recovery of the related pistillate variants of N type viz., N 145-4 and CNES-1 (Claassen and Hoffman,1950) indicated promise for stability of the recessive genotype for sex expression to be used as female parent in hybrid combination. Nevertheless, the high cost of rouging in the variable population and consequent recovery of low frequency of genuine recessive pistillate plants were major issues in this case.

The NES pistillate type which is homozygous for pistillate gene (f) also contains environmentally sensitive genes (s) for interspersed staminate flowers which express in summer (Zimmerman and Smith 1966). This system offers advantages in breeding for transferring the single recessive gene for femaleness and at the same time use the environmentally sensitive genes that promote maleness in summer for maintenance of the pistillate system (Ankineedu and Rao 1973).

A stable pistillate line TSP-10R (Texas S-Pistillate-10) was also released in USA in 1962 (Ramachandram and Prasad1995).

In India the first stable S type pistillate line VP-1was developed from the crosses involving TSP-10R and used in hybrid breeding program (Ramchandram and Prasad 1995) to develop the first released castor hybrid GCH3 from Gujarat.

Subsequently several pistillate lines of both S type (SKP 1, 2, 4, 6, 15, 18, 23, LRES 17, DPC10 and DPC 14), NES type [M571, M574, M619, M587 which are mutants of VP1 and NES 6, DPC 15, 16, JP 65 derivatives of the cross involving 240 pistillate x (625-4 x HC6), NES19 from H86 x Bhagya, DPC 11] were developed (Lavanya and Solanki 2010).

In addition to the above new pistillate lines viz., *Geeta* (from 48-1), DCP 18 (from 163-1-10-2x 48-1) and DCP 19 (from M619 x Ji 225F4) were developed creating a wider base for castor pistillate attribute (Lavanya and Solanki 2010). The disadvantage of VP1 was its susceptibility to *Fusarium* wilt and difficulty in its maintenance due sex revertents. These problems were overcome in the new pistillate lines of NES type and the pistillate derivatives of 48-1 which is a wilt resistant genotype.

Efficiency in Successful seed production technology for stable pistillate parents to be used in hybrid development has been standardised at the Directorate of Oilseeds Research exploiting environmentally sensitive interspersed male character (Ramachandram and Ranga Rao 1988 and 1990 ; Prabhakaran *et al.*, 2009; Reddy and Prabhakaran 2010).

The conventional method followed earlier for maintenance of pistillate parent (VP1) relied upon maintaining 20 to 25% monoecious revertents as pollen source. This caused a high proportion (40-65%)of monoecious plants in certified pistillate parental seed production plots leading to high cost of seed production due to roguing and also warranting rejection of seed production plots due to low genetic purity of the pistillate parental materials.

The modified method, however, involves environmentally sensitive interspersed staminate flowers (ISF) as pollen source in summer season with an isolation distance of 1000 meters. This has resulted in spectacular improvement in seed production programmes for certified seed production (Zaveri 2009).

Hybrids of Castor: Another problem of heterosis breeding in castor was the absence of perceptible levels of heterosis in F1 following hybridization between parents considered to be diverse genetically and absence of any inbreeding depression following selfing of any open pollinated variety (Swarnalata *et al* 1984b).

Nevertheless, some notable success has been achieved so far in castor hybrid breeding through the choice of certain selected parental combinations involving diverse pistillate lines and high yielding male parental varieties possessing positive variability for attributes such as maturity, branching habit, spike density, seed weight and oil content leading to the development of productive hybrids which are indicated in the succeeding paragraph.

Secondly unlike other crops, where three-line system of hybrid breeding is practised, in castor the hybrid development involves two-line system involving a pistillate line and a male parental variety.

The first castor hybrid viz., GCH3 originated from the cross involving TSP10R and JI 15 was popular by virtue of its drought resistance high yield and high oil content over the standard varieties like S-20. Among the fourteen castor hybrids released so far, GCH4 developed by using wilt resistant and high yielding 48-1 (*Jwala*) as the male parent is still popular. Among the hybrids released later DCH 177 and DCH 519 are resistant to *Fusarium* wilt and very high yielding with productivity levels of 1.7 to 2.0 tons/ha under dry land conditions and more than 2 tons/ha under irrigated conditions. The details pertaining to some important recommended castor hybrids are given below in the Table 4.

Table 4: Salient Features of Castor Hybrids Recommended for different Regions in India (*Source*: Lavanya and Solanki 2010 & Anonymous 2018-2019)

Hybrid	Parentage	Mean seed yield (kg/ha)	Areas recommended	Salient features
GCH 3	TSP10R x JI15	1540 (I)	Gujarat	Heavy branching, dark brown seed with black spot
GAUCH-1	VP-1 x GAUC1	1520 (I)	Irrigated areas of Gujarat, rainfed areas of southern India	Early maturing hybrid with drought escape mechanism
GCH 2	VP-1 x JI35	1750 (I)	Irrigated areas of Gujarat	Tolerant to root rot
GCH 4	VP-1 x 48-1	1200 (R) 2200 (I)	Both rainfed and irrigate areas, all over country	Resistant to leafhoppers, tolerant to Fusarium wilt
GCH 5	Geeta x SH 72	1800(R) 2800 (I)	Rainfed and irrigated areas of Gujarat	Red, double bloom, medium duration (120-180 days), semi spiny, wilt tolerant.
DCH 32	LRES17x DCS5	1030 (R) 2460 (I)	Rainfed areas of Andhra Pradesh, Karnataka, Tamil Nadu, Maharashtra and Orissa	Red, triple bloom, spiny, early duration (90-150 days), resistant to jassids
GCH 6	JP65 X JI96	1300 (R) 2300 (I)	Rainfed and irrigated late kharif regions of Gujarat, Rajasthan and Maharashtra	Red, single bloom, spiny, medium duration (120-180 days), Tolerant to Macrophomina root rot
TMVCH 1	LRES17 x TMV 5	1180 (R)	Rainfed areas of Tamil Nadu	Red, triple bloom, spiny, resistant to jassids
DCH 177	DCP9 x DCS9	1550 (R) 2130 (I)	Rainfed areas of Andhra Pradesh, Karnataka, Tamil Nadu, Maharashtra and Orissa	Red single bloom, spiny, early duration (90-150 days), Resistant to Fusarium wilt, both parents are resistant to wilt
RHC 1	VP-1 x TMV5-1	900-1200 (R) 3000-3200 (I)	Rainfed and irrigated areas of Rajasthan	Mahogany, triple bloom, spiny, medium duration (100-180 days), resistant to jassids
PCH 1	VP-1 x PCS 124	1500 (R) 2000 (I)	Rainfed areas of Andhra Pradesh	Tolerant to wilt, resistant to jassids.
DCH 519	M574 x DCS78	1740(R) 2130 (I)	Both rainfed and irrigated areas	Green, triple bloom, spiny, resistant to Fusarium wilt, leaf hoppers and both parents are resistant to wilt.
GCH 7	SKP 84 x SKI 215	3000 (I)	Irrigated areas of Gujarat	Resistant to nematode-wilt complex
DCH-177	DPC 9 x DCS9	1550 (R) 2130 (I)	AP Telangana, Karnataka TN and Maharashtra	Resistant to Fusarium wilt and leaf hoppers
ICH 66	SKP84 x ICS164	1570(R) 3375(I)	AP, Telangana, Karnataka, TN & Odisha	Resistant to Fusarium wilt

R=yield under Dry land conditions; I=yield under Irrigated conditions

Strengthening hybrid castor programme through development of genetically diverse parental populations

Despite notable success that has been achieved in breeding castor hybrids adopting two-line parental system, it must be noted that there is a need to diversify the genetic base of parental materials for recovering higher levels of heterosis and to sustain the castor hybrid breeding programmes over longer periods. This assumes relevance in the face of the restricted genetic diversity within *Ricinus communis* which is a monotypic genus, characterised by very low inbreeding depression and low levels of heterosis (Lavanya and Solanki 2010). The levels of heterosis observed in castor are not as substantially as high as in other cross-pollinated crops (Lavanya and Varaprasad 2013). The criticism about slow pace of development of heterotic gene pools and exclusive dependence (Siddiq *et al.,* 2004) on restricted genetic diversity needs to be attended to expeditiously on sound scientific lines. There is also a caution that whatever productivity advantage that has been achieved so far in castor hybrids is more a function of the genetic recombination rather than that of the real heterosis per se.

Broadening of genetic base through enrichment of source nursery with germplasm for yield vigour is important keeping in view the fact that available genetic variability in castor is inadequate to realize the desired levels of yield heterosis.

- One of the methods could be to pool up the available variable genotypes and extract inbreds out of such population, which are crossed with two diverse heterotic testers to estimate their heterotic response in the combination of the testers. Based on their response with the testers the heterotic parental lines are inter-mated in all possible combinations and the resulting segregating populations are further used to develop genetically diverse inbred parental materials for use in hybrid program (Lonnquist 1964).
- Second important method is based on the use of multivariate analysis /D square statistic to identify genetically diverse clusters in castor populations to come up with heterotic parental materials (Costa *et al.,* 2006). The parental lines in each diverse group for heterosis could be identified for use in hybrid breeding program.
- The third option could be development of Multi-Parent Advanced Generation Inter-Cross (MAGIC) populations for the founder lines selected to maximize the number of diverse favourable genes pooled in each population as suggested by Bandillo *et al* (2010).
- It should not be out of context to indicate that there is a need to collect more diverse genetic material from Africa with specific reference to Ethiopia, which represents one of the important centres of diversity (Vavilov, 1951), in order to diversify the genetic base adequately. In this context a mention

may be made of castor material resistant to semi-looper (*Achoea Janata* L.) collected from Ethiopia (Prasad, 2014 unpublished) which is being used in castor breeding at the Indian Institute of Oilseeds Research, Hyderabad.

Drought Resistance in Castor: Castor being a crop grown under conditions of dryland agriculture, it is assumed generally that the crop is inherently resistant to drought. However, different genotypes of castor have shown variable response to drought under rain-fed conditions (Lakshmamma *et al.,* 2004; Lakshmamma and Lakshmi, 2010)

Drought being non-specific, it becomes imperative to incorporate considerable levels of tolerance or resistance to long spells of drought that occur under conditions of dryland agriculture.

Drought resistance is a polygenic attribute in which a number of characters come into play. Among wide ranging characters, root structure and density is the most reliable index in this context. Screening procedures for root attributes related to drought resistance viz., root length, root volume, root penetrability, root thickness and root density have been standardized (Lakshmi Prayaga and Lakshmamma 2009), which can be useful indices in this regard. Additionally, shoot attributes viz., Chlorophyll index as indicated by high Spad Chlorophyll Meter Reading (SCMR), Low Specific Leaf Area (SLA) and high Total Dry Matter (TDM) are found to confer tolerance to drought (Lakshmamma and Lakshmi 2010). The above attributes should be combined with yield attributes including harvest index in a crop variety to realize high level of performance under conditions of dryland agriculture.

Drought-resistant plants share a mechanism known as Crassulacean Acid Metabolism (CAM), which allows them to survive despite low levels of water. According to Dr Xiaohan Yang, a plant biologist at the US Department of Energy's Oak Ridge National Laboratory CAM is a proven mechanism for increasing water-use efficiency in plants. The research paper (Yang *et al.*,2017) brings out that CAM is essentially a form of photosynthesis in which the pores in a plant's leaves only open to let in carbon dioxide at night. During the day, when the sun is out, the pores remain closed in order to prevent water escaping through them. This means they are better able to tolerate dry conditions.

(https://www.independent.co.uk/news/science/drought-resistant-crops-plants-genetic-engineering-a8091816.html)

Some castor genotypes that showed promise for drought resistance are indicated in the Table 5.

Table 5: Promising Genotypes of Castor Identified for abiotic stresses at the IIOR Hyderabad by Anjani et al., (Lavanya and Solanki, 2010)

Abiotic stresses	Tolerant accessions
Drought	RG-72, RG-82, RG-89, RG-111, RG-248, RG-272, RG-289, RG-298, RG-415, RG-1494, RG-2048, RG- 2139,RG-2474, RG-3013, RG-3063, RG-3116
Temperature	RG-72, RG-89, RG-111, RG-211, RG-282, RG-941, RG-1618, RG-1661, RG-1826, RG-1941, RG-2048, RG-2059, RG-2094, RG-2153, RG-2439, RG-3063
Salinity	RG-72, RG-289, RG-298, RG-539, RG-941,RG-1463, RG-1582, RG-2149
Drought & salinity	RG-72, RG-289, RG-298
Drought & temperature	RG-72, RG-89, RG-111, RG-2048, RG-3063
Drought, temperature & salinity	RG-72
Drought, temperature, wilt & root rot	RG-111
Temperature and salinity	RG-941

Note: All the abiotic tolerant accessions need further confirmation under real natural conditions that vary with site specific situations.

Biotechnology of Castor

The toxic proteins ricin and hyperallergenic agglutinin (RCA) in the endosperm of Castor plant (*Ricinus communis*) have been major handicap in the use of de-oiled seed cake as cattle feed. Due to the monotypic nature of *Ricinus communis*, availability of genetic variability for such quality attributes is scant to take up any recombination breeding for the amelioration of the above traits. Hence, the option of exploiting biotechnological tools, including genetic engineering could be one of the promising options to introduce resistance to biotic and abiotic stresses as well as seed quality traits. Castor oil is gaining tremendous importance in various industrial applications including in pharmaceutical and bioenergy sectors (http://www.castoroil.in/uses/uses.html). Because of a growing interest in castor oil as a biofuel and the presence of the powerful toxin ricin in its seed, metabolic pathways and regulatory genes involved in both oil and ricin production have been analyzed and characterized. Genetic engineering of castor plant offers new avenues to increasing oil yield and oxidative stability, confer stress tolerance, and improve other agronomics traits, such as reduced plant height to facilitate mechanical harvesting. However, difficulties in tissue culture-based regeneration and poor reproducibility of results are major bottlenecks for genetic transformation of castor bean. Despite advances in tissue culture research over the past four decades, direct or callus-mediated adventitious shoot regeneration systems that are genotype-independent remain a much sought-after goal in castor bean. Genetic transformation attempts to develop insect-resistant and ricin-free transgenic castor bean lines have been based on shoot proliferation from meristematic tissues (Sujatha and Tarakeswari, 2018).

Jatropha cucrus belonging to the family Euphorbiaceae offers wider genetic variability for several attributes of productivity and oil quality under tropical conditions (Chhetri *et al*., 2008; Prasad, 1994c).Hence the hybridization involving castor and Jatropha was planned with the use of modern biotechnological tools (Sujatha *et al*., 2008). Genetic diversity of castor was assessed using protein based markers while use of molecular markers is at infancy. *In vitro* studies in castor have been successful in shoot proliferation from meristematic explants, but not callus-mediated regeneration (Sujatha *et al*., 2008). Genetic transformation experiments have been initiated for development of insect resistant and ricin-free transgenics which resulted in a very low transformation frequency (Sujatha *et al*., 2008) Reliable and highly efficient tissue culture protocols for direct and callus-mediated shoot regeneration and somatic embryogenesis have been achieved in respect of *Jatropha* which indicates potential for widening the genetic base through biotechnological tools. Assessment of genetic diversity using molecular markers disclosed low interaccessional variability in local *Jatropha curcas* germplasm (Sujatha *et al.*, 2008)

Response of castor tissues to *in vitro* culture has been poor, which necessitated a study on understanding the molecular basis of organogenesis in cultured tissues of castor, through *de novo* transcriptome analysis, by comparing with two other crops (*Jatropha* and sunflower) with good regeneration ability (Puvvala *et al*., 2019). Puvvala *et al* (2019) carried out RNA-seq. analysis with hypocotyl explants from castor, *Jatropha* and cotyledons from sunflower cultured on MS media supplemented with different concentrations of hormones. Genes that showed strong differential expression analysis during dedifferentiation and organogenic differentiation stages of callus included components of auxin and cytokinin signalling, secondary metabolite synthesis, genes encoding transcription factors, receptor kinases and protein kinases. In castor, many genes involved in auxin biosynthesis and homeostasis like WAT1 (Wall associated thinness), vacuolar transporter genes, transcription factors like short root like protein were down-regulated while genes like DELLA were upregulated accounting for regeneration recalcitrance. Validation of 62 differentially expressed genes through qRT-PCR showed a consensus of 77.4% with the RNA-Seq analysis (Puvval *et al*., 2019). As a result the information on the set of genes involved in the process of organogenesis in Castor and Jatropha is now available (Puvvala *et al.*, 2019) which forms a basis for understanding and improving the efficiency of plant regeneration and genetic transformation in castor.

Biotic Stresses of Castor

Insect Pests of Castor

Over one hundred insects are known to feed on castor at various developmental stages of the plant; but not all of them cause serious damage to yield. Among the insects known to inflict significant yield losses, the following are found to be important, which need to be managed to cut down economic losses.

1. Red hairy caterpillar - *Amsecta albistriga* Wk., A. moorie Butler
2. Castor Semilooper - *Achoea Janata* L.
3. Tobacco caterpillar - Spodoptera litura Fabr.
4. Capsule borer - Conogethes punctiferalis Guen
5. Leaf hopper - Empoasca flavescence Fabr.
6. Leaf miner - Liriomyza trifolii Burgess
7. White fly - Trialeurodes ricini Misra and
8. Thrips - Retithrips syriacus Mayet

Among the above the first three are defoliators and are considered important in terms of the damage caused to the crop. While Red hairy caterpillar and tobacco caterpillar are polyphagous, castor semilooper is highly specific to castor, which causes serious losses in the months of August and September in the Deccan Plateau. The pest of the next importance in terms of economic losses is the capsule borer since it directly attacks the developing capsule leading severe yield losses. The infestation of capsule borer is reported to be on increase in Andhra Pradesh, Telangana, Karnataka and Tamil Nadu, causing around 50% damage; but low in Gujarat and Rajsathan (Lakshminarayana 2010).

Red hairy caterpillar, a migratory insect even in larval stage is a major pest causing severe defoliation of the young seedlings soon after germination in Telangana and Andhra Pradesh in the monsoon season. As the crop picks up growth the damage by this pest gets abated. Tobacco caterpillar too has assumed the status of a major pest and its incidence is on increase in the recent years (Lakshminarayana 2010). Other defoliators of sporadic incidence are Bihar hairy caterpillar *(Spilosoma obliqua)* and smaller size hairy caterpillar *Euproctis* spp, which sometimes cause some damage in pockets of castor growing area.

Among the insects listed above the ones at serial numbers 5,6, 7 and 8 are the sap-sucking pests, which are noticed throughout the castor growing zones of the country. Incidence of leaf hoppers are found to be more localized in the dryland areas and the whitefly incidence is found to be more intense in irrigated belt of castor. Till recently attack of thrips was not very much intense; but off late this group of sucking pests is reported to be causing perceptible damage. Based on the severity of damage, the leafhopper is reported to be the most important (Lakshminaranayana 2010) among the sucking pests, followed by thrips.

The leaf miner *Liriomyza trifolii* belonging to the order of Diptera has made its presence in 1991; but is of polyphagous nature and is spreading faster.

Yield losses: As per the reports gathered from 1985 to 1992 (Anonymous 2006), the yield losses due to the pest complex although considerable, appear to vary greatly depending upon the region and the season. The yield losses due to the total pest complex put together is reported to vary from 14 to 20% in the zone of North Gujarat, where sucking pests alone are reported to have caused 13 to 20% losses in yield . In Telangana semilooper and leafhopper have reported to have inflicted a damage ranging from 36 to 41%. The capsule borer alone caused 49 to 52% in Raichur zone of Karnataka, while semilooper and tobacco caterpillar have caused damage of the tune of 31 to 41% in this zone.

The mean avoidable yield losses due to all insect pests on castor cultivars ranged from 17.2 to 63.3% during *kharif* monsoon season, while the loss was higher (22.5 to 89.4%) during *rabi* post monsoon season. Among the cultivars, DCH-519 revealed minimum avoidable yield losses due to insect pests in both the above seasons (Lakshminarayana and Duraimurugan 2014).

1. Red hairy caterpillar (RHC) *Amsecta albistriga* wlk., *A. moorei* Butler (*Arctiidae: Lepidoptera)*

This is a very serious polyphagous pest in all the castor growing tracts in Peninsular India particularly in Telangana and Andhra Pradesh. *A.moorei* is common in the central and north India, while *A. albistriga* is southern India. The red hairy caterpillar (Fig. 5) causes severe damage by defoliation. The moths (Fig.6) emerging following soaking rains are attracted to light, since peak emergence takes place after the dusk from 7.30 PM. The moths emerge in 2 to 5 peaks with 1 to 3 days' duration of each peak. The population of female moths is high in the initial peaks, while in the later peaks males out-number restricting the females to less than 10% (Lakshminarayana and Raoof 2005). Each female moth lays as many as 150 to 2,000 eggs following the copulation, which happens only once in its life. The eggs are laid in masses on the foliage of crops, weeds and sometimes on soil surface or even on stony surface. The gregarious larvae emerge in 3 to 4days and feed on vegetation for a week to ten days before they move on migrating to other sources of vegetation. Many a time, the migratory larvae feed on young seedlings and kill the emerging vegetation. The migratory larvae are covered with dense reddish brown hair and destroy

Fig. 5: Amsecta albistriga larva

Fig. 6: Adult moth of Amsecta albisrtiga

a vast stretches of crops (Qayum and Sanghi 1994). Sometimes larvae indulge in cannibalism too. The larvae after 6 instars, go in for pupation usually from August in soil of open fields under trees, bushes etcat adepth of 2.5 to 6.5 cm based on the soil texture and moisture (Gupta *et al.,* 1966). The new flush of moths get released with the onset of monsoon mostly during May to October in the southern state of Tamil Nadu. In Telangana and Andhra Pradesh the new generation of moths occur in June – July.

There are several predatory natural enemies of red hairy caterpillar out of which the egg parasitoid viz., *Trichogramma* spp. *Telenomos manolus*, the larval parasitoid *Apanteles oblique,* predatory bug *Canthoconodea* spp. Tiger beetle, nematodes and Nuclear Polyhedrosis Virus (NPV) are important in checking the proliferation of the pest. There are instances of parasitization to the tune of 35% eggs by *Trichogramma brasiliensis* (Lakshminarayana 2010). Maheswari and Rao (2000) reported around 10% mortality of the pest by NPV.

2. Castor Semilooper: *Achaea Janata* L. (*Noctuidae : Lepidoptera*)

Castor Semiloopeer (Fig.7) is a major pest of castor inflicting serious losses during the period of July to September in Deccan Plateau of Peninsular India. In Saurashtra region of Gujarat, the pest is reported to cause serious crop damages during September to November. The pest is wide spread from India to Hawaii and from Taiwan to New Zealand (Salvador *et al.,* 1986). In India, semilooper is reported to feed on the leaves of roses, pomegranate, *Ziziphus* spp and *Euphorbia* spp., (Khan and Rao 1948; Singh MP 1982.)

Fig. 7: Castor Semilooper Achaea janata larva

The perennial or sporadically located castor plants are harboured by the adult moth for egg laying with the onset of monsoon. The female moth lays around 400 to 500 bluish green coloured eggs on leaves of the main *kharif* sow castor crop.The larvae emerge after 3 to 4 days. The earlier instars of the larva feed voraciously destroying the crop. Of the total of 5 instars the larval stage the full grown grey caterpillar measures around 35 mm, possessing lateral red and brown stripes, black head and anal tubercles. The subsequent pupal period is of 10-14 days. The pupal case is chitinous and shows ashy cover. Out of the 5 to 6 generations the pest goes through in a year, the pest populations emerging out of the second and third generations are said to cause very severe damage to castor crop (Khan 1946; Pandey *et al.,*

Fig. 8: Achaea janata Adult moth

1967). In the north Indian state of U.P the pupae lie dormant in the soil during the winter months of November and December before the adult moths come out emerging in the month of June (Pandey *et al.,* 1967)

There are many natural parasitoids on castor semilooper, out of which the local indigenous parasitoid, *Trichogramma chilonis* and the larval parasitoid *Microplitis maculipennis* arethe most effective ones in bringing down the population of castor semilooper (Kundu *et al.,*1966, Prabhakar and Prasad 2004). Microplitis is an internal larval parasitoid affecting mostly the 3rd instar larvae. The parasitized semilooper larvae may be noticed in the filed carrying at the posterior end a brown pupa of the parasitoid. The affected larvae stop feeding and die of starvation. Somasekhar *et al.,* (1993) reported that around 60 to 90% larval parasitization by *Microplitis.* The rest of the parasitoids are sporadic in their action without much significant effect.

There are non-insect agents too which can be used to check the proliferation of semilooper. Among them the well-known pathogenic bacterium *Bacillus thuringiensis* var. *kurstaki* is found to be effective against semilooper (Devi *et al.,* 1996). Also granulosus virus too has been found to cause mortality of the pest in the field. A species of nematode *Mermis* sp., has been found to parasitize castor semilooper (Chatterjee *et al.,* 1968). Also several species of birds viz., *myna* (*Acridotheres tristis*), house crow (*Corvus spendens*), black drongo (*Dicrururs adsinilis),* rosy-paster (*Stemus roseus)* and large gray babble (*Trudoides Malcolm)*, are some of the avian species that predate on semilooper (Laksminarayana 2010). Hence it is advised to have some bird perches placed at different locations in castor fields to provide anchor for predatory birds which can contribute to controlling the proliferation of castor semilooper.

3. Tobacco caterpillar: *Spodoptera litura* (Fabr) (*Noctuidae: Lepidopera*)

Spodoptera litura is one of the most polyphagous larval pests of several crops, with a wide range of alternate hosts. The larval stage of this species is known to be feeding on 40 known crop plants and several weeds (Thobbi 1961). The pest has exhibited its preference for castor leaves over the other crops viz., okra, cotton, pigeon pea, mung bean and groundnut (Patel *et al.,* 1965). In Telangana, Andhra Pradesh and Tamil Nadu, the pest makes its presence in castor in August and continues causing very severe damage in October and November.

Fig. 9: Larva and adult moth of *Spodoptera litura*

Fig. 10: Spodoptera larvae of initial instars on castor leaf.

The adult female moth lays its eggs numbering over 500 to 600 covered by fine woolly mass on the lower surface of leaves. The eggs hatch out in 4 to 5 days and the young larvae feed gregariously leaving only the network of veins on the leaves. Later the larvae migrate to wide range of preferred hosts. When the attack is severe one can see the larvae feeding on the outer coat of capsules and also boring the stem leading to wilting of the plant. The larvae are also nocturnal causing damage to the crop and during the day they seek shelter in the soil crevices. After the five larval instars (Lal and Nayak 1963) the pupation takes place in the soil. The insect completes 7 to 8 generations in a year (Narayanan 1959), which gives an account of rapid proliferation of the pest.

Among the biological agents that bring about mortality of tobacco caterpillar, *Trichogramma evanescens minutum, Telonemous remus, Apanteles* spp. *Campoletis chloridae* are considered important (Satyanarayana *et al.,* 2005). Other predatory mechanisms available are spiders, lacewing *(chrsoperia),* predatory bug *(Cantheconidia furcellata)* (David and Basheer, 1961) and insectivorous birds are found to check the proliferation of the pest population. The other biological agents include *Spodpotera*-NPV, entomopathogenic fungus *Nomuraea rileyi,* the pathogenic bacteria *Bacillus thuringiensis* and *B. cereus* too are reported to bring about killing of larvae. (Lakshminarayana 2010).

The other larval pests like Bihar hairy caterpillar and *Euproctis* cause damage to castor crop during the period October to December. *Euproctis lunata* was reported from Delhi (Anonymous, 1953) and Rajasthan (Kushwaha and Bhardwaj, 1967). The other species viz., *E. guttata* Wlk., *E.scintillans* and *E. fratema* have been found to infect castor in Telangana and Andhra Pradesh (Khan and Rao 1948).

4. Shoot and Capsule Borer [*Conogethes (Dichocrosis) punctiferalis* (Guen) (Pyralidae: Lepidoptera)

Fig. 11: Capsule borer attack on castor, its larva and adult moth

The incidence of capsule borer was reported in 1962 (Anonymous 1962); but the pest has assumed serious proportions in recent years particularly in Peninsular India. The pest is found to infest castor starting from peak flowering and subsequently continues its damage till the crop maturity. At times it turns into a shoot borer too. The peak damage of the pest is found during October to December on dryland castor.

The pest is also found on ginger, turmeric, cardamom, soap-nut, guava, peaches, cacao, pear, jack fruit, citrus etc. (Misra and Teotia, 1965). The female moth lays pinkish oval eggs singly or in small groups on the fully developed inflorescence and tender capsules and tender terminal shoots.

The eggs hatch out in 6 to 7 days releasing pale greenish larvae with pinkish tinge, fine hairs and dark head. The larval phase lasts for about two weeks with five instars (David *et al.,* 1964) and pupates in the damaged tissue of host plant.

The young caterpillars eat away the green epidermis of castor capsule and pushes through the capsule boring the pedicellar and stigmatic end. As the larva enters the capsule and silken gallery is woven in which the excreta and frass get accumulated. Once the damage to crop advances, webbing of the capsules along with the frass is observed in which more than one larva could be found. A single larva can damage many capsules (Lakshminarayana, 2010).

David *et al.*, (1964) reported the occurrence of icheumonid larval parasitoid *Diadegma ricini*, *Theronia spp,* the brancoid parasitoids *Apanteles* spp., *Bracon hebetor* and pupal parasitoid *Brachymeria euploege* as natural enemies of shoot and capsule borer of castor.

5. Leafhopper –*Empoasca flavescens* (Fabr) (Cicadellidae : Hemiptera)

Among the sap sucking pest complex leaf hoppers are important. Nymphs and adult insects suck sap from the lower surface of leaf and also inject toxin which causes curling of leaf edges and the leaves turn brown and brittle (Jayaraj, 1967). The damage caused results in hopper burn which renders the attacked leaves dry, uneven, curl downward in the shape of an inverted boat, margins turn brown and eventually death of the plant. This causes premature drying of leaves and consequent loss of photosynthetic surface resulting in stunting leading to poor capsule formation (Rai, 1976). The leaves, thus become unfit for rearing Eri silk worms. Economic thresh-hold level of the pest is estimated at 2 to 3 adults/leaf.

The pest is active in cold and humid weather. The peak infestation is seen in September to December. The female lays 15 – 25 eggs inside the veins of the lower leaf surface and the nymphs emerge out in weeks' time and turn into adults in 17 to 18 days (Jayaraj and Basheer, 1964)

According to Subba Rao *et al.,* (1968), lygaied bug (*Geocoris tricolor*) feeds on nymphs and adults of leafhopper. The larvae of the family *Erythreidae* and few spiders are also reported to feed on nymphs and adults of leafhopper (Lakshminarayana 2010)

6. Serpentine leaf miner– *Liriomyza trifolii* Burges (*Agromyzidae : Diptera).*

This pest was noticed initially in the year1991 on castor and other crops (Lakshminarayana *et al.,* 1992). The insect is reported to be a major pest of vegetables, ornamentals and fruit crops in the west. The number of the wide range of hosts of this insect is over 70 ranging from fibre crops, legumes, vegetables, ornamentals, green manures, narcotics and weedy plants (Srinivasan *et al.,* 1995).

The maggots feed on the green part (chlorophyll) of the leaf tissue between the epidermal layers and create a serpentine mined tract of feeding. The pupation of the developed yellow coloured maggots of the age of 7 to 9 days happens in the soil when they dropdown from the host. The pupation period is around a week when the brown coloured pupae release the adults. When the attack is severe, the dried leaves dropped down through out in the castor field (Lakshminarayana, 2010).

The initial damage by the pest starts from the lower most leaves including cotyledonary leaves and moves upwards. Thus the younger leaves are less damaged as compared to the older ones. The castor genotypic response to this pest is highly variable with those with cup-shaped and broad leaves are damaged most. Normally no yield losses are observed with the kind of damage in the form of leaf mines by the pest. Plants exposed to the leaf-miner one month after planting exhibited plant height and biomass at par with the normal plants (Patil *et al.*, 1997). No satisfactory control of the pest was possible with the traditional insecticides. The host genotypic variation with regard to pest incidence is observed (Kapadia 1995). Castor accession with Papaya leaf type (Fig.4) from the DOR Germplasm exhibited resistance to the serpentine leaf-miner (Anjani 2010). Also Purple colured leaf type of castor germplasm too exhibited resistance against the serpentine leaf miner of castor. (Fig. 13)

Fig. 13: Purple coloured Leaf type of Castor germplam exhibiting resitance against the leaf-miner of castor. Leaf of the susceptible host is shown sideby. (Anjani et al 2010)

7. Whitefly – *Trialeurodes ricini* (Misra) (Aleyrodide : Hemiptera)

Castor whitefly is unlike the other sap sucking pests in the sense that it occurs spring – summer season (March to June) and hence the rainfed castor grown from June to November is not affected by the pest. However, the pest impinges upon the crop grown in the post rainy season under irrigated conditions.

The nymphs and adults damage the crop by sucking mostly from the lower surface of the leaf, resulting in yellowing of leaves. When the infestation is severe the whole plant gets stunted. One could observe suity mould developing on the honey secreted by the pest. The insects are often found in thick crowds on the undersides of leaves. When infested plants are disturbed, great clouds of the winged adults fly into the air.

Adult whitefly looks like tiny moth with a white coating on its body. The female lays its eggs in singles numbering around 85 to 90 on the lower having waxy projections. The nymphs remain adhered to the lower surface of leaves.

Fig. 14: Whiteflies feeding on leaf surface

Among the predators of whiteflies, the coccinellid *Menochiles sexmaculata,* the lygaeid bug *Geocoris triolor,* the drosophilid fly *Acletoxenus indica,* the syrprid *Ischiodon scutellaris* and thrips *Tenothrips ricini* are noticed to be important (Lakshminarayana 2010).

Thrips: *Retithrips syriacus* (*Thripidae : Thysanoptera*)

Thrips are wide spread throughout India. In recent years the damage due to thrips is on increase. The crop damage is caused by the tiny pinkish nymphs and brown coloured adults, as they feed on the upper and lower surfaces of leaves causing wrinkling of leaves. Thrips damage mostly confined to the most tender leaves and the leaves which are not fully open. Also the tender emerging spikes are attacked resulting in its withering.

Fig. 15: Thrips on castor foliage

Apart from castor, thrips are found to attack coffee plants in India (Kumar *et al.,* 1984).

The female of the species lays around 50 to 60 eggs at the rate of 6 to 12 per day inside the soft tissues of the host plant, especially the portions of leaf veins. The larvae emerge out in 4 to 5 days, leaving circular holes which distort the leaves at the damaged part. The second instars of 7 to 9 days fall from the host and enter the soil to pupate. The life cycle of thrips is confined to a duration of 15 to 20 days (Lakshminarayana, 2010).The thrip population is found to decline with rains (Ananthakrishnan, 1956). *Scolothrips indicus* is reported to feed on immature stages of thrips. A predatory mite is also found to reduce the thrip population during July and August. The biochemical profiles of the host plant are reported to influence the infestation levels of thrips The interaction of *Retithrips syriacus* with the host viz., castor is governed essentially by the biochemical profiles of its host, which tend to be altered subsequent to infestation, thus manifesting induced resistance through enhanced production of phenolics (Ananthakrishnan *et al.,* 1992)

Non-insect pests of castor

In this category there are few pests which are not serious in terms of regular crop losses; but they do attack the crop at times necessitating immediate action to check their proliferation and damage to the crop.

1. Reniform nematode- *Rotylenchus reniformis* : This is a polyphagous pest, amphimictic, semi-endoparasite feeding on phloem cells of the root system of castor in India (Jain *et al.*, 2007). The root knot nematodes *Meloidogyne incognita* and *M.javanica* have also been found to attack castor (Varaprasad and Swaroop, 1986). The roots of the host are found to be covered with the egg masses of nematodes. The shoots of the affected plants exhibit chlorosis and stunted look. The yield losses depend upon the nematode density in the soil and may go up to 40% at times if the nematode density in the soil is very high. The reniform nematode is also reported to be associated with wilt complex in castor since the nematode incidence predisposes the castor plant to wilt causing pathogen.
2. Red spider mite – *Tetranychus telarius:* All active stages of the mite suck-up the sap from leaves, causing leaf blotching. The infested leaves yellowish and then brown becoming brittle getting detached from the plant. The pest might turn serious during hot summer, when the crop is grown with irrigatioin (Lakshminarayana, 2010).

Pest Management in Castor: The biotic stresses the crop faces are to be managed with a view to enhancing not only the productivity but also to safe guard and sustain the ecosystem. Hence, efforts are on the way towards developing eco-friendly technologies that would serve the purpose of checking the pest proliferation and yield losses and also to ensure minimum or no damage to the ecosystem (Lakshnimnaryana 2010).

(i) **Deep summer ploughing**: This practice would be useful in checking the wilt disease of castor and also pests like root grub and other pests that pupate in the soil including nematodes. This, however, may not be effective against red hairy caterpillar (RHC) since the pest pupates and hibernates on waste and barren lands, beneath the bushes where ploughing may not be possible.

(ii) **Community participation:** This approach of involving the vigilance of local communities will lead to the effective management of red hairy caterpillar (RHC). Preparation of village maps indicating the places of initial pest incidence and the usual direction of migration of the pest based on the past local experiences shall be useful in coming-up with suitable pest management strategy. Satisfactory control of RHC was achieved during 1990 to 1992 in Telangana zone with this approach. The control of RHC requires special management strategies due to the peculiar features of the pest viz., high fecundity, egg laying on weeds, ability to survive on different hosts, larval migration across vast areas and less vulnerability of the pest to chemical insecticides (Qayum and Sanghi 1994).

(iii) **Bonfires:** These are reported to be effective in the control of RHC. A massive programme launched by the Centre for World Solidarity, Hyderabad and other voluntary organizations in collaboration with the Department of Agriculture and Transfer of Technology Wing of the Directorate of Oilseeds Research, Hyderabad over 2,000 ha during 1988-89 yielded promising results. Nevertheless, the difficulties and health hazards associated with the use of rubber wastes for burning, the emphasis has been shifted to employing light traps (Lakshminanarayana 2010)

(iv) **Light traps**: The device involving light traps for the control of RHC has paid good dividends. During the years from 1990 to 1992, over an area of 5,000 ha consisting of five pest-endemic districts of the erstwhile Andhra Pradesh was covered with the joint participation of certain voluntary organizations and the departments of the government. In order to cover an effective unit area of 10 ha one 200 mercury lamp was used. The area covered with light traps that attracted the adult moths in the reproductive. No incidence of RHC was observed and no re-sowings due to pest attack were observed. In the other surrounding areas not covered by light traps, re-plantings to the tune of 20 to 50% had to be carried out due to the RHC incidence (Qayum and Sanghi 1994). In the areas where installation of mercury electric lamps was not possible, petromax lights each of 200 candle power per 6 acres were employed with effective results (Narayanan 1953).

(v) **Trap cropping and vegetative traps for RHC management**: The well known trap crops viz., cucumber or cowpea may be raised all along the borders of the field. The larvae of RHC shall be attracted readily by the trap crops in large numbers., which can be killed by jerking the larvae into the container of kerosoinized water. The trap crops should be sown much earlier to the sowing of the main crop of castor. The twigs of *Jatropha, Ipomoea* or *Calotropis* may be placed on the filed boarders coinciding with the migration of RHC to which the pest gets attracted. The larvae thus collected may be destroyed (Qayum and Sanghi 1994).

(vi) **Resistant varieties of castor:**There are no any clear and confirmed sources of host genetic resistance to foliage feeding caterpillar pests. However, some preference or non-preference of the pest to certain genotypes has been observed. A case of preference of the foliage feeding pests to bloomed varieties of castor was noticed as compared to the non-bloomed varieties. However, in the absence of bloomed varieties, the non-bloomed green varieties were attacked severely (AICORPO 1974). Lowest oviposition of *Spodoptera* was observed on GCH 4 and GCH 5, while VI-9, GAUCH-1 and SKI-73 were preferred more (Thanki *et al.*, 2001).Castor genotypes with deep red stem (RG 103 and RG99) were found to be more susceptible to Bihar hairy caterpillar (AICORPO 1992). Most of the castor varieties are not resistant to leaf-miner. It has been observed that castor genotypes with broad leaves and cup shaped leaves are the most susceptible to castor leaf-miner. On the other-hand a castor genotype RG 1930 with pink foliage was found to be highly resistant to leaf-miner due to its high phenol content (Anjani *et al.*, 2007). Castor genotypes with loose spike, non-spiny small capsules are found to be less damaged by capsule borer. The strains of castor with compact spikes and larger capsules were more vulnerable to caspule borer (Lakshminarayana 2005) Among the germplasm accessions at IIOR viz. RG 1934, 2546,2770,2543 and 2786 and among the released varieties 48-1, Geeta, Kiran and GCH-4 exhibited low infestation of capsule borer (Lakshminarayana 2010).

Considering some of the specific characters of castor in this context, it was found that the castor varieties with single bloom or no bloom are more susceptible to leaf hopper; but are found resistant to white fly. The castor types with double or triple bloom are resistant to leaf hopper; but are susceptible to white fly. The reaction of castor genotypes to thrips is similar to that of leaf hopper. Among sap sucking pests leaf hopper is economically more important followed by thrips. Most of the released castor varieties and hybrids viz., GCH-4, GCH-7, DCH-32, DCH 519 and 48-1 are of double and triple bloom, showing field tolerance to leafhoppers (Lakshminarayana 2010).

The intensive screening of germplasm accessions at different Oilseed research centres in the country over two decades resulted in the identification of different sources of resistance against leaf hopper, whitefly, capsule borer and leaf-miner. There are 63 accessions resistant to leaf hopper, 4 against leaf miner and 5 against capsule borer. There are also 33 accessions exhibiting multiple resistance against leaf hopper and white fly and one accession showing resistance against wilt and leaf hopper (Lakshminarayana 2010).

(vii) **Sowing time in relation pest management :** Castor crop sown during early to mid-June on drylands is likely to suffer the infestation of the foliage feeding pests viz., red hairy caterpillar and semilooper and also that of whitefly and caspule borer. The rainfed crop sown after the first week of July is found to escape the heavy incidence of the above pests. Since the yield of the late sown crop is likely to decline, a balance has to be struck between these options viz., (1) managing the pest in the early sown crop to harvest good yield and (2) boosting the yield of the late sown crop with appropriate agronomic management techniques. The cost effectiveness of the suggested options revealed that the first option is preferable to the second one (Lakshminarayana 2010).

(viii) **Spacing and fertilizer dose in relation to pest management**: Different spacings of castor did not seem to influence the incidence sucking pests. With regard to the application of crop nutrients, the incidence of semilooper and capsule borer was very low in plots where no nitrogen was applied. Also it appeared that with low levels of nitrogen the incidence of leaf hopper, foliage feeding pests and capsule borer was low. However, their infestation increased with sulphur application. Application of high doses of nitrogen (80 kg/ha) to dryland castor favoured the incidence of semilooper, capsule borer and leaf hopper. While the infestation of all sucking pests increased with the application of higher doses of nitrogen, it did not influence the incidence of leaf miner (Lakshminarayana 2010).

(ix) **Intercropping for pest management:** Castor intercropped with pigeon pea exhibited 1 low incidence of semilooper, while the infestation of the pest increased when castor was intercropped with sesame. Incidence of *Spodptera* was high when castor was intercropped with cowpeas (Lakshminarayana 2010).

(x) **Manual management of castor pests**: One of the most effective and eco-friendly methods of pest management is manual control of pest proliferation. The feasibility of manual picking up of the larvae has been demonstrated in the case of successful control of older larvae of semilooper, gregarioius stages of *Spodptera* , *Euproctis* spp and Bihar hairy caterpillar. Lakshminarayana (2005) has brought out that practicing manual picking up of the above larvae twice a week had checked the defoliation of castor to the bare minimum of less than 25% resulting is saving of extra costs involved in other operations. The manual management of semilooper has in fact enhanced the parsitization of the pest by *Microplitis* under the field conditions, while any chemical control of the pest would have been harmful to the natural parasitiod *Microplitis,* often leading to the resurgence of the pest. Despite the fact that hand picking of egg masses and gregarious stages of larvae was recommended in some cases, this practice may not be feasible since egg laying is usually done on weeds and live grassy materials in vast uncultivated barren lands.

(xi) **Bird perches**: The feasibility of using avian species as bio-control agents for the management of the defoliating larval pests depends on the frequency of occurrence of the predatory birds in the vicinity of the castor growing areas. Installation of bird perches in castor fields adds to the cost with no certainty about the success of the operation. Castor plant itself could be a suitable perch for predatory birds in the locality. The experience of some farmers has shown that erection of bird perches over vast areas has not yielded the desired results that commensurate with the cost involved.

Pheromone traps: Use of pheromones of castor semilooper and *Spodoptera* have since been identified and often recommended for monitoring the pest situation to take up the needed control measures in respect of the above pests. Gedia *et al.*, (2007) reported that maximum temperatures had positive correlation with the male moth catches of *Spodoptera,* while wind speed and rainfall exhibit a negative correlation between male moth catches.

(xii) **Biological Control**: In the earlier paragraphs various natural parasitoids, predators and microbial agents for each of the pests have been presented. Among all the natural enemies, the larval parasitoid of castor semilooper *Microplitis maculipennis* is by far the most effective in terms of the control of the castor semilooper pest. Under field conditions the activity of this parasitoid has resultedin almost 100% control of the semilooper, practically obviating the use of insecticides when the parasitoid is active in the field. The parasitoid of next importance is *Trichogramma chilonis,* which is reported to bring about 50% parasitization of semilooper eggs. The rest of the parasitoids, microbial agents, spiders, coccinellids and chrysopids add to the parasitisation; but the individual contribution of each one of them may not be significant.

Telenomus proditor, an exotic egg parasitoid was multiplied used successfully and released in Mahabubnagar district of Telangana region during 1960s and 1970s. The parasitisation of the eggs of semilooper ranged from 40 to 90%, which indicated the adaptation of the introduced parasitoid to the local conditions of the zone (Thobbi and Srihari1968). The indigenous parasitoid *Trichogramma chilonis* too played a significant role in the control of semilooper. The flight range of the parasitoid was around 10 meters from the point of release. The inundative field releases of *T. chilonis* was at the rate of 10-1 cc/ha in split doses at weekly intervals, which caused the semilooper egg parasitisation of the order of 90%. The indigenous *Trichogramma chilonis* was also highly competitive and effective over the introduced *Telenomus proditor* in parasitisation of the semilooper eggs (Lakshminarayana 1992). Under the wider field conditions, the indigenous parasitoid seemed to have dominated over the exotic *Telenomus proditor.*

The entomophagous fungus, *Nomouraea riley* is also reported to be effective against *Spodoptera.* Nevertheless, the fungus requires high humid conditions and relatively lower temperature for its proliferation in the field, the conditions favourable for the incidence of *Botrytis*, the grey mould rot.

Among viruses granulosus virus (GV) against semilooper and nuclear poly hydrosis virus against *Spodoptera* were found effective. The use of viruses for control of castor pests has not gained popularity among farming communities due constraints such as non-availability of viruses on large scale for field applications over vast areas, their photo-degradability and lower field efficiency.

(xiii) **Botanical and microbial pesticides**: There has been a considerable research on the use of botanical and microbial pesticides and to minimise the use of chemicals.

Among the wide range of products that were tested and evaluated for their role in managing the pest complex of castor, the following were found promising.

- The extracts of neem *(Azadirachta indica),* karanj *(Pongamia pinnata),* mahua *(Madhuca longifolia)* (Jain *et al.,* 2005) exhibited inhibition of growth of castor semilooper(Prakash and Rao 2018). Neem seed kernel extract (NSK) (Chari and Muralidharan, 1985; Basappa and Lingappa 2004) and custard apple seed extract (Raman *et al.,* 2000*)* were found effective against castor semilooper. NSK 5% was found to be superior in checking *Spodoptera* than the commercial neem formulations (AICOP0 2003).
- Singaravelu and Ramakrishanan (1998) characterised a granulosus virus causing mortality of castor semilooper.

Bacillus thuringiensis sub.sp.*kurstaki* was found to be highly effective in causing feeding cessation and mortality of castor semilooper (Devi *et al.,* 1996). Spraying DOR Bt-1 strain of *Bacillus thuringiensis* sub.Sp. *kurstaki* could effectively and economically control castor semilooper (Devi and Sudhakar 2006). Cost-effective mass production technology of Bt based principle of solid state fermentation has been achieved (Devi *et al.,* 2005).

Another Bt isolate DOR4 was also found to be toxic to castor semilooper (Madhusudan *et al*., 2008).

Commercial formulations of *B. thuringiensis* var. *kurstaki* viz., Dipel and Delfin were found to be effective against castor semilooper. (AICORPO 1995). Nuclear Polyhedrosis Virus (NPV) of *Spodoptera* @ 500 LE/ha was found effective in castor and many other crops (Lakshminarayana 2010).

While the effectiveness, low cost and eco-friendly nature of several botanical and microbial pesticides have been established, it is necessary to enhance their stability under field conditions.

There are wide range of chemical insecticides which are very effective against almost all insect pests of castor. Nevertheless, the recent experiences from experimental fields as farmers' fields in wide range of crops bear ample testimony that it is indeed not possible to achieve a comprehensive and effective control of pest complex solely depending on chemicals. Additionally, continuous and exclusive dependence on chemicals for pest control is fraught with dangers of environmental pollution and food poisoning. Hence it must be realised that the chemical pesticides must be used with caution and as per the recommendations of the public research organizations. It must be remembered that the chemicals might provide with a moderate control of the pest (Lakshminarayana 2010).

Many a time the use of chemical insecticides might become inevitable for management of different pests, particularly when other means of pest control do not yield satisfactory results that might lead to considerable crop losses.

- In the case of red hairy caterpillar (RHC) certain insecticides have been recommended. If the economic threshold for chemical treatment is reached (eight egg masses per 100m length or 10% leaf damage), dusting of insecticides might not control young larvae. In such cases, spraying fenvalerate (0.02%), Quinalphos 2ml, or Methyl Parathion (0.02%) or Methyl Parathion (0.02%) might be effective against the early stages of the pest (Lakshminarayana and Raoof, 2005). The later grown up instars of the pest are migratory with a with dense hair cover, which render the caterpillar resistant to insecticides.
- Castor semilooper is easily vulnerable to several insecticides. Sprays of cholorpyrifos 0.04%, Quinalphos 0.05%, or carbaryl 0.15% are very effective against the pest (AICORPO 1993; Singh 1982). Care must be taken not to spray against castor semilooper when the parasitoid *Microplitis* is active.
- In the case of *Spodoptera,* cypermethrin 0.03%, acephate0.075%, chloropyrifos 0.05% and carbaryl 0.05% were found to be effective (AICORPO 1992).
- Management of castor capsule borer with chemicals is more difficult. While most of the chemicals were ineffective in controlling the capsule borer, Profenophos 0.05% was effective against the borer (AICORPO, 2003).
- The castor leaf miner could be checked with the sprays of methyl oxydematon 0.05% and dimethoate 0.05% (AICORPO 1992). It may be noted the sprays of other recommended pesticides against other pests, have accentuated the incidence of leaf miner.
- Leaf-hopper and white-fly the sap-sucking pests may be controlled by sprays of dimethoate 0.05% (AICORPO 1989), methyl-o-demeton, and imidacloprid (Patel, MM*et al.*, 1986, Paramar *et al.*, 2004).
- Among some of the recent chemicals, thiodicarb 0.075%, spinosad 0.018% and indoxcarb 0.015% were found to be toxic to semilooper, *Spodoptera* and capsule borer. The chemical lefenuron was found to be relatively safer to the natural enemies of the pests (AICORPO 2005). The chitin inhibitor, diflubenzuron adversely affected the growth of castor semilooper (Kadam *et al.*,1995). The other chemical lufexuron was effective against *Spodoptera*

(Sudhakaran 2002). Penfluron was more effective than diflubenzuron or triflumuron in inhibiting the larval and pupal moulting of semilooper (Gujar and Mehrotra, 1989)

- Transgenic castor for pest management: *Bacillus thuringiensis* var. *kurstaki* protein *Cry 1Aa* was found to be more toxic to semilooper (Sujatha and Lakshminarayana, 2005). Castor variety DCS-9 has been transformed through *Agrobacterium* mediated particle gun bombardment employing appropriate vectors with *Bt* genes *Cry 1 Aa* (Sujatha *et al.,* 2009).

Few events *viz.,* PCP 201 AMT1, PCP202 AMT 18, which conferred resistance to semilooper and *Spodoptera* were identified during field bioassays conducted under net confined conditions and high pressure (Lakshminarayana *et al.*, 2009). Semilooper resistant transgenic castor cultivars VP-1 and Jyoti expressing *Cry 1Ab* protein were also developed (Malathi *et al.*, 2006)

Diseases of Castor

On lighter soils with low moisture holding capacity in Peninsular India, castor is a traditional crop. Straight Castor varieties are recommended for cultivation for dryland agriculture of this region. Nevertheless, the area under castor in this zone is declining due to the increasing incidence of two major diseases viz., (1). Botrytis or gray mold and (2). Wilt disease causing immense yiled losses.

1.Botrytis disease or **Gray mold disease** is caused by *Botryotinia ricini,* Godfrey-Whetzel (Ramanjaneyulu *et al* 2014). The symptoms of the disease are characterised by the appearance of initial pale, blotch like spots on the capsulesfrom which drops of yellow liquid exudes. Later the fungal threads grow on these spots and transform into dense wooly growth (Fig.16), which may vary from pale to olive gray in colour. Severely infected capsules rot and fall. The seeds turn hollow, exhibiting discloured seed coat. Black sclerotia develop on the infected seeds. Circular or irregular lesions develop on leaves. Water soaked blackish lesions develop on slender shoots with the contact of infected spikes falling down, causing breakage. The disease is more pronounced causing devastation in the monsoon season.

Consequently, castor production is getting shifted to *rabi,* the post rainy season under irrigation, when Botrytis is not a serious problem (Anonymous, 2020)

Management:

(i) Adopt wider spacing between the crop rows.

(i) Prophylactic spray ofcarbendazim @ 1g /litre or Propiconazole @ 1 ml/lite before the onset of cyclonic weather based on weather forecast. This should be followed by another spray if the disease appears

(ii) Remove the infected spikes and destroy

(iii) Top-dress with Urea 20 kg/ha after cessation of rains to encourage formation of new productive spikes.

Fig. 16: Severe incidence of Botrytis disease of Castor

2. Wilt disease is caused by *Fusarium oxysporum* f. sp. ricini Nanda & Prasad. The symptoms of the disease include discolouration of hypocoyl, loss of turgidity of top leaves. Marginal necrosis at the seedling stage. The adult plants exhibit gradual yellowing and sickly apperarence of foliage followed by necrosis, browning of xylem tissues, formation of dark stripe and pinkish over growth on the stem upto the infected leaves, Irrivesible wilting and with drooping of the apical leaves and branches. Necrotic streaks at thebase of the stem, root degeneration, chlorosis of leaves and necrosis of the affected tissues causing death of the plant.

Management of Botrytis disease:

(i) Burning and destroying of the cropdebris.

(ii) Crop rotation with cereal crops

(iii) Grow wilt resistant varieties and hybrids viz., DCH-177, DCH_519, , GCH-7,DCS-107, DCH-519, DCS-107, 48-1 (Jwala), ICH-66 and GCH-8

(iv) Avoid water-logging in the field

(v) Seed treatment with Trichoderma 10g / kg or thirum/captan @ 3 g/kg or Carbendazim @ 1g/kg

(vi) Soil drenching with Carbendazim @ 10 g/litre, two to three times at 15 days interval

The efficacy of biocontrol agents and chemicals under *in vitro* condition was evaluated against *Fusarium oxysporum* f. sp. *ricini*. Pot culture studies were conducted with biocontrol agents and chemicals for the management of wilt disease complex of castor caused by *F. oxysporum* f. sp. ricini and nematode, *Rotylenchulus reniformis*. All the treatments significantly decreased wilt incidence and improved plant growth compared to untreated inoculated control. Combination treatment of carbofuran @ 2g/kg soil + carbendazim @ 1g/kg soil was found most effective in reducing wilt incidence and reniform nematode population followed by *Trichoderma viride* @ 4g/kg seed + *Pseudomonas fluorescens* @ 10g/kg seed (Shalini *et al*., 2014)The diseases of next importance that inflict castor crop are as follows. The names of the causative pathogens are given in the parenthesis.

3. Seedling blight (*Phytophthora nicotiane*):Dull green patches on both surfaces of cotyledonary leaves leading to the appearance of dark brown concentric zones. This results in leaf rot and death of leaf tissue causing drying of the seedlings. The disease may be managed by avoiding planting in low lying aeas and providing good drainage to the crop. Seed treatment with captan or thiram@ 3g/kg would prevent the disease. In the case of severity of the disease Metlaxyl may be sprayed @ 2g/ litre twice, each a 10 days' interval

4. Root rot and Dieback (*Macrophomina phaseolina*): Dark black lesions develop on the stem near grond level.Tap root shows signs of drying and the root bark gets peeled-off. The damage to root results in withering of leaves and and branches. Wilting of leaves start from top and proceeds down words. The nodes exhibit brown depressed lesions.

The disease can be checked by avoiding low-lying areas and water-logged conditions, adopting crop rotation with cereal crops, cultivating disease resistant cultivards viz., GCH-4 and ICH-66. Seed treatment with *Trichoderma* @ 10g/kg or thiram or captan @2g/kg. Burn the crop debris. Soil drenching with Carbendazim @ 1g/litre 2 to 3 times each at 15 days interval was found effective. The infected plants must be uprooted and destroyed. Crop sanitation must be maintained by burning and destroying the crop debirs.

5. Alternaria blight (*Alternria ricini*): Light brown circular spots appear on the cotyedonary leaves and irregular spots with concentric rings surrounded by yellow hallos mark their presence on mature leaves.The spots coalesce, causing leaf blight. The immature capsules turn brown and fall due to collapse of the pedicel. Sunken spots form on mature capsules and enlarge gradually covering the entire capsule with balck sooty fungal growth.

The disease can be controlled by seed treatment with thiram or captan @ 2 to 3 g/kg or spray of copper oxychloride @ 3 g/litre twice or thrice, each at 15 days interval with effect from 90 days of crop growth

Bacterial leaf spot (*Xanthomonas compestris):* Circular water soaked spots with pale green centre and dark margin appear on leaves turning into angular shape with dark brown colour . The affected leaves become brittle and dry-up. Stems and petioles exhibit dark and elongated lesions.

This disease may be managed by seed treatment with hot water of 50 to 60^{0}C for 10 minutes. At the field level a spray of streptocycline @ 1 g/10 litres of water or a combination of paushamyin @ 0.25 ml / litre + copperoxychloride @ 3 g/ litre given twice each at 10 to 15 days interval.

6. Powdery mildew (*Leveillula taurica)* : The under surface of leaf presents a whitish powdery mass. The upper surface of leaf exhibits light green patches corresponding to the diseased patces of the lower surface. The foliage later turn brown with necrotic patches and drop-off.

As a control mesure, a spray of wettable sulphur @ 3 g/litr given at 15 days' interval with effect from 30 days of crop growth with the onset of symptoms. Also spray of hexaconzolle @ 1 ml/ litre or dinoccap @ 2ml/litre at 15 days intervals may be resorted to when the disease assumes severity.

The diseases of Fusarium wilt, Macrophomina root rot and Botrytis are the foremost diseases of castor casuing heavy yield looses in India. Several sources of resistance to these diseases have been identified in castor germplasm and developed parental lines with resistance (Anjani *et al.*, 2014; Anjani *et al.*, 2017).

Insect pests such as leafhopper, capsule borer, *Spodoptera* and leafminer and semilooper are the major pests causing substantial yield losses in castor. Sources of resistance for leafhopper, leafminer and capsule borer have been identified in castor repository (Prasad and Anjani 2008; Anjani *et al.*, 2010; Lakshminarayan and Anjani 2010; Anjani *et al.*, 2017; Anjnai *et al.*, 2018). Several souces of genetic resistance to these pests have been registered with Plant Germplasm Registration committee, India (Anjani *et al.*, 2017). These sources of resistnance have served as donors in resistance brbeeding programmes across India for developing resistant castor lines and varieties (Desai *et al.*, 2001; Lavanya and Solanki, 2010; Santhalakshmi Prasad *et al.*, 2019).

Castor Crop Production: Under the current scenario of variable weather and climate change, castor is one crop which can be used successfully to obtain remunerative returns provided an appropriate production technologies are implemented depending upon the location specific conditions.

Constraints in castor production

(i) In dryland areas, the notable constraint is the drought due to inadequate and erratic rainfall. Also most of the drylands of castor production are characterised by shallow depth and poor soil fertility including lack of organic matter.

(ii) The constraints limiting the production in irrigated areas with specific reference to the zones of Gujarat, Rajasthan and Haryana are salinization of soils, accentuated diseases such as wilt and root-rot due to continuous cultivation of castor without crop rotation.

(iii) The general constraints common to all the castor growing areas are incidence of pests and diseases viz., *Fusarium* wilt causing serious losses, Botrytis and incidence of pests viz., Tobacco caterpillar (*Spodopteralitura*) castor semi-looper (*Achoea janata*), capsule borer (*Conogathus punctiferalis)* and leaf hoppers (*Empoasca flavescence*)

(iv) The biotic stresses afflicting castor may be managed following appropriate integrated plant protection measures as well as the choice of resistant and tolerant varieties.

In this section crop production management technologies are presented.

1. Land preparation in relation to moisture conservation on drylands

Notwithstanding the generality that castor is drought resistant, castor seed yields can decline adversely due to long spells of soil drought.

Castor normally yields well with a moderate and well distributed rainfall of 600 to 700 mm. The crop can perform reasonably well even under conditions of low rainfall of 350 to 500 mm provided it is well distributed.

There are instances of drastic reduction of the average yields when soil drought persists for more than 40 days.

This problem could, however, be managed with the adoption of appropriate tillage and moisture conservation practise.

- Summer ploughing of the filed with the receipt of summer showers helps in better infiltration and conservation of water received from subsequent showers, in addition to checking the proliferation of weeds. (Raghavaiah and Suresh2010)

- It is always recommended to apply Farm Yard Manure (FYM) @ 5 to 6 tons/ ha with the preparatory tillage or even with summer ploughing to ensure availability of minimum organic matter in the soil for good crop growth.(Prasad and Sudhakar Babu 1997)

- Preparation of field into ridges and furrows across the slope facilitates moisture conservation in the furrows and checks soil erosion. (Prasad and Sudhakar Babu 1997).
- The ridges may be spaced at 60 to 90 cm on drylands depending upon the soil type and its moisture holding capacity, soils. Lighter soils with low moisture holding capacity need wider spacing of 90 cm. Seed is planted on the ridge with intra-row spacing varying as described below.

2. Seed rate and plant population

Seed rate of castor varies depending upon the soil type and conditions, under which the crop would be grown. In the case of castor hybrids which produce more prolific growth the seed rate however, should be adjusted accordingly.

- Under conditions of dryland agriculture of totally un-irrigated situation, a plant population of 28,000 to 30,000 plants /ha with a seed rate of 8 to 9 kg at a spacing of 90 cm x 45 cm is recommended.
- If castor hybrids are to be grown on drylands with better moisture holding capacity a plant population of 20,000/ha with a seed rate of 6 to 7 kg may be adopted with a spacing of 90cm x 60 cm to provide for higher degree of branching etc.
- In the case of irrigated hybrid castor production of North Gujarat, a spacing of 150 cm x 75 cm was recommended with a seed rate of 3 to 4 kg /ha, to facilitate luxurious growth of crop with proliferating spikes. In Sourashtra region of Gujarat (Junagarh) a spacing of 120 cm x 60 cm and 90 cm x 60 cm is recommended depending upon the situation for GCH6 hybrid castor (Raghavaiah and Suresh 2010).
- Planting with a seed drill along with the plough shall help placing the seed in moisture-rich zone of the soil and thus facilitates speedy and vigorous germination.
- The general recommendations regarding seed rate and spacing for castor varieties and hybrids in respect of Peninsular India are given in Table below. (Anonymous 2019)

Table 6 : Seed rate and Spacing (Anonymous 2019)

Season	Variety/hybrid	Seed rate(kg/ha)	Spacing (cm)
Rainy/Kharif			
Heavy soils	Hybrids	5 to 6	90 x 60
Light /red soils	Varieties	8 to 9	90 x 45
	Hybrids	5 to 6	90 x 60
Post rainy Irrigated			
Heavy soils	Hybrids	5 to 6	120 x 90
Lighter soils	Hybrids	5 to 6	90 x 90

3. Seed treatment

Pre-sowing seed treatment of castor with recommended fungicides may be necessary to ward off seed-borne diseases.

The following seed treatment schedules are recommended to ensure good plantstand. (Raghavaiah and Suresh 2010)

Table.7: Seed treatment schedules for castor (Raghavaiah and Suresh 2010)

Fungicide	Dose	Disease to be prevented
Thiram / Captan	3 g /kg seed	Alternaria blight & Seedling blight
Carbendazim	2 g/kg seed	Wilt
Trichodermaviride	10 g/ kg	Wilt
Trichoderma viride (soil application)	2.5 kg in 125 kg FYM/ha to be applied to soil.	Wilt

4. Castor Crop Nutrition

Castor is a highly nutrient responsive crop. It is estimated that castor crop yielding 2 tons/ha would remove 80 kg nitrogen, 20 kg phosphorous and 35 kg potash (Raghavaiah and Suresh 2010). According to Pasricha and Tandon (1993) castor crop removes 40kg N, 9 kgP2O5 and 16 kg K2O per ton of seed yield. The low productivity levels of dry land castor in Telangana are attributed to very low input of crop nutrients (Raghavaiah and Suresh 2010). Basal application of FYM @ 5 to 6 t/ha at preparatory tillage helps in good nutrient uptake by the crop in addition to facilitating moisture conservation.

- Following are the crop nutrient recommendations for castor under varying management situations in Peninsular India (Anonymous 2019)

Table 8: Major Nutrient Recommendations for Castor Production (Anonymous 2019)

Situation	Kind	Recommended N:P:K kg/ha
Dryland	Varieties	60:40:30
	Hybrids	70:45:30
Irrigated (Post rainy season)	Varieties	70:45:30
	Hybrids	90:45:45

- Apply 12.5 kg ZnSO4 ha-1 (if the soil available Zn is < 1.2 ppm) and 25 kg FeSO 4 ha-1 (if the soil available Fe is < 3.7 ppm for non-calcareous soil and < 6.3 ppm for calcareous soil)
- While the total quantities of P and K could be applied as basal just before sowing, one third of N dose may be given at the time of sowing all along the seed zone. The rest of the quantities of N could be applied as spilt doses at flowering initiation and peak flowering to capsule formation stage of the crop. These developmental stages are continuous process and hence can't be demarcated separately.

The recommendations for nutrient management should take into consideration the soil test values of the locations.

5. Integrated Nutrient Management

The investigations conducted at different oilseed research centres of India on the Integrated Nutrient Management of castor have revealed the following results (Raghavaiah and Suresh 2010).

- In the dryland region of Telangana (Palem), substituting 25% of Recommended Dose of Fertilizer (RDF) with neem cake was found be equally effective as 100% RDF and resulted in lower cost of production.
- In Rajasthan (Mandore) inoculation of castor seed with Phosphorous Solubilizing Bacteria (PSB) + application of 5 tons of FYM along with 50% RDF was found to be productive and economical as compared to that of 100% RDF.
- In North Gujarat and Saurashtra zone, substituting 25% Nitrogen (amounting to 30 kg of N) through FYM, castor cake and green manuring amounting to 75%N substitution was useful in getting good yields.
- In lighter soils of Gujarat. Seed treatment with *Azospirillum* or PSB @50g /kg of seed together with castor cake @1 ton/ha is advocated for obtaining good yields.

6. Weed management in Castor crop

Castor crop grown on drylands without any extraneous irrigation is more vulnerable to weed competition. Weed infestation of dryland castor will adversely affect the growth and yield of castor by (1) robbing the precious moisture and nutrients from the soil and (2) harbouring pests and diseases acting as alternate hosts. Due to slow growth of crop in the initial stages following germination, the crop is adversely smothered by weeds. It becomes imperative to keep the field weed free particularly in the case of rainfed castor in the initial stages of crop growth. The critical phase of weed competition for castor is 45 to 50 days following sowing.

The most effective simple weed management procedure for rainfed castor is to carry out two to three inter cultural operations with bullock drawn blade harrow starting from 25 days' age of the crop at weekly intervals and followed by one hand weeding. In the case of dryland castor, square planting with a spacing of 60 cm x 60 cm or 90 cm x 90 cm will be more convenient to carryout intercultural operations for weed management. As compared to the above mentioned practices of two harrowings followed by hand weeding, harrowings alone gave better cost-benefit ratio (Subba Reddy *et al* 1999). In the case of dryland castor, good and timely land preparation including summer ploughing, sowing across the slope, working with blade harrow, laying the field in ridges across the slope at 45 days of emergence helps a lot in effective weed control and also in maintaining the moisture level of the soil.

Pre-emergence application of Butachlor @ 1 kg active ingredient/ ha followed by hoeing at 30 days resulted in least weed density and also better uptake of nutrients by castor crop (Sreedevi *et al.,* 1997). Also, pre-emergence application of Fluchloralin @ 0.9 to 1.35 kg active ingredient / ha was effective in achieving good control.

7. Efficient irrigation systems

A vast area of castor is under dryland agriculture. Nevertheless, castor is also being grown as a commercial crop under irrigated conditions principally in Gujarat and Rajasthan (Fig. 16). The area under irrigated castor particularly in the post-rainy season following the advent of castor hybrids has increased even in Peninsular India.

Since this system of castor production has resulted in remunerative income levels. Increasing levels of irrigation, however, have sparked up the increased incidence of root rot-wilt complex causing build-up of higher levels of fungal inoculum in soils apart from declining levels of castor productivity. Additionally, it must be remembered that castor is also sensitive to water-logging which should be avoided at all costs in irrigated castor management.

The number and intensity of irrigation depends upon the type and its water holding capacity of the soil on which the crop is grown.

- Since the castor tends to become more vegetative under irrigated conditions, it is advisable to with-hold irrigation initially following germination for about three weeks to check excessive vegetative growth, facilitate condensation of internodes and promote early flowering (Raghavaiah and Suresh 2010).
- Lighter soils on which the castor crop is grown under rain-fed conditions in monsoon season, often encounter drought of varying intensities. Under such conditions one to two supplemental irrigations depending upon the kind of drought would augment castor yields (Anonymous 2019).
- During the post-rainy *rabi* season, the number of irrigations may normally reach 6 to 8 and in summer the irrigation frequency may increase to meet the crop's demand to 12 to 14 at the intervals of 7 to 10 days depending upon the soil type and prevailing weather conditions.
- The first irrigation being given at pre-sowing phase, the second irrigation is normally recommended to be given at 3 to 4 leaf stage of the crop and the subsequent third irrigation may be given at 6 to 8 leaf stage. The subsequent irrigations may be effected at 15 to 20 days' interval depending upon the soil and prevailing weather conditions.
- The most critical phase of the crop needing the irrigation is from the terminal bud initiation to full flowering of the primary spike. The crop

should never be subjected to moisture stress at the above developmental stage (Raghavaiah and Suresh 2010).

- Irrigation given in furrows is appropriate when the crop is raised on the ridges in ridge-furrow system. The consumptive use in medium red soils of Telangana region is reported to be 500 mm (Raghavaiah and Suresh 2010)
- Micro irrigation adopting the drip system of water management (0.2, 0.4. 0.6 and 0.8 of Cumulative Pan Evaporation, i.e., CPE) has resulted in significantly higher yield levels than the normal surface irrigation in North Gujarat employing the castor hybrid GCH-4. This system is reported to have resulted in 24% saving of irrigation water and at the same time brought about 36% higher seed yield (Raghavaiah and Suresh, 2010). The drip system should be laid out at a lateral spacing of 90 cm and dripper spacing of 60 cm and operated at a pressure of 1.2 kg/sq.cm using 4LPH dripper (Raghavaiah and Suresh 2010). The schedules of operation of the system should be 41 minutes in November – December, 35 minutes in December –February and 65 minutes in March-April (Malavia *et al.*, 1995).
- The investigations carried out on Fertigation through micro-irrigation in *kharif* season castor at Junagarh, Gujarat have indicated that drip irrigation scheduled at 0.6 CPE and fertigation of 100% dose of nitrogen had significant beneficial effect on seed and stalk yield of castor with a significant higher water use efficiency and net monetary returns (Lakkad *et al.,* 2005). In North Gujarat, adoption of drip system and application of 150 kg N/ha as fertigation has resulted in 17% higher income (Patel *et al.,* 2006)

8. Castor to combat aberrant weather

Delayed onset of monsoon and incidence of drought are common features of the semi-arid tracts of India. Whenever monsoon showers do not occur before the first week of July, the production of dryland crops more particularly cereals gets adversely affected. In the consequent event of delay of *kharif* plantings, castor crop comes up as handy tool to manage the aberrant weather. The investigations carried out in the All India coordinated Research Project on Oilseeds (AICORPO) and the Central Research Institute for Dryland Agriculture (CRIDA) have indicated the following recommendations for production of a contingent crop of castor for the above situation (Raghavaiah and Suresh 2010).

- The short duration and early maturing varieties of castor such as Jyothi (DCS 9), Kranti (PCS 4) and GC 2 are recommended to be grown with a closer spacing of 60 cm x 30 cm to curtail the excessive and unwanted vegetative growth.
- Provide a dust mulch using a blade harrow following each shower, which would check evaporation losses of soil moisture.

- In the event of occurrence of intermittent drought spells preceding the flowering phase or at flowering of the main spike, apply 20 kg of N/ha to encourage emergence of secondary and tertiary order spikes to compensate the losses due to failure of the development of primary spike.
- If irrigation water is available, give one or two protective irrigations at 50 to 80 days of crop cycle matching with the development of later order spikes
- In the event of occurrence of continuous cloudy weather for 40 to 90 days, the consequent problem of blasting of male flowers leading to the abortion of female flowers should be managed with application of 20 kg of N /ha to encourage emergence of subsequent order of spikes.
- In the face of problems of impending failure of rice production due to scanty water availability, castor comes up as a contingent crop which can be raised with two to three irrigations during the months of August or September to offset the production losses and boost the farmers' income levels.

9. Castor based Cropping Systems

Castor is grown as a sole crop as well as an inter-crop on drylands by virtue of its adaptability to the conditions of low moisture availability. Castor becomes a sole crop under the rain-fed situations with better soil type and better moisture availability. In light red or *chalka* soils of Telangana region, castor is also grown as dryland crop in association with other monsoon season crops like pigeon pea, cotton, horse gram and cowpeas as inter crops.

With the availability of high yielding varieties, particularly hybrids, castor has come to be grown as a sole irrigated commercial crop particularly in the post monsoon season under high input production systems.

(i). Castor based crop sequences: In Peninsular India where the castor is a popular dryland crop, a two year, two crop sequences are adopted. Castor followed by cotton or sorghum or pigeon pea are the common crop sequences.

Under irrigated conditions obtainable in North Gujarat, it was observed that castor preceded or succeeded by pearl millet or sorghum or mung bean resulted in higher monetary returns than that of sole crop of castor without any other sequence crop. Also mustard- castor rotation was found to result in higher castor yields as compared to continuous cultivation of castor hybrids viz., GAUCH-1 or GCH-4 hybrid. Planting castor early by10 to 15 day in advance as a relay crop in the fields of cluster beans, pearl millet and cowpeas, before their harvest has resulted in higher yields of castor due to better utilization of soil moisture as compared to that sown later by two weeks.

Studies carried out at the Directorate of Oilseeds Research (DOR 2003) have revealed that the five-year castor based crop sequences with sole castor, cluster beans, castor + pigeon pea (1:1), castor + groundnut (1:5), castor + cluster beans (1:2) and castor + horse gram (1:2) as preceding crops followed by sorghum, pigeon pea, sorghum + pigeon pea (2:1) and soybean as succeeding crops showed that castor + cluster bean, castor+ pigeon pea, castor + groundnut resulted in highest castor equivalent yields and gross returns than castor + horse gram and sole castor.

Castor was observed to have utilized the residual soil fertility effectively in succession with sunflower. At Junagadh, post rainy season *(rabi)* castor grown after *kharif* fodder sorghum gave higher castor yields than that grown after black gram, bunch groundnut and fallow.

Mavarkar *et al.,* (2008) evaluated economic viability of castor based sequential realy inter-cropping systems under rainfed conditions. They found that the castor crop equivalent yield was significanty higher in paired row of castor inter-cropped with baby corn followed by field bean over different relay inter-cropping systems.

Fig. 16: Irrigated Castor Crop in Gujarath

(ii). Castor based Intercropping Systems: In the preceding paragraph of this section, the traditional castor based inter-cropping systems such as castor inter-cropped with pigeon pea, cotton, horse gram and cowpeas prevalent on drylands are indicated. However, the data gathered in recent years by Raghavaiah and Suresh (2010) have indicated that due to vagaries of monsoon if rain-fed crops could not be sown, the inter-cropping system involving black gram + castor (6:1) could become a viable alternative, which would result in income comparable to that obtained from groundnut.

At Junagarh, planting castor 30 days after groundnut sowings in 2:1 ratio (groundnut+ castor) resulted in highest yield of castor and remunerative returns from the inter cropping system. With the application of 100% Recommended Dose of Fertilizers (RDF) for both groundnut + castor planted 3:1 ratio the farmers can reap the highest returns.

Studies carried out in Hyderabad showed cowpeas grown in between two rows of castor and later incorporated as green manure (10 t/ha) at around45 days added 30 kg N/ha to the soil, thereby obtaining N from green manure at least cost.

Castor based intercropping systems with modified crop geometries such as castor spaced at 90 cm, paired with 60/120 cm and at 60 cm to include 2, 3 and 6 rows of intercrops viz., green gram, cluster beans and sesame under two systems viz., irrigated and unirrigated-rainfed conditions in Western Rajasthan (Mandore) resulted in higher returns as compared to sole castor under irrigation and dryland systems (Raghavaiah and Suresh 2010).

Castor based intercropping with grain legumes such as groundnut, green gram and black gram in paired row systems of intercrop (i.e., one row castor intercropped with two rows of legume/pulse crop) seemed to fetch better monetary returns, with higher relative net returns indices (Srilatha *et al.,* 2000) Similar trends of profitability have been reported by Agarwal and Porwal (2006).

10. Harvesting of Castor crop

The developmental rhythm of castor plant is of a perennial, despite the crop being treated as an annual. The old traditional varieties of castor show a longer life-cycle with crop maturity much staggered. Hence, harvesting was done earlier in phases extending over a longer period with the first picking of mature capsules starting after 160 days. However, most of the currently recommended varieties exhibit more or less a life cycle of around 150 to 160 days with the maturity of the different spike orders falling with in the above duration. The early maturing varieties reach harvest within 120-150 days from emergence. With the modern varieties of castor, the harvesting may be carried-out with 4 pickings at the most. Stage of harvesting of spikes is the most important factor. The spikes should be harvested at physiological maturity when the leaves show senescence and capsules turning orange yellow in colour, thereby avoiding shattering that occurs when the capsules get totally dried and start splitting (Fig. 17) The first harvest or picking of spikes, which starts with the principal or primary spike getting ready after 90 to 100 days in early maturing varieties and around 120 days in the case of medium early to medium late varieties. This is normally followed by subsequent three pickings at fortnightly or 20 days' intervals depending upon the varietal maturity pattern.

Fig.17: Dry castor capsules on the spike.

In the tropics most harvesting is done manually, by cutting of the spikes, that reach maturity. The capsules are stripped off into a wagon or sled, or into containers strapped on the workers. The capsules must be spread-out to dry quickly. Harvesting of unripe and green capsules, however, must be avoided, since such an operation will have an adverse effect on the yield and oil content of the seed.

When the whole crop is gathered, it is dried in the sun for a few days and the threshing is done either by beating the dried capsules with sticks or by treading them under the feet of bullocks. Winnowing is done in the usual manner.

After harvest, the remaining stalks in the field may be cut mechanically and incorporated into the soil. The stalks decompose rapidly and furnish organic matter. Castor bean hulls, which are scattered over the field during harvest, are about equal to farm yard manure in fertilizer value.

Harvesting of castor must be carried out before reaching this extreme case of shattering

Castor Oil and Other Products of Castor

Castor oil is a viscous colourless to pale yellow coloured liquid (Fig. 18).

This oil is the only vegetable oil that contains up to 85% of the unique hydroxy fatty acid viz., ricinoleic acid, which confers distinctive industrial properties to the oil. Chemically, castor oil is a triglyceride (ester) of fatty acids. Up to 90 % of the fatty acid content of the oil is ricinoleic acid (12-hydroxyoleic acid), an 18-carbon acid having a double bond in the 9-10 position and a hydroxyl group on the 12th carbon. Additional characteristics include relatively high viscosity, non-drying, faint but characteristic odour slightly acrid taste and an aftertaste taste. Many derivatives can be produced which have a similar chemical composition to petroleum based oils. (Naughton, 2011)

Fig.18. Castor oil with castor seeds and castor capsules on the panicle. (www.paramountcolors.com)

Blown Castor oil is also known as oxidised castor oil. It is a derivative that has a higher viscosity and specific gravity compared to natural castor oil. These properties are induced by bubbling air through it at elevated temperatures. Its main use is as a plasticizer for inks, lacquers and adhesives.

Hydrogenated castor oil (HCO) or castor wax is a hard, brittle wax that is insoluble. It is produced by adding hydrogen in the presence of a nickel catalyst. It is mainly used for coatings and greases where resistance to moisture, oils and other petrochemical products is required. As a result, castor oil and products derived therefrom are used for numerous industrial products, including bio-based lubricants, fuel, paints and coatings, plastics, anti-fungal compounds, and cosmetics. (Thomas, 2005; Howarth, 2003).

The growing importance of castor oil business internationally is vividly reflected in the imports and exports of castor oil, which were estimated of the order of 268,500 tonnes and 289,500 tonnes respectively in 2009 (www.oilworld.biz). The international industry's demands for castor oil are of the order of 3–5% per annum. The global castor oil and its derivatives market will register a CAGR of more than 5% by 2022 (Cision, 2018)

The bulk of industrial use of castor oil lies with the European Union, USA and Japan. The traditional castor producing countries viz., China and India are also gearing up their technologies towards exploiting castor oil in their Industry. According to the report (Cision 2018), one of the major drivers for this market is the increasing demand for castor oil and its derivatives in major end-use in industries.

Further, the report states that one of the major factors hindering the growth of this market is the fluctuations in castor oil prices caused by extreme dependence on climate. Fluctuations in raw material prices, due to the availability of inexpensive substitutes as well as the widening supply-demand gap demand of long storage and hoarding period pose a challenge in the market.

Castor oil can be broken down into other chemical compounds that have numerous applications (Hatice, *et al.*,2010). Transesterification followed by steam cracking gives undecylenic acid, a precursor to specialized polymer nylon 11, and heptanal, a component in fragrances (Ashford, 2011). Breakdown of castor oil in strong base gives 2-octanol, both a fragrance component and a specialized solvent, and the dicarboxylic acid sebacic acid. Hydrogenation of castor oil saturates the alkenes, giving a waxy lubricant (Thomas, 2005).

Castor oil may be epoxidized by reacting the OH groups with Epichlorohydrin to make the triglycidyl ether of castor oil which is useful in epoxy technology (Hermansen, 2017). Castor oil is also being used in food industry with its conversion as food grade castor oil. (Wilson *et al.*, 1998)

Castor oil could be an ideal candidate feedstock for the emerging biodiesel industry (Auld *et al*., 2009), but it is currently it may be a far cry because the international price of castor oil is too high to be used for biodiesel production. Hence it becomes too expensive for castor oil to compete even with petroleum based diesel. Increasing castor areas and production all over the world might pave way for producing castor oil based biodiesel at more competitive rates in future.

Castor wax: This is nothing but hydrogenated castor oil. Castor wax is opaque and white vegetable wax. It is produced by the hydrogenation of pure castor oil often in the presence of a nickel catalyst to increase the rate of reaction. The hydrogenation of castor oil forms saturated molecules of castor wax; this saturation is responsible for the hard, brittle and insoluble nature of the wax. Castor wax is used in polishes, cosmetics, electrical capacitors, carbon paper, lubrication and coatings and greases where resistance to moisture, oils and petrochemical products is required. Castor wax is also useful in polyurethane coating formulation, as it contains three secondary hydroxyl groups. These coating-compositions are useful as a top coat varnish for leather, wood and rubber. Castor wax can also be added to beeswax for encaustic painting. (BBC - GCSE Bitesize 2018)

Castor Cake: Castor de-oiled cake is the solid residue obtained following crushing of castor seed for extraction of castor oil. It forms as much as 60% of the crushed castor seed. At the current production of castor seed at more than 1.123 million tons, the annual production of castor cake is around 0.67 tons of Castor cake contains around 4.5 to 6.0% of N, 2.5% of P2O5 and 2.5% of K2O. Additionally, castor cake is rich in protein (25%), sugar (25%0 and minerals (20%) (Kumar and Ashfaq 2010).

It forms a good organic manure particularly to check the incidence of termites and nematodes. In general, the fields that received soil incorporation of castor cake @ 3tons/ha 15 days before plating are stated to be relatively free from incidence of insect pests. The castor meal, however, is mineralized extremely fast, and, a large quantity of N is released in a short period of time (Lima *et al.,* 2011). The mineralization of castor meal measured by microbial evolution of CO_2 is 7 times faster than bovine manure and 15 times faster than sugarcane bagasse (Severino *et al*., 2004). It also has traces of nutrients like Manganese, Zinc and Copper, thus making it a balanced fertilizer. Moreover, it helps to neutralize the detrimental effects of chemical fertilizers. Apart from their contribution to Nutrients, it has number of other benefits in agriculture, which none of the synthetic fertilizers or pesticides can offer. The castor cake brings in the wonderful molecules that nature has designed to help the plants flourish naturally. It also provides slow and steady nourishment, stimulation, protection from soil nematodes and insects; improve yields, and quality of product like taste, flavour, amino acid composition etc.100 kg of Castor de-oilcd cake brings to the floor as much nitrogen (N) as 1200 kg of cow dung. It is can be used in organic farming (Sunray International 2019).

However, castor cake can't be used as cattle feed due to the presence of toxic elements viz., ricin and RCA at 125.5 g per 100 g of the cake (Kumar and Ashfaq 2010).

Castor Genome and Ricin: Ricin is a very toxic lectin (a carbohydrate-binding protein), produced in the seeds of the castor oil plant. Ricin is synthesized in the endosperm of castor seeds (Lord and Roberts, 2005). Castor bean genomics is relevant to biosecurity as the seeds contain high levels of ricin, a highly toxic, ribosome-inactivating protein. Chan *et al*., (2010) reported the draft genome sequence of castor bean (4.6-fold coverage), the first for a member of the *Euphorbiaceae*. Whereas most of the key genes involved in oil synthesis and turnover are single copy, the number of members of the ricin gene family is larger than previously thought. Comparative genomics analysis suggests the presence of an ancient hexaploidization event that is conserved across the dicotyledonous lineage (Chan *et al*., 2010).

11

Oil Bearing Trees

Introduction

Indian oilseeds scenario has been able to surmount the crisis of prolonged stagnation with increased production in the decades of 1980s and 1990s (Prasad 1991). Nevertheless, in the face of growing demands of vegetable oil both for human consumption and industrial use, there is every need to augment the oilseed production to new heights of sustainability (Prasad 1994a).

Apart from the traditional sources of vegetable oils from annual crops like groundnut, rapeseed-mustard, sesame, sunflower, safflower, niger, linseed, castor and soybean besides cotton seed and rice bran, more than 100 species of domesticated or wild plant species of India are known to contain oil in considerable quantities (Prasad, 1994c). In addition to the production from nine traditional oilseeds mentioned above, 3 million tonnes of vegetable oil is being harnessed from secondary sources like cottonseed, rice bran, coconut, Tree Borne Oilseeds (TBOs) and Oil Palm. Subramanian *et al.,* (2005) reported that there are over 300 different species of trees which produce oil bearing seeds. Thus, there is a significant potential for non-edible oil source from different plants for biodiesel production as an alternative to petrodiesel.

It is estimated that TBOs have the potential to contribute more than 3 million tons of oil; but only a fraction of the potential is realised currently. Seeds of many of the oil bearing trees have been used by the ethnic communities for different purposes. Apart from seed oil, the trees yield valuable by-products including cake, which have industrial applications (Hocking, 1993; Chakravarty *et al*., 2012).

There are many species of plants found wild or cultivated inside as well as outside the forest areas containing sizeable amount of vegetable oil having domestic and industrial utility. Out of these species.The country has enormous potential of oilseeds of tree origin like Mahua (*Madhuca indica* and *M. longifolia*), Sal (*Shorea robusta*), Neem (*Azadirachta indica*), Simarouba (*Simarouba glauca*), Karanj (*Pongamia pinnata*), Ratanjyot /physic nut (*Jatropha curcas),* Jojoba (*Simmondsia chinesis),* Cheura (*Diploknema butyracea*), Kokum (*Garcinia indica*), wild Apricot *(Prunus armeniaca*), wild Walnut (*Aleurites molucana*), Kusum (*Schleichera oleosa*), Undi (*Calophyllum inophyllum),* Tung *(Verniciafordii)* etc.

which can be grown and established on the wastelands and under varied agro-climatic conditions.

Azam *et al.* (2005) studied the prospects of fatty acid methyl esters (FAME) of some 26 non-traditional oil seed plants including *Jatropha* for use as potential biodiesel feed stocks in India. Among them, *Azadirachta indica, Calophyllum inophyllum, J. curcas* and *Pongamia pinnata* were found most suitable for use as feed-stocks for biodiesel production, since they meet the major specification of biodiesel for use in diesel engine. *J. curcas* is being considered as one of the most promising potential oil source to produce biodiesel in Asia, Europe and Africa (Chhetri *et al.,* 2008).

Paradoxically, there is no much emphasis in literature on oilseed bearing *Moringa oleifera* popularly known as 'drumstick tree' about its potential as an oil bearing plant species.

Nagaraj and Mukata (2004) analysing the oil content and fatty acid profiles of some oil bearing tree species reported that the oil content of their seeds ranged from 4.6% in *Parkinsonia aculeate* to 37.4 % in *Pongamia pinnata.* Oleic and linoleic acids were the major fatty acids in seven species and *Jatropha curcus* had these fatty acids in equal proportions.

The data on fatty acid composition of oil of some oil bearing trees including *Melia* neem (*Azadirachta indica*), karanj *(Pongamia pinnata*), mahua (*Madhuca indica*) and kapok (*Ceiba pentandra*) presented in the Table 1, which indicates a wide variability among the oil bearing tree species offering a scope for selection of oil quality. The indicated tree species facilitate intercropping of annual dryland grown cereal crops, pulses and oilseeds, thereby adding to the income of farmers.

Table1: Fatty acid composition of oils of some oil bearing tree species (Prasad, 1994)

Oil bearing Tree	16:0	18:0	18:1	18:2
Neem	15	14	62	8
Melia	19	11	37	32
Mahua	24	23	38	13
Kapok	23	1	39	33
Karanj	6	9	46	138

As per the estimates of Solvent Extractors' Association (SEA), India can produce about 11-15 lakh tonnes of oil for edible and industrial uses from TBOs. About 5 lakh tonnes of seeds are being collected from oil bearing tree species and crushed to produce approximately 1 lakh (100,000) tonne of oil. As per the survey report (2012) conducted by Indian Council of Forestry Research & Education (ICFRE), Dehradun, the above mentioned 11 TBOs are grown in about 14.7 lakh hectare area under Government or Private lands. Presently 7.5 to 10% of potential is being harnessed. Many of these TBOs possess 20-60% oil content in seeds, which are of edible and non-edible fat for industrial uses. TBOs offer domestic and industrial utility like agriculture, cosmetic, pharmaceutical, diesel and substitute, cocoa-butter substitute etc.

The enormous potential of TBOs is not being harnessed fully due to lack of awareness of their uses, collection, proper processing facilities, organized marketing sector and others. Besides, lack of proper facility for storage, seed collection, long gestation period and their fruiting/maturity season coinciding with rainy season are main constraints limiting collection and utilization of above TBOs. Once they are planted, they continue to yield for about 60-150 years starting after gestation period without requiring much input in the later stage. TBOs have disadvantages namely long gestation period (2-8 years), harvesting coinciding with rainy season, very tall stature, involvement of higher number of man-days in harvesting, non-synchronized maturity, non-availability of manual/mechanical harvester etc. Besides, non-availability of superior genotypes/varieties/ hybrids with high seed yield and oil content are the other limiting factors which are essential requirements for augmenting potential of TBOs. The tree borne oilseeds (TBO's) at present, contribute an insignificant portion of vegetable oil production in the country mainly due to lack of improved varieties, elite planting material and agronomic practices.

In order to strengthen the production of TBOs, the following steps need to be put in place without delay

(i.) Integrated development of nurseries and plantation on the wastelands,

(ii.) Maintenance of TBOs plantation and their improvement to produce superior and dependable planting material,

(iii.) Incentives for undertaking intercropping with TBOs,

(iv.) Distribution of pre-processing, processing and oil extraction equipment

(v.) Training of extension workers and

(vi.) Encourage and strengthen the basic and applied research on TBOs.

Authentic estimates of the potential of TBOs are not available because of their scattered occurrence over vast and diverse areas. No detailed surveys of their enumeration have been undertaken so far. Largest number of trees of these oilseed species are found scattered in forests, homesteads as well as in cultivated and uncultivated lands. Besides these are crushed in several hundreds of oil mills spread all over the country, due to which correct estimates of their production potential are not available. However, based on the investigations conducted by the Directorate of Oilseeds Development, and the Directorate of Khadi Village Industries and the information gathered by some trade and Industrial organizations involved in their collection, certain estimates are prepared, which indicate that a potential of 17 lakh tonnes of oil is available annually (Prasad,1994c). Among the

TBOs, sal, mahua, karanj and neem are important and constitute 80%of total tree oilseeds production.

Among TBOs, Sal (*Shorea robusta*) forms the major component occupying 111,500 sq.km. of area in the country mainly in the central zone consisting of Madhya Pradesh, Bihar and Orissa as well as Himalayan foothills. The next major area is under mahua (*Madhuca indica)* distributed in the zones of Uttar Pradesh, Madhya Pradesh, Gujarat and Andhra Pradesh.

Another major component of TBOs is karanj (*Pongamia pinnata*) which is spread out throughout the country along the banks of streams, rivers or near sea coast. It is a very hardy tree, drought resistant, tolerant to salinity and fast growing (Bringi and Mukherjee, 1987). Karanj is reported to have undergone initial phase of domestication in the areas of current zones of erstwhile Andhra Pradesh, Karnataka and Tamil Nadu (Bringi, 1987). Neem (*Azadirachta indica*) is widely distributed particularly in dry areas (Bringi, 1987).

While the above mentioned tree species are native to India, there is another oil bearing perennial plant species viz., *Jatropha* spp.(Ratanjyot/ physic nut) introduced from American continent and as such not native of India. According to Patel *et al.,* (1988) *Jatropha* is adapted to arid and semi-arid conditions and can grow on stony, gravelly, shallow and gravelly soils. *Jatropha* is mainly distributed in Gujarat, Rajasthan, Madhya Pradesh, Maharashtra and to a limited extent in other states of India.

Thumba (*Citrullus colocynthis)* the creeper adapted to arid ecosystem, resembles the common water melon, but bears fruits which are small and hard characterised by bitter taste.*C. colocynthis* is a desert viny plant that grows in sandy, arid soils. It is native to the Mediterranean Basin and Asia, and is distributed among the west coast of northern Africa, eastward through the Sahara, Egypt until India. On the island of Cyprus, it is cultivated on a small scale; it has been an income source since the 14th century and is still exported today. It is a wild annual or a perennial plant in Indian arid zones and has a great survival rate under extreme xeric conditions (Lloyd, 1898). In fact, it can adapt itself to annual precipitation of 250 to 1500 mm and an annual temperature of 14.8 to 27.8 °C. It grows from sea level up to 1500 meters above sea level on sandy loam, sub-desert soils, and sandy sea coasts with a pH range between 5.0 and 7.8. Though it has a production potential of one million tonnes, the present collection is only one lakh tonnes (Prasad, 1994c). It is grown in the crop stands of pearl-millet (*bajra),*in the western districts of Rajasthan viz., Jodhpur, Bikaner, Barmer and Jaisalmer. The data onoil content in respect of some of the known oil bearing tree species and the corresponding fatty acid composition are given in Tables.2 and 4 respectively.. The estimated potential and production of the tree species is given in Table 3.

Table 2: Oil content and nature of oil of major non-traditional oil bearing tree species (Prasad, 1994)

Tree species	Oil content (%)	Nature of oil
Neem (Azaadirachta indica)	20	Semi-hard
Mahua (Madhuca indica)	35	Hard
Sal (Shorea robusta)	12	Hard
Karanj (Pongamia pinnata)	27	Semi-hard
Kusum (Schelichera oleosa)	33	Hard
Khakan (Salvadora oleoides)	33	Nut oil
Nahor ((Mesua ferrea)	40	Smi-hard to Soft
Kokum (Garcinia indica)	40	Hard
Undi (Calophyllum inophyllum)	60	Semi-hard to Soft

Table 3: Estimated Potential and actual production of seeds and oil from oil bearing plant species of forest origin (in lakh tonnes)
(Suri and Mathur, 1988)

Wild oilseed species	Potential		Atual Production	
	seeds	oils	seeds	oils
Sal (Shorea robusta)	60 – 70	6 -8	2.0 -3.0	o.25
Mahua (Madhuca indica)	4-30	2-4	0.70-0.85	0,25 -0.30
Neem (Azadriacta indica)	4 -30	1-3	1.0 -1.5	0.20 -0.30
Karanj (Pongamia pinnata)	1 - 4	0.3 – 1.0	0.25 – 0.30	0.07- 0.1
Kusum (Schleichera oleosa)	0.9 – 1.9	0.29	0.20	0.06
Khakan (Salvodora oleoides)	0.46	0.15	-	-
Undi (Calophyllum inophyllu)	0.11	0.07	-	-
Nahor (Mesua ferrea)	0.09	0.03	-	-
Kokum (Garcinia indica)	0.02	0.01	-	-
Total	70 – 115	11 - 17	4.92	1.02

Table 4: Fatty Acid Composition (%) of different oil bearing tree species
(Prasad, 1994)

Tree species	Palmitic Acid (16: 0)	Stearic Acid (18:0)	Oleic Acid (18:1)	Linoleic Acid (18:2)_
Neem (Azadirachta indica)	15	14	62	8
Melia (Melia azedarach)	0719	11	37	32
Mahua (Madhuca indica)	24	23	38	13
Kapok (Ceiba pentandra)	23	01	30	33
Karanj (Pongamia pinnata)	06	09	46	138
Jatropha curcas	16.3	4.2	38.8	29.0

Several other tree species including soap nut (*Sapindus mukorossi)* (Chhetri *et al*., 2008)wild Apricot *(Prunus armeniaca*), Cheura (*Diploknema butyracea*), Simarouba (*Simarouba glauca*), Jojoba (*Simmondsia chinensis),* Tung *(Aleurites fordii)* andOlive (*Olea europaea*) (Dhyani *et al.,* 2015) have been reported to be promising oil yielders in addition to the tree / creeper species described above of many exotic species introduced into India. *Simmondsia chinensis* (Jojoba) and an ever green tree *Simaruba glauca* (paradise tree) have shown promise for arid and semi-arid zones. There are oil bearing new annual species too viz., *Cuphea, Euphorbia lathysis, L., Limathes* spp., *Crambe abyssinica and Losquerella* spp. (Prasad, 1994)

Two plant species, soapnut (*Sapindus mukoross*i) and jatropha (*jatropha curcas, L.*) are presented by Chhetri *et al*., (2008) as newer sources of oil for biodiesel production. According to them both oils of *Sapindus* and *Jtropha* have great potential to be used as feedstock for biodiesel production. Fatty acid methyl ester (FAME) from cold pressed seed oil was envisaged as biodiesel source for the first time. Soapnut oil was found to have average of 9.1% free Fatty Acid (FA), 84.43% triglycerides, 4.88% sterol and 1.59% others. *Jatropha* oil contains approximately 14% free FA, approximately 5% higher than soapnut oil. Soapnut oil biodiesel contains approximately 85% unsaturated FA while biodiesel from *Jatropha* oil was found to have approximately 80% unsaturated FA. Oleic acid was found to be the dominant FA in both soapnut and *Jatropha* biodiesel. Over 97% conversion to FAME was achieved for both soapnut and *Jatropha* oil (Chhetri *et al*., 2008).

***Jatropha* spp**: Jatropha species have originated from Mexico and adjoining Central America. In India it is called by different names viz. *Ratan Jyot* , hedge castor, wild castor or *Junglee Arandi, Adavi Amudamu* and *Nepalam* in different regions and languages of India. In English it is called physic nut. From the Caribbean, *Jatropha spp* was probably introduced by Portuguese seafarers into India via the Cape Verde Islands and former Portuguese Guinea (now Guinea Bissau) (Malhotra *et al*., 2020). In Asia it is grown as a hedge or a fence and is commonly known as hedge castor or wild castor; but it is different from castor although they belong to the same family. *J. curcas* is a semi-evergreen shrub or a small tree reaching to a height of 6m.

The genus name *Jatropha* derives from the Greek *jatrós* (doctor), *trophé* (food), which implies medicinal uses. The plant is planted as a hedge (living fence) by farmers all over the world around homesteads, gardens and fields, because it is not browsed by animals. *Jatropha curcas* L. (JCL) seems to be the more productive of the other species of the genus in terms of seed yield, based on the evaluation carried out at the ICAR-Directorate of Oilseeds Research Hyderabad during 1985-1995.

Description: Jatropha is tropical season plant and cannot withstand frost.

It is an evergreen shrub tending towards a small tree habit with soft wood at times hollow in the centre. The plant sheds leaves once fruiting terminates. *J. curcas* is a tall bush or small tree (up to 5 m height) and belongs to the Euphorbiaceae family. The genus *Jatropha* contains approximately 170 known species. *Jatropha curcas* L., or physic nut, has thick glabrous branchlets. The tree has a straight trunk and gray or reddish bark, masked by large white patches. It has green leaves with a length and width of 6 to 15 cm, with 5 to 7 shallow lobes. the leaves are green to pale green, alternate to sub-opposite, and three to five-lobed with a spiral phyllotaxi, the leaves are arranged alternately.

Male and female flowers are produced on the same inflorescence, averaging 15 to 20 male flowers to each female flower. The inflorescence can be formed in the leaf axil. Plants occasionally present hermaphroditic flowers in very small number. Hence, *Jatropha* plant is monoecious with male (staminate) and female (pistillate) arranged on the same spike, as shown in Fig.1.

Capsules or fruits are produced in winter, or there may be several flushes during the year if soil moisture is adequate and temperatures are sufficiently high. Most fruit production is concentrated from midsummer to late October with variations in production peaks where some plants have two or three harvests and some produce continuously through the season.

Seeds are mature when the capsule changes colour from green to yellow. The seeds contain 25–40% oil by weight with around 20% saturated and 80% unsaturated fatty acids. In addition, seeds contain other chemical compounds, such as saccharose, raffinose, stachyose, glucose, fructose, galactose, and protein. The oil is largely made up of oleic and linoleic acids.

Furthermore, the plant also contains curcasin, arachidic, myristic, palmitic, and stearic acids and curcin.

Jatropha species have 2n=2x=22 chromosomes, corresponding to the diploid level. The chromosomes are small, from ca. 1 to 2 μm with median or sub-median centromeres (Dahmer *et al.*, 2009). The *Jatropha* genome was sequenced by *Kazusa DNA Research Institute*, Chiba Japan in October 2010 (Sato *et al.*, 2010).

Today the members of the species are being grown in almost all tropical and subtropical countries as protection hedges around homesteads, gardens and fields, since it is not browsed by animals.

Propagation: *Jatropha* may be propagated by seeds and /or stem cuttings. Propagation through seed (sexual propagation) leads to a lot of genetic variability in terms of growth, biomass, seed yield and oil content. Clonal techniques can help in overcoming these problems. Vegetative propagation has been achieved by stem techniques. Cuttings should be taken preferably from juvenile plants and treated with 200 micro gram per litre of IBA (rooting hormone) to ensure the highest level of rooting in stem cuttings. However, cuttings planted on soil strike root easily without use of hormones. Cuttings and air layers used for propagation result in shallow fibrous root system which does not offer good anchorage to the plant, which is a disadvantage under dryland conditions (Gadekar, 2006).

Processing: Seed extraction and processing generally needs specialized facilities.

Oil content varies from 28% to 30% and the extraction efficiency may be 80%, one hectare of plantation may be expected to yield 400 to 600 litres of oil if the crop is grown on average soil with good standard crop management (*Jatropha Curcas* Plant". *www.svlele.com).*

The Straight Vegetable Oil extracted from seeds is reported to fuel combustion engines or may be subjected to transesterification to produce biodiesel (Wikipedia).

Jatropha oil is not suitable for human consumption, as it induces strong vomiting and diarrhoea and could be toxic.

Evaluation of different species of *Jatropha*: A block of *Jatropha* species was established at the Directorate of Oilseeds Research Hyderabad with the collections from Gujarat, Orissa, former Andhra Pradesh including Telangana state and Karnataka. The species of *Jatropha* viz., *J. podagarica, J. multifidia, J. pandurifolia, J. curcus* and *J. gassipifolia, were studied.* Selections were made in *J. gossipifolia* for agronomic attributes such as productivity and maturity. A substantial degree of variability was observed for fatty acid composition of oil among different species of *Jatropha.* (Table 5)

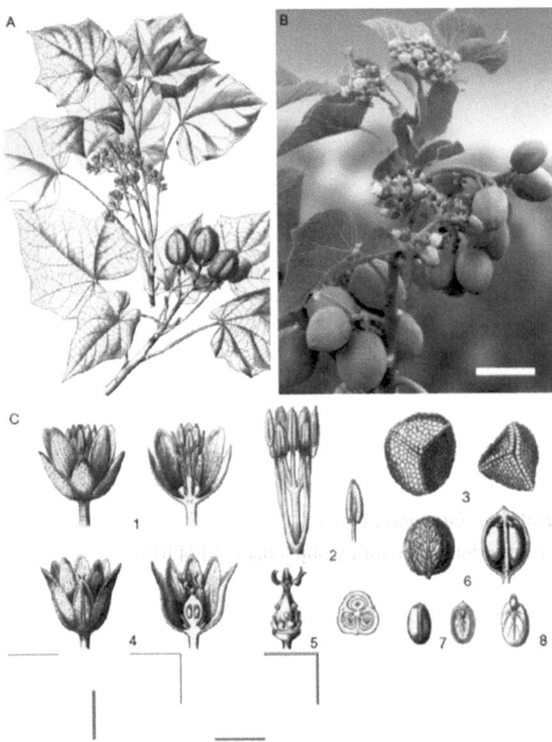

Fig. 1: Aerial Parts of Jatropha curcas Plant: [Source: Nicolas Carels (2013) - Academia.edu_ files\] A &B. Branch with leaves, flowers and capsules (fruits);; C: 1. Staminate (male) flower & it longitudinal section (LS) 2, Stamens showinganthers with fused filaments & a single anther with short filament, 3.Sections of anther ; 4. Pistillate (female) flower and its LS, 5.gynoecium with stigmas, & ovary and cross section of ovary, 6. Capsule (fruit) and its LS. 7. Seed and its LS; 8. Seed endosperm with cotyledon and embryonic structure

Table 5: Variability for fatty acid composition among different species of Jatropha (Prasad, 1994c)

Jatropha species	16:0 Plamitic Acid	18:0 Stearic acid	18:1 Oleic acid	18:2 Linoleic acid	18:3 α-Linolenic acid
J. curcas	16.3	4.2	38.8	29.0	4.2
J. gossipifolia	13.2	3.9	15.6	58.4	8.5
J. pondurifolia	7.8	4.4	13.5	68.2	1.9
J. podagarica	17.9	8.2	18.6	41.0	11.6
J. multifida	13.4	3.5	24.9	36.	15.1

A detailed analysis of growth parameters of the different species of Jatropha (Prasad 1994c) indicated that *J. gossipifolia* flowered and fruited precociously offering a better yield per plant as compared to *J. curcus* (Prasad, 1994c). Despite its better yield potential *J. gossipifolia* had a serious limitation of fruit shattering. Among the above species of *Jatropha,* the choice automatically falls on *J. curcus.*

Fig. 2: Leaf canopy with capsules (fruits) of *Jatropha curcas*
(CC BY-SA 2.5, https://commons.wiki- media.org/w/index.php?curid=1424405)

Fig. 3: *J. curcus* green fruits and seeds

Fig. 4: Well managed and fully grown bush of Jatropha curcas

Advantages of *Jatropha*: The Planning Commission, of India has given priority to Jatropha due to its crude oil, which has high free fatty acids content, more similar to fossil diesel in chemical composition and easy conversion into biodiesel with conversion efficiency is about 95%. (Malhotra *et al.*, 2020)

Jatropha has certain advantageous characteristic features. It is (i) best hedge plant, (ii) has less gestation period (three years), (iii) capable to grow and establish in various biotic and abiotic stress conditions, (iv) has high oil content (30-42% in seed), which has multiple uses as biodiesel, direct fuel, lubricant, medicine, besides other industrial uses. (v) *Jatropha curcas* crop is precocious and as such dry capsules may be harvested in 3 to 5 years. The dry capsules can be plucked with ease from 1 to 2 meters' height, cutting down the use of excess labour. Its leaves are not palatable to animals and as such no fencing is required to protect the crop. Dry capsules can be opened just by pressing by hand. Jatropha oil does not have much wax, which could be an advantage for manufacture of Biodiesel. (http://svlele.com/jatropha_why.htm) (vi) The by-products of biodiesel are also quite useful for industrial application such as glycerine and bio fertilizers. The residue is a good substrate for biogas production (Dhyani *et al.*, 2015)

Jatropha oil contains approximately 14% free FA. (Chhetri *et al.,* 2008). It was found that over 97% ester conversion was achieved after acid catalysed transesterification. The FA composition of biodiesel produced from *Jatropha* oil is presented in Table 6

Table 6: Fatty Acid (FA) analysis of Jatropha oil biodiesel. (Chhetri *et al.*, 2008)

FA	Content	Amount (%)	% as reported by Gubitz *et al.*, (1999)
Lauric acid	12:0	0.31	
Palmitic acid	16:0	13.38	14.1–15.3
Palmitoleic acid	16:1	0.88	0–1.3
Stearic acid	18:0	5.44	3.7–9.8
Oleic acid	18:1	45.79	34.3–45.8
Linoleic acid	18:2	32.27	29.0–44.2
Others		1.93	others
Total		100	

Jatropha is considered drought resistant. But the recent experiences with *Jatropha curcus* (Prasad, 2013) under rainfed conditions of Peninsular India and Deccan plateau indicated that *Jatropha curcas* could not withstand drought periods of more than 25 days and failed on drylands during monsoon season.

According to Nicolas Carels (2013) *Jatropha curcas* remains a semi-wild species and it is far from being an industrial crop. It should go through the domain of scientific integration towards complete domestication for profitable biodiesel production (Nicolas Carels, 2013).

The percentage of survival on degraded soils was significantly higher in *Pongamia pinnata*, than *Jatropha curucas.* This may be due to the fact that *P. pinnata* is a leguminous plant having a very deep root system of the depth of more than 10m and lateral spread of 9 m and the inherent N2 fixing symbiotic associations with 'rhizobia' which enables it to cope-up with adverse soil situations. *Pongamia. pinnata* showed better accumulation potential with regard to Cr, Cu and Mn than *J.curcas* at 95% confidence level (Singh, 2013), thereby demonstrating *Pongamia*'s efficacy in amelioration of polluted and toxic soils.

After the glass house studies, both *J. curcas* and *P. pinnta* were introduced in the fly ash dumps by Singh (2013) and nearby contaminated soil. The plantations were established at a spacing interval of 1 m x 1m. *P. pinnata* showed comparatively better survival rate in both fly ash and fly ash contaminated soil rather than *J. curcas* and showed 78% survival value for initial 6 months which gradually became constant to 70% after 18 months, respectively. *J. curcas* on the other hand showed an initial survival of 92.8% for 6 months which gradually fell to 44.6% after 24 months. (Singh, 2013)

Methyl esters from the biodiesel of *Pongamia* are non-corrosive and are produced at low pressure and low temperature conditions, whereas biodiesel produced from *jatropha* has corrosive effect on the piston-liner (Singh, 2013).

The seed propagation of *Jatropha,* which is the common method followed is the least successful method of propagation, resulting in lower survival rate of seedlings and poor uniformity of growth leading to variable productivity of the progeny (Brittaine and Lutaladio, 2010). Vegetative propagation of *Jatropha* by cuttings (Fig.5) is better than seed propagation since clones give more uniform productivity and potentially higher yields per hectare. Nevertheless, sufficient mass of cuttings of good plant materials may be difficult to obtain and costly to source. Lack of a tap-root system of clones results in poor soil anchorage and less capacity to extract water and nutrients and as such less suited to dryland agriculture with intercropping (Brittaine and Lutaladio, 2010)

The growth and development of *Jatropha* gets a setback under conditions of salinity and drought (Niu *et al.,* 2012; dos Anjos *et al.*, 2017) Gmunder *et al.,* (2012) observed that although *J. curcas* has the potential to grow rain-fed in water scarce areas, irrigation may be applied for increased yields, and consequently use relatively large amounts of water compared to other oil yielding tree species.. Such practice would also lead to relatively high stress on water resources. It is believed that optimal rainfall for its growth and development is between 1 000 and 1500 mm (FACT, 2007; Brittaine and Lutaladio, 2010)). Lama *et al.*, (2019) found that the highest reduction of mean yield (63 %) was recorded under severe drought stress to which *Jatropha* plants were subjected to. Also, stem dry biomass and root/shoot ratio were adversely affected by the interaction of drought stress.

Another drawback encountered in the case of *Jatropha* is with regard to the proportion of male and female flowers on its panicles. Most of *J. curcas* strains had typical inflorescences with separate sexes (monoecious); meanwhile, some were atypical (gynoecious, androecious, andromonoecious) (Anaya *et al.*,2016). By and large *Jatropha* being monoecious, it bears male and female flowers on the same panicle. With rising temperature and drought which are common under conditions of dryland agriculture, the proportion of male flowers get augmented thereby pulling down the seed yields. Also as observed by Anaya *et al.,* (2016) *Jatropha* often exhibits unstable sex phenotype, which reflects adversely on its performance.

Fig. 5: Rooted cutting of Jatropa curas following treatment with IBA200µg/l
(*Source*: By Kumarsukhadeo at English Wikipedia)

Ranga Rao (2006) studied and analysed Jatropha production in different regions in India and concluded that notwithstanding the reported versatility of Jatropha as a promising oilseed yielding tree for biodiesel production, any plans of its large scale commercial production are unlikely to succeed without back up of appropriate relevant and situation specific tested planting materials and agro-production techniques. He further stated that in the face of the limitations associated with the current nondescript planting materials and half-baked and assembled production practices and information disseminated, any hasty expansion of Jatropha production might cause irreparable losses and damage to the farmers' economy.

Fig. 6: A single plant of Jtropha curcas from a five years' old plantation on a dryland near Hyderabad, India.

According to Jongschaap *et al.,* (2007), at lower rainfall levels of semi-arid tropics of India, the precipitation is not sufficient to provide the required amount of PET (potential evapotranspiration) of *Jatropha curcas* and most probably the soil characteristics prevent the efficient storage of water and release of available water in respect of *Jatropha.* (Fig.6). Jongschaap *et al*., (2007) observed that the positive claims of high oil yields of *Jatropha curcas* seem to have emerged from incorrect combinations of unrelated observations often based on singular and older trees of *J. curcas*. Oil yields per hectare in respect of *Jatropha* are over estimated and extrapolation of such measurements ignores the growth reduction in such systems occurring from the competition of natural resources such as radiation, water and nutrients.Concepts for reclamation of marginal soils by *Jatropha curcas* are reported (Spaan *et al*.,2004) although its oil production is not proven to be economically subsistent level (Francis *et al.,* 2005). In monoculture, *Jatropha curcas* may lead to soil exhaustion if not produced in a sustainable way (Jongschaap *et al.,* 2007).

Toxicity of *J. curcus* is based on its inherent components viz. phorbol esters, curcains, trypsin inhibitors and other. These are present in considerable amounts in all plant parts and components including oil. This makes the detoxification a complicated process, which has remained a laboratory process and does not seem to be suited for application for a small scale unit and local application (Jongschaap *et al.,* 2007).

Dehue and Hettinga (2008) propose the changes to the production system and management technology of *Jatropha curcus* to achieve a production of 6 tons/ha in the place of current >3 tons/ha through the adoption of following.

- Applying 125 kg/ha/yr of Urea currently, results in direct emissions and, much larger, indirect soil emissions and lead to a significant worsening of the Green House Gases (GHG)-performance to 60% GHG-emission saving.
- Applying 12.5 kg/ha/yr of Triple Super Phosphate (TSP) and Gypsum (*Quantity not indicated*) results in almost no increase in GHG-intensity.
- Applying 1200 kg/ha of Lime (once at plantation establishment) results is almost no increase in GHG-intensity.

They Conclude that by applying lime or TSP there is a net beneficial effect of higher yields. If Urea is applied, this beneficial yield-effect is outbalanced by resulting GHG-emissions from Urea production and soil emissions.

In a significant implication for any country's biofuel policy, a specialized arm of the United Nations, has warned (Anonymous, 2010 A) that the developing countries should not buy blindly into the 'Jatropha' for biodiesel argument. Warning against the hype and half-truths around *Jatropha curcas*, an oil seed plant touted as a major potential source of biofuels, the United Nations Food and Agriculture Organization has warned in a special report that yields need to improve significantly for the crop to give an adequate return.

"Although there have been increasing investments and policy decisions concerning the use of Jatropha as an oil crop, they have been based on little evidence-based information," the report said, adding that identifying the true potential of Jatropha requires separating the evidence from the hyped claims and half-truths."

The report comes two weeks after two researchers at Belgium's University of Leuven said that the crop requires more water than had been thought, and was best suited for small-scale farming in remote areas, where alternative fuel supplies are erratic and expensive.

The cautioning report is also a pointer to several giant corporate houses worldwide such as GM that have invested in the crop. US automobile giant GM was one of the companies that invested in Jatropha following a surge of interest five years ago in the potential for biofuels.

In April, it announced a five-year programme with the US Department of Energy and India's Central Salt and Marine Chemicals Research Institute to demonstrate the oilseed's commercial potential. A good chunk of India's biofuel programme rests on the success of Jatropha as a green diesel alternative, but the FAO report has said that it is unrealistic to expect Jatropha to substitute oil imports significantly in the developing countries where it is grown.

The FAO has also disproved the argument that growing Jatropha utilizes marginal lands effectively. The level of economic returns needed to secure private sector investment "may not be attainable on degraded land", FAO said, noting considerably better gross margins which can be gained on sugar cane and oil palm plantations. The UN organization, however, does not rule out the oilseed completely, but has flagged an urgent need to multiply the yields.

The FAO acknowledged that while there was limited information available on private sector work in Jatropha, "it may be assumed that advances have been made".

A detailed analysis carried out by Kant and Wu (2011) on Jatropha production, led to the release of the following findings on Jatropha curcas:

(i). GM Biofuel crop *Jatropha curcas* under-performs in Madgascar: In the United Kingdom, GEM Biofuels reports that their Jatropha plantations have not fared as well as hoped. An internal review has made it clear that despite having planted 55,737 hectares of Jatropha in Madagascar between 2007 and 2009, a lack of resources has resulted in significantly less success than had been hoped. Low intervention and maintenance following planting has resulted in a lower than anticipated number of plants reaching maturity and producing oil-bearing seeds. The Board has commissioned external agronomic consultants to conduct a full review of both the Company's plantations in Madagascar and to assess the potential of its land bank for other crops.

(ii). Greenleaf Global, closed down following Government investigation: Greenleaf Global plc ran a Jatropha Plantation in West Africa and misled investors by promising high returns on investment while knowing this was not true. Greenleaf Togo acted as the main leaseholder of the Jatropha plantation while Greenleaf UK acted as an agent in marketing the scheme to the public. The amounts taken from investors amounted to £8.2million. Plots of land in Togo were leased for the cultivation of Jatropha trees from seedlings. Greenleaf UK had also recently began marketing a scheme in Ghana for the commercial production of maize. Investors were told that they could expect to receive a return of up to 20% in the first year. These expected rates of return were not evidence based. Greenleaf UK made, in uniform distribution to various investors, on-account payments at a time when no commercial or other harvest had taken place in respect of the plantation.

Reasons for failure of Jatropha

- Lack of planting materials with certain yield potential
- Peculiar floral biology with predominance of male flowers under high temperature conditions causing drastic decline in yield.
- Needs nutrient application in the form of chemical fertilizers.
- Can't withstand prolonged spells of acute drought even on heavier soils of high moisture retention. Needs irrigation to get reasonable level of yields.
- Prone to attack of wide range of pests and diseases
- Fragile stems leading to canopy damage under conditions of high wind velocity common under semi-arid and arid conditions
- Higher labour requirement for pruning and harvesting
- Low yields per unit of moisture and nutrients and higher total cost of production
- Not sustainable for production on marginal lands
- Not really carbon negative &
- Seed cake loaded with a dangerous toxin viz., curcin

Fig. 7: A well maintained and irrigated block of Jatropha curcas

The well-known defects and drawbacks of *Jatropha curcas* are poor seed viability, low germination, unstable pistillate flower expression, unstable and low seed yield levels and scanty and delayed rooting of seedlings etc. The plants propagated by cuttings show a lower longevity and lack of drought and disease resistance. In the context of achieving a faster rate of the progress in the genetic improvement of Jatropha, Plant biotechnology serves as a powerful tool.

New hope for *Jatropha* Improvement: Thanks to advances in molecular genetics and DNA sequencing technology, the San Diego startup SGBiofuels (SGB) has, in a few years claims to have succeeded in domesticating *Jatropha*, a process that once took decades (Anonymous, 2008) (Fig.8).

SGB is growing hybrid strains of the plant that produce biofuel in quantities that it says are competitive with petroleum priced at $99 a barrel.SGB's technology allows its scientists to identify potentially productive hybrids in the laboratory at the molecular level before the plants are crossbred. Given that this crop has somewhat of a checkered past, ultimately getting growers to plant it is going to be the key hurdle," says Michael Cox BSE

Fig. 8: Scientists of SG Biofuels with Jatropha plants showing green capsules

In March 2010, SG Biofuels launched its first JMax100 cultivar, a proprietary cultivar of *Jatropha* optimized for growing conditions in Guatemala with yields averaging approximately 100 percent greater than existing varieties

Read more at:

(1) http://economictimes.indiatimes.com/articleshow/27928278.cms?utm_source=contentofinterest&utm_medium=text&utm_campaign=cppst

(2) A comparative study of biochemical traits and molecular marker s for assessment of genetic relationships between *Jatropha curcas* L. germplasm from different countries *Plant Science*, 24 March 2009

Three protocols have been developed that will serve as an ideal system for future transgenic research (Mukherjee and Jha, 2009). Also, successful *in vitro* micro-propagation in *J. curcas*, has been achieved by Mukherjee and Jha (2009), using different explants from nodal meristem culture from field grown plant. Complete plant regeneration of *J. curcus* through somatic embryogenesis has since been accomplished. Plant propagation by somatic embryogenesis not only helps to obtain a large number of plants year round, but also can act as a powerful tool for further transgenic research. In this context, plant production from excised immature zygotic embryos has also been achieved (Mukerjee and Jha, 2009).

Fig. 9: A four to five years' old plantation of Jatropha curcas on a marginal dryland at Kapurthala, Punjab, India (2013)

Medicinal Value of *Jatropha curcas*

Reddy Prasad *et al.,* (2012) reported that *J. curcas* has been used in traditional medicine to cure various infections. Researchers had isolated and characterized numerous biologically active compounds from all parts of this plant. The plant typically contains mixtures of different chemical compounds that may act individually, additively or in synergy to improve health. In addition, the mechanisms of action of these active compounds have been studied in relation to the applications in traditional medicine. Before exploiting any plant for medicinal application, it is crucial to have complete information about the medicinal uses of each part of the plant. The medicinal uses of the leaves, fruit, seed, stem bark, branches, twigs, latex and root of *J. curcas* are discussed by Reddy Prasad *et al.*, (2012). It is suggested that much more research is required to develop herbal medicinal value using modern science and technology. The role of *J. curcas* in medicinal uses should be taken into consideration as it shows promising potentials in the pharmaceutical field. Commercializing on the medicinal product derived from *J. curcas* may turn out to be more profitable than using Jatropha as fuel substitution.

Table 6: Bio chemicals isolated from different parts of J. curcas plant.
(Reddy Prasad *et al.*, 2012)

Plant parts	Beneficial Chemical compounds isolated
Leaves	Flavanoids, apigenin, vitexin, isovitexin, sterol stigmasterol. β-D-sitosterol, β-D-glucoside, sapogenins, alkaloids, triterpenae alcohol and 1-triacontanol • Also contain a dimer of a tripene alcohol •A complex of 5-hydrxypyrrolidin-2-one and pyrimidine-2,4-dione isolated from the leave • Stigmast-5-en-3β, 7 β-diol, stigmast-5-en-3β, 7α-diol, cholest-5-en-3β, 7β-diol, cholest-5-en-3fl, campesterol and 7-keto-β-sitosterol
Stem bark	• Saponins, steroids, tannins, glycosides, alkaloids and flavonoids • B-amyrin, β-sitosterol and taraxero
Latex	• Tannins, saponins, wax and resin • Curcacycline A (a cyclic octapeptide) • Curcacycline B (a cyclic nonapeptide) • Curcain (a protease)
Seeds	• Curcin (a lectin) • Phorbol esters • Esterases (JEA and JEB) and lipase (JL)
Roots	• Steroids, alkaloids, saponins • Jatropholone A and B (diterpenoids), Jatropholol (diterpenoid) • β-Sitosterol and its β-d-glucoside, marmesin, propacin, the curculathyranes A and B and the curcusones A–D. the coumarin tomentin, the coumarino-lignan jatrophin as well as taraxerol

Fig. 10: Jatropha Oil and green and dry capsules (Image:Shutterstock)

Jatropha offers wide range of other uses, particularly as a medicinal plant. Several healing and remedial properties of Jatropha have been on record.

Some of the medicinal uses of *Jaropha* are as under

Fresh latex of Jatropha can cure the wounds and skin diseases.

The decoction of Jatropha leaves can be used as mouth wash for curing gum diseases.

The leaf decoction can be used as skin wash to cure scabies.

The paste of leaves is used for curing abscess and

Heated leaves bound in a cloth to be used as poultice or heated leaves used as bandage for enhancing breast health and milk secretion.

(https://herbpathy.com/Uses-and-Benefits-of-Jatropha-Cid1144)

According to https://drhealthbenefits.com/herbal/herbal-plant/ benefits- of-jatropha-leaves, following health benefits of Jatropha are reported.

Boiled leaves with cups of water, if taken in orally regularly, the herbal drink would cure rheumatic disease.

Consuming Jatropha leaves raw will be much better and fast to cure constipation.

The slices of jatropha leaves are covered children's forehead and left for a while to cure fever.

The oil extract called KUFI in Hausa has been used a s rubefacient for rheumatic conditions and curing itches and parasitic skin diseases. *Jatropha curcus* has been used in the woollen industry in England and in the preparation of Turkey Red.

(https://nimedhealth.com.ng/2019/08/24/the-medicinal-uses-of-jatropha-curcas/)

Non-toxic *Jatropha*: It is reported that *Jatropha* seeds are edible once the embryo is removed (Levingston and Zamora, 2006) Also non-toxic variety of Jatropha have been reported from Mexico. Birgit Schmook (cited by Henning) reports that the seeds in the zone around Misantla, Veracruz of Mexico are very appreciated by the population as food once they have been boiled and roasted. Such genetic material could be a potential source of oil for human consumption, and such seed cake can be a good protein source for humans as well as for livestock. (Levingston and Zamora, 2006).Ochse (cited by Duke) reports, "the young leaves may be safely eaten, steamed or stewed." They are favoured for cooking with goat meat, said to counteract the peculiar smell. (Levingston and Zamora, 2006) The availability of edible non-toxic *J. curcas* without toxic phorbol esters content has been demonstrated (Valdes-Rodríguez, *et al.*, 2013).

Cooked nuts of non-toxic Jatropha are eaten in certain parts of Mexico. *Xuta, chuta, aishte* or *piñón manso* (among others) are some of the names given in Mexico to edible non-toxic *Jatropha curcas* (Gomez-Pompa *et al.,*2009).

Biotic Stresses of Jatropha: Some of the important pests and diseases of *Jatropha* which cause considerable damage and yield losses are presented below. Appropriate management strategies to control the under mentioned biotic stresses have to be put in place to avoid serious yield losses.

Pests and diseases of *Jatropha* (https://jatropha.pro/diseases/)

Insect pests

Polyphagotarsonemus latus -Spider mite (The leaves become leathery and leaf tips get dried quickly leading to death of the pant)

Scutellaria nobilis - sap sucking plant bug

Ferrisia virgata-Mealy bug

Salubriya–Leaf webber

Aphtonia dilutipes– Beetle feeding on leaves

Termites – cause serious damage

Diseases

***Lasiodiplodia* spp** - causes rotting and dieback and stem-end rot.

Colletotrichumssp – Anthracnose disease

Macrophomiaspp – Collar rot disease

Fusarium – wilt disease

A comparative account of *Jatropha* and *Pongamia* in respect of other parameters (VAYUGRID, 2013) is presented in Table 4.

Among the oil bearing trees tested for consistent productivity, *Pongamia pinnata* (karanj) proved to the most promising (Prasad *et al*., 1994; Prasad, 2009).

Table 8: Comparative Account of Pongamia and Jatropha (VAYUGRID 2014)

S.No.	Factor	Jatropha curcas	Pongamia(Elite genotype)
1	Minimum Rainfall (mm) needed	600-2000	250-500
2	Oil Yield in 3yrs (t/ha)	3	10
3	Oil Yield in 10 years (t/ha)	9.5	18.0
4	Oil Yield in 15 years (t/ha)	10.0	28.0
5	Mechanized harvest	Not possible	Possible
6	Presence of toxin	Yes. Curcin & Phorbol	NIL
7	Drought Resistance	Fails with +60 days drought	Highly drought resistant. No failures.
8	Depth of root system	Less than 2.0 m	+10 m
9	Labour requirement	Very high	low
10	Planting material	Non descriptive seedlings	Elite grafted saplings
11	Pests and diseases	Wide range causing yield losses	Relatively Negligible. Very few biotic stresses with no yield loss.
12	Tolerance to problem soils	Not Tolerant	Resists salinity, water logging, alkalinity and comes up rocky soil
13	Carbon credits	No	Yes sequesters 20 t of C/ha
14	Nitrogen fixation	No	Yes (up to 1100 kg/ha)
15	Fertilizer use	Yes to boost yield	No need
16	Effect on soil	Neutral for soil fertility. Positive for soil structure.	Enhances soil organic matter, soil fertility, soil structure and soil biology.
17	Floral biology & yield	Male and female flowers on the same plant. Higher proportion of male flowers reduces yield	Uniformly all bisexual flowers, Stability of yield performance.

The advantages of *Jatropha* over *Pongamia* are only with regard to its precocity and oil quality. Otherwise in respect of all other features including yield and resistance to biotic and abiotic factors, *Pongmia pinnata* is a better choice.

While there exists a wide variability and choice, the performance of the species of oil bearing trees varies depending upon the agro-ecological factors prevailing in different zones. Within each species too there exists perceptible genetic variability for wide range of attributes including seed yield and oil content. It is therefore necessary that the tree species and their varieties have to be carefully evaluated taking into consideration above factors.

Pongamia pinnata L.: (Karanj)

Pongamia pinnata belongs to the family Fabaceae (Papilionaceae) with chromosome number 2n=2x=22. It is also called *Derris indica, Millettia pinnata* and *Pongamia glabra.* It is a medium sized evergreen oil seed bearing tree with a spreading crown and a short bole. The tree is known for its shade and yield of non-edible oil of medicinal value and is grown as ornamental tree.

The tree is commonly known as Indian beech (English), *Karanj* (Hindi), *Kanuga / Ganuga* (Telugu), *Pongam* (Tamil), *Honge* (Kannada) and *Panigrahi* (Odissi). The tree is native to eastern and tropical Asia. India possesses a wide genetic diversity for *Pongamia.*

Pongamia is a medium-sized tree that generally attains a height of about 8.0 to 12.0 m and a trunk diameter of more than 50-75 cm. The bark is thin, grey to greyish-brown and yellow on the inside. The alternate, compound pinnate leaves of the tree consist of 5 or 7 leaflets, which are arranged in 2 or 3 pairs, and a single terminal leaflet. Pods are elliptical, 3-6 cm long and 2-3 cm wide, thick walled, and usually contain a single seed. Seeds are 10-20 cm long, flat, oblong, and light brown in colour. Seeds yield a non-edible oil and a cake is rich in protein.

Pongamia thrives in areas having an annual rainfall ranging from 400 to 2500 mm. In its natural habitat, the maximum temperature exceeds 38°C and the minimum can be as low as 1°C.The root system of *Pongamia* is thick and well developed deep tap-root of 10m depth and a spread of later roots to around 9m. *Pongamia pinnata* is known for its ability to fix atmospheric Nitrogen in association with soil *Rhizobia.* Mature tree apart from its inherent drought resistance, can withstand water logging and slight frost. This species grows at elevations of 1200 m, but in the Himalayan foothills it is not found above 600 m. It can grow on most soil types ranging from stony to sandy to clayey, including dry sands and saline soils. (Duke, 1983; Daniel 1997; Orwa *et al.*, 2009). The natural distribution of this species is along coasts and river banks in India and Myanmar.

Pongamiapinnata has a broad distribution across Asia including the Indian subcontinent, Africa, Pacific and America including the Caribbean zone. Dense tree populations of *Pongamia* occur in India, Bangladesh, Pakistan, Nepal, Sri Lanka, Myanmar, Thailand, Vietnam, Malaysia, Brunei, Java,, Sumatra, Singapore, Philippines, China (Fujian, Guangdong), Taiwan and Japan (Kyushu, Ryukyu Islands).

Acreage and Production in India: Published reports on the acreage and production of *Pongamiapinnta* in India are too scanty to mention. It is estimated that the annual production of *Pongamia* kernels in India is of the order of 130,000 tons (Rabab and Dixit- Madhurima1997).

An effort was made (Anonymous, 2012A) to gather the area under *Pongamia* in India through discussions with various organizations pertaining to Forestry, Agroforestry and Dryland Agriculture.The discussions have revealed that the area under *Pongamia* in India is of the order of 500,000 ha in the form of unorganized tree stands, with an estimated domesticated tree population of 37.5 million all along the country's rural belt. Assuming an average minimum yield of 10 kg of pods or 4 kg of kernels per tree, the total estimated pod production should be of the order of 370,000 tons resulting in an approximate total kernel / seed production of 148,000 to 150,000 tons per annum. It may be noted that the trees are not growing under any system of tree management; but are totally neglected without any care. (Sources: Agro-forestry Department of ANGRAU 2012, Hyderabad and Andhra Pradesh & Telangana Forest departments, Hyderabad 2012 & ICAR Central Research Institute for Dryland Agriculture, Hyderabad 2012)

Taxonomy of *Pongamia pinnata*

Kingdom:	Plantae
(unranked):	Angiosperms
(unranked):	Eudicots
(unranked):	Rosids
Order:	Fabales
Family:	Fabaceae
Genus:	Millettia
Species:	M. pinnata
Binomial name	
Millettia pinnata	
(L.) Panigrahi	
Pongamia pinnata (L.) Pierre	

Potential of *Pongamia* apart from oil yield:*Pongamia pinnata* is an important nitrogen-fixing evergreen tree with multifarious uses. Sangwan *et al.,* (2010) have reported wide ranging uses uses of *Pongamia pinnata* in pharmacology. It was suggested that the tree possesses anti-inflammatory, anti-diarrhoeal, anti-oxidant and anti-hyper ammonemic, anti-ulcer, anti-hyper glycemic and anti-lipid peroxidative properties in addition to the fact that the oil of the tree forms a good feed-stock for bio-energy. According to Yadav *et al* (2011)the roots of *Pongamia pinnata* are good for not only controlling insect pests effectively, but also for cleaning foul ulcers, cleaning teeth, strengthening gums and gonorrhoea. The root paste is used for local application in scrofulous enlargement. The fresh bark of *Pongamia pinnata* is sweet and mucilaginous to taste, soon becomes bitter and acrid. It is antihelmintic and useful in beri-beri, ophthalmology, dermatopathy, vaginopathy, and ulcers. Pongamia leaf, oil and seed extracts form good tools in crop protection programmes due to presence of insect toxic product karanjin (Kumar *et al.,* 2006).

It is indigenous to the Indian subcontinent and South-East Asia and it has recently been recognised as a viable feedstock of oil for the growing biofuel industry

(Karmee & Chadha 2005). The source of its oil at present is naturally growing and non-descript tree complex and very few young plantations (Divakara and Das 2011). The leaves and twigs of Pongamia are incorporated as green manure in rice fields of South India.

Studies carried out by Al Muqarrabun*et al.*, (2013) indicated that several different classes of flavonoid derivatives, such as flavones, flavans, and chalcones, and several types of compounds including terpenes, steroid, and fatty acids have been isolated from all parts of this plant. The pharmacological studies revealed that various types of preparations, extracts, and single compounds of this species exhibited a broad spectrum of biological activities such as antioxidant, antimicrobial, anti-inflammatory, and anti-diabetic activities. The results of several toxicity studies indicated (Al Muqarrabun *et al* 2013) that extracts and single compounds isolated from this species did not show any significant toxicity and did not cause abnormality on some rats' organs. Thus, this plant has a potential to be used as an effective therapeutic remedy due to its low toxicity towards mammalian cells.

Pongamia has been reported to be efficient in remediating problem soils containing heavy metals and other pollutants that impede plant growth (Prasad 2007, Tulod *et al.*, 2012 ; Kumar *et al.,* 2017; Prasad, 2019a). The phytoremediation potential of *Karanj* has been examined in few comparative studies; in pot experiments, on coal mine spoil, fly-ash amended soils and on landfill wastes (Pandey 2017).

Fig. 11: A part of canopy of Pongamia pinnata showing green pods

Growth and Development: *Pongamia* is a fast growing, medium sized evergreen tree attaining a height of 7—10 m and Stem diameter of 50 -80 cm many times with drooping canopy. The tree at times can grow to a height of even 20-25 m with a trunk diameter of 60cm. It has smooth grey-brown bark with vertical fissuring. Leaves compound, pinnate and alternate. *Pongamia* tree has the typical feature of producing sprouts from the base of the trunk and more particularly root-suckers. Mature leaves are glossy and dark green above and pale below. New leaves are pinkish-red. Flowers are white, pink or lavender pea-like blossom. Bloom occurs in late spring/early summer. Seeds are 1.5 cm long, light brown, oval and contained in clusters of brown, eye-shaped pods. The tree reproduces from viable seeds but asexual root suckers too grow into trees. Yield ranges from 9 to 90 kg seed/tree. The seeds possess oil content ranging from 25 to 40%. A major component of the seed oil is Oleic Acid, which ranges from 45 to 60%.

Floral Biology: Flowers are fragrant, white to pinkish, paired along rachis in axillary, pendent, long racemes or panicles. Calyx is cup-shaped. Tree flowers in 4th to 5th year. The flowers are paplinaceous with one Standard, two wings and keels. Stamens are free at the base, but joined with others into a closed tube. Ovary is short-stalked, pubescent with generally two ovules but rarely three. Style is filiform with upper half incurved and glabrous. Stigma is small, terminal, knob shaped white and papillate that remains receptive during 06.00 to 11.30 AM following flower opening. Completion of anthesis is indicated by the closure of the flower. *Pongamia* is dry season bloomer and produces flowers with copious amount of pollen and additionally secrets nectar which form droplets on wing and keel petals that attract bees (Raju and Rao 2006). Pollination in Pongamia is effected by several species of bees of which *Apis dorsata, A. millifera, A. cernaindica, Amegilla spp, Megachile* and *Xylocorpa spp* are important.There is a substantial out-crossing in *Pongamia* due to insect pollination. Pollen viability is of the order of 83%. After 72 hours of pollination, pollen tubes are found to traverse through stylar tissue. Visit of the honey bee viz. *Apis mellifera* is reported to increase seed yields (Arpiwi *et al.,* 2014) . Setting up of bee hives in Pongamia plantation for honey production could be a practical proposition (Daisy *et al.,* 2001).

Pods and Seeds: Pods are short stalked, oblique to oblong, flat, smooth, thickly leathery to sub-woody and indehiscent. Pods are normally 1-seeded; but occasionally contains two seeded (Fig.18). Seeds are thick, reniform and viable for approximately one year and contain 28 to 40% of oil. Seeds germinate in 10 to 16 days after sowing. Pod shells of *Pongamia* form a good material for making briquettes, which could be an ideal kitchen fuel (VAYUGRID 2013) in the rural areas.

Wood: Wood is beautifully grained and medium to coarse textured, but is not durable. Therefore, it is used as fuel-wood and cheap timber.

The leaves are not readily eaten by animals, but it has some fodder value in dry areas. Leaves are also used as insect repellent in stored grains. The oilcake has use as poultry feed, as manure with insecticidal and nematicidal value. The cake can also be used to produce biogas.

Seed Oil: *Pongamia* oil is extracted from the seeds by expeller pressing, cold pressing, or solvent extraction (Berk, 2013). The oil is yellowish-orange to brown in colour. Pongamia oil is known for its insect repellent properties (Pavela and Herda 2007). However, theoil can be used to produce bio-diesel through the process of trans-esterification. The average seed oil content ranges from 25-40%, but higher percentages are claimed in elite genotypes. The yield of methyl esters from the oil under the optimal conditions was 97–98%. The oil is thick and yellow-orange to brown in colour. Traditionally, besides the cooking and lighting uses in rural areas, it was used as a lubricant, water-paint binder, pesticide, and in soap making and tanning industries. The oil is known to have value in folk medicine for the treatment of rheumatism as well as human and animal skin diseases. Seed oil is also used in scabies, leprosy, piles, ulcers, chronic fever, lever pain and lumbago. (Yadav *et al.*,2011).

Straight vegetable oil too can be used as source of energy, particularly to run the village pump-sets, tractors and old heavy vehicles. The cake left out after extraction of oil forms valuable organic manure. The seed also contains protein (>17%) of good quality. The seed cake can be used to produce bio-gas.

Pongam oil is used in tanning industry for dressing of leathers. In leather industry it is used for making fat liquors. After neutralization, the oil is used as lubricant for heavy lathes, chains, bearings of small gas engines, enclosed gears and heavy engines.

A study by RANWA (Research and Action in Natural Wealth Administration) says the Pune Municipal Transport could run its 800-bus fleet on karanja oil if the tree is planted on 68.6 sq. km. (Annonymous 2002) The karanj oil has been used in 7.5 KVA gensets and has electrified three villages in Adilabad district of Andhra Pradesh and in six villages in Jadol area of Rajasthan (Annonymous 2002). The utility of the seed oil of *Pongamia* has been well established (Shrinivasa 2001; Vivek and Gupta 2004). Karikalan and Chandrasekaran, (2015) studied the fuel characteristics of vegetable oils and found karanja oil can be used as an efficient and cheap raw material for synthesis of Biodiesel. The various fuel properties of karaja oil in relation to neem oil and diesel are given in Tables 8 and 9.

Table 9: Comparison of the Properties of Pongamia oil and Neem Oil (Arvind and Bekal, 2018)

Properties	Pongamia oil	Neem oil
Acid value	1-1.5 mg of KOHgm	-
Viscosity	44 mm: sec	37.42 mm- sec
Flash point	19S°C	>200eC
Fire point	210-212°C	-
Cloud point	22°C	20eC

Properties	Pongamia oil	Neem oil
Iodine value	S2-S4	S4-94
Density	925 kg m-5	919 kg m:'
Calorific Value	34.2 MJ kg	29.97 MJ kg

Arvind and Bekal (2018) reported that fire power is found to be better for *Pongamia* oil as compared to other vegetable oils including Neem oil (Table 9)

Fig. 12: Pongamia cake

Pongamia has a production potential of 135,500 metric tons per year in India, with 30-40% content of seed oil.A single average tree gives yield 9-90 kg seed per year, indicating a yield potential of 900-9000 kg seed/ha. It is possible to extract 25% of oil using a mechanical expeller (Arvind and Bekal, 2018)

The Pongamia oil is reported to exhibit following fatty acid profile (Anonymous 2012B)

Palmitic acid: 3.7 to 7.9 %

Stearic acid: 2.4 -8.9%

Oleic acid: 44.5 – 71.3%

Linoleic acid: 10.8 -18.3%

Linolenic acid: 2.6%

Arachidic acid: 2.2 -4.7%

Eicisenoic acid: 9.5 – 12.4%

Behenic acid: 4.2 – 5.3% and

Lignoceric acid: 1.1 -3.5%

***Pongamia* cake:** The cake is not used as a source of human or animal nutrition due to the presence of some irritants and anti-nutritional substances such as karnjin, saponins and phytates, although not being toxic. Nevertheless, it has been conclusively demonstrated by CFTRI that protein of *Pongamia* cake is comparable to that of soybean in terms of quality following removal of the above mentioned irritants through simple acid hydrolysis.

Feeding trials carried out by Soren and Sastry (2009) on male lambs with processed karanj cake on digestibility, nutrient balances and its effects on microbial crude protein synthesis, indicated that it is comparable to soybean cake in terms of nutritional value. It is also concluded that solvent extracted karanj cake after water washing could replace 50% of soybean meal nitrogen in protein supplementation (Soren *et al.*, 2009) in the feed for lambs.

The karanj seed cake is also recommended to be used as an alternative protein source in animal feed formulation after it has been properly detoxified (Babitha *et al.*, 2007).

Raj *et al.*, (2016) found that detoxified neem cake and detoxified karanja cake can be included in total mixed rations of medium producing dairy cattle (yielding 5-8 litres of milk per day) replacing standard soybean meal without adversely affecting milk composition and milk production efficiency.

It is worth attempting converting Pongamia seed cake into animal feed in view of its promising nutritional qualities.

Pongamia cake could be used as an effective organic fertilizer.A perceptible increase in cotton yield has been reported due to application of 100% Pongamia cake, which is higher as compared to the application of 100% of inorganic fertilizers (Osman *et al.,* 2005) (Table 10).

Table 10: Plant Nutrient Composition of Pongamia Cake (Osman *et al.*,2005)

Plant Nutrent	Content
Nitrogen %	4.28
Phosphorous%	0.4
Potassium%	0.74
Calcium%	0.25
Magnesium%	0.17
Zinc (ppm)	59
Iron (ppm)	1000
Copper (ppm)	22
Manganese (ppm)	74
Boron (ppm)	19
Sulphur (ppm)	1894

Karanjin & Pongamol: Pongamia produces natural chemical compounds, such as karanjin and pongamol, which impart bitter taste to the foliage and other parts of the tree. These compounds deter consumption of leaves by herbivores and insect infestations, resulting in lower management costs and enhanced sustainability. In addition, these compounds are valued for use in crop sprays, cosmetics, and sunscreen products.Pongamol and karanjin isolated from the pods of *Pongamia pinnata* exhibited significant antihyperglycemic activity in Streptozotocin-induced diabetic rats and type 2 diabetic db/db mice and protein tyrosine phosphatase-1B may be the possible target for their activity (Vismaya *et al*., 2010a).

(i) **Karanjin,** a white crystalline powder is furano-flavanol (3-methoxy furano-2,3,7,8-flavone) in quantities of 2% by seed weight and 4-5% by weight of oil.

The extraction of karanjin from seeds of Pongamia has been reported by Vismaya *et al.,* (2010)

Karanjin is an acaricide and insecticide. Acaricide means a product that is used to kill mites (Acarina). Karanjin is used as bio pesticide / bio insecticide and is reported to have nitrification inhibitory properties (Majumdar *et al.,* 2004). The resistance of Pongamia trees to termites and wide range of other pests is due to karajin and pongamol in the plant system. Karanjin could be used as an efficient pesticide / pest –repellent in organic agricultural programmes where chemical pest control is not allowed. (Lale and Kulkarni 2010).

Vismaya *et al*., (2010A) reported the gastro-protective properties of karanjin in addition to its anti-oxidant nature, there by suggesting its use in medicine.

Saini *et al*., (2017) isolated *Karanjin* from *Pongamia pinnata* (L.) pierre and *embelin* from *Embelia ribes* Burm.f. and their purity was confirmed by ultraviolet spectrophotometric and Thin layer chromatography based study. Anti-Alzheimer's activity of isolated compounds were evaluated through elevated plus maze and Morris water maze model on Swiss albino mice. Both isolated compounds and standard significantly reversed the amnesia induced by diazepam and improved learning and memory of mice in dose and time dependent manner

(ii)Pongamol is 1-(4-Methoxy-5-benzofuranyl)-3-phenyl-1,3-propanedione. It is a pale yellow to cream coloured powder. Extraction of karanjin and pongamol from Pongamia oil by reverse phase HPLC has been reported by Gore and Satyamoorthy (2000). The quantity of Pongamol in Pongamia oil is much lesser than that of karanjin.

In view of its low recovery it is stated that the cost of Pongamol is always much higher than that of Karanjin.

The University of Queensland, node of the Australian Research Council Centre for Excellence in Legume Research, under the directorship of Professor Peter Gresshoff, in conjunction with Pacific Renewable Energy are currently working on *Pongamia pinnata* for commercial use for the production of biofuel (Gresshoff, 2014; Indrasumunar *et al.,* 2017). Emphasis is given to analyzing carbon sequestration (in relation to carbon credits) and nitrogen gain. Research has also been carried out on using the material leftover from the oil extraction as a feed supplement for cattle, sheep and poultry as this byproduct contains up to 30% protein. Other studies have shown some potential for biocidal activity *Pongamia* against *V. cholerae* and *E. coli,* as well an anti-inflammatory, antinociceptive (reduction in sensitivity to painful stimuli) and antipyretic (reduction in fever) properties. There is also research indicating that *Pongamia pinnata* can be used as natural insecticides.

Carbon sequestration: The Carbon sequestration potential of *Pongamia pinnata* during the 10 to 15 years of its growth was found to be many folds higher than that of several other tree species. Pongamia was found to sequester around 45 to 50 kg of C per tree per annum as against 28 to 35 kg of Neem (*Azadirachta indica),* 23 to 26kg of Mahua (*Madhuca latifolia)* and 11 to 15 kg in respect of Tendu (*Diospyros melanoxylon).* (Chaturvedi *et al.,* 2011; Jesse More, 2009; Gupta, 2009; Kamarkar *et al.,* 2012; Gera and Chauhan, 2010; Reddy,N.S. *et al.,* 2009)

A conservative estimate of carbon sequestration by Pongamia is as follows: A population of 200 trees of Pongamia (aged 20 years) in an area of one acre can sequester 8,200 lbs or 6118.2 kg of Carbon. Some estimates arrived at by CSIR indicate that the above figure can go up to 25 tons /ha (Prasad, 2019a).

A 2006 study estimates that over the course of a 25-year period, one Pongamia tree has the potential to sequester 767 kg of carbon. The carbon sequestration ability of Pongamia was calculated for 3,600 trees planted in the Powerguda village in Adilabad district of Telangana State in India. Over the course of seven years, the trees are estimated to sequester 147 MT of carbon equivalent and yield about

51,000 kg of oil, resulting in a total value for the village of about $845 (Prasad, 2019a)

The current trends of global warming are of great concern. It is estimated that an increase of 2.5% of global temperature would reduce agricultural productivity in USA by 6%; but by 38% in India (Sengupta 2013). Added to the problems of climate change leading to increasing temperatures, the world is facing the problems of soil deterioration due to pollution and mining activities. It is therefore necessary to come up with appropriate technologies involving tree species that are efficient in not only cutting down global warming and bring about perceptible levels of C sequestration; but also ameliorate the problem soils. *Pongamia pinnata* is one such species that is capable of carrying out the soil amelioration as well as enhancement of upper environment (Prasad, 2019a).

In India, pongamia is used in land reclamation and as a soil stabiliser. It is being tried for 'phytocapping' landfill waste in Rockhampton (Venkatraman and Ashwath 2009).

Arpiwi *et al.,* (2013) found that salinity tolerance of *Pongamia* saplings was 200 mM NaCl under saline drained and 150 mM NaCl under waterlogged conditions. According to them *Pongamia pinnata* diversity could be exploited for selection of superior genotypes for oil production on marginal land.

Symbioitc Niirogen fixation and Soil amelioration by *Pongamia*

Pongamia produces its own nitrogen through Symbiotic Nitrogen Fixation, thereby displacing approximately $200 / ha/year of nitrates applied as compound fertiliser. Soils under *Pongamia*stands of +15 years' age accumulate almost 1100 to 1300 kg of Nitrogen per hectare per annum, apart from enhancing the soil organic matter and consequent positive soil biological activity (Elevitch and Wilkinson, 1999).

Table 11: Data on analysis of soils with and without Pongamia (Vayugrid 2014)

Parameter	Under Pongamia (6yr old trees)	Barren soil
pH	6.5	6.7
Bulk density of soil (g/cubic cm)	4.75	4.87
Oxidisable C (mg/g of Soil)	1.6	0.45
Organic matter (%)	3.27	1.16
N (%)	0.091	0.027
P (%)	0.0093	0.0030
K (%)	0.086	0.032

The soil under six-year-old trees of *Pongamia* exhibited marked improvements not only soil organic matter and Nitrogen, but also in other plant nutrients viz., Phosphorous and Potash.

Propagation: Pongamia can be propagated by seed as well as by vegetative means. The techniques of propagation and their evaluation in relation to the improvement of *Pongamia pinnata* have been well reviewed by Mukta and Sreevalli (2010)

(i) **Seed propagation** is widely adopted in afforestation programmes due to ease and feasibility (Allen and Allen 1981, Duke 981 and 1983). In order to obtain healthy and good quality seeds, their collection from selected mother trees, extraction and storage in a scientific manner needs attention. Therefore, principles for seed-testing for the determination of purity, germination, moisture content and freedom from pathogens according to rules laid down by ISTA (Anonymous,1966) need to be applied. Seeds lose their viability with in a period of six to eight months if not stored properly under optimum temperature conditions. Seed viability, however, may be enhanced to one year following harvesting, if stored under temperatures lower than 20*C. (Rout-Sandeep and Naik-Saswat, 2015) The ideal storage conditions are always those which reduce the growth processes of respiration and transpiration to the lowest possible degree, without impairing the inherent vitality and strength of the seed embryo (Phartyal *et al.*,2002). Seeds of *P. pinnata* stored in the middle layer of glass-stoppered containers retained higher viability even after one year and exhibited higher germination rate (Athaya, 1985). Also Seed viability may be better ensured if the seeds are stored as whole unshelled, sun-dried, uninfected, bold and healthy pods.

Seed germination could be enhanced by steeping the fresh seeds of *Pongamia* in hot water of around 50°C for 20 minutes before sowing. Seeds are sown in seed beds / polythene bags / sand trays with the micro-pile facing downwards not deeper than 3 cm to the top of seed and 5 cm to the bottom. Mixture consisting of 1part of sand,1 part of red soil and 1 part of well decomposed and grounded cattle manure has been found to yield good results for filling-up of polythene nursery bags and preparation of seed beds for sowing. Seed germinates within two weeks of sowing.

Seed propagated tree populations exhibit a wide variation both in terms of growth and development, branching, seed and pod and seed characters. Some seeds throw out seedlings with less or no chlorophyll termed as *albina, xantha, chlorina* etc., if the seeds happen to be derived from self-pollination. Raising of healthy seedlings in well managed nursery is absolutely necessary for the production of healthy and vigorous root-stocks for grafting.

Transplanting the seedlings into the field should occur at the beginning of the rainy season when seedlings are 60 cm in height with well-developed root system. Soil mass should be retained around the roots during transplanting.

(ii) **Propagation by stem cuttings** is successful and is practised to certain extent (Palanisami and Kumar, 1997). Stem cuttings, however, result in fibrous root system of the saplings, which renders them vulnerable to drought under rainfed conditions. Also, the fibrous root system does not offer good anchorage to the growing sapling under fierce windy conditions which are very common.

Additionally, with stem cuttings it would not be feasible to produce vast planting materials for raising plantations in large areas. Despite the limitations associated with the stem cutting propagation, the ease of doing it encourages many to resort to this process. In such cases, adhering to the following recommendations (Mukta and Sreevalli 2010) would be useful.

- IBA at 1000ppm as an alcoholic dip induces best rooting in cuttings.
- Cuttings root better in the months when cambium is in active phase.

(iii) **Root suckers** are rather plentiful as well. It is a rapid-growing coppice species that can be cloned.

(iv) ***Pongamia* can be propagated by grafting**, by far the best method of propagating valuable genotypes and high yielding elite trees.

(v) **Tissue culture technology for propagation of elite trees**: A critical examination of the work carried out in this area indicates some success in regeneration and micro-propagation (Sujata and Harzra 2006, Sugla *et al.,* 2007, Sujatha *et al.,* 2008 and Biswas *et al.*, 2013). However, micro-propagation technology has not yet come to the stage of developing plantlets /clones for commercial plantations on large scale.

One of the major limitations associated the tissue cultured plants of *Pongamia* are week root system that can't withstand the rigours of dryland rain grown conditions (VAYUGID 2014).

Through research efforts initiated by the Hyderabad unit of the National Bureau of Plant Genetic Resources (NBPGR) the following improved accessions of *Pongamia pinnata* were identified (Varaprasad, 2014 ; Varaprasad and Sudhakara Babu, 2015) Table. 12.

Table.12: Imporved accessions of Pongamia identified in different regions of India (Varaprasad, 2014 ; Varaprasad and Sudhakara Babu, 2015)

Agrocliinatic zone	Improved varieties accessions
Eastern Plateaus & Hills Region	Baramunda . Mancheswar
Southern Plateau & Hills Region	KNRI-(29.31 %). KSR1(27.34%) and TCR-MD(27.2 %). TCR-C1-1, PT 1 and PT 2 of Dharwad.
Central Plateau & Hills Region	NRCP-13(eariy flowering). NRCP 24. NRCP 26 (high yielding). IC 527952 (34 %)
Eastern Plateaus & Hills Region	KKVPP-02(31.7%). KKVPP-17(42%)
Central Plateau & Hills Region	RPK 14
Western Plateau & Hills Region	RHRAK 50. RHRAK 44

% Oil content in parenthesis

Production of Elite Clones of *Pongamia* through Grafting

The grafting techniques involve the use of root stock seedlings over which the vegetative scions are grafted by several methods viz., top grafting, veneer grafting, bud-grafting or budding and wedge-cleft grafting etc., which are normally employed in the vegetative propagation of trees (USAID 2007). Wedge-grafting has been found to be most successful (95%)using3-month-oldseedlings raised in poly-bags as the stock and semi-hardwood scions of 12–15cm length (Karoshi and Hegde,2002). However, it has been observed that higher rate of success could be achieved in *Pongamia* using root-stocks with soft-wood (75 to 80 days old vigorous root-stocks) employing cleft-wedge grafting (Fig. 13); as has been standardized for vegetative propagation of cashew (Prasad *et al.*, 2000, BhaskaraRao *et al.,* 1993)

Under the aegis of VAYUGRID, grafted saplings were developed by adopting a special grafting technology consisting of appropriate root-stock age and vigour resulting in a true copy of its genetically elite mother *Pongamia* tree The following steps are involved in the development of elite grafted saplings of *Pongamia pinnata* (Prasad 2013)

1. Identification and selection of genetically elite *Pongamia* trees based on the selection criteria for stable characters (Prasad, 2013) and confirming the tree performance in the two to three subsequent years before propagating it (Figs. 15 and 16)
2. Growing of vigorous root-stocks adopting appropriate nursery technologies,
3. Grafting dormant scion stem tissues of the elite mother trees on to the root-stocks possessing soft wood, employing a special and efficient grafting technology (Prasad *et al.*, 2000) and
4. Nurturing of the fresh grafts in a nursery to a suitable stage of development for their transplantation in the main filed during the rainy season (Fig. 17).

Planting and initial care of planted grafted saplings: The investigations carried out at the ICAR-Central Institute of Dryland Agriculture, Hyderabad have come out with the following recommendations.

- It is recommended that pits of the size of 45-50 cubic cm be dug before the rainy season.
- Under semi-arid rainfed conditions and on light soils of low fertility, a spacing of 5m X 4m is recommended for planting, leading to a density of 500 trees per hectare.

- The excavated top soil is mixed with sand and organic manure / FYM in the proportion 1(sand):2(soil): 1 (FYM) and 100 g of neem cake with which the pits are filled at the time of planting with the receipt of rainfall of at least 50 to 60mm or before surface soil moisture gets exhausted.
- It is desirable to make a basin at the base of each planted sapling to conserve rainwater and support the sapling with a stake till it gets established. In order to keep the seedling stable, the soil around it should be pressed by hand and the seedling covered with grass straw after planting.

Fig.13. Cleft-wedge grafting technique

- Weeding may be carried out around each sapling after 60 days of panting and till the plant gets established with firm and deep root-system in the soil
- No irrigation may be needed normally as Pongamia saplings can withstand even moderate to severe soil drought.
- During the first summer / rainless hot period following planting, watering of saplings with 30 to 40 litres of water / plant either from water harvested from run-off or other source would ensure proper canopy development.

In areas with exposed rock, artificial explosion can be used for pit preparation for planting high value elite grafts. Powder is used to explode a pit to a depth of 1 m and a diameter of 2-3 m. Then good top soil and organic matter are put into the pit into which an elite graft is planted and watered. The nitrogen from the explosion powder remains in the pit after explosion thus increasing the nutrients for tree growth (VAYUGRID, 2014).

Following two important steps should be implemented to ensure proper seedling establishment and zero plant mortality in the first year of the plantation.

- Undertake planting of elite grafts only when the soil profile is saturated with rain water to a depth of more than 60 cm.
- Mix the dugout soil with 100 g of neem cake in addition to the FYM-sand-soil mixture suggested above and fill the pit at planting to ward of damage to root by any soil born biotic agents.

Plant Nutrition

As flowering season approaches in October to December, trees need to be replenished with nutrients in order to have enough food for the developing pods. The decision for supplementary nutrition may be taken depending upon the growth and development of canopy. If the growth of sapling is stunted after the first year, the following procedure may be followed. For this a small trench of 30 cm deep, 120 to 150 cm long and 40 to 60 cm wide should be made around the tree canopy. Around 2 kg of FYM + 50 g of SSP (Single Super Phosphate) may be applied in the trench dug around the plant and the trench closed with the dugout top soil.

Nipping of root-stock sprouts: It is necessary tocarry out nipping / removal of sprouts arising from the root-stock zone of the grafted sapling. Also eliminate all root sprouts and maintain the principal / main trunk clean devoid of side growth up to a height of 1 meter.

Irrigation: No artificial irrigation may be necessary, as the grafted saplings are highly drought resistant by virtue of the deep rooted root-stock and as such can thrive well with rainfall moisture. Nevertheless, it should be ensured that the adverse impact of drought is minimised to recover tangible levels of productivity.

In the event of acute drought due to total or near total failure of seasonal rainfall, periodical sprays of canopy with 2% Urea solution once in ten days or fortnight depending upon the soil moisture level would protect the plantation from yield decline.

Inter-croppingof drought resistant annual crops viz., castor, pigeon pea, cowpeas, green gram, sesame, cluster beans, horse gram niger, sunflower, fodder grasses etc., is recommended depending upon the local conditions in inter-row spaces of the plantation in the first three years.

Inter-cropping may not be possible after three years of establishment of plantation.

Fig. 14: Flowering on a branch of Pongamia tree

Biotic stresses of Pongamia: Despite some insects and some fungi harboured by *Pongamia* tree, there was no economic damage caused by the insects and fungi. Pongamia is basically resists many biotic stresses due to the presence of karanjin and pongomol. *Pongamia* is totally resistant to termites, which can't penetrate the stem or root.

Eriophyd mite (*Aceria Pongamiae*) induces leaf galls in India. *Aceria pongamiae* is a highly host specific eriophyid mite, producing varying numbers of small finger-like leaf galls on *P. pinnata.* The number of galls on infested leaf varies, quite often individual galls fused to form complex, irregularly shaped, massive structures, covering entire laminar area including the midrib, vein and vein lets. Each gall carries hundreds of mites in different stages of development, namely, the egg, 1st nymph, 1st quiescent stage, 2nd nymph, 2nd quiescent stage and adult male and female\ on the leaves of *P. pinnata.*(Nasareen *et al.,* 2012).No plant protection measure is recommended as the mite induced leaf galls are not known to adversely affect the pod and seed yield. However, the colonization of mite affects the leaf shape and the leaves give a crumpled appearance with no adverse effect on pod and seed yield.

Only one dipteran fly *Aspondylia pongamiae* (Fig. 21) was found to cause yield losses in which the maggots eat away the developing ovules of Pongamia flowers leaving sterile rounded bead like structure. The adult fly itself doesn't seem to cause any direct damage; but it lays eggs in the developing flower buds from which the maggots emerge to cause the damage (Devaraj and Sundararaj 2014). The pest was observed in the districts Southern Karnataka in the fag-end of the rainy season associated with elevated levels of atmospheric humidity. On the brighter side the hymenopteran parasitoids viz., *Eurytoma dentata, Megastigmus albizziae, Neanastatus proximus and Ormyrus kama.* are very effective in checking the proliferation of the pest. The Combined parasitisation, amounting to 37.56% in 2008-2009 and 44.54% in 2007-2008 was observed indicating that these parasitoids played significant role in keeping the population of *A. pongamiae* under control (Devaraj and Sundararaj 2014)

Fig. 15: Pod bearing on the branches of an elite strain of *Pongamia pinnata* (Tumkur Krnataka, India)

Fig. 16: An elite tree of *Pongamia pinnata* (Tumkur distt. Karnataka, India)

Fig. 17: Pongamia nursery with elite grafted saplings

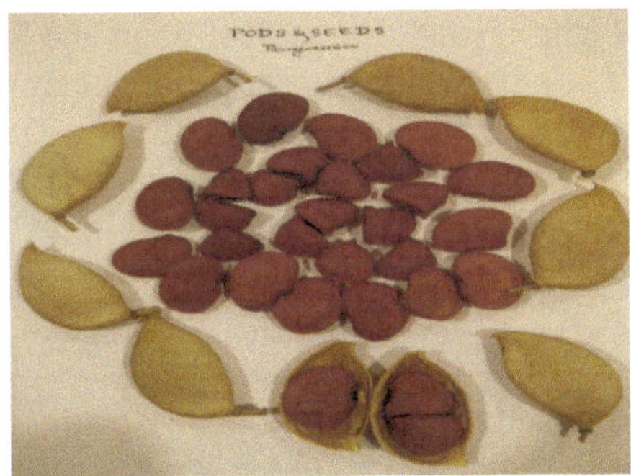

Fig. 18: Pods and seeds of *Pongamia pinnata*

Fig.19: A view of the three and a half years' old young elite clonal plantation of *Pongamia pinnata* raised on typical marginal dryland (rainfall 350-400 mm) near Turkapally- Hyderabad, India

Fig. 20: Three years' old elite Pongamia trees growing on copper mine toxic dumps in Zambia (*Courtesy*: Dr Benjamin Warr, Zambia)

A Phanerogamic parasite viz., *Dendrophthoe falcata* (L.f) Ett. belonging to the family Loranthaceae (Selvi and Kadamban, 2009) was observed on one tree of *Pongamia pinnata* near Hyderabad for the first time in October 2019 (Prasad, unpublished) in the zone of Telangana (Fig.22). Selvi and Kadamban (2009) however, reported the occurrence *D. falcata* on a wide range of tree species including *Pongamia pinnata* in southern Tamil Nadu.

Fig. 21: Adult fly of Aspondylia pongamiae

Chromosome Number: The somatic chromosome number of *Pongamia pinnata* is 2n=2x=22 (Patel and Narayana 1937; Raghavan and Arora 1958; Bhatt and Sanjappa 1976; Sarbhoy 1977). Singhal *et al* (1990) reported existence of B chromosomes to the tune of 7 numbers in the somatic cells of *Pongamia pinnata.* DNA fragment lengths ranged from 250 to 1000 bp with a variable number of polymorphic bands, ranging between 13-24 (Aminah *et al.,* 2017)

Fig. 21: *Dendrophthoe falcata* parasitic plant on one tree of *Pongamia pinnata* near Hyderabad Oct- 2019

A higher recovery and better quality of oil of *Pongamia pinnata* as compared to other crops, its non-competition with food crops as it is a non-edible source of fuel, non-toxicity, medicinal value and its suitability to be grown on degraded and marginal lands render it a highly desirable not only in ameliorating the farmers' economy, but also enhancing the eco-system and environment (Prasad 2012a).

Melia Azedarach (**Persian Lilac / China berry tree)**: *Melia azedarach* was first described by Linnaeus n (1753) (Spec. Plant. 384) a belongs to the family

Meliaceae. The generic name is derived from the Greek word melia - manna ash, referring to the resemblance of the leaves to *Fraxinus ornus (an* avenue tree *Viz.* Flowering Ash, found in Hungary). The specific name is from the Persian *'azad-darakht'*, meaning 'noble tree'. The type was formally described from an Indian cultivar.

Members of the familyMeliaceae are a woody family widely distributed throughout the tropics and subtropics, with only slight penetration into temperate zones; they occur in a variety of habitats from rain forests and mangrove swamps to semi-deserts. The timbers of certain Meliaceae are some of the most sought after in the world, such that natural stands have been much depleted.Other uses of Meliaceae comprise shade and street trees, fruit trees, and, last but not the least, sources of biologically active compounds (Mabberley *et al.*, 1995). Among other secondary metabolites, that Meliaceae synthesizes and accumulates are bitter and biologically active nortriterpenoids, which are also known as limonoids and meliacins. These and other compounds have aroused considerable commercial interest due to their insect-antifeedant and insecticidal, molluscicidal, antifungal, bactericidal and plant-antiviral activities as well as their numerous medicinal effects in humans and animals. Together with *Melia azedarach*, another more prominent and important species that belongs to family of Meliaceae is *Azadirachta indica* , neem tree with known medicinal, curative and pesticidal properties.Mabberley (1984) and Mabberley *et al.* (1995) reviewed the taxonomy and botanical history of *M. azedarach* sens. Lat. The authors argue that the whole complex be described as a single species *Melia azedarach* and suggest an informal infraspecific classification under three groups: wild trees, Chinese cultivars and Indian cultivars, each containing many forms with different synonyms or varietal names. They treat as synonyms or varieties *Melia sempervirens* (L.) Sw., *Melia dubia* Cav., *Melia composita* Willd., *Melia australasica* A. Juss., *Melia bukayun* Royle and *Melia japonica* G. Don. However, Sasidharan and Sivarajan (1996) regard the Indian *Melia dubia* as a separate species because of its distinct morphological and field characteristics.

Fig. 23: Melia azedarach in bearing (Picture: Emmanuel Douzery)

Fig. 24: A branch of Melia azedarach with inflorescence and berries

Description: The adult tree has a rounded crown, and commonly measures 7–12 metres (20–40 ft) tall, The leaves are up to 50 centimetres (20 in) long, alternate, long-petioled, two or three times compound (odd-pinnate); the leaflets are dark green above and lighter green below, with serrate margins. (Figs 23 and 24) The flowers are numerous, small, purple and fragrant, with five pale purple or lilac petals, growing in clusters.(Fig. 24) The opposite, pinnate leaves are 20–40 centimetres (7.9–15.7 in) long, with 20 to 30 medium to dark green leaflets about 3–8 centimetres (1.2–3.1 in) long. The terminal leaflet often is missing. The petioles are short, staminal tube deep purple blue brown.0.6 cm long. Fruit or berries are small, yellow drupe, nearly round, about 15 mm in diameter, smooth and hard as a stone, containing 4 to 5 black seeds. Seed are oblongoid, 3.5 mm x 1.6 mm, smooth, brown and surrounded by pulp. The berries are yellow at maturity, hanging on the tree all winter, and gradually becoming wrinkled and almost white. The seed contains up to 40% of a drying oil, which is used for lighting, varnishes and in traditional *Ayurvedic* medicine.

As the stem ages and grows, changes occur that transform the surface of the stem into bark. The fruits have evolved to be eaten by animals which eat the flesh surrounding the hard endocarp or ingest the entire fruit and later vent the endocarp. Fruits are poisonous or narcotic to humans if eaten in quantity. However, the toxins of the fruit are not harmful to birds, who gorge themselves on the fruit, eventually reaching a “drunken” state. The birds that are able to eat the fruit spread the seeds in their droppings. The toxins are neurotoxins and unidentified resins are found mainly in the fruits. The first symptoms of poisoning appear a few hours after ingestion. *Melia* seed oil can be utilized successfully for the synthesis of *styrenated poly* (*ester-amide*) *resins*, characterized through physicochemical analysis and spectroscopic studies. (Ahamad *et al*.,2015)

Nativity and Distribution: *Melia azedarach* is reported to be native to Asia probably from Baluchistan, (Pakistan) and Kashmir (India) (Troup and Joshi 1981); but has long been cultivated throughout the Middle East, the Indian subcontinent and China.

Mabberley *et al.* (1995) reports the distribution of ‘wild trees’ from India, Nepal, Sri Lanka and tropical China, south and east through Malesia: Sumatra, Java, Philippines (Luzon, Negros, Mindanao), the Lesser Sunda Islands (Flores, Timor, Wetar), Papua Barat and Papua New Guinea, Australia and the Solomon Islands. It is widely scattered in China, through Malesia to the Solomon Islands and Australia (Ahmed and Idris, 1997). In Australia it extends from northern Queensland to southern New South Wales, usually within 100 km of the coastline (Doran and Turnbull, 1997), and it also occurs in the Kimberley region in northern Western Australia (Wheeler *et al.*, 1992), northern areas of the Northern Territory and the Gulf of Carpentaria in north western Queensland. The latitudinal range covering both natural and naturalized populations is from approximately 35°N-35°S indicating a purely tropical distribution.

Special attributes: *Melia azedarach*'s special attributes are (a) fast growing habit, (b) good and very easy establishment in the field (c) early flowering and bearing (d) more or less uniform fruit development and synchronous maturity, (e) moderate levels of average yield on drylands ranging from 4.5 to 5.0 tonnes / ha and (f) seed oil content of 39-40% with a balance of fatty acid composition with higher content of oleic acid (Prasad and Ranganatha,1992)

Melia is reported to be feed stock for biodiesel production (Akhtar *et al.*, 2016) and used for synthesis of styrenated poly(ester-amide) resins (Ahamad *et al.*, 2015)

In view of its attributes viz., easy establishment of saplings, precocity and synchrony of flowering and fruiting and fair degree of seed and oil yield, *Melia azedarach* should do well under dryland conditions characterised by low rainfall.

Production: Management of *Melia azedarach* is simple and easy. The saplings of 6 to 9 moths' age raised from the seed in the nursery may be used. Also, the root sprouts of coppices arising from the adult trees may be used for planting since they represent the productive genotype of the selected mother tree. Saplings may be planted during the monsoon season in the pits of 45cubic cm size in a well prepared field. Prior to planting two weeks before, each pit may be filled with 4 kg of FYM, 2 kg of silt or mixture of red-soil and sand and 500 g of SSP. The saplings establish very quickly without any hurdles. The plantation may be totally rain-fed and no need for any extraneous irrigation. There are no any special recommendations except for pruning of the spent branches after harvest of berries or drupes to encourage new vegetative growth. The trees are very hardy and easy to manage with less financial inputs,

Tissue culture: The *in vitro* multiplication technology of *Melia* through tissue culture of nodal segments has been standardized (Sujatha *et al.,* 1994). Among various growth regulators tested at different concentrations, Benzyl amino purine (BA) at 0.5 mg/l was found to be the best for induction of shoot proliferation. Auxiliary branching along with rapid proliferation from the base lead to higher rate of multiplication, which was ten-fold for every two weeks. Solidified half strength MS medium incorporated with 1.0 mg/l NAA induced good rooting in one week with 100% success (Sujatha *et al.,* 1994). The tissue cultured plants started bearing fruits earlier compared to plants raised from seeds. (Prasad, 1994c).

Chromosomes: *Melia azedarach* is a diploid species with chromosome number 2n = 2x= 28 (Khosla and Styles, 1975). There were clear 14 bivalents at Meiotic Metaphase with regular and normal cell division. The Chromosomes were observed with median and submedian constrictions varying in length from 1.45μ to 0.50μ. The four types identified in the complement are 2Lsm (1.45μ to 1.40μ) + 3M sm (1.15μ to 0.95μ) + 3Mm (0.95μ to 0.80μ) +6S nearly m to sm (0.60μ to 0.35μ) (Khosla and Styles, 1975)

Selection of superior genotypes: In *Melia,* five superior genotypes were identified based on the attributes of fruiting, branching, canopy spread, stem girth and bearing habit (Prasad *et al.*, 1993). These progenies grew to 1.5 m height each and initiated fruiting in less than a year (Prasad, 1994c).

Evaluation studies: An evaluation of 55 different oil bearing trees carried out at the Directorate of Oilseeds Research Hyderabad indicated that *Melia azedarach* is precocious and efficient utilizer of soil moisture (Prasad 1994c) its growth matching with the rainfall pattern of Deccan Plateau. The trees of *Melia start* yielding from the second year of establishment of the plantation and the range of yield over 5 years was of the order of 10 to 15 kg/tree.

Cultivars: Mabberley (1984) indicated that China berry tree (*Melia azedarach*) cultivars throughout the world are variable in form and are morphologically different from wild types growing in Asia. It is likely that all China berry tree trees introduced into the American continent descended from some form of cultivar. In the United States, two cultivars occur that are morphologically distinct from the common type (see above Description) in North America. *'Umbraculiformis'* is distinguished from the common type by its multiple stems, umbrella-like canopy and shorter stature (20 to 25 feet (6-7.6 m) tall). *'Jade Snowflake'* cultivar is distinguished by its variegated leaves as per the review by Dirr and Michael (1998.)

Pollination and breeding system: China berry trees have bisexual or perfect flowers or a combination of perfect and staminate (male) flowers (Zomlefer, 1994). China berry tree flowers are fragrant which attract insect pollinators. The trees of mahogany family (Meliaceae) as a whole use nectar and fragrance to attract pollinators, most often moths and bees. Flowers of the China berry tree are self-pollinated or cross-pollinated and it is likely both systems occur in wild populations of *Melia* (Zomlefer, 1994.)

Biotic stresses: Despite the incidence of bacterial and fungal diseases, no serious damage is reported to the growth and production of *Melia* (Ahmed and Idris, 1997). In India there are no reports of serious incidence of pests and diseases on *Melia azaedarach.*

Nevertheless, *M. azedarach* var. *australasica* is reported to be subjected to periodic defoliation by the larvae of the white cedar moth (*Leptocneria reducta*) and a spider mite (Doran and Turnbull, 1997).

Uses: According to Milimo (1994). *M. azedarach* is one of the few species that are multipurpose and used by small farmers in arid and semi-arid parts of the world where it is not easy to grow promising trees successfully. In Vietnam it is one of the most common trees found in home gardens and in India it is frequently cultivated on roadsides. It is also suitable for use in windbreaks and in avenue plantings. In the Philippines it is planted as a shade tree in coffee and abaca (*Musa textilis*) crops (Ahmed and Idris, 1997). In the rural afforestation programme of Thailand, the tree is used in the *taunga* reforestation method in which combinations of trees and crops grown together. *M. azedarach* was one of the main species grown in combination with cassava, maize, sorghum, coffee, cashew nuts, banana, pineapples and other crops (Watanabe *et al*., 1988). In India, *M. azedarach* is used for planting dry, eroded hill slopes between 900-1500 m elevation in the Himalayas (Troup and Joshi, 1981). In Jammu and Kashmir, it has been successfully used in the afforestation of bare south and south westerly slopes near Srinagar.

Oil content of *Melia azedarach* is reported to be of the order of 39 to 40% (Bachheti *et al.,* 2012). The specific gravity, refractive index, saponification value, and unsaponifiable matter of *Melia azedarach* seed oil were found to be 0.80 g/ cm3, 1.36 40 D n , 84.15 (mg KOH/g) and 0.71 (%w/w) respectively. Elemental analysis of oil by AAS' titration method shows that it contains Ca (1230 mg/100g), Mg (990 mg/100g), K (121 mg/100 g), Zn (3.12 mg/100 g), Mn (3.4 mg/100g), Fe (19.52 mg/100g) and P (213 mg/100g). The extracted oil was methyl-esterified and analysed by GC and GC-MS. The major fatty acids reported here, in *Melia azedarach* seed oil, are Palmitic acid (5.68%), Linoleic acid (74.57%), Oleic acid (16.39%), Stearic acid (3.33%). (Bachheti *et al.*,2012).

A handsomely marked wood, the heartwood of *Melia azedarach* is red to reddish-brown. It is clearly demarcated from the whitish-yellow sapwood. The texture is coarse; the grain is straight; lustre is bright; there is a musk-like aroma. The wood is tough, moderately heavy, somewhat brittle, somewhat durable. It is easy to work; takes an excellent polish. It is used for making furniture, packing cases, construction, decorative veneer etc. (Gamble, 1972; Vines,1987; Genders,1994; Loewe.2009) The timber is soft, pinkish to yellow-brown resembling mahogany, with prominent growth rings which give it a fairly decorative appearance (Anderson, 1993)

M. azedarach has been widely known for medicinal qualities and contains compounds with anti-feedant and growth-disruption properties in insects. The main bioactive compounds are azadirachtin, salannin and meliantriol, which also occur in *A. indica* (Milimo, 1994). However, Ahmed and Idris (1997) reported information that indicates that azadirachtin, the most important compound in *A. indica*, is absent in *M. azedarach*. This conflicts with a paper by Huang *et al.* (1995) which reports that a new limonoid, azedarachin C, isolated from the bark of *M. azedarach*, and the work of Cabral *et al.* (1995), reporting the extraction of azadirachtin-A from *M. azedarach* seed collected in Brazil. The author states that this seed source is unlike some geographical strains which appear to lack azadirachtin-A. This highlights that chemotypes are present in *M. azedarach* and that there is a need to undertake studies into variation in the species.

Extracts from the fruit have long been used as an insecticide on crops. Gupta (1993) reports that the leaves, bark and fruits are placed inside books and woollen garments to repel insects. In Ghana, it is used to protect cocoa beans against infestation by *Ephesia.* Chen *et al.* (1996) provides information on the toxic benefits of *M. azedarach* fruits extracts on the larvae of the diamond back moth, *Plutella xylostella.* Extract from *M. azedarach* fruits was used for control of cotton leaf worm *Spodoptera* in Egypt (Salam and Ahmed. 1997). Chen *et al.*, (1996) reported on the antifeedant and growth inhibitory effects of *sendanin*, a limonoid extracted from the bark of *M. azedarach* and *M. toosendan* on *Peridroma saucia*, a variegated cutworm. Extracts from the leaves of *M. azedarach andA. indica* are reported to control pests including locusts. Ramakrishnan *et al.* (1995) notes that the application of leaf extracts from six species, *A. indica* followed by *M. azedarach* were the most effective in reducing levels of root-knot nematodes (*Meloidogyne incognita*).

Melia azedarach is found to possess medicinal value (Mabberley *et al.*, 1995). The bark is reported to produce a drug that expels intestinal worms. The root is recorded as of medicinal value in *'Cortex Meliae azedarach'* in pharmacopoeias, but it is generally recorded that the bark is more productive. Bark, leaves and roots have been used for treatment of rheumatism, fever, swellings and inflammation (Lassak and McCarthy, 1983). A glycopeptide called *meliacin* isolated from the leaves and roots is reported to inhibit in-vitro replication of various DNA and RNA viruses, e.g. polio virus (Mabberley *et al.*, 1995).

The seeds of *M. azedarach* yield oil high in linoleic acid and oleic acid. They also yield a greenish-yellow bitter oil (Troup and Joshi, 1981) and the trunk yields gum, which has been little utilized.Anderson (1993) records that the Aborigines of the Tully River area of north Queensland used the bruised bark and leaves as a fish poison, which was reported to act fairly rapidly. Gupta (1993) reports that a form of whisky is made from the fruits. The stone of the fruits is used as a bead in necklaces and rosaries. *M. azedarach* leaves are reported to be palatable and nutritious for stock with *in vivo* dry matter digestibility of 77% (Vercoe, 1989).

Among its pesticidal properties, a special mention may be made with of nematicidal properties as reported by Ramakrishanan *et al.*, (1995), which revealed that the extract *M. azedarach* checks the proliferation of root knot nematode *Meloidogyne incognita.*

Neem (*Azadirachta indica*): *Azadirachta indica*, commonly known as neem, nim-tree or Indian lilac is a tree in the mahogany family Meliaceae. It is one of the two species in the genus Azadirachta, and is native to the India. It is typically grown in tropical and semi-tropical regions(USDA 2017). Neem trees also grow in islands located in the southern part of Iran. Its fruits and seeds are the source of neem oil.

The "*divine tree*" neem (*Azadirachta indica*) is mainly cultivated in the Indian subcontinent. Neem has been used extensively by humankind to treat various ailments before the availability of written records which recorded the beginning of history. The world health organization estimates that 80% of the population living in the developing countries relies exclusively on traditional medicine for their primary health care. More than half of the world's population still relies entirely on plants like neem for medicines, and plants supply the active ingredients of most traditional medical products (Kumar and Navaratnam, 2013).

The opposite, pinnate leaves are 20–40 centimetres (7.9–15.7 in) long, with 20 to 30 medium to dark green leaflets about 3–8 centimetres (1.2–3.1 in) long. The terminal leaflet often is missing. The petioles are short.The white and fragrant flowers are arranged in more-or-less drooping axillary panicles which are up to 25 centimetres (9.8 in) long. The inflorescences, which branch up to the third degree, bear from 250 to 300 flowers. An individual flower is 5–6 millimetres (0.20–0.24 in) long and 8–11 millimetres (0.31–0.43 in) wide. Protandrous, bisexual flowers and male flowers exist on the same individual tree.

Neem (*Azadirachta indica* A. Juss. Meliaceae) is an attractive, evergreen tree that can grow up to 20 m tall. Its spreading branches form rounded crowns up to 10 m in diameter. The short, usually straight trunk has a moderately thick, strongly furrowed bark that has a unique odour and a bitter, astringent taste. Its leaves are imparipinate, 20-38 cm long, crowded near the branch end, oblique, lanceolate and deeply and sharply serrate. Neem is rarely leafless and is usually in full foliage even during months of prolonged drought. Its small, white bisexual and staminate (functionally male) flowers are borne in axillary clusters. They have a honey-like scent and attract many bees which act as pollinators. The fruit is a smooth, ellipsoidal drupe, 1-2 cm long that is yellow when mature and comprises a sweet pulp enclosing a seed. The seed is composed of a shell and a kernel (sometimes two or three kernels), the latter having a high oil content. Neem will begin bearing fruit after 3-5 years, becomes fully productive in 10 years and can produce up to 50 kg of fruits annually (NRC, 1992; Tewari, 1992; Gunasena and Marambe, 1998).

The family includes 50 genera and 1400 species. In India, it is represented by 20 genera and 70 species. However, the chemical composition of these species exhibits high variability. There is need to study to understanding the genetic distance of neem tree in relation to other family members (*Azadirachta indica, Amoora lawii, Dysoxylum malabaricum, Khaaya, Melia Azadirachta, Melia dubia, Swietenia macrophylla, Sweitenia mahogany, Soymida fabrifuga, Toona ciliate, Walsuria trifoliate*). Additionally, species were also analysed for Azadirachtin concentration in the leaf with the aim to finding clues about the evolution of Azadirachtin pathway and the family members who share it (Kuravadi and Gowda, 2019). Neem tree synthesises a large number of unique bioactive compounds. Chemical synthesis of many of these compounds has been tried in the laboratory; complex synthesis process and molecular complexity involved makes the process not a viable option for commercial production (Veitch *et al.* 2008). However, to increase the natural production or engineering new production methods require a deeper understanding of the synthesis process and also the genes/enzymes involved in their biosynthesis. Sequencing of neem genome is aimed at elucidating all the genes that are present in neem and understanding pathways of synthesis for many of the compounds that are produced by neem tree. (Kuravadi and Gowda, 2019).

The neem tree is often confused with a similar looking *Melia azedarach* or Bakain or China berry. Bakain also has toothed leaflets and similar looking fruit. One difference is that neem leaves are pinnate but bakain leaves are twice- and thrice-pinnate.

There are two varieties acknowledged in neem: *A. indica* ('Indian neem') and *A. indica* var. *siamensis* ('Thai neem') . *A. indica* var. *siamensis* has a wide natural distribution in Thailand at altitudes below 200 m (Willan *et al.*, 1990). It extends south on the Thai peninsula where the variety is reported to hybridise with Indian neem to give progeny of intermediate characteristics. *A indica* var. *siamensis* has leaves which are larger, less bitter and darker in colour than the Indian neem, and is also reported to grow faster, typically with a single, straight stem. The two varieties can also be distinguished by differences in their pollen morphology and isozyme patterns (Lauridsen, 1991).

Fig. 25: Neem branches with flowers

Phenology of Flowering and Fruiting: Neem will normally begin flowering and producing seed after 3-5 years of age, but flowering and fruiting can be extremely variable within the species (Kandaswamy and Raveendaran, 1988; Mahadevan, 1991; Gogate and Gujar, 1993; Gupta *et al.*, 1995). For example, Gogate and Gujar (1993) found flowering and fruiting in a 50-year old plantation of neem in Maharashtra, India, to be highly variable. Out of 331 trees under study, a total of 292 flowered and fruited in the 1992 season, and fruit production varied from 7.5 to 17.3 kg (dry weight) per tree. Studies on the fructal phenology of neem in a three-year-old provenance trial at Jodhpur in India revealed that 25.52% of the plants in the trial were fruiting in the third year, although

Fig. 26: Neem branches with fruits (drupes/berries)

this varied between provenances from 0 to 65% (Gupta *et al.,* 1995). The majority of the trees were fruiting synchronously.

In India, neem flowers (Fig. 25) in April and the fruits are ready for harvesting in July. Late flowering genotypes have, however, been identified from Tirupur, Tamil Nadu, India (Shanti *et al.*, 1996). Seed germination was reported to get reduced in the late-flowering type, and a higher proportion of abnormal seedlings was produced. These abnormalities may be due to an increased level of selfing in the late-flowering type (Solanki, 1998).

Breeding Systems: Neem has bisexual and staminate (functionally male) flowers, that occur on the same tree (Singh, V.P. *et al.,* 1996). This kind of sex expression is known andromonecious. The anthers start to dehisce around 08.00 hrs in the morning in the closed flower, and the pollen is mature before the stigma becomes receptive (protandry) (Gupta *et al.,* 1995). Pollination is effected by bees and other small insects, and the ovary is trilocular, i.e. it has three chambers each with two ovules. Generally, the endocarp encloses one seed, sometimes two and rarely three, a phenomenon termed polycarpy (Singh, V.P. *et al.,* 1996). The seeds are dispersed mainly by birds and mammals. Neem is out-breeding. Selfing has, however, been reported in the species and viable seed is produced on self-pollination (Solanki, 1998). The information on genetic resources of *Azadirachta indica* could be accessed from Germplasm Resources Information Network of US department of Agriculture *(*GRIN, 2012).

Chromosomes: The chromosomal complement of *A. indica* consists of 2n = 2x =28, indicating the diploid nature of the species with basic chromosome number of x=14 as in the case of *Melia azedarach.* The average size of the chromosomes being 1.10 μm average size of chromosomes, size of long and small chromosomes within the metaphase plates varies from 0.66 to 1.67 μm. Total chromatin length in Neem is calculated as 30.80 μm (Jha, 2014). There are clear 14 bivalents at the meiotic metaphase leading to normal cell division (Khosla and Styles, 1975). One pair of chromosomes is larger than the remaining pairs. The chromosome size is reported to vary in length from 1.60μ to 0.55μ. Both median and sub-median types of chromosomes are identified as IL sm (1.60 μ) + 3L m to sm (1.25μ to 1.00μ) +4M nearly sm (0.75μ to 0.70μ) + 6sm (0.60μ to 0.50μ) (Khosla and Styles, 1975).

Ecology and Distribution: Neem is thought to have originated in Assam and Myanmar where it is common throughout the central dry zone and the Siwalik hills (NRC, 1992). However, the exact origin is uncertain and some authors suggest it is native to the dry forest of south and south-east Asia, including Pakistan, Sri Lanka, Thailand, Malaysia and Indonesia (Ahmed and Graingc, 1985). In the 19th century, neem was introduced by Indian immigrants to the Caribbean (i.e. Trinidad and Tobago, Jamaica and Barbados), South America and the South Pacific (NRC, 1992). The cultivation of neem spread to Africa in the 1920□s when it was introduced to Ghana, Nigeria and the Sudan, and the species is now well established in more than 30 countries (NRC, 1992).

Habitat: Neem can grow in tropical and subtropical regions with semi-arid to humid climates. Neem will typically experience a mean annual rainfall of 450-1200 mm, mean temperatures of 25-35°C and grow at altitudes up to 800 m.a.s.l.

Normally it thrives in areas with sub-arid to sub-humid conditions, with an annual rainfall of 400–1,200 millimetres (16–47 in). It can grow in regions with an annual rainfall below 400 mm, but in such cases it depends largely on ground water levels. It is a typical tropical to subtropical tree and exists at annual mean temperatures of 21–32 °C (70–90 °F). It can tolerate high to very high temperatures and does not tolerate temperature below 4 °C (39 °F). Neem is one of very few shade-giving trees that thrive in drought-prone areas e.g. the dry coastal, southern districts of India, and Pakistan.

The species is drought-tolerant, and thrives in many of the drier areas of the world. There is, therefore, considerable interest in neem as a means to prevent the spread of deserts and ameliorate desert environments, e.g. in Saudi Arabia (Ahmed *et al.*, 1989), sub-Saharan Africa (NRC, 1992) and western India (Gupta, 1994). In Saudi Arabia, several thousand neem trees have been planted on the Plains of Arafat near Mecca to provide shade and relief from the intense desert heat for Muslim pilgrims (Ahmed *et al.*, 1989)

Neem grows on all types of soils, including clay, saline and alkaline soils, but does well on relatively heavier soils like black cotton soils (NRC, 1992). It can tolerate dry, stony, shallow soil with a waterless sub-soil, or places where there is a hard calcareous or clay pan near the surface. In Niger, neem was found to utilise groundwater at depths of 6-8 m, whereas an adjacent crop of millet extracted water from the top two metres of soil (Smith *et al.*, 1997A). The species is also thought to be an efficient water user in comparison to some other tree species. Neem does not tolerate water-logging, is fire-resistant and has a unique property of calcium mining which neutralises acidic soils (Gunasena and Marambe, 1998). The species is a light demander and when juvenile, neem will push up vigorously through scrubby vegetation. It is hardy, but and does not withstand excessive cold, especially in the seedling and sapling stages (NRC, 1992; Chaturvedi, 1993).

The tree height, collar diameter, clear bole length and canopy diameter of 13-year-old trees grown at CAZRI, Jodhpur in India were also assessed for variation in growth characteristics (Jindal *et al.*, 1999). All characters were found to be highly variable, and tree height was positively and significantly correlated with collar diameter, dbh, canopy diameter and fruit yield per tree.

Fruit and seed characteristics, such as weight, length, width, diameter, yield and azadirachtin and oil content are highly variable both within and between provenances of neem (Kumaran *et al.,* 1993; Rengasamy *et al*., 1993; Rengasamy *et al.,* 1995; Veerendra, 1995; Kundu, 1999; Rathore *et al.,* 1998; Sridharan *et al.,* 1998; Jindal *et al.,* 1999). The weight, length, diameter and form (diameter/length) of neem seeds from four populations (two from Bangladesh, one from Thailand and one from Kenya) were measured and analysed (Kundu, 1999). There were significant differences between populations for all seed parameters, and the populations from Thailand and Kenya were differentiated both from one another and the material from Bangladesh. This study had similar results to that carried out by Kumaran *et al*., (1993) who also found significant differences between populations of neem from Tamil Nadu, India. The authors also found seed length and seed oil content to be highly heritable and the weight of 100 seeds to be robust selection index by virtue of its high genotypic coefficient of variance, high genetic advance and moderate heritability. The variation in fruit yield per tree, 100-fruit weight, 100-seed weight and 100-kernel weight and oil content in 13-year-old neem trees from Jodhpur, was measured and analysed (Jindal *et al*., 1999). Fruit yield per tree was highly variable and positively correlated with tree height, collar diameter, dbh, and canopy diameter. 100-fruit weight was positively correlated with 100-seed weight and 100-kernel weight, but there were no correlations between kernel oil content and other fruit characteristics.

Variation in oil and protein content of neem seeds from Laos, Nepal, Ghana, Bangladesh, Myanmar and three areas of India was found to be highly significant (Rathore *et al.,* 1998). Neem seeds were collected from 12 different locations in Tamil Nadu and their azadirachtin and oil contents determined (Sridharan *et al.,* 1998). Both seed characters were highly variable, but showed some correlation with climatic factors. There was a significant positive correlation between oil content and the number of sunshine hours from September to March, and a negative correlation between azadirachtin content and the total number of rainy days during the fruiting season (April to August).

Azadirachtin content within-samples, and within and between-regions varied substantially from 0.14 to 1.66% (Rengasamy *et al.,* 1993) The results suggested that the environment had a strong effect on seed azadirachtin content with trees grown in regions with red and black soils at altitudes above 500 m.a.s.l. being rich in azadirachtin. The properties of the oil were also highly variable, but the results suggested that there was a positive correlation between azadirachtin content and the saponification and acid values of the oil. A later study on seeds from the same 21 zones (Rengasamy *et al.,* 1995) suggested that trees growing in coastal, arid or semi-arid regions had a high azadirachtin content (>0.72%), and that those from sub-humid regions showed low (mean 0.27%) azadirachtin content. Neem fruits collected from nine agro-ecological zones of Rajasthan, India were also analysed for azadirachtin content and were found to vary from 0.19 to 0.67% azadirachtin by seed kernel weight (Gupta and Prabhu, 1997).

Molecular Variation: Indian subcontinent being the home of *Azadiracta indica*, it is normally presumed that the neem tree populations of India would be characterized by considerable diversity. However, the RAPD profiles of seventeen accessions of Indian genotypes of neem generated using 49 random DNA primers (Farooqui *et al.,* 1998) revealed the RAPD profiles were very similar of the studied genotypes. The observations suggested that there was less variation than expected within neem tree populations of India. In addition, the pattern of RAPD similarities obtained did not correspond to the pattern of geographical variation in neem, i.e. accessions from a particular region were not always grouped together in the dendrogram of similarities. In contrast, a study on isozyme variation in four provenances of neem (two from Bangladesh, one from Thailand and one from Kenya) found substantial variation both within- and between provenances of neem, as expected in the case of widely distributed and outcrossing tree species (Kundu, 1999). The provenance from Thailand was shown to be distinctly different from the provenances from Bangladesh and Kenya, a result that supports the status of the material from Thailand as a distinct variety of neem. Wickremasinghe and Simons (1994) also demonstrated the similarities between neem from India, Sri Lanka and Nepal on the basis of isozyme variation, and their differentiation from *A. indica* var. *siamensis* from Thailand.

Azadirachtin content in seeds: Azadirachtin and oil content in neem seeds is reported to vary with seed maturity, hence time of seed harvesting. The fresh flowers and fruits of a 10-year-old neem tree in India were analysed for their azadirachtin concentration by Rengasamy and Parmar (1994). The compound was not detectable in the flowers and in green fruits collected at 20, 30 and 40 days after anthesis (DAA). Green fruits analysed at 50 DAA and yellow fruits analysed at 60 DAA contained 0.051 and 0.112 % azadirachtin respectively. Johnson *et al.*, (1996), however, found the concentration of five major limonoids in neem to vary only slightly with maturity, but the azadirachtin concentration was highest (10 mg/g seed kernels) in green newly ripening fruits.

Seed germination: Fresh neem seed germinates seven days after sowing and germination is complete after 25 days (Singh, B.G. *et al.,* 1995). Bharathi *et al.,* (1996) found that greenish-yellow drupes had the highest viability and germination percentage when compared to both younger (green) and more mature (yellow) seeds. The seeds germinate readily without pre-treatment, although Kumaran *et al.,* (1993) found that soaking in 2% KH2PO4 gave maximum germination, shoot length and leaf number, and 1% KH2PO4 gave maximum root length. Pre-soaking de-pulped seeds for 30 minutes (Hegde, 1993) to three days (Fagoonee, 1984) has also been found to improve germination. Radwanski and Wickens (1981) suggested that germination improves when the inner shell is removed to expose the embryo

Procedure to be adopted for good germination.: Soak the seed for 24 hrs in cold water if the seeds have been previously dried. If the seed is fresh, sow the seed within two weeks of fruit collection. Fresh seed should have a germination rate of 85%.

Propagation of neem occurs in nature by birds, squirrels, bats and rodents that consume the fruit and vent the seeds at different locations, thereby assisting in regeneration of natural vegetation.

Neem can be propagated by seed readily in the nursery. In the last 20 years, however, considerable research vegetative propagation of neem has been conducted, to facilitate propagation and multiplication of new and improved genotypes for various characters.

Processing of Fruits for Seed Propagation: Nagaveni *et al.*, (1987) observed that the fruits of neem were normally collected from the ground, de-pulped and sown immediately because of the risk of low seed viability. It is recommended that the fruits be collected from the tree when they are greenish-yellow, rather than waiting for the fruits shed on the ground below the tree for making collections for seed extraction. The collected fruits may be de-pulped and shade-dried for two days (Nagaveni *et al.,* 1987; Chaturvedi, 1993; Gunasena and Marambe, 1998). The procedure of picking the fruits at the right stage from the tree avoids the risk of fermentation of the unopened cotyledons, which enhances viability and germination percentages up to several months after harvest.

Vegetative Propagation: Vegetative propagation of neem allows for the mass production of genetically identical clones. Traditional methods of vegetative propagation of neem include grafting, root cuttings, stem cuttings and stump cuttings.

Stump cuttings generally consist of a stem cutting (2.5 cm in length) and a root mass (approx. 23 cm in length). The stumps prepared from two year-old plants have a better survival rate than younger plants in the monsoon season.

Sivagnanam *et al*., (1989) reported that stem cuttings of neem dipped into IAA and IBA rooted effectively in a mist propagator in 135 days. Coppice shoots gave better results than the cuttings from the main woody stem.

The use of IBA at 1000ppm increased the rooting percentage and root biomass. Kamaluddin and Ali (1996) recommended that treatment with 0.2% and 0.4% IBA could bring about significant root development to sustain subsequent growth of the cuttings. Palanisamy and Kumar (1997) suggested that rooting was dependent on the endogenous auxin levels of the shoot, and that longer cuttings (25 cm long) rooted more readily than shorter ones (5cm).

Gupta *et al*., (1998) found that using IBA (800ppm) for air layering is useful in promoting faster root development in air-layered branches.

Root rot disease of neem caused by *Ganoderma applantum* (Chakraborty and Konger, 1995) appears to be more serious in the case of vegetatively propagated stem cuttings and air-layers with fibrous root system. The adventitious root system doesn't provide enough anchorage to plant and renders it more sensitive to drought particularly under dryland conditions (Khan *et al.*, 2016).

Micro-propagation: Callus induction from young leaves and cotyledons has been standardised (Narayan and Jasiwal, 1985), and tissue cultured seedlings have been obtained from cotyledons, young leaves and stem segments (Rao *et al.,* 1988).

Joshi and Thangane (1996) described a procedure for the clonal propagation of neem from woody explants, generating between two and three shoots per explant. Drew (1993) reported on clonal propagation from stem nodal sections.

The tissue cultured shoots successfully developed roots and the regenerated plants established well in a soil medium. Plantlets regenerated from embryos (Thiagarajan and Murali, 1993), and through axillary bud culture (Joarder *et al.,* 1993) have also been successfully established under field conditions. Venkateswarlu *et al., (*1999) described a procedure for the selection of plus trees and their mass micropropagation. Micropropagated plants flowered 25 months after being transferred to the field and exhibited the attributes of the original mother tree. They concluded that mass production of neem seedlings through micropropagation could produce trees with known azadirachtin content in their seeds. In spite of the seeming success of tissue cultured plants getting established under field conditions, the long term field performance of micropropagated plants has yet to be ascertained (Bahuguna, 1997).

Substantial tissue culture research has explored the possibility of isolating limonoids from neem cultures (Sarkar and Datta, 1986; Ramesh Kumar and Padhya, 1988; Stephen *et al*., 1998). Allan *et al*., (1994) showed that tissue cultures could produce azadirachtin. A project funded by the Europen Union investigated the possibility of producing azadirachtin on a commercial basis (van der Esch, 1999) under tissue culture. It was observed that the available technology, however, is not perfect enough for tissue cultured production of azadirachtin economically.

Production of azadirchtin and other compounds through induction of hairy root cultures of Neem: Allan *et al.,* (2002) established hairy root cultures from stem and leaf explants of *Azadirachta indica* A. Juss (neem), following infection with *Agrobacterium rhizogenes*. Transformation was confirmed using polymerase chain reaction analysis. The cultures had a relatively fast growth rate (0.134g dry weight / day) and showed a 100-fold increase in biomass over a 4-week culture period. Both the roots and cultivation medium were antifeedant to insect pests as determined by a no-choice bioassay using the desert locust, *Schistocerca gregaria*. Azadirachtin, nimbin, salannin, 3-acetyl-1-tigloylazadirachtinin and 3-tigloylazadirachtol were detected in the hairy roots by supercritical fluid chromatography. The cultivation medium contained 3-acetyl-1-tigloylazadirachtinin, 3-tigloylazadirachtol and desacetylsalannin

Management of Neem Plantation

Neem seeds are commonly germinated in plastic bags in the nursery, although direct sowing is said to be successful where there is adequate rainfall (Benge, 1988). For example, a group of Indian villagers have been successfully carrying out direct sowing neem seed for about 30 years (Vivekanandan, 1998). Seeds should be planted at a depth of 2.5 cm to avoid predation by rodents and birds (Bahuguna, 1997). Research by Habte *et al.*, (1993) concluded that neem could be a highly vascular mycorrhiza (VAM)-dependent species. The dry matter yield of neem grown in association with mycorrhiza at a soil P concentration of 0.02mg/l was shown to be comparable to neem grown with mycorrhiza at a soil P concentration of 0.2mg/l. It would therefore seem that improved growth of seedlings in the nursery could be achieved through sowing in inoculated soils.

De Joussieu, (1963) and Radwanski (1977) recommend transplanting of 12 weeks' old seedlings obtained from a well-managed nursery. On the other hand, Bahuguna (1997) suggest that transplanting of more than one-year-old saplings selected with well-developed tap root for transplanting in dry areas characterised by water deficit.

The transplanted seedling being very sensitive, excessive drought or waterlogging should be avoided to promote their better growth. Young neem trees cannot tolerate frost or excessive cold conditions (Chaturevedi, 1993; Hegde, 1993). Shade may be provided by the way of live plantation since the saplings grow faster under shade (Bahuguna 1997; Vivekanandan, 1998)

Two-year-old neem seedlings may be transplanted into a pit of cubic meter size, containing sand mixed with river silt to improve water retention, and the planted trees irrigated until the plants were well established. In western India, low and erratic rainfall, high potential evapotranspiration, high temperatures and high wind speeds prompted research into water harvesting and conservation techniques for the establishment of neem (Gupta, 1994; 1995). The establishment of neem in troughs using ridge and furrow method and in 1.5 m diameter pits covered with surface mulch were the most effective techniques of harvesting and maintaining water, and corresponding increases in plant biomass were observed. Both techniques were found to be cost-effective, but the ridge and furrow method was more labour intensive. A picture of neem plantation is given in Fig. 27.

Fig. 27: Neem Plantation

Seed storage: Viability of neem seeds can be extended through drying (Mohan *et al.*, 1995) and at to low temperature storage conditions (Singh *et al.*, 1997). A 5-6 days' drying period in shade conditions caused a drop in seed moisture content to 11% (Singh *et al.*, 1997). Drying was found not to adversely affect germination rates, and seeds could be stored under ambient or refrigerated conditions for up to six months. Seed samples were stored under ambient temperature and humidity, and samples collected and analysed over a period of four weeks. Azadirachtin content was unstable and reduced to 68% of the original level in the dark, and to 55% in daylight. Johnson *et al.*, (1996) found losses in both azadirachtin and salannin under similar storage conditions over a period of six months.

Optimal harvesting of neem seed to improve viability: Fruit should be collected ripe from the tree. A tarpaulin may be spread underneath the tree, and the tree shaken so drupes of optimum maturity fall. Grade fruits to remove undeveloped, immature and damaged fruits. Store it in thin layers to avoid damage and the risk of fermentation.

De-pulp the fruit within 24 hrs after harvesting.

Remove the kernels from the seed if they are to be used immediately.

Storage: Dry the seed under shade for 3-5 days. Store the seeds in clean, cloth bags or airtight containers at low temperature ~10 C.

Constraints: Hand picking of fruits may ensure better quality seeds. However, it is often too laborious and expensive.

Lack of water in remote zones for de-pulping fruits is another practical problem.

Lack of shelter or space for spreading and drying neem seed.

Drying of neem seed is done during rainy season when the relative humidity may be in excess of 75%.

Neem Products for Farming: Several neem products are used on farms in developing countries. They include neem seed kernel solution (NSKS), neem oil, powdered neem seed, neem leaves and neem seed cake (NSC). All these products have been used for pest control, but the most well-known are the incorporation of neem leaves into stored grain, or the coating of stored grain with neem oil, to protect against storage pests post-harvest, and the use of NSC as a fertiliser and pesticide against soil nematodes (NRC, 1992). Aqueous neem leaf extracts have been traditionally prepared by soaking chopped or ground leaves, which are reported to contain limonoids but not azadirachtin, in water (Hellpap and Dreyer, 1995). The aqueous neem extract can then be sprayed onto crops (IRRI, 1989). However, the production and use of aqueous NSKS has gained significance in many developing countries in recent years, and is a simple, effective means of producing neem pesticides for application on crops in cultivation.

Moser (1996) reported that neem based pesticides are used by a wide range of farmers for the reasons of low cost and effectiveness. (Marz, 1989). Highly commercial cardamom farmers in southern India use neem seed cake to control nematodes. Commercial tobacco growers in Andhra Pradesh, India, used an aqueous NSKS to control tobacco leaf worm and transmission of the tobacco mosaic virus (Ahmed, 1988). In Ghana, the use of neem was also found to be related to the intensity of farming (Childs, 1999). A group of farmers stated that in their village neem was used to control pests of cowpea, but that other villages in the area used no method of pest and disease control as lower cropping intensities resulted in fewer pest problems. Neem does not directly kill insects on the crop. It acts as an anti-feedant, repellent, and egg-laying deterrent and thus protect the crop from damage. The insects starve and die within a few days. Neem also suppresses the hatching of pest insects from their eggs. Neem can help control more than 200 pest species through the possession of a complex group of compounds called limonoids (Ahmed, 1995). Neem oil has been shown to avert termite attack as an eco-friendly and economical agent (YashRoy and Gupta, 2000).

Neem oil for polymeric resins

Many of these limonoids have been isolated and extracted from neem and have demonstrated the ability to block insect growth. The most well-known and intensively studied of these limonoids are azadirachtin, meliantriol, salannin and nimbin (NRC, 1992). Azadirachtin is found in the highest concentration in the seeds of neem, and has been the main focus of research on neem and the development of commercial pesticide formulations. The other limonoids found in neem have also received substantial research attention and their modes of action are sometimes known. It is thought that there is a synergistic effect between the various limonoids in neem that gives the species its potent pesticidal effect. However, this synergy is not currently understood, and may be of interest both from a research perspective and the development of new commercial products.

Neem has been used for medicinal purposes for many centuries, but its pesticidal properties were not reported until the early part of this century. It was not until 1962 when Indian scientists ground up the seeds of neem in water and sprayed the resulting suspension over different crops that the potential of neem as a pesticide was envisaged (Pradhan *et al.,* 1962). The research that followed this initial investigation went in two directions. The potential for safe, cheap, environmentally benign pesticides that could be produced on-farm using existing farm, or using home utensils was recognised, and has been the focus of research in many developing countries (Moser, 1996). Neem-based fertilizers have been effective against the pest southern armyworm.

Neem cake is often sold as a fertilizer. Simultaneously, the potential for pesticides that could break the resistance of pests to synthetic pesticides and reduce negative impacts on the environment was of great interest to commercial pesticide companies, especially those in developed countries

Neem Composition: A. Neem Fruit (fresh):

1. Greenish Brown Kernels - 30%

2. Other Shell, Pulp etc., -70%

B. Neem Seed: Shell - 55.3% Kernel - 44.7%

C. Neem Kernel Oil Content: - 46-48%

D. Other Ingredients:

1. Azadirachtin - 0.3%

2. Nimbidin - 1.2 – 1.6%

3. Nimbin – 1%

4. Nimbinin - 0.01%

5. Vepinin - 0.15%

Neem Oil: The Neem oil varies in colour depending on the number of days the seeds are soaked in water; it can be golden yellow, yellowish brown, reddish brown, dark brown, greenish brown, or bright red It has a strong odour that is said to combine the odours of peanut and garlic. It is composed mainly of triglycerides and contains many triterpenoid compounds, which are responsible for the bitter taste. Azadirachtin is the most well-known and studied triterpenoid in neem oil. Nimbin is another triterpenoid which has been credited with some of neem oil's properties as an antiseptic, antifungal,antipyretic and antihistamine (Kraus. 1995).

Method of processing is likely to affect the composition of the oil, since the methods used, such as pressing (expelling) or solvent extraction or standard cold pressed method are unlikely to remove exactly the same mix of components in the same proportions. The neem oil yield that can be obtained from neem seed kernels also varies widely in literature from 25% to 45%.

Neem seed oil can also be obtained by solvent extraction of the neem seed, fruit, oil, cake or kernel in addition to standard procedures of cold press and normal crushing in a cursher. A large industry in India extracts the oil remaining in the seed cake using hexane. This solvent-extracted neem oil is of a lower quality as compared to the standard cold pressed neem oil and is mostly used for soap manufacturing. Neem cake is a by-product obtained in the solvent extraction process for neem oil.

It is hydrophobic in nature; in order to emulsify it in water for application purposes, it is formulated with surfactants. The fatty acid composition of neem oil is given in Table 17.

Table. 17: Fatty acid composition of Neem oil (Sadanandasamy *et al.*, 2013)

Fatty acid	Systematic Name	Formula	Structure	%
Linoleic acid	9,12-octadecadienoic acid	C18H32O2	C18:2	34.69
Oleic acid	9-octadecenoic acid	C18H34O2	C18:1	20.46
Stearic acid	octadecanoic acid	C18H36O2	C18:0	20.42
Palmitic acid	hexadecanoic acid	C16H32O2	C16:0	18.66
Arachidic acid	eicosanoic acid	C20H40O2	C20:0	3.59
Behenic acid	docosanoic acid	C22H44O	C22:0	0.80
Lignoceric acid	tetracosanoic acid	C22H48O2	C24:0	0.55
Palmiticoleic acid	9- hexadecenoic acid	C16H30O2	C16:1	0.17

The juice of this plant is a potent ingredient for a mixture of wall plaster, according to the *Samarā□ga□a Sūtradhāra*, which is a Sanskrit treatise dealing with *Śilpaśāstra* (Vedic science of art and construction)

Other uses (*Wikipedia)*

- Hair Comb: Wood of neem tree is used to handcraft hair combs and it is believed that regular use can control hair loss, dandruff and other scalp problems.
- Toiletries: Neem oil is used for preparing cosmetics such as soap, shampoo, balms, and creams as well as toothpaste
- Animal Treatment: Used to treat sweet itch and mud fever in horses

- Toothbrush: Traditionally, slender neem twigs (called *datun)* are first chewed as a toothbrush and then split as a tongue cleaner. This practice has been in use in India, Africa, and the Middle East for centuries. It is still used in India's rural areas. Neem twigs are still collected and sold in rural markets for this use. It has been found to be as effective as a toothbrush in reducing plaque and gingival inflammation (Bhambal *et al.*, 2011)
- Tree: Besides its use in traditional Indian medicine, the neem tree is of great importance for its anti-desertification properties and possibly as a good carbon sink (Paul, 1992; Chavan and Rasal, 2010).
- Neem gum is used as a bulking agent and for the preparation of special purpose foods.
- Neem blossoms are used in Andhra Pradesh, Tamil Nadu, and Karnataka to prepare *Ugadi pachhadi*. A mixture of neem flowers and jaggery (or unrefined brown sugar) is prepared and offered to friends and relatives, symbolic of sweet and bitter events in the upcoming new year festival, Ugadi. "*Bevina hoovina gojju*" (a type of curry prepared with neem blossoms) is common in Karnataka throughout the year. Dried blossoms are used when fresh blossoms are not available. In Tamil Nadu, a *rasam* (*veppam poo rasam*) made with neem blossoms is a culinary speciality.
- Cosmetics: Neem is perceived in India as a beauty aid. Powdered leaves are a major component of at least one widely used facial cream. Purified neem oil is also used in nail polish and other cosmetics.
- Bird repellent: Neem leaf boiled in water can be used as a very cost-effective bird repellent, especially for sparrows.
- Lubricant: Neem oil is non-drying and it resists degradation better than most vegetable oils. In rural India, it is commonly used to grease cart wheels.
- Fertilizer: Neem extract is added to fertilizers (urea) as a nitrification inhibitor.
- Crop protection: In Karnataka, people grow the tree mainly for its green leaves and twigs, which they puddle into flooded rice fields before the rice seedlings are transplanted. This operation is useful in repelling the pests of rice crop.
- Resin: An exudate can be tapped from the trunk by wounding the bark. This high protein material is not a substitute for polysaccharide gum, such as gum arabic. It may, however, have a potential as a food additive, and it is widely used in South Asia as "Neem glue".
- Bark: Neem bark contains 14% tannin, an amount similar to that in conventional tannin-yielding trees (such as *Acacia decurrens*). Moreover, it yields a strong, coarse fibre commonly woven into ropes in the villages of India.

- Honey: In parts of Asia neem honey commands premium prices, and people promote apiculture by planting neem trees.
- Soap: 80% of India's supply of neem oil now is used by neem oil soap manufacturers.
- Neem oil is used in Ayurvedic and Unani medicine to treat a wide range of afflictions such as skin diseases, inflammations and fevers, and more recently rheumatic disorders.

A comparative account of Neem and Pongamia is given in Table. 18.

Table. 18: Comparative account of Neem and Pongamia (VAYUGRID 2014)

Parameter	Neem	Pongamia
Genetic Diversity	Only 20 million trees in India http://www.neemfoundation.org/neem-articles/using-neem/greening-india-with-neem.html)	Tree population 37 to 38 million trees in India.
Seed viability	<3weeks (poor)	Around 6 to 8 months
Temperature tolerance	0-48*C	1-50*C
Rainfall	450-1,200 mm	400-2,500 mm
Soil requirement	Black, silt, loamy with good moisture holding capacity and drainage. Desert sandy soils are not suitable	All kinds of soils including sandy and rocky types
Tolerance to drought	Good, but seedlings are vulnerable	Very Good
Tolerance to water-logging	Bad. Vulnerable to water-logging	Good. Tolerates water logging
Biomass at 10 yrs (fresh weight)	15.3 t/ha	125 t/ha
Spacing	6m X 5m	5m X 4m
No. of trees /hectare	333	500
Full growth at	12-15 years	8-10 years
Seed yield/ tree at 10 yrs	30-40 kg (?)	40-60 kg (elite Pongamia)
Oil content in seed	20%-40%	33 - 40%
Initial Seed yield	4 to 5 years	3 – 4 Years (Elite grafted Pongamia)
Method of propagation	Seed, Seed propagation is most popular in Neem	Seed, cuttings & Veg. Propagation by grafting
Tolerance to soil problems	Tolerates alkaline soils	Tolerant to salinity and alkalinity
% of Oleic and Linoleic Acids in oil	low	high
Ingredients of commercial interest	Azadirachtin & Nimbin	Karanjin & Pongamol

Mahua (*Madhuca indica* var. *latifolia*)

Madhuca indica var. latifolia / Madhuca longifolia commonly known as *Mahua* is found largely in the central and north Indian plains and forests. The other common names prevalent in India are *mahuwa, mahwa, mohulo,* or *Iluppai* or *vippa chettu*in the regional Indian languages. The tree is fast-growing reaching up to approximately 20 meters in height with evergreen or semi-evergreen foliage, *Madhuca* is of the family Sapotaceae.

Synonyms

Madhuca longifolia J. F. Gmel, *Madhuca latifolia* (Roxb.) J. F. *Macbr, Illipe latifolia* (Roxb.) *F. Muell., Bassia longifolia* L., *Bassia latifolia* Ro

Mahua is sometimes substituted for *Diploknema butyracea* (Roxb.) H. J. Lam, a related species with similar properties that was formerly known as *Madhuca butyracea* (Roxb.) J. F. Macbr.

Mahua (*Madhuca longifolia* (L.) J. F. Macbr.) is a multipurpose tropical tree mainly cultivated or harvested in the wild in Southern Asia for its edible flowers and oil seeds (CJP, 2007; Fern, 2014; Orwa *et al.*, 2009).

It is adapted to low rainfall environments including arid zones. Mahua forms a prominent tree in tropical mixed deciduous forests in India in the states of Odisha, Chhattisgarh, Jharkhand,

Uttar Pradesh, Bihar, Maharashtra, Gujarat, Telangana, Andhra Pradesh, Madhya Pradesh, Kerala, Karnataka and Tamil Nadu.

Mahua tree produces around 20 to 200 kg of oleaginous seeds per year depending upon its maturity and various other factors. The fat extracted out of the seeds is used in skin care and dermatology, in the manufacture of vegetable butter, detergents and soaps. The fat / oil may be used as fuel oil too. Mahua flowers, fruits and leaves are edible and used as vegetables in India and other Southern Asian countries. The sweet, fleshy flowers are eaten fresh or dried, powdered and cooked with flour, used as a sweetener or fermented to make alcohol (Fern, 2014). The flowers are used to produce an alcoholic drink in tropical India.

Several parts of the tree, including the bark, are used for medicinal properties. *Madhuca* tree considered holy by many tribal communities because of its multifarious uses in the day to day lives of tribal populations of India.

Fig. 28: A portion of Madhuca indica tree showing fruiting branches with leaves

The leaves of *Madhuca indica* (*M. longifolia*) form the feed of the larvae of the moth *Antheraea paphia*, which produces tassar silk of commercial importance in India (Vantomme, 2002). Leaves and fruits (Fig. 28) are also lopped to feed goats and sheep.

The cake resulting from oil extraction is used as a fertilizer, and could be used to control root-knot nematode and fungal infections. Because of its high saponin content the incorporation of the cake reduces nematodes and phytopathogenic fungi (Gupta, 2013; Orwa *et al.*, 2009). Mahua wood is also used for its hard, strong, dense and reddish timber (Orwa *et al.*, 2009). Mahua flowers produce a nectar that is very valuable to honey bees in periods of scarcity (Singh *et al.*, 2008). Mahua is reported to have many applications in traditional medicine, and to provide several environmental benefits. Mahua is an oil bearing tree whose seeds yield between 35 and 47% oil (CJP, 2007; Ratnabhargavi, 2013).

Candidate Plus Trees (CPTs) of *Madhuca* spp., variation and inter-relationship among important characters: Kumaran (2015) identified 30 candidate plus tree (CPTs) of mahua, which included 29 from Tamil Nadu and one from Chhattisgarh.

Regarding the fruit characteristics (Kumaran, 2015), length of fruit varied from 30.3 mm (Thentirupathy -1) to 51 mm (Bellathy), while the fruit diameter varied from 23 mm (Bellathy) to 31 mm (M.G. Puddur). The fruit weight ranged from 10.2 g (Bellathy) to 19 g (Thentirupathy -1), while the number of seeds/fruit ranged from one (Bellathy) to five (Thentirupathy -1). Kumaran (2015) also reported the variation for 100 seed weight ranging from 160 g (Villupuram) to 270 g (Karaikal). Variation for oil content ranged from 30.0 percent to 52.8 percent. The highest seed oil percentage was recorded from the source TNML-12 (Poiyanallur of Ariyalur district).

Seed oil content was found to be positively correlated with 100 seed weight (Kumaran, 2015). Also, oil content showed a positive and significant association with other traits viz. kernel weight, seed weight, seed length and seed width. The kernel weight in the above study exhibited a highly positive correlation with germination percentage.

Flowering: The flowering season starts in February and extends up to April. The copious fall of succulent, corollas create a cream coloured carpet on the ground. Mahua flower is rich in sugar (73%) and next to cane molasses constitute the most important raw material for alcohol fermentation. The yield of 95% alcohol is 405 litres from one ton of dried flowers. (http://svlele.com/mahuae.htm)

Fruiting: The mature fruits are shed from the tree in May and July in the North India and August and September in the South India. The orange brown ripe fleshy berry is 25 to 50 mm long and contains one to four shining seeds. The seeds can be separated from the fruit wall by pressing. Drying and decortication yield 70% kernels by the seed weight.((http://svlele.com/mahuae.htm)

Kernel: Each seed contains two kernels. Seed collection of mahua is not easy due to logistic problems and prevalence of abiotic and biotic problems involved in the forest ecosystem. Birds and animals eat the fleshy fruits. Collection of seeds and kernel separation is done by tribal populations who inhabit the forest and rural people in a very unorganized manner. Hence, there is considerable loss during drying and storage of kernels. Kernels contain 50% oil.

Storage of Kernels: The storage condition determines the quality of expelled oil, as the kernels are susceptible to both fungus and insect attack during the storage. *Aspergillus flavus* and *Rhizophus* sp. are chief among the fungi and *Oryzaephilus surinamensis*, an insect, is known to cause maximum damage. Storage of kernels at 5 to 7% moisture level in Gunny bags with aluminium phosphide, would preserve the kernels well. Fresh oil from properly stored seeds is yellow in colour, while commercial oils are generally greenish yellow with a disagreeable odour and taste. The FFA of oil from fresh kernels is 1 to 2 percent. Poorly stored kernels yield oil of 30% FFA or even higher.

(http://svlele.com/mahuae.htm)

Oil: Mahua seed contains 35% oil and 16% protein.

Mahua oil is reported to have potential use in biodiesel production. In India, mahua oil produced contributing to biodiesel production of the order of 60 million t/year (CJP, 2007). The high saponin content of mahua cake reduces nematodes (Yadav *et al.*, 2005) and phytopathogenic fungi, when applied as a soil amendment of fertilizer value. (Gupta, 2013; Orwa *et al.*, 2009).

Mahua oil is used also to make soaps and candles and is also used as a seed preservative against pests (Orwa *et al.*, 2009). The characteristic features of mahua oil and its fatty acid composition are given Tables 19 and 20 respectively.

Mahua is reported to have many applications in traditional medicine, and to provide several environmental benefits. The tree has pronounced association with soil mycorrhizae which results in fixation of atmospheric nitrogen. The dual kind of root system consisting of vast and intense network of superficial root system in addition to the profound tap root system, brings about conservation of soil by binding the soil, thereby checking erosion. Leaves of mahua falling on the soil, contribute to soil organic matter (Bargali *et al.*, 2015; Manna *et al.*, 2004)

Table. 19: Characteristic features of fat / oil of Mahua are as under (http://svlele.com/mahuae.htm)

Characteristic feature	Value / observaion
Colour	Pale yellow
Consistency	Plastic
Refrctive index @40*C	1.452 -1.462
Specific gravity @ 15*C	0.856 -0.870
Iodine Value	58 to 70
Specification value	187 to 196
Un saponification value	1 to 3

Table 20: The fatty acid profile of Mahua fat is as given below (http://svlele.com/mahuae.htm)

Fatty acid	Percentage
Palmitic acid	16.0 – 28.2
Stearic acid	20.0 – 25.1
Arachidic acid	0.0 – 3.3
Oleic acid	41.0 – 51.0
Linoleic acid	8.9 – 13.7

Management: Propagation by grafting of the scions of elite mother trees on to one-year-old root stocks is a successful method (Singh, 2016). The grafted saplings are planted in September at 8 m spacing intervals. Mahua saplings are reported to respond to application of N, P, K fertilizers. Fruit bearing takes almost 10 years following planting. The trees yield up to 60 years. Old trees are more productive than young trees. Although the younger trees respond to irrigation, it may be avoided at flowering. Irrigation may be useful during fruit development. Inter-cropping of annual crops may be carried out in a plantation of mahua.

In India, fruits are harvested in May, June and July. Seed yield ranges from 60 to 80 kg/tree/year. Mahua trees also yield 100 to 150 kg of flower dry matter/tree/year (Singh, 2016) (Fig.29).

The fruit has one to two kidney-shaped kernels (Mohibbeazam *et al.*, 2005). The oil content of dried Mahua seeds is about 50 %. The oil is characterized by free fatty acid (FFA) content of around 20 wt% and a comparatively high percentage of saturated fatty acids such as stearic (14.0 wt%) and palmitic (17.8 wt%) acids (Ghadge and Raheman, 2006). The remaining fatty acids are mostly spread among unsaturated components such as linoleic (17.9 wt%) and oleic (46.3 wt%) acids (Mekala and Gupta, 2014). The relatively high percentage of saturated fatty acids (35.8% wt) found in Mahua oil results in relatively poor low-temperature properties of the parallel methyl esters, as evidenced by PP value of 6 °C (Ghadge and Raheman, 2006). The refractive index of the oil is1.452 (Wikipedia)

Propagation: The tree reproduces through seed and coppices. The suitable sowing time in India is July to August. Natural reproduction takes place by seeds which germinate early in the rainy season, soon after falling from a tree. Forest fires, trampling and browsing by animals, however, take a heavy toll of the seedlings under natural forest conditions.

Germination and growth of seedlings originating from seeds is slow. Seed scarification with hot water and concentrated H2SO4 and, pre-sowing treatment with GA3 significantly increase percentage germination (Dhillon *et al.*, 2014) Seeds of fruits of round, round, tapering, oval and oval tapering shapes produce robust and healthier seedlings and such saplings form better rootstocks for grafting of scions of superior plus trees.(Kumaran, 2015)

In general, woody trees like mahua are difficult to regenerate under in vitro conditions.

The success of stem cuttings depends upon the physiological status of mother stock plant, its nutrient status and general growth conditions; type of cutting, pre-planting treatments and environmental conditions including season (Dhillon *et al.*, 2014).

The success rate of sprouting of cuttings is very low. However, grafting (cleft grafting) was found to be quite successful method of vegetative propagation of Mahua tree (Kumaran, 2015). According to Kumaran (2015) using the rootstocks arising from the same tree contributing scions, gave better success and is recommended for rapid production of grafted clones of superior mahua trees.

Other uses: The alkaloids in the press cake of *mahua* seeds is reportedly used in killing fishes in aquaculture ponds in some parts of India. The cake serves to fertilise the pond, which can be drained, sun dried, refilled with water and restocked with fish fingerlings.

The *mahua* flower is edible and is a food item for tribal populations. They use it to make syrup for medicinal purposes (Trifed, 2009).

Processing of flowers and Use as food material: The flowers are also fermented to produce country liquor. Tribals of Bastar in Chhattisgarh and peoples of Western Orissa, Santhals of Santhal Paraganas of Jharkhand state, Koya tribals of North-

East and Andhra Pradesh, Bhil tribal of western Madhya Pradesh and tribals of North Maharashtra consider the tree and the *mahua* drink as part of their cultural heritage. *Mahua* drink is considered an essential for tribal men and women during celebrations (India9.com 2014)

The main ingredients used for making it are *chhowa gud* (granular molasses) and dried *mahua* flowers.

Mahula fruit is an essential food of Western Odisha people. The tree has a great cultural significance. There are so many varieties of food prepared with the help of fruit and flower. Also Western Odisha people pray this tree during festivals. The liquor produced from the flowers is largely colourless, opaque and not very strong. The taste is reminiscent of sake with a distinctive smell of *mahua* flowers. It is inexpensive and the production is largely done in home stills.

Mahua flowers are also used to manufacture jam, which is made by tribal co-operatives in the Gadchiroli district of Maharashtra (Times of India, 2012)

Fig. 29: Sun dried mahua flowers and Normal flowers of mahua tree

In many parts of Bihar, such as villages in the district of Siwan, the flowers of *mahua* tree are sun-dried; these sun-dried flowers are ground to flour and used to make various kinds of breads.

Trifed (2009) of the Ministry of Tribal Affairs, Government of India reports: "Mahua oil has soothing properties and is used in the treatment of skin disease, rheumatism and headache. It is also a laxative and considered useful in habitual constipation, piles and haemorrhoids and as an emetic. Native tribes also use it as an illuminant and hair fixer and it has also been used as biodiesel" (Hindu, 2014).

Oil extraction: Mahua seeds contain up to 50% oil. In the industrial extraction process, the seeds are first broken and flaked, the resulting flakes are then steam-cooked. The cooked flakes are crushed and solvent-extracted with hexane at 63°C. The resulting mahua oil meal contains less than 1% oil. In smallholder farms, the seeds are only crushed, resulting in an energy-rich mahua seed cake containing up to 17% oil (Ratnabhargavi, 2013; Singhal *et al.*, 1986).

Detoxification: Palatability of mahua seed cake is reported to get affected due to certain degree of its toxicity. However, the quality of cake can be enhanced by a process of detoxification involving washing with water, which was tested with some success for ruminants and poultry (Singhal *et al.*, 1986).

Slow growth and long gestation period of *Madhuca indica* are perhaps responsible for the lack of more basic scientific data on physiology, genetics and breeding of the tree. Nevertheless, with the availability of modern tools of biotechnology and genetic engineering it should be possible to over the limitations cited above.

SAL (*Shorea robusta*)

Fig. 30: Canopies of Sal tree

Fig. 31: Seeds of Sal Tree- some seeds with winged calyx

Shorea. robusta (Sal) is a deciduous tree that grows up to 50 m and 5 m girth. The tree belongs to the family of Dipterocarpaceae.– Meranti family. Under normal conditions they reach 18–32 m with girths of 1.5–2 m. The trunk is clean, straight and cylindrical, often bearing epicormic branches. The crown is spreading and spherical. The bark is dark brown and thick, with longitudinal fissures deep in poles, becoming shallow in mature trees, and provides effective fire protection. The tree develops a long taproot at a young age. The tree grows at 100–1,500 m altitude. The mean annual temperature required is between 22–27 °C and 34–47 °C The tree requires mean annual rainfall between 1,000–3,000 mm and maximum of 6,600 mm *S. robusta* flourishes best in deep, well-drained, moist, slightly acid, sandy to clayey soils. It does not tolerate waterlogging. The most favourable soil is a moist sandy loam with good subsoil drainage. Soil moisture is essential.

Leaves are simple, shiny, about 10–25 cm (3.9–9.8 in) long and broadly oval at the base, with the apex tapering into a long point. New leaves are reddish, soon becoming delicate green.

Flowers are yellowish-white, arranged in large terminal or axillary racemose panicles.

Fruit at full size is about 1.3–1.5 cm (0.51–0.59 in) long and 1 cm (0.39 in) in diameter; it is surrounded by segments of the calyx enlarged into 5 rather unequal wings about 5.0–7.5 cm (2.0–3.0 in) long. Fruit content is 66.4% kernel and pod, 33.6% is shell and calyx. The fruits generally ripen in May.

The seed contains 14-15% fat. It has calyx and wings. The de-winged seeds contain a thin, brittle seed pod. The kernel has 5 segments covering the embryo. 2 kg (4.4 lb) of seeds give 1 kg (2.2 lb) of kernel. The seeds contain 10.8% water, 8% protein, 62.7% carbohydrate, 14.8% oil, 1.4% fiber and 2.3% ash (World Agroforestry, 2013).

Habitat

The habitat of *Sorea robusta* is situated in ranges south of the Himalaya, from Myanmar in the east to Nepal, India and Bangladesh (Oudhia and Ganguli, 1998). The tree is widely distributed in tropical regions and covers about 13.3% of the forested area in the country. Sal (*Shorea robusta*) tree occurs either gregariously or mixed with other trees in Himalayan foot hills and central India. In the Himalayan foothill belt it extends up to Assam valley (including Meghalaya and Tripura) in the east to foothills of north-west Bengal, Uttar Pradesh, Uttaranchal, Kangra region of Himachala Pradesh.

Livelihood to Tribal Populations

Sal tree is the major source of livelihood to forest dwellers in the Central Indian states of Orissa, Chhattisgarh and Madhya Pradesh, which form India's major sal belt, spread over 45% of their forested areas. Around 20-30 million forest dwellers in the above regions depend on collection of sal seeds, leaves and resins (NTFP, 2019). The proper storage of seeds before processing is crucial. Excess moisture (>6-8%) damages oil quality via hydrolysis in the seed fat, with resulting high free fatty acid oil output. Sal fat is extracted by three methods. Traditionally fat is extracted by water rendering. Secondly, the mechanical system of extraction is carried out by oil expeller and rotary mills. The third method involves solvent extraction in which the seed is pressed as flakes first in a flaker mill and exposed to solvent extraction.

Characteristics of Sal fat/ oil

The extracted crude sal oil/fat is greenish-brown and has a characteristic odour. Due to the presence of more saturated fatty acids, it is solid at room temperature. Because of this, it is known as sal fat or sal butter. The oil is used as cooking oil after refining. The oil contains 40-50% stearic acid and 35-45 % oleic acid (Table 21). The refined oil is used as substitute for cocoa butter in chocolate manufacturing. (Orwa *et al.*, 2009)

Table. 21: Physical Properties and faty acid composition of Sal oil

Property	Range/limit
Appearance	white colour solid
Odor	characteristic odour
Taste	Typical taste
Specific gravity	.88-0.914
Slip melting point	30-35 °C
Peroxide value	4.0
Iodine value	38-42
Saponification value	187-193
Unsaponifiable matter	1.2% max
Refractive Index at 40 °C	1.4500-1.4600[12]
Titer °C	46-53[12]

Fatty acid	Percentage
Palmitic acid (C16:0)	4-8
Stearic acid (C18:0)	40-50
Arachidic acid (C20:0)	5-8
Oleic acid (C18:1)	35-45
Linoleic (C18:2)	1-4

Applications

Sal oil or butter is used for cooking locally and used for soap making up to 30%. Refined, modified fat is a substitute for cocoa butter and used in confectionery industry. Sal butter is used in the manufacturing of hydrogenated edible ghee (*vanaspati*), paints and pigments, lubricants, auto oil, etc.

Processing of seeds: Calyx segments surround the seed (Fig.31). These take the form of five unequal wings about 5–7.5 cm long (Orwa *et al.*, 2009) and provide the seed with dispersal mechanism by allowing the seed to be blown up to 100 m away from the tree to regenerate into future seedlings.

After collection, the seeds are decorticated (the "wings" removed) (Fig.31) and they are dried. Sometimes they are decorticated by setting fire to them. This can be quite risky as this can reduce the oil content of the seed. Alternatively, the seeds are decorticated mechanically. The moisture content of the seeds is important in terms of keeping a low level of FFA. It is particularly important to collect the seeds, decorticate them, and dry them as quickly as possible because the collection season is followed by the monsoon season. It is also important to keep the time gap between collection and seed crushing to a minimum to ensure a good quality oil. A maximum of 72 hrs is said to be ideal but this can often be up to 4 or 5 months.Oil is removed from the seeds either by expelling or solvent extraction (or a combination of the two processes). The kernels typically contain between 14% and 18% oil (Talbot 2015). In its crude state, sal oil has a deep green color; even when fully refined a slight green tinge still remains. Typical fatty acid compositions of sal oil indicates (Table 21) that major fatty acids are stearic and oleic acids found in roughly equal quantities together with an appreciable level (typically about 7%) of arachidic acid. Like illipe butter, this fatty acid configuration will give an oil rich in SOS triglycerides, although the higher level of oleic acid will also give a significant amount of SOO triglycerides. The oil contains between 55% and 75% total SOS and between 20% and 30% more unsaturated triglycerides. This means that, as far as its use is concerned, it could, in some circumstances, be used unfractionated but would most likely need some form of fractionation to concentrate the SOS triglycerides (Talbot, 2015).

Illipe butter is the fat from the nuts of the *Shorea stenoptera*, which is a wild crop in the Borneo jungle of Southeast Asia. It flowers from October through January and the nuts are collected once they fall to the ground. It becomes a pale yellow, solid fat after extraction with a chemical composition that parallels cocoa butter with a lightly higher melting point. Illipe butter contains oleic, palmitic, steric, and linoleic acids. It is a hard butter that melts when in contact with skin (https://www.naturallycurly.com/curlreading/home/what-is-illipe-butter).

Reproductive Biology

Shorea robusta is bisexual or hermaphroditic, self-incompatible species. Pollen vectors or pollinators in its natural habitat are insects of the family *Thysanoptera*. (Orwa *et al.*, 2009). Flowering of *Shorea* is correlated with the previous drought period. Beginning at about age 15 years , *S. robusta* bears fruit regularly every 2 years or so, and higher levels of seed yield are harvested once in 3-5 years. Sal seed is dispersed by wind and water.

Ambiental factors needed for *Shorea robusta*

Sal tree has specific requirements unlike many other forest trees as given below.

Altitude: 100-1500 m,

Mean annual temperature: (min. 1-7) 22-27 (max. 34-47) deg. C,

Mean annual rainfall: 1000- 3000 (max. 6600) mm

Soil type: *S. robusta* flourishes best in deep, well-drained, moist, slightly acid sandy to clayey soils. It does not tolerate waterlogging. The most favourable soil is a moist sandy loam with good subsoil drainage. Availability of soil moisture is an important factor determining the occurrence of *S. robusta*. Sal accounts for about 14% of the total forest area in India. For example, southwest Bengal harbours luxuriant Sal forests. Fire is normally responsible for its frequent occurrence in pure stands or as the dominant species of mixed stands, as *S. robusta* is better equipped to survive conflagrations than other tree species

Intercropping /Mixed cropping: In Assam India, artificial regeneration of *S. robusta* is practised in combination with crops such as upland rice, maize, sesame and mustard. Good results have also been achieved with mixed plantations in which *S. robusta* is cultivated together with *Tectona grandis*. (Orwa *et al*., 2009)

Food products from Sal Tree: In India seeds are boiled into a porridge with flowers of *Madhuca latifolia* and fruits (green pods) of *Dolichos biflorus*.

In Tamil Nadu India, seeds are ground into a coarse flour used to make bread, and the plant is used as a famine food.

Composition of seeds: Seeds consist of 10.8% water, 8% protein, 62.7% carbohydrate, 14.8% oil, 1.4% fibre and 2.3% ash. *S. robusta* butter, used in cooking, is derived from the seeds. A de-fatted kernel powder, popularly known as sal seed cake, contains about 50% starch, in addition to proteins, tannins and minerals. The physico-chemical property of the starch can be exploited for preparing canned food products (Orwa *et al.*, 2009)

Fodder: In India, S. robusta is lopped for fodder, but the leaf fodder is considered to be of medium to poor quality. The oil cake, though rich in tannins (5-8%), has been used without detrimental effects in concentrates for cattle in proportions of up to 20%.

Sal-seed cake may be used for animal feed. It constitutes up to 10% of poultry and pig feeds .

The leaves can be used as roughage for cattle. Leaves are also fed to the larvae of *Antheraea mylitta*, a tasar silk-producing worm.

Lipids: *S. robusta* fat is being exported thereby contributing to the foreign exchange earnings of India. (Orwa *et al.,* 2009)

Timber: The dark, reddish brown, hard and heavy heartwood (specific gravity of 0.83-0.93 cm^3) is very durable and highly resistant to termite attack; grain is strongly spiralled and rather coarsely structured; seasoning also presents problems. Wood is hardwood, which is well suited for structures subject to heavy stress in house construction, hydraulic engineering, ships and railway cars. It is also used for poles, railway ties and posts, simple interior finishing such as window frames and floors, and many other applications

Other products: *S. robusta* leaves are widely used for making leaf plates and cups for household use and for community festivity events. When tapped, the tree exudes large quantities of a whitish, aromatic, transparent resin known as *'lal dhuna'*. It is used to caulk boats and ships and as incense. In some places in the Upper Tista forests of the Darjeeling District, large pieces, often 450 - 600 cubic centimetres in size, are found in the ground at the foot of the trees (Gamble, 1972). The leaves are widely used for making plates, cups and for wrapping (Manandhar, 2002). An oil obtained from the seed is used for illumination (Manandhar,2002). The bark is a source of tannins (Manadhar, 2002)

Medicinal(https://www.bimbima.com/ayurveda/sal-tree-shorea-robusta-information-and-uses/49/): The resin obtained from the tree is stomachic, astringent, antidysenteric, styptic, antiseptic, and antigonorrhoeic. The bark, and leaves are cooling, astringent, acrid, anthelmintic, pain-relieving, constipating, urinary astringent, Tonic, and purifying.

- Analgesic (resin): Acting to relieve pain.
- Antipyretic (resin): Effective against fever.
- Anti–inflammatory (leaves decoction): Reducing inflammation by acting on body mechanisms.
- Antinociceptive (leaves): Inhibits nociception, the sensation of pain
- Antibacterial (flower decoction):
- Anti–Obesity (leaves): Reduces obesity
- Antiulcer (resin): Tending to prevent or heal ulcers.
- Immunomodulatory (bark): Modifies the immune response or the functioning of the immune system.
- Wound healing: Heals the wound.

Tree Management: Young plants grow quickly, attaining top heights of up to 6 m after 6 years. The first thinning is usually performed after 5 years, and thereafter the trees are thinned every 5-10 years. Rotations of 30-40 years are used when coppice regeneration is practised, and 80-160 years for high forest regeneration. Sal tree requires a lot of light, *S. robusta* coppices well. Both coppice regeneration and planting of seedlings are used in plantation systems. It is also ideally used for growing under Taungya systems, which consists of land preparation for tree plantation, growing agricultural crops for 1 to 3 years after the tree plantation and moving on to another area to repeat the cycle. It is a method of raising forest plantations in which cultivators are allowed to raise agricultural crops for initial periods of a few years and in return they are made to raise forest plantations.

Sal Seeds: One kilogram consists of 400 to 1000 seeds. Seed Storage is beset with difficulties, since seed viability is lost within 2 weeks in open storage at room temperature. Around 10% germination is observed on desiccation at 33-36*(dig) C to 9.6% moisture content.

Biotic Stresses

Among the insect pests in India, *Hoplocerambyx spinicornis* is the most destructive. Its larvae tunnel through the bark, sapwood and finally to the heartwood, causing death of the tree. The population of the pest keeps building up and assumes an epidemic level. Other insect pests include *Diacavus furtivus* and *Xyleborus* spp. The major fungal diseases include those caused by *Polyporus shorea* and *Polyporus gilvus*. The phanerogamic parasite vz., *Loranthus scurrula* can also cause increment losses (Orwa *et al.*, 2009)

Drumstick Tree (*Moringa oleifera)*

Moringa oleifera (drumstick tree) is a fast-growing, drought-resistant tree of the family Moringaceae (Morton, 1991) native to tropical and subtropical regions of South Asia(Olson, 2002).Common names include Moringa (GRIN, 2017), drumstick tree (from the long, slender, triangular seed-pods), horseradish tree(GRIN, 21017) (from the taste of the roots, which resembles horseradish), and ben oil tree or benzolive tree (GRIN, 2017) (from the oil which is derived from the seeds).

Moringa derives from a Tamil word, *murungai*, meaning "twisted pod", alluding to the young fruit.(Olson, 2002). The species name is derived from the Latin words *oleum* "oil" and *ferre* "to bear" ("*Moringa oleifera*". Flora of Australia Online. Department of the Environment and Heritage, Australian Government.*)* A combined analysis of morphology and DNA shows that *M. oleifera* is most closely related to *M. concanensis*, and the common ancestor of these two diverged from the lineage of *M. peregrine* (Olson, 2002)

Description of the Tree: The tree can reach a height of 10–12 m (32–40 ft) and trunk diameter of 45 cm (1.5 ft) (Parotta and John,1993). The bark has a whitish-grey colour and is surrounded by thick cork. Young shoots have purplish or greenish-white, hairy bark. The tree has an open crown of drooping, fragile branches and the leaves build up a feathery foliage of tripinnate leaves.

The flowers are fragrant and hermaphroditic, surrounded by five unequal, thinly veined, yellowish-white petals. (Fig.32). The flowers are about 1.0–1.5 cm (1/2") long and 2.0 cm (3/4") broad. They grow on slender, hairy stalks in spreading or drooping flower clusters which have a length of 10–25 cm (Parotta and John, 1993). The evergreen or deciduous foliage has leaflets 1 to 2 cm in diameter. flowers are bisexual and white or cream coloured.

Flowering begins within the first six months after planting. In seasonally cool regions, flowering only occurs once a year between April and June. In more constant seasonal temperatures and with constant rainfall, flowering can happen twice or even all year-round (Parotta and John, 1993)

The fruit is long, hanging, three-sided brown capsule of 20–45 cm size which holds dark brown, globular seeds with a diameter around 1 cm. (Fig.33). The seeds have three whitish papery wings and are dispersed by wind and water (Parotta and John, 1993).

The tree tolerates a wide range of soil conditions, but prefers a neutral to slightly acidic (pH 6.3 to 7.0), well-drained sandy or loamy soil (Radovich, 2011). In waterlogged soil, the roots have a tendency to rot (Radovich, 2011). Moringa is a sun and heat-loving plant, and does not tolerate freezing or frost. Moringa is particularly suitable for dry regions, as it can be grown on rainwater without expensive irrigation techniques.The tree is often cut back annually to 1–2 m and allowed to regrow so the pods and leaves remain within arm's reach (Parotta and John, 1993)

India is the largest producer of moringa, with an annual production of 1.2 million tonnes of fruits from an area of 380 km^2 (Radovich, 2011).Moringa is grown in home gardens and as living fences in South Asia and Southeast Asia, where it is commonly sold in local markets. In the Philippines and Indonesia, it is commonly grown for its leaves which are used as food. Moringa is also actively cultivated by the World Vegetable Centre in Taiwan, a Centre for Vegetable Research.

More generally, moringa grows in the wild or is cultivated in Central America and the Caribbean, northern countries of South America, Africa, Southeast Asia and various countries of Oceania.

Propagation

Moringa can be propagated from seed or cuttings. Direct seeding is possible because the germination rate of *M. oleifera* is high. Moringa seeds can be germinated year-round in well-draining soil. Cuttings of 1 m length and at least 4 cm diameter can be used for vegetative propagation.

Planting and cultivation

For intensive leaf production, "the spacing of plants should be 15 x 15 cm or 20 x 10 cm, with conveniently spaced alleys (for example: every 4 m) to facilitate plantation management and harvests (Amaglo, 2006). Weeding and disease prevention are difficult because of the high density. In a semi-intensive production, the plants are spaced 50 cm to 1 m apart. This gives good results with less maintenance.

Moringa trees can also be cultivated in alleys, as natural fences and associated with other crops. The distance between moringa rows in an agroforestry cultivation is usually between 2 and 4 m. Moringa can also be grown as an intercrop with other crops. Drumstick plants are grown as live fences and windbreaks. Moringa plants are grown as live fences and windbreaks. Moringa is also used as wood source after coppicing (cutting back the main stem to encourage side shoots). It is also grown as an intercrop with other annual and perennial crops.

In the tropics, it is used as forage for livestock. The press cake, obtained following oil extraction, is useful as a soil conditioner

Cytology:*Moringa oleifera* possesses 2n = 28 chromosomes, paired as regular bivalents at Metaphase-I of meiosis (Silva *et al.,* 2011). Meiotic abnormalities were rare. Tripolar spindles were observed in Metaphase-II, leading to the formation of unreduced gametes, which were recorded in some plants at a low frequency. Pollen viability is higher than 88% (Silva *et al.,* 2011)

Plant Breeding

In India, from where moringa most likely originated (Olson, 2002), the diversity of wild types is large. This gives a good basis for breeding programs. In countries where moringa has been introduced, the diversity is usually much smaller among the cultivar types. Locally well-adapted wild types, though, can be found in most regions.

Because moringa is cultivated and used in different ways, there are different breeding aims. The breeding aims for an annual or a perennial plant are obviously different. The yield stability of fruits is an important breeding aim for the commercial cultivation in India. On less favourable locations, perennial cultivation has big advantages. Erosion is much smaller with perennial cultivation. In India selections are effected for a higher number of pods and dwarf or semi dwarf varieties. Breeders in Tanzania, though, are selecting for higher oil content.

Yield and harvest

M. oleifera can be cultivated for its leaves, pods, and/or its kernels for oil extraction and water purification. The yields vary widely, depending on season, variety, fertilization, and irrigation regimen. Moringa yields best under warm, dry conditions with some supplemental fertilizer and irrigation (Radovich,

2011). Harvest is done manually with knives, sickles, and stabs with hooks attached to adequately longer pole to reach the height of the tree (Radovich, 2011). At harvest pruning is recommended to promote branching, increase production, and facilitate cost effective harvesting (Grubben, 2004).

Fruits (Pods) and Seeds

When the plant is grown from cuttings, the first harvest can take place 6–8 months after planting. Often, the fruits are not produced in the first year, and the yield is generally low during the first few years. By year two, it produces around 300 pods, by year three around 400–500. A good tree can yield 1000 or more pods (Booth and Wickens, 1988). In India, a hectare can produce 31 tons of pods per year (Radovich, 2011). Under North Indian conditions, the fruits ripen during the summer. Sometimes, particularly in South India, flowers and fruit appear twice a year resulting in two harvests in July to September and March to April (Ramachandran *et al.,* 1980). It also contains oleic acid-type oil. According to Sengupta and Gupta (2000), *M. oleifera* seeds contain between 33 and 41% w/w of vegetable oil. Anwar *et al*., (2005) also investigated the composition of *M. oleifera,* including its fatty acid profile and showed that *M. oleifera* oil is high in oleic acid (>70%). Additionally the green pods are rich in vitamins and mineral (Table 22).

Leaves

The leaf of Moringa is a valuable source of protein and can be used effectively to combat malnutrition among rural children (Prasad, 2017). Senegal (Africa) has taken a significant step to produce Moringa leaf material exclusively to supply the leaf powder to rural children and pregnant mothers to overcome the problem of malnutrition in rural areas (Prasad, 2017).

One rounded spoon (8g) of leaf powder will satisfy about 14% of the protein, 40% of calcium, 23% of iron and nearly all the vitamin-A needs of a child aged 1 to 3 years. Six rounded spoonful of leaf powder will satisfy nearly all of a woman's daily iron and calcium needs during pregnancy and breast-feeding (Fuglie, 2001b).

Average leaf yields of the order of 6 tons/ha/year in fresh matter can be achieved. The harvest differs strongly between the rainy and dry seasons, with 1120 kg/ha per harvest and 690 kg/ha per harvest, respectively. The leaves and stems can be harvested from the young plants 60 days after seeding and then another seven times in the year. At every harvest, the plants are cut back to within 60 cm of the ground (Sogbo, 2006a and 2006b). In some production systems, the leaves are harvested every 2 weeks. The leaf juice has a stabilizing effect on blood pressure. The leaf juice controls glucose levels in diabetic patients. Fresh leaves and leaf powder are recommended for tuberculosis patients because of the availability of vitamin A that boosts the immune system. If leaf juice is used as diuretic, it increases urine flow and cures gonorrhoea. Leaf juice mixed with honey treats diarrhoea, dysentery and colitis (colon inflammation). Fresh leaves are good for pregnant and lactating mothers; they improve milk production and are prescribed for anaemia. *Moringa* leaves contain more Vitamin A than carrots, more calcium than milk, more iron than spinach, more Vitamin C than oranges, and more potassium than bananas," and that the protein quality of Moringa leaves rivals that of milk and eggs. Also, the leaves and stem of *M. oleifera* are known to have large amounts of their calcium bound in calcium oxalate crystals (Olsen *et al.*, 2007)

Production of Leaf material

Taking into account various factors influencing the overall efficiency such as the cost of seed and the cost of soil preparation, it was concluded that for the production of green leafy matter, the optimal planting density in a sandy soil is with10cm x 10cm spacing leading to one million plants per hectare under conditions of Northern Senegal (Prasad, 2017). The following leaf yields were reported by Fuglie (2001a) at first cutting.

Table 22: Leaf yields of Moringa under different plant densities.(Fugile, 2001a)

Planting Density (Plants/ha)	Fresh matter (tons/ha)	Dry matter (tons/ha)
95,000	19.6	3.30
350, 000	29.7	5.05
900,000	52.6	8.98
100,000	78.0	13.26

Water soluble fertilizer 21/7/20(NPK)was mixed with the water through the feeding tank attached to the drip-irrigation system which facilitated judicious nutrient management. Additionally, urea was applied @ 8 kg/ha every two weeks (Prasad, 2017).

The cultivation of *M. oleifera* can also be done intensively with irrigation and fertilization with suitable varieties (Amaglo, 2006). Trials in Nicaragua with 1 million plants per hectare and 9 cuttings/year over 4 years gave an average fresh matter production of 580 metric tons/ha/year, equivalent to about 174 metric tons of fresh leaves (Amaglo, 2006). The quantity of water utilized was of the order of 72,000 litres/day/ha @ one-hour watering with one bar pressure in the rainy season. During the dry season, 108,000 litres of water was used per hectare per day @ one hour and a half watering with one bar pressure (Prasad, 2017).

The harvested leaves were washed carefully, cleaned and dried in a spacious room with good ventilation before they were crushed and sieved to make consumable powder.

Varieties of Moringa: Moringa varieties may be broadly classified into two groups: perennial and annual(Agriculture forum, 2012).

Perennial types have probably been in cultivation for thousands of years. In India perennial types are typically propagated from cuttings. The perennial varieties exhibit long growing period before reaching maturity for production of pods, and they are more vulnerable to the attack of pests and diseases. Also the planting material in the form of the stem cuttings is rather limited to undertake plantations on large scale. Due the prolonged duration, their water requirement is much higher and as such they are not suitable drylands which suffer drought and shorter spell of rainy season.

Annual types such as Periyakulam-1 (PKM-1) and PKM-2, are largely the products of recent plant breeding research and have now replaced most the perennial varieities that previously dominated commercial production in India. They are seed propagated and they attain maturity rapidly with a wider degree of adaptability resulting in higher yields. Annual types may have significant variation in some cultural characteristics. Suthanthirapandian *et al* (1989) studied the variability of nine traits in seedling populations of annual moringa. Among the traits studied, the number of flowers per inflorescence (19.0-126.0), fruit weight (25.0-231.5g) and yield by number of fruit per plant (1.0-155.0), showed widest variability.

Varieties of *Moringa oleifera* from India: (https://sustainablebioresources.com/plants/plant-families/moringaceae/moringa-oleifera/varieties-ecotypes-moringa-oleifera/#_edn15)

The land races include Chemmurungai, Durga, , K.M.1., Kodikkal, Palmurungai, PAVM, Punamurungai, Saragva, Yalpanam, Yazhpanam.

Kodikkal murungai is a perennial ecotype grown predominantly in the betel vine growing areas of the Tiruchirapalli district of Tamil Nadu. The trees are short statured with small leaves.

Punamurungai is a perennial ecotype grown in home gardens of Tirunelveli and Kanyakumari districts[2]. It is preferred for its thick pulp and taste but other references describe it as bitter and the variety unpopular.

Palmurungai is a perennial ecotype preferred for its thick pulp and taste[2] but other references describe it as bitter and the variety unpopular.PAVM. The PAVM variety was recently developed by a farmer in the Dindigul district of Tamil Nadu and is in wide use. It is a high yielding drumstick variety that starts yielding pods from the 5th or 6th month of planting. The variety is propagated from by cuttings made by air layering (Prabu, 2009)

Among the drumstick varieties developed outside India, the following two varieties appear to be promising.

Mbololo variety developed by the Kenya Forestry Research Institute (KFRI) of Nairobi, Kenya. Oil from the seeds of this variety has been characterized (Tsaknis *et al*., 1999). It has high resistance to oxidation and a high ratio of monounsaturated to saturated fatty acids that may make it an acceptable substitute for olive oil in the diet.

Jaffna (Yazhpanam). A drumstick variety from Sri Lanka introduced into Southern India and cultivated commercially in the Tirunelveli and Tuticorin districts of Tamil Nadu. The natural antioxidants present in the seed oil of this variety has been determined and compared with that of other vegetable oils(Bhatnagar and Krishna, 2013.).

Anupama is a drumstick production variety released from KAU, India.

Coimbatore 1 is widely available in India and considered superior for drumstick production and quality. Drumsticks are 45-60cm long with two harvests per year. Tree yields product for eight to ten years. For additional information on this variety refer to the AgriFarming Drumstick Farming Guide(Agrifarming, 2017).

Coimbatore 2 is a more popular variety than Combiatore 1 in Tamil Nadu with shorter drumsticks (25-35cm long). A high yielding drumstick production variety with bulky pods. Production life from three to four years. The trees can be maintained up to 15 years without pruning(TNAU, 2017)

Valayapatti is a perennial ecotype cultivated in and around Usilampatti and Andipatti. It yields 1000-1200 pods per tree(TNAU, 2017)

Improved Indian Drumstick Varieties Developed by the Public Sector

These include KM 1, PKM 1, PKM 2, GKVK 1, 2, 3, Dhanaraj, Bhagya (KDM 1), Konkan Ruchira, Anupama and Rohit (Agriculture forum, 2012).

Bhagya (KDM-1) is a drumstick variety developed by the University of Horticultural Sciences, Bagalkot (UHS Bhagalkot, 2017) It has recently been shown to be pest resistant and well adapted to growing in the hot, dry region of Chikkodi Taluk. The variety is long-lived with a lifespan of 15-20 years (Patil, 2016)

Dhanaraj (S 6/4) was developed by TNAU and is the only dwarf variety described in the literature. It has a canopy of only 2.00 – 2.5m in height. It flowers early in 7-8 months, is high yielding (250 -300 pods/plant/year) and of good cooking quality.

Seeds of this variety may be available from the seed bank.

G.K.V.K. -1, G.K.V.K. -2 and G.K.V.K. -3 are improved drumstick varieties developed In Karnataka, Gandhi Krishi Vigyana Kendra Bangalore.

Periyakulam 1 and 2 (PKM 1 and PKM 2) are the two commercially viable annual varieties for drumstick production developed by the Horticultural Research Station of Tamil Nadu Agricultural University (TNAU). As annuals they may have to be replanted more often but may be most suited for areas with short growing seasons and climates that are too cold for the trees to survive through the winter. In India the variety can be in production for four or five years before replacement is required. Refer to the Agricultural Forum reference[1] for summary characteristics and comparison of these two varieties.

PKM1 was developed through pure line selection and is only propagated from seed. It is an early variety of medium or dwarf stature, growing fast in the first year after planting and can produce flowers and pods just six months after sowing. PKM1 may be the most widely planted variety for large scale plantation of drumstick production. It has a yield potential of 50-54 tonnes/ha and is suitable for growing in the stubble of other crops previously harvested on the same land (a ratoon crop). A full characterization of the oil from this variety has been reported (Lalas and Tasaknis, 2002.)

In addition to its rapid growth and maturity the other characteristics of the variety are as follows:

- Shorter stature
- Higher pod yield
- Uniform pod length with green colour
- Non-bitter flavour
- The pods are fleshy, non-fibrous and soft even when harvested late.
- The stem/ trunk is tensile and more flexible avoiding breakage due to heavy winds etc., and
- Pods and leaves exhibit extended shelf life

PKM 2 is a high yielding hybrid derived from cross involving MP 31 and MP 38. It produces an average of 240 fruits per tree with an average yield of 98 tons/ ha. According to a TNAU website optimal planting densities for higher yield are not uniform for the two varieties. In the case of PKM1 the recommended spacing is 1.5m X 1.0 m with two plants/hill; while a closer spacing of 1.2 x 1.2 m is recommended for PKM2. PKM1, PKM2 and KM 1 are preferred annual varieties in southern India. (Agriculture forum 2012). The water requirement of PKM2 is higher than that of PKM1. However, PKM2 is recommended for intercropping as an intermediate crop with coconut and tropical fruit orchards.

KM 1 is another promising variety recommended by the Tamil Nadu Agricultural University (TNAU 2017).

Rohit 1 is a long duration variety with early bearing habit for drumstick production. The variety yields the first flush of pods at four to six months following planting. (Agrifarming, 2017).

Oil

One estimate for yield of oil from kernels is 250 l/ha (Radovich, 2011).

According to Orhevba *et al.*, (2013) the seeds contain up to 40% of oil by weight which is used for cooking, soap manufacture, cosmetic base and in lamps. Application of enzymes for oil extraction, though giving lower yields compared to solvent-assisted extraction, may allow labelling of the oil as organic oil. The high oleic acid content of the oil confers on it a functional food property, as oils containing high oleic content have been shown to reduce the risk of coronary heart disease. The oil is also an excellent frying medium, with a low degradation rate when applied at high cooking temperatures. The use of *M. oleifera* for the prevention and treatment of various ailments has been reported in many folk medicines throughout the world. The oil is reported to be of therapeutic and nutritional value and is popularly used to cure the cases of fungal infections, constipation, prostate function, and bladder function, apart from its value as an antioxidant. The unrefined oil contains certain minor components, including tocopherols and other metabolites, that have led to its use in treating other medical disorders as well (Ghazali *et al.*, 2011)

The physicochemical properties of Moringa oil were reported by Orhevba *et al.*, (2013). The values of moisture content, ash, crude protein, crude fat and carbohydrate were 0.60%, 1.50%, 2.19%, 39.3% and 56.42% respectively. While the values obtained for pH, saponification value, iodine value, free fatty acid and specific gravity were 5.96, 164.09mg/g, 68.23g/mol, 8.27mgKOH/g and 0.86 respectively. The results showed that *Moringa oleifera* seed oil has suitable saponification and iodine values for industrial use. Ben oil is pressed from the seeds of the tree. The oil is characterized by an unusually long shelf life and a mild, but pleasant taste. The name of the oil is derived from the presence of behenic acid. The fatty acid composition of the tree oil is as follows:

Component	Percentage
Oleic acid	65.7%
Palmitic acid	9.3%
Stearic acid	7.4%
Behenic acid	8.6%

Ben oil has been used for thousands of years as a perfume base, and continues to be used in that capacity today. The oil can also be used as a fuel. Burkill (1966) reports: It burns with a clear light and without smoke. It is an excellent salad oil, and gives a good soap... It can be used for oiling machinery, and indeed has a reputation for this purpose as watch oil, but is now superseded by sperm oil

Cake, Woody-powder, Bark etc.: The press cake, obtained following oil extraction, is useful as a soil conditioner. The wood pulp may be used for paper-making. In the tropics, it is used as forage for livestock; in many countries, Moringa is used as a micronutrient powder to treat diseases. The green pods, fresh and dried leaves are used as vegetable.

The superficial layers of bark of the moringa root should be scraped off because of its toxicity and the flesh of the root may be eaten sparingly (Bever, 1986).

Medicinal Value: Apart from their use as vegetables, the green pods and leaves are widely used in traditional herbal medicine. It is also used for water purification. (Kalibbala *et al.,* 2009; Kalibbala., 2012). The tree's bark, roots, fruit, flowers, leaves, seeds, and gum are also used medicinally. The uses include as an antiseptic and in treating rheumatism, venomous bites, and other conditions. The flowers, leaves and roots are widely used as remedies for several ailments. Paste made from bark treats boils. Paste from ground bark can be applied to relieve pain caused by snake, scorpion and insect bites. Oil is sometimes applied externally for skin diseases (Maroyi, 2007)

Moringa seeds are effective against skin-infecting bacteria *Staphylococcus aureus* and *Pseudomonas aeruginosa* (Bever, 1986) They contain the potent antibiotic and fungicide *terygospemin.* Moringa seem to have most of the food nutrients required by the body to replenish its defensive mechanisms, it is considered useful against asthma, gout, rheumatism and enlarged spleen or liver. It also helps in the removal of wind from the stomach and as a snuff and can be used to alleviate ear and toothache (Bever, 1986). All parts of the plant are used in a variety of traditional medicines.

Healthy Snack: When mature, the seeds from the pods can be extracted and treated like green peas and can be fried or roasted and eaten like peanuts. The seeds contain up to 40% of healthy oil by weight which is used for cooking, soap manufacture, cosmetic base and in lamps.

Moringa tree's various parts can be used to develop wide ranging products (Sandeep *et al.,* 2019) useful for augmenting health of people and use in diet of populations

Table 23 : Nutritional Value of green pods of Moringa oleifera(per 100 g) (USDA Nutrient Database)

Item	Value
Energy	37 kcal (150 kJ)
Carbohydrates	8.53 g
Dietary fibre	3.2 g
Fat	0.2 g
Protein	2.10 g
Vitamins	Quantity % DV
Vitamin A equiv.	1%

Item	Value
Thiamine (B1)	5% 0.0530 mg
Riboflavin (B2)	6% 0.074 mg
Niacin (B3)	4% 0.620 mg
Pantothenic acid (B5)	16% 0.794 mg
Vitamin B6	9% 0.120 mg
Folate (B9)	11% **44 μg**
Vitamin C	170% 141.0 mg
Minerals	**Quantity%DV†**
Calcium	3% 30 mg
Iron	3% 0.36 mg
Magnesium	13% 45 mg
Manganese	12% 0.259 mg
Phosphorus	7% 50 mg
Potassium	10% 461 mg
Sodium	3% 42 mg
Zinc	5% 0.45 mg

Biotic stresses

Moringa tree is not affected by any serious diseases in its native or introduced ranges. In India, several insect pests are seen, including various caterpillars such as the bark-eating caterpillar, the hairy caterpillar or the green leaf caterpillar.

The major pests of Moringa in India are as follows (Mridha, and Barakah, 2017).

(i) Bud worm (*Noorda moringae*), (ii) Leaf caterpillar (*Noorda blitealis*), (iii) Hairy caterpillars (*Eupterote mollifera, Pericallia ricini, Metanastria hyrtaca and Streblote (Taragama) siva*), (iv) bark borer (*Indarbela tetraonis*) and (v) Long horn beetles (*Batocera rubus*).

There are other insect pests of minor nature viz., aphids (*Aphis gossypii*) scale insects (*Ceroplastodes cajani*), bud midge (*Stictodiplosis moringae*) and leaf eating weevils (*Myllocerus* spp.)

Moringa tree is vulnerable to termite attack. If termites are numerous in soils, insect management costs may be higher (Parotta and John, 1993). Hence the soil should be rid of termites before taking up planting of *Moringa.*

The moringa tree is a host to *Leveillula taurica*, a powdery mildew which causes damage in papaya crops in south India.

Fig. 32: Blossoming flowers of Moringa oleifera With one small and tender developing pod By Harvey McDaniel from Naalehu, HI - Flickr - image description page, CC BY 2.0, https://commons.wikimedia. org/w/index.php?curid=346366

Fig. 33: Moringa oleifera Tree showing enlarged green pods and foliage with tripinnate leaves. By Krish Dulal - Own work, CC BY-SA 3.0, https://commons.wikimedia. org/w/index.php?curid=26452783

Jojoba (*Simmondsia chinensis*)

The popular common names of the tree are Goat nut, Coffee berry, Quinine nut, Wild hazel, Pig nut etc.The plant is a native shrub of the Sonoran Desert (Gentry and Scot, 1958) Colorado Desert, Baja California Desert, and California chaparral and woodlands habitats in the Peninsular Ranges and San Jacinto Mountains. It is found in southern California, Arizona, and Utah (U.S.), and Baja California state (Mexico).

The seeds of jojoba are one of the only known sustainable sources of liquid wax esters. They have been used as an eco-friendly replacement for similar oils that were once harvested from the spermaceti organ of the sperm whale (*Physeter macrocephalus*), which nearly drove it to extinction. "Jojoba is the only plant to store wax in its seeds. Such a vegetable oil has heretofore been unavailable," says Dr Eberhard Munz (research group AAN).

Description: Jojoba (*Simmondsia chinenesis*) is only single species in the Simmondiaceae, and possibly belongs to the subclass-Hamamelididae in the order Caryophyllales. (Angiosperm Group, 2016).

Jojoba is a dioecious wind-pollinated shrub, reaching a height of 1-5 meters and having a long life span (100-200 years) (Matthews, 1994). The plant appears as a small tree or a big shrub, typically grows to 1–2 meters tall, with a broad, dense crown, but there have been reports of plants as tall as 3 meters. Genetic variability in morphology, anatomy and physiology within the species is very large and enable selection of clones for high yield and other agricultural attributes. The leaves are opposite, oval in shape, 2–4 centimeters (0.79–1.57 in) long and 1.5–3 centimeters (0.59–1.18 in) broad, xerophytic with a thick cuticle and sunken stomata (Fig. 35). They contain special tissue with a high concentration of phenol compounds. The flowers are small and greenish-yellow, with 5–6 sepals and no petals. The female flowers are usually solitary (Figs. 34 and 35), one per two nodes although flowers that occur at every node or in clusters are not rare. The male flowers are clustered. Flower buds form in the axels of leaves solely on the new vegetative growth occurring during the warm season under favourable temperatures and water regime. New flower buds are dormant and will open only after a cool season with enough cold units for the fulfilment of their chilling requirements. Anthesis occurs in the spring when the soil and air temperature rise to above 15°C. Severe water stress prevents opening of flowers. The plant typically blooms from March to May (Benzoini, 1997)

The fruit is a capsule containing one to three dark brown seeds that normally ranges in their dry weight between 0.5-1.1 g and contain 44-56% wax. Fruits ripens during the spring and early summer and seeds fall to the ground in late summer.

Fig. 34. Female flower of Jojoba
(http://agritech.tnau.ac.in/bio-fuels/Biofuel_Jojoba.html)

Fig. 35. Female Jojoba Bush, CC BYKenneth Bosma 2.0,
https://commons.wikimedia.org/w/index.php?curid=11908393

Liquid wax (Oil) of Jojoba: Jojoba is grown for the liquid wax, commonly called jojoba oil, in its seeds. This oil is rare since it is an extremely long (C36–C46) straight-chain wax ester and not a triglyceride, and this renders jojoba oil and its derivative viz., jojoba esters more similar to whale oil than to traditional vegetable oils. Hydrolysed Jojoba Esters (HJE's) are developed, adopting the saponification reaction which liberates more than 12 natural long chain fatty alcohols making them available for anti-viral purposes. Jojoba oil has also been suggested as a possible feed stock for biodiesel fuel (Al-Hamamre, 2013) Jojoba wax is made up of straight-chain esters of mono-unsaturated long chain fatty acids and fatty alcohols and has an average total carbon chain length of 42 carbons. At room temperature, it is a light yellow liquid and does not become rancid or damaged by prolonged exposure to high temperatures. The wax can be isomerized, hydrogenated, sulphurized, chlorinated or trans-esterified.

Ecology: Jojoba is endemic to North America, and occupies an area of approximately 260,000 square kilometers (100,000 sq mi) between latitudes 25° and 31° North and between longitudes 109° and 117° West (Gentry and Scot,1958). Jojoba is a long day plant and thrives on coarsely textured soils with good drainage. The plant can't tolerate water-logging. It is tolerant to soil salinity and nutrient-poor soils with pH between 5 and 8 (www.calscape.org). High temperatures are tolerated by jojoba, but frost can damage or kill plants (Borlaug, 1985). Currently Jojoba is being grown in Argentina, Australia, Israel, Mexico, Peru and the United States. It is considered the second most economically valuable native plant in Mexico's Sonoran Desert.

Propagation: Jojobai is dioecious and hermaphrodite plants are extremely rare. Half of the seedlings are males which should be rouged-out as only 8-10% males are sufficient for pollination.

The fruit is an acorn-shaped ovoid, three-angled capsule 1–2 centimeters (0.39–0.79 in) long, partly enclosed at the base by the sepals of the female flower (Fig. 34) The mature seed obtained from female plants (Fig. 35) is hard oval that is dark brown and contains an oil (liquid wax) content of approximately 54%. An average-sized bush produces 1 kilogram of pollen, to which few humans are allergic (Steven *et al.*, 2000)

The female plants produce seed from flowers pollinated by the male plants. Jojoba leaves have an aerodynamic shape, creating a spiral effect, which brings wind-born pollen from the male flower to the female flower. In the Northern Hemisphere, pollination occurs during February and March. In the Southern Hemisphere, pollination occurs during August and September.

Jojoba may be propagated by seed. The seed progenies, however, are heterogeneous with lot of variation with in the progenies.

Selection may be effected for potentially promising genotypes based on their fruiting and maturity. The selected plants have to undergo testing after vegetative propagation, which facilitates improvement of the crop performance. Vegetative propagation enables the establishment of plantations with the desired proportion of male to female plants of preselected superior clones. Preparation of rooted cuttings of jojoba is done in a greenhouse with substrate temperatures of 30°C and mist applied for 10 seconds every 8-10 min. Three-node young branches are held for 48 hours in a refrigerator, dipped in rooting powder, and placed in peat containers (Jiffys). Rooting occurs within three to five weeks. The rooted cuttings are transferred to final pots after about four to five weeks and gradually hardened by transferring to unheated tables, and then to a net house (30% shade) (Benzioni, 1997).

Plants are ready for transplantation to the field after three to six months. Rooting percentage range from 15 to 95%, depending on the clone and the season.

***In vitro* propagation** is possible, and rooted plantlets may be produced. Hardening and transfer to the greenhouse is yet not very successful (Benzioni, 1997).

According to Agarwal *et al.,* (2002), nodal explants excised from 18 to 20-year-old female plants of *Simmondsia chinensis* if cultured on Murashige and Skoog's medium supplemented with 20 μM N^6-benzyladenine (BA) differentiated an average of 2.7 ± 0.4 shoots in 11.5% explants. The percentage of nodal explants inducing multiple shoots enhanced significantly if *in vitro* raised shoots were used as source of explants. Nearly 100% cultures differentiated an average of 4.7 ± 2.0 shoots per explant on the same medium. Nearly 85% of the shoots induced roots when a pulse treatment of 50 μM indole-3-butyric acid (IBA) was given prior to their transfer to semi-solid MS medium containing 10 μM IBA + 0.5% activated charcoal + 1 μM BA. Plantlets were gradually hardened in *Soilrite* and acclimatized to soil (Agarwal *et al.,* 2002).

Chromosome number: Somatic cells of jojoba are tetraploid; the number of chromosomes is $2n = 4x = 52$ (Hiroshi *et al.,* 1992)

Genomic studies:Seeds of the desert shrub, jojoba (*Simmondsia chinensis*), are an abundant, renewable source of liquid wax esters, which are valued additives in cosmetic products and industrial lubricants. Jojoba is relegated to its own taxonomic family, and there is little genetic information available to elucidate its phylogeny. Sturtevant *et al.*, (2020) reported the high-quality, 887-Mb genome of jojoba assembled into 26 chromosomes, the basic chromosome number of this tetraploid species, with 23,490 protein-coding genes. The jojoba genome has only the whole-genome triplication (γ) shared among eudicots and no recent duplications. These genomic resources coupled with extensive transcriptome, proteome, and lipidome data helped to define heterogeneous pathways and machinery for lipid synthesis and storage, provided missing evolutionary history information for this taxonomically segregated dioecious plant species, and will support efforts to improve the agronomic properties of jojoba.

Production practices of Jojoba: Giving an Irrigation about three weeks before planting facilitates germination of weeds, which are then treated with contact herbicides to bring about weed control. The plantlets may be planted in rows on soil saturated with moisture, 4.5 m apart and with 2 m of intra-row spacing from plant to plant within a row (1,110 plants/ha). In order to ensure availability of pollen for pollination, it is necessary to plant seeds from particular clones after every16th row. The direction of the rows depends on the wind direction and the topography (to prevent soil erosion). Planting is carried out in wet soil, with each plant being placed directly adjacent to a dripper. The plants should be handled very carefully so that the delicate root system remains intact (Benzioni, 1997).

Drip irrigation is recommended, which may be extended daily in the first month and then at weekly intervals in sandy soil and once in two to three weeks during the summer.

Jojoba plantation does not require any chemical fertilizers except in the first year when nitrogen may be needed to increases the plant growth (Nelson, 2001). Fertilizers are applied to provide N, P &K with the irrigation water in the first year. The amount of water for a fully developed plantation is about 0.35 of evaporation as measured with a pan class A. Optimal planting time is spring and early summer (Benzioni, 1997).

Supplemental irrigation could maximize production if rainfall as less than 400 mm (www.calscape.org). Jojoba is normally harvested by hand, since the plant does not exhibit any kind of synchronous maturity of capsules. Yield is around 3.5 t/ha depending on the age of the plantation (Yermaros, 1979). Selective breeding is being practised to develop plants that produce more beans with higher wax content, as well as other characteristics that will facilitate harvesting (Steven *et al.*, 2000)

The plants' ability to withstand high salinity (up to 12 dS m^{-1} at pH 9) and the high value of jojoba products make jojoba an interesting plant for the use of desertification control. It has been used to combat and prevent desertification in the Thar Desert in Rajasthan of India (Alsharhan *et al.,* 2003)

Biotic Stresses: It has been found that during the propagation of cuttings the dominant disease in the nursery is caused by two species of *Fusarium* viz., *F. solani* and *F. oxysporum*. In addition, a number of other pathogens viz., *Alternaria* spp., *Pythium* spp., *Phytophthora* spp., *Rhizoctoniasolani*, *Cylindrocladium* sp., *Diplodia* spp., *Colletotrichum* spp., *Phoma* spp. and *Erwinia* spp.

In jojoba plantations, the major problems are due to infection by *F. oxysporum* and but also by *F. solani*. The symptoms of the disease by *Fusarium* are wilt and defoliation of leaves, which develop into desiccation and death of the plant. The infection may lie dormant and the disease can take hold in quite well-developed plants three and four years after the infection sets in.

Usesof Jojoba: Jojoba foliage provides year-round food feed many animals, including deer, bighorn sheep (*Ovis canadensis*) and livestock of Mexico. Its nuts are eaten by squirrels, rabbits, other rodents, and larger birds. In large quantities, jojoba seed meal is toxic to many mammals. Later this effect was found to be due to *simmondsin*, which inhibits hunger. The wax acts as a laxative in humans. The Mexican indigenous peoples heated jojoba seeds to soften them. They then used a mortar and pestle to create a salve or buttery substance. The latter was applied to the skin and hair. The O'odham people of the Sonoran Desert treated burns with an antioxidant salve made from a paste of the jojoba nut. Native Americans also used the salve to soften and preserve animal hides. Their pregnant women ate jojoba seeds, believing they assisted during childbirth. Hunters and raiders ate jojoba on the trail to keep hunger at bay. The Seri tribe of Mexico who utilize nearly every edible plant in their domain, do not regard the beans as real food and in the past ate it only in emergencies (Steven *et al.*, 2000)

The seed oil is used for lubrication as a substitute for sperm whale oil, in heavy machinery. It can also be used with little or no refining at high temperatures and pressures. It is a good source material for pharmaceuticals, cosmetics, food related industries. (Mukta and Varaprasad, 2013).

The liquid wax of jojoba and its derivatives have a wide range of industrial uses, mainly in cosmetics in which it is incorporated in formulations for skin care preparations such as lotions, moisturizers, massage oils, and soothing creams. It is also widely used in hair care products, such as shampoos, gels and mousses, and is a very good base for lipstick, makeup and nail products. There are many potential uses in pharmaceuticals, and in industry (as extenders for plastics, printer's ink, gear-oil additives, various lubricants etc., apart from its touted potential for the production of biodiesel.

Performance of Jojoba in Rajasthan state of India: Jojoba was introduced in India around 1965, but it was given importance only during 80s due its novel product. (Mukta and Varaprasad, 2013). According to CJP (2007) Jojoba generally does not produce an economically useful yield until the fourth or fifth year after planting. Seed yields in natural stands of jojoba range from a few seeds to as much as 5 Kg of clean, dry seed per plant. Production of seed varies greatly from plant to plant in a stand and from year to year for a particular plant. Currently, the average yield of commercial Plantations that were established with selected higher yielding clones is 1 to 11 tons from 7th to 15th year per hectare (Fig.36).

Fig.36. Jojoba Plantation in Rajasthan& Seeds shown in a palm
Source: Rajasthan Direct (2013)

Association of the Rajasthan Jojoba Plantation and Research Project (**Ajorp**), has been set up by Rajasthan State Government in 1995 for Jojoba promotion in the state.

Fourteen patents have been filed by the Indian Institute of Petroleum, Dehradun, India for using jojoba oil as a lubricant in various automotive segments (Kumar, 2007). Association of the Rajasthan Jojoba Plantation and Research Project (**Ajorp**), has been set up by the state government in 1995. **Ajorp** has already established and developed two model jojoba farms in Rajasthan viz., one at Fatehpur (Sikar district) and the other 37ha farm at village Dhand (Jaipur district) in association with Haigud, an NGO based in Israel. The third model jojoba farm, spread over 50ha, has been established in collaboration with the Gujarat State Rural Development Corp. at Mujpur in Patan district (Kumar, K.P.N., 2007)

The development of the jojoba market is very positive as jojoba oil became a recognized raw material mainly in the cosmetics industry. Global consumption of jojoba oil is increasing year over year. Alongside the US, major jojoba oil importers include Germany, followed by Japan. China has increased its jojoba imports significantly. Global consumption in the industry is said to have increased to over million metric tons per year (Rajasthan Direct, 2013*)*

Simarouba (*Simarouba glauca)* : *Simarouba glauca* is a species of flowering tree that is native to Florida, South America, and the Lesser Antilles. Thepopularvernacular names of the tree are Simarouba, Paradise tree, Aceituno etc.

Fig. 37: Foliage of Simarouba (Simarouba glauca) with fruits

Taxonomy*:Simarouba* is a genus of trees and shrubs in the family Simaroubaceae, native to the neotropics. It has been grouped in the subtribe Simaroubina along with the *Simaba* and *Quassia* genera. They have compound leaves, (Fig.37) with alternate pinnate leaflets. Their flowers are unisexual, relatively small (around 1 cm long) and arranged in large panicles. The individual flowers have between 4 and 6 sepals and petals and between 8 and 12 stamens. The fruit is a carpophore, (Fig.37) and has up to 5 drupaceous mericarps.

Habitat and plant's features: The tree is well suited for warm, humid, tropical regions. Its cultivation depends on rainfall distribution, water holding capacity of the soil and sub-soil moisture. It is suited for temperature range of 10 to 40 °C. It can grow at elevations from sea level to 1,000 m. It grows 12 to 15 m tall and has a span of 7.6 to 9.1 m. It bears yellow flowers and oval elongated purple coloured fleshy fruits. The plants are polygamodioecious with varying percentage of staminate, pistillate and few bisexual flowers in the population. It can grow well in tropical climate and can withstand scanty to high rainfall. All types of well-drained soils are suitable for simarouba cultivation.

Uses: It is both a source of edible oil from its seeds and could also from a feed-stock for biofuels. The oil can be used as cocoa butter substitute/extenders in confectionary and bakery industry. The oil cake is valued organic manure. The extracts from parts of the tree, demostrated potent anticancer properties. However, no systematic research, using phytochemicals isolated from *Simarouba glauca*, has been carried out to explore the molecular mechanisms leading to cancer cell death (Jose *et al.*, 2019)

Improved accessions available are given in Table 24 (Mukta and Varaprsasd, 2013). Simarouba can be propagated by seeds, grafting and tissue culture technology. Fruits are collected in the month of April / May, when they are ripe and then dried in sun for about a week. Skin is separated and seeds are germinated in seed bags to produce saplings. Saplings of 2 to 3 months' age can be transplanted to a plantation.

Table.24: Improved varieties/accessions

Agroclimatic zone	Improved varieties/accessions
Southern Plateau & Hills Region	Palem-1. 2. 3.4
Central Plateau & Hills Region	HAUP-09. 13, 22, 28 .
Western Plateau & Hills Region	PDKV SG-23, 25, 27, 30

Environmental Impact: The tree forms a well-developed root system and dense evergreen canopy that efficiently checks soil erosion, supports soil microbial life, and improves groundwater position. Besides converting solar energy into biochemical energy all-round the year, it checks overheating of the soil surface all through the year and particularly during summer. Large-scale planting in wastelands facilitates wasteland reclamation, converts the accumulated atmospheric carbon dioxide into oxygen and contributes to the reduction of greenhouse effect or global warming.

Kokum (*Garcinia indica*): The popular vernacular names of *Garcinia indica* tree are Kokum, Mangosteen, Kokum butter tree, Aamsol etc. This is an important minor fruit, besides its seed oil being a source for edible and non-edible preparations.The genus *Garcinia*, belonging to the family Clusiaceae, includes about 200 species found in the Old World tropics, mostly in Asia and Africa. *Garcinia indica* is indigenous to the Western Ghats region of India located along the western coast of the country. Of the 35 species found in India, 17 are endemic. Of these, seven are endemic to the Western Ghats, six in the Andaman and Nicobar Islands and four in the north-eastern region of India. The kokum variety from the Ratnagiri and Sindhudurg districts from the coastal Konkan region of the state of Maharashtra in India has received the GI (Geographical Indication) tag (Wikipedia).

Garcinia indica is an evergreen, monoecious tree (Asinelli *et al.*, 2011) The tree can grow as high as 18 meters. On maturity the trees attain a pyramid shape. The fruit is a berry of the size of orange with purple colour (Fig. 38) with fleshy endocarp (Raju and Reni, 2001). The fruit of the plant contains five to eight large seeds which account for 20-23% of the fruit's weight. The kernels account for 61 percent of the weight of the seed, while the oil content of the kernel accounts for about 44%. The seeds are compressed and embedded in an acidic pulp. The fat contains large amount of stearic and oleic acids (Prasad, 1994). It is reported to have been introduced from Zanzibar to India. It is found to be grown widely in tropical rain forests of Western-Ghats in Konkan, Goa, Southern Karnataka and Kerala. It is a slender ever green tree with drooping branches.

Fig. 38: Branches of Garcinia indica tree with fruits

Kokum fruit is a promising industrial raw material for commercial exploitation in view of its interesting chemical constituents. Kokum seeds yield a valuable edible solid fat known as kokum butter, used in chocolate and confectionary preparations and also in manufacture of soap, candle and ointments. It could be a source of feed stock for biodiesel, provided the cultivation of the tree is taken up systematically with high yielding strains. Improved accessions available for cultivation in west coast plains and hills are *Konkan Amrita, Loanwala kokum*, *Khola kokum* and *Khane kokum* (Mukta and Varaprasad, 2013).

12

Designer Oilseed Crops

In recent years, oilseed crops are being increasingly exploited for production of pharmaceuticals, surfactants, plasticisers, emulsifiers, detergents, lubricants, adhesives, cosmetics, oleo-chemicals, biofuels apart from their demand for their healthy vegetable oils and livestock feeds (Metzger and Bornscheuer, 2006). The increased interest resulted in an 82% expansion of oilseed crop cultivation areas and about a 240% increase in total world production over the last 30 years (Rahman and Jimenez., 2016).

Notwithstanding the overall enhancement in oilseeds production in the past three decades, the growth and production of oilseed crops exhibit fluctuation without any consistency. While, the trends of production of soybean and rapeseed – mustard have been encouraging, production of linseed and safflower declined (Rahman and Jimenez., 2016). Despite the perceptible increases in productivity of linseed, ironically the total area and production of the crop registered a marked decline as compared to other oilseed crops. Also, the area of safflower crop has come down. A significant rise in the seed yield per unit area could be noticed in respect of castor seed, oil-palm, soybean and sunflower globally. On the other-hand moderate increases in area, production and productivity have been witnessed in the cases of sesame, groundnut, cotton seed and castor bean. In the above cited backdrop of oilseeds production scenario, there has been a consistent enhancement in the consumption pattern of world population. Vegetable oil consumption has drastically increased in the past decade and a half along with increase in the production of biodiesel for which the vegetable oil, largely that of rapeseed-mustard is the main feed-stock. For example, during the years from 2000 to 2010, there was a considerable increase from 1 to 7 million tons in non-food utilization of rapeseed oil, largely for the production of biodiesel (Lu *et al.*, 2011)

In order to satisfy the increasing world demand, sustainable oil production through classic breeding efforts needs to be coupled with biotechnological approaches in order to expand oil yield per unit area. The expansion of oilseed growing areas can be another approach utilized to meet this increased demand. Nevertheless, in countries like India there is little scope for area expansion under oilseed crops (Mohapatra, 2017). Hence, it becomes absolutely necessary to take recourse to science and technology to enhance productivity and diversify the production of oilseeds.

Vegetable oils are made of triglycerides (TAG), arising through the esterification of diverse fatty acids in three different positions of the glycerol backbone. Thus, the quality of vegetable oil is totally depends upon the composition of fatty acids in TAG. The predominant fatty acids in the vegetable oils are palmitic acid (C16:0), stearic acid (C18:0), Oleic acid (C18:1), linoleic acid (C18:2), linolenic acid (C18:3), arachidic acid (C20:0), eicosenoic acid (C20:1) and erucic acid (C22:1). The two kinds of fatty acids are saturated fatty acids, which do not have double bonds between carbon atoms and the unsaturated fatty acids which have at least one double bond with hydrogen. Depending upon the number of such double bonds, the fatty acids are known as mono-unsaturated or poly-unsaturated fatty acids.

The major value based component of oilseed crop being oil, the first objective of biotechnological approach should be to maximise the oil content of oilseed crops. Additionally, it is necessary to manipulate the oil quality to suit the various downstream applications. Also, there is a need to enhance the quality of by-products such as seed proteins along with the elimination of undesirable constituents such as erucic acid and glucosinolates of rapeseed-mustard and gossypol of cotton of cotton seed. As an example it may be cited that in UK and Germany high oleic rapeseed is grown for edible oil, while high erucic rapeseed is grown to produce industrial oil often on adjacent farms (Murphy, 1994). The problem of out crossing is very much checked by the self-incompatibility of the crop (Murphy, 1994).

Many genes in TAG (Tri-acyl-glycerol) biosynthesis pathways have been identified and studied well. New biotechnology methods allow insertion or modification of genes involved in the biosynthesis of a desired fatty acid, in order to accumulate a higher level of fatty acid or even to produce a novel fatty acid. (Maisonneuve *et al.*,2009; Westlake *et al.*, 2009)

Modification of Fatty Acid Profile to suit the needs of consumers and industry

The case which deserve attention is that of utility of certain specific fatty acids associated with vegetable oils such as stearic acid. A higher content of stearic acid in vegetable oil would be of advantage for solid fat applications due to its cooking utility, in addition to its health value (Rahman and Jimenez, 2016). Soybean genotypes with stearic acid levels of 9 to 23% were developed through mutagenesis (Pantalone *et al.*, 2002). Also, a novel mutation in a 9-steroyl-ACP-desaturate of soybean has been reported (Ruddle *et al.*, 2013) as a new source of high stearic acid resulting in molecular markers useful in breeding. Nevertheless, most of the new high stearic acid lines exhibit problems such as lower seed yield, poor germination and reduced seedling vigour and growth rate (Wang *et al.*, 2001). In the case of cottonseed, hairpin RNA – mediated post-transcriptional gene slicing has been utilized to modify stearic and oleic acid content (Liu *et al.*, 2002).

The biotechnology company viz., Calgene has come up with the first genetically engineered lauric acid rapeseed with a view cater to the needs of industry involved in the manufacture of soaps and detergents (Kumar, 1998a). It may be noted that rapeseed originally does not possess lauric acid, which is a constituent of coconut oil. It is interesting to note that high level of lauric acid in rapeseed was achieved through the introduction of gene encoding specific C12 thioesterase (Kumar 1998a). Since coconut cannot be grown in temperate areas to harness lauric acid, the new rapeseed with lauric acid gives a novel opportunity to produce lauric acid in temperate regions too.

It was possible to clone the genes of the fatty acid biosynthesis viz., transcylases, thioesterases and desaturase-related enzymes and transform into oil bearing crops (Lessire, 2005), which formed a novel step towards designing oilseed crops.

Rapeseed (canola) oil with oleic acid of the order of 75%and 4% linolenic acid by McVetty in 1999 (Rahman and Jimenez, 2016) not only enhanced the quality of it as the edible oil but also provided for its use in oleo-chemical industry.

Enhancing Protein quantity and quality in oilseeds

Apart from increasing the quantum and quality of vegetable oils, the important objective of increasing protein content and quality of the meal or cake is to be achieved along with the decreasing or eliminating the anti-nutritional factors. Since Oilseeds are rich in protein, the by-products such as cake or meal are being increasingly used as cattle, poultry and fish feed (Ramachandran *et al.*, 2007). Also, the oil seed meals are rich in plant macro and micronutrients. Meal proteins are highly decomposable and as such they promote nutrient uptake by plants, thus qualifying to be good organic fertilizers for agriculture (Moore, 2001).

A classical example which needs attention in this context is that of soybean. Soybean is well known source of protein that varies from 35 to 50% (Krishnan, 2005). However, the nutritional quality of soybean protein is stated to be not optimal due to low level of sulphur containing amino acids, such as methionine and cysteine, which cannot be synthesised by monogastric animals (Krishnan, 2005). Genetic engineering approaches, however, could increase sulphur-rich proteins in soybean by a small amount, using the expression of heterogeneous proteins such as methionine-rich zeins proteins and Brazil Nut Albumin (BNA). Nevertheless, with the increase of heterologous sulphur amino acids in transgenic soybean, there was decrease in the endogenous sulphur rich proteins (Streit *et al.,* 2001), indicating the impracticality of introduction of sulphur-rich proteins to enhance the nutritional quality of soybean. It is further stated that in order to increase the cysteine level, or any other amino acid, the key enzymes involved in their biosynthesis pathway have to be engineered to attain the expression in the developing seeds (Krishnan, 2005).

Genetic modification of carbohydrate biosynthetic pathway

Scientists at DuPont have successfully used two separate methods viz., conventional breeding together with germplasm screening on one hand and chemical mutagenesis on the other to modify the soluble carbohydrate biosynthetic pathway and developed soybean strains with low raffinose saccharide content or with high sucrose and low raffinose saccharide contents.

Soya meal prepared from these novel varieties display improved metabolisable energy value.

Thus the above varieties provide a means of developing soya flour with low raffinose saccharide for human consumption (Kinney, 1996 ; Kerr, 1996).

Further experiments with the mutants with low raffinose saccharide have helped the scientists to define their targets for genetic engineering approaches. One molecular strategy to decrease oligosaccharides may involve blocking the expression of GS gene for the production of galactinol in the seed by antisense techniques because GS is considered to be the key enzyme in the biosynthesis of oligosaccharides (de Lumen, 1992). Mutant phenotypes have been generated in soybean plants specified by antisense RNA technique (Mol *et al.*, 1990)

Designer Vegetable Oils for Biodiesel: Currently rapeseed and soybean oils world over and palm oil in particular in south-eastern countries of Malaysia, Indonesia and Thailand are being used for the production of biodiesel (Pleanai and Gheewala, 2009 ; Demibras, 2002 ; Yin *et al.*, 2008). Soybean oil appears to be a favourite choice for biodiesel production for the reasons of recovery of more usable energy and less greenhouse gasses (Hill, *et al.*, 2006). Nevertheless, biodiesel from soybean oil suffers from low storage life due to the oils high oxidative reactivity, which results in formation of compounds that can clog fuel filters (Canakci *et al.*, 1999; Mittelbach and Gangl, 2001). Since higher degree of oleic acid and low levels of linoleic acid can enhance the quality of soybean oil for biodiesel such genotypes with perceptible reduction in PUFA oils together with significant enhancement of oleic acid have been developed incorporating the gene Fad2-1 through micro-projectile bombardment (Kinney and Knowlton, 1997).

The transgenic soybean with four different transgenes resulted in over 10-fold increase in vitamin E-type molecules indicated as tocotrienols (Karunanandaa *et al.*, 2005).

Since ricinoleic acid facilitates better lubricity in oils for biodiesel production (Kinney and Clemente, 2005), the enzyme oleate-12-hydroxylase (Bafor *et al.,* 1991) was targeted and transgenic soybean carrying castor bean oleate12-hydoxyase gene has been developed (Kinney and Clemente, 2005).

Soybean oil and rapeseed oil are edible oils and their use in enormously large quantities world over for production of biodiesel is jeopardizing the availability of edible oils for human consumption. Hence it is advisable to switch over to non-edible industrial oils such as castor oil and oils from tree species such as *Pongamia pinnata*, *Calophyllum* etc., for production of biodiesel. *Pongamia* for example is highly drought resistant suited for growing under wide ranging agro-ecological situations and even on wastelands and soils of poor quality where regular crop production is not possible. Oil of *Pongamia* is reported to be well suited for production of biodiesel (De and Bhattacharya 1999; Karikalan and Chandrasekaran,2015; Prasad 2019a). Among non-edible oil bearing annual plants castor (*Ricinus communis*) which possess ricinoleic acid is another drought hardy plant that could be exploited for biodiesel production (Dias *et al.,* 2013).

It is expected that the varietal outcome of biotech research would incorporate pest and disease resistance too, thereby reducing the need to apply chemicals for control of biotic stresses. All the above would go to rationalise the design of oilseed crop plant architecture in order to optimise the kind of oilseed yield, growing time, flowering time, desiccation rates, harvesting potential and other useful agronomic attributes.

Biotechnology is assisting the improving oilseed crops in two ways. Firstly, the potential and efficiency of classical breeding programmes is being enhanced by increasing the genetic diversity within the breeding lines and by using marker assisted selection to transfer useful genes into a desired agronomic background in a relatively shorter period of time. Secondly, genetic engineering is being used to isolate genes directly from distantly related or unrelated plant species. The two approaches should be seen as complimentary.

Genetic engineering and genomics started a new era for designer oil crops and has created opportunities for sustainable oilseed crop production around the world.

The direct study of DNA sequences known as genomics are being increasingly applied for improvement of oilseed crops (Lai *et al.*, 2012). Genetic manipulation of plant genome and production of genetically modified plants by means of metabolic engineering and genomics is promoted as the most efficient route to improve oilseed crops (Maheshwari and Kovalchuk, 2014). Improved genome-wide technologies, supported by precise plant phenotyping and field monitoring have resulted in rapid progress (Nelissen *et al.*, 2014).

The components of genomic-tool box include gene discovery, trait analysis through genome-wide studies (GWAS), global gene expression analyses, genomic selection and predictive breeding strategies (Snowdon and Luy, 2012). Based on genome-sequencing projects of oilseed crops like *Brassica,* the genomic tool box offers much more promise for the discovery of favourable allelic combinations (Rahman, 2013) for more efficient plant breeding. Also the new genome based technologies will accelerate the domestication of new crops such as *Diplotaxis tenufolia, Cochlearia officinalis (*Sedbrook *et al.*, 2014*)*.

References

Abate, M. and Mekbib, F. (2015)Assessment of Genetic Diversity in Ethiopian Sesame (*Sesamumindicum* L.) **Germplasm using Random Amplified Polymorphic DNA(RAPD)** Markers. *Journal of Advances in Agriculture* **5** (2): 639. ISSN 2349-0837

Abbadi *et al.,* (2004) Biosynthesis of very-long-chain polyunsaturated fatty acids in transgenic oilseeds: constraints on their accumulation. *Plant Cell.* **16(**10):2734-48

Abdellatef, E., Ahmed, M.M.M., Daffalla, H.M. and Khala-falla, M.M. (2010). Enhancement of adventitious shoot regeneration in sesame (*Sesamum indicum* L.) cultivar promo KY using ethylene inhibitors. *Journal of Phytology***2**: 61-67.

Abdellatef, E., Sirelkhatem, R., Ahmed, M.M.M., Radwan, K.H. and Khalafalla, M.M. (2008). Study of genetic diversity in Sudanese sesame (*Sesamum indicum* L.) germplasm using random amplifed polymorphic DNA (RAPD) markers. *African Journal of Biotechnology* **7:** 4423-4427

Abdul Nizar, M. and Mulani, R.M. (2015). Genetic diversity in indigenous and exotic linseed germplasm (*Linum usitatissimum* L.) *Electronic Journal of Plant Breeding*, **6**(3): 848-854

Abel, G.H. and Lorance, D.G. (1975). Registration of 'Dart' safflower. *Crop Sci.* **15**: 100

Abraham (2011) Flax Protein .*Supplements in Review*https://supplementsinreview.com/protein/ **flax-protein/#fn-13524-23**

Abraham, V. (1994). Rate of outcrossing in Indian mustard. *Cruciferae Newsletter***. 16** :69-70.

Abraham, E. V, Natarjan K. and Murugaesan M. (1977) Damage by pests and Phyllody to *Sesamum Indicum* in relation in time sowing. *Madras Agriculture Journal, 1977:* **64**:298-301.

Abreu, C.A. de, Raij, B.van and Tanaka, R.T. (1996) Sources of manganese for soyabean and their effects on soil analysis. *Revista Brasileira de Ciencia do Solo.***20**(1):91-97

Abrol, D.P. and Shankar, U.(2011) All India Coordinated Research Project on Linseed. Annual Reports. -2010-2011. SKUAT Jammu.

Abrol, D.P. and Uma Shankar (2016). Integrated Pest Management. In: Gupta, S.K. (Ed) *Breeding Oilseed Crops for Sustainable Production*, Elsevier /Academic Press, London / San Francisco. Pp 523-545

Ackman, R.G. (1983) Chemical composition of rapeseed oil, in *High and Low Erucic Acid Rapeseed Oils*, eds. J.K.G. Kramer, F.D. Sauer and W.J. Pigden, Academic Press, New York, pp. 85-129

Adda, S., Reddy, T.P. and Kavi Kishor, P.B. 1993. Somatic Embryogenesis and Organogenesis in *Guizotia abyssinica. In Vitro Cellular & Developmental Biology - Plant* **30**: 104-107.

Adda, S., Reddy, T.P. and Kavi Kishor, P.B. 1994. Andro-clonal variation in niger (*Guizotia abyssinica* Cass). *Euphytica***79**: 59-64.

Adhikary, N.K., Chowdhury, M.R., Begum, T. and Mallick, R. (2019). Integrated Management of Stem and Root Rot of Sesame (Sesamum indicum L.) caused by *Macrophomina phaseolina* (Tassi) Goid. *Int.J.Curr.Microbiol.App.Sci.***8(**4): 804-808. doi: https://doi.org/10.20546/ijcmas.2019.804.089

Adriana, A., Rosa, B., Jao, G. and Stefano, P. (2014) Agricultural Biodiversity in Southern Brazil: Integrating Efforts for Conservation and Use of Neglected Underutilized Species, *Sustainability* Feb. 2014DOI: 10.3390/su6020741

Adda, S., Reddy, T.P. and Kavi Kishor, P.B. 1993. Somatic Embryogenesis and Organogenesis in Guizotia abyssi-Adhikary, N.K., Chowdhury, M.R., Begum, T. and Mallick, R.(2019)Integrated Management of Stem and Root Rot of Sesame (*Sesamum indicum* L.) caused by *Macrophomina phaseolina* (Tassi) Goid. *International Journal of Current Microbiology and Applied Sciences ISSN: 2319-7706* (**8**)4Adugna, W and Labuschagne, MT (2003) Association of linseed characters and its variability in different environments. *Journal of Agricultural Science, Cambridge* 140, 285–296

Adugna, W and Labuschagne, MT (2008) Genotype-environment interactions and phenotypic stability analyses of linseed in Ethiopia. *Plant Breeding* **121**(1): 66–71

Agarwal, S.K. and Porwal, M.K. (2006) Growth dynamics of intercropping systems in castor (*Ricinus communis* L.) under irrigated ecosystem. *Annals of Agricultural Research,* **27**(3):265-270

Agrawal, R., Tsujimoto, H., Tandon, R., Rama Rao, S., and Raina, S. N. (2013). Species-genomic relationships among the tribasic diploid and polyploid *Carthamus* taxa based on physical mapping of active and inactive 18S–5.8S–26S and 5S ribosomal RNA gene families, and the two tandemly repeated DNA sequences. *Genetics*, **52**:136–144.

Agrawal, V., Prakash, S. and Gupta, S. (2002) Effective Protocol for *in Vitro* Shoot Production Through Nodal Explants of *Simmondsia Chinensis*. *Biologia Plantarum* **45:** 449–453 (2002). https://doi.org/10.1023/A:1016238205522

Agribusiness (2018) India's edible oil consumption to exceed 34 million tonnes by 2030: Report. *BusinessLine., The Hindu. 2018*

Agriculture Forum (2012). All about *Moringa.* http://greenagrow.blogspot.com/2012/11/all-about-moringa.html

Agrifarming. (2017). Drumstick farming: detailed information guide. http://www.agrifarming.in/drumstick-farming/

Agropedia, (2012). http://agropedia.iitk.ac.in/content/linseed-varieties-and-its-charectristics

Ahir, K.C., Saini, A. and Rana, B.S. (2018)Estimation of yield losses due to major insectpests of groundnut *(Arachis hypogaea* L.)*Journal of Entomology and Zoology Studies.***6(2):** 312-314

Ahloowalia, B.S., Nichterlein, K. and Maluszynski, M., (2004). Global impact of mutation-derived varieties. *Euphytica*: *Int. J. Plant Breed.* **135**: 187–204.

Ahmad, M.S., Ahmad, M.U., Ansar, A.A. and Osman, S.M. (1978) Studies on herbaceous seed oils V. *Fette Seifen Anstrichm*. **80**, 353–354.Google Scholar

Ahmadi, R.and Omidi, A.H. (1997) Studyand determination of natural crossing in winter safflower (*Carthamus tinctorius*L.). *Sesame and Safflower Newsletter,* **12** : 94-97.

Ahamad, S., Ahmad, S.A. and Hasnat, A.(2015)Synthesis and Characterization of Styrenated Poly(ester-amide) Resin from *Melia Azedarach* Seed Oil -An Eco-friendly Resource. *Chemical Science Transactions* 2015, **4**(4) :1047-1053 DOI:10.7598/cst2015.1105

Ahmed, S. (1988). Farmer pest-control practices in India: Some observations. *Final Workshop of IRRI-ADB-EWC Botanical Pest Control in Rice Based Cropping Systems, IRRI,* December 12-16, 1988, IRRI, Los Banos, Philippines

Ahmed, S. (1995). The multiple uses of the pesticide tree. *Ceres***155** (5): 4-6

Ahmed, S., Bamofleh, S. and Munshi, M. (1989). Cultivation of neem (*Azadirachta indica,* Meliaceae) in Saudi Arabia. *Economic Botany***43**: 35-38.

Ahmed, S. and Grainge, M. (1985). Use of indigenous plant resources in rural development: potential of the neem tree. *International Journal of Development Technology***3**: 123-130. A

Ahmed, S. and Idris, S.(1997). *Melia azedarach.* In: Hanum IF, Maesen LJB van der, (Eds). *Plant Resources of South-East Asia.* No. 11 Auxiliary plants. Prosea Foundation, Bogor, Indonesia. Leiden, Holland: Backhuys Publishers, 187-190

AICORPO (1974) Progress Report of the All India Coordinated Research Project on Oilseeds 1974, Directorate of Oilseeds Research, Hyderabad

AICORPO (1989)Progress Report of the All India Coordinated Research Project on Oilseeds 1989, Directorate of Oilseeds Research, Hyderabad

AICORPO (1992) Progress Report of the All India Coordinated Research Project on Oilseeds 1992, Directorate of Oilseeds Research, Hyderabad

AICORPO (1993) Progress Report of the All India Coordinated Research Project on Oilseeds 1993, Directorate of Oilseeds Research, Hyderabad

AICORPO (1995) Progress Report of the All India Coordinated Research Project on Oilseeds 1995. Directorate of Oilseeds Research, Hyderabad

AICORPO (2003). Progress Report of the All India Coordinated Research Project on Oilseeds 2003, Directorate of Oilseeds Research, Hyderabad

AICRIP-DA (2004) Districtwise Promising Technologies for Rainfed Groundnut based Production System in India, *All India Coordinated Research Project for Dryland Agriculture ICAR,* Central research Institue for Dryland Agriculture - ICAR, Hyderabad-500 059, India.Pp: 92

AICRP-RM (2000). Annual Progress Report of National Research Centre on Rapeseed-mustard. 1999-2000, pp. 24.

AICRP-RM (2004). Annual Progress Report of National Research Centre on Rapeseed-mustard. 2004-05, pp. 9–11, 2005.

AICRP-RM (2005). Annual Progress Report of All India Coordinated Research Project on Rapeseed-Mustard, pp. A1–28, 2005.

Ajith Prasad, H. N., (2004) Transmission and serological diagnosis of sunflower necrosis virus from various sources and screening for resistance. *M.Sc. (Agri.) Thesis, Univ. Agri. Sci., Bangalore*, 99 pp.

Akhtar K.P, Dickinson M, Sawar G, Jamil FF and Haq, MA. (2008) First report on the association of a 16S r II phytoplasma with sesame phyllody in Pakistan. *Plant Pathology*, 2008; **57**:771. 5

Akhtar, T., Tariq, M.I. and Iqbal, S. (2016) Production of Biodiesel from *Melia azedarach* seed oil: A nonedible feedstock for biodiesel. *Science International* **28:** 3805-3807

Alagirisamy, M. (2016) Groundnut, In: S. K. Gupta (Ed) *Breeding Oilseed Crops for Sutainable Production: Opportuities and Constraints, Academic Press, Elsivier,* Oxford U.K., Sa Diego, USA. Pp: 89-134

Al-Amery, M. M., Hamza, J. H. and Fuller, M. P., (2011), Effect of Boron foliar application on reproductive growth of sunflower (*Helianthus annuus* L.*). Int. J. Agron.* (**230712**): 1- 5.

Alemaw, G and Wold, A.T. (1995) An agronomic and seed quality evaluation of *noug* (*Guizotia abyssinica*, Cass.) germplasm in Ethiopia. *Plant Breeding.* **114** (4):375-376.

Alemu,H. (2016). Review Paper on Mutation Breeding as applied in Groundnut (*Arachis Hypogea* L.) Improvement. *Gene and Cell Therapy.***1** (5): 35-40. doi: 10.11648/j. gct.20160105.11

Al-Hamamre, Z. (July 2013). "Jojoba is a Possible Alternative Green Fuel for Jordan". *Energy Sources, Part B: Economics, Planning, and Policy.* **8** (3): 217–226. doi:10.1080/15567240903330442

Ali, F., Abdurrahim Yil maz, A., Chaudhary, H.J., Nadeem, M.A., Malik Ashiq Rabbani, M.A., et al., (2020) Investigation of morphoagronomic performance and selection indices in the international safflower panel for breeding perspectives. *Turkish Journal of Agriculture and Forestry* (2020) **44**: 103-120, http://journals.tubitak.gov.tr/agriculture/ Turk

Ali GM, Yasumoto S. and Katsuta MS. (2007). Assessment of genetic diversity in sesame (*Sesamum indicum* L.) detected by amplified fragment length polymorphism markers. *Electron. J. Biotechnol.* **10**: 12–23

Ali, H.M.A. and Shah, S.A. (2013) Evaluation and selection of rapeseed (*Brassica napus* L.) mutant lines for yield performance using augmented design. *J. Animal and. Plant Sci.***23:** 1125-1130

Allaby, R. G., Peterson, G. W., Andrew, D. M. and Fu, Y. B. (2005). Evidence ofthe domesticationhistory of flax (*Linum usitatissimum*) from genetic diversity of the Sad2 locus. *Theor. Appl. Genet.*, **112**: 58–65

Allan, E.J., Eeswara, J.P., Johnson, S., Mordue, A.J., Morgan, E.D. and Stuchbury, T. (1994.) The production of azadirachtin by in-vitro tissue-cultures of neem, *Azadirachta indica. Pesticide Science***42** (3): 147-152.

Allan, E.J., Eeswara, J., Jarvis, A.P. and Luntz, A.M. (2002) Induction of hairy root cultures of *Azadirachta indica* A. Juss. and their production of azadirachtin and other important insect bioactive metabolites. *Plant Cell Reports* **21**(4):374-379. DOI: 10.1007/s00299-002-0523-3

Allen CAW. (1988) Prediction of biodiesel fuel atomization characteristics based on measured properties. *Ph.D. Thesis* Faculty of Engineering, Dalhousie University, Canada.1998. pp. 200.. [Google Scholar]

Allen, O.N. and Allen, E.K. 1981. The Leguminosae. The University of Wisconsin Press. 812 p

Al-Muhsen S, Clarke AE and Kagan RS (2003). "Peanut allergy: an overview". *CMAJ.* **168** (10): 1279–1285.

Al Muqarrabun, L.M.R., Ahmat., Ruzaina, S.A.S., Ismail, N.H. and Sahidin,I (2013) Medicinal uses, phytochemistry and pharmacology of *Pongamia pinnata* (L.) Pierre: A review, *Journal of Ethnopharmacology Review*http://dx.doi.org/10.1016/j.jep.2013.08.041

Alsharhan, Abdulrahman S., Fowler, Abdulrahman, Goudie Andrew S.; Abdellatif, Eissa M. and Wood, Warren W. (2003*). **Desertification in the third millennium***. Lisse: Balkema. pp. 151–172. ISBN 978-0-415-88943-8.

Alves, D. M. T., Pereira, R. W., Leal-Bertioli, S. C. M., Moretzsohn, M. C., Guimarães, P. M., and Bertioli, D. J. (2008). Development and use of single nucleotide polymorphism markers for candidate resistance genes in wild peanuts (*Arachis* spp). *Genetics and Molecular Research,***7:** 631-642 https://doi.org/10.4238/Vol7-3gmr453

Amaglo, N. (2006). **"How to Produce Moringa Leaves Efficiently?"**(PDF). Retrieved 19 November 2013.

Ambreen,H., Kumar, S. S., Kumar, A., Agarwal, M., Jagannath, A. and Shailendra Goel, S.(2018). **Association Mapping for Important Agronomic Traits in Safflower** (*Carthamus tinctorius* L.) Core Collection Using Microsatellite Markers. Frontier *Plant Science***9**: 402. doi: 10.3389/fpls.2018.00402

Ambreen, H., Kumar, S., Variath, M.T., Joshi, G., Bali, S., Agarwal, M., *et al.* (2015) Development of genomic microsatellite markers in *Carthamus tinctorius* L. (Safflower) using next generation sequencing and assessment of their cross-species transferability and utility for diversity analysis. *PLoS ONE* **10**(8): e0135443. doi:10.1371/journal.

Ambuel, J., Ninomiya, S. and Takahashi, N. (1997) Fuzzy logic evaluation of soybean plant shape. *Breeding Science,* **47** (3): 253-257

Amin, P.W. (1985). Resistance of wild species of groundnut to insect and mite pests. *International Workshop on Cytogenetics of Arachis.* 31 Oct-2 Nov.1983. Patancheru, A.P. India.

Amin. P.W., Singh, K.N. Dwivedi, S.L. and Rao, V.R. (1985) Sources of resistance to jassids (*Empoasca kerri* Pruthi), thrips (*Frankliniella schultzei*) and termite (*Odonto terrnes*sp.) in groundnut. *Peanut Sci.***12**: 58-60.

Aminah, A., Suprianto, Suryani, A. and Siregar, I.Z. (2017) Genetic diversity of *Pongamia pinnata (Millettia pinnata*, aka. malapari) populations in Java Island, Indonesia, *B IO D I V E R S IT A S* ISSN: 1412-033X **18** (2):677-681., April 2017 E-ISSN: 2085-4722 Pages: 677-681 DOI: 10.13057/biodiv/d180233

Amrithadevarathinam, A., Arunachalam, V. and Murty, B.R.(1976). A quantitative evaluation of intervarietal hybrids of *Brassica campestris. Theoretical and Applied Genetics*. **48**: 1-8.

Anand, I.J. 1987. Breeding hybrids in rapeseed and mustard, In: *Proc. 7th International Rapeseed Conference,* Poland. pp. 79-85.

Anand, I.J. and Rawat, D.S. (1978) Exploiing cytoplasmic male sterility for heterosis breeding in Indian Mustard *Brassica juncea* L. (Czern)*Proc. Intern. Symp. On Cytogenetics and Crop Improvement*. BHU, Varanasi India 1978.

Anand, I.J., D.S. Rawat and P.K. Mishra. (1985). Mechanism of male sterility in *Brassica juncea* IV. Commercial exploitation of hybrid vigour. *Cruciferae Newsletter*. **10**: 52-54.

Ananthakrishnan, T.N. (1956). On the incidence of *Retithrips syriacus* on castor in Madras. *Zool. Anz.***157**: 33-35

Ananthakrishnan, T.N. (1971a) Mycophagous Thysanoptera - III. *Oriental Insects* **5**: 189–207.

Ananthakrishnan, T.N. (1971b) Thrips (Thysanoptera) in agriculture, horticulture and forestry-diagnosis, bionomics and control. *Journal of Scientific andIndustrial Research, India,* **30**(3):113-46

Ananthakrishnanan, T.N., Gopichandran, R. and Gurusubrahmanian, G. (1992) Influence of chemical profiles of host plants on the infestation diversity of *Retithrips syriacus. Journal of Biosciences,***17** (4): 483-389.

Anaya, M.L.A., Castillo, E.P. *et al.,* (2016) Sex expression and floral diversity in *Jatropha curcas*: a population study in its center of origin.*Peer J – Life & Environment* PubMed 27257548. May 24, 2016. https://peerj.com/

Anderson E, (1993). Plants of central Queensland - their identification and uses. *Department of Primary industries. Brisbane*: Queensland Government Printer

Anderson, W.F., Holbrook, C.C. and Wynne, J.C. (1991). Hertiability and early generation selection for resistance to early and late leaf sopt in peanut. *Crop. Sci.,***31**: 588-593

Andrade, P.B.D., Freitas, B.M., Epifânia., Rocha, E.E. M; Lima, J.A.D. and Rufino, L.L. (2014) Floral biology and pollination requirements of sesame (*Sesamum indicum* L.)*Acta Sci., Anim. Sci.* **36** (1) Maringá Jan./Mar. 2014http://dx.doi.org/10.4025/actascianimsci.v36i1.21838

Angiosperm Group (2016). Angiosperm Phylogeny Group; An update of the Angiosperm Phylogeny Group classification for the orders and families of flowering plants: *APG IV. Bot. J. Linn. Soc.* **181**, 1–20 (2016)

Anil Kumar, H. R., (1999) Studies on sunflower necrosis virus disease. *M.Sc. (Agri.) Thesis, Univ. Agri. Sci., Bangalore, India*. 102 pp

Anjani, K. (1997a) Feasibility of recessive genetic male sterile lines in safflower hybrid seed production. In: Corleto, A., Mundel, H.H. (Eds) Safflower: a multipurpose species with unexploited potential and world adaptability. *IVthInternational Safflower Conf,* Bari, Italy. 278-280

Anjani K .(1997b). Safflower hybrids developed using recessive genetic male sterility system. In: Corleto A, Mundel HH (Eds.) Safflower: a multipurpose species with unexploited potential and world adaptability. *Proc of IVth International Safflower Conf,* Bari, Italy. 281-282

Anjani K. (1998). DSH 129: The first safflower hybrid released for commercial cultivation. In: Heterosis: its commercial exploitation. *Proc. Vasantarao Naik Memorial National Seminar, Dr. Punjabrao Deshmukh Krishi Vidyapeeth, Nagpur,* India. 32.

Anjani, K. (2000).Components of seed yield in safflower (*Carthamus tinctorius* L.). *Indian J. of Agricultural Sciences* **70** (12): 873-875.

Anjani, K., (2000). Catalogue of Castor, vol. 1. *Directorate of Oilseeds Research*, Hyderabad, India.

Anjani, K. (2001) Exploration of castor (*Ricinus communis* L.) germplasm in Andaman and Nicobar Islands. *IBPGR News Letter of Asia, the Pacific and Oceania*. **34**: pp10

Anjani, K. and Ashoka Vardhana Reddy, P., (2002). Genetic divergence and potentiality of castor (*Ricinus communis* L.) germplasm collected from northeast India. *Indian J Plant Genetic Resources* **15**(2), 157-159.

Anjani,K. (2003) Occurrence of dwarf fertile plants in safflower hybrid. *DOR Newsletter***9**:4.

Anjani, K., (2005). Pistillate selections from castor germplasm. *IPGRI newsletter for Asia, The Pacific and Oceania.* **47**: 26-27.

Anjani, K. (2005a). Stabilization and maintenance of male sterility percent in recessive genetic **male sterile lines of safflower** (*Carthamus tinctorius* L.). *Indian J Genetics and Plant Breeding,* **65**(2): 141-142.

Anjani K, (2005b) Development of cytoplasmic genic male-sterility in safflower. *Plant Breeding*, **124** (3): 310-312

Anjani, K. (2010) Genetic resources in castor: A primary gene pool for exploitation. *Research and Development in Castor: Present Status and Future Strategies*. (Ed) Hegde, D.M. *Indian Society of Oilseeds Research*. Directorate of Oilseeds Research Hydearabad. Pp 26-35

Anjani, K, (2012)Castor Genetic resources-A primary gene pool for exploitation. *Industrial Crops and Products -10.1016/j.indcrop.2011.06.011*.**35**(1). 1-14

Anjani, K. (2013) Genetic linkage between male sterility and non-spiny trait in safflower (*Carthamus tinctorius* L.). *Plant Breeding*.**132**: 180-184.

Anjani, K. (2014)A re-evaluation of castor (*Ricinus communis* L.) as a crop plant. CAB Reviews 2014 **9**(38): 1-20 ,doi: 10.1079/PAVSNNR20149038

Anjani K., Bhavna, Mishra, D. and Prasad R.D. (2018). Identification of simple sequence repeat markers linked to Fusarium wilt (*Fusarium oxysporum* f.sp. *carthami*) resistance **and marker assisted selection for wilt resistance in safflower** (*Carthamus tinctorius* L.) **interspecific offsprings.** *Plant Breed*ing **137**: 895-902.).

Anjani, K., Chakravarty, S.K. and Prasad, M.V.R.1(994). Collecting castor (*Ricinus communis* L.) germplasm in North-Eastern Hill Province of India. *IBPGR Newslett. Asia Pacific Oceania* 17, 13.

Anjani. K., Harvir Singh and Prasad, R. D. (2005). Performance of multiple resistant line of **safflower** (*Carthamus tinctorius* L.). Indian J Agricultural Science, 75(3), 178-179

Anjani, K. and Jawaharlal, J. (2015). Differential longevity of castor (*Ricinus communis* L.) germplasm conserved under uncontrolled storage conditions across extended periods. *J Crop Improvement*, **29**: 706-716.

Anjani, K., and Mukta, N. (2008). Varieties and Hybrids of Safflower. *Directorate of Oilseeds Research, Hyderabad*, India. P95.

Anjani, K., Pallavi, M. and Babu, S.N.S. (2007) Uniparental inheritance of purple leaf and associated resistance to leaf-miner in castor bean. *Plant Breeding* **126** (5): 515-520

Anjani, K., Pallavi, M., Bhavna, P., DurgaPrasad, R. and Sarada, C. (2019) Introgression of resistance to *Alternaria* **leaf spot from wild species into susceptible cultivated safflower.** *Plant Breeding,* **139**(2). https://doi.org/10.111/pbr.12775

Anjani, K., Pallavi,M., Prasad, R. D., Dinesh Kumar, V., Madanamohan, B. and Janaki **Ramayya, P. (2012) Identification of RAPD markers flanked to** *Fusarium* wilt resistance **gene in safflower** (*Carthamus tinctorius* L.). *J Oilseeds Res* **29** (Spl.Issue): 4-6

Anjani, K., Pallavi, M. and Sudhakara Babau, S.N. (2010). Biochemical basis of resistance to leafminer in castor (*Ricinus communis* L.). *Industrial Crops and Products,* **31** (1): 192-196.

Anjani, K. and Praduman Yadav. (2017). High yielding-high oleic non-genetically modified Indian safflower cultivars. *Industrial Crops and Products.* **104**: 7-12.

Anjani, K., Raoof, M.A and Desai, A.G. (2014). Evaluation of world castor (*Ricinus communis* L.) germplasm for resistance to Fusarium wilt (*Fusarium oxysporum* f. sp. *ricini*). *European J Plant Pathology.* **139**:567–578.

Anjani, K., Raoof, M.A., Santha Lakshmi Prasad, M., Duraimurugan, P., Lucose, C., Praduman Yadav, Prasad, R.D., Jawahar Lal, J. and Sarada, C. (2018) Trait-specific accessions in global castor (*Ricinus communis* L.) germplasm core set for utilization in castor improvement. Industrial Crops and Products. **112**: 766-774

Anjani, K., Raoof, M.A., SanthalakshmiPrasad, M.,Lakshminarayana, M., Duraimurugan, P., Praduman Yadav, Lakshmamma, P., Jawaharlal, J. and Vishnuvardhan Reddy, A. (2017). Castor Genetic Resources: Status and Utilization in Castor Genetic Improvement. *ICAR-Indian Institute of Oilseeds Research, Hyderabad, India.*

Anjani, K., Sudhakara Babu, S. N., Padmavathi, P., Srinivas, P.S., Prasad, R. D. and Vishnuvardhan Reddy, A. (2020). Hybrid Seed Production in Safflower.*ICAR-Indian Institute of Oilseeds Research,* Hyderabad, India. P25

Anjula, N. and Nagaraju (2002) Seed transmission studies of sunflower necrosis virus disease. In: Nagaraju *et al* (eds) *National symposium on plant disease Scenario in Southern India,* pp 21–22, 19–21 December. IPS (S Zone)/University of Agricultural Science, Bangalore, pp 74

Ankineedu, G., Sharma, K.D. and Kulkarni, L.G. (1968) Effect of fast-neutrons and gamma rays on castor. *Indian J Genet.* **28**:31–39

Ankineedu, G. and Rao, N.G.P. (1973). Development of pistillate castor. *Indian. J Genet.* **33**: 416–422

Anonymous (1953). An unusual epidemic of castor hairy caterpillar *Euproctis lunata* in Delhi. *Plant Protection Bulletin.* **5**: 50-51

Anonymous (1962) Report of the Indian Central Oilseeds Committee on pests of oilseeds, New Delhi, India.

Anonymous, (1966.) International rules for seedtesting procedure. *International Seed Testing Association,* **31**:1–132.

Anonymous (1976) Progress Report of Dryland Research Main Centre Kharif 1976, Central Arid Zone Research Institute, ICAR, Jodhpur, India,

Anonymous(1989). Evaluation of *Taramira* under rain-fed situation. *Annual Report. 1988-89. All India Coordinated Research Project for Dryland Agriculture Main Centre Jodhpur.* PP 59-61

Anonymous (1997). Promotion of research and development efforts on hybrids in rapeseed mustard. *Annual Report 1995-96.*ICAR-*National Research Centre on Rapeseed-mustard, Bharatpur-321 303, India.*

Anonymous (2001)http://agris.fao.org/agris-search/search.do?recordID=ET2003000105 / GKVK University of Agricultural Sciences, Banglaore, India

Anonymous (2002) *The Indian Express* dated 2/6/2002.

Anonymous (2006). Research Achievements in Castor: All India CoordinateResearch Project on Castor. Directorate of Oilseeds Research, Hyderabad. India pp111.

Anonymous (2006A) AICRIP on Sunflower DOR 2006. Research Achievements in Sunflower. All India Coordinated Research Project on Sunflower. Directorate of Oilseeds Research, Hyderabad, India. Pp117.

Anonymous (2008) http://economictimes.indiatimes.com/articleshow/27928278.cms?utm_source=contentofinterest&utm_medium=text&utm_campaign=cppst

Anonymous (2010) https://iisrindore.icar.gov.in/CROPPD1.htm. Indian Institute of Soyabean Research. Indore, India

Anonymous (2010A) Don't Fall for Jatropha, *Economic Times –India Times,* New Delhi.http://articles.economictimes.indiatimes.com/2010-07-26/news/27574470_1_jatropha-biofuels-marine-chemicals-research-institute

Anonymous (2012) Australian Oilseeds Federation (2012), Better Sunflowers Agronomy Training Package (Big Yellow Sunflower Pack}, https:// www.bettersunflowers.com ; https://grdc.com.au/__data/assets/pdf_file/0019/370612/GrowNote-Sunflower-North-04-Physiology.pdf

Anonymous (2012A) Agroforestry Depatment of ANGRAU, Hyderabad and A.P. Forest Department, Hyderabad.

Anonymous (2012B) https://www.easyayurveda.com/2012/12/21/karanja-pongamia-pinnata-benefits-usage-ayurveda-details/

Anonymous (2013) TNAU Agritech Portal, Agriculture-Stress Management: Drought (http://agritech.tnau.ac.in/agriculture/agri_drought_effect_on_crops.html)

Anonymous (2014-15). Annual Report of the ICAR-Indian Institute of Oilseeds Research, Hyderabad for 2014-2015http://icar-iior.org.in/index.php/annual-report-new-

Anonymous (2015) TNAU Agritech Portal. Crop Improvement: Sesame. http://agritech.tnau.ac.in/crop_improvement/crop_imprv_emasculation_oilseeds.html

Anonymous (2018)DSH-185: New CGMS-based Safflower Hybrid Indian council of Agricultural Research (ICAR) New Delhi.*https://www.icar.gov.in/content/dsh-185-new-cgms-based-safflower-hybrid*

Anonymous (2018A) TNAU Agritech Portal (2018)

Anonymous (2018-2019) Annual Report of ICAR-IIOR for 2018-2019 http://icar-iior.org.in/media/docs/ann-rep/iior-annrep-2018-19.pdf

Anonymous (2019) *Vyavasaaya Panchaangam* (Agricultural Encyclopaedia) inTelugu. 2017-2018 Prof. Jaisankar Telangana State Agricultural University, Hyderabad – 500 030.Pp 440. http://te.vikaspedia.in/agriculture/c35c4dc2fc35c38c3ec2f-c2ac02c1ac3ec02c17c02-1/amudam

Anonymous (2020). Castor. Production Technology, ICAR-Indian Institute of Oilseeds Research, Hyderabad. P 26

Anwar F, Ashraf M, Bhanger MI. Interprovenance variation in the composition of *Moringa oleifera* oilseeds from Pakistan. *J Am Oil Chem Soc*. 2005;**82**:45–51.

Appaji, S., Raoof, M.A., Chattopadhyay, C. and Prasad, M.V.R. (1996) Sunflower Downy mildew – athreatening disease in India. *Indian Farming.* **46** (5): 6-8.

Appelqvist, L. A and Ohlson, R. (ed). (1972) Rapeseed – Cultivation, Composition. Processing and Utilization. *Elsevier Publ.Comp.* Amsterdam. 197

Arant, F.S. *et al.,* (1951) *The Peanut, the Unpredictable Legume*. National Fertilizer Association, Washington D.C. Pp.333

Aravind S. M. (2002) Transmission, vector relationship and weed hosts of sunflower necrosis virus disease. *M.Sc. (Agri.) Thesis, Univ. Agri. Sci., Bangalore*, India. 2002, 96 pp

Arora R K. (1998). The Indian gene center – priorities and prospects for collection. In: Paroda, R. S., Arora, R. K. and Chandel, K. P. S. (Eds) *Plant Genetic Resources*: *Indian Perspective*,Indian Society of Plant Genetic Resources, National Bureau ofPlant Genetic Resources, New Delhi. pp 66–75.,

Arvind and Bekal, S.(2018) Experimental Investigation on the performance of Novel Stove for Use with Vegetable Oil, *Energy and Power***8**(2): 46-50;doi:10.5923/j.ep.20180802.02

Arpiwi, N. L., Yan G, Barbour EL, Plummer JA (2013) Genetic diversity, seed traits and salinity tolerance of *Millettia pinnata* (L.) Panigrahi, a biodiesel tree. *Genetic Resources and Crop Evolution***60** (2): 677 – 692. doi: 10.1007/s10722-012-9866-y.

Arpiwi, N.L., Guijun Yan, Elizabeth L Barbour and Julie A Plummer (2014). Phenology, pollination and seed production of *Millettia pinnata,* In: Kununurra (Ed.) NorthWesternAustralia. *Jurnal Biologi*XVIII(1): 19 – 23

Arunachalam, V., Bandyopadhyay, A., Nigam, S.N. and Gibbons, R.W. (1982) Heterotic potential of single crosses in groundnut (*Arachis hypogaea* L.,) *Oleaginneux***37** :415-420

Arunachalam, V., Bandyopadhyay, A., Nigam, S.N. and Gibbons, R.W. (1980) Some basic results of applied value in groundnut breeding. *Proc. National Seminar on the Appliction of Genetics to the Improvement of Groundnut.* Tamil Nadu Agricultural University, Coimbatore.16-17 July, 1980 pp: 1-9

Arunachalam, V., Bandyopadhyay, A., Nigam, S.N. and Gibbons, R.W. (1984) Heterosis in relation to genetic divergence and specific combining ability in groundnut (*Arachis hypogaea,* L.) *Euphytica*,**33:**33-39.

Arvalis, (2016). Niger, nyger, *Guizotia abyssinica.* Les fiches couverts, espèce pure

Asad, A., Blame, P.C. and Edwards, D.G. (2003) Effects of Boron foliar applications on vegetative and reproductive growth of sunflower. *Ann. of Botany*, **92**: 565-570.

Asanome, N. and Ikeda, T. (1998) Effect of branch direction's arrangement on soybean yield and yield components. *Journal of Agronomy and Crop Science,* **181** (2):95-102

Ashford (2011). Ashford's Dictionary of Industrial Chemicals, Third edition, 2011, page 6162

Ashish, J., Nadaf, H.L. and Gangadhara, K. (2014) Genetic analysis of rust and late leaf spot in advanced generation recombinant inbred lines of groundnut (*Arachis hypogaea* L.) *Int. J. Genet. Eng. Biotechnol*. **5**(2): 109-114

Ashley, J.M. (1984) Groundnut. In: P.R. Goldgworthy and N.M. Fisher (Eds)*The Physiology of tropical field crops*. Johnwiley & Sons Ltd. Pp. 453-493

Ashoka Vardhana Reddy, P., Anjni, K. and Manikyam, M. (2002) Collecting castor (*Ricinus communis* L.) land races from Tamil Nadu, India. *Plant Genetic Resources Newsletter.* **132**: 60-62

Ashri A. (1971) Evaluation of the world collection of safflower, *Carthamus tinctorius* L. I. Reaction to several diseases and associations with morphological characters in Israel. *Crop Science***11**: 253–257.

Ashri A (1975) Evaluation of the germplasm collection of safflower (*Carthamus tinctorius*): Distribution and regional divergence for morphological characters. *Euphytica* **24:**651–659.

Ashri, A. (1981) Increased genetic variability for sesame improvement by hybridization and induced mutations. In: A. Ashri (ed) *Sesame: Status and Improvement. FAO Plant Production and Protection, Paper 29,* Rome. Pp141-145.

Ashri, A. (1988) Sesame breeding — objectives and approaches, p. 152-164. In: A. Omran (ed.), *oil crops: sunflower, linseed and sesame*. *IDRC-MR205e, IDRC*, Ottawa, Canada.

Ashri, A. (1994) Genetic Resources of Sesame In: R.K. Arora and K.W. Riley(Eds). *Sesame Biodiversity in Asia: Conservation, Evaluation and Improvement Proceedings of the IBPGR-ICAR/NBPGR. Asian Regional Workshop on 'Sesame Evaluation and Improvement' held at Nagpur & Akola in India* 28-30 September, 1993.PP25-39

Ashri, A. (1995) Sesame research overview: Current status, perspectives and priorities, p. 1-17. In: M.R. Bennett and I.M. Wood (eds.), *Proc. 1st Austral. Sesame Workshop. NT Dept. Primary Industry and Fisheries, Darwin.*

Ashri, A. (2001) Induced Mutations in Sesame Breeding. Faculty of Agriculture. *Sesame Improvement by Induced Mutations. IAEA, VIENNA, 2001 IAEA-TECDOC-1195 ISSN 1011–4289 © IAEA, 2001 XA0100273: 32007909*.pdf.pp1-8

Ashri, A. (2007) Sesame (*Sesamum indicum* L.) Chapter 8. In: SINGH, R. J. (Ed*.). Genetic resources, chromosome engineering and crop improvement, oilseed crops.* Boca Raton: CRC Press, 2007. v. 4, p. 231-289.

Ashri, A. (2010) Sesame Breeding. In:*Plant Breeding Reviews* Janick, J (Ed) John Wiley and Sons. **51**: pp179-226

Ashri, A. and Knowles, P.F. (1960) Cytogenetics of safflower (*Carthamus* L.) species and their hybrids. *Agronomy Journal,* **52** (1): 11-17.

Asinelli, M.E.C.; Souza, M.C.o.d and Mourao, K.t.S.M. (2011). Fruit ontogeny of *Garcinia gardneriana* (Planch. & Triana) Zappi (Clusiaceae). *Acta Botanica Brasilica.* **25** (43–52).

Atlagić, J and Korić, D.S. (2008) Cytogenetic study of *Helianthus laevigatus* and its F_1 and BC_1F_1 hybrids with cultivated sunflower, *Helianthus annuus. Plant Breeding,* **118** (6):555-559 https://doi.org/10.1046/j.1439-0523.1999.00423.x

Athaya, C.D. (1985). Ecological studies of some forest tree seeds. II. Seed storage and viability. *Indian J. For.* **8:**137–140.

Auld, D.L., Zanotto, M.D., McKeon, T.A., Morris, J.B. (2009). Castor. In: Vollmann, J., Rajcan, I. (Eds.), *Oil Crops, Handbook of Plant Breeding* **4**, pp. 317–332

Avidov Amit and Learner, Alon.(2008). Polyploid castor plants.Patent: Publ.of the Int.Appl. with Int.search report Polyploid Castor Plants, Compositions Derived Therefrom and Uses of the same. WO2009060455A1.WIPO (PCT) *International Application Published with International Search Report.* https://patents.google.com/patent/WO2009060455A1/en

Ayad, T. S. and Saad, A. A., (2011) Effect of different levels and timing of Boron foliar application on growth, yieldand quality of sunflower genotypes. *Mesopotamia J. of Agric.***39** (3).

Ayad, T.S and Saad, A.A. (2012) Response of some sunflower hybrids to zinc foliar spraying and phosphorous fertilizer levels under sandy soils. *J. Tikrit Univ. for Agri. Sci.,* **12**(4): 174-182.

Azam, M.M., Waris, A. and Nahar, N.M. (2005) Prospects and potential of fatty acid methyl esters of some non-traditional seed oils for use as biodiesel in India. *Biomass and Bioenergy.***29**:293–302. [Google Scholar]

Babitha. B., Amrutha, P., Fashina-Bnau., Akshatha. B.S. and Bhavana. M. (2007) Evaluation of Protein and amino acid nutritional components of Pongamia seed cake. *Project Reference No. 38S_B_MSC_007, MaharaniLakshmi Ammanni College for Women,* Bangalore (India).

Babu, B.K., Saikia, R. and Arora, D.K. (2010) Molecular characterization and diagnosis of *Macrophomina phaseolina*: Acharcal Rot Fungus. In: Gherbawy.Y, Voigt, K. (Eds). *Springer Berlin Heidelberg,* 2010, 179-193.

Bachheti, R.K., Dwivedi, H., Rana, V. and Rai, I. (2012) Characterization of fatty acids in *Melia azedarach* L. seed oil. *Internatioal Journal of Current Research and Reviews.***4:** 108-114

Badri J., Yepuri V., Ghanta A., Siva S., Siddiq E. A. (2014). Development of microsatellite markers in sesame (*Sesamum indicum* L.). *Turk. J. Agric. For.* **38:** 603–614. 10.3906/tar-1312-104

Badwal, S.S. and Gupta, V.P. (1968) Correlation of quantitative traits and selection indices for improving pod yield in groundnut. *Journal of Research, Punjab Agricultural University.* **5** (2): 20-23

Badwal, S. S. and Harbans Singh. 1973. Effect of growth habit on correlations and path coefficients in groundnut. *Indian J. Genet*. **33:** 101-111

Badwal, S.S. and. Labana, K.S (1983). Male sterility in Indian Mustard In: 15th International conference Genetics, New Delhi. pp. 514

Bafor, M., Smith, M.A., Johnson, Stobart, A.K. and Stymme, S. (1991) acylRicinoleic acid biosysnthesis and triacylglycerol assembly in microsomal preparations from developing castor-bean (*R.commnis)*endosperm. *Biochem. J.***280**: 507-514

Baghel, K., Salam, J.L., Kanwar, R. and Bhanwar, R.R. (2018) Genetic Variability Analysis of Yield and its Components in Niger [(*Guizotia abyssinica* (L. f.) Cass.]*International Journal of Current Microbiology and Applied Sciences* ISSN: 2319-7706;7(08): 4266-4276

Baghel, S. and Bansal, Y.K. (2017) In vitro regeneration of *Guizotia abyssinica* Cass. and evaluation of genetic fidelity through RAPD markers. *South African Journal of Botany*, **109**: 294-307

Baghyalakshmi, K., Shaik, M., Manohar rao, M., Shaw, R., Lavanya, C., Mnajunath, T., Senthilvel, S. (2020). Development and characterization of tetraploid castor plants. *Plant Gentic Resources: Characterization and Utilization*, 1-7.doi:10.1017/S1479262120000039

Bahuguna., V.K. (1997). Silviculture and management practices for cultivation of *Azadirachta indica* (neem). *Indian Forester* **123**: 379-386

Bains S.S. and Jhooty J.S. (1978) Increased susceptibility of virus infected *Brassica juncea* to *Peronospora parasitica. Plant Dis. Reptr.* **62:** 1043 -1046

Bains, S.S. and Jhooty, J.S. (1979) Mixed infection by *Albugo candida* and *Peronospora parasitica* on *Brassica juncea* inflorescence and their control. *Indian phytopath.* **32**: 268-271 5.

Bains S.S. and Jhooty J.S (1980) Studies on *Peronospora parasitica* causing downey mildew of Crucifers. *Indian J. Mycol. Pl. pathol.* **10**: LIV (Anstr.)

Bajaj, Y.P.S., Mahajan, S.K. and Labana, K.S. (1986) Interspecific hybridization of *Brassica napus* and *B. juncea* through ovary, ovule and embryo culture. *Euphytica* **35**: 103–109

Bakhetia, D.R.C. and Sekhon, B.S. (1984) Review of the research work on insect pests of rapeseed-mustard in India. *Paper presented at All India Annual Rabi Oilseeds Workshop held at Sukhadia University R.R.S. Durgapura (Jaipur) India.* Aug.6-10. 1984

Bakhetia, D.R.C. and Singh, J. (1992) Technological Advances in Pest Management in Rapeseed-Mustard.,In: Kumar. D. and Rai. M. (Eds). *Adv. in Oilseeds research.* Vol.1Scientific Publishers, Jodhpur, India. Pp 120-147

Balaiah, C., Reddy, P.S. and Reddi, M.V. (1977) Genetic analysis in groundnut. I. Inheritance studies on 18 morphological characters in crosses with 'Gujrat Narrow Leaf Mutant'. *Proceedings of the Indian Academy of Sciences.* **85B:** 340-350

Bamber, C.J. (1916). Plants of the Punjab: a descriptive key to the flora of the Punjab, North-west Frontier province and Kashmir. *Superintendent Government Printing, Punjab, Lahore*. Google Scholar

Bancroft, I., Morgan, C., Fraser, F. *et al.* Dissecting the genome of the polyploid crop oilseed rape by transcriptome sequencing. *Nat Biotechnol* **29:**762–766 (2011) https://doi.org/10.1038/nbt.1926

Bandillo N., Muyco P. A., Caspillo C., Laza M., Sajise A. G., Singh R. K. *et al., (*2010) Development of multiparent advanced generation intercross (MAGIC) populations for gene discovery in rice *(Oryza sativa L.). Philipp. J. Crop Sci.* ***35****, suppl 1, 96.*

Banga, S. S. and Banga, S. K. (1997). *Enarthrocarpus lyratus* cytoplasmcauses male sterility in oilseed rape, *International Symposium on Heterosis in Crops, Mexico,*pp 120–121

Banga, S.S. andLabana. K.S. (1983a). Male sterility in Indian mustard. II Genetics and Cytology of MS-1. *Proc. 6th International Rapeseed Conference,* Paris. pp. 349-353

Banga, S.S. and Labana, K.S.(1983b)Production of F_1 hybrids using ethrel-induced male sterility in Indian mustard (*Brassica juncea* (L.) Coss.)*J.Agric. Sci. (Camb)* **101**:453-455

Banga, S.S. and Labana.K.S.(1984). Heterosis in Indian mustard. *Zeitscriftfiir Pflanzenjiichtung.* **92**(1): 61-70.

Banga, S.S. and Labana, K.S. (1985). Male sterility in Indian mustard IV. Genetics of MS4. *Canadian Journal of Genetics and Cytology.* **27**: 487-490.)

Bange, M.P., Hammer, G.L., and Rickett, K.G. (1997) Environmental control of potential yield of sunflower in the sub-tropics. *Australian Journal of Agricultural Research.* **48** (2): 231-240.

Bannerot. H. Bnoulidard, L. Canderon, Y. and Tempe, J. (1974) Cytoplasmic male sterility transfer from *Raphanus* to *Brassica* : *Proc. Eucarpia Crop Sect. Cruciferae* **25**: 52-54

Bannerot H., Boulidard L. and Cauderon Y. (1977) Unexpected difficulties met with the radish cytoplasm. Eucarpia *Cruciferae* Newsletter **2:** 16 Google Scholar

Bansal, K.N. and Katara, G.S. (1993) Response of safflower (*Carthamus tincorius* L) to nitrogen and irrigation. *Annals of rid zone*, **32** (3) :193-194

Bargali, S. S., Kiran Shukla, Lalji Singh, Lekha Ghosh and Lakhera, M. L., (2015). Leaf litter decomposition and nutrient dynamics in four tree species of dry deciduous forest. *Trop. Ecol.*,**56** (2): 191-200

Barakat, M.N., and E.C. Cockling. (1983). Plant regeneration from protoplast derived tissues of *Linum usitatissimum* L. (fl ax). *Plant Cell Rep.* **2**:314–317

Barakat, M.N., and E.C. Cockling. (1985). An assessment of the cultural capabilities of protoplasts of some wild species of Linum. *Plant Cell Rep.* **4**:164–167

Barwale, R.B. and Panchabhaye, K.M. (1994) Commercial production of sesame hybrids with special reference to seed production technologies. *Hybrid Sesame, Proceedings of Group Discussion*, June 14, 1994, *Directorate of Oilseeds Research*, Hyderabad: pp35-37

Bas, N., Pavelek, M., Maggioni, L. and Lipman, E. (2006) Report on working group of fibre crops (Flax and Hemp) ECP GR *European Cooperative Programme for Plant Genetic Resource, IPGRI and INIBAP Supported by the CGIAR* 14-16 June 2006, Wageningen, The Netherlands.

Basappa, H. and Lingappa, S. (2004). Evaluation of various neem preparations against castor semilooper *Achoea janata L.* under field conditions. *Journal of Research of ANGRAU***32** (4): 25-28.

Basappa H and Prasad M.S.L. (2005). In: Hedge DM (ed) Insect pests and diseases of sunflower and their management. Directorate of Oilseeds Research, Hyderabad, 80pp

Basappa, H and Singh, V. (1990) Record of insect pests on niger (*Guizotia abyssinnica*)*: Journal of Oilseeds Research* 1990 **7**(2) :86-87

Baskaran, R.K and Mahadevan, N.R. (1991). Effect of sowing time on shoot webber *Antigastra catalaunalis* and*phyllody* incidence and on the yield of sesame *Sesamum indicum, Indian Journal of Agricultural Sciences***61**(1): 70-72

Bassett, C. M., Rodriguez-Leyva, D. and Pierce, G. N. (2009). Experimental andclinical research findings on the cardiovascular benefits of consuming flaxseed. *Appl. Physiol. Nutr. Metab.*, **34**: 965–74

Bassil, E. S. and Kaffka, S. R. (2002). Response of Safflower (*Carthamus tinctorius* L.) to Saline Soils and Irrigation. I. Consumptive Water Use. *Agric. Water Manage.*, **54**: 67-80).

Bastaki N.K. and Cullis C.A. (2014). Floral-dip transformation of flax (*Linum usitatissimum*) to generate transgenic progenies with a high transformation rate. *Journal of Visualised Experiments*, **94**: e52189

Bateman A.J. (1955) Self-incompatibility systems in angiosperms. III. Cruciferae. *Heredity***9**: 52–68

Batkal, B.G. (1994). Inaugural Address,Sesame Biodiversity in Asia: In: R.K. Arora and K.W. Riley(Eds.) *Conservation, Evaluation and Improvement.Proceedings of the IBPGR-ICAR/ NBPGR Asian Regional Workshop on 'Sesame Evaluation and Improvement' held at Nagpur & Akola, India.*Pp 28-30 September, 1993: pp9-11.

Battacharya, D.K. (2002) Lesser known plant sources for fats and oils. *Inform.***13** :151-157

Baud, S and Lepiniec, L (2010) Physiological and developmental regulation of seed oil production. *Progress in Lipid Research* 49, 235–249.

Bavbiskar, A.P. (1994). Note on Production of Hybrid Seed in Sesame. *Hybrid Sesame, Proceedings of Group Discussion*, June 14, 1994, Directorate of Oilseeds Research,Hyderabad: pp 50-52

Baydar, H. (2005)Breeding for the improvement of the ideal plant type of sesame.*Plant Breeding***124**(3):263 – 267

Baydar, H., Gokmen, O.Y. and Friedt, W. (2003). Hybrid seed production in safflower (*Carthamus tinctorius* L.) following the induction of male sterility by gibberellic acid. *Plant Breeding,* **122** (5): 459–461.

BBC - GCSE Bitesize (2018)*: Double bonds and hydrogenation". Retrieved 2018-09-1*

Beard, B. H. and Comstock, V. E. (1980). Flax. In: W.R. Fehr and H.H. Hadley (eds.), *Hybridization of Crop Plants* American Society of Agronomy - Crop Science Society of America, Madison, WI.

Beech, D.f. and Norman, M.J.T. (1963) The effect of time of planting on yield attributes of varieties of safflower. Australian Journal of Experimental Agriculture and Animal husbandry. **3** (9): 140-148.

Bedigian, D. (1984)*Sesamum indicum*, L. Crop origin, Diversity, Chemistry and Ethnobotany. *Ph.D. Dissertation, University of Illinois, Urbana – Champaign*. Illinois, USA

Bedigian, D. (2003). Evolution of sesame revisited: domestication, diversity and prospects. *Genetic Resources and Crop Evolution*, **7** (50) :779-787, 2003.

Bedigian, D. and Harlan, J.R. (1986) Evidence of cultivation of sesame in the ancient world. *Econ.Bot.***40:** 137-154

Bedigian, D., Seigler, D.S. and Harlan, J.R. (1985). Sesamin, sesamolin and the origin of sesame, *Biochemical Systematics and Ecology*, **13** (2): 133-139

Bedigian,D., Smyth,,C.A., and Harlan,J.R. (1986) Patterns of morphological variation in *Sesamum indicum . Economic Botany***40** (3): 353–365 198

Begum, T. and Dasgupta, T. (2015). Amelioration of Seed yield, Oil content and Oil quality through Induced Mutagenesis in Sesame(*Seamum indicum,* L.) *Bangladesh J. Bot.* **44**(1): 15-22, 2015 (March)

Bekele, E., Geleta, M., Dagne K., Jones, A.L., Barnes, I., Bradman N. and Thomas, M.G. (2007). Molecular phylogeny of genus *Guizotia* (Asteraceae) using DNA sequences derived from ITS. *Genetic Resources and Crop. Evolution***54**: 1419–1427.

Bleho J., Obert B., Takáč T., Petrovská B., Heym C., Menzel D. and Šamaj J. (2011): ER disruption and GFP degradation during non-regenerable transformation of flax with *Agrobacterium tumefaciens*. *Protoplasma,***249**: 53–63.

Bell, M.J. and Wright, G.C. (1998) Groundnut growth and Development in Contrasting Environments. Growth and Plant Density Responses. *Experimental Agriculture*, **34** (99): 112-124

Belonogova, M.A., and G.N. Raldugina. (2006). Shoot regeneration from cotyledon explants of fi ber fl ax (Linum usitatissimum) and their subsequent rooting. *Russ. J. Plant Physiol.* **53**:501–506

Benerjee, P.P and Kole, PC. (2011). Heterosis, Inbreeding Depression and Genotypic Divergence for Some Physiological Parameters in Sesame (Sesamun indicum L.)Journal of Crop Improvement **25** (1) : 11-25

Benelli, V.G. (2015)Comparison of Seed Yield, Oil and Phenotypic Traits Among Selected Parents and Crosses of Niger. *Thesis submitted for the degree of Master of Science,*The University of Tennessee, Knoxville, USA. Pp 149.

Benga, S.H. and Angadi, S.V. (2016) Effect of plating date on winter canola growth and yield in South western U.S. *American J. Plant Sciences,***7:**201-217.

Benge, M.D. (1988). Cultivation and Propagation of the Neem Tree. In: M. Jacobson(Ed). *Focus on Phytochemical Pesticides (1) The Neem Tree, CRC Press Inc.*, Boca Raton, Florida. pp. 2-17.

Bennet, M.R. (1996) Sesame Production in Australia. *Ses. Staff. Newsletteer,***11**: 4-9

Bennett, R. N., Mellon, F.A., Botting, N.P., Eaglcs, J., Rosa, E.A. and Williamson G. (2002) Identification of the major glucosinolate (4-mercaptobutyl glucosinolate) in leaves of *Eruca sativa*l. (salad rocket) *Phytochemistry.* 2002; **61**:25–30

Bentham, G. (1841). On the structure and affinities of *Arachis* and *Voandzeia. Transactions of the Linnaen Society of London***18**:155-162

Benzioni, A. (1997) Jojoba, *NewCrop Fact Sheet, University ofPurdue, USA.https://hort.purdue.edu/newcrop/CropFactSheets/jojoba.html#Botany*

Bera, S.K., Paria, P. and Radhakrishnan, T. (2002). Wide hybridization in *Arachis hypogaea*: Genetic improvement and species relationship. *Int. Arachis Newsletter*, **22**: 37-39.

Berenguer, M.A. and Joseph, E. (2019) Siderophores: From natural roles to potential applications. *Advances in Applied Microbiology*, 2019

Bergman, J.W., D.E. Baldridge, P.L. Brown, A.L. Dubbs, G.D. Kushnak, and N.R. Riveland. (1987). Registration of 'Hartman' safflower. *Crop Sci*. **27**: 1090–1091.

Bergman, J.W., G. Carlson, G. Kushnak, N.R. Riveland, and G. Stallknecht. (1985). Registration of 'Oker' safflower. *Crop Sci*. **25**: 1127–1128.

Bergman, J.W., G. Carlson, G. Kushnak, N.R. Riveland, G. Stallknecht, L.E. Welty, and D. Wichman. (1989a). Registration of 'Girard' safflower. *Crop Sci*. 29: 828–829.

Bergman, J.W., G. Carlson, G. Kushnak, N.R. Riveland, G. Stallknecht, L.E. Welty, and D. Wichman. (1989b). Registration of 'Finch' safflower. *Crop Sci*. **29**: 829.

Bergman, J.W. and N.R. Riveland. (1983). Registration of 'Sidwill' safflower. *Crop Sci*. **23**: 1012–1013.

Berk, Zeki (2013). "Food Process Engineering and Technology (Second Edition)". *Advances in Molecular Toxicology, 2014*. ISBN 978-0-12-415923-5.

Berry, P.M. and Spink, J.H. (2006). A physiological analysis of oilseed rape yields: Past and future, *The Journal of Agricultural Science,* **144** **(**5) . 381-392

Berlinger, M. J. 1986. Host Plant Resistance to Bemisia tabaci. Agric. *Ecosystems Environ.***17**: 69-82.

Bernardo R.(2010)Breeding for Quantitative Traits in Plants. Stemma Press, Woodbury, MN.Google Scholar

Berry, P.M and Spink, J.H. (2006). A physiological analysis of oilseed rape yields: Past and future, *The Journal of Agricultural Science,***144** **(**5) . 381-392

Bertioli, D.J and Cannon, S.B. (2016) The genome sequences of *Arachis duranensis* and *Arachis ipaensis*, the diploid ancestors of cultivated peanut. *Nature Genetics* **48**: pp 438–446 (2016)

Bever B.O. (1986) Medicinal Plants in Tropical West Africa. *Cambridge University Press*, 1986, Cambridge.

Beversdorf, W.D. and Knott, L.S. (1987) An *in-vitro* mutagenesis / selection system for *Brassica napus.Iowa State J. Res*. **61**:435-443

Bezdenezhoykh, S.M.(1956) Pollination of fibre flax and training of bees. In: Kishchunas, I.V. and Gubin, A.F. (Eds) *Pollinationof Agricultural Plants. Gos Izd-vo. Selkhoz Litry, Moscova* (Moscow). pp. 21-24

Bhagya, S. and Shamanthaka Sastry, M. C. (2003). Chemical, functional and nutritional properties of wet dehulled niger (*Guizotia abyssinica* Cass.) seed flour. *Food Sci. Technol.,***35** (7): 703-708

Bhambal, A., Kothari, S., Saxena, S. and Jain, M. (2011). "Comparative effect of neemstick and toothbrush on plaque removal and gingival health – A clinical trial" (PDF). *Journal of Advanced Oral Research*. **2**(3): 51–56. ISSN 2229-4120.

Bhambure C.S. (1958) Effect of honeybee activity on Niger seeds production, *Indian Bee J.* 1958;**20**:189-195.

Bharathi, A., Umarani, R., Karivaratharaju, T.V., Vanangamudi, K. and Manonmani, V. (1996). Effect of drupe maturity on seed germination and seedling vigour in neem. *Journal of Tropical Forest Science* **9** (2): 147-150.

Bhardwaj, S.P. and Gupta, R.K (1977) Tilangi (Niger) a potential high yielding oilseed crop. Indian Farming **27** (6) : 18-19

Bhargav, D.K and Meena, H.K. (2014) Sunflower Necrosis Disease: A New Threat for Sunflower (*Helainthus annuus* L.) Production in Southern India, *Popular Kheti***2** (1).

Bhaskaran, P. and Jayabalan, N. (2006) *In-vitro* mass propagation and diverse callus orientation on *Sesamum indicum* L.- an important oil plant. *Journal of Agricultural Technology* **2**: 259-269

Bhaskaran and Mahadevan (1991)Effect of sowing time on shoot webber Antigastra catalaunalis and Phyllody incidence on yield of sesame *Indian J. Agric. Sci.* **61**(1):70-72

Bhaskara Rao, E.V.V., Swamy, K.R.M., Yadukumar, N. and Sreenath Dixit (1993) Cashew Production Technology,*National Research Centre for Cashe*w, Puttur,-574202, Karnataka, India.pp.36.

Bhat, B. (2011) Epidemiology and management of sunflower necrosis disease. *Ph.D. thesis, Acharya N.G. Ranga Agril. University, Hyderabad, 289 pp*

Bhat B.N. and Reddy D.R.R. (2016) Status of Viruses Infecting Sunflower and Strategies for their Management. In: Gaur R., Petrov N., Patil B., Stoyanova M. (eds) *Plant Viruses: Evolution and Management.* Springer, Singapore. DOI: https://doi.org/10.1007/978-981-10-1406-2_16

Bhat, J.G. and Murthy, H.N. (2007). Factors affecting in-vitro gynogenic haploid production in niger [*Guizotia abyssinica* (L. f.) Cass.)].*Journal of Plant Growth Regulator,* **52**: 24

Bhat, J.G. and Murthy, H.N. (2008). Haploid plant regenera-tion from unpollinated ovule cultures of *niger (Guizotia abyssinica (L. f.) Cass.). Russian Journal of Plant Physiology* **55***:* 241-245

Bhat K.V., Babrekar P.P. and Lakhanpaul S. (1999) Study of genetic diversity inIndian andexotic sesame (*Sesamum indicum*L.) germplasm using random amplified polymorphic DNA (RAPD) markers, *Euphytica,* **110:** 21-33 (1999)

Bhat, R.V., Vasanthi, S., Rao, B.S., Rao, R.N., Rao, V.S. *et al.* (1996) Aflatoxin B1 contamination in groundnut samples collected from different geographical regions of India: a multicentre study. *Food Additives and Contaminants* **13***:* 325-331

Bhat, S. R., Vijayan, P., Ashutosh, K., Dwivedi, K. and Prakash, S. (2006). *Diplotaxis erucoides* induced cytoplasmic male sterility in *Brassica juncea* is rescued by the *Moricandia arvensis* restorer-genetic and molecular analysis. *Plant Breeding,* **125**: 150–5

Bhateria, S., Sood, S.P. and Pathania, A. (2006) Genetic analysis of quantitative traits across environments in linseed (*Linum usitatissimum* L.). *Euphytica,* **150:**185-194

Bhatia, V.S., Piara Singh, Wani, S.P., Kesava Rao, A.V.R. and Srinivas, K. (2006) Yield Gap Analysis of Soybean, Groundnut, Pigeonpea and Chickpea in India Using Simulation Modeling,*Global Theme on Agroecosystems Report no. 31, ICRISAT International Crops Research Institute for the Semi-Arid Tropics* Patancheru 502 324, Andhra Pradesh, India. 156pp.

Bhatnagar, A.S. and Krishna, A.G.G. (2013). Natural antioxidants of the Jaffna variety of *Moringa Oleifera* seed oil of Indian origin as compared to other vegetable oils. *Grasas Y Aceites.* **64** (5):537-545

Bhatt, R. P. and M. Sanjappa. (1976) Cytology of pongam oil tree *Derris indica* (Lamk.) Bennett. *Curr. Sci.* 45: 388–389.

Bhatty R.S. (1995) Nutritional composition of whole flaxseed and flaxseed mIn: Flaxseed in HumanNutrition. In: *Flaxseed in HumanNutrition.*, p. 22–45.

Bhavsar, G.J., Syed, H.M. and Andhale, R.R. (2017) Characterization and quality assessment of mechanically and solvent extracted Niger (*Guizotia abyssinica*) Seed oil. *Journal of Pharmacognosy and Phytochemistry* 2017; **6**(2): 17-21

Bhuyan, J. and Sarma, M.K.,(2003), Identification of heterotic crosses involving cytoplasmic genic male sterile lines in sesame (*Sesamum indicum* L.) *Sesame and Safflower Newsletter* 2003 No.**18** pp.7-11 ref.5

Bidney, D. C. Scelonge, J. Martich, M. Burrus, L. Sims, and G. Huffman, (1992). Microprojectile bombardment of plant tissue increases transformation frequency by Agrobacteriumtumefaciens. *Plant Mol. Biol.* **18**: 301—313. B

Bidney, D.L. and Scelonge, C.J., (1997). Sunflower biotechnology. In: Schneiter, A.A. (ed.): *Sunflower Technology and Production*. ASA-CSSA-SSSA Madison, Wisconsin, USA, pp. 559-593.

Bilgen B. (2016). "Characterization of sunflower inbred lines with high oleic acid content by DNA markers," in *Proceedings of the 19th International Sunflower Conference* Edirne: 662–668.

Billore, S.D.,Ramesh, A.,Joshi, O.P. and Vyas, A. K. (2010) Energy budgeting of soybean based cropping system under various tillage and fertility management. [2009]*Indian Journal of Agricultural Sciences (India)*79: 827-830

Boateng, L., Ansong, R., Owusu, W.B. and Steiner-Asiedu, M. (2016) Coconut oil and palm oil's role in nutrition, health and national development: A review. *Ghana Med J.*, **50**(3): 189–196.

Biswas, B., Kazakoff, S.H., Jiang, Q., Samuel, S., Gresshoff, P.M., and Scott, P.T. (2013) Genetic and genomic analysis of the tree legume *Pongamia pinnata* as a feedstock for biofuel. *Plant Genome*6., doi:10.3835/plantgenome2013.05.0015

Blackman, B.B., Scascitelli, M., Kane,N.C., Luton,H.H., Rasmussen, D.A., Bye, R.A. Lentz, D.L and Rieseberg.L.H. 2011.Sunflower domestication alleles support single domestication center in eastern North America. *Proc. Natl. Acad. Sci. USA*108:14360–14365.

Bleho, J., Obert, B., Takáč, T. *et al.* ER disruption and GFP degradation during non-regenerable transformation of flax with *Agrobacterium tumefaciens* . *Protoplasma* 249, 53–63 (2012). https://doi.org/10.1007/s00709-010-0261-2

Bloem, A.A. and Bernard, R.O. (1995) Effect of residual pus applied nitrogen on grain yield of maize in the Highveld Region. *South African Journal of Plant and Soil.*12(30: 99-104

Bobade S.N.and Khyade V.B. (2012) Detail study on the Properties of *Pongamia Pinnata* (Karanja) for the Production of Biofuel. *Research Journal of Chemical Sciences*, **2**(7) : 16-20

Bockelman, H. E., Knowles, P. F. and Klisiewicz, J.M. (1973) Inheritance of resistance to *Fusarium* wilt in cultivated safflower, *Carthamus tinctorius* L. *Agron. Abstr*. **3**

Bolhuis, G.G. (1951) Naturlijke bastaardering bij de aardnoot (*Arachis hypogaea*L.) *landbouwkundig tijdschrift* 63:447-55

Bonfim-Silva, E.M., Pacheco, A.B., da Silva, T.J.A., Santo, E.S.S., Lemes. C.S. and Sousa, H.F. (2016) Biometric characteristics and visual diagnosis of safflower plants under macro and micronutrient omission. *International Journal of Current Research***8** (05): 31071-31076

Booth, F.E.M.; Wickens, G.E., (1988): Non-timber Uses of Selected Arid Zone Trees and Shrubs in Africa, p.98, *FAO,* Rome

Booth, I., Harwood, R. J., Wyatt, J. L. and Grishanov, S. (2004). A comparative study of the characteristics of fibre-flax (*Linum usitatissimum*). *Ind. Crops Prod.*, **20:** 89–95

Boothe, J., Nykiforuk, C.Y., Zaplachnski, S. and Snarka, S. (2010). Seed basedexpression systems for plant molecular farming. *Plant Biotechnology Journal.***8**: 588-606

Boraiah,K.M., I. Shanker Goud, I ., Kotreshi Gejli, Konda, C. R. and Prashanth babu, H. (2012) Heterosis for yield and yield attributing traits in groundnut (*Arachis hypogaea* L.) *Legume Researchhttp://krishi.icar.gov.in/PDF/ICAR_Data_Use_Licence.pdf*

Borchani C., Besbes S., Blecker C. H. and Attia H. (2010). Chemical Characteristics and Oxidative Stability of Sesame Seed, Sesame Paste, and Olive Oils. *Journal of Agriculture, Science and Technology*. 12: 585-596

Borlaug, N. (1985). Jojoba. New Crop for Arid Lands, *New Raw Material for Industry.* National Academy Press.

Boureima, S., Eyeletters, M., Diouf, M., Diop, T.A. and Van Damme, P. (2011) Sensitivity of seed germination and seedling radicle growth to drought stress in sesame (*Sesamum indicum* L.). Res. J. Environ. Sci., 5 : 557-564

Boureima, S., Oukarroum, A., Diouf, M., Cisse, N. and Damme, P.V. (2012) Screening for drought tolerance in mutant germplasm of sesame (*Sesamum indicum* L). Environ. Exp. Bot. 81: 37-43

Boussama, N., Ouariti, O., Suzuki, A. and Ghorbal, M.H. (1999) Cadmium stress on nitrogen assimilation. *J. Plant Physiol.*, **155**: 310-317.

Boyce, J.A., Assa'ad, A., Burks, A.W., *et al.*, (2010). "Guidelines for the diagnosis and management of food allergy in the United States: report of the NIAID-sponsored expert panel". *J. Allergy Clin. Immunol.* **126** (6 Suppl): S1–58.

Boye, J.I., (2012) Food allergies in developing and emerging economies: need for comprehensive data on prevalence rates. *Clin Transl Allergy.* **2**: 225. Published online 2012 Dec 2012. doi: 10.1186/2045-7022-2-25

Bradbear, N. and Clement, A.(1991). Beekeeping for Rural Development. *IBRA Publications, UK.*, pp: 6.

Bradley, V.L. and Johnson, R.C. (2001). Managing the U.S. safflower collection. In *Proceedings of the 5th International Safflower Conference, Williston, ND, and Sidney,* MT, July 23–27, 2001. Bergman, J.W. and H.H. Mundel, Eds.,pp. 143–147.

Branch. W.D. (1987) Inheritance of curly leaf shape in peanut. *The Journal of Heredity***78**: 125

Branch, W.D. (1995) Inheritance of peanut testa colors involved in market acceptability. Crop Science. 35 (10: 270-271.

Branch, W. D. (2002). Variability among advanced gammairradiation induced large-seeded mutant breeding lines in the 'Georgia Brown' peanut cultivar. *Plant Breeding*, **121**: 275.

Brar, G.S. and Ahuja.K.L. (1979). Sesame: its culture, genetics, breeding and biochemistry pp 245-313. In: Malik, C.P. (ed.). *Annual. Rev. of Plant Sci.* Kalyani publishers, New Delhi.

Bridges, M.; Jones, A. M. E.; Bones, A. M.; Hodgson, C.; Cole, R.; Bartlet, E.; Wallsgrove, R.; Karapapa, V. K.; Watts, N.; Rossiter, J. T. (2002). Spatial organization of the glucosinolate-myrosinase system in brassica specialist aphids is similar to that of the host plant. *Proceedings of the Royal Society* B. 269

Brigham, R.D. (1985) Status of sesame research and production in Texas and the USA. : In **Ashri, A. (Ed) Sesame and Safflower. Status and Potentials. Publ.No.66, FAO Rome.**

Brim, C.A.(1966). A modified pedigree method of selection in soybeans. *Crop Sci.***6**:220.

Bringi, N.V. (1987) Non-traditional oilseeds and oils of India. *Oxford & IBH.* New Delhi.

Bringi, N.V. and Mukherjee, S.K. (1987) Karanja seed Oil, In: Bringi, N.V. (Ed) Non-traditional oilseeds and oils of India. Oxford and IBH. New Delhi.pp 143-166.

Brittaine, R. and Lutaladio, N. (2010) Jatropha: A Smallholder Bioenergy Crop. The Potential for Pro-Poor Development, *Integrated Crop Management* Vol. **8**. 2010; FAO,UN. Rome, 2010 pp.27-53.

Brock, R.D. (1971). The role of Induced Mutations in Plant breeding. *Radiation Botany***11:**181

Broyles, S.S.*et al.,* (2014)The wild side of a major crop: soybean's perennial cousins from Down Under.*American journal of botany***101**(10):1651-65DOI:10.3732/ajb.1400121

Bruckner, G. (1992) Biological effects of polyunsaturated fatty acids, in *Fatty Acids in Foods and Their Health Implications*, ed. C.K. Chow, Marcel Dekker, New York, pp. 631-646. Google Scholar

Brutch, N. B. (2002). The flax genetic resources collection held at the N.I. Vavilov Institute, Russian Federation. In: Maggioni, L.M., Pavelek, M., van Soest, L. J. M., Lipman, E. (eds) *Flax genetic Resourcesin Europe. IPGRI,* Maccarese Rome, Pp 61–65

Bubeck, D.M., Fehr, W.R. and Hammond, E.G. (1989) Inheritance of palmitate and stearic acid mutants of soybean. *Crop Sci.* **29**, 652–656.Google Scholar

Bulcha, W., (2007). *Guizotia abyssinica* (L.f.) Cass. In: van der Vossen, H.A.M. & Mkamilo, G.S. (Eds). Record from PROTA4U. PROTA (*Plant Resources of Tropical Africa / Ressources végétales de l'Afrique tropicale),* Wageningen, Netherlandshttp://uses.plantnet-project.org/en/Guizotia_abyssinica_(PROTA)

Bunting, A.H. (1955) Classification of cultivated groundnut. *Empire Journal of Experimental Agriculture* **23:** 158-170

Burdon, J. J. and Leather, S. R. (1990). *Pests, Pathogens and Plant Communities. Oxford*: Blackwell. Pp. 344

Burke, J.M.; Tang, S.; Knapp, S. and Rieseberg, L.H. (2002) Genetic analysis of sunflower domestication. *Genetics*, **162**: 1257–1267

Burton, J.W. (1997) Soyabean (*Glycine max* L. Merril.) *Field Crops Research.* **53** (1/3) 171-86

Butler, E. J. (1918) Fungi and diseases in plants. *Thatcher, Spink Co. Calcutta,* 1918; 547 pp.

Butler, G. D. Jr. and Wilson, F.D.(1984). Activity of Adult Whiteflies (Homoptera: Aleyrodidae) Within Plantings of Different Cotton Strains and Cultivars as Determined by Sticky-Trap Catches. J. Econ. Ent. 77(5): 1137-1140.

Butterworth, M. H. and Mosi, A.(1986). The intake and digestibility by sheep of oat straw and maize stover offered with different levels of *noug* (*Guizotia abyssinica*) meal. *Anim. Feed Sci. Technol.*, **16** (1-2): 99-107

Caballero B and Allen LH. (2012), In: Prentice, A.(Ed).. *Encyclopedia of human nutrition. Oxford: Academic press; 2012. [Google Scholar]*

Cabral,M.M.O., Garcia, E. S. and Kelecom, A. (1995) Lignanes from the Brazilian *Melia azedarach*, and their activity in *Rhodnius prolixus* (Hemiptera, Reduviidae). *Memorias do Instituto Oswaldo Cruz,* **90**(6):759-763.

Cagirran, M.I. (2006)Selection and morphological characterization of induced determinate mutants in sesame. *Field Crops Research,* **96**(1): 19-24

Cai, H.W. and Morishima, H. (2002) QTL clusters reflect character associations in wild and cultivated rice. *Theor. Appl. Genet.,* **104** :1217–1228.

Cahener, A. and Ashri, A. (1974) Vegetative and reproductive development of *Virginia* type peanut varieties in different stand densities. *Crop. Sci.* **23:**412-416

Camacho-Villa, T., Maxted, N., Scholten, M. and Ford-Lloyd, B. (2005) Defining and identifying crop landraces. *Plant Genetic Resources: Characterization and Utilization* **3**: 373–384.

Campos, A.P., Boica Junnior, A.L. and Ribeiro, Z.A. (2010) Nao preferencia para oviposicao e alimentaco de *Spodoptera frugiperda* por cultivares de amendoim. *Arquivos de Instituto Biologico* **77**: 251-258(Portuguese)

Campbell, E.J. (1983) Sunflower oil. *J. Am. Oil Chem. Soc.* **60**, 387–392.Google Scholar

Canakci, M., Moneyem, A. and Van Gerpen. J. (1999) Accelerated oxidation process in biodiesel. *Trans.ASAE,* **42**: 1565 - 1572

Caron, D.M and Connor, L.J. (2013) Honey bee biology and bee keeping.Kalamazoo, MI:*Wicwas Press.*

Carapetian, J. (1994) Effects of safflower sterility genes on inflorescence and pollen grains. *Australian Journal of Botany.* **42** (3): 325-332

CAZRI, (1976). Improved Dryland Agriculture for Western Rajasthan. Technical Bulletin-1 CAZRI, Jodhpur, 28pp.

Chae, YA, Park SK and Anand IJ (1987). Selection *in vitro* for herbicide tolerant cell lines of *Sesamum indicum* L. 2. Selection of herbicide tolerant calli and plant regeneration. *Korean Journal of Beeding* **19**: 75-80.

Chahal, A.S. (1981) Seed borne infection of *Alternaria brassicae*in Indian Mustard and its elimination during storage. *Curr.Sci.* **50**: 621-623

Chakraborty, R. and Konger, G. (1995) Root rot disease of neem caused by *Ganoderma applantum. : Indian Forester,* **121** (11) :1081-1083

Chakravarty, S., Shukla, G. and Dey, A.N. (2012) Tree-borne Oilseeds Species: Tree Borne Oilseed Species (Ecological and Economic Benefits) Paperback – December 12, 2012. LAMBERT Academic publishing

Chamberlain, J.R., Childs, F.J., and Harris, P.J.C. (2000) An Introduction to Neem and its Use and Genetic Improvement of Neem (*Azadirachta indica*) and its potential benefits to poor farmers in developing countries. *Forestry Research Programme of the Renewable Natural Resources Knowledge Strategy Department for International Development (DFID)*Pp 52. https://assets.publishing.service.gov.uk/media/57a08d82e5274a31e00018d6/R7348_report.pdf

Chan, A.P., Crabtree, J., Zhao, Q., Lorenzi, H., Otvos, J., Puiu, D., Berhan, A.M., Jones, K.M., Redman, J., Chen, G., Cahoon, E.B., Gedi, M., *et al.* (2010)Draft genome sequence of the oilseed species *Ricinus communis*. *Nature Biotechnology* **28**, pp 951-956 (2010)

Chandan, K. and Pandya, S.M. (2009) Morphological characterization of *Arachis* species of section *Arachis*. *PGR Nwesletter-FAO-Biodiversity.* No.**121**: 38-41

Chandler, M. R. (1978)Some Observations on Infection of *Arachis hypogaea* L. by Rhizobium. *Journal of Experimental Botany*,**29**(3): 749–755, https://doi.org/10.1093/jxb/29.3.749

Chandra Mohan, J., Mohamed Ali, A. and Subrahmanian, C. (1967) Correlation studies in groundnut. Correlation of certain quantitative characters with yield in the strain TMV2. *Madras Agricultural journal.***54**: 482-484

Chandrawati, Singh, N., Kumar, R., Kumar, S., Singh, P.K., Yadav, V.K., Ranade, S.A. and Yadav, H.K. (2017) Genetic diversity, population structure and association analysis in linseed (Linum usitatissimum L.) *Physiol Mol Biol Plants***23.** DOI 10.1007/s12298-016-0408-

Chang, Z., Chen, Z., Wang, N., Xie, G., Lu, J., Yan, W., Zhou, J., Tang, X. and Deng, X.W. (2016) Construction of a male sterility system for hybrid rice breeding and seed production using a nuclear male sterility gene *PNAS* December 6, 2016 **113 (49)**: 14145-14150; first published November 18, 2016 https://doi.org/10.1073/pnas.1613792113

Chapin, C. (2008) Product Development Solutions: It's a soy world, *Industry News*, United Soybean Board, USA

Chapkin, R.S. (1992) Reappraisal of the essential fatty acids, in *Fatty Acids in Foods and Their Health Implications*, ed. C.K. Chow, Marcel Dekker, New York, pp. 429.

Chapman, J.F., Daniels, D.W. and Scarisbrick, D.H. (1984). Field studies on 14C assimilate fixation and movement in oilseed rape *(Brassica napus*). *J. Agric. Sei., Camb.*,**102**:23-31.

Chapman, M. A., and Burke, J. M. (2007). DNA sequence diversity and the origin of cultivated safflower (*Carthamus tinctorius* L.; Asteraceae).*BMC Plant Biology*, **7**:1–9.

Chapman, M. A., Hvala, J., Strever, J., Matvienko, M., Kozik, A., Michelmore, R. W., *et al.* (2009). Development, polymorphism, and cross-taxon utility of EST–SSR markers from safflower (*Carthamus tinctorius* L.). *Theor. Appl. Genet.* **120,** 85–91.

Chari, M.S. and Muralidharan, C.M. (1985) Neem (*Azadirachta indica* L.) as feeding deterrent of castor semilooper (*Achoea Janata* L.) *Journal of Entomological Research,* **9** (2): 243-245

Chatterjee, A.K. and Singh, H.P. (1993) Plant regeneration from leaf calli of safflower In: Proceedings of the 3rd International Safflower Conference (Dajue, L and Yunzhou, H. eds.) Beijing, 139-143

Chatterjee, P.N., Singh, P. and Shivaramakrishna, V.R. (1968) Further records of insect hosts of *Mermis* sp (*Mermithidae: Nematoda*). *Indian Forester*.**94**: 251-252

Chattopadhyaya, B., Banerjee, J., Basu, A., Sen, S.K. and Maiti, M.K. (2010). Shoot induction and regeneration using internodal transverse thin cell layer culture in*Sesamum indicum* L. *Plant Biotechnology Reports* 4: 173-178.

Chaturevedi, A.N. (1993). Silviculture in Neem Research and Development, In: N.S. Randhawa and B.S. Parmar(Eds). *Society of Pesticide Science, India.* pp. 38-49.

Chaturvedi R.K., RaghubanshiA.S.and Singh, J.S.(2011) Carbon density and accumulation in woody species of tropical dry forest in India, *Forest Ecology and Management,* **262**(8): 1576–1588

Chaudhary B. A, Patel M. P, Dharajiya D. T, Tiwari K. K. (2019). Assessment of Genetic Diversity in Castor (*Ricinus Communis* L.) Using Microsatellite Markers. Biosci Biotech Res Asia 2019;16(1). Available from: https://bit.ly/2UXEbWh

Chaudhary, R.I. and Patel, C.C. (2016). Screening of Brassica germplasm for resistance to mustard aphid, *Lipaphis erysimi* (Kalt.). *Internat. J. Plant Protec.*, **9**(1): 62-67.

Chaudhari, S., Khare D., Patil, S.C., Sundravadana, S.,Variath, M.T., Hari K. Sudini, H.K., Surendra Manohar, S.S., Bhat, R.S. and Janila. P. (2019) Genotype × Environment Studies on Resistance to Late Leaf Spot and Rust in Genomic Selection Training Population of Peanut (*Arachis hypogaea* L.).*Front. Plant Sci.,* 04 December 2019 https://doi.org/10.3389/fpls.2019.01338

Chauhan, J.S., Singh, K.H., Singh, V.V. and Kumar, S. (2011) Hundred years of rapeseed-mustard breeding in India: Accomplishments and future strategies. *Indian Journal of Agricultural Sciences* **81**(12):1093-1109

Chauhan, J.S., Singh, C.V. and Singh, R.K. (1996). Analysis of growth parameters and yield in sesame (*Sesamum indicum* L.). *Indian Journal of Agricultural Sciences***66**(1): 55-58.

Chauhan, Y.S. and Kumar, K. (1986). Gamma ray induced chocolate B seeded mutant in *Brassica campestris*. *Curr. Sci*. **55 :**410

Chavan, B. L. and Rasal, G.B. (2010). Sequestered standing carbon stock in selective tree species grown in University campus at Aurangabad, Maharashtra, India. *International Journal of Engineering Science and Technology* **2**(7): 3003–3007.

Chavan, V.M. (1961). Niger and Safflower.*Monograph published bythe Indian Central Oilseeds Committee,* Hyderabad, India.

Chen CC, Chang SJ, Cheng LL and Hou RF, (1996). Effects of chinaberry fruit extract on feeding, growth and fecundity of the diamondback moth, *Plutella xylostella* L. (Lep., Yponomeutidae). *Journal of Applied Entomology,* **120**(6):341-345.

Chen, C.Y. and Wan, H. (1968) .Variety X environment interactions in regional trials of soybeans and groundnut and their importance in breeding. *Journal of Agricultural Association, China. N.S.* **64:** 1-12.

Chen, M., Welch, M. and Laubach, S. (2018). "Preventing Peanut Allergy". *Pediatric Allergy, Immunology, and Pulmonology*. **31** (1): 2–8. doi:10.1089/ped.2017.0826

Chen, Z., Wang, M. L., Barkley, N. A., and Pittman, R. N. (2010). A simple allele-specific PCR assay for detecting FAD2 alleles in both A and B genomes of the cultivated peanut for high-oleate trait selection. *Plant Mol. Biol. Rep.* **28:** 542–548. doi: 10.1007/s11105-010-0181-5

Chen Y., Zhou X.R., Zhang Z.J., Dribnenki P., Singh S. and Green A. (2015). Development of **high oleic oil crop platform in flax through RNAi-mediated multiple FAD2 gene silencing.** *Plant Cell Reports*, **34**: 643–653

Cheng, F.C., Jinn, T. R., Hou R.C.W. and Jason T. C. Tzen (2006) Neuroprotective Effects of Sesamin and Sesamolin on Gerbil Brain in Cerebral Ischemia, *Int J Biomed Sci.* 2006 Sep; **2**(3): 284–288.

Cherry, J.P. (1983) Cottonseed oil. *J. Am. Oil Chem. Soc*. **60**, 360–367.Google Scholar

Chetia, P. (2020) Excessive Pesticides in Agriculture Contaminating Dairy Products; Can Cause Cancer in Humans: Study. *AGRICULTURE WORLD/Krishijagaran.com.* Tuesday, January 14, 2020.

Chhetri, A.B., Tango, M.S., Budge, S.M., Watts, K.C. and Islam M. R. (2005) Non-Edible Plant Oils as New Sources for Biodiesel Production, *Int J Mol Sci.* **9**(2): 169–180. doi: 10.3390/ijms9020169

Chhonkar, A.K. and Kumar, A. (1987). Correlation andregression studies between different physiological attributesand pod yield ingroundnut. *J. Oilseeds Res*. **4**: 132-135

Childs, F. 1999. Improvement of neem (*Azadirachta indica*) and its potential benefits to poor farmers in developing countries. *Report on a visit to Ghana. HDRA, Coventry,* UK. 16 pp.

Choopanya D (1973) Mycoplasma- like bodies associated with sesame phyllody in Thailand. *Phytopathol.***63**:1536-1537

Choudhary, V. and Jambhulkar, S. J. (2016) Genetic Improvement of Rapeseed Mustard through Induced Mutations. In: Gupta, S.K. (Ed).. *Breeding Oilseed Crops for Sustainable Production. Academic Press/ Elsevier*. Pp 377-390

Chowdhury, J.B. and Das.K.(1966). Male sterility in *Brassica campestris* var. yellow sarson. *Indian Journal of Genetics and Plant Breeding*. 26 : 374-380.436.Google Scholar

Chowdhury, J.B. and Das, K(1967a) Functional male sterility in brown sarson. *Indian Journal of Genetics and Plant Breeding*. **27**: 143-147.

Chowdhury, J.B. and Das. K.(1967b). Male sterility genes in brown sarson. *Indian Journal of Genetics and Plant Breeding*. **27**: 284-288.

Chowdhury, J.B. and Das, K. 1968. Cytomorphological studied on male sterility in *B. campestris*. *Cytologia*. **38**: 195-199.

Chowdhury K., Banu, L. A., Khan, S. and Latif, A. (2007) Studies on the fatty acid composition of edible oil. *Bangladesh J of Sci and Ind Res*. 2007;**42**(3):311–316. [Google Scholar]

Chu, Y., Wu, C. L., Holbrook, C. C., Tillman, B. L., Person, G., and Ozias-Akins, P. (2011). Marker-assisted selection to pyramid nematode resistance and the high oleic trait in peanut. *Plant Genome* **4**: 110–117. doi: 10.3835/plantgenome2011.01.0001

Chun, J.A., Lee, J.W., Yi, Y.B., Park, G.Y. and Chung, C.H. (2009) Induction of hairy roots and characterization ofperoxidase expression as a potential root growth marker in sesame. *Preparative Biochemistryand Biotechnology,* **39:** 345-359

Cision (2018). Cision PR Newswire 2018:Global Castor Oil and its Derivatives Market to 2022 - High Potential for Producing Bioplastics & Biopolymers. News Provided by *Research and Markets* Dec 18, 2018, 15:15 ET

CJP(2007). *Madhuca indica* the new source of biodiesel & *Agronomy for Jojoba Farming Technology,Centre Jatropha Promotion Biodiesel*, Churu, Rajasthan, Indiahttp://www.jatrophabiodiesel.org/Madhuca%20Indica/index.php

Classen, C.E. (1950) Natural and controlled crossing in safflower (*Carthamus tinctorius* L.) *Agronomy Journal***42**: 381-384.

Claassen, C.E., Hoffman, A. (1950). The inheritance of the pistillate character in castor and its possible utilization in the production of commercial hybrid seed. *Agron. J.* **42**: 79–82.

Claassen, C.E., Ekdahl, W.G. and Severson GM (1950). The estimation of oil percentage in safflower seed and the association of oil percentage with hull and nitrogen percentages, seed size, and degree of spininess of the plant. *Agronomy Journal* **42**: 478–482.).

Cloreto, A. (2008)A note on safflower plant ideotype suitable for Mediterranean environments. *Proceedngs of the 7th* ***International Safflower Conference,*** Wagga Wagga Australia. https://safflower.wsu.edu/7th-international-safflower-conference-germplasm/

Cloutier, S. (2016) Linseed. *Science Direct* 2016. Elsivier

Cloutier, S., Ragupathy, R., Miranda, E., Radovanovic, N., Reimer, E., Walichnowski, A., Ward, K., Rowland, G., Duguid, S. and Banik, M., (2012). Integrated consensus genetic and physical maps of flax (*Linum usitatissimum* L.). *Theor. Appl. Genet*. **125** (8):1783–1795.

Cloutier, S., Ragupathy, R., Niu, Z. and Duguid, S. (2011) SSR-based linkage map of flax (*Linum usitatissimum* L.) and mapping of QTLs underlying fatty acid composition traits. Mol. Breed., 28: 437–451

Coffelt, T.A. (1983) Stability of Florigiant peanut variety and its component lines in six environments. *Agronomy Abstracts. American Society of Agronomy.* P.59.

Coffelt, T.A. (1989) In: *Oil Crops of the World*. Robbelen, G., Downey, R. Ashri, A. (eds) McGraw-Hill Publ. Comp., New York. 1989. Pp.319-338

Coffelt, T. A. and Hammons, R. O. (1971). Inheritance of an albino seedling character in *Arachis hypogaea* L. *Crop Sci.* **11** (5): 753-755.

Coffelt, T.A and Hammons, R.O. (1974) Correlation and heritability studies of nine characters in parental and infraspecific cross populations of *Arachis hypogaea* L. *Oleagineux*, **29**.: 23-27

Cohen, S. and V. Melamed-Madjar. (1978). Prevention by soil mulching of the spread of tomato yellow leaf curl virus transmitted by *Bemisia tabaci* (Gennadius) (Hemiptera: Aleyrodidae) in Israel. *Bull. Ent. Res.* **68**(3): 465-470.

Collard, B.C.Y. and Mackill, D.J. (208) Marker-assisted selection: an approach for precision plant breeding in the twenty-first century. *Philos Trans R Soc Lond B Biol Sci.* 2008 Feb 12; **363**(1491): 557–572.

Cook RB.(1981)The biogeochemistry of sulphur in two small lakes. *Ph.D. Dissertation. Columbia University*, NY. 1981.

Cooper, W.E. and Gregory, W.C. (1960) Radiation induced leaf spot resistant mutnts in peanut (*Arachis hypogaea* L.) *Agronomy Jornal* **52**: 1- 4.

Costa, M. N da., Pereira, W. E., Bruno, L.A. de., Freire, R., Mnobrega, E. C. de., Milani, M.B., Oliveira, M. A. P. de. 2006. Genetic divergence on castor bean accesses and cultivars through multivariate analysis. *Pesq. agropec. bras.* 41:1617-1622.

Craig-Schmidt, M.C. (1992) Fatty acid isomers in foods, in *Fatty Acids in Foods and Their Health Implications*, ed. C.K. Chow, Marcel Dekker, New York, pp. 363-398.Google Scholar

Craufurd, P. Q., Prasad, P. V., Kakani, G. V., Wheeler, T. R., and Nigam, S. N. (2003). Heat tolerance in groundnut. *Field Crops Res.* **80**: 63–77. doi: 10.1016/S0378-4290(02)00155-7

Crozier. M.N. (1950) Micropscopic examination of the sectoned stemof linen flax (*Linum usitatissimum* L). Part 1. Microsopic analysis as a means of predetermining yield of scotched fibre. *The N. Z. Journal of Science and Technology* **31**: 17-30

Culbertson, J.O. (1954). Breeding flax. *Adv. Agron.* **6**: 174–178.

Cullis C. (2007) "Oilseeds" *Springer*, p. 275

Cullis, C. (2011). Linum. In: *Wild Crop Relatives: Genomic and Breeding Resources Oilseeds*. Kole C, (ed.) Springer, New York. pp 177–189

Cunha, A. and Ferreira, M.F. (1997) Differences in free sterols content and composition associated with somatic embryogenesis, shoot organogenesis and calli growth of flax. *Plant Sci* **124**:97–105

Cunha, A.C. and Ferreira, M. F. (1999) Influence of medium parameters on somatic embryogenesis from hypocotyl explants of flax (*Linum usitatissimum* L.) – effect of carbon source, total inorganic nitrogen and balance between ionic forms and interaction between calcium and zeatin. *J Plant Physiol* **155**:591–597

Czuj T., Zuk M., Starzycki M., Amir R., and Szopa J. (2009): Engineering increases in sulfur amino acid contents in flax by overexpressing the yeast Met25 gene. *Plant Science*, **177**: 584–592

D'Antuono, L.F. and Rossini, F., (1995). Experimental estimation of flax (Linum usitatissimum L.) crop parameters. *Ind. Crops Prod.* **3**: 261–271.

Dagne, K. (1994) Meiosis in interspecific hybrids and genome interrelationships in *Guizotia* Cass. (Compositae). *Hereditas (Lundskorna)*, **121** (2): 119-129.

Dagne, K. and Jonsson, A. (1999) Oil content and fatty acid composition of seeds of *Guizotia* Cass (Compositae). *Journal of the Science of Food and Agriculture.* **73** (3):274-278.

Dahake (2009) Mutation breeding in Groundnut (*Arachis hypogaea*, L.) *M.Sc.Thesis* Junagadh. Pp185http://krishikosh.egranth.ac.in/handle/1/5810012599

Dahmer, N., Wittmann,M.T.S., and dos Santos Dias, L.A. (2009). Chromosome numbers of *Jatropha curcas* L.: an important agrofuel plant, *Crop Breeding and Applied Biotechnology* **9:** 386-389 Brazilian Society of Plant Breeding.

Daisy, Thomas., Naresh Pal., Chawda S.S. and Subba Rao, K (2001) Bee Flora and Migratory Routes in India.*Proc. 37th Int. Apic. Congr.,* 28 Oct – 1 Nov 2001, Durban, South Africa

Dajuae, L. and H.H. Mundel. (1996). Safflower, *Carthamus tinctorius L. Promoting the Conservation and Use of Underutilized and Neglected Crops. 7. Institute of Plant Genetics and Crop Plant Research, Gatersleben/International Plant Genetic Resources Institute, Rome,* 83 pp

Daniel, J.N. (1997) *Pongamia pinnata* a nitrogen fixing tree for oilseed. *FACT 97-03,Forest Farm and Community Tree Network,*USA.

Daniell, H., Kumar, S. and Dufourmantel, N.(2005). Breakthrough in chloroplast genetic engineering of agronomically important crops. *Trends Biotechnol,* **23**(5): 238–245.

Daniels, R.W., Scarisbrick, D.H. and Smith, L.J., 1986. Oilseed rape physiology. In: D.H. Scarisbrick and R.W. Daniels (Eds), *Oilseed rape.* Collins, London, pp. 83-126.

Dantata, I.J (2014) Effect of Legume-Based Intercropping on Crop Yield: A Review. *Asian Journal of Agricultural and Food sciences* **2** (4).

Darby, H., Cummings, E., Madden, R. and Monahan, S. (2013). Sunflower population and nitrogen rate trial. *Univ. of Vermont Ext. Publ.* Burlington, Vt.USA.

Darlington, C. D. and Wylei, A.P. (1956) Chromosome atlas of flowering plants, New York, Macmillan, 1956.

Das, K. and Pandey, B.D. (1961). Male sterility in brown sarson. *Indian Journal of Genetics and Plant Breeding*. **21**: 185-190.

Dash, P. K., Cao, Y., Jailani, A. K., Gupta, P., Venglat, P., Xiang, D., Rai, R, Sharma, R., Thirunavukkarasu, N., Abdin, M. Z., Yadava, D. K., Singh, N. K., Singh, J., Selvaraj, G. Deyholos, M., Kumar, P. A. and Datla R. (2014). Genome-wide analysis of drought induced gene expression changes in flax (*Linum usitatissimum) GMCrops & Food*, **5**(2): 106–119

Das, R. and Jha, S. (2019) Insect Pollinators of Sesame and the Effect of Entomophyllous Pollination on Seed Production in New Alluvial Zone of West Bengal. *International Journal of Current Microbiology and Applied Sciences* **8** (3):1400-1409

Das, S., Tyagi, K. and Kaur, H. (2004). "Evaluation of taramira oil-cake and reduction of its glucosinolate content by different treatments". *Indian Journal of Animal Sciences.* **73** (6): 687–691.

Dasthagiriah, P. and Nagaraj, G. (1993). Seed and oil quality characteristics of some niger genotypes. *J. Oil. Tech. Ass. India.,* **25** (2): 42-44.

Daun, J.K. (1983) The introduction of low erucic acid rapeseed varieties in Canada prior to 1970, in *High and Low Erucic Acid Rapeseed Oils*, eds. J.K.G. Kramer, F.D. Sauer and W.J. Pigden, Academic Press, New York, pp. 161-180.Google Scholar

Da Via, D. J., Knowles, P.F. and Klisiewicz, J.M. (1979). Evaluation of the safflower world collection for resistance to *Phytophthora. Crop Sci.* **21**: 226-229.

David, B.V. and Basheer, M. (1961) Mass occurrence of predatory stink bug, *Cantheconidia furcellata* (Wolff) on *Amsecta albistiga* Wlk. In South India. *Journal of National Historical Society,*Bombay, **58**: 817-819.

David, B.V., Narayanaswamy, P.S. and Murugesan, M. (1964) Bionomics and control of castor shoot and capsule borer *Dichocrosis punctiferalis* Guen (*Lepidoptera: Pyralidae*) in Madras state. *Indian Oilseeds Journal*.**8**: 146-158

Davies, C.S. and Nielsen, N.C. (1986) Genetic analysis of a null-allele for lipoxygenase-2 in soybean. *Crop Sci.* **26**, 460–463.Google Scholar

Davies, C.S. and Nielsen, N.C. (1987) Registration soybean germplasm that lacks lipoxygenase isozymes. *Crop Sci.* **27**, 370–371.Google Scholar

Davies, C.S., Nielsen, S.S. and Nielsen, N.C. (1987) Flavor improvement of soybean preparations by genetic removal of lipoxygenase-2. *J. Am. Oil Cheelgado, N.,Layrissem. Soc*. **64**, 1428–1433.Google Scholar

De, B.K. and Bhattacharya, D.K. (1999) Biodiesel from minor vegetable oils like karanja oil and nahor oil. *Lipid/Fett.***101**(10):404-406.DOI:10.1002/(SICI)1521-4133(199910)101:10<404:AID-LIPI404>3.0.CO;2-K

Debeake, P. and Raffailac, D. (1996) Light interception as an indicator of leaf area index and risk of diseases in sunflower. *Helia,* **19** (24): 1-16

De Bruin J. L. and Pedersen P. (2008)Effect of row spacing and seeding rate on soybean yield. *Agron. J.* **100***:* 704–710.CrossRefWeb of ScienceGoogle Scholar

Dehmer K. J. and Friedt W. (1998). Development of molecular markers for high oleic acid content in sunflower (*Helianthus annuus* L.). *Ind. Crops Prod.* **7:** 311–315. 10.1016/ S0926-6690(97)00063-0

Dehue, B. and Hettinga, W. (2008) GHG Performance *Jatropha* Biodiesel. ***Ecofys*** 2 June 2008 Commissioned by: D1 Oils Copyright Ecofys 2008 Ecofys reference: PBIONL073010 www.ecofys.nl pp 1- 44

De Jussieu, A. 1963. *Azadirachta indica* and *Melia azeedarach* Linne: silvicultural characteristics and planting methods. *Rev. Bois Fôrets. Trop.***88**: 23.

Delaplane, K. S. *et al.,* (2013)I. Standard methods for pollination research with *Apis mellifera.* In: V. Dietemann., J.D. Ellis. and P. Neumann. (Eds.). The Colossus Bee-book, volume I: standard methods for *Apis mellifera* research. *Journal of Apicultural Research,* v. **52** (4) : 1-28, 2013.

Delgado, M. (1972) Yield components of sesame under different plant population densities. Thesis for M.S. degree, University of California (Quoted from Sharma, S.M. 1994)

Delgado, N.,Layrisse, A and Quijada, P. (1994) Iheritance of indehiscent capsule in sesame (*Sesamum indicum,* L.). *Agronomia Tropical (Maracy)* **44**(3):499-512

de Lumen, B.O. (1992) Molecular Strategies to Improve Protein Quality and Reduce Flatulence in Legumes: A Review, *Food Structure* **11:** (1) Article 4 1992 , Pp 33-46.

Demibras, A. (2002) Biodiesel from vegetable oils via transesterification in supercritical methanol. *Energ. Convers. Manage.* **43**: 2349-2356

Demurin, Y., Skoric, D. and Karlovic, D. (1996) Genetic variability for tocopherol composition in sunflower seeds as basic of breeding for improved oil quality. *Plant Breed.* **115**: 33-36.

Denis, M., Delourme, R., Gourret, J.P., Mariani, C. and Renard, M. (1993) Expression of Engineered Nuclear Male Sterility in *Brassica napus* (Genetics, Morphology, Cytology, and Sensitivity to Temperature). *Molecular BIolgy and Gene Regulation DOI: https://doi.org/10.1104/pp.101.4.1295*

Deokar, A.B. and. Patil, F.B. (1980.) Combining ability analysis in a diallel cross of safflower. *J. Maharashtra Agric. Univ.***5:** 214–216.

de Oliveira, I.J., Maurício Dutra Zanotto, M.D., Krieger, M. and Vencovsky, R. (2012) Inbreeding depression in castor bean (*Ricinus communis* L.) progenies, *Crop Breed. Appl. Biotechnol.* **12(**4) Viçosa Dec. 2012 https://doi.org/10.1590/S1984-70332012000400006

Deotale, R.D. *et al.,* (1996). Physiological studies in safflower as influenced by TIBA. *PKVV Research Journal*, **20** (2) :133-135

Desai, A.G., Dange, S.R.S. and Pathak, H.C. (2001) Genetics of resistance to wilt in castor caused by *Fusarium oxysporum* ff. sp *ricini*. Nanda and Prasad. *J Mycol Plant Pathol***31**:322–326

Deshpande, R. B. (1940) A sterile mutant in safflower (*Carthamus tinctorius* L.). *Current Science*, **9:** 370-371

Deshpande, R. B. (1952). Wild safflower (*Carthamus oxyacantha* Bieb.) — a possible oil crop for the desert and arid regions. *Indian J. Genet. Plant Breed.* **12**: 10-14 Google Scholar.

Deshpande, S.L., Saxena, M.K., Bharaj, G.S. and Sing, J. (2005). Extra early safflower variety is the need of non-traditional region. Enver Esendal (Editor), *Proceedings VIth International Safflower Conference,* Istanbul - Turkey, June 6-10, 374.

Desmae, H., Janila, P., Okori, P., Pandey, M.K. *et al*., (2017) Genetics, genomics and breeding of groundnut (*Arachis hypogaea* L.) *Plant Breeding.* 2018:1–20. https://onlinelibrary.wiley.com/doi/full/10.1111/pbr.12645

Devaraj, R. and Sundararaj, R (2014).Parasitoids of *Asphondylia pongamiae* (Diptera: Cecidomyiidae), the flower gall inducer of *Pongamia pinnata* and their roles in biological control. *Journal of Tropical Forest Science* 2014 Vol. 26 No. 2 pp. 173-177

Devarathinam, A.A. and Sundaresan, N. (1990) New wild variety of *Sesamum-Sesamum indicum* (L) var. *sencottai* ADR & NS compared with *S. indicum* (L) var. *yanamalai ADR&* MS and *S. indicum* L. *Journal of Oilseeds research.* **7:** 121-12

Devi, P. and Rani, S. (2002) *Agrobacterium rhizogenes* induced rooting of in vitro regenerated shoots of the hybrid *Helinathus annuus* x *H. tuberosus.Scientia Horticulturae* **93** (2) :179-186

Devi, P.S.V., Prasad, Y.G. and Rajeswari, B. (1996) Effect of *Bacillus thuringiensis* var. *Kurstaki* and neem on castor defoliators-*Achoea janata* (L) and *Spodoptera litura* (Fabr.) *Journal of Biological Control.***10** (1/2): 67-71

Devi, P.S.V., Ravinder, T. and Jayadev, C. (2005) Cost-effectiveness of *Bacillus thuringiensis* by solid state fermentation. *Journal of Invertebrate Pathology.***88** (2): 163-168

Devi, P.S.V. and Sudhakar, R. (2006) Effectiveness of local strainof *Bacillus thuringiensis*(Bt.) in the management of castor semilooper *Achoea janata* on castor (*Ricinus cummunis* L.). *Indian Journal of Agricultural Sciences* **76** (7):447-449

Devi, P.S.V., Ravinder, T. and Jayadev, C. (2005) Cost-effectiveness of *Bacillus thuringiensis* by solid state fermentation. *Journal of Invertebrate Pathology.***88** (2): 163-168

Dexter J.E., Jhala A.J., Hills M.J., Yang R., Topinka, K.C., Weselake R.J., Hall, L.M. (2010): **Quantification and mitigation of adventitious presence of volunteer flax** (*Linum usitatissimum*) in wheat. *Weed Science*, **58**: 80–88

Deyholos, M.K., (2006). Bast fiber of flax (*Linum usitatissimum* L.): Biological foundations of its ancient and modern uses. *Isr. J. Plant Sci.* **54**: 273–280

Dhawan, K., Yadava, T.P., Kaushik, C.D. and Thakral, S.K. (1982) Changes in phenolic compounds and sugars in relation white rust Indian mustard. *Crop. Improv*. **9**: 95-97

Dhillon, B.S. Banga, S.S., Mangat,B.K., Allah-Rang, Randhawa,L.S., Bharaj, T.S., andRang, A.(1996.) Hybrid breeding in crop plants. *Journal of Research, Punjab Agricultural University.* **33:**1-4,

Dhillon, H.S., Kumar Thakur, A.K. and Singh, K.J. (2014). Growth and propagation aspects of some medicinally important trees in Chandigarh, India: a review *Journal of Medicinal Plants Studies* 2014; **2**(5): 29-35

Dhillon, S.S. and Labana, K.S. (1988). Outcrossing in Indian mustard. *Annals of Biology.* **4:**100-102.

Dhillon, S.S., Labana,K.S., and Banga,S.K.(1990). Studies on heterosis and combining ability in Indian mustard. *Journal of Research*, PAU. **27**: 1-8.

Dhyani, S. K., Vimala Devi, S. and Handa A. K. (2015): Tree Borne Oilseeds for Oil and Biofuel. *Technical Bulletin 2/2015. ICAR-CAFRI*, Jhansi. India. pp: 50.

Dias J.M.,Araújo J.M., Costa, J.F.,Alvim-Ferraz,M.C.M.and Almeida F.M. (2013). Biodiesel production from raw castor oil, *Energy* 53 : 58-66

Diederichsen, A. (2007) Ex-situ collections of cultivated flax (*Linum usitatissimum* L.) and other species of the genus *Linum* L. *Genet. Resour. Crop Evol.*, **54**: 661–678

Diederichsen, A. and Fu, Y.B. (2008). Flax genetic diversity as the raw material for future success. In: ***Proceedings of the international conference on flax and other bast plants***, Saskatoon, Canada, Pp. 270–280

Diederichsen, A., Kusters, P. M., Kessler, D., Bainas, Z. and Gugel, R. K. (2013). Assembling a core collection from the flax world collection maintainedby Plant Gene Resources of Canada. Genet. Resour. *Crop Evol.*, **60**: 1479–1485

Diederichsen, A. and Raney, J.P. (2006) Seed colour, seed weight and seed oil content in *Linum usitatissimum* accessions held by Plant Gene Resources of Canada. *Plant Breed.* 2006;**125**(4):372–377. doi: 10.1111/j.1439-0523.2006.01231.x. [CrossRef] [Google Scholar

Diederichsen, A. and Raney, J.P. (2008) Pure-lining of flax (Linum usitatissimum L.) gene bank accessions for efficiently exploiting and assessing seed character diversity. *Euphytica* **164**:255–273

Diederichsen, A. and Richards, K.(2003). Cultivated flax and the genus *Linum* L.: Taxonomy and germplasm conservation. In: Muir, A.D., Westcott, N.D. (Eds.), Flax: The Genus Linum. Taylor and Francis Ltd, New York, NY, pp. 22–54

Diederichsen, A., Rozhmina, T. A., Zhuchenko, A. A. and Richards, K. W. (2006). Screening for broad adaptation in 96 flax (*Linum usitatissimum* L.) accessions under dry and warm conditions in Canada and Russia. *Plant Genet. Resour. Newsl.*, **146**: 9–16

Diederichsen, A. and Ulrich, A.(2009). Variability in stem fibre content and its association with other characteristics in 1177 flax (*Linum usitatissimum* L.) genebank accessions. *Ind. Crops Prod.* **30**: 33–39

DiPietro, C.M. and Lierner (1989) Soybean Protease inhibitors in Foods. Food Science, **54** (3):606-609

Dillman, A.C. (1938) Natural crossing in flax. J. American Soc. Agron. **30**: 279-286.

Dillman, A.C. (1953). Classification of flax varieties, 1946. *USDA Technical Bulletin No. 1054. United States Department of Agriculture, Washington, DC*, Pp 56.

Dikshit, N. and Sivaraj, N. (2015). Analysis of agro-morphological diversity and oil content in Indian linseed germplasm. *Grasas Aceites*, **66(**1): e060. doi: http://dx.doi.org/10.3989/gya.068914

Dimitrijevic A., Imerovski I., Miladinovic D., Jockovic M., Cvejic S., Jocic S., *et al.* (2016). "Screening of the presence of *ol* gene in NS sunflower collection," in *Proceedings of the 19th International Sunflower Conference* Edirne: 661–667.

Dimitrijevic A., Imerovski I., Miladinovic D., Cvejic S., Jocic S., Zeremski T., *et al.* (2017). Oleic acid variation and marker-assisted detection of Pervenets mutation in high- and low-oleic sunflower cross. *Crop Breed. Appl. Biotechnol.* **17**: 229–235. 10.1590/1984-70332017v17n3a3

Dinesh Kumar, V. and NarasimhaRao, N.(2008)Development ofgene constructs for induction of male sterility and restoration of fertility in safflower. In *Proceedings of the 7th International Safflower Conference, Wagga Wagga, New South Wales, Australia* from 3rd-7th November 2008

Dirr and Michael A. (1998). Manual of woody landscape plants: Their identification, ornamental characteristics, culture, propagation and uses. 5th ed. Champaign, IL: Stipes Publishing. 1187 p. [74836]

Disasa, T., Feyissa, T. and Dagne, K (2011) *In vitro* regeneration of niger (*Guizotia abyssinica* L.F. Cass.), *International Journal of Biosciences* (IJB), **1**(6): 110-118, 2011 ISSN: 2220-6655 (Print) 2222-5234 (Online)

Divakara, B.N. and Das, R. (2011) Variability and divergence in *Pongamia pinnata* for further use in tree improvement. *J. Forest Res.* **22**(2): 193-200

Diwakar, M.C. and Singh,A.K. (1993). Combining ability for oil content and yield attributes in yellow seeded Indian mustard. *Annals of Agricultural Research.* **14**(2): 194-198.

Dixit A., Jin M. H., Chung J. W., Yu J. W., Chung H. K., Ma K. H., et al. (2005). Development of polymorphic microsatellite markers in sesame (*Sesamum indicum* L.). *Mol. Ecol. Notes* **5** :736–738. 10.1111/j.1471-8286.2005.01048.

Dixit, R.K. (1976) Heterosis and inbreeding depression in sesame. *Indian Journal of Agricultural Sciences.***46**: 514-517

Dobarganes, M.C., Marquez, R.G. Perez, C.M.C (1993) Thermal stability and frying performance of genetically modified suflower seed.

Donald, C.M., (1968). The breeding of crop ideotypes. *Euphytica*, **17**: 385-403.

Dong J.Z. and McHughen A. (1993): An improved procedure for production of transgenic flax plants using *Agrobacterium tumefaciens*. *Plant Science*, **88**: 61–71

Dolle U.V. (1984) Studies on leaf blight of sesame (*Sesamum inducum* L.) caused by *Alternaria sesame*. *Mysore J. Agric. Sci.* **18**(1):89-90.

DOR (1987) Report on Production of Oilseeds. *Directorate of Oilseeds Research*, Hyderabad, India

Dorairaj, M.S. (1962) Preliminary steps for the formation of selection of index for yield in groundnut. *Madras Agricultural journal.* **49:** 12-27

Doran. J. C. and Turnbull, J.W. (1997). Australian trees and shrubs: species for land rehabilitation and farm planting in the tropics. *Australian trees and shrubs: species for land rehabilitation and farm planting in the tropics.*, viii + 384 pp

dos Anjos, R.A.R., Santos, L.C.S. *et al.*, (2017) Initial growth of *Jatropha curcas* plants subjected to drought stress and silicon (Si) fertilization, *Australian Journal of Cop Science,* **11**(04):479-484 (2017) ISSN:1835-2707 doi: 10.21475/ajcs.17.11.04.377

Dossa K., *et al.*, (2016) Analysis of Genetic Diversity and Population Structure of Sesame accessions from Africa and Asia as Major Centers of Its Cultivation. *Genes (Basel). 2016 Apr 12*; **7**(4)

Dossa, K. *et al.*, (2017) The emerging oilseed crop *Sesamum indicum* enters the Omics era. *Frontiers in Plant sciences* **8:** DOI 10.3389/fpls.2017.01154

Downey, R.K. (1964) A selection of *Brassica campestris* L. containing no erucic acid in its seed oil. *Can. J. Plant Sci.* **44**: 295.Google Scholar

Downey, R.K. and Rôbbelen, G. (1989) *Brassica* species, in *Oil Crops of the World*, eds. G. Röbbelen, R.K. Downey and A. Ashri, McGraw-Hill, New York, pp. 339-362.Google Scholar

Down To Earth (2015) Intercropping soybean with sugar cane not only is more profitable but also benefits the latter crop. *Deep-rooted gains. Down To Earth 04July 2015Kisan World,* **22**(6).

Drew, R.A. (1993). Clonal propagation of neem by tissue culture. *Proceedings of the World Neem Conference, Bangalore,* India

Du, H., Zeng, X., Zhao, M., Cui, X., Wang, Q., Yang, H., et al. (2016). Efficient targeted mutagenesis in soybean by TALENs and CRISPR/Cas9., J. *Biotechnol.* **217**: 90–97. doi: 10.1016/j.jbiotec.2015.11.005

Dubey, S.K. (1996) Combined effect of *Bradyrhizobium japonicum* and phosphate-solubilizing *Pseudomonas striata* on nodulation, yield attributes and yield of rainfed soybean (*Glycine max*) under different sources of phosphorus in Vertisols. *Ind J Microbiol* **33:**61–65

Duhoon, S.S. (2004) Exploitation of heterosis for raising productivity in sesame. *Paper presented at International Crop Science Congress, Brisbane, Australia,* Sep 26–Oct 1, 2004

Duke, J.A. (1981). Handbook of legumes of world economic importance. *Plenum Press.* New York and London 1981. Size 21 × 27 cm, 345 pages, with 146 figs, and 7 Tables, cloth £ 45.

Duke, J.A. (1983) Handbook of Energy Crops. unpublished. *Wikipedia*

Duncan, W.G., McCloud, D, W., McGrow, R.L. and Botee, K.J. (1978) Physiological aspects of peanut yield improvement. *Crop Science*. **18:**1015-20

Dunwell, J.M. (1996). Microspore culture, In: Jain, S.M.,Sopory, S.K. and Veilleux, R.E. (Eds). *In-vitro haploid production of higher planats,*1. *Kluwer Academic Publishers,* Dordrecht, Dordrect. Pp. 205-216

Durga, K.K., Ragnunatham, G., Ranganatha, A.R. and Sharma, P.S. 1997. Genetic variability for physiological attributes in sesame {*Sesamum indicum* L.). *Crop Research Hisar* **13**(2): 349-359.

Durrant, A. (1971). Induction and growth of flax genotrophs. *Heredity* **27:** 277–298.

Dutta, P.C. *et al.,* (1994) Variation in lipid composition of niger seed (*Guizotiz abyssinica* Cass.) samples collected in different regions of Ethiopia. *Journal of American Oil Chemists Society***71** (8): 838-843.

Dwivedi A, andSharma G.N. (2014) A Review on Heliotropism Plant: *Helianthus annuus* L.*The Journal of Phytopharmacology* 2014; **3**(2): 149-155

Dwivedi, C., Natrajan, K. and Mathees, D. (2005). Chemo-preventive Effects of Dietary Flaxseed Oil on Colon Tumor Development, *Nutrition and Cancer* **51**(1): 52-58

Dwivedi, G., Jain, S. and Sharma, M.P. (2011) Pongamia as a Source of Biodiesel in India, *Smart Grid and Renewable Energy* .**2** (2):184-189. Article ID:6600,6 pages DOI:10.4236/sgre.2011.23022

Dwivedi, S.L., Amin, P.W., *et al.,* (1986) Genetic association of trichome characters associated with resistance to jassids (*Empoasca kerri* Pruthi) in Peanut. *Peanut Sci*.,**13:**15-18.

Dwivedi, S.L., Jambunathan, R., Nigam, S.N., Raghunath, K., Ravi Shankar,K. and Nagabhushanam, G.V.S. (1990) Relationship of Seed Mass to Oil and Protein Contents in Peanut (*Arachis hypogaea* L.)*Peanut Science,* **17**(2)**:** 48-52.

Dwivedi, S.L., Nigam, S.N., Nageswara Rao, R.C. *et al.,* (1996) Effect of drought on oil, fatty acids and protein content of groundnut (*Arachis hypogaea*, L.) seeds. *Field Crops Res*. **48:** 125-133

Dwivedi, S.L., Nigam, S.N. and Prasad, M.V.R. (1998) Induced genetic variation for seed quality traits in groundnut *Arachis hypogaea* L. *I A N* **18**: 44 -46 http://oar.icrisat.org/2006/1/IAN18_44-46_1998.pdf

Dwivedi, S.L., Pande, S., Rao, J.N. and Nigam, S.N. (2002) Components of resistance to late leafspot and rust among interspecific derivatives and their significance in foliar disease resistance breeding in groundnut. (*Arachis hypogaea* L.) *Euphytica,* **125:** 81-88

Dwivedi, S.L., Thendapani, K. and Nigam, S.N. (1989) Heterosis and Combining Ability Studies and Relationship Among Fruit and Seed Characters in Peanut. *Peanut Science:* **16** (1) :14-20.

Eapen S. (2003) Regeneration and Genetic Transformation in Peanut: Current Status and Future Prospects. In: Jaiwal P.K., Singh R.P. (Eds.) *Applied Genetics of Leguminosae Biotechnology. Focus on Biotechnology*, vol.**10B.** Springer, Dordrechthttps://doi.org/10.1007/978-94-017-0139-6_11

Eaves, P.H., Dupuy, H.P., Holzenthal, L.L., Rayner, E.T. and Brown, L.E (1968) Elimination of the Halphen response of cottonseed oils in conjunction with deodorization. *Cancer Res*. **45**: 293–295.Google Scholar

Ebrahimian, E., Ahmad, B. and Bahman, P. E.(2010) Efficiency of zinc and iron application methods on sunflower. *J. Food Sci. Agril. & Envt.*, **8** (3&4): 783-789.

Ecoport, (2010). Ecoport database. Ecoport. Web:http://www.ecoport.org

Egonyu, J., Kyamanywa, S., Anyanga, W. and Ssekabembe, C. 2005. Review of pests and diseases of sesame in Uganda. *Africa Crop Science Conference Proceedings***7**:1411-1416.

Elayaraja, D. (2018) Effect of Micronutrients and Organic Manures on Sesame.*Journal of Ecobiotechnology* **10**: 12-15

El-Beltagi, H., Salama, Z., and El Hariri, D. (2008) Some biochemical markers for evaluation of flaxcultivars under salt stress conditions. *J Nat Fibers***5**:316–330

Elevitch, C.R. and Wilkinson, K.M. (1999) Nitrogen fixing Tree Startup Guide. Sustainable Agricultural Research. USDA. Permanent Agricultural Resources, Holualoa, HI96726, USA.

Elfadl, E., Reinbrecht, C. and Claupein, W. (2010) Evaluation of phenotypic variation in a worldwide germplasm collection of safflower (*Carthamus tinctorius* L.) grown under organic farming conditions in Germany. *Genet Resour Crop Evol* **57**:155–170

Elkan, G.H., Wynne, J.C., Schneeweis, T.J. and Isleib, T.G. (1980) Nodulation and Nitrogenase Activity of Peanuts Inoculated with Single Strain Isolates of *Rhizobium. Peanut Science* (1980) **7:** 95-97

Elnaz, E., Ahmad, B. and Bahman, P. E.(2010) Efficiency of zinc and iron application methods on sunflower. *J. Food, Agric. & Envi.* **8** (3&4): 783-789.

El Tinay, A.H., Khattab, A.H. and Khidir, M.O. (1976) Protein and oil composition of sesame seed. *J. Am. Oil Chem. Soc.* **53**, 648–653.Google Scholar

Enikuomehin, O.A., Peters, O.T. (2002) Evaluation of crude extracts from some Nigerian plants for the control of field diseases of sesame (*Sesamum indicum* L.) *Trop. Oilseeds J.,* 2002: 84-93.

Enyi, B.A.C. (1977) Physiology of grain yield in groundnut. *Experimental Agriculture* **13**: 101-110

EPPO (2015)*EPPO Bulletin* **45**(3): 410-444

Erickson, E.A., Wilcox, J.R. and Cavins, J.F. (1988) Inheritance of altered palmitic acid percentage in two soybean mutants. *J. Hered.* **79**: 465–468.Google Scholar

Espina M.J., Sabbir Ahmed,C.M., Adeleke E., Yadegari,Z.,Prakash Arelli, P., Vince Pantalone, V.and Taheri, A. (2018)Development and Phenotypic Screening of an Ethyl Methane Sulfonate Mutant Population in Soybean*Front. Plant Sci.,* **29** (March 2018):| https://doi.org/10.3389/fpls.2018.00394

Estilai, A. 1977. Interspecific hybrids between *Carthamus tinctorius* and C. alexandrinus. *Crop Sci.***17**: 800–802.

Estilai, A. and Knowles, P.F. (1978) Relationship of *Carthamus leucocaulos* to other *Carthamus* species. *Canadian Journal of Genetics.* **20:** 221-233.

Estilai, A. and Knowles, P.F. (1980). Aneuploids in safflower. *Crop Sci.***20**: 516–518.

Fabiyi, Oluwatoyin AdenikeAtolani, and Olubunmi (2013) Varietal response of sunflower *Helianthus annus* (L) to root-knot nematode *Meloidogyne* spp and other field pathogens. *Agriculture and Biology Journal of North America* ISSN Print: 2151-7517, ISSN Online: 2151-7525, doi:10.5251/abjna.2013.4.1.14.18 © 2013, ScienceHuβ, http://www.scihub.org/ABJNA

FACT (2007). Position Paper on *Jatropha curcas* L. State of the art, small and large scale project development. *Fuels from Agriculture in Communal Technology* (available at http://www.fact-fuels.org).[c.f.:

Fagoonee, I. 1(984). Germination tests with neem seeds. In: *Natural Pesticides from the Neem Tree and Other Tropical Plants,* (eds.) H. Schmutterer and K.R.S. Ascher. GTZ Press, Ecshborn, Germany.

Falke,A.B., Hamidou, F., Halilou, O. and Harou A. (2019)Assessment of Groundnut Elite Lines under Drought Conditions and Selection of Tolerance Associated Traits. *Advances in Agriculture*, Volume 2019, Article ID 3034278, 10 pages https://doi.org/10.<1155/2019/3034278

FAO (1989) Protein Quality Evaluation Report of the Joint FAO/WHO Expert Consultation. Rome (1989*): FAO Food and Nutrition Paper No. 51*

FAOSTAT. (2016) Production of Crops: Linseed: Area Harvested and Production (tonnes). Retrieved July, 2016 from http://faostat3.fao.org/ home/index.html

FAOSTAT (2019) Production of Crops: Castor bean.https://www.scirp.org/journal/Articles.aspx?search Code = Castor+ crop+world+ production& searchField= All&page =1& SKID=59221709

Farooq, M.A., Ali, B., Gill. R.A., Islam, F., Cui, P. and Zhou, W. (2016) Breeding Oilseed Crops for Sustainable Production : Heavy Metal Tolerance, In: Gupta, S.K. (Ed). *Breeding Oilseed Crops for Sustainable Production-Opportunities andConstraints. Academic Press/ Elsevier*. Pp. 19-31

Farooqui, N., Ranade, S.A., and Sane, P.V. (1998). RAPD profile variation amongst provenances of neem. *Biochemistry and Molecular Biology International* **45**: 931-939.

Fasahat, P., Rajabi, A., Rad, J.M., *et al.* (2016) Principles and utilization of combining ability in plant breeding. *Biom Biostat Int J.* 2016;4(1):1-22. DOI: 10.15406/bbij.2016.04.00085

Fatemi, S.H. and Hammond, E.G. (1977) Glyceride structure variation in soybean varieties. I. Stereospecific analysis. *Lipids* **12**, 1032–1036.Google Scholar

Fatemi, S.H. and Hammond, E.G. (1980) Analysis of oleate, linoleate and linolenate hydroperoxides in oxidized ester mixtures. *Lipids* **15**, 379–385.Google Scholar

Fattahi, M., Janmohammadi, M., Dashti, S., Nouraein, M. and Sabaghnia, N. (2018)Effects of nitrogen and micronutrients on the growth of safflower under limited water conditions in a high-elevation region. *Biologia*, **64** (3): 235-248 © Lietuvos mokslų akademija, 2018

Favero, A.P., Moraes, S.A., Gracia, A.A.F., Valls, J.F.M.and Vello, N.A. ((2009) Characterisation of rust, early and late leafspot resistance in wild and cultivated peanut germplasm. *Scientia Agricola* (*Piracicaba, Brazil*) **66** (1): 110-117

Favero, A.P., Simpson, C.E., Valls, J.F.M. and Vello, N.A (2006) Study and evolution of cultivated peanut thrughncrossability studies among *Arachis ipaensis*, *A.duranensis* &*A. hypogaea*. *Crop Sci.***46:**1546-1552

Fazli, I.S., Ahmad, S., Jamal, A. and Abdin, M.Z. (2005) Changes in oil and fatty acid composition of developing seeds of Taramira (*Eruca sativa* MILL.) as influenced by Suphur and Nitrogen nutrition. *Indian J. Plant Physiol.*, **10** (4) :354-361 (Oct.-Dec., 2005)

Fehr, W.R., Welke, G.A., Hammond, E.G., Duvick, D.N. and Cianzio, S.R. (1991a) Inheritance of reduced palmitic acid content in seed oil of soybean. *Crop Sci.* **31**, 88–89.Google Scholar

Fehr, W.R., Welke, G.A., Hammond, E.G., Duvick, D.N. and Cianzio, S.R. (1991b) Inheritance of elevated palmitic acid content in soybean seed oil. *Crop Sci.* **31**, 1522–1524.Google Scholar

Fehr, W.R., Welke, G.A., Hammond, E.G., Duvick, D.N. and Cianzio, S.R. (1992) Inheritance of reduced linolenic acid content in the soybean genotype A16 and A17. *Crop Sci.* **31**, 1522–1524.Google Scholar

Fehr, W.R., Welke, G.A., Hammond, E.G., Duvick, D.N. and Cianzio, S.R. (1993) Expression of combinations of mutant alleles for fatty acid composition of soybean. *Crop Sci.*

Felker, P., Bunch, R. and Leung, A.M. (2016) Concentrations of thiocyanate and goitrin in human plasma, their precursor concentrations in brassica vegetables, and associated potential risk for hypothyroidism. *Nutr Rev.* 2016 Apr; **74**(4): 248–258.

Fenwick, G.R., Heany, R.K. and Mullin, W.J. (1983)CRC Critical Rev. *Food nutrition.* **18**:123-201.

Ferfuia C., Turi M. and Vannozzi G. P. (2015). Variability of seed fatty acid composition to growing degree-days in high oleic acid sunflower genotypes. *Helia* **38**: 61–78. 10.1515/helia-2014-0022

Fern, K., (2014). *Madhuca longifolia.* Useful Tropical Plants Database 2014http://tropical.theferns.info/viewtropical.php

Fernandez, P., Peniego, N., Lew, S., Hopp, H.E. and Heinz, R.A. (2003) Differential representation of sunflower ESTs in enriched organic specific cDNA libraries in small scale sequencing project.*BMC Genomics* **4:** 40

Fernandez, J.M. Perez-vich, B., Velasco, L. and Dominguez, J (2007) Breeding for specality oil types in sunflower. *Helia,* **30** (46): 75-84.

Ferrie, A.M.R. and Keller, W.A. (2002) Appliction of double haploidy and mutagenesis in *Brassica. 13th Cruciferae Genetics Workshop.University of California, Davis. CA.* March 23-26, 2002.

Ferry, Edwards, M.G. and Gatehouse, A.M.R. (2004) Plant-insect interaction: Molecular approaches to insect resistnce. In: T. Sasaki and P. Christou (Eds).*Biotechnology.* **15:**155-161 (20004)

Fick.G.N. (1989) Sunflower, in *Oil Crops of the World*, eds. G. Röbbelen, R.K. Downey and A. Ashri, McGraw-Hill, New York, pp. 301-318.Google Scholar

Fick, G.N and Miller, J.F (1997) Sunflower Breeding, Chapter 8 -395 In: *Sunflower Technology and Production, Agronomy Monograph no. 35* Access DL, Alliace of Crop, Soil and Environmental Societies, American Society of Agronomy, Crop Science Society of America and Soil Science Society of America. Madison, WI 53711, US

Fieldes, M.A., (1994). Heritable effects of 5-azacytidine treatments on the growth and development of flax (*Linum usitatissimum*) genotrophs and genotypes. **Genome 37**: 1–11.

Fila,G., Bagatta, M., Maestrini, C. and Potenza, E. (2018) Linseed as a dual-purpose crop: evaluation of cultivar suitability and analysis of yield determinants. *The Journal of Agricultural Sciences* **156** (2): 162-176

Finnegan E.J., Lawrence G.J., Dennis E.S., Ellis J.G. (1993): Behaviour of modified Ac elements in flax callus and regenerated plants. *Plant Molecular Biology*, **22:** 625–633.

Fiore, M.C., Trabase, T. and Sunseri, F. (1997) High frequency of plant regeneration in sunflower from cotyledons via somatic embryogenesis. *Plant Cell Rep.* **16** :295-298

Firdous, S.S., Asghar, R and Murtaza, G. (2014) Histopathology of leafspot of Sesame (*Sesamum indicum* L.) caused by *Psedomonos syringaepv. sesami. The Journal of Animal & Plant Sciences,* **24**(3): 2014, Page: 814-819 ISSN: 1018-7081

Flor, H.H.1(955). Host-parasite interaction in flax rust—its genetics and other implications. *Phytopathology***45**: 680–685.

Fontes, R.L.F. and Cox, F.R. (1998) Iron deficiency and zinc toxicity in soyabean grown in nutrient solution with different l Soilevels of sulphur. *Journal* of *Plant Nutrition,* **21** (8) : 1715-22

Foster, P.L. and Rosche, W.A. (2001) Aflatoxins. In: *Encyclopedia of Genetics*, 2001https://www.sciencedirect.com/topics/pharmacology-toxicology-and-pharmaceutical-science/aflatoxin

Fouilloux, G. (1989) Breeding Falx Methods. In: G. Marshall (Ed). FLAX: Breeding and Utilizaion © 1989 *ECSC, EEC, EAEC, Brussels and Luxemberg.* Pp. 14-25

Francis, G., Edinger, R. and Becker, K. (2005) A concept of simultaneous wasteland reclamation, fuel production and socioeconomic development in degraded area of India: need, potential and perspectives of *Jatropha* plantations. *Natural Resource Forum*, **29**:12-24.

Frankel, E.N., Warner, K. and Klein, B.P. (1988) Flavor and oxidative stability of oil processed from null lipoxygenase-1 soybeans. *J. Am. Oil Chem. Soc.* **65**, 147–150.Google Scholar

Frankel, O.H. (1984) Genetic perspectives in germplasm conservation. In: Arber, W., K. Limensee, W.J. Peacock and P. Starlinger, (Eds). *Genetic Manipulation: Impact on Man and Society*. pp.161-170.CUP, Cambridge., UK.

Freeman, T.P. (1995) Structure of Flaxseed. In: Cunnane, S. C. and Thompson, L. U.(Eds). *Flaxseed in Human Nutrition, Champaign Illinois*: AOCS Pres.Pp. 11-21

Frey, K.J. and Hammond, E.G. (1975) Genetics, characteristics and utilization of oil in caryopses of oat species. *J. Am. Oil Chem. Soc.* 52, 358–362.Google Scholar

Friedt, W. (1992) Present state and future prospects of biotechnology in sunflower breeding *Field Crops Research***30** (3–4): 425-442

Friedt, W., Nichterlein, K. and Nickel, M. (1989) Biotechnology in Breeding of Flax (*Linum usitatissimum*) In: G. Marshall (Ed) *FLAX: Breeding and Utilization.* © 1989 ECSC, EEC, EAEC, Brussels and Luxembourg. Kulwer Academic Publishers. PP 5-13

Friedt, W., Bickert, C. and Schaub, H. (1995) *In vitro* breeding of high-linolenic, doubled haploid lines of linseed *(Linum usitatissimum* L.) via androgenesis. *Plant Breed***144**:322–6

Friedt W., Nurhidayah T., Röcher T., Köhler H., Bergmann R. and Horn R. (1997) Haploid production and application of molecular methods in sunflower (*Helianthus annuus* L.). In: Jain S.M., Sopory S.K., Veilleux R.E. (eds) In Vitro Haploid Production in Higher Plants. *Current Plant Science and Biotechnology in Agriculture*, **29**. Springer, Dordrecht

Fu, T.D. (1981) Production and Research of rapeseed in the People's Republic of China. In: *Proceedings of the International Rapeseed Symposium,* Shanghai, China. April 23-May 2, 1981. China

Fu, T.D. (1981a) Breeding of maintainer and restorer of self-compatible lines in *B. napus*. *Eucarpia CruciferaeNews Letter.***6**: 39-42

Fu, T., Yang, G.and Yang, X. (1990) Studies on "three line" *Polima* cytoplasmic male sterility developed in *Brassica napus*. *Plant Breed.***104:** 115–120.

Fu, Y.B., Diederichsen, A., Richards, K. *et al.* (2002). Genetic diversity within a range of cultivars and landraces of flax (*Linum usitatissimum* L.) as revealed by RAPDs. *Genetic Resources and Crop Evolution* **49:** 167–174 (2002). https://doi.org/10.1023/A:1014716031095

Fu, Y. B. (2011). Genetic evidence for early flax domestication with capsular dehiscence. *Genet. Resour. Crop Evol.*, **58**: 1119–1128

Fuglie, L, J. (2001a) Combattre la malnutrition avec le *Moringa* in L'arbre de la vie, Les multiples usages du *Moringa.CTA et CWS, Dakar, Senegal*.pp 119-139 (French)

Fuglie, L.J. (200b). The Miracle Tree: *Moring oleifera*: Natural Nutrition for Tropics. *Training Manual. Chuch World Services,* Dakar, Senegal

Fukushima, A. *et al.,* (1997) On colour change from yellow to red in floral tissues of a dyer's saffron cultivar. *Journal of Plant Physiology,* **150** (6):697-706.

Gadekar, D.A. and Jambhale, N.D.(2003). Inheritance of yield and yield components in safflower (*Carthamus tinctorius* L.*). J. Oilseeds Res*. **20**: 63–65.

Gadekar Kumarsukhadeo Prakash. (2006), Vegetative propagation of Jatropha, Karanj and Mahua by Stem cuttings, Grafting, Budding and Air layering, *M.Sc. Thesis,Department of Forestry, Indira Gandhi Agricultural University* Raipur (C.G.)

Gamble. J. S. (1972) *A Manual of Indian Timbers*. PublisherBishen Singh Mahendra Pal Singh, Neew York.

Gan, Y., Basnyat, P., McDonald, C.L., Campbell, C.A., and Liu, L., 2009. Water use and distribution profile under pulse and oilseed crops in semiarid northern high latitude areas. *Agric. Water Manage*. **96**: 337–348

Ganapathy, M. (2016). Plants as bioreactors- A review. *Advanced Techniques in Biology& Medicine***4**:1 DOI: 10.4172/2379-1764.1000161

Gandhi A.P. (2009) Simplified process for the production of sesame seed (*Sesamum indicum* L) butter and its nutritional profile. *Asian Journal of Food and Agro-Industry.* **2**(01): 24-27

Ganesan, J. (1995) Induction of genic male sterility system in sesame.*Crop Environment*.**22** (2): 167-169.

Gangwar, S.K. *et al.,* (1994) Influence of intercropping on infestation by pests of crops at medium high altitude of Meghalaya. *Indian Journal of Agricultural Science.* **64**(2):137-140

Gangaprasad, S, Sreedhar, R.V., Salimath, P.M. and Ravikumar, R.L. (2004) Induction of Male Sterility in Niger (*Guizotia abyssinica* Cass.). Paper presented at the Poster Session of *4th International Crop Science Congress, Brisbane, Australia,* September, 2004. www.cropscience.org.au/icsc2004

Gangopadhyay, G., Poddar, R. and Gupta, S. (1998) Micro-propagation of sesame (*Sesamum indicum* L.) by *in vitro multiple* shoot production from nodal explants. *Phytomorphology* **48**: 83-90

Gantait S. and Mondal, S. (2018) Transgenic approaches for genetic improvement in groundnut (*Arachis hypogaea* L.) against major biotic and abiotic stress factors, *J Genet Eng Biotechnol.* 2018 Dec; **16**(2): 537–544. doi: 10.1016/j.jgeb.2018.08.005

García-Moreno, M. J., Fernández-Martínez, J. M., Velasco, L., and Pérez-Vich, B. (2011). Molecular tagging and candidate gene analysis of the high gamma-tocopherol trait in safflower (*Carthamus tinctorius* L.). *Mol. Breed.* **28**, 367–379.

Gardner, F.P.and Auma, E.O. (1989) Canopy structure, light interception, and yield and market quality of peanut genotypes as influenced by planting pattern and planting date.*Field Crops Research* **20**(1):13-29 ·

Garnatje, T., Garcia, S., Vilatersana, R. and Valles, J. (2006) Genomic size variation in the genus *Carthamus;*Systematic implications and additive changes during allopolyploidisation. *Annals of Botany*, **97**: 461-467

GCP (2011). CGIAR generation challenge programme. (2011) *Project updates (incorporating projects completed in 2010 and 2009). Texcoco, Mexico: Generation Challenge Programme.*

Gebremedhin, B.; Hirpa, A.; Berhe, K., (2009). Feed marketing in Ethiopia: Results of rapid market appraisal. *Improving Productivity and Market Success (IPMS) of Ethiopian farmers' project Working Paper 15.* ILRI (International Livestock Research Institute), Nairobi, Kenya. 64 p

Gebremedhn, H. and Tadesse, A. (2014)Diversity of *Guizotia abyssinica* pollinators and their effect on seed quality in Mekelle, Northern Ethiopia.*Livestock Research for Rural Development. Volume* **26,** *Article #192.* Retrieved November 1, 2019, from http://www.lrrd.org/lrrd26/11/hgeb26192.html

Gedia, M.V., Vyas, H.J. and Acharya, M.F. (2007) Influence of weather on *Spodoptera litura* male moth catches in pheromone traps and oviposition in castor. *Indian Journal of Plant Protection.* **35** (1):118-120

Gehring, M. and Henikoff, S. (2007). DNA methylation dynamics in plant genomes. *Biochim. Biophys. Acta***1769**: 276–286.

Geleta, M. and Bryngelsson, T.(2010). Population genetics of self-incompatibility and developing self-compatible genotypes in niger (*Guizotia abyssinica*). *Euphytica* **176** (3):417-430.

Geleta, M., Bryngelsson, T. Bekele, E. and Dagne, K. (2007)Genetic diversity of *Guizotia abyssinica* (L. f.) Cass. (Asteraceae) from Ethiopia as revealed by random amplified polymorphic DNA (RAPD) .*Genetic Resources and Crop Evolution* **54**(3):601-614

Geleta, M. and Ortiz, R. (2013) The importance of *Guizotia abyssinica* (niger) for sustainable food security in Ethiopia.*Genetic Resources and Crop Evolution.* **60** (5) :1763-1770

Geleta, M., Stymne, S., and Bryngelsson, T. (2011). Variation and inheritance of oil content and fatty acid composition in niger (*Guizotia abyssinica*). *J Food Comp. Anal***24**(7): 95- 1003

Genders. R. (1994) Scented Flora of the World.*Robert Hale. London.*ISBN0-7090-5440-8

Gentry and Scot .H (1958). "The natural history of Jojoba (*Simmondsia chinensis*) and its cultural aspects". *Economic Botany*. **12** (3): 261–295. *doi:10.1007/BF02859772.*

Geoffrey A. Mueller, Soheila J. Malekiand Lars C. Pedersen (2014) The Molecular Basis of Peanut Allergy, (U.S. Government Work) *Curr. Allergy Asthma Rep.*(2014) **14:**429

George, L., Bapat, A. and Rao, P.S. (1987). *In vitro* multiplication of Sesame (*Sesamum indicum*) through tissue Culture. *Annals of Botany* **60**: 17-21

George, L. and P.S. Rao. (1982). *In vitro* multiplication of safflower (*Carthamustinctorius* L.) through tissue culture. *Proc. Ind. Natl. Sci. Acad.* **B48**: 791–794.

Gera, M and Chauhan, S. (2010) Opportunities for Carbon sequestration benefits from growing trees of medicinal importance on farm lands of Haryana. *Indian Forester*, **136** (3): 287-300

Getinet, A. and Sharma, S.M. (1996) Niger. *Guizotia abyssinica* (L. f.) Cass. Promoting the conservation and use of underutilized and neglected crops. 5. *Institute of Plant Genetics and Crop Plant Research, Gatersleben/International Plant Genetic Resources Institute, Rome.*ISBN 92-9043-292-6. Pp: 59

Ghadge, S. V. and Raheman, H. (2006). Process optimization for biodiesel production from mahua (*Madhuca indica*) oil using response surface methodology. *Bioresource Technology*, **97**(3): 379–384

Ghazali, H.M., Abdulkarim and Mohammed, S. (2011) Moringa (*Moringa oleifera*) Seed Oil: Composition, Nutritional Aspects, and Health Attributes,Chapter 93 : InVictor R. Preedy, Ronald Ross Watson and Vinood B. Patel(Eds) *Nuts and Seeds in Health and Disease Prevention Academic Press* Pp 1226. ISBN978-0-12-375688-https://doi.org/10.1016/B978-0-12-375688-6.10093-3

Ghorpade, P.B. and Wandhare, M.R.(2001). Application of simplified triple test cross and combining ability analysis to determine the gene action in safflower. In *Proceedings of the 5th International Safflower Conference, Williston, ND, and Sidney, MT,* July 23–27, 2001. Bergman, J.W. and H.H. Mundel, Eds., pp. 79–82.

Ghosh, R. K., Bandopadhyay, P. and Mukhopadhyay, N. (1994) "Performance of Rapeseed-mustard cultivars under various moisture regimes on the Gangetic Alluvial Plain of West Bengal," *Journal of Agronomy and Crop Sciences,***173**(1): 5–1

Gibbons, R.W., Bunting, A.H. and Smartt, J. (1972) The classification of varieties of groundnut, *Arachis hypogaea* L. *Euphytica*, **21**, 78–85,

Gibbons, R.W. and Tattersfield, J.R. (1969) Outcrossing trials in Groundnuts (*Arachis hypogaea* L.) *Rhodesia Journal of Agricultural Reasearch*7: 71-85

Gill KS. (1987) Linseed. *Publications and Information Division. Indian Council of Agricultural Research; 1987* New Delhi Inida. p. 386.

Gill, R.A., Zang, L., Ali. B., Farooq, M.A., Cui, P., Yang, S., Ali. and Zhou, W. (2015). Chromium induced physico-chemical andultra structural changes in four cultivars of *Brassica napus* L. *Chemosphere*, **120** :154-164

Giller, K. E. and Dashiell, K. E., (2007). *Glycine max* (L.) Merr. Record from Protabase. In: van der Vossen, H.A.M. & Mkamilo, G.S. (Eds). *PROTA (Plant Resources of Tropical Africa), Wageningen, Netherlands*

Gizachew, D.; Szonyi, B.; Tegegne, A.; Hanson, J. and Grace, D.(2016) Aflatoxin contamination of milk and dairy feeds in the Greater Addis Ababa milk shed, Ethiopia. *Food Control,***59**: 773-779

Glenn, E.P., O'Leary, J.W., Watson, M.C., Thompson, T.L. and Kuehl, R.O. (1991) *Salicornia bigelovii* Torr.: an oilseed halophyte for seawater irrigation. *Science* **251**, 1065–1067. Google Scholar

Gmunder, S., Singh, R., Pfister, S., Adheloya, A. and Zah R. (2012) Environmental Impacts of *Jatropha curcas* Biodiesel in India, *BioMed Research International.*Volume **2012**/ Ten pages.https://doi.org/10.1155/2012/623070

Gogate, M.G. and Gujar, D.R. (1993). An assay of genetic variability through phenological studies on a neem (*Azadirachta indica* A. Juss.) plantation. In: *Proceedings of the World Neem Conference 1993*, Bangalore, India. pp. 132-137.

Goldmon, D. L. (1991) Relay intercropping soybean into winter wheat: genetic and environmental factors (1991*). Retrospective Theses and Dissertations.* Iowa State University, USA. 10034. https://lib.dr.iastate.edu/rtd/10034

Golkar and Mokhtari, N. (2018). Molecular diversity assessment of the world collection of safflower genotypes by SRAP and SCoT molecular markers. *Physiology and Molecular Biology of Plants*. **24**(6): 1261–1271.

Gómez-Pompa, A., Marín-Andrade, A.I., Campo-García, J., Domínguez-Landa, J.H., Cano-Asseleih, L.M., Segura-Juárez, L.M., Cuéllar-Martínez, M., Fernández-Sánchez, M.J., Sánchez-Sánchez, O. and Lozoya, X. (2009). La Xuta se come. *Universidad Veracruzana,* México. 72 p.

Gopal K, Jagadeswar R, Prasad G. Evaluation of sesame (*Sesamum indicum*) genotypes for their reaction to powdery mildew and phyllody diseases. *Plant Disease Research.* 2005; **20**(2):126-13

Golkar, P., Eshan, S. and Nouraein, M. (2017) Combining ability × environment interaction and genetic analysis for agronomic traits in safflower (*Carthamus tinctorius* L.): biplot as a tool for diallel data. *Acta agriculturae Slovenica* **109**(2):229 DOI: 10.14720/aas.2017.109.2.07

Gopani, D.D. and Vaishnani, N.L. (1970) Two Mutant forms of groundnut (*Arachis hypogaea* L) *Indian Journal of Agricultural Sciences,***40**: 43-437

Gopinath, K.A., Venkateswarlu, B., Venkateswarlu, S., Yadav, S.K., Balloli, S.S., Srinivasa Rao, Ch., Prasad, Y.G. and Maheswari, M., (2011). *Organic Sesame Production, Technical Bulletin*, Central Research Institute for Dryland Agriculture, Santoshnagar, Hyderabad, Andhra Pradesh, India. 34p.

Gor. H.K., Dhaduk, L.K. and Lata Raval (2012) Heterosis and inbreeding depression for pod yield and its components in groundnut (*Arachis hypogaea* L.).*Electronic Journal of Plant Breeding,***3**(3): 868-874 (Sep 2012)

Gore, V.K and Satyamoorthy, P. (2000) Determination of Pongamol and Karanjin in Karanja Oil By Reverse Phase Hplc, *Analytical Letters*, 33:2, 337-346, DOI: 10.1080/00032710008543056 https://doi.org/10.1080/00032710008543056

Goulden, C.H. (1939). Problems in plant selection. Conference information. In: *Proc. 7th Int. Genet. Congr.* August 23–30, 1939, pp. 132–133. G

Gowda, J.and Seetharam, A. (2008) Response to Mass and SI Selection for Autogamy, Seed Yield and Oil Content in Sunflower Populations (*Helianthus annuus* L.) *HELIA,* **31** (48): 101-110

Graef, G.A., Fehr, W.R. and Hammond, E.G. (1985) Inheritance of three stearic acid mutants of soybean. *Crop Sci.* **25**, 1076–1079.Google Scholar

Grafius, J.E. (1956) Components of yield in oats- Ageometrical interpretation. *Agronomy journal.,* **48**(9): 419-423

Graham, S.A. (1989) *Cuphea*: a new plant source of medium-chain fatty acids. *Crit. Rev. Food Sci. Nutr.* **28**:139–173.Google Scholar

Green, A.G. (1986) A mutant genotype of flax (*Linum usitatissimum* L.) containing very low levels of linolenic acid in its oil. *Can. J. Plant Sci.* **66**, 499–503.Google Scholar

Green, A. G., Chen, Y., Singh, S. P.and Dribnenki, J. C. P. (2008). Flax. In :*Compendium of transgenic crop plants*. Kole C and Hall, T.C. (eds.) Oxford, Blackwell Publishing Ltd. Pp. 199–226

Green, A. G., and Marshall, D. R. (1981). Variation for oil quantity and quality in linseed (*Linum usitatissimum* L.*). Australian J. Agric. Res.,***32**(4), 599-607. http://dx.doi.org/10.1071/AR9810599

Green, A.G. and Marshall, D.R. (1984) Isolation of induced mutants in linseed (*Linum usitatissimum*) having reduced linolenic acid content. *Euphytica* **33**, 321–328.Google Scholar

Green, C.C. and Wynne, J.C. (1986) Diallel and generation mean analysis for the components of resistance to *Cercospora arachidicola* in Peanut. *Theor. Appl. Genet.*, **73**: 228-235

Greer, F.R., Sicherer, S.H. and Burks, A.W. (2019). "The Effects of Early Nutritional Interventions on the Development of Atopic Disease in Infants and Children: The Role of Maternal Dietary Restriction, Breastfeeding, Hydrolyzed Formulas, and Timing of Introduction of Allergenic Complementary Foods". *Pediatrics.* **143** (4): e20190281. doi:10.1542/peds.2019-0281

Gregory, W.C. (1955) X-ray breeding of peanuts (*Arachis hypgaea* L.) *Agronomy Journal* **47**: 396-399.

Gregory, W.C. and Gregory, M.P. (1976) Groundnut, Arachis hypogaea In: Simmonds, N.W.(Ed) *Evolution of Crop plants*pp.151-154. Longman, London.

Gregory, W.C., Gregory, M.P., Krapovickas, A. *et al.* (1973) Structures and genetic resources of peanuts, In: C.T. Wilson(Ed).*Peanuts: Culture and Uses, American Peanut Research and Education Association,* Stillwater, Oklahoma, pp. 47–133.

Gregory, W.C., Krapovickas, A. and Gregory, M.P. (1980) Structure, variation and classification in *Arachis,***In: Summerfield, R.J. and Bunting, A.H.(Eds)***Advances in Legume Science.*pp 469-481 Royal Botanic Gardens, Kew, England UK.

Gregory, W.C., Smith, B.W. and Yarbrough, J.A. (1951) Morphology, Genetics and breeding In *The Peanut- Unpredictable Legume* pp 28-85 National Fertilizer Association, Washington DC USA

Greiner, C.A. (1990) ***Economie Implication of Modified Soybean Traits***, Iowa State University Experiment Station, Ames, IA.Google Scholar

Gresshoff, P.M. (2014) The Contrasting Need for Food and Biofuel: can we afford biofuel? A Love of Ideas, *Melbourne University Press*. pp 144-152.

GRIN (2012).*Germplasm Resources Information Network - (GRIN)* [Online Database]. National Germplasm Resources Laboratory, Beltsville, Maryland.USDA, ARS, National Genetic Resources Program.URL: http://www.ars-grin.gov/cgi-bin/npgs/html/genus.pl?3179 (10 October 2012)

GRIN (2017)."*Moringa oleifera*". *Germplasm Resources Information Network(GRIN).* Agricultural Research Service (ARS), United States Department of Agriculture (USDA).

Grosso, N.R. and Guzman, C.A. (1995) Chemical composition of aboriginal peanut (*Arachis hypogaea* L.) seeds from Peru. *Journal of Agriculture and Food Chemistry.* **43** (1) :102-105

Grubben, G (2004). Plant resources of tropical Africa In: Grubben, G. J. H. (ed.). *Vegetables.* **2**: 394. ISBN 978-9057821479.

Guan, L.L. (2007) The correlation and path analysis of the seed yield per plant and main **agronomic traits in oil-type safflower**(*Carthamus tinctorius* L.).*Southwest China Journal of Agricultural Sciences* 2007**, 03**.

Guan, L. L., Xu, Y. W., Wang, Y. B., Chen, L., Shao, J. F., and Wu, W. (2012). Isolation and characterization of a temperature-regulated micro-somaloleate desaturase gene (*ctFAD2-1*) **from safflower** (*Carthamustinctorius* L.). *Plant Molecular Biology Reporter.* **30**: 391–402

Gubitz GM, Mittelbach M and Trabi M. (1999) Exploitation of the tropical oil seed plan*t. Jatropha curcas L. Bioresource Technology*. **67:**73–82.

Guggari, A.K. *et al.,* (1995) Response of niger to application of macro and micro nutrients in combination with farm yard manure. *Farm systems* **11** (3/4): 51-53

Gujar, G.T. and Mehrotra, K.N. (1989) Effect of chitin synthesis inhibitors on castor semilooper, *Achoea Janata. National Academy Science Letters,* **12** (8): 289-291

Gulya, T.J., Rashid, K.Y. and Masirevic, S.M. (1997) Sunflower diseases. In: Schneither AA, Seiler G (Eds.) ***Sunflower technology and production. Amer Soc Agron,*** Madison, pp 263–379

Gunstone, F.D. and Norris, F.A. (1983) *Lipids in Foods. Chemistry, Biochemistry and Technology*, Pergamon Press, Oxford, pp. 147-155.Google Scholar

Gunasena, H.P.M. and Marambe, B. (1998). Neem in Sri Lanka: *A Monograph. University of Peradeniya, Peradeniya*, Sri Lanka. 62 pp.

Gupta, A.(2013). An integrated approach for mahua seed cake utilization. In: *PhD Thesis, Centre Rur. Dev. Technol., Indian Inst. Technol.* Delhi, India

Gupta, H.S. (2009) Forest as Carbon Sink: Temporal Analysis for Ranchi district. *Indian Forester***, 32** (1): 7-11

Gupta, G.N. (1994.) Influence of rain-water harvesting and conservation practices on growth and biomass production of *Azadirachta indica* in the Indian desert. *Forest Ecology and Management* **70:** 329-339.

Gupta, G.N. (1995) Rain-water management for tree planting in the Indian Desert. *Journal of Arid Environments* **31**: 219-235.

Gupta, K., Prem, D. and Agnihotri, A. (2004). Role of Biotechnology for Incorporating White Rust Resistance in *Brassica* Species In:P.S. Srivastava., Narula, A. and Srivastava, S. (Eds.) *Plant Biotechnology and Molecular Markers*, Copyright © 2004 Anamaya Publishers, New Delhi, India. Pp.156-168

Gupta, K.N., Naik, K.R. and Bisen, R. (2018)Status of sesame diseases and their integrated management using indigenous practices.*International Journal of Chemical Studies*. **6**(2): 1945-1952

Gupta, P.K., Pal, R.S. and Emmanuel, C.J.S.K. (1995). Initial flowering and fruiting of the Neem National Provenance Trial. *Indian Forester* **121:** 1063-1068

Gupta, P.K. and Prabhu, V.V. (1997). On the high performance liquid chromatographic determination and monitoring of azadirachtin in neem ecotypes. *Indian Forester***123**: 1067-1071.

Gupta, R.D. and Gupta, S.K. (2016) Strategies for Increasing the Production of Oilseed on a Sustainable Basis. In: S.K. Gupta (Ed) *Breeding Oilseed Crops for Sustainable Production: Opportunities and Constraints.* Academic Press/ Elsivier, London/ San Diego. Pp 1-18

Gupta, R.K.,(1993). Multipurpose trees for agroforestry and wasteland utilization.: *Oxford & IBH*, New Delhi, India

Gupta, R.L., Pandey, N.D., Ram, S. and Sukhani, T.R. (1966) Bionomics of *Amsecta moorie* Butler (*Lepidoptera*: *Arctiidae*). *Allahabad Farming* **40** (2): 65-70

Gupta, S.K. (2016) Brassicas, In: Gupta, S.K. (Ed). *Breeding Oilseed Crops for Sustainable Production: Opportunities and Constraints. Academic Press/ Elsevier*. Pp.33-53

Gupta, S. K., Kamal Dhawan, Parkash Kumar and Yadava, T.P. (1982). Note on the chemical composition of some groundnut strains. *Indian J. Agric. Sci.* **52**:343-344.Gupta, S.K., Sharma, T.R., Singh, Harbhajan and Singh, H. (1997). Structural male sterility in *Brassica rapa*. *Cruciferae Newsletter*. **19**: 75-77

Gupta, S.K., Yadava, T.P., Saharan, G.S. and Kumar, P. (1984) Changes in phenolic compounds, sugars and total nitrogen in relation to *Alternaria* leaf blight in Indian mustard. *Haryana Agric. Univ. J. Res.***14**(4): 535-537

Gupta, V.K., Solanki, K.R., Gupta, R., Kumar, R.V. and Datta, A. (1996). Reproductive biology of neem (*Azadirachta indica* A. Juss). *Range Management and Agroforestry* **17** (2): 187-192.

Gupta, V.K., Solanki, K.R., Kumar, R.V. and Datta, A. (1998). Propagating neem (*Azadirachta indica)* by air layering. Forest, Farm and Community Tree Research Reports **3**:29-31.

Gürel, E. and Kazan, K. (1999) Evaluation of various sunflower (*Helianthus annuus* L.) genotypes for *Agrobacterium tumefaciens*-mediated gene transfer. *Turk J Bot.***23**:171-177.

Guth, H. and Grosch, W. (1991) Detection of furanoid fatty acids in soya-bean oil — cause for the light induced off-flavour. *Fat Sci. Technol*. **93**: 249–255.Google Scholar

Habibuddin, A.K. (1987) Genetic analysis of certain economic and drought reisitance characters in wheat (*Triticum aestivum* L.) *Ph.D. Thesis, Andhra Pradesh Agricultural University*, Hyderabad.

Habte, M., Muruleedhara, B.N. and Ikawa, H. (1993). Response of neem (*Azadirachta indica)* to soil P concentration and mycorrhizal concentration. *Arid Soil Research and Rehabilitation*7 (4): 327-333.

Haile, D. *et al.*, (2018) Genetics, genomics and breeding of groundnut (*Arachis hypogaea* L.) *Plant Breeding*. 2018;1–20.

Hajare, S.S., Hajare, S. and Sharma, A. (2005) Aflatoxin Inactivation Using Aqueous Extract of Ajowan (*Trachyspermum ammi*) Seeds. *Journal of Food Science* **70**(1):C29 - C34 .DOI: 10.1111/j.1365-2621.2005.tb09016.x

Hajika, M., Igita, K. and Kitamura, K. (1991) A line lacking all the seed lipoxygenase isozymes in soybean [*Glycine max* (L.) Merrill] induced by gamma-ray irradiation. *Jpn. J. Breed.* **41**, 507–509.Google Scholar

Hajjar, R.and Hodgkin, T. (2007) The use of wild relatives in crop improvement and survey of the developments over the last twenty years. *Euphytica*, **156**: 1-13

Hakansson A (1956) Seed development of *Brassica oleracea* and *B. rapa* after certain reciprocal pollination. *Hereditas,***42**: 373–395

Halakeri, A.V. (1999). Studies on sunflower necrosis disease. *M.Sc. (Agri.) Thesis, University of Agricultural Sciences, Dharwad,* Karnataka, India.

Halder, J., Mohapatra, A.K. and Rao, S.V. (1997) Response of niger (*Guizotia abyssinica*) to *Azotobacter*and nitrogen. *Indian Journal of Agronomy*, **42** (2):375-376.

Hall, L.M. ... Randall J. Weselake, R. J. (2016).Flax (*Linum usitatissimum* L.)In:Thomas A. McKeon, Douglas G. Hayes, ... Randall J. Wese (Eds) *Industrial Oil Crops, Academic Press and AOCS Press* ISBN 978-1-893997-98-1; pp 157-194

Hamidou, F., Halilou, O. and Vadez, V. (2012),Assessment of Groundnut under Combined Heat and Drought Stress.*Journal of Agronomy and Crop Science*.**199**: 1-11 DOI: http://dx.doi.org/10.1111/j.1439-037X.2012.00518.x Open Access Reoitory, ICRISAT: pp1-38

Hamilton, R.J. (1987) In: Recent Advances in Chemistry and Technology of Fats and Oils, Hamilton, R.J and Bhati, A. (Eds) Elsevier *Applied Science, New York.* 1987. Pp109-166

Hammond, E.G. (1985) Stability of soybean oil to oxidation, in *World Soybean Research Conference III: Proceedings*, ed. R. Shibles, Westview Press, Boulder, CO, pp. 251-258. Google Scholar

Hammond, E.G. (1990) The raw materials of the fats and oils industry, in *Fat and Oil Technology*, ed. P. Wan, American Oil Chemists' Society, Champaign, IL, pp. 1-15.Google Scholar

Hammond, E.G. (1991) Organization of rapid analysis of lipids in many individual plants, in *Modem Methods of Plant Analysis. New Series*, Vol. 12, eds. H.F. Linskens and J.F. Jackson, Springer-Verlag, Berlin, pp. 321-330.Google Scholar

Hammond, E.G. (1992a) Genetic alteration of food fats and oils, in *Fatty Acids in Foods and Their Health Implications*, ed. C.K. Chow, Marcel Dekker, New York, pp. 313-327.Google Scholar

Hammond, E.G. (1992b) The seeds on new technology. *Inform* **3**, 1268–1292.Google Scholar

Hammond, E.G. (2000) Genetic alteration of food fats and oils in fatty acids. In: Chow, C.K. (ed) Foods and their health implications. Dekker. NewYork. Pp 357-374.

Hammond, E.G. and Fehr, W.R. (1983) Registration of A5 germplasm line of soybean. *Crop Sci.* **23**, 192.Google Scholar

Hammond, E.G. and Fehr, W.R. (1984) Progress in breeding for low-linolenic acid soybean oil, InC. Ratledge, P. Dawson and J. Rattray(Eds): *Biotechnology for the Oils and Fats Industry*,American Oil Chemists' Society, Champaign, IL, pp. 89-96.Google Scholar

Hammond, E.G., Duvick, D.N., Fehr, W.R., Hildebrand, D.F., Lacefield, C. and Pfeiffer, T.W. (1992) Rapid screening techniques for lipoxygenases in soybean seeds. *Crop Sci.* **32**, 820–821.Google Scholar

Hammons, R.O. (1963) Artificial crosspollination of peanut with bee collected pollen. *CropScience* **3**: 562-563

Hammons, R. O. (1964) Krinkle, a dominant leaf marker in peanut. *Arachis hypogea* L. *Crop Science,***4**: 22-24

Hammons, R.O. (1973). Genetics of *Arachis hypogaea* L. (In) *Peanuts- Culture and Uses pp 135-173. American Peanut Rsearch and Education Association.* Stillwater Oklahoma USA.

Hammons. R.O. (1976) Peanuts: Genetic vulnerability and breeding strategy. *Crop. Sci.***16:** 577-630

Hammons, R.O., Herman, D. and Stalker, H.T. (2016) Chapter 1 - Origin and Early History of the Peanut, In: *Peanuts-Genetics, Processing, and Utilization.* AOCS Press. Pp 1-26

Han, XiaoZeng and Xu, YanLi (1988) Study of the main factors causing yield decrease in continuously cropped soyabeans, *Soybean Science,* **17**(3): 207-212

Hanelt, P.(1961) Systematic study of genus *Carthamus* L. – *A Monographic Review. Ph.D. Thesis.* Martin Luther University. Halle. Witterburg. Germany.

Hansen, G. and Wright (1999) Reent Advances in transformation of plants. *Trends Plant Sci.***4**: 226-231

Hansen, R. (2011) Sesame Profile. 19/08/2011. Available at http://www.agmrc.org/commodities __products/grains__oilseeds/sesame_profile. cfm

Hanson,M.R., Gray,B.N. and Ahner, B.A. (2013) Chloroplast transformation for engineering of photosynthesis,*Journal of Experimental Botany,* **64** (3) :731–742, https://doi.org/10.1093/jxb/ers325

Hanumantharaya, L., Balikai, R.A., Mallapur, C.P., Venkateshalu and Kumar, C.J. (2007) Integrated Pest Management Strategies against safflower aphid, *Uroleucon compositae* (Theobald), *Proceedings of the 7th Safflower Conference, Wagga Wagga,* Australia.

Hao-jie, L.I., Xiao-bin, P.U., Jin-fang, Z., Qi-xing, Z. and Liang-cai, J. (2005) Primary study on the radiation induced mutation in *Brassica npus. Southwest China J. Agric. Sci.* **18**: 542-546

Haque, E., AnuPeter.V. and Jairath, M.S. (2018) Tackling aflatoxin in groundnut, *BusinessLine, (The Hindu)* dt. Jan.24, 2018.

Harrigan, E.K.S. 1987. Safflower registration of cv. Sironaria. *Sesame Safflower Newsl.* **3**: 47–49.

Harrigan, E.K.S. 1989. Review of research of safflower in Australia.In: V. Ranga Rao, and M. Ramachandram, (Eds.)*Proceedings of the 2nd International Safflower Conference*, Hyderabad, India, January 9–13, 1989. ISOR, Directorate of Oilseeds Research, pp. 97–100.

Harris, D., Matthews, R.B., Nageswara Rao, R.C. and J.H. Williams (1988) The Physiological Basis for Yield Differences between Four Genotypes of Groundnut (*Arachis hypogaea*) in Response to Drought. III. Developmental Processes· *Experimental Agriculture.* **24** (2): 215-226

Harris, H.C. (1949) The Effect on the Growth of Peanuts of Nutrient Deficiencies in the root and pegging zone. Published by *www.plantphysiol.orgCopyright © 1949 American Society of Plant Biologists.* pp 150-160

Hartley, Matt (2008). "Grain Farmer Claims Moral Victory in Seed Battle Against Monsanto". *The Globe and Mail.* Retrieved 30 August 2011

Hartley R.W.(1989) Barnase and barstar: two small proteins to fold and fit together.*Trends Biochem Sci. 1989 Nov;14(11):450-4.*

Harvey, B.L. and Knowles, P.F. (1965) Natural and artificial alloploids with 22 pairs of chromosomes in the genus *Carthamus* (Compositae). *Can J Genet Cytol.,***7**:126–139

Harvir Singh (2005) Thrips incidence nd necrosis disease in sunflower. *Journal of Oilseeds Research,* **22** (1): 90-92.

Hasan, S.Q., Ahmad, I., Sherwani, M.R.K., Ansari, A.A. and Osman, S.M. (1980) Studies on herbaceous seed oils X. *Fette Seifen Anstrichm.* **82**, 204–205.Google Scholar

Hasselquist, F. (1762) *Reise nach Palastina.* Rostock, USSR

Hatice, M., Meier and Michael, A. R. (2010). "Castor oil as a renewable resource for the chemical industry". *European Journal of Lipid Science and Technology.* **112**: 10–30. doi:10.1002/ejlt.200900138.

Haumann, B.F. (1990) Low-linolenate flax: variation on familiar oilseed. *Inform* **1**, 934–941. Google Scholar

He, FongFa*et al.*, (1994) Karyotype of sesame (*Sesamum* L.) as related to its phylogenesis. *Journal Southwest Agricultural University (Sichuan)***16**(6): 573-576

He, Z.H.(1996) The eco-geographical distribution of chemical composition of soybean seeds. *Journal of Northeast Agricultural University*, **3** (2): 94-102

Heaton, T.C.and Klsiewicz, J.M. (1981). A disease-resistant safflower alloploid from*Carthamus tinctorius* L. X *C. lanatus* L.*Can. J. Plant Sci.* **61**: 219-224

Heaton, T.C. and Knowles, P.F. (1982) Inheritance of male sterility in safflower.CropScience,**22:** 520-522

Heenan,P. B. and Dawson,M. I. (2005) Spontaneous hybrids between naturalised populations of pak choi (*Brassica rapa* var.*chinensis*) and wild turnip (*B. rapa* var.*oleifera*) from near Ashburton, Canterbury, New Zealand. *New Zealand Journal of Botany* **43**(4):pp 817-824, DOI: 10.1080/0028825X.2005.9512992

Hegde, D.M. (2000). Oilseed scenarion in India-past,present and futurewith special reference to rapeseed-mustard, In: *Advances in Rapeseed-Mustard Research Technology for Sustainable Production of Oilseeds.Compendium of Winter School Lecture Notes,* NRCRM: pp. 1–15384pp+vii.

Hegde, D.M. (2012) In: K.V. Peter (Ed): *Handbook of Herbs and Spices* (Second Edition**), 2.***Woodhead Publishing* Series in Food Science, Technology and Nutrition Book.2012. ISBN:978-0-85709-040-9

Hegde, D.M. (2020) Enhancing profitability of oilseeds production through resource conservation technologies and efficient cropping systems. *Souvenir, "National Seminar -2020 Technological Innovations in Oilseed Crops for Enhancing Productivity, Profitability and Nutritional Security"*, Feb. 7-8, 2020Indian Society of Oilseeds Research, ICAR-Indian Institute of Oilseeds Research, Hyderabad, India. pp 22-35

Hegde, I.C. (1976) A systematic and geographical survey of the old world *Cruciferae*, In: Vaughan, J.G., Macleod, A.J. and Jones, B.M.G.(Eds) *"Biology and Chemistry of Cruciferae"*: 1-45pp. *Academic Press*, New York

Hegde, M. (2019) Unleashing the Power of Linseed to improve Public Health and Combat Non-Communicable Disease Menace in the Country. *Indian J Agric Biochem***32** (1): 1-9, 2019 doi 1 0.59581097 4-4479.201 9.00001 .7

Hegde, N.G. (1993) Improving the Productivity of Neem Trees. *World Neem Conference,* Bangalore, India: 69-79

Heiser, C.B. (1976) ***The Sunflower***, University of Oklahoma Press. USA (cf. Weiss 2000)

Hellpap, C. and Dreyer, M. (1995). The smallholder□s neem products. In: H. Schmutterer (Ed) *The Neem Tree: Source of Unique Natural Products for Integrated Pest Management, Medicine, Industry and other Purposes,*Weinheim, Germany.

Hema, B.P. and Murthy, N.H. (2008) Improvement of *in vitro* androgenesis in niger using amino acids and polyamines. *Biologia Plantarum* **52**(1):121-125 ·

Hemingway J S. (1976). Mustard: *Brassica* spp. and *Sinapis alba*(Cruciferae) In: Simmonds (Ed.)*Evolution of Crop Plants*. Longmans, London pp 56-59

Hepburn A.G., Clarke L.E., Blundy K.S. and White J. (1983): Nopaline Ti-plasmid, pTiT37, T-DNA insertions into a flax genome. *Journal of Molecular and Applied Genetics*, **2**: 211–24.

Herb, S.F. and Martin, V.G. (1970) How good are analyses of oils by GLC? *J. Am. Oil Chem. Soc.* **47**, 415–421.Google Scholar

Hereman, S.I. (1868). Plaxton's Botanical Dictionary. *Bradbury Evans and Co. London*. pp 225

Hermansen, R. D. (2017) Polymeric Thermosetting Compounds: Innovative Aspects of Their Formulation Technology. *CRC Press.* ISBN 9781771883153. *(32)*

Hewezi, T., Jardinaud, F., Alibert, G. *et al.* (2003) A new approach for efficient regeneration of a recalcitrant genotype of sunflower (*Helianthus annuus*) by organogenesis induction on split embryonic axes. *Plant Cell, Tissue and Organ Culture* **73:** 81–86 (2003). https://doi.org/10.1023/A:1022689229547

Heyn F.W. (1976) Transfer of restorer genes from *Raphanus* to cytoplasmic male sterile *Brassica napus*. *Cruciferae Newslett.* **1**: 15–16 [Google Scholar]

Heyn, F.W. (1978) Cytoplasmic male sterility in *Brassica napus*. *Cruciferae* Newletter, **3:** 34-35

Higgins, B.B. (1938) Peanut breeding, pp. 57-58. *Proceedings of the 39th Annual Convention Association of South Agricultural Workers.* Atlanata, Georgia. USA.

Hildebrand, D.F. and Hymowitz, T. (1982) Inheritance of lipoxygenase-1 activity in soybean seeds. *Crop Sci.* **22**, 851–853.Google Scholar

Hildebrandt, V.M. (1931) The Plant Resources of the Worldas Initial Material in Plant Breeding.: The Sesame,*Lenin. Acad Agric Sci Inst Plant Indust, Moscow and Leningrad,*

Hilditch, T.P. (1954) The fats: a story of nature's art. *J. Sci. Food Agric.* **5**, 557–566.Google Scholar

Hill, A.B.(1989) Hybrid safflower breeding. *Proceedings of the 2nd International Safflower Conference, Hyderabad*, India, Jan. 9-13, 1989, pp. 69-170

Hill, J., Nelson, E., Tilman, D., Olasky, S. and Tiffany, D. (2006) Environmental, economic and energetic costs and benefits of biodiesel and ethanol biofuels. *Proc. Natl. Acad. Sci. USA.* **103**: 11206-11210.

Hinata, K. and Konno, N. (1976). Cytoplasmic male sterile strain of Yukina (*Brassica campestris)* produced by nuclear substitution. *Japan. J. Breed.* **26:** 127-128

Hindu (2014) "Farm Query - Mahua oil". *The Hindu.* Chennai, India. 2014-01-22.

Hiremath, R.V. (1976) Studies on diseases of Sesame: Incidence of powdery mildew on some varieties. *Oilseeds Journal* **6**(3): 18.

Hiroshi, T., Sachiko, Y. and Kazuo, O. (1992). Seed coat anatomy, karyomorphology, and relationships of *Simmondsia*(Simmondsiaceae). *The Botanical Magazine Tokyo.* **105** (4): 529–538. doi:10.1007/BF02489427.

Hocking, D. (1993). Trees for Drylands. Oxford and IBH Publishing Company. New Delhi. Pp 1-41

Hondelmann, W. (1985) Das Vorkommen einer ungewöhnlichen Fettsäure, der Petroselinsinsäure, in der Familie der Doldengewächse als Ausgangspunkt für die Entwicklung neuer Ölfrüchte — Bestandsaufnahme und kritische Würdigung. *Landbauforschung Völkenrode* **35**, 185–190.Google Scholar

Hondelmann, W. and Dambroth, M. (1987) Identifizierung und Evaluierung von Samenölhaltigen Wildarten der Krautflora als potentielle Nutzpflanzen für die Gewinnung von Industriegrundstoffen. *Landbauforschung Völkenrode* **37**, 189–184.Google Scholar

Hondelmann, W. and Radatz, W. (1984) Zur Evaluierung ölsamentragender Wildarten. *Landbauforschung Völkenrode* **34**, 145–154.Google Scholar

Hooker. J.D. (1872). The flora of British India. Part I. *L. Reew and Co., Covent Garden*, London. pp 158.

Hopkins MS, Casa AM, Wang T, Mitchell SE, Dean RE, Kochert GD, *et al*., (1999) Discovery and characterization of polymorphic simple sequence repeats (SSRs) in peanut *Crop Sci.,* 1999; **39:** 1243–1247. 6

Horn, R., Kusterer, B., Lazarescu, E., Prüfe, M. and Friedt, W. (2003) Molecular mapping of the Rf1 gene restoring pollen fertility in PET1-based F1 hybrids in sunflower (*Helianthus annuus* L.). *Theor Appl Genet. 2003 Feb;* **106**(4):599-606.

Hosoda, T., Namai, H. and Gotoh, J. (1963) On the breeding of *Brassica napus* obtained from artificially induced amphidiploids. III. On the breeding of synthetic rutabaga (*B. napus* var. *rapifera)*. Part 1. *Jpn J Breed* **13**: 99–106

Howarth, G.A. (2003). "Polyurethanes, polyurethane dispersions and polyureas: Past, present and future". *Surface Coatings International Part B: Coatings Transactions.* **86** (2): 111–118. doi:10.1007/bf02699621. ISSN 1476-4865.

Howell, R.W. and Collins, F.I. (1957) Factors affecting linolenic and linoleic acid content of soybean oil. *Agron. J.* **49**, 593–597.Google Scholar

Huang. (1980). Oil crops of China No.**12**:1-5

Huang, L., He, H., Chen, W. *et al.* (2015) Quantitative trait locus analysis of agronomic and quality-related traits in cultivated peanut (*Arachis hypogaea* L.). *Theor Appl Genet* **128:** 1103–1115 (2015) doi:10.1007/s00122-015-2493-1

Huang, R.C., Okamura, H., Iwagawa, T., Tadera, K. and Nakatani M, (1995). *Azedarachin* C, a limonoid antifeedant from *Melia azedarach. Phytochemistry*, **38**(3):593-594

Hubner, S. *et al*.,(23 associates) (2019) Sunflower pan-genome analysis shows that hybridization altered gene content and disease resistance.*Nat Plants*. 2019 Jan;**5**(1):54-62. doi: 10.1038/s41477-018-0329-0 (https://www.ncbi.nlm.nih.gov/pubmed/30598532)

Hughes, H.J., Ryan, D.J., Mukherjea, R and Schasteen, C.S, (2011) Protein Digestibility Corrected Aminoi Acid Scores (PDCAAS) for Soy Protein Isolateand Soy Protein Concentrate: Criteria for Evaluation.*Journal of Agriculture and Food Chemistry*: **59(**23), 12707 – 12712.

Hull and Fred H. 1952. Recurrent selection and over dominance.. In: John W. Gowen, (ed.), *Heterosis. Hafner Publishing Co.,* N ew York, London. P.451-473

Hunter, J.E. (1992) Safety and health effects of isomeric fatty acids, in *Fatty Acids in Foods and Their Health Implications*, ed. C.K. Chow, Marcel Dekker, New York, pp. 857-868. Google Scholar

Hurburgh, C.R. (1988) Moisture and compositional analysis in the corn and soybean market. *Cereal Food World* **33**, 503–505.Google Scholar

Husain, S.K., Ahmad, M.U., Sinha, S., Ansari, A.A. and Osman, S.M. (1978) Studies on herbaceous seed oils III. *Fette Seifen Anstrichm*. **80**, 225–227.Google Scholar

Hussain, M.M., Rauf, S., Riaz, M.A., Al-Khayari, J.M and Monneveux, P.(2017) Determination of drought tolerance related traits in *Helianthus argophyllus*, *Helianthus annuus*, and their hybrids, *Breed Sci*. 2017 Jun; **67**(3): 257–267.

Hosseinian, F. S., Rowland, G. G., Bhirud, P. R., Dych, J. H. and Tyler, R. T. (2004). Chemical compositionand physicochemical and hydrogenation characteristics of high-palmitic acid solin (low linolenic acid flaxseed) oil. *Journal of the American Oil Chemists' Society*, **81**: 185–18

Hymowitz, T. (1976) Soybeans, In: *Evolution of Crop Plants (*Ed) Simmonds, N.W. *Longman Publishers*, London (from Weiss, 2000)

Ibrahim, Y.M. (1994) Response of safflower to different nitrogen levels in United Arab Emirates. *Annals of Arid Zone*, **33** (1): 77-78

ICAR (1992). Niger: Package of practices for increasing production. Extension Bulletin No. VII, Directorate of Oilseeds Research, ICAR. Hyderabad.Pp12

ICAR-DGR (2019) ICAR-Directorate of Groundut Research, Junagadh Research Highlights (http://www.dgr.org.in/index.php/crop-improvement/)

ICAR-DRMR (2019) Indian Council of Agricultural Research (ICAR) Highlights of technologies developed in rapeseed-mustard. ICAR-Directorate of Rapeseed-Mustard Research, Sewar, Bharatpur.

ICRISAT (2017) http://exploreit.icrisat.org/profile/Groundnut/250

ICRISAT (2018) ICRISAT Happenings News Letter 2018 https://www.icrisat.org/groundnut-cultivation-improves-incomes-for-tribal-farmers-in-india/

ICRISAT (2019) Advanced Breeding Techniques Produce High-Oleic Groundnut Lines. ICRISAT Happenings Newsletter. https://www.icrisat.org/advanced-breeding-techniques-produce-high-oleic-groundnut-lines/

Imery, B.C and Knowles, P.F. (1970).**Inheritance studies in interspecific hybrids** between*Carthamus flavescens* and *C. tinctorius. Crop Sci.* **10**:349-352.

Imrie, B.C. and Knowles, P.F.(1971) Genetics studies of self-incompatibility *in Carthamus tinctorius* Spreng Proceeding. *Crop Science*, **11**: 6-9

Indi, D.V., Lukade, G.M. and Patil, P.S. (1986) Influece of *Alertnaria* leaf spot on growth and yield of safflower. *Curr. Res. Rep.* **2**: 137-139

India9.com (2014) https://www.india9.com/i9show/Mahuwa-59485.htm

Indrasumunar, A., Gresshoff, P. M. and Scott, P. T. (2017).Prospect of the legume tree *Pongamia pinnata* as a clean and sustainable biodiesel feedstock. In: Mohammad G. Rasul, Abdul Kalam Azad and Subhas C. Sharma (Eds.), *Clean energy for sustainable development* (pp. 399-417) Amsterdam, The Netherlands: Elsevier. doi:10.1016/B978-0-12-805423-9.00012-0

Indi, D.V., Lukade, G.M., Patil. P.S. and Shambharakr, D.A. (1988) Estimation of yield losses in safflower due to *Aletrnaria* leafspot under dryland conditions. *Pesticides,* **22** :41-43

Inomata, N. (1976) Culture *in vitro* of excised ovaries in *Brassica campestris* L. I. Development of excised ovaries in culture media, temperature and light. *Jpn J Breed* **26**: 229–236

IPGRI (1994) Ecotypes in safflower. *IPGRI Newsletter for Asia-Pacific and Oceania,* **15**: 5-7.

IRRI. 1989. Research on botanical pest control in rice-based cropping systems. *Project Completion Report.* Los Banos, Philippines.

Irvine, R.B., McConnell, J., Lafond, G.P., May, W.E., Hultgreen, G., Ulrich, A., Stonehouse, K., Chalmers, S. and Stevenson, F.C., (2010). Impact of production practices on fiber yield of oilseed flax under Canadian prairie conditions. *Can. J. Plant Sci.* **90**: 61–70.

Iselib, T.G. Wynne, J.C., Elkan, G.H. and Schneewis, T.J. (1978) Quantitative Genetic aspects ofnitrogen fixation in peanut. *Proceedings of the American Research Education Association,* USA pp71.

Ishag, H.M. (1970) Growth and yield of irrigated groundnut (*Arachis hypogaea* L.) grown in different spacings in Sudan, Gezira. I. flowering, yield and yield components. *Journal of Agricultural Sciences Camp.* **74** (3): 533-537

ISOR (2020) Introductory Bulletin, National Oilseeds Seminar, 2020, Feb. 7-8. Indian Society of Oilseeds Research & Indian Institute of Oilseeds Research. ICAR. Hyderabad.

Jacob, K.M., (1957). Secondary association in the castor oil plant. *Cytologia* **22:** 380–392.

Jacques, S., Bacon, R.K., and Parsch, L.d. (1997) Comparison of single cropping, relay cropping and double cropping of soya beans with wheat, using cultivar blends. *Experimental Agriculture,* **33** (4): 477-486

Jadhav, G. D. and Shinde, N. N. (1979) Genetic studies in groundnut (*Arachis hypogaea* L.) *Indian Journal of Agricultural Research.* **13**: 93-96

Jadhav, S.A., Dhuppe, M.V. and Salunke, P.M. (2018) Correlation Coefficient and Path Analysis in Safflower (*Carthamus tincorius* L.). *Int.J. Curr.Microbiol.App.Sci* (2018)

Jadimath, V.G., Murthy, H.N., Pyati, A.N., Kumar, H.G.A.and Ravishankar, B.V. (1998). Plant regeneration fromleaf cultures of*Guizotia abssinica* (Niger) and*Guizotiascabra. Phytomorphologia* **48**: 131-135*Special Issue*-**6:** 1234-1241(Journal homepage: http://www.ijcmas.com)

Jagannath, A., Arumugam, N., Gupta, V., Pradhan, A., Verma, P. K. and Pental, D. (2002). Development of transgenic *barstar* lines and identification of a male sterile *(barnase)* / restorer (*barstar*) combination for heterosis breeding in Indian oilseed mustard (*Brassica juncea*). *Current Science* **82**: 46–52.

Jain, K.L., Jain, H.K. and Sekkawat, K.S. (2005) Growth inhibiting effect of some non-edible oils on *Achoea janata* L., *Journal of Applied Zoological Research* **16** (2): 192-194

Jain, R.K.,Mathur, K.N. and Singh, R.V. (2007) Estimation of losses due to parasitic nematodes on different crops in India. *Indian Journal of Nematology.* **37**: 219-221

Jaiswal, N., Yadav, P.P., Maurya, R., Srivastava, A.K. and Tamrakar, A.K. (2011) Karanjin from *Pongamia pinnata* induces GLUT4 translocation in skeletal muscle cells in a phosphatidylinositol-3-kinase-independent manner.*Eur J Pharmacol.* 2011 Nov **16**:670(1):22-8. doi: 10.1016/j.ejphar.2011.08.049. Epub 2011 Sep 14.PMID: 21939653

Jambulkar, S.J. (1995) Rapid cycling through immature embryo culture in sunflower (*Helianthus annuus*, L.) *Helia***18**: 45-49

Jana T, Sharma TR. Singh NK. (2005) SSR –based detection of genetic variability in the charcoal root rot pathogen *Macrophomina phaseolina. Mycol Res.,* 2005: **109**:81- 86.

Janila, P., Manohar, S. S., Patne, N., Variath, M. T., & Nigam, S. N. (2016a). Genotype × environment interactions for oil content in peanut and stable high-oil-yielding sources. Crop Science, 56, 01– 10.

Janila, P., Manohar, S. S., Rathore, A., & Nigam, S. N. (2015). Inheritance of SPAD chlorophyll **meter reading and specific leaf area in four crosses of groundnut (Arachis hypogaea L.).** Indian Journal of Genetics, 75, 408– 412.

Janila, P., and Nigam, S. N. (2012). "Phenotyping for groundnut (*Arachis hypogaea* L.) improvement," in Phenotyping for *Plant Breeding*, eds S. K. Panguluri and A. A. Kumar (New York, NY: Springer Publishing), 129–167

Janila, P., Nigam, S.N., Pandey, M.K., Nagesh, P. and Varshney, R.K. (2013a)Groundnut improvement: use of genetic and genomic tools. *Front Plant Sci.*; **4**: 2

Janila, P., Nigam, S. N., Abhishek, R., Kumar, V. A., Manohar, S. S., and Venuprasad, R. (2014). Iron and zinc concentrations in peanut (*Arachis hypogaea* L.) seeds and their relationship with other nutritional and yield parameters. *J. Agric. Sci.* 153, 975–994. doi: 10.1017/S002185961400052

Janila, P., Nigam, S.N., Pandey, M.K., Nagesh, P. and Varshney, R.K. (2013a) Groundnut Improvement: Use of Genetic and Genomic tools, *Frontiers in Plant Science,* **4** (23) :1-16

Janila, P., Pandey, M. K., Shasidhar, Y., Variatha, M. T., Sriswathi, M., Khera, P., *et al.* (2016b). Molecular breeding for introgression of fatty acid desaturase mutant alleles (*ahFAD2A* and *ahFAD2B*) enhances oil quality in high and low oil containing peanut genotypes. *Plant Sci.* **242**: 203–213. doi: 10.1016/j.plantsci.2015.08.013

Janila, P., Ramaiah, V., Rathore, A., Upakula, A., Reddy, R. K., Waliyar, F., and Nigam, S. N. **(2013b). Genetic analysis of resistance to late leaf spot in interspecific groundnuts.** *Euphytica,* **193:** 13– 25.

Janila. P, Variath, M.T., Pandey, M. K., Desmae, H., Motagi,B.N., Okori, P.Manohar ,S.S., Rathnakumar,A.L., Radhakrishnan,T., Liao.B. and Varshney, R.K. (2016c)Genomic Tools in Groundnut Breeding Program: Status and Perspectives, *Front. Plant Sci.,* 17 March 2016 https://doi.org/10.3389/fpls.2016.00289

Jaradat, A.A. (2016) Breeding Oilseed crops for Climate Change In: S.K. Gupta (Ed.) 2016 *Breeding Oilseed Crops for Sustainable Production: Opportunities and Constraints.* Academic Press/ Elsivier, London/ San Diego. Pp 421-445

Jarl, C.L., Ljungberg, U.K. and Borhman, C.H. (1988) Correction of chlorophyll 11-defective male sterile winter oilseed rape (*Brassica napus*) through organelle exchange: **Characterisation of the chlorophyll deficiency.** *Physiol. Plantarum***72:** 505-510.

Jaswal, S.V. and Gupta, V.P. (1967) Selection criterion in improving erect types of groundnuts. *Journal of Research, Punjab Agricultural University***4:** 188-191.

Jat, N.L., Keshwa, G.L. and Singh, G.D. (1987) Efficacy of herbicidal weed control in taramira (*Eruca sativa* Mills.) *Haryana J. Agronomy*. **3**: 92-93

Jat. R, Naga, S, R., ChoudharyR. S and Mohammad. (2017a)Effect of Potassium and Sulphur on Quality of Sesame (*Sesamum indicum* L.) *Int.J. Curr.Microbiol. App.Sci* (2017) **6**(4): 1876-1878

Jat. R, Naga, S, R., Choudhary R.S and Jat, S. (2017b).Effect of Potassium and Sulphur on economics and optimum does of Sesame (*Sesamum indicum* L.*)Journal of Pharmacognosy and Phytochemistry* 2017; **6**(3): 168-170.

Jatothu, J., Dangi, K. and Kumar, S. (2013) Evaluation of Sesame crosses for heterosis in yield and yield attributing traits. Journal of Tropical Agriculture **51:** 84 – 91.

Javed, M.A., Siddiqui, M.A. Khan, M.K.R. *et al.*, (2003). Development of high yielding mutants of *Brassica campestris* L. cv Toria selection through gamma rays irradiation. *Asian J. Plant Sci.* **2**: 192-195

Javed, N. (1989) Self-incompatibility and autogamy of sunflower (*Helianthus annuus* L.) cultivars *agris.fao.org*

Jaworski, J. and Cahnoon, E.B. (2003) Industrial oils from transgenic plants. Curr.Opin. Plant Biol.**6** :178-84

Jayalakshmi, V., Reddy, C. R. and Reddy, G. L. K. (2000) Heterosis in groundnut (*Arachis hypogaea* L.). *Legume Research* **23** (3) :155-158

Jayan, T.V. (2018) Regulator defers decision on GM mustard trials.Agri Business,*Business Line*. September 03, 2018

Jayant, M and Sinha, S.K. (1981) Control of leaf spot disease of brown *sarson* caused by *Alternaria brassicae* and *Alternaria brassicicola* by the antibiotic substance produced by a strain of *Streptomyces hygroscopicus. Third Int. Symp. Pl. Path.* IARI, New Delhi. Dec. 14-18 (Abst.) P.71

Jayaraj, S. (1967) Hopper burn disease of castor bean varieties caused by ***Empoasca flavescens*** (F) in relation to the history of leaves. *Phytopathology* **58**: 397-406

Jayaraj, S. and Basheer, M. (1964) Biological observations on the castor leaf hopper, *Empoasca flavescens* (F). *Madras Agricultural Journal***51**: 89-90.

Jayaraman, K.S. (2017)GM mustard: Far from the kitchen. *Nature India.*doi:10.1038/nindia.2017.101 Published online 1 August 2017 https://www.natureasia.com/en/nindia/article/10.1038/nindia.2017.101

Jayewar, N.E., Gosalwad, S.S. and Sonkambke, M.M. (2017) Estimation of yield losses caused by defoliators in sunflower. *Agric. Update*, **12** (*TECHSEAR-1*) 2017: 54-57

Jefers, D.L., Triplett, G.B Jr. and Lafever, H.N. (1977) Relay Intercropping Wheat and Soybeans. *Research Circular,***233.** Ohio Agricultural Research and Development, Ohio, USA.

Jesse More (2009) Carbon Sequestration Potential of the MillionTrees.NYC Initiative, Biofuels and Bio-Based Carbon MitigationCenter U. S. 250 and OHIO 83 South Wooster, OHI

Jeyaramraja,P.R., Nithya Meenakshi,S. and Woldesenbet, F. (2018)Relationship between drought and preharvest aflatoxin contamination in groundnut (*Arachis hypogaea* L.) *ISSN 0029-568Xorld, Mycotoxin Journal* **11** (2): 187 – 199 https://doi.org/10.3920/WMJ2017.2248

Jha, T.B. (2014) Somatic chromosomes of Aegle marmelos and *Azadirachta indica* through EMAmethod, *The Nucleus*, **57** (3):185-188. DOI 10.1007/s13237-014-0124-

Jhala A.J., Bhatt H., Topinka K., Hall L.M. (2011): Pollen mediated gene flow in flax (*Linum usitatissimum* L.): can genetically engineered and organic flax coexist? *Heredity*, **106**: 557–566

Jin, U.H., Chun, J.A., Han, M.O., Lee, J.W., Yi, Y.B., Lee,S.W. and Chung, C.H. (2005) Sesame hairy root cultures for extra-cellular production of a recombinant fungal phytase. *Prog Biochem***40**:3754–3762

Jindal, S.K., Vir, S. and Pancholy, A. (1999). Variability and associations for seed yield, oil content and tree morphological traits in neem (*Azadirachta indica*). *Journal of Tropical Forest Science* **11**: 320-322

Joarder, O.I., Naderuzzaman, A.T.M., Islam, R., Hossain, M., Joarder, N. and Biswas, B.K. (1993). Micropropagation of neem through axillary bud culture. *World Neem Conference,* Bangalore, India, 1993.

Jogloy. S., Wynne, J.C. and Beute, M.K. (1987) Inheritance of late leaf spot resistance and agronomic traits in peanut. *Peanut Sci.*, **14**:86-90

Jogloy, S., Pensuk, V., Patanothai, A., Wongkaew, S. (1999) Combining ability of resistance to late leaf spot and rust of peanut. *Thai. J. Agric. Sci.* **32:** 281-287.

John, C. M., Naryana, G. V. and Seshadri, C. R. (1950) The wild gingelly of Malabar. *Madras Agric J.*,**37**:47–50

Johnson, R. C., Kisha, T. J., and Evans, M. A. (2007). Characterizing safflower germplasm with AFLP molecular markers. *Crop Sci.* **47**: 728–1736.

Johnson, S., Morgan, E.D. and Peiris, C.N. (1996)Development of the major triterpenoids and oilin the fruit and seeds of neem (*Azadirachta indica* A. Juss). *Ann. Bot.*,**78: 383–388**

Jongschaap, R.E.E., Corre, W.G., Bindraban, P.S. and Brandenburg, W.A. (2007) Claims and Facts on *Jatropha curcus* L.-*Global Jatroha curcus Evaluation, Breeding and Propagation Programme. Plant Research International B.V. Wageningen* UR Stichting Het Woudt, Laren, Netherlands, October 2007. Report **158**, Pp 42 + Appendices I to IV

Jordan, M.C., and A. McHughen. (1988) Glyphosate tolerant flax plants from Agrobacterium mediated gene transfer. *Plant Cell Rep.*7:281–284

Jorgensen, R.B. and Andersen, B. (1994) Spontaneous hybridization between oilseed rape (*Brassica napus*) and weedy *Brassica campestris* (*Brassicaceae*): a risk of growing **genetically modified oilseed rape.** *Am J Bot* **81**:1620–1626

Jose, A., Kannan, E., Ramakrishnan, P., Kumar, A.V., Subbarao, V.M. (2019) "Therapeutic Potential of Phytochemicals Isolated from Simarouba glauca for Inhibiting Cancers: A Review". ***Systematic Reviews in Pharmacy***. **10** (1): 73–80. doi:10.5530/srp.2019.1.12. Retrieved 17 January 2020.

Joshi, A.B. (1961) Sesamum: *A Monograph. Indian Central Oil Seeds Committee,* Hyderabad, India.

Joshi, B. M., Nerker, Y.S. and Jambhale, N.D. (1983) Induced male sterility in safflower. *JournalofMaharashtraAgriulturalUniversities* **8**: 194-196 NT

Joshi, M.S. and Thangane, S.R. (1996). In vitro propagation of *Azadirachta indica* A. Juss. (neem) by shoot proliferation. *Indian Journal of Experimental Biology***34**: 480-482.

Kadam, N.V., Dalvi., C.S.and Dumbre, R.B. (1995) Efficacy of diflurobenzuron against castor semilooper. *Journal of Maharashtra Agricultural Universities.* **20** (1): 20-23.

Kale, D. M., Chandra Mouli, Murty, G.S.S. and Rao, M.V. P (1997) Development of a new Groundnut Variety 'TG-26' byusing induced mutants in cross breeding. *Mutation Breeding Newsletter, Joint FAO/IAEA Division, International Atomic Energy Agency, Vienna.***43**: 25-27 ISSN 1011-260X

Kalibbala, M.G. (2012) Removal of Natural Organic Matter and Control of Trihalomethanes Formation in Water Treatment. *Water and Sewage Technology Department of Land and Water Resources Engineering Royal Institute of Technology (KTH) SE-100 44 STOCKHOLM, Sweden TRITA LWR PhD Thesis 1064*, ISSN 1650-8602 ISRN KTH/LWR/ PHD 1064-SE ISBN 978-91-7501-323-7

Kalibbala H. M, Wahlberg O. and Hawumba T. J., (2009). The impact of *Moringa oleifera* as a coagulant aid on the removal of trihalomethane (THM) precursors and iron from drinking water. *Water Science & Technology – Water Supply***9**(6): 707 - 714.

Kallamadi P.R. and Sujatha M. (2018) Genetic Diversity in Castor Bean. In: Kole C., Rabinowicz P. (Eds) *The Castor Bean Genome. Compendium of Plant Genomes. Springer.*

Kalyan, R.K. and Ameta, O.P. (2015)Management of major insect pests in soybean. *Ph.D. Thesis.Maharana Pratap University of Agriculture and Technology,* Udaipur, India.

Kamal-Eldin, A. and Appleqvist, L.A. (1994) Variation in fatty acid composition of the different acyl lipidsin seed oils from *Sesamum* species. *Journal of the American Oil Chemists Society,* **71** (2):135-139

Kamaluddin, M. and Ali, M. (1996). Effects of leaf area and auxin on rooting and growth of rooted stem cuttings of neem. *New Forests***12** (1): 11-18.

Kamara, A.Y., Tof, A.,Ademulegun, T and Solomon, R. (2017) Maize-soyabean inter cropping **for sustainable intensification of Cereal-legume cropping systems in Northern Nigeria.** *Experimental Agriculture.*10.1017/S0014479717000564

Kamarkar, A., Kamarkar,S and Mukherjee,S. (2012)Biodiesel production from neem towards feed stock diversification,*Renewable and Sustainable Energy Reviews,* **16**:1050–1060

Kambiranda, D.M., Vasanthaiah, H.K.N., Ananga, R.K.A., Basha, S.M. and Naik, K (2011) Impact of drought stress on peanut (*Arachis hypogaea L.*), In: *Productivity and Food Safety. Plants and Environment*, pp. 249–272, InTech, 2011. View at Google Scholar

Kamel, S.M., Bal, A.H., Mahfouz, H.M. and Said, M. (2013) The most common insect pollinator species on sesame crop (*Sesamum indicum* L.) In: Ismailia Governorate Egypt. *Arthropods*, **2.** (2013).

Kammili, A., Mandalapu, P., Ponukumatla, B., Ruvulapalli, D.P.and Sarada, C.(2020) Introgression of resistance to *Alternaria* leaf spot from wild species into susceptible cultivated safflower. *Plant Breed.* 2020; 139: 368– 374. https://doi.org/10.1111/pbr.12775

Kan, Chia-Cheng; Chung, Tsui-Yun; Juo, Yan-An; Hsieh and Ming-Hsiun (2015). "Glutamine rapidly induces the expression of key transcription factor genes involved in nitrogen and stress responses in rice roots". *BMC Genomics. **16** (1). doi:10.1186/s12864-015-1892-7. ISSN 1471-2164. PMID 26407850.*

Kandaswamy, A. and Raveendaran, N. 1988. Productivity of neem trees in Tamil Nadu, India – an observational study. *Paper presented at the Final Workshop of the IRRI-ADB-EWC Project on Botanical Pest Control in Rice-based Cropping Systems.* IRRI, Los Banos, Philippines. 10 pp.

Kanehira, T. *et al.,* (2003) Anovel and potent biological anti-oxidant, kinobeon A, from cell cultures of safflower. *Life Science*, **74**: 87-97

Kang, C.W. (1994) Sesame breeding and agronomy in Korea. In: R.K. Arora and K.W. Riley(Eds). *Sesame Biodiversity in Asia: Conservation, Evaluation and Improvement* International Plant Genetic Resources Institute (IPGRI) Office of South Asia, New delhi. Pp 65-90

Kang, C.W., Lee, J.I. and Choi. B.H. (1995). Mutation breeding for disease resistance and high yield of sesame in Korea. *Sesame and Safflower Newsletter***10**: 21-36. (South Korea)

Kang,L. Li,P., Wang,A., Ge,X. and Li·Z.(2017)A Novel Cytoplasmic Male Sterility in *Brassica napus* (inap CMS) with Carpelloid Stamens via Protoplast Fusion with Chinese Woad. *Front Plant Sci.* (2017)**8**: 529.

Kannaiyan, S. (2009) *Agrobiodiversity Hotspots*. ISBN: 978-81-7319-916-5, *Narosa Publishing House,* New Delhi. Pp322.

Kansara, S.S. and Sabalpara, A.N. (2016) Assessment of Yield Loss due to Niger (*Guizotia abyssinica* L.F. Cass) leafspot caused by *Alternaria alternata* (FR) Keissl. *The Bioscan International quarterly journal of life sciences***11** (4): 2873 -2875 www.thebioscan.com 2873.

Kant, P. and Wu, S. (2011). The Extraordinary Collapse of *Jatropha* as a Global Biofuel Feed-stock. *Environmental Science & Technology***45**: 7114-7115.

Kapadia, M.N. (1995) Varietal preference of CastorLeafminer, *Liriomyza trifolii* (Burges) and its parasitoids. *International Journal of TropicalAgriculture*. **13** (1-4): 269-271.

Kapadiya, H.J., Gohil, V.N. and Monpara, B.A. (2015).A Black Sesame Variety Gujarat Til 10 (GT 10) Field Resistance to Phytophthora Blight Disease. *Int. J. Curr. Res. Biosci. Plant Biol.* 2015, **2**(12): 53-63

Karikalan, L. and Chandrasekaran, M. (2015) Karanja oil biodiesel: A potential substitution for diesel fuel in diesel engine without alteration, *ARPN Journal of Engineering and Applied Sciences*,**10** (1): 152-161www.arpnjournals.com

Karmee, S.K. and Chadha, A. (2005) Preparation of biodiesel from crude oil of *Pongamia pinnata.Bioresour Technol.* 2005. **96(**13):1425-9. Epub 2005 Feb 17.

Karoshi, V.R. and Hegde, G.V. (2002) Vegetative Propagation of *Pongamia pinnata* (L) Pierre: Hitherto a Neglected Species. *Indian Forester,* **128** (3).

Karunanandaa, B., Qia, Q. *et al.,* (2005) Metabolically engineered oilseed crops with enhanced seed tocopherol. *Metab. Eng.*7: 384 - 400

Karve, A.D., A.K. Deshmukh, and S.M.H. Qadri. (1981) Breeding disease resistant safflower for cultivation in the Deccan peninsula of India.*Proceedings of the 1st International Safflower Conference, Davis, CA, July 12–16, 1981,* pp. 103–107.

Kataria, V.P., Rao, S.K. and Kushwaha, J.S. (1984) Yield components in buch tpe of groundnut. *Mysore Journal of Agricultural Sciences.***18**: 13-16

Katiyar, R.K. (1983)*Current Science***52**: 1067-1068

Katz, F. (1997) The move towards genetically improved oils. *Food Tech.* **51**:66

Katiyar, R.K. (1983). Male sterility in turnip rape. *Current Science.* **52**(22): 1067-1068.

Katiyar, R. K. and Chopra, V. L. (1995) A somaclone of *Brassica juncea* is processed in to a variety and is released for commercial cultivation in India. *Cruciferae Newsletter***17**: 92–9

Kaul, S.L. and Prasad, M.V.R. (1983) Genetic basis of yield heterosis in Castor, In: *Procedings of the Pre-Congress Scientific Meeting on Genetic Improvement of Heterotic Systems,* TNAU Agri. University, Coimbatore, India. **54:** 15

Kaur, V., Kumar, V.S., Yadav, R., Wankhede, D.P., Aravind, J., Radhamani, J., Rana, J.C. and Kumar, A. (2018) Analysis of genetic diversity in Indian and exotic linseed germplasm and identification of trait specific superior accessions. *Journal of Environmental Biology* **39**(5):702-709 · September 2018 :

Kavitha, M., Ramalingam, R.S., Raveendran, T.S. and Punitha, D., (2000) Cytoplasmic-genic male sterile lines in sesame. Effect of environmental factors. *Crop Research,* **19**: 162-164.

Kaya, Y. (2016) Sunflower, Chapter-4, In: Gupta, S.K. (Ed) *Breeding Oilseed Crops for Sustainable Production: Opportunities and Constraints. Academic Press, Elsevier Inc.* London, SanDeigo, Oxford: pp 58-88.

Kaya, Y. (2016) Sunflower In: Gupta, S.K.(Ed) *Breeding Olseed Crops for Sustainable Production- Opportunities and Constraints.*http://dx.doi.org/10.1016/B978-0-12801309-0.00004-5 Copyright © Elsivier Inc. Pp 55-88

Kaya, Y., Evci, G., Pekcan, V. and Yilmaz, M.I. (2014) Broom rape resistance breeding in sunflower: A case study in Turkey. Proceedings of thzenzuchte *Third International Symposium on Broomrape (Orabanche Spp.) in Sunflower*, June 3-6, 2014. Cordoba, New York, pp 194-199.

Kebede, A.; Tegegne, F.; Mekuriaw, Z. and Tegegne, A. (2009). On-farm evaluation of the effect of concentrate and urea treated wheat straw supplementation on milk yield and milk composition of local cows. In: *Zelalem Yilma and Aynalem Haile (*Eds). *Proceedings of the 17th Annual conference of the Ethiopian Society of Animal Production (ESAP) held in Addis Ababa*, Ethiopia, September 24 to 26, 2009. ESAP, Addis Ababa

Keiller, D.R. and Morgan, D.G.(1988). Effect of pod removal and plant growth regulators on the growth, development and carbon assimilate distribution in oilseed rape *(Brassica napusLei L.). J. Agric. Sei., Camb.,* **Ill**: 357-362.

Keijzer, P. (1989) Synchronization of fibre and grain maturation of flax (*Linum usitatissimum* L). In: G. Marshall (Ed). FLAX: Breeding and Utilizaion © 1989 *ECSC, EEC, EAEC, Brussels and Luxemberg*Pp 26-36.

Keijzer, P. and Metz, P. (1992) Breeding of flax for fibre production in Western Europe. In: Sharma, H. and Van Sumere,.C (eds) *The biology and processing of flax.M Publications, Belfast,* NorthernIreland, pp 33–66

Keller, W.A. and Armstrong, K.C. ((1978) High frequency production of microspore derived plants from *Brassica napus* anther cultures. *Z. Pflanzenzucht* **80** :100-108

Kempthorne, O and Nordskog, A. W. (1959). Restricted selection indices. *Biometrics.***15:** 10-19.

Kenaschuk, E.O., (1975). In: Harpiak, J.T. (Ed.), Oilseed and Pulse Crops in Western Canada—A Symposium. *Western Co-operative Fertilizers, Calgary, Alberta*, pp. 203–221.

Kerr, P. (1996) Utilization and quality of identity preserved oilseed co-products. *Presented at the Institute of Food Technologists 1996 Symposium*: Identity Preserved Oils, New Orleans, LA, June 21–22.Google Scholar

Kessavalou, A. and Walters, D.T. (1997) Winter rye as cover crop following soybean under conservation tillage. *Agronomy Journal,* **89:** 68-74

Khalil, M.K., Ahh, A. and Farrag, A.M. (2002) Nitrate accumulation, growth, yield, and chemical composition of rocket (*Eruca sativa*) plant as affected by npk fertilization, kinetin and salicylic acid. *Ann Agric Sci (Cairo)* **47**:1–26

Khan A. R., D. A. Emery and J. A. Singleton, (1974). Refractive index as a basis for assessing fatty acid composition in segregating population derived from intraspecific crosses of cultivated peanuts. *Crop Sci.* **14**: 464–468.

Khan, M.A., Dorcus C. G. and Villordon, A. (2016) Root System Architecture and Abiotic Stress Tolerance: Current Knowledge in Root and Tuber Crops, *Front Plant Sci.* 2016; **7**: 1584. Published online 2016 Nov 1. doi: 10.3389/fpls.2016.01584

Khan, M. A., von Witzke-Ehbrecht, S., Maass, B. L., and Becker, H. C. (2009). Relationships among different geographical groups, agro-morphology, fatty acid composition and RAPD marker diversity in safflower (*Carthamus tinctorius* L.). *Genet. Resour. Crop Evol.* **56**: 19–30.

Khan, M.M., Manaf, A., Hassan, F. and Ahmed, M.S. (2017) Micronutrients (Zn, B and Mn) Effects on Sesame (*Sesamum indicum* L.)*Int. J. Agron. Agri. Res.* **10**(5) : 92-101, May 2017. https://innspub.net/ijaar/micronutrients-zn-b-mn-effects-sesame-sesamum-indicum-l/

Khan, M.Q. (1946) Life history and bionomics of castor semilooper in Hyderabad (Deccan). *Indian Journal of Entomology***8**: 111-115.

Khan, M.Q. and Rao, A.S. (1948) *Annual Report of the Scheme for Research on Pests and Diseases of Castor and other Oilseeds in Hyderabad State* (1943-1945) Department of Agriculture. H.E.H. Nizam of Hyderabad Government, Hyderabad Division. HEH Nizam Government Press. Pp87.

Khatri, A., Khan, I.A, Siddiqui, M.A., Raza, S. and Nizamani, G.S. (2005) Evaluation of high yielding mutants of *Brassica juncea*Cv. S-9 developed through gamma rays and EMS. *Pak. J. Bot.***37**: 279-284.

Khidir, O.M (1969) Evolution of the Genetic Systems of Safflower *(Carthamus,* L.) *Gentica.***40**:84-88

Khidir, O.M. and P.F. Knowles. (1970). Cytogenetic studies of *Carthamus* species (Compositae) with 32 pairs of chromosomes. II. Intersectional hybridization. *Can. J. Genet. Cytol.* **12**:90-99.

Khosla, P.K. and Styles, B.T. (1975) Karyological studies and chromosomal evolution in Meliaceae. *Silvae Genetica,* **24**(2-3) : 73-82

Khatri, A., Khan, I.A., Siddiqui, M.A., Raza, S. and Nizamani, G.S. (2005) Evaluation of high yielding mutants of *Brassica juncea* Cv. S-9 developed through gamma rays and EMS. *Pak.J. Bot.* **37**:279-284

Kim D. H., Zur G., Danin-Poleg Y., Lee S. W., Shim K. B., Kang C. W.*et al.,* (2002). Genetic relationships of sesame germplasm collection as revealed by inter-simple sequence repeats. *Plant Breed.* 121 259–262. 10.1046/j.1439-0523.2002. 00700.x

Kim, M.K., Park, S.K. and Chae, Y.A. (1987). Selection *in vitro* for herbicide tolerant cell lines of *Sesamum indicum* L: Effects of explants and hormone combinations on callus induction. *Korean Journal of Plant Breeding* **19:** *70-94.*

Kim, K.S., Park, S.H. and Jenks, M.A. (2007) Changes in the leaf cuticular waxes of sesame (*Sesamum indicum,* L.)plants exposed to water deficit. *Journal of Plant Physiology.* 164: 1134-1143

Kim Y. H.(2001), Effects of BA, NAA, 2-4, D and AgNO3 treatments on the callus induction and shoot regeneration from hypocotyls and cotyledon of sesame *(Sesamum indicum L.). The Korean Soc. Hort. Sci.,***42**(1): 70-74.

Kinman, M. L. (1970) Key to hybrid sunflower found. *Crops and Soils,* **23**: 21

Kiniry, J.R., R. Blanchet, R., Williams, J.R., Texier, V.C., Jones, A. and Cabelguenne, M. (1992) Sunflower simulation using the EPIC and ALMANAC models, *Field Crops Research,* **30** (3–4) :403-423

Kinney, A.J. 1996. Soybean biotechnology: improving soybean seed quality by genetic engineering. *Paper No. 21-B, presented at 87th American Oil Chemists' Society Annual Meeting and Expo, Indianapolis, IN,* April 28-May 1.

Kinney, A.J. and Clemente, T.E. (2005) Modifying soybean oil for enhanced performance in biodiesel blends. *Fuel. Process Technol.***86**: 1137-1147

Kinney, A.J., Knowlton, S. (1997) Designer oils for food applications. In: Harlander, S. and Roller, S. (Eds). Genetic Engineering for food Industry: *A strategy for food quality improvement. Blakie Academic, London.* pp.193-213

Kiran, B.U., Mukta, N., Kadirvel,P. and Alivelu, K (2017) Genetic diversity of safflower (*Carthamus tinctorius* L.) germplasm as revealed by SSR markers, *Plant Genetic Resources,* **15** (1): 1-11

Kirk, J.T.O and Oram, R.N. (1981)Isolation of erucic acid-free lines of *Brassica juncea:* Indian mustard now a potential oilseed crop in Australia [selections] [1981]*J. Australian Inst. Agric. Sci.***47**: 51-52

Kirti, P. B. and Chopra, V. L. (1989). Rapid plant regeneration fromorganogenesis and embryogenesis from protoplasts of *Brassica juncea. Plant Cell, Tissue and Organ Culture* **20** (1): 65–7

Kirti, P. B., Mohapatra, T., Prakash, S. and Chopra, V. L. (1995a). Development of a stable cytoplasmic male sterile line of *Brassica juncea* from the somatic hybrid *Trachystoma ballii + B. juncea.Plant Breeding***113**: 21–24

Kirti P B., Mohapatra T, Khanna H, Prakash S, and Chopra V L. (1995b*). Diplotaxis catholica*plus *Brassica juncea* somatichybrids: molecular and cytogenetic characterization. *Plant Cell Reporter,***14**: 593–7

Kirti, P. B., Prakash, S. and Chopra. V. L. (1991). Interspecific hybridization between *B. juncea* and *B. spinescenes* through protoplast fusion. *Plant Cell Report***9**: 639–42

Kirti, P.B., Prakash, S., Gaikwad, K., Dinesh Kumar, V., Bhat, S.R. and Chopra, V.L. (1998). Chloroplast substitution overcomes leaf chlorosis in a *Moricandia arvensis-* based cytoplasmic male sterile *B. juncea. Theoretical and Applied Genetics* **97:** 1179–82

Kisha, T. J. and Johnson, R. C. (2012) "Chapter 6: Safflower" (2012). *Publications from USDA-ARS / UNL Faculty. 1307.* https://digitalcommons.unl.edu/usdaarsfacpub/1307

Kishor P.B.K., Reddy T.P., Sarvesh A. and Venkatesham G. (1997) Haploidy in niger (*Guizotia abyssinica* Cass). In: Jain S.M., Sopory S.K., Veilleux R.E. (Eds) In Vitro Haploid Production in Higher Plants. *Current Plant Science and Biotechnology in Agriculture,* **29**. *Springer, Dordrecht*

Kitamura, K., Cavies, C.S., Kaizuma, N. and Nielsen, N.C. (1983) Genetic analysis of a null-allele for lipoxygenase-3 in soybean seeds. *Crop Sci.* **23**, 924–927.Google Scholar

Kitashiba,H. and Nasrallah,J.B. (2014)Self-incompatibility in *Brassicaceae* crops: lessons for interspecific incompatibility.*Breed Sci.* 2014 May; **64**(1): 23–37.

Klepper, B., Belford, R.K. and Rickman.R.W.(1984). Root and Shoot Development in Winter Wheat. *Agronomy Journal***76**: 117–122.

Kleiman, R. (1990) Chemistry of new industrial oilseed crops, in *Advances in New CrCrop Physiology and Ecology*J. Janick and J.E. Simon (Eds). Timber Press, Portland, OR, pp. 196-203

Klisiewicz, J.M. (1975) Race 4 of *Fusarium oxysporum* f.sp. *carhmai*. Plant Disease Reporter 59: 712

Klisiewicz, J.M. and Thomas, C.A. (1970) Pathogen races of *Fusarium oxysporum* f.sp. *carhmai*. *Phytopathol***60**:83

Klisiewicz, J.M. and Urie. A.L. (1982). Registration of Fusrium resistant safflower germplasm, *Crop Science* **22**: 165.

Knauft, D.A., Norden, A.J., Gorbet, D.W. (1987) In:Fehr (Ed.)*Principles of Cultivar Development-Crop Species, Macmillan Publ.Comp.*, NewYork. 1987, pp346-384

Knoll, J.E., Ramos, M. L., Zeng, Y., Holbrook, C.C. *et al.*, (2011). TILLING for allergen reduction and improvementof quality traits in peanut (*Arachis hypogaea*L.)BMC *Plant Biology 2011*,**11.**

Knowles, P.F. (1968). Registration of 'UC-1' safflower. Crop Sci. 8: 641. Knowles, P.F. 1969. Centers of plant diversity and conservation of crop germplasm: safflower. Econ. Bot. 23: 324–329.

Knowels, P. F., (1969) Modification of quantity and quality of safflower oil through plant breeding. *J. Am. OIL Chem. Society*. **46**: 139-142

Knowles, P. F. (1983) Genetics and Breeding of Oilseed Crops. *Economic Botany,* **37** (4): 423-433

Knowles, P.F. (1988). *Carthamus* Species Relationships. *Lecture presented at Beijing Botanical Garden, Institute of Botany,Chinese Academy of Sciences,* Beijing.

Knowles, P.F. (1989). Safflower. In: Downey, R.K., G. Robellen, and A. Ashri, (Eds).*Oil Crops of the World*. McGraw-Hill, New York, pp. 363–374.

Knowles, P.F. (1991) Global Perspective of Safflower. In: Ranga Rao, V. and Ramachandram, M. (Eds) ***Proceedings of the Second International Conference on Safflower, Hyderabad,*** *India, 9-13, January 1989.Indian Society of Oilseeds Research, ICAR-Directorate of Oilseeds Research,* Hyderabad. Pp 13-16.

Knowles, P.F. and Schank.S.C. (1964). Artificial hybrids of *Carthamus nitidus* Boiss. and *C. tinctorius* L. (Compositae). *Crop Sci.***4:** 596–599

Kobayashi, T. (1958) Radiation genetics of sesame. III. Morphological changes and mutants induced by ionizing radiations. *Japan J. Genet*. **33** (1958): 239-261

Kobayashi, T (1965). Radiation-induced beneficial mutants, p. 399-403. In: *The use of induced mutations in plant breeding. Pergamon Press, Oxford (1965).*

Kobayashi, T. (1991) Cytogenetics of Sesame (*Sesamum indicum*) p 581-582 In: T. Tsuchya and P.K. Gupta. (Eds). *Chromosome Engineering in Plants: Genetics, Breeding and Evolution.* B. Elsevier. Amsterdam. (https://doi.org/10.1016/B978-0-444-88260-8.50036-7)

Kolkman, J.M. *et al*., (2007) Single nucleotide polymorphisms and linkage disequilibrium in **sunflower.** *Genetics,***177**: 457–468

Kolte, S.J. (1985) Diseases of Annual Edible Oilseed Crops, Vol. III: Sunflower, safflower, and niger seed diseases. *CRC Press, Boca Raton, FL.* 118

Koprna, R., Kucera, V.,Kolovrati, O., Vyvadilova, Klima, M (2005).Development of Self-incompatible Lines with Improved Seed Quality in Winter Oilseed Rape (*Brassica napus* L.) for Hybrid Breeding. *Czech J. Genet. Plant Breed.,* **41** (3): 105–111

Korayem, A.M., Mohamed, M. M. M. and El- Ashry, S.M. (2016) Yield and oil quality of **sunflower infected with the root-knot nematode,** *Meloidogyne arenaria. International Journal of Chem. Tech. Research CODEN (USA):* IJCRGG ISSN: 0974-4290 **9**(3): 207-214, 2016

Korshunova, Y. O., Eide, D., Clark, W. G., Guerinot, M. L., and Pakrasi, H. B. (1999). The IRT1 protein from *Arabidopsis thaliana* is a metal transporter with a broad substrate range. *Plant Molecular Biology,* **40**(1): 37–44. https://doi.org/10.1023/A:1026438615520

Kottle SG (1985) Diseases of edible oil seed crops. Vol-II. *Rapeseed- Mustard and sesame diseases*. P. 83-122 *CRC press, Inc.* Boca Raton, Florida. 1985

Kolte SG (1985A) Diseases of Annual Edible Oilseed Crops, Vol. III: ***Sunflower, safflower, and*** *niger seed diseases.CRC Press Inc*. Boca Raton, FL. 118

Koutroubas, S. D. and Papakosta, D. K.(2010) Seed filling patterns of safflower: Genotypic and seasonal variations and association with other agronomic traits. *Industrial Crops Products***31**: 71–76.

Kozik, A., Chan, B. and Michelmore, R. (2003) Lettuce, Sunflower EST CGPBD Project: DNA analysis assembly visualisation and validation. *Plant Animal Genome Conference.***9** :886

Kozin, R. B. (1954) Effect of bees on the seed crop of flax. Pchelowdstvo. Mosk. 6: 41-43

Kramer, J.K.G., Sauer, F.D. and Pigden, W.J. (1983) High and low Erucic Acid Rapeseed Oils. *Academic Press, New York.* 1983

Krapovickas, A. (1968*)*Origen, varibilidad y diffusion de mani (*Arachis hypogaea* L) (In) *Actas Memorias XXXVII Congreso Internacional Americanas II* pp517-53

Krapovickas, A. (1973). Evolution of the genus *Arachis*. Pp. 135-151 In: R. Moav, (ed.). *Agriculture Genetics: Selected Topics NRCO*, Jerusalem, Israel.

Krapovickas, A. and V. A. Rigoni. (1960). La nomenclatura de las subespecies y variedades de Arachis hypogaea L. *Revista de Investigaciones Agrícolas* **14**: 197– 228.

Kraus, W. (1995) "Biologically active ingredients-azadirachtin and other triterpenoids", In: H. Schutterer (Ed.), *The Neem Tree Azadirachta indica A. Juss and Other Meliaceous Plants*, Weinheim, New York, 1995, p 35-88

Krause, R., Mankowska, G., Marcin, L. and S. Jan (2003) Regeneration of flax (*Linum usitatissimum* L.) plants from anther culture and somatic tissue with increased resistance to *Fusarium oxysporum. Plant cell reports* **22.** DO10.1007/s00299-003-0662-1

Krishna, K. R., (2013). Agroecosystems: Soils, Climate, Crops, Nutrients Dynamics, and Productivity. *Apple Academic Press, CRC Press, Taylor & Francis Group*https://books.google.fr/books?id=P6FiAgAAQBAJ&pg=PA209#v=onepage&q&f=false

Krishnan, H.B. (2005). Engineering soybean for enhanced sulphur amino acid content. *Crop Sci.***45:** 454-461

Krishnaswamy, S., Appadurai, R. and Sree Rangaswamy, S.R. (1985) Report on Oil crops, **Sesame and Safflower,** *Proceedings of the Second Oilseeds Network Workshop held in Hyderabad* Feb.5-9, 1985. Pp 78-89

Kučera,V., Chytilová, V., Vyvadilová, M.and Klíma, M. (2006)Hybrid breeding of cauliflower using self-incompatibility and cytoplasmic male sterility, *Hort. Sci.* (Prague), **33** (4): 148–152

Kulkarni, L.G., Ankineedu, G.(1966). A new sex phenotype in castor. *Indian. J. Agric. Sci.* **36**: 255–257.

Kulkarni, L.G. and Ramanamurthy, G.V. (1977*) Castor.* Monograph. Pub. *Indian Council of Agricultural Research* (ICAR), New Delhi. Pp105

Kulkarni, V. V. (2006)Studies on Interspecific Hybridization with Particular Reference to Development of Male Sterility in Sesame (*Sesamum indicum* L.). Ph.D. Thesis. University of Agricultural Sciences, Dharwad -580 005, India.

Kulkarni, V.V., Shankergoud, I. and Govindappa, M.R. (2015) Identification of sunflower powdery mildew resistant sources under artificial screening. *SABRAO Journal of Breeding and Genetics* **47** (4): 502-509, 2015

Kumar, A. (1992) Production technology for yield enhancement of Indian mustard under irrigated conditions, In: Kumar, D. and Rai, M. (Eds). *Adv.Oilseeds Res. Vol.1* **Scientific** Publishers, Jodhpur India. Pp 72-

Kumar, D. (1998) *Managing Saline and Alakali Situations for Optimizing Oilseeds Production.* ICAR-Central Arid Zone Research Institute, Jodhpur.

Kumar, D and Rai, M.(Eds) (1992) Advances in Oilseeds Research Vol.1, Rapeseed and Mustard. ISBN: 81-7233-046-4, Scientific Publishers, Jodhpur, India. Pp399.

Kumar, D., Singh, B. and Sharma, Y.C. (2017) Bioenergy and Phytoremediation Potential of *Millettia pinnata,* Chapter 6 In: *Phytoremediation of Bioenergy plants* Baudh, K., Singh, B. and Korstad (Ed.), *Springer* pp169-188

Kumar, D. and Yadav, I.S. (1986a). Genetics of yield and its components in taramira. *Indian J. Agric. Sci.,* **56**(3): 167-70.

Kumar, D. and Yadav, I.S. (1986b*).* Combining ability in traramira under rainfed conditions. *Indian J. Agric. Sci.,***56**(4): 229-33.

Kumar, D and Yadav, I.S. (1992) Taramia (*Eruca sativa* Mill.) Research in India-Present Status and Future Programme of Yield Enhancement. In: Kumar, D. and Rai. M. (Eds).*Advances in Oilseeds Research,* Vlo.1 1992Scientific Publishers, Jodhpur, India. Pp 330-358

Kumar, D., Yadav, I.S. and Gupta, B. S. (1988) Combining ability analysis for quantitative traits of rocket –salad *(Eruca vesicaria* subsp. *sativa) Indian J. Agric. Sci.,* **58**(1): 11-15.

Kumar, H. (1996) Influence of chromosome doubling on seed filling in safflower (*Carthamus tinctorius* L.). *Sesame and Safflower Newsletter.***11**: 95-104

Kumar, H., Pillai, R.S.N.and Singh, R.B. (1981). Cytogenetic studies in safflower. *Proceedings of the 1st International Safflower Conference, Davis, CA,* July 12–16, 1981,pp. 126–136.

Kumar, H.D. (1998A) Modern Concepts of Biotechnology, Vikas Publishing House Pvt. Ltd. New Delhi.

Kumar, K.P.N. (2007) Rajasthan farmers hope to strike it rich with jojoba plantation. *live **mint***03 Sep 2007, 12:29 AM IST

Kumar, P.K.V., Balakrishanan, M.M. and Govindarajan, T.S. (1984) *Retitthrips syriacus* (Mayet)- a new record on coffee from South India. *Journal of Coffee Research* **14** (3): 131-132.

Kumar, P. and Mishra, U.S. (1992) Diseases of *Sesamum indicum* in Rohilkhand: intensity and yield loss.*Indian Phytopathology* **45** (1):121-122

Kumar, P., Yadav, T.P., Lekh Raj, Gupta, S.K., Thakral, N.K., Kumar, P. and Raj, L. (1997). Combining ability and heterosis for oil content in toria (*B. campestris*). *Cruciferae Newsletter*. **19**: 87-88

Kumar, P.R. (1994) Genetic improvement in rapeseed –mustard In: Prasad, M.V.R. *et al.,* (Ed) *Sustainability in Oilseeds Research, Indian Society of Oilseeds Research,* Directorate of Oilseeds Research, Hyderabad. Pp39-47

Kumar, P.R.,Arora, R.K., Singh, N.P., Yadav, R.C. and Kumar, P.(1990). A study of heterosis in Indian mustard. *Acta Agronomica Hungarica.***39**(1-2): 137-143.

Kumar- Ravi (2018) "Genetically Modified Mustard: Dhara Mustard Hybrid-11". *www. biotecharticles.com*. Archived from the original on 2018-10-2019. Retrieved 2018-10-05 (cited Wikipedia)

Kumar S. (2002) Identification of trap crop for reducing broomrape infestation in the succeeding mustard. *Agronomy Digest*. **2**:99–101.

Kumar, S., Kerkhi, S.A., Gangwar, L.K., Chand, P. and Kumar, M. (2012): Improvement in the genetic architecture through study of variability, heritability and genetic advance in linseed crop (*Linum usitatissimum* L.). *Int. J. Res. Eng. IT Soc. Sci.***2:** 58-65 (2012)

Kumar, S., Prasad, Y., Ram, S. and Chakraborty, M. (2019) Role of genetic variability for seed yield and its attributes in linseed (*Linum usitatissimum* L.) improvement. *Journal of Pharmacognosy and Phytochemistry***2**: 266-268

Kumar, V., Chandrasekhar, K. and Siddhu, O.P. (2006) Efficacy of karangin and different extracts of *Pongamia pinnata*against selected insect pests. *J. Ent. Res.* **30** (2): 103-108.

Kumar, V.D. and Ashfaq, A. (2010). Rcin free Castor: Can it be a Reality? In: Hegde, D.M. (Ed) 2010. *Research and Development in Castor: Present Status and Future Strategies*. Indian Society of Oilseeds Research, Hyderabad. Pp 69-90

Kumar, V.S. and Navaratnam, V. (2013) Neem (*Azadirachta indica*): Prehistory to contemporary medicinal uses to humankind *Asian Pac J Trop Biomed.* 2013 Jul; **3**(7): 505–514. doi: 10.1016/S2221-1691(13)60105-7

Kumaran, K. (2015) Mahua (*Bassia latifolia*. Roxb.) *Paper presented at the National Workshop on TBOs at CARI, Jhansi* - 15th -16th October 2015

Kumaran, K., Surendran, C and Rai, R.S.V. (1993). Variation studies and heritable components of seed parameters in neem (*Azadirachta indica* A. Juss.). In: *Proceedings of the World Neem Conference 1993*, Bangalore, India. pp. 167-173.

Kumari A. (2008). Floral biology, pollen germination, pollen viability and pollination ecology of safflower (*Carthamus tinctorius* Linn.). *Research and Reviews in Biosciences*, **2**, 2-6).

Kumari, V., Singh, A., Kumar, A., Prasad, R., Jambulkar, S. and Sanju, S. (2018) Identification of *Phytopthera* resistant mutants through induced mutagenesis in sesame (*Sesamum indicum*, L.). *Indian Phytopathology*, **72.**

Kundu, S.K. (1999). Comparative analysis of seed morphometric and allozyme data among four populations of neem (*Azadirachta indica*). *Genetic Resources and Crop Evolution*:**46**: 569–577. https://doi.org/10.1023/A:1008761503500

Kundu.G.G.,Anand, R.K., Sharma, V.K. and Rai, S. (1966) New host records of some *Hymenopterous* parasites. *Indian Journal of Entomology* **28** :560-581

Kurt, C. (2018) Variation in oil content and fatty acid composition of sesame accessions from different origins.*GRASAS Y ACEITES***69** (1):January–March 2018, e241

Kuravadi N.A., Gowda M. (2019) Phylogeny of Neem and Related Species in the Meliaceae Family. In: Gowda M., Sheetal A., Kole C. (eds) *The Neem Genome. Compendium of Plant Genomes. Springer, Cham.* DOI https://doi.org/10.1007/978-3-030-16122-4_5

Kurita, M.(1946). Secondary association of chromosomes in castor oil plant. *Japn. J. Genet.* **21:** 63

Kurt, O. and Bozkurt, D. (2006). Effectof temperature and photoperiod on seedling emergence of flax (*Linum usitatissimum* L.). *Journal of Agronomy*, **5**: 541–54

Kushwaha, K.S. and Bhardwaj, S.C. (1967) Biology and external morphology of forage pests. *Indian Journal of Agricultural Sciences*. **39**: 93-107

Kushi, K. K. and M. N. Khare (1979). Seed borne fungi of sesame and their significance. *Seed research***17**(1): 48-53

Kutuzova, S. N. (1998). "Genetika lna," [Genetics of flax], In:Dragavcev, V.A. and Fadeeva, T.S. (**Eds**.), Genetika kulturnych rastenij (len, kartofel, morkov, zelennye kultury, gladiolus, jablona, ljucerna), *[Genetics of cultivated plants (flax, potato, carrot, leafy vegetables, gladiolas, apple, alfalfa)]*, VIR, St. Petersburg, Pp. 6–52

Kutuzova, S. N. (2000). Katalog mirovoj kolekcii VIR, Vypusk 714, Donory chozjajstvenno cennych priznakov dja selekcii lna-dolgunca. [Catalogue of the world collection at the VIR, *Volume 714, Donors of economically important characters for breeding of fibre flax*]. VIR, *St. Petersburg*, Pp. 50

Labana, K.S. and. Banga, S.S. (1984). Floral biology in Indian mustard. *Genetic Agreria.*: **38:** 131-138.

Labana, K.S., Jindal, S.K. and. Mehan, D.K.(1978). Heterosis and combining ability in yellow sarson. *Crop Improvement*. **5**: 50-55.

Labana, K. S., Singh, M., Sangha, A.S. and Jaswal, V.S. (1980) Variability and inter-relations among characters in F2 progeny of groundnut. *J. of Res. Punjab Agrl. University* **17**:107-114.

Lacombe S., Leger S., Kaan F., Bervillé A. and Sas M. (2002). Genetic and molecular expression features of the *Pervenets* mutant leading to high oleic acid content of seed oil in sunflower. *OCL* **9**: 17–23. 10.1051/ocl.2002.0017

Lacombe, S., Souyris, I. and Bervillé, A. J. (2009) An insertion of oleate desaturase homologous sequence silences via siRNA the functional gene leading to high oleic acid content in **sunflower seed oil.** *Mol Genet Genomics*. 2009 Jan; **281**(1):43-54.

Lai, K., Lorence, M. and Edwards, D. (2012) Genomic database for crop improvement. *Agronomy,* **2** :62-73

Lakkad, L.V., Asgdaria, K.B., Mathukia, R.K.,Sagarka, B.K. and Khanpara, V.D. (2005) **Efficient use of water and fertilizers through drip system for maximizing castor production.** *Sustainable Management of Water Resources*. Pp 146-147

Lakra, B.S. and Saharan, G.S. (1988) Morphological and physiological variations in *Albugo candida. Indian J. Mycol. Pl. Pathol.* **18**: 149-156

Lakshmamma, P. and Lakshmi, P. (2010) Variability for water use efficiency traits and drought tolerance in castor (*Ricinus communis*L) germplasm lines. *Journal of Oilseeds Research,* **27** (1): 81-84

Lakshmamma, P., Lakshmi, P. and Anjani, K. (2004) Screening of castor germplasm accessions for drought tolerance. *Indian Journal of Plant hysiology,***9**(1):104-106

Lakshminarayana, M. (1992). Competitive ability of *Trichogramma chilonis* and *Telenomous proditor* in prasatizing the eggs of castor semilooper (*Achoea Janata). Journal of Oilseeds Research,* **9** (2): 351-352

Lakshminarayana, M. (2005) Effect of mechanical control of defoliators in the pest management of castor. *Journal of Oilseeds Research***22** (2):331-334

Lakshminarayana, M.(2010) Eco-friendly management of Insect pests of Castor. In: Hegde, D.M. (Ed) 2010. *Research and Development in Castor: Present Status and Future Strategies*. Indian Society of Oilseeds Research, Hyderabad. Pp36-55

Lakshminarayana M. and Anjani K. (2010). Sources of resistance in castor against capsule borer, *Conogathes* (*Dichocrocis*) *punctifaralis* Guen. *J Oilseeds Research,***27** (SpecialIssue): 273-274

Lakshminarayana, M. Basappa, H. and Vijay Singh (1992) Report on the incidence of hitherto unknown leaf-miner *Liriomyza trifolii* Burges on castor. *Journal of Oilseeds Research,* **9** (1): 175-176.

Lakshminarayana, M. and Duraimurugan, P. (2014) Assessment of avoidable yield losses due to insect pests in castor (*Ricinus communis* L.) *J. Oilseeds Res.,***31**(2) : 140-144,

Lakshminaryana, M., Kalpana Sastry, R. and Prasad, M.V.R. (1999) Breeding for pest resistance in oilseed crops. In: Mukherji, K.G. and Rajak, R.L.(Eds.) *OilSeeds -:IPM Systems in Agriculture*, **5:** 114 Pub. Vedam Books Ltd. (India)

Lakshminarayana, M. and Raoof, M.A. (2005) Insect pests and diseases of castor and their management. *Directorate of Oilseeds Research,* Hyderabad. p 7

Lakshminarayana, M., Tarakeswari, M. and Sujatha, M. (2009) Castor trangenics harbouring, *Cry 1Ec, Cry 1Aa* and their reaction to major defoliators. *Journal of Oilseeds Research,* **26** (Special Issue): 177-178

Lakshmamma, P., Lakshmi, P. and Anjani (2004) Screening of castor germplasm accessions for drought tolerance. *Indian J. Plant Physiology.***9**(4): 447-450

Lakshmamma, P. and Lakshmi, P. (2010) Variability for water use efficiency traits and drought tolerance in castor (*Ricinus communis* L.) germplasm lines. *Journal of Oilseeds Research,***27** (1): 81-84

Lakshmi Prayaga and Lakshmamma, P (2009) New technique for root studies. DOR Newsletter, **14** (4): 1-2.

Lal, R. and Nayak, G.N. (1963) Effect of plants on the development of caterpillars of *Prodenia litura* Fab and their susceptibility to different insecticides. *Indian Journal of Entomology***25:** 299-306

Lalas, S. and Tasaknis, J. (2002). Characterization of *Moringa oleifera* seed oil variety "Periyakulum 1". *J Food Composition and Analysis*. **15**: 65-67. DOI: 1006/jfca.2001.1042.

Lale, A. and Kulkarni, D.K. (2010) A mosquito repellent karanja kunapa from *Pongamia pinnata.* Asian Agri.History. 14(2):207-211.

Lama, A.D., Klemola, T., Tyystjärvi, E., Niemelä, P., Vuorisalo, T. (2019) Physiological and compensatory growth responses of *Jatropha curcus* (L.) seedlings to simulated herbivory and drought stress, *South African Journal of Botany*, **121** : 486-493

Langham, D.G. (1946) Genetics of sesame.III. *Journal of Heredity*, **37** (5):149-152

Langham, D.R. (2007) Phenology of Sesame. *Issues in new crops and new uses. 2007.*In: J. Janick and A. Whipkey (eds.). ASHS Press, Alexandria, VA. Pp.144-182.

Lassak, E.V. and McCarthy, T.(1983). *Australian Medicinal Plants*. Sydney, Australia: Methuen

Laurentin H.E. and Karlovsky P (2006) Genetic relationships in sesame (*Sesamum indicum* L.) germplasm collection using amplified fragment length polymorphism (AFLP). *BMC Genet. 2006 Feb 16;* **7**:10.

Laurentin, Herna and Karlovsky, P. (2007) AFLP fingerprinting of sesame (*Sesamum indicum* L.) cultivars: identification, genetic relationship and comparison of AFLP informativeness parameters, *Genet Resour Crop Evol* (2007) **54**:1437–1446 DOI 10.1007/s10722-006-9128-y

Lauridsen, E.B., Kanchanabufagura, C. and Boonsermusuk, S. (1991). Neem (*Azadirachta indica* A. Juss.) in Thailand. *Forest Genetic Resources Information***19**: 25-33.

Lavanya, C. and Solanki, S.S (2010) Crop Improvement of Castor: Challenges ahead. In: Hegde, D.M. (Ed) 2010. *Research and Development in Castor: Present Status and Future Strategies*. Indian Society of Oilseeds Research, Hyderabad. Pp36-55

Lavanya, C. and Varaprasad, K.S. (2013) Castor Hybrids in India: A Success story. National Association of Seed Industry (NASI) May 7, 2013, *Seed Times*, **5** (4): 111-117

Lavanya C., Vishnuvardhan Reddy A., Dutta B. and Bandopadhyay R. (2018) Classical Genetics, Cytogenetics, and Traditional Breeding in Castor Bean. In: Kole C., Rabinowicz P. (eds) *The Castor Bean Genome. Compendium of Plant Genomes. Springer,* Cham DOIhttps://doi.org/10.1007/978-3-319-97280-0_3

Lawrence, R.C.N. (1974) Population and spaing studies in Malawian groundnut cultivars. *Experimental Agriculture* **10:** 177-184.

Lay, C. L. and Dybing, C. D. (1989). Linseed. In: G. Röbbelen, R. K. Downey and A. Ashri (Eds).*Oil Crops of the World* McGraw-Hill, New York.pp. 416 - 430

Layrisse, A., Wynne, J.C. and Isleib, T.G.(1980). Combining ability for yield, protein and oil of peanut lines from South American Centers of diversity. *Euphyhca* **29:**561-570.

Lazzeri, L., Errani, M., Leoni, O. and Venturi,G. (2004) Eruca *sativa*spp. *Oleifera*: A new non-food crop. *Ind Crop Pro.* **20**:67–73

Lecrlerc, P. (1969) Une sterelite male cytoplasmque chez le trounesol. *Annuls. Amelior. Plantes.***19:** 99-106.

Lee, J.I., Park, Y.H., Park, Y.S. and Kim, B.G. (1985). Propagation of sesame (*Sesamum indicum L.*) through shoot tip culture. *Korean Journal of Plant Breeding* **17**: 367-372

Lessire, R. (2005) Genomics applied to oilseed crops, any benefit? *European J. Lipid Sci. Tech.***107:** 211-212

Leterme, P. (1985). Modélisation de la croissance et de la production des siliques chez lecolza d'hiver. *Thèse de Doctorat, Institut National Agronomique Paris-Grignon, 1*12 pp.

Levingston and Zamora (2006) (cited by IPGRI), in "Assessment of the potential of *Jatropha curcas*, (biodiesel tree) for energy production and other uses in developing countries." Mike Benge (bengemike@aol.com), Senior Agroforestry Officer, USAID (Ret.), July 2006 and updated August 2006.

Levy, A. and Ashri, A, (1978) Induced plasmon mutations affecting the growth habit of peanuts *A. hypogaea* L.*Mutation Research/Fundamental and Molecular Mechanisms of Mutagenesis***51** (3): 374-360. https://doi.org/10.1016/0027-5107(78)90123-9

Lewis L, Onsongo M, Njapau H, Schurz-Rogers H, Luber G et al. (2005) Aflatoxin contamination of commercial maize products during an outbreak of acute aflatoxicosis in eastern and central Kenya. *Environmental Health Perspectives* **113:** 1763–1767.

Li C, Miao H, Wei L, Zhang T, Han X, Zhang H (2014). Association mapping of seed oil and protein content in Sesamum indicum L. using SSR markers, *PLoS One*. 2014; **9**(8): e105757.

Li, C.Q. (1995) A comparative study of structure of soyabeans. *Acta Agricultural Boreali-Sininca.***10** (Supplemet). 64-9.

Li, G.B., Hu Yingand Cui, Zhuan (1996) The study of the characteristics of K uptake by sesame (*Sesamum indicum* L.) plants and application techniques of K fertilizers. *Oil Crops of China* **18** (4): 49-51

Li, L.Y. *et al.,* (1995). Ontogeny of *Carthamus tinctorius. Acta Agronomica Sinica* **21** (6) :740-745.

Li, Q, Y., Wu, Z. and Stefansson, B. R. (1988) Genetic Male Sterility in Rape (*Brassica napus* L.) Conditioned by Interaction of Genes at two loci.*Canadian Journal of Plant Science*, 1988,68(4):11151118.https://doi.org/10.4141/cjps88-132

Li, Y. (1997) The induced mutations of sesame male sterility and heterosis breeding, p. 18-23 In: *Proc. 2 nd FAO/IAEA, Res. Coord. Mtg, Induced mutations for sesame improvement. IAEA,* Vienna (1997).

Li, Y., Li, Z.,Cai, Q., Yang, G., He, Q., Liu,P. (2011) Identification of a SSR marker linked to the fertility-restoring gene for *polima* cytoplasmic male sterile line in *Brassica napus. African Journal of Biotechnology***10** (47) DOI: 10.5897/AJB11.1009

Li, Xiu-Qing, Jean, M, Landry, B.S. and Brown, G (1998) Restorer genes for different forms of *Brassica* cytoplasmic male sterility map to a single nuclear locus that modifies transcripts of several mitochondrial genes. PNAS August 18, 1998. **95** (17): 10032-10037; https://doi.org/10.1073/pnas.95.17.10032

Lichter, R. (1982) Induction of haploid plants from isolated pollen of *Brassica napus. Z. Pflanzenphysiol.***103**: 229-237

Lim, E.S. and Gumpil, J.S. (1984) The flowering, pollination and hybridization of groundnuts (*Arachis hypogaea* L.) *Pertanika*7(2): 61-66 (1984)

Lima, R.L.S.,Ligia, L.S., Sampaio,R.,Sofiatti, V.,A.Gomes, J.A. andBeltrão, N,E.M.(2011) Blends of castor meal and castor husks for optimized use as organic fertilizer. *Industrial Crops and Products,***33** (2). Pp364-368

Lin, Edo. (2005) Production and Processing of Small Seeds for Birds. *Agricultural and Food Engineering Technical Report 1. Food and Agriculture Organization of the United Nations, Rome*.http://www.fao.org/docrep/008/y5831e/y5831e00.htm#C ontents.

Ling H.Q. and Binding H. (1997): Transformation in protoplast cultures of *Linum usitatissimum* and *L. suffruticosum* mediated with PEG and with *Agrobacterium tumefaciens*. *Journal of Plant Physiology*, **151**: 479–48

Linnaeus, C.V. (1753). Species Plantarum. *Laurentii Salviae*. Holmiae.

Liu, A.Z. and Burke, J.M. (2006). Patterns of nucleotide diversity in wild and cultivated sunflower.*Genetics,***173**:321–330

Liu, H.L. (1975) Science and Technology of OilCrops No.**4**:77-85

Liu H. Y., Zhou X. A., Wu K., Yang M. M., Zhao Y. Z. (2015). Inheritance and molecular mapping of a novel dominant genic male-sterile gene in *Sesamum indicum* L. *Mol. Breed.* **35**:9 10.1007/s11032-015-0189-5

Liu, K.S. (1997) Soybean Improvements through Plant Breeding and Genetic Engineering In: KeShun Liu (Ed) *Soybeans, 1997. Springer Link*: Pp 478 -523 ISBN : **978-1-4613-5711-7**

Liu, Q and Zhou, L. (2014) "*Linum*". *Flora of China*. **11**. Archived from the original on 5 April 2015. Retrieved 2 October 2014. (cf. *Wikipedia*)

Liu, Q., Cao, S., Zhou, X., Wood, C., Green, A., and Singh, S. (2013). Non sense- mediated mRNA degradation of *CtFAD2-1* and development of a perfect molecular marker for *ol* mutation in high oleic safflower(*Carthamustinctorius* L.). *Theoretical and Applied Genetics* **126**: 2219–2231.

Liu, Q., Singh, S.P. and Green, A.G. (2002) High stearic and oleic acid cottonseed oils produced by hairpin RNA- mediated post-transcriptional gene slicing. *Plant physiol.,* **129**: 1732-1734.

Liu, Y.Y. *et al.,* (1998) Effect of selenium on activity of glutathione peroxidase in soyabean. *Soyabean Science,* **17** (2): 157-161

Liu, Z. and Jan, C.C. (2012) Molecular techniques for sunflower breeding. In: Kovacevic, Z., Skoric, D.and Sakac, Z (Eds.) *International Monongram on Sunflower Genetics and Breeding. Serbian Academy of Science*. Belgrade, Serbia.pp 355-441.

Liu, Z.,Wang,D., Feng, J., Seiler, G.J., Cai,X.and Jan, C. C.(2013) Diversifying Sunflower Germplasm by Integration and Mapping of a Novel Male Fertility Restoration Gene. *Genetics*. 2013 Mar; **193(**3): 727–737.

Lloyd, J. U. (1898). "Citrullus Colocynthis". The Western Druggist. Chicago, USA

Loewe M., V. (2009) Usos mercados de la madera de paraíso *Chile Forestal* 2009 No.347 pp.46-49

Lokesh B.K, Maraddi GN, Agnal MB and Upperi SN (2008) Survey for the incidence of necrosis virus disease and thrips in sunflower growing areas of Southern Karnataka. *Int J Plant Prot***1**(2):58–60

Lonnquist,J.H. (1964) Modification of the ear to row procedure for the improvement of maize populations. *Crop Sci.,***4:** 227-228

Lopez-Gonzalez, G. 1990. Acerca de la clasificacion natural del genero *Carthamus L., s.1. Anales Jardin Bot.* Madrid **47**: 11–34.

Lord, M. J. and Roberts, L. M. (2005). "Ricin: structure, synthesis, and mode of action". In Raffael S, Schmitt M (eds.). *Microbial Protein Toxins*. Topics in Current Genetics. **11**. Berlin: Springer. pp. 215–33. doi:10.1007/b100198. ISBN 978-3-540-23562-0

Lorenc-Kukula K., Wróbel-Kwiatkowska M., Starzycki M., Szopa J. (2007): Engineering flax with increased flavonoid content and thus *Fusarium* resistance. *Physiological and Plant Molecular Pathology*, **70**: 38–48.

Lorenc-Kukuła K., Zuk M., Kulma A., Czemplik M., Kostyn. K., Skala J., Starzycki M., Szopa J. (2009): Engineering flax with the GT family 1 *solanum sogarandinum* glycosyltransferase SsGT1 confers increased resistance to *Fusarium* infection. *Journal of Agricultural and Food Chemistry*, **57**: 6698–6705

Lou, Q.Y.(1996) The main component of sunflower head pectin. *Chemistry and Industry of Forest Products (China)* **16** (3) : 45-50

Loumouamou B., Silou T.H. and Desobry S. (2010). Characterization of Seeds and Oil of Sesame (*Sesamum indicum* L.) and the Kinetics of Degradation of the Oil During Heating. *Research Journal of Applied Sciences, Engineering and Technology,* **2**(3): *227- 232.*

Love, H.K.,Rakow, G., Raney, J. P and Downey, R.K. (1990) Development of low glucosinolate mustard.*Canadian Journal of Plant Science,* 1990, 70(2): 419-424, https://doi.org/10.4141/cjps90-049

Lovelli, S., Perniola, M., Ferrara. A. and Di Tommaso, T. (2007) Yield response factor to water (Ky) and water use efficiency of *Carthamus tinctorius* L. and *Solanum melongena* L. *Agric Water Manage* **92**(1–2): 73–80, http:// dx.doi.org/ 10.1016/j.agwat.2007.05.005

Low, D., Pistillio, P. and Mackay, J. (1975) Oilseed breeding*: Sunflower. Annual Report,* CSIRO Div.IRR. Res. Griffith. NSW. Australia.

Loza, C. and Brostoff J (1995). "Peanut allergy". *Clin. Exp. Allergy*. **25** (6): 493–502. doi:10.1111/j.1365-2222.1995.tb01086.x. PMID 7648456.

Lu, C., Napier, J.A., Clemente, T.E. and Cahoon, E.B. (2011) New frontiers in oilseed biotechnology; meeting the global demand of vegetable oils for food, feed and biofuel and industrial applications. *Curr. Opin. Biotechnol.*, **22**: 252-259.

Lu, J., Mayer, M. and Pickersgill, B. (1990) Stigma morpoology and pollination in *Arachis* L.(Leguminosae). Ann. Bot. **66: 73-82.**

Ludvíková, M.and Griga, M. (2015). Transgenic Flax/Linseed (*Linum usitatissimum* L.) Expectations and Reality. *Czech J. Genet. Plant Breed.* **51** (4): 123–141

Luis, J.M. Ozias-Akins,P., Holbrook,C.C.,Kemerait, Jr. R.C., Snider, and Vasilis Liakos (2016) Phenotyping Peanut Genotypes for Drought Tolerance, *Peanut Science* **43**(1) https://www.peanutscience.com/doi/pdf/10.3146/PS15-5.1

Lynch, R.E, Branch, W.D., Garner, J.W. (1981) Resistance of *Arachis* spp to fall army worm *Spodoptera frugipedra. Peanut Sci.* **8:** 106-109

Mabberley, D. J. (1984). A monograph of *Melia* in Asia and the Pacific: The history of white cedar and persian lilac. *Gardens' Bulletin Singapore*. **37**: 49-64. [75909]

Mabberley, D.J., Pannell, C.M., Singh, A.M. (1995). Meliaceae. Flora Malesiana: Series I, *Spermatophyta.* Volume **12** (1), ii + 407 pp

Madhavi, K.S. (1988) Studies on some morphological and physiological characters with specific reference to their inter-relationship and heterosis in *Arachis hypogaea* L. *M.Sc. (Ag). Thesis, Andhra Pradesh Agricultural University, Hyderabad,* India.

Madhavi, K.S., Nagabhushanam, G.V.S. and Prasad, M.V.R. (1993). Efficient groundnut (*Arachis hypogaea*, L.) breeding through the use of induced mutants for intermediate canopy development. *ISOR National Seminar on Oilseeds Research and Development in India: Status and Strategies*. Aug. 2-5, 1993. Extended Summaries: 92-93, Hyderabad, India.

Madhusudhan, B., Meur, G., Devi, P.S.V., Kirti, P.B. and Gupta, A.D. (2008) Characterisation of *Bacillus thurungiensis* strain DOR 4 toxic to semilooper *Achoea janata*. Proteolyitc Processing and building toins to Receptors. *Current Microbiology,* **57** (1): 72-77

Madusudhan, K.T. and Singh, N.: Studies on linseed proteins. *Journal of Agricultural and Food Chemistry*., **31**: 959–963 (1983

Maeda, K. (1972) Growth analysis on the plant type in peanut varieties. *Arachis hypogaea,* L. IV Relationship between the varietal differences of the progress of leaf emergence on the main stem during pre-flowering period and the degree of morphological differentiate on of leaf primordia in embryo. *Proceedings of the Crop Science Society of Japan.* **41:**179-186

Mahadevan, N.P. (1991). Phenological observations of some forest tree species as an aid to seed collection. *Journal of Tropical Forestry* 7 (3): 243-247.

Mahapatra, L.N. (1966) Preliminary studies on correlation of yield with other characters of groundnut. *Indian Agriculture.* **10:** 97-106

Maheshwari, P. and Kovalchuk, I. (2014) Genetic engineering of oilseed crops. *Biocatal.*Agric. *Biotechnol*, **3:**31-37.

Maheshwari, T.U. and Ro, N.V. (2000) NPV on red hairy caterpillar, *Amsecta albistriga. Insect Environment* **6**(2):69

Maisonneuve, S., Bessoule, J.J., Lessireet, R., Delseny, M. and Roscoe, T.J. (2009) Expression of rapeseed microsomal lysophosphatidic acid acyltransferaseisozyme enhances seed oil content in *Arabidopsis.* Plant Physiol. **152**, 670-684.

Maiti, S., Hegde, M. R. and Chattopadhyay, S. B. (1988). Handbook of Annual Oilseed Crops. Ltd. *Oxford and IBH Publ. Co. Pvt.,* New Delhi.

Majumdar, D., Pandya, B., Arora.A. and Dhara, S. (2004) Potential use of karanjin (3-methoxy furano-2,3,7,8-flavone) as nitrification inhibitor in different soil types. *Archives of Agronomy and Soil science*. **50** (4-5): 455-465.

Makkar, H. P. S.; Singh, B. and Negi, S. S., (1990). Tannin levels and their degree of polymerisation and specific activity in some agro-industrial by-products. *Biological Wastes*, **31**: 137-144

Makne, V.G. and Choudhari., V.P. (1980). Combining ability in safflower (*Carthamus tinctorius* L.). *J. Maharashtra Agric. Univ*. **5**: 128–13

Makne, V.G., Patil, V.D.and Choudhari, V.P. (1979). Genetic variability and character association in safflower. *Indian J. Agric. Sci.***49**: 766–768.

Malathi, B., Rmesh, S., Rao, K.V. and Reddy, V.D. (2006) *Agrobacterium* mediated genetic transformation and production of semilooper resistant transgenic castor (*Ricinus communis* L) *Euphytica,* **147**(3): 441-449.

Malavia, D.D., Khanpara, V.D., Shobhana, H.K., Golakiya, B.A. and Lamm, F.R. (1995) A comparison of irrigation methods in arid and semi-arid western Gujarat: Micro irrigation for Changing World, conserving resources & Preserving the Environment. *Proceedings of the 5th International Micro Irrigation Congress. Orlando Florida, USA.* April 02-06, 1995 PP 464-469

Malek, M.A., Ismail, M.R., Monshi, F.I., Mondal, M.M.A. and Alam, M.N. (2012) Selection of promising rapeseed mutants through multi-location trials. Bngladesh *J. Bot.* **41**: 111-114

Maliwal, P.L. (1985) Effect of dates of sowing on yield and quality of taramira (*Eruca sativa)* varieities. *Indian J.Agron.***30**: 112-118

Malhotra, K., Kumar, D., Pandey, V.D. and Singh, S. (2020) Diversity of *Jatropha* in India. In*: Book: Biodiversity: An overview, I.K. International Publishing House Pvt. Ltd. New Delhi,* pp.235-245

Malik, Y.P. and Singh, S. V. (1997) Evaluation of linseed germplasm against budfly *Dasyneura lini* Barnes and some important diseases. *J. Insect Sci.,***10:** 70-71

Maliwal, P.L., Jangir, R.P., and Sharma, S.L. (1984) Effect of date of sowing on yield and yield attributes of taramira varieties. *J. Oilseeds Res*. **1**: 1-9

Mallikarjuna, N. (2005) Production of hybrids between *Arachis hypogaea* and *A. chiquitana* (Section *Procumbentes*). *Peanut Science*, **32**: 148-152

Mallikarjuna, N. and Hoisington, D. (2009) Peanut improvement: Production of fertile hybrids and back cross progeny between *Arachis hypogaea* and *A. kretschimeri*. *Food Sec*. **1:** 457-462

Mallikarjuna, N. and Sastri, D.C. (2002) Morphological, cytological and disease resistance studies of inter sectional hybrids between *Arachis hypogaea* L. and *A. glabrata*, Benth. *Euphytica*, **126** (2): 161-167

Mallikarjuna, N. Pande, S., Jadhav, D.R., Sastri, D.C. and Rao, J.N. (2004) Introgression of disease resistance genes from *Arachis kempffmercadoi* into cultivated groundnut. Plant Breed. **123** :573-576

Maluszynski, M., Kasha, K.J., Forster, B.P.and Szarejko, I. (2003) Double haploid production in crop plants*: A Manual. Kluwer Academic Publisher, Dordrecht*, The Netherlands.428pp.

Manandhar. N. P. (2002) Plants and People of Nepal, *Timber Press, Oregon. USA*. ISBN: 0-88192-527-6

Mandal, A.B. (1990). Variability, correlation and path coefficient analysis of safflower *(Carthamus tinctorius* L.). *Phytobreedon***6**: 9–18.

Mandal, A.K.A., Chatterj'ee, A.K. and Das Gupta, S. (1995) Direct somatic embryo genesis and plantlet regeneration from cotyledonary leaves of safflower. *Plant Cell, Tissue and Organ Culture,***43**: 287-289

Mandal, K.G., Saha, K.P., Ghosh, P.K., Hati, K.M and Bandyopadhayay, K.K.(2002)Bioenergy and economic analysis of soybean-based crop production systems in central India. *Biomass and Bioenergy*, **23** (5): 337-345

Mandal, S. *et al.,* (1993)Sulphur boosts the yield of summer sesamum. *Indian Farming,***43** (5):17-18

Mandel, J.R.; Dechaine, J.M.; Marek, L.F. and Burke, J.M. (2011) Genetic diversity and population structure in cultivated sunflower and a comparison to its wild progenitor, *Helianthus annuus L. Theor. Appl. Genet.,* **123**: 693–704

Manivannan, N. and Ganesan, J. (1995) An Improved Hybridization Technique for Commercial Hybrid Seed Production in Sesame. *Plant Breeding Newsletter***4** (3): Feb-April 1995

Manjunath, K.G. and Sannappa, B. (2014). Evaluation of Castor Ecotypes of Selected Regions of the Western Ghats of Karnataka, India through Bio-Chemical Assay. *International Journal of Science and Research (IJSR*). **3**(11): 2045-205

Manna, M. C.; Jha, S.; Ghosh, P. K.; Ganguly, T. K. and Singh, K. N., (2004). Litter decomposition in three plantation species in semi-arid and sub-humid regions of central India: impact of climate, substrate quality and earthworm activity. *J. Sustain. Forestry,* **18** (1): 47-62,

Manokari,M., Latha,R., Priyadharshini, S., Cokul, R.M. *et al*., (2019).Green synthesis of zinc oxide nanoparticles from aqueous extracts of *Sesamum indicum* L. and their characterization *WNOFNS***23** (2019) 200-210 ; EISSN 2543-5426 http://www.worldnewsnaturalsciences.com/wp-content/uploads/2019/01/WNOFNS-23-2019-200-210.pdf]

Maria, S., Daniela, D., Daniela, M., Carmen, S., Adriana, G., Ioana, O., Cioroianu T. and Nicoleta, V. (2009) Influence of foliar fertilizers on the penetration, uptake and the distribution of the micronutrients in different organs of sunflower plants. *Scient. Pap. USAMV Bucharest, Series A.,* **LII** (ISSN): 1222-5339.

Mariani, C., Beuckeleer, M.D., Truettner, J., Leemans, J. and Godberg, R.B. (1990) Induction of male sterility in plants by chimeric ribonuclease gene. Nature, **347**: 737-741

MarianiC.,Gossele, V., Beuckleer, M.D., Bwick, M.D., Goldberg, R.B., Greef, W.D. and Leemans, J. (1992). A chimeric ribonuclease inhibitor gene restorer of fertility to male sterile plants. *Nature*, **357**: 284-387

Mariotti, M.and Masoni, A. (1996)Spectral properties of iron-deficient corn and sunflower leaves.*Remote Sensing of Environment* **58** (3): 282-288

Maroyi A (2007). The Utilization of *Moringaoleifera* in Zimbabwe: A Sustainable Livelihood Approach. *J Bot Sci.* 2007 **23**:172 – 185.

Mary, R.J. and Jayabalan, N. (1997). Influence of growth regulators on somatic embryogenesis in sesame. *Plant Cell Tissue and Organ Culture* **49**: 67-70.

Marz, U. 1989. The economics of neem production and its use in pest control. *Farming Systems and Resource Economics in the Tropics* 5.

Masuo, Y. (1958) On natural crossing in flax. *Crop Science Society* Japanese *Proceedings***27:** 321-323(In Japanese). Abstract in *Plant Breed.* **29**(4):817-818

Matlock, R.S., Banks, D.J., Tai, Y.P. and Kirby, J.S. (1970) Inheritance of narrow leaf character in peanut, *Arachis hypogaea*, L. *Agron. Abstr.* **1970:**15

Matsuzawa, Y. and Sarashima, M. (1986). *Croeilerae News*. No.**11**: 17.

Matthews, R. F., 1994. *Simmondsia chinensis*. In: Fire Effects Info. Syst., [Online]. USDA, *Forest Service, Rocky Mountain Res. Station, Fire Sci. Lab.*

Mau, R.F.L. and Martin, J.L. (2007) *Bemisia tabaci* (Gennadius)http://www.extento.hawaii.edu/kbase/Crop/Type/b_tabaci.htm

Mavarkar, N.S., Shetty, T.K.P., Sridhara, S., Naik, T.B. and Sridhara, C.J. (2008) Economic viability of castor based sequential relay inter-cropping systems under rainfed conditions. *Environmnet and Ecology.***26**(2A): 883-887

Mawson. R., Heaney, R.K., Zdunczyk, Z and Kozlowska, H. (1993) Rapeseed meal-glucosinolates and their antinutritional effects. Part II. Flavour and palatability,*Molecular Nutrition,***37**(4): 336-344

Mayerhofer, R., Archbald, C., Bowles, V. and Good, A.G. (2010). Development of molecular markers and linkage maps for *Carthamus* species *C. tinctorius* and*C. oxyacanthus. Genome***53**: 266-276.

Mayeux, A. H., and Ntare, R. B. (2001). Accessions with resistance to foliar diseases, *A. flavus/* Aflatoxin contamination and rosette disease. *Groundnu*Mayeux, A.H., Waliyar, F., & Ntare, B. R. (2003). Groundnut varieties recommended by the Groundnut Germplasm Project (GGP) for West and Central Africa (in En, Fr.). *ICRISAT, Patancheru* 502 324, Andhra Pradesh, India. 80 pp

Mazhar, H., Quayle, R., Fido, R.J., Stobart, A.K., Napier, J.A. and Shewry, P.R. (1998). Synthesis of storage reserves in developing seeds of sunflower. *Phytochemistry* **48**: 429–432.

McCloud, D., Duncan, W.G., McGraw, R.L., Sibale, P.K., Ingram, K.T., Dreyer, J. and Campbell, I.S. (1980) Physiological basis for increased yield potential in peanuts. *Proceedings of the International Workshop on Groundnuts. 13-17 Oct. 1980, ICRISAT,* Patancheru, Andhra Pradesh, India. Pp. 125-132

McDonald, M.V., Ahgmad, I., Meneten, J.O.M and Ingram, D. S. (1991). Haploid culture and *in-vitro* mutagenesis (UV light, X-rays and gamma rays) of rapid cycling *Brassica napus*for improved resistance todisease. *Plant Mutation Breeding for Crop Improvement* vol. **2,** *IAEA, Vienna.*pp 129-138

McGuire, P.E., Damania, A.B., and Qualset, C.O. (eds.) (2012). Safflower in California. *The Paulden F. Knowles personal history of plant exploration and research on evolution, genetics, and breeding. Agronomy Progress Report No. 313,* Dept. of Plant Sciences. University of California. Davis CA USA.

McHughen, A. (1987). Salt tolerance through increased vigor in a flax line (STS-II) selected for salt tolerance in vitro. *Theor. Appl. Genet.* **74**:727–732

McHughen, A. (1989.)*Agrobacterium* mediated transfer of chlorsulfuron resistance to commercial flax cultivars. *Plant Cell Rep.* **8**:445–449.

McHughen A. (2000): Transgenic linseed flax (*Linum usitatissimum*). In: Bajaj Y.P.S (ed.): Transgenic Crops I. *Biotechnology in Agriculture and Forestry*, **46:** 338–351.

McHughen, A. (2002). Transgenic linseed flax. *Marcel Dekker*, New York.

McHughen, A.and Holm, F.A.(1995). Transgenic flax with environmentally and agronomically sustainable attributes. *Transgenic Res*. **4:** 3–11.

McKenzie, R.R.(2011). Genetic and Hormonal Regulation of Stem Vascular Tissue Development in Flax (*Linum usitatissimum* L.) *M.Sc. thesis. University of Alberta*, Edmonton, Alberta, pp. 1–220

McPherson, M.A., Good, A.G., Keith, C.T. and Hall, L.M. (2004) Theoretical Potential of transgenic safflower (*Carthamus tinctorius* L) with weedy relatives in New World. *Canadian Journal of Plant Sciences,* **84** : 923-934

MEDLINE, PubMed, and PMC (PubMed Central) (2019). "How are they different?". www.nlm.nih.gov. 9 September 2019.

Meena, H.P., Dudhe, M.Y., Mukta, N. and Anjani, K. (2012)Heterosis breeding in safflower: present status and future prospectsunder Indian scenario, *Journal of Oilseeds Research*, **29** (Special issue):164-167

Meena, K., Anjani K.,and Venkat Ramya, K. (2014). Molecular diversity in castor germplasm collection originated from North-Eastern Hill Province of India. *International Journal of Research and Scientific Innovation.* **I,** (VI),http://www.rsisinternational.org/IJRSI.html.

Meena K., Kammili, A., Usha kiran, B.and Vivekananda. K. (2015). Agro-morphological and molecular diversity in castor (*Ricinus communis* L.) germplasm collected from Andaman and Nicobar Islands, India. *Czech J. Genet. Plant Breed.,***51**, 2015 (3): 96–109.

Mehta P.P. and Prasad R.N. (1976) Investigation on leaf blight of ***til*** caused by *Alternaria sesami. Proceeding, Bihar Academy Agri. Sci.,***24**:104-109

Mehta, P.R. (1951) Observations on new and known diseases. *Plant Protection Bulletin,***3** :7-12.

Mehtre, S.S., Ghatge, R.D. and Lad, S.K. (1994). Wild *Sesamum mulayanum,* a source of multiple disease resistance. *Annals of Agricultural Research.***15** (2):243-244.

Mehtre, S.S., Bhondve, T.S. and Dalvi, N.D. (1997). Intercropping of soyabean in different crops. -a review.*Crop Research (Hissar)***13**(2): 519-531

Mekala, N.K. and Gupta, V.K. (2014) Current Bioenergy Researches In: Gupta V., Tuohy, M., Kubicek,C. and Feng Xu, J.S.(Eds) .*Bioenergy Research: Advances and Applications,Book ISBN: 9780444595645 Imprint: Elsevier* Pp 500.

Melaku, E. T., 2013. Evaluation of Ethiopian nigerseed (*Guizotia abyssinica* Cass) production, seed storage and virgin oil expression. *Msc Dissertation, Landwirtschaftlich-Gärtnerische Fakultät der Humboldt-Universität zu Berlin*

Mendham, N.J, Russell, J. and Buzza, G.C., (1984). The contribution of seed survival to yield in new Australian cultivars of oil-seed rape (*Brassica napus* L.). *J. Agric. Sei., Camb*., **103**: 303-316.

Mendham, N.J., Shipway, P.A. and Scott, R.K., (1981). The effects of delayed sowing and weather on growth, development and yield of winter oilseed rape (*Brassica napus* L.) *Agric. Sei., Camb*., **96:** 389- 416.

Merrill, S.D., Tanaka, D.L., Krupinsky, J.M., and Ries. R.E.(2004). Water Use and Depletion by Diverse Crop Species on Haplustoll Soil in the Northern Great Plains. *Journal of Soil and Water Conservation***59:** 176–183.

Metz., G.L., Green, D.E. and Shibbles, R.M. (1984) Relationships between soybean yield in narrow rows and leaflet canopy and developmental characters. *Crop Science*, **24**:257-462.

Metzger, J.O. and Bornscheuer, U. (2006). Lipids as renewable resources: Current state of chemical and biotechnological conversion and diversification. *Appl. Microbiol. Biotechnol.***71** :13-22

Mezzarobba, A. and Jonard, R. (1986) Physiologic végétale *CR Acad Sci Paris*, **303** III:181–186

Mhiret, W.N., Harrison, J.S.H. (2018) Biodiversity in Ethiopian linseed (*Linum usitatissimum* L.): molecular characterization of landraces and some wild species. *Genet Resour Crop Evol.,* **65,** 1603–1614 (2018). https://doi.org/10.1007/s10722-018-0636-3

Mihail, J. D. and Taylor, S. J. (1995). Interpreting variability among isolates of *Macrophomina phaseolina* in pathogenicity, pycnidium production,and chlorate utilization. *Can. J. Bot.***73**: 1596-1603.

Milimo, P. B. (1994). Mechanisms of drought resistance in *Melia volkensii* and *M. azedarach. PhD thesis, Department of Forestry, Australian National University, Canberra*

Millam S, Obert B, Preťová A (2005) Plant cell and biotechnology studies in *Linum usitatissimum* – a review. Plant Cell Tissue Organ Cult 82:93–103

Min, Y. Y., and Toyota, K. (2019). Occurrence of Different Kinds of Diseases in Sesame Cultivation in Myanmar and Their Impact to Sesame Yield. *Journal of Experimental Agriculture International,* **38***(4), 1-9 https://doi.org/10.9734/jeai/2019/v38i430309*

Misra, S.C. and Teotia, T.P.S (1965) Studies on the biology of castor shoot and capsule borer *IndianOilseedsJournal* **9** :184-187

Mishra, S., Singh, R., Kumar, S. and Kumar, Sudheer (2017) Morphological, Biochemical and Molecular Characterization of *Alternaria brassicicola* and *Alternaria brassicae*: A Comparative Overview. *Am. J. Biochem. Mol. Biol*., **7** (2): 53-64, 2017

Mitkowski, N.A. and Abawi.G.S. (2003). Root-knot nematodes. *The Plant Health Instructor*. DOI:10.1094/PHI-I-2003-0917-01

Mittapalli, O. and Rowland, G.G. (2003). Inheritance of seed color in flax. *Crop Sci.* **43**: 1945–1951 Mittelbach, M. and Gangl, S. (2001) Long storage stability od biodiesel made from rapeseed and used frying oil. *J. Am. Oil. Chem. Soc.,***78**: 573-577

Miyazawa M, Maehara, T. and Kurose K. (2002) Composition of the essential oil from the leaves of *Eruca sativa. Flav Frag J.* 2002, **17:**187–90

Mizushima. U. (1950). *TohokuJ. Agric. Res.,***1**: 1-14.

MNOOP (2018), National Mission on Oilseeds and Oil Palm (NMOOP),Ministry of Agriculture and Farmers Welfare, Government of India Report, July 2018 (http://nmoop.gov.in/)

Mohamed, S., Binsfeld, P.C., Cerboncini, C. and Schnabl, H. (2003) Regenration systems at high frequency from high oleic *Helianthus annuus*, L. genotypes. *J.Appl. Bot.***77**: 85-89

Mohammadi, A.A., Saeidi, G. and Arzani, A. (2010). Genetic analysis of some agronomic traits in flax (*Linum usitatissimum* L.). *Aust. J. Crop Sci.* **4**: 343–352. M

Mohan, J., Chaudhari, L.D. and Jha, M. (1995.) A note on increasing the viability of neem seed. Indian-Forester **121** (11): 1085-1086.

Mohankumar, B.N., Basavegouda, Vyakaranahal, B. and Deshpande, V.K. (2011)Influence of sowing dates on production of seed yield in niger (*Guizotiaabyssinica* Cass.) *Karnataka Journal of Agricultural Sciences* **24**(3) · September 2011

Mohanti, N.N. and Behera, B.C. (1958) Blight of sesame caused by *Alternaria sesame*. New Comb. *Current Sci.* 1958; **27**:492-493.

Mohanty, R.N. (1964). Seed setting of niger under controlled environmental conditions. *Indian Oilseeds J.* **8**:158.

Mohapatra, D. and Bajaj, Y.P.S. (1984) *In vitro* hybridization in an incompatible cross–*Brassica juncea* x*Brassica hirta.Curr Sci.,***53**: 489–490

Mohapatra, T. (2017) Strategies for Increasing Oilseeds Production in India. ICAR-11OR *Foundation Day Lecture. ICAR-Indian Institute of Oilseeds Research,* Hyderabad. Pp10

MohibbeAzam, M.; Amtul W.and Nahar, N.M. (2005) Prospectsand potential of fatty acid methyl estersof some non-traditional seed oils for use as biodiesel in India. *Biomass Bioenerg*., 2005, 29:293–302.

Mol, J.N.M., van der Krol, A.R., van Tunen, A.J., van Bfokland, R., de Lange, P. and Stuitje, A.R. (1990) Regulation of plant gene expression by antisense RNA, *FEBS 08632.* **268**(2): 427-430.

Molla, A. and Muhie, K. (2011). Tef (*Eragrostis tef*) based cropping systems in the hot to warm moist valleys of North Shewa, Ethiopia. ***Scientific Research and Essays*** **6(**6):1411-1416, 18 March, 2011. http://www.academicjournals.org/SRE DOI: 10.5897/SRE09.090 ISSN 1992-2248 ©2011 Academic Journals, Full Length Research Paper

Mondal, S. and Badigannavar, A.M. (2013) A narrow leaf groundnut mutant, TMV2-NLM has a G to A mutation in AhFAD2A gene for high oleate trait. *Indian J. Genet.,***73(**1): 105-109 (2013) DOI: 10.5958/j.0019-5200.73.1.016

Monson A (2007) An early ptolemaic land survey in Demotic: *P. Cair* II:31073.

Monyo, E. S., Njoroge, S. M. C., Coe, R., Osiru, M., Madinda, F., Waliyar, F. and Anitha, S. (2012). **Occurrence and distribution of aflatoxin contamination in groundnuts (***Arachis hypogaea***) and population densities of aflatoxigenic aspergilli.** *Crop Protection,***42:** 149–155. https://doi.org/10.1016/j.cropro.2012.07.004

Moore, A. (2011). Fertilzer potential of biofuel byproducts. In: Dos Santos Bernardes, M.A. (Ed)., Biofuel Production- Recent developments and prospects. *InTech Rijeka*, Croatia.

Moore, K.M. Knauft, D.A. (1989) The Inheritance of High Oleic Acid in Peanut*Journal of Heredity*, 1989, **80**: 252-253.

Morton J.F. (1991) The horse-radish tree, *Moringapterygosperma* (Moringaceae)—a boon to arid lands? *Economic botany*. 1991; **45**(3):318-333

Morinaga T, Fukushima E, Kano T, Yamasaki Y (1929) Chromosome numbers of the cultivated plants II. *Botany Magazine***43:**589.

Moser, G. (1996). Status Report on Global Neem Usage. *Pesticide Service Project,* PN 86.2588.1. GTZ, Griesheim, Germany. 39 pp.

Moshkin, V.A. (1986). Castor. *Amerind Publishing Co. PVT Ltd*, New Delhi.

Moshkin, V.A. and Dvoryadinka. A.G. (1986) Castor Genetics. In: V.A. Moshkin(Ed) *Castor. Amerind Publ. Co., New Delhi*. 1986. pp. 93-102.

Mridha, M.A.U. and Barakah, F.N. (2017) Diseases and pests of Moringa: a mini review *Proc. International Symposium on Moringa* Eds.: A.W. Ebert and M.C. Palada. *Acta Hortic. 1158. ISHS 2017*. DOI 10.17660/ActaHortic.2017.1158.14

Muddanuru, T., Vasavi, S., Vijay, S. and Sujatha, M. (2018) tsv-cp article pdf. *Plant Cell Tissue and Organ Culture* https://www.researchgate.net/publication/326508016_tsv-cp_article_pdf

Muir, A.D., and. Westcott, N.D. (2003). Mammalian metabolism of flax lignans. In A. D. Muir and N. D. Westcott (ed.) *Flax: The genus Linum. Culinary and Hospitality Industry Publications Services (CHIPS), Weimar, T.* p. 230–242

Mukherjee, P. and Jha, T.B. (2009) Biotechnological Improvement of a Biofuel crop *Jatropha curcus,*In: Claude Ponterio *et al.*(Eds). *Jatropha curcas as a Premier* ISBN: 978-1-60876-003-9. *Nova Science Publishers, Inc.*

Mukta, N.(2008) Safflower genetic resources in India - an overview. *7th International Saflower Conference,* Wagga Wagga, Australia

Mukta, N. and Sreevalli, Y. (2010) Propagation techniques, evaluation and improvement of the biodiesel plant, *Pongamia pinnata* (L.) Pierre—a review. *Industrial Crops and Products* **31**:1–12

Mukta, N. and Varaprasad, K.S. (2013). Tree borne oilseeds for agroforestry. In: Rao, G.R., Kumar, N.R., Prasad, J.V.N.S., Prabhakar, M., Venkatesh, G., Srinivas, I., Ramana, D.B.V. and Venkateswarlu, B. (Eds*) Agroforestry as a Strategy for Adaptation and Mitigation of Climate Change in Rainfed Areas Central Research Institute for Dryland Agriculture, Hyderabad*, pp.102-110.

Mundel, H.H. (2008) Major Achievements in Safflower Breeding and Future Challenges In: Knights, S.E., Pottar, T.D (Eds)Unexploited Potential and World Adaptability. Proceedings of the 7th International Safflower, Wagga Wgga, New Soouth Wales, Conference, Australia, Nov. 3-7, 2008.

Mundel, H.H., and Bergman, J.W. (2009). "Safflower," In:J. Vollmann and I. Rajcan (Eds) *OilCrops, Handbook ofPlantBreeding4*, Berlin: *Springer-Verlag*, 423–447.

Mundel, H.H. and H.C. Huang. (2003). Control of major diseases of safflower by breeding for resistance and using cultural practices. In:*Advances in Plant Disease Management.* Huang, H.C. and S.N. Acharya, (Eds). *Research Signpost,Trivandrum*, Kerala, India, pp. 293–310.

Mundel, H.H., Huang, H.C., Burch, L.D. and Kiehn, F. (1985). Saffire safflower. *Can. J. Plant Sci.* **65**: 1079–1081

Murashige,T. and Skoog, F. (1962). "A Revised Medium for Rapid Growth and Bio Assays with Tobacco Tissue Cultures". *Physiologia Plantarum.* **15** (3): 473–497. doi:10.1111/j.1399-3054.1962.tb08052.x.

Muravenko, O.V., Lemesh, V.A., Samatadze, T.E., Amosova, A.V., Grushetskaya, Z.E., Popov, V.K., Semenova, O.Y., Khotyuleva, V.L. and Zelenin, A.V., (2003). Genome comparisons with chromosomal and molecular markers for three closely related flax species and their hybrids. *Russ. J. Genet.***39:** 414–421

Muravenko, O.V., Amosova, A.V., Samatadze, T.E. *et al.* (2004) Chromosome Localization of 5S and 45S Ribosomal DNA in the Genomes of *Linum* L. Species of the Section *Linum* (Syn. *Protolinum* and *Adenolinum*). *Russian Journal of Genetics* **40,** 193–196 (2004). https://doi.org/10.1023/B:RUGE.0000016994.89043.2b

Murdock D, editor (2002). Encyclopedia of foods: *A guide to healthy nutrition. California: Academic Press*; 2002. [Google Scholar]

Murphy, J.D. (1994)Designer Oil Crops *VCH Verlagsgesellschaft mbH. D*-69451(Germany) pp317

Murty, B.R., Arunachalam, V. and Anand, I.J. (1967) Diallel and partial diallel analysis of some yield factors in *Linum usitatissimum. Heredity* **22**:35-41

Murthy, G.S. (1979). Heterosis in inter mutant hybrids of *Sesamum indicum* L. *Curr.Sci.,* **48:** 825-827

Murthy, G.G.S., Salunkhe, C.K. and Joshua, D.C. (1995A) Inheritance of induced stiff stem mutant in sesame (*Sesamum indicum* L.) ***Sesame Safflower Newsletter*, 10: 62-66.**

Murthy, G.S.S. (1997) Induced mutants for the improvement of sesame and hybrid seed production. p. 31-38. In: *Proc. 2"d FAO/IAEA Res. Coord. Mtg, Induced mutations for sesame improvement. IAEA,* Vienna (1997).

Murthy, H.N., Hiremath, S.C., and Pyati, A.N. (1995) Genomic classification in *Guizotia* (Asteraceae). *Cytologia*, **60** (1): 67-73

Murthy, H. N.; Hiremath, S.C. and Salimath, S.S. (1993). Origin, evolution and genome differentiation in *Guizotia abyssinica* and its wild species. *Theor. Appl. Genet.* **87**(5): 587-592

Murthy, H.N., Jeong, J., Choi, Y.E. and Paek, K. Y. (2003) *Agrobacterium*-mediated transformation of niger [*Guizotia abyssinica* (L. f.) Cass.] using seedling explants.*Plant Cell Reports* **21**(12):1183-7. DOI: 10.1007/s00299-003-0573-1

Murthy, I.Y.L.N. (2011) Recent advances in secondary and micronutrient management in safflower (*Carthamus tinctorius* L), *Journal of Oilseeds Research***28**(2): 94-104.

Musialak M., Wróbel-Kwiatkowska M., Kulma A., Starzyc-ka E. and Szopa J. (2008): Improving **retting of fibre through genetic modification of flax to express pectinases.** *Trans-genic Research,* **17**: 133–147

Mustafa, H.S.B. *et al.,* (2015). Selection Criteria for Improvement in Sesame *(Sesamum indicum*). *American Journal of Experimental Agriculture*, **9**(4): XX-XX, 2015, Article no. AJEA.17524 ISSN: 2231-0606

Myles, S., Peiffer,J., Brown, P.J., Ersoz, E.S., Zhang, Z., Costich, D.E. and Buckler ES (2009) Association mapping: critical considerations shift from genotyping to experimental design. *Plant Cell* **21**:2194–2202

NAAS (2017) Resolution and Report on GM Mustard. National Academy of Agricultural Sciences. naasindia.org/Documents/GMmustard.pdf

Nabloussi, A., Velasco, L. and Fernandez-Mrtinez, J.M. (2013) Cross Pollination of Safflower (*Carthamus tinctorius* L.) under Moroccan Environmental Conditions.*International Journal of Plant Breeding***7**(2): 145-147©2013 Global Science Books

Nag, S., Mitra, J. and Karmakar, P. (2015) An over view of Flax (*Linum usitatissimum* L) and its Genetic Diversity. *International Journal of Agriculture, Environment and Biotechnology.***8:** 805 DOI: 10.5958/2230-732X.2015.00089.3

Nagabhushanam, G.V.S. (1989) Studies on Genetic manipulation of canopy and reproductive characters in Groundnut (*Arachis hypogaea* L.) *Ph.D. Thesis, Andhra Pradesh Agricultural University,* Hyderabad. Pp515.

Nagabhushanam, G.V.S., Prasad, M.V.R. and Jagadesh, C.A. (1990). A note on aerial pod bearing variant in groundnut. *J. Oilseeds Res.***7**(2):84-85.

Nagabhushanam, G.V.S. and Prasad, M.V.R. (1992a). Catergorization of groundnut genotypes based on canopy development for stability and yield. *Indian J. Agric. Sci.* **62**(11):725-730

Nagabhushanam, G.V.S. and Prasad, M.V.R. (1992b). Selection criteria for groundnut (*Arachis hypogaea,* L.) breeding. *Oleagineaux (*Paris) :**47**(1):23-28.

Nagabhushanam, G.V.S. and Prasad, M.V.R. (1992c). Canopy development as a basis for parental choice in breeding. *Indian J. Agric. Sci.***63**(9):537-541.

Nagabhushanam, G.V.S., Prasad, M.V.R. and Jagadeesh, C.A. (1992a). Studies on association **among canopy and reproductive attributes in F1 and F2 generations of intra-specific and inter-subspecific crosses in groundnut** (*Arachis hypogaea, L.).J. Genetics and Breeding(Rome)* **46:**155-162.

Nagabhushanam, G.V.S., Prasad, M.V.R. and Jagadeesh, C.A. (1992b). Mutagenic effectiveness and efficiency of Sodium Azide in groundnut (*Arachis hypogaea, L.)* in relation to canopy development. *J. Oilseeds Res.* **9**(1):28-32.

Nagaraj, G. (1995) Quality and Utility of Oilseeds. *Directorate of Oilseeds Research.* Rajendranagar, Hyderabad.

Nagaraj, G. and Mukta, N. (2004) Seed composition and fatty acid profile of some tree borne oilseeds. *J. Oilseeds Res.***21** (1): 117-120

Nagaraju and Hanumantha Rao, C. (1999) Information on Sunflower necrosis virus, *Directorate of Oilseeds Research*, Hyderabad, 1999, **4.**

Nagaraju, Muniyappa, V., Singh, S.J. and Virupakshappa, K.(1997) Occurrence of mosaic disease on Sunflower in Karnataka, *Indian Phytopathology*, 1997, **50:** 277-281.

Nagarathna T. K., Shadakshari, Y.G. and Ramanappa T. M. (2011). Molecular analysis of sunflower (*Helianthus annuus* L.) genotypes for high oleic acid using microsatellite markers. *Helia* **34**:63–68. 10.2298/hel1155063

Nagaveni, H.C., Ananthapadmanabha, H.S. and Rai, S.N. (1987). Note on the extension of viability of *Azadirachta indica. Myforest.* **23**: 245.

Nagella, P., Hosakatte, N. M., Ravishankar, K. V., Hahn, E.J. and Paek, K.Y. (2008) Analysis of genetic diversity among Indian niger [*Guizotia abyssinica (*L. f.) Cass.] cultivars based on randomly amplified polymorphic DNA markers *Electronic Journal of Biotechnology*,**11**(1) (2008) doi: 10.2225/vol11-issue1-fulltext-14

Nageswara Rao, R.C. and Singh, A.K. (1994) Prospects of utilization of genetic vriability for yiled improvement in Groundnut. In: Prasad, M.V.R. *et al.*, (Ed) *Sustainability in oilseeds. Indian Society of Oilseeds Research.* Hyderabad. pp. 38.

Nageswara Rao, R.C., Williams, J.H., Sivakumar, M.V.K., Wadia and K.D.R. (1988) Effect of water deficit at different growth phases of Peanut. II. Response to drought during pre-flowering phase. *Agron.J.***80**: 431-438

Nagy, E. D., Guo, Y., Tang, S., Bowers, J. E., Okashah, R. A., Taylor, C. A., and Knapp, S. J. (2012). A high-density genetic map of *Arachis duranensis*, a diploid ancestor of cultivated peanut. *BMC Genomics*,**13**: 469. https://doi.org/10.1186/1471-2164-13-469

Naidu, L.K. *et al.*, (2006) Manual Soil-Site Suitability, Criteria for Major Crops. National Bureau of Soil Survey and Land Use Patterns, *Indian Council of Agricultural Research (ICAR),* Nagpur, India.pp118

Naik, P.M. and Murthy, H.N. Somatic embryogenesis and plant regeneration from cell suspension culture of niger (*Guizotia abyssinica* Cass.). *Acta Physiol Plant* **32:** 75 (2010). https://doi.org/10.1007/s11738-009-0380-6

Nalawade, B.B. and Chavan, P.D. (1990) Changes in the mineral nutrients during growth of niger (*Guizotia abyssinica* Cass). *Biovigyanam(India).* **16** (2): 83-89

Nambiar, P.T.C. and Dart, P.J. (1980) Studies on Nitrogen fixation by groundnut at ICRISAT. *Proceedings of the International Workshop on Groundnuts held at ICRISAT Centre.*pp110-124 Patancheru, Andhra Pradesh. `

Narain, A., Singh, P.(1968) Haploid meiosis and its bearing on the constitution of the castor oil plant. *J. Heredity***59**, 287–288

Narangalkar, A. L.; Shivpuje, P. R. (1990)Estimation of crop losses due to aphid in safflower. *Journal of Maharashtra Agricultural Universities* 1990. **15** (2) :239-240

Narasimham, N.V., Von Oppen, M. and Parthsarathy Rao, P. (1987) World markets for handpicked and selected (HPS)groundnuts. *Resource Management Program Economics Group Progress Report- 82,International Crops Research Institute for the Semi- Arid Tropics (ICRISAT).* Patancheru P.O. 502 324, India. October 1987: 34pp.

Narsimhulu, B. S. and Chopra, V. L. (1989). Shoot regeneration from cotyledons of three diploid species of Brassica. *Plant Breeding* **99**: 49–55.

Narayanan A and Chand K K (1986). The nature of leaf-senescence in groundnut (*Arachis hypogaea* L.) genotypes. *Indian J. Plant Physiol.* **29**: 125-132

Narayanan, E.S. (1953) The red hairy caterpillar (*Amsecta moorei*) and its control. *Indian Farming***3** (6): 8-9

Narayanan, E.S. (1959*) Insect Pests of Castor*. Indian Central Oilseeds Committee

Narayan, P. and Jasiwal, V.S. (1985). Plantlet regeneration from leaflet callus of *Azadirachta indica* A. Juss. *Tree Science* **4** (2): 65-68.

Narkhede, B. N. and Patil, A.M. (1987) Heterosis and inbreeding depression in safflower. *Journal of Maharashtra Agricultural University* **12**: 337–340

Narkhede, B.N., Patil, A.M and Deokar.A.B. (1992). Gene action of some characters in safflower. *J. Maharashtra Agric. Univ.* **17**: 4–6.

Nasrallah M.E. and Wallace D.H. (1967) Immunogenetics of self-incompatibility in *Brassica oleracea* L. *Heredity* **22**: 519–527

Nasareen P.N.M., Vardhanan Y.S. and Ramani N. (2012) Damage Assessment of the Gall Mite *Aceria pongamiae* Keifer 1966 (Acari: Eriophyidae) on *Pongamia pinnata* (L.) Pierre. In: Sabu A. and Augustine A. (Eds) Prospects in Bioscience: Addressing the Issues. Springer, Indiahttps://doi.org/10.1007/978-81-322-0810-5_38

Nasirullah, D., Mallika, T.; Rajalakshmi, S.; Pashupathi, K. S.; Ankaiah, K. N.; Vibhakar, S.; Krishnamurthy, M. N.; Nagaraja, K. V.; Kapur, O. P. (1982). Studies on niger *(Guizotia abyssinica*) seed oil. *J. Food Sci. Technol.*, **19** (4): 147-149

Naughton, F.C. (2011) Castor Oil, Published online: 4 December 2000, *Wiley Online Library.* DOI: 10.1002/0471238961.0301192014012107.a01https://doi.org/10.1002/0471238961.0301192014012107.a01.pub2

Nautiyal, P. C., Nageswara Rao, R. C and Joshi, Y. C. (2002). Moisture Deficit–Induced Changes in Leaf Water Content, Leaf Carbon Exchange Rate and Biomass Production in Groundnut Cultivars Differing in Specific Leaf Area. *Field Crops Research*, **74**: 67-79

Nayr, N.M and Mehra, K.L. (1970). Sesame: Its uses, botany, cytogenetics and origin. *Econ. Bot.***24**:20-31

Naz, R.M.A. and Islam (1979) Effect of irradiation of *Brassica trilocularis* (yellow *sarson*). Pak.J. Agril.Res. **17:** 87-93.

Nelissen, H., Moloney, M. and Inze, D. (2014) Translational research: From pot to plot. *Plant Biotechnol.J.* **12**:277-285

Narkhede, B. N. and Patil, A. M. (1987) Heterosis and inbreeding depression in safflower. *Journal of Maharashtra Agricultural University***12**: 337–340.

Nelson, M. (2001). "Nitrogen fertilization effects on jojoba seed production". *Industrial Crops and Products.* **13**: 145–154. doi:10.1016/s0926-6690(00)00061-3.

Nevil, D.J. (1980) Studies on resistance to foliar diseases in groundnut. *Proceddings of the International Workshop on Groundnuts,* 1980. Hyderabad., India.pp. 199-202.

Nicolas Carels (2013) Birth of New Energy Crop. In: Nicolas Carels, Mulpuri Sujatha and Bir Bahadur(Eds) (2013*) Jatropha, Challenges for New Energy Crop.*2013. Vol **1:** *Farming, Economics and Biofuel.* 10.1007/978-1-4614-4806-8_199© Springer, New York.

NICRA team of Groundnut Pest Surveillance, (2011). Manual for Groundnut Pest Surveillance. Jointly published by National Centre for Integrated Pest Management, New Delhi, Central Research Institute for Dryland Agriculture, Hyderabad and Directorate of Groundnut Research, Gujarat. 29 pp.

Nie, Z., Chen, F. and Shi, X.C. 1988. A study on selection indices of seed yield in safflower. *Oil Crops China***1**: 15–19

Nigam, S. N., Chandra, S., Sridevi, K. R., Bhukta, M., Reddy, A. G. S., Rachaputi, N. R., et al. (2005). Efficiency of physiological trait-based and empirical selection approaches for drought tolerance in groundnut. *Ann. Appl. Biol.* **146**: 433–439. doi: 10.1111/j.1744-7348.2005. 040076.x

Nigam, S.N., Dwivedi, S.L., Gibbons, R.W. (1991) Groundnut breeding: constraints, achievements and future possibilities. *Plant Breeding Abstracts* **1056**:1127-1136

Nigam, S.N., Nageswara Rao, R.C. and Wynne, J.C. (1998) Effects of Temperature and Photoperiod on Vegetative and Reproductive Growth of Groundnut (*Arachis hypogaea* L.), *J. Agronomy & Crop Science* **181**: 117—124 (1998)

Nigam, S.N., Nageswara Rao, R.C. and G.C. Wright (2002) Field screening for drought tolerance in groundnut. In: N.P. Saxena,O'Toole, C. John (Eds). Field Screeningfor drought tolerance in crop plants with emphasis on rice. *Proceedings of the InternationaolmWorkshop for Field screedning for drought tolerance in rice.* Dec.11-14.2000. ICRISAT Pattancehru. India and Rockfellar Foundation, New York, pp. 147-151.

Nigam, S.N., Upadyaya, H.D., Chandra, S., Nageswara Rao, R.C.N., Wright, G.C. and Reddy.,A.G.S. (2005) Gene effects for specific leaf area and harvest index in three crosses of groundnut (*Arachis hypogaea*), *Annals of Applied Biology*https://doi.org/10.1111/j.1744-7348.2001.tb00143

Nigam, S. N., Waliyar, F., Aruna, F. R., Reddy, S. V., Lava Kumar, P., Craufurd, P. Q., and Upadhyaya, H. D. (2009). Breeding peanut for resistance to aflatoxin contamination at ICRISAT. *Peanut Science*,**36:** 42–49. https://doi.org/10.3146/AT07-008.1

Nikam, T.D. and Shitole.M. G (1999). In vitro culture of safflower cv. Bhima: initiation, growth optimization and organogenesis. *Plant Cell Tiss. Org. Cult.* **55**: 15–22.

Nimbkar, N. (2008). Issues in safflower production in India. In: *Safflower Unexploited and World Adaptability. Proceedings of the 7th International Safflower Conference*, Wagga Wagga, New South Wales, Australia Nov. 3 to 6, 2008. ISBN: 978-0-646-50329-5

Nimbkar, N. (2019) Safflower rediscovered, *Times Agricultural Journal,* November 25, 2019

Nimje, P.M. (2015) Soybean Production technology, *AISECT/Agriculture Skill Council of India (ASCI) /National Skill Development Corporation (NSDC)* New Delhi. India. Pp12

NIPHM (2014) ASEA Based IPM Package-Safflower, National Institute of Plant Health Management, Directorate of Plant Prtotection, Quarantine & Storage, Ministry of Agriculture, New Delhi. Pp34.

Nishi, S., Kawata, J. and Toda, M. (1959) On the breeding of interspecific hybrids between two genomes "c" and "a" of *Brassica* through the application of embryo culture techniques. *Jpn. J. Breed.***18:** 215–222 https://www.nfsm.gov.in/StatusPaper/NMOOP2018.pdf

Nishiyama, I. and Inomata, N. (1966) Embryological studies on cross-incompatibility between 2x and 4x in *Brassica. Jpn. J. Genet.,***41**: 27–42

Niu, G., Rodriguez, D., Mendoza, M., Jifon, J. and Ganjegunte, G. (2012) Responses of *Jatropha curcas* to Salt and Drought Stresses, *International Journal of Agronomy.* Volume **2012** |Article ID 632026 | 7 pages | *https://doi.org/10.1155/2012/632026*

NMOOP (2018) Present Status of Oilseed crops and vegetable oils in India, *National Mission on Oilseeds and Oil Palm.**Status Paper.*** pp 1-71

Norden, A.J. (1982) Breeding of cultivated Peanut. In: *Peanut Science and Technology.*American Peanut Research and Education Society., Yoakum, Texas, USA. Pp 95-122.

NRC (1992)National Research Council. 1992. Neem: A tree for solving global problems. *National Academy Press,* Washington D.C., USA. 141 pp.

NTFP (2019) Sal Seed., NTFP –Enterprise for Forest Governance, Andhra Pradesh, Madhya Pradesh and Orissa.*Centre for Peoples' Forestry*ISBN 978 - 81-906691-7-7(SET)https://pubs.iied.org/pdfs/G02282.pdf

Ntiamoah, C. and Rowland, G.G. (1997). Inheritance and characterization of two low linolenic acid EMS-induced McGregor flax (*Linum usitatissimum*). *Can J. Plant Sci.* **77**: 353–358.

Nuwanyakpa, M. and Butterworth, M., 1(987). Effects of urea, molasses, molasses-urea, noug cake and legume hay on the intake and digestibility of teff straw by highland sheep. *Proceedings of a workshop held at Ryall's Hotel, Blantyre, Malawi, September 1986 Ed. Little, D. A. and Said, N. A.,*

Nybom, H. (2004). Comparison of different nuclear DNA markers for estimating intraspecific genetic diversity in plants. *Molecular Ecology***13:** 1143-115

Nykiforuk, Cory L., Shen, Yin, Murray, Elizabeth, W., Boothe, Joseph G., Busseuil, David, Rhéaume, Eric, Tardif, Jean-Claude, Reid, Alexandra, Moloney and Maurice, M. (2011). Expression and recovery of biologically active recombinant Apolipoprotein AIMilano from transgenic safflower (*Carthamus tinctorius*) seeds. *Plant Bbiotechnology Journal*,**9**(2): 250-263.

Obert, B., Bartošová, Z. and Preťová, A. (2004a) Dihaploid production in flax by anther and ovary cultures. *J Nat Fibres***1**:1–14

Obert, B., Dedicova, B., Hricova, A., Samaj, J. and Preťová, A. (2004b) Plant regeneration from flax anther culture: genotype, pretreatment and media effect. *Plant Cell Tissue Organ Cult***79**:233–238

O'Connor, B. J., Robertson, A. J., Gusta, L.V. (1991). Differential stress tolerance and cross adaptation in a somaclonal variant of flax. *J. Plant Physiol.*,**139**: 32–36

Oelke, E.A., E.S. Oplinger, E.S., Teynor, T.M., Putnam, D.H., Doll, J.D., K Kelling, K.A., Durgan, B.R. and Noetzel, D.M.(1992) Safflower, *Alternative Field Crops Manual. University of Wisconsin & University of Minnesota*, USA

Ogasawara,T., Chiba K. and Tada, M. (1993) Production and high yield ofnapthoquinone by a hairy root culture of *Sesamum indicum.Phytochemistry* 33:1095–1098

Ogura, H. (1968). Studies on the new male sterility in Japanese radish with special reference to the utilization of this sterility towards practical raising of hybrid seeds. *Mem. Fac. Agric. Kogoshima Univ*. 639-678

Oh, E. *et al*., (2018)An Improved Screening Method for Evaluation of Resistance to *Phytophthora* Blight in Sesame Cultivars and Accessions.*Plant Breed. Biotech.* 2018 (September) **6**(3):305~308

Ojiambo P.S., Ayicco PO, Nyabundi JD. (1999) Effect of plant age on sesame infection by *Alternaria* leaf spot. *African crop Science Journal*. 1999; **7**:91-96.

Oladosu, Y., Rafii, M.Y. *et al*., (2016) Principle and application of plant mutagenesis in crop improvement: a review. Journal *Biotechnology & Biotechnological Equipment* **30** (1) : 1-16.

Olivera, M.W. de. *et al.,* (1997) Growth of soybean cultivars in soils with in soils with low manganese availability. 11. Manganese content and distribution *Revesta Ceres,* **44** (251): 43-52.

Olson,M. E. (2010). Moringaceae Martinov; Drumstick Tree Family; In: Flora of North America, North of Mexico, Volume **7***: Magnoliophyta: Dilleniidae, Part 2*. Oxford University Press. p. 168. ISBN 978-0195318227.

Olson, M. E. and Carlquist, S. (2008). Nutritional Quality of *Moringa oleifera* leaves. *Bot J Linnean Soc.* 2001; **135**(4): 315-348

Oomah, D. B. (2001) Flaxseed as a functional food source. *J Sci Food Agr.*,**81**:889-894

Oplinger, E., Oeleke, E., Doll, J., Bundy, L., and Schuler, R. (1989). Alternative Field Crops Manual: Flax. Online. *University of Wisconsin Cooperative Extension Service, University of Minnesota Extension Service, and the Center for Alternative Plant and Animal Products. Republished online by Purdue University Center for New Crops and Plants Products.*

Oplinger, E.S., Putnam, D.H., Kaminski, A.R., Hanson, C.V., Oelke, E.A., Schulte, E.E., and. Doll, J.D. (1990). "Sesame". *Alternative Field Crops Manual, University of Wisconsin and University of Minnesota, USA.*

Orf, J.H. (2008) Breeding, Genetics, and Production of Soybeans. - *10.1016/B978-1-893997-64-6.50005-6*

Orhevba, B.A., Sunmonu, M.O. and Iwunze, HI (2013) Extraction and Characterization of *Moringa oleifera* Seed Oil Research & Reviews: *Journal of Food and Dairy Technology,* **1**

(1): 22-27 |July – September, 2013http://www.rroij.com/open-access/extraction-and-characterization-of-moringa-oleifera-seed-oil-22-27.pdf

Orlikowska, T.K. and Dyer, W.E. (1993) *In-vitro* reneration and multiplication of safflower. (*Carthamus tinctorius* L.) *Plant Sciences* **93:**151-157

Ortiz-Burgos S. (2016) Shannon-Weaver Diversity Index. In: Kennish M.J. (eds) Encyclopedia of Estuaries. *Encyclopedia of Earth Sciences Series. Springer, DE.S. ordrecht.*

Orwa, C., Mutua, A., Kindt, R., Jamnadass, R. and Anthony, S. (2009). Agroforestree Database: a tree reference and selection guide version 4.0. *World Agroforestry Centre, Kenya*

Osman, H.E. (1981) Genetic male sterility in sesame (*Sesamum indicum*, L.) reproductive characteristic and possible use in hybrid vigour. ***Seame Safflower Newsletter,*** 7: 36-41.

Osman, H.E. and Yermanos, D.M. (1982)Genetic Male Sterility in Sesame: Reproductive Characteristics and Possible Use in Hybrid Seed Production*Crop Sci.* **22**:492-498.

Osman, M., Wani, S.P., Balloli, T., Sreedevi, K., Srinivasa Rao, CH and D'Silva, E. (2005). Pongamia Seed Cake as a valuable source of plant nutrients for sustainable agriculture. *Indian. J. Fert.* **5** (2):25-26&29-32.

Oudhia P. and Ganguali R.N. (1998). Is *Lantana camara* responsible for Sal-borer infestation in M.P?*Insect Environment*. **4** (1): 5.

Padmavathy, N. and Vijayaraghavan R. R.(2008) Enhanced bioactivity of ZnO nanoparticles-an antimicrobial study. *Sci Technol Adv Mater* **9** (2008) 035-040

Pal, B.P. (1945) Studies in hybrid vigour. I. Notes on the manifestation of hybrid vigour in Gram, *Sesamum,* Chilli and Maize. *Indian journal of genetics and plant breeding,* **5**: 106-121

Palanisamy, K. and Kumar, P. (1997). Effect of position, size of cuttings and environmental factors on adventitious rooting in neem (*Azadirachta indica* A. Juss.). *Forest Ecology and Management* **98** (3): 277-280. P

Pallavi, M., Prasad, R.D. and K. Anjani. (2007). Novel sources of resistance to *Fusarium* wilt in *Carthamus* genus. *In*: Extended Summary: *National Seminar on Changing Global Vegetable Oils Scenario: Issues and Challenges Before India. 29-31 January 2007. Indian Society of Oilseeds Research, Hyderabad,* India. Pp59-60.

Palmer, C.E. and Keller, W.A. (1994) *In-vitro*culture of oilseeds. In: *Plant Cell and Tissue Culture*. Kluwer Academic Publishers. Dordrecht. Netherlands.

Palter, R. and Lundin, R.E. (1970) A bitter principle of safflower: matairesinol monoglucoside. *Phytochemistry,***9** (11): 2407-2409

Panda, B. S. and V. J. Rao, (1976). The genetics of self incompatibility in *Guizotia abyssinica* Cass. *J. Res. Orissa UA&T.***6**: 17 – 25.

Pandey, A.K. and Kumari, A. (2008) Pollination ecology of safflower (*Carthamus tinctorius* linn.) 7th International Safflower Conference Wagga Wagga Australia.

Pandey, D.D. (1997) Standing crop biomass and net primary productivity of *Guizotia abyssinica* Cass. *Environment and Ecology,***15**(3): 564-566.

Pandey I.D., Singh, B. and Sachan, J.N.*Brassica* Hybrid Research in India: Status and Prospects. *The Regional Institute Online Publlishing.* http://www.regional.org.au/au/gcirc/4/263.htm

Pandey, M. K., Monyo, E., Ozias-Akins, P., Liang, X., Guimarães, P., Nigam, S. N., *et al.* (2012). Advances in *Arachis* genomics for peanut improvement. *Biotechnol. Adv*. **30:**639–651. doi: 10.1016/j.biotechadv.2011.11.001

Pandey, N.D., Sukhani, T.R. and Gupta, R.L. (1967) Bionomics of *Achoea janata* Linn. *Labdev Journal of Science and Technology***5** (2): 127-128

Pandey, S. and Gardner, C.O. (1992) Recurrent Selection for Population, Variety, and Hybrid Improvement in Tropical Maize. *Advances in Agronomy*, **48** :1-

Pandey, S.K., Majumdar, E. and Dasgupta, T. (2017) Genotypic Variation of Microelements Concentration in Sesame (*Sesamum indicum* L.) Mini Core Collection*Agric Res* DOI 10.1007/s40003-017-0252-z

Pandey, S.K., Dasgupta,T., Rathore, A and Vemula, A. (2018) Relationship of Parental Genetic Distance with Heterosis and Specific Combining Ability in Sesame (*Sesamum indicum* L.) Based on Phenotypic and Molecular Marker Analysis. *Biochem Genet*. 2018 Jun;**56**(3):188-209. doi: 10.1007/s10528-017-9837-2. Epub 2018 Jan 10

Pandey, V.C. (2017) "Managing waste dumpsites through energy plantations," In: K. Bauddh, S. Bhaskar, and J. Korstad, (Eds)*Phytoremediation Potential of Bioenergy Plants*, 1st ed., Springer, 2017, p. 15

Pantalone, V.R., Wilson, R.F., Novitzky, W.P. and Burton, J.W. (2002) Genetic regulation of elevated stearic acid concentration in soybean oil. J. Am. Oil Chem. Soc. **79**: 549-553

Paramar, P.D., Vyas, H.J. and Rathod, R.R. (2004). Efficacy and economics of insecticides against castor leaf hopper, *Jacobiasca furcastylus (*Ramakrishnan and Menon). *Indian Journal of Plant Protection.* **32** (1):135-137

Parmar K. S., Siddiq E. A. and Swaminathan M. S. (1979)Chemical Induction of Male Sterility in Rice *Indian Journal of Genetics and Plant Breeding*, **39** (3): 529-541

Parameshwarappa, K.G., B.T. Ninganur, V. Rudranaik, G.G. Gulaganji, U.K. Hulihalli, and N.K.B. Patil. 1995. Combining ability analysis of seed yield, percent oil and other yield attributes in safflower. *J. Oilseeds Res.* **12***:* 258–26

Parotta and John, A. (1993). "*Moringa oleifera* Lam. Reseda, horseradish tree. Moringaceae. Horseradish tree family" (PDF). USDA Forest Service, *International Institute of Tropical Forestry*. Retrieved 20 November 2013.

Pascale, A.J. and Damario, E.A. (1997) Changes in Agroclimaticconditions for sunflower cultivation and other oilseeds. In: H.L.S. Tandon (Ed). *.Fertilizer Management in Commercial Crops*. FDCO New Delhi.

Pasricha, N.S. and Tondon, H.L.S. (1993) Fertilizer management in Argentina during the period 1961-90. *Revista Facultad Agronomia (Universidad de Buenos Aires)* **16** (1/2) : 119-125.

Patel, B.D., Maliwal, G.L., Dudhatra, N.P., Kaswala, R.R., Patel, P.U., Raman, S. and Patel, M.L. (1998) Effect of drip irrigation on yield and water expenses efficiency of castor (*Ricinus communis* L.*) Gujarat Agricultural University Research Journal*. **24** (1): 86-88.

Patel, I.D.,Patel, P.G., Patel, P.S. and Patel, V.I. (1988) *Jatropha:* A new oilseed crop. In: Ranga Rao, V. and Prasad, M.V.R. (Ed.) Proceedings of the National Seminar on "Strategies for making India self relientin vegetable oils. Indian Society of Oilseeds Research, Directorate of Oilseeds Research. Hyderabad. Pp.342-343.

Patel, H.K., Patel, R.M. and Patel, V.C. (1965) Effect of food and temperature on the duration of various stages of tobacco leaf eating caterpillar, *Prodenia litura* F. *Indian Tobaco,* **15**: 174-176.

Patel, J.S and Narayana, G.V. (1937) Chromosome number in *Pongamia pinnata.Curr.Sci.***5:** 479

Patel, K.S.,Patel, M.K. Patel, G.N.and Pathak, H.C. (2006) Fertigation study in castor, *Ricinus communis* L. Journal of Oilseeds Research, **23** (1) 122-123

Patel, M.M., Naik, N.M., Vyas, H.N. and Patel, A.T. (1986) Evaluation of insecticides against white-fly (*Thialeurodes ricini.* Misra) and jassid (*Empoasca kerri.* Pruthi) infesting castor. *Indian Journal of Plant Protection,* **14** (1): 81-82.

Patel, V.R., Dumancas, G.G., Viswanath, L.C.K., Maples.R. and Subong, B.J.J. (2016) Castor Oil: Properties, Uses, and Optimization of Processing Parameters in Commercial Production *Lipid Insights.* 2016; **9**: 1–12. (Sept. 2016) doi: 10.4137/LPI.S40233

Pathak, A.N and Pathak, A.R. (1972) Effect of sulphur component of some fertilizers on groundnut (*Arachis hypogaea* L). *Indian Journal of Agricultural Research***6** (1): 23-26.

Pathirana, R. (1992). Gamma ray induced field tolerance to *Phytopthora* blight in sesame. *Plant Breed.***108**: 314-319

Pathirana, R. (1994). Natural cross pollination in sesame. (*Sesamum indicum* L.) *Plant Breed.* **112:** 167-170.

Pathirana, R. (1995) Comparison of selection procedures in breeding for seed yield in segregating sesame populations. *Euphytica,* **82**: 73-78

Patil, B.V., Somasekhar, Asiram, K.T., Muzibur Rahman and Sushila, N. (1997) Vegetative growth loss estimation due to serpentine leafminer. *Liriomyza trifolii.* On cotton and castor. *Indian Journal of Entomology* **59** (4): 351-353

Patil, C.B. and Joshi, B.P. (1978) Niger yields can be doubled. *Indian Farming.* March. 1978.

Patil, C.B. (1979) Economics of niger oil. cake and seed Yield and yield attributes of niger as affected by different agro-techniques. *Indian Journal of Agricultural Science* **49** (8):633-666

Patil, H. S., Ranganatha, A.R.G., Paroha, S. and Tripathi, A. (2012). Niger (*Guizotia abyssinica*). In; Breeding of field crops. *Agrobios (India)* pp 567-590.

Patil, M.B., Y.M. Shinde, and K.A. Attarde. (1993). Evaluation of safflower cultures for resistance to Alternaria leaf spot (*Alternaria carthami*) and management strategies. *Proceedings of the 3rd International Safflower Conference, Beijing*, June 14–18, 1993. Li, D. and H. Yuanzhou, Eds., pp. 269–278

Patil, P.S., Deokar, A.B.and Katule, B.K.(1987) Safflower-Breeding (Eds P. S. Patil *et al.*), Mathama Phule Agril. Univ., Solapur (Zonal Research Station), Maharashtra, India, pp. 7-60

Patil, P.S., Patil, A.M and Deokar, A.B. (1992). Line X tester analysis for combining ability in safflower. *J. Maharashtra Agric. Univ*. **17**: 64–66.

Patil V. (2016). Drumstick, the game changer for him. The Hindu. April 3, 2016. http://www.thehindu.com/news/national/karnataka/drumstick-the-game-changer-for-him/article8427905.ece [accessed May 30, 2017].

Patra, G.J. and Mohanty (1987) Stability of cross derivatives of groundnut. *Indian Journal of Agricultural Sceinces,* **57** (5): 294-298

Paul, S. (1992) Carbon storage potential of short rotation tropical tree plantations." *Forest Ecology and Management* **50** (1–2): 31–41. doi:10.1016/0378-1127(92)90312-W

Pavela, R. and Herda, G. (2007) Repellent effects of *Pongamia* oil on settlement and oviposition of the common green house whitefly *Trialeurodes vaparariorum* on Crysanthemum. *Insect Sci.,* **14**(3): 219-224.

Pavelek M., Tejklová E., Bjelková M. (2015) Flax and Linseed. In: Cruz V.M.V., Dierig D.A. (eds) *Industrial Crops.* Handbook of Plant Breeding, vol **9.** Springer, New York, NY

Pawar, D. V. (2017) Studies on sesamum leaf blight caused by *Alternaria sesami* (kawamura) mohanty and behera. *Ph.D. Thesis.*Vasantrao Naik Marathwada Krishi Vidyapeeth, Parbhani, India.

Peanut Base (2019) Changes between the *Arachis hypogaea var. Tifrunner* genome assemblies version 1 and version 2.https://peanutbase.org/peanut_genome_v1_v2.

Pearson, O.H. (1972) Cytoplasmically inherited male sterility characters and flavour component from the species cross *Brassica nigra X Brassica olracea* L. *J. Amr. Soc. Hortic. Sci.*97: 398-402

Pelletier, G. (1988)Cybrids in oilseed *Brassica* crops through protoplast fusion. In. Bajaj, Y.P.S. (Eds) *Biotechnology in Agriculture and Forestry. Legumes and Oilseed Crops –I*10: 418-433

Pelletier, G., Primard, C., Vedel, F., Chetrit, P., Remy Rousselle, R.and Renard, M. (1983) Intergeneric cytoplasmic hybridization in *cruciferae* by protoplast fusion. *Molecular and General GeneticsMGG* 191(2): 244–250

Peng, S., Feng, N., Guo, M., Chen, Y. and Guo, Q. (2008) Genetic variation of *Carthamus tinctorius* L. and related species revealed by SRAP analysis. *Biochem Syst Ecol* 36: 531–538.

Peng, Z., Gallo,M., Tillman, B.L., Rowland, D., and Wang, J. (2016) Molecular marker development from transcript sequences and germplasm evaluation for cultivated peanut (*Arachis hypogaea* L.*). Mol Genet Genomics.* 2016 Feb;291(1):363-81. doi: 10.1007/s00438-015-1115-6. Epub 2015 Sep 11.

Pérez-Vich, B., Fernández-Martínez, J.M., Grondona, M., Knapp, S.J. and Berry, S.T. (2002) Stearoyl-ACP and oleoyl-PC desaturase genes cosegregate with quantitative trait loci underlying high stearic and high oleic acid mutant phenotypes in sunflower.*Theor Appl Genet.*104(2-3):338-349.

Peter, Böger (2002). Herbicide Classes in Development: Mode of Action, Targets, Genetic Engineering, Chemistry. Wakabayashi, Ko., Hirai, Kenji. Berlin, Heidelberg: Springer Berlin Heidelberg. ISBN 9783642594168.

Petraitiene, E. and Brazauskiene, I. (2011) Complex of major fungal diseases in winter and spring oilseed rape in Lithuania, *Thirteenth (13th) International Rapeseed Congress,* June 5-9, 2011. Prague, Czech Republic.

PGRC(2011). Plant Gene Resources of Canada. http://pgrc3.agr.ca/about-propos_e.html (accessed 06.11.11.). www.phytozome.org/flax.ph

Pham, T.,Bui, T., Werlemark, G. Bui, T., Merek, A. andCarlsso, A. (2009) A study of Genetic Diversity in Sesame (*Sesamum indicum,* L.) in Vietnam and Cambodia estimated by RAPD markers.*Genetic Resources and Crop Evolution* 56 (5): 679-690

Phartyal, S.S., Thapliyal, R.C., Koedam, N., Godfroid, S., (2002). *Ex situ* conservation of rare and valuable forest tree species through seed-gene bank. *Curr. Sci.*83, 1351–1357.

Phillips, O. L., and Meilleur, B.A.(1998.) Usefulness and economic potential of the rare plants of the United States: A statistical survey. *Econ Bot.* 52(1):57-68

Pillai, R. S. N., Kumar, H. and Singh, R. B. (1981a). Karyotype analysis of safflower. *Crop Science*21: 809–811.

Pillai, R. S. N., Kumar, H. and Singh RB. (1981b). Translocation homozygotes in safflower identification and frequency of chromosomes involved in the interchanges. *Crop Science*21: 815-818.

Pimratch, S., Jogloy, S., Toomsan, B., Jaisil, P., Sikhinarum, J., Kesmala, T. and Patanothai, P. (2004a). Evaluation of seven peanut genotypes for nitrogen fixation and agronomic traits. *Songklanakarin J. Sci. Technol*., 26: 295-304.

Pimratch, S., Jogloy, S., Toomsan, B., Jaisil, P., Kesmala, T. and Patanothai, A. (2004b) Heritability and correlation for nitrogen fixation and agronomic traits of peanut (*Arachis hypogaea* L.) *Songklanakarin J. Sci. Technol.*, 2004, 26(3): 305-315

Pleanjai, S. and Gheewala, S.H. (2009) Full chain energy analysis of biodiesel productionfrom palm oil in Thailand. *Appl. Energy,* 86 : S209-S214

Poehlman J.M. (1987) Breeding Soybeans. In: *Breeding Field Crops.* Springer, Dordrechthttps://doi.org/10.1007/978-94-015-7271-2_17

Pokharia, A.K. (2008) Record of macrobotanical remains from the Aravalli Hill, Ojiyana, Rajasthan: evidence for agriculture-based subsistence economy. *Curr Sci*94:612–62

Polhill, R.M., Raven, P.H. and Stirton, C.H. (1981) Eolution and systematics of the *Leguminosae* In: R.M. Polhill and P.H. Raven (eds) *Advances in Legume Systematics, Part 1.* pp 1-26. Royal Botanic Gardens, Kew.

Polsoni, L., Knott, LS. and Beversdorf, W.D. (1988). Large scale microspore culture technique for mutation-selection studies in *Brassica napus. Canadian J. Botany.***66**: 1681-1685

Pompeu, A.S. (1983) Cruzamentos entre *Arachis hypogaeae* as especies *Arachis diogoi* e A.spp. (30006, 30005). *Brangantia* **42** (6): 261-265

Pope, S. (2020) Nutritional benefits of Peanut Oil, *The healthy Home Economist.*https://www.thehealthyhomeeconomist.com/peanut-oil/

Potter, T.I., Zanewich, K.P.and Rood, S.B. (1993) Gibberellin physiology of safflower: Endogenous gibberellins and resonse to gibberellic acid. *Plant Growth Regulation.* 12 (1-2): 133-140

Prabhakar, M. and Prasad, Y.G. (2004) Natural parasitisation of semilooper *Achoea janata* L. infesting castor, *Ricinus communis*. L. *Indian Journal of Entomology*66 (4): 301-307

Prabhakaran, K.A.A.J. (1992) Idetification of male sterile sources through wide hybridization and induced mutagenesis in sesame (*Sesamum indicum* L.). *Final Report.IDRC Sesamum improvement Project.School of Genetics, Ac & RI, TNAU*, Coimbatore 641003, India.

Prabhakaran, K.A.A.J., Lavanya, C., Suresh, G. and Dinesh Kumar.V.(2009) Guidelines for quality seed production in castor. *Directorate of Oilseeds Research*, Hyderabad.pp55.

Prabakaran, A. J. and Rangasamy, S.R. (1995) Observations on interspecific hybrids between *S. indicum* and *S. malabaricum*. I. Qualitative characters. *Sesame Safflower Newsl*10:6–10

Prabhu, K.V.(2017)Pental's Transgenic Method Enables Perfect Mustard Hybrid Development. *Smart Indian Agriculture*. Agri-biotechnology • Agriculture Policy • **Briefing**, February 24th, 2017.

Prabhu, K. V., Somens, D. J., Rakow, G. and Gugel, R. K. (1997). Molecular markers linked to white rust resistance. *Theoretical and Applied Genetics* 97: 865–70

Prabu, M. J. (2009) A farmer's experimentation leads to a highly popular drumstick variety. The Hindu. January 29, 2009. Updated September 16, 2010. http://www.thehindu.com/todays-paper/tp-features/tp-sci-tech-and-agri/A-farmerrsquos-experimentation-leads-to-a-highly-popular-drumstick-variety/article15951226.ece

Pradhan, S., Jotwani, M.G. and Rai, B.K. (1962.) The neem seed deterrent to locusts. *Indian Farming*12: 7-11.

Praduman Yadav and K. Anjani. 2017. Assessment of variation in castor genetic resources for oil characteristics. *J Am Oil Chem Soc*. 94:611–617

Prajapat, K *et al.,* (2018). Energy input-output based soybean based cropping systems under different nutrient supply conditions. *Journal of Environmental Biology*, 39: 93-101

Prakash, A. and Rao,J. (2018) Botanical Pesticides in Agriculture. BOOK. Chapter C: 9 pages. *Taylor& Francis Group*. Pp46110.1201/9781315138572.

Prakash, S., Ahuja, I., Upreti, H. C., Kumar, D. V., Bhat, S.R., Kirti, P. B. and Chopra, V. L. (2001). Expression of male sterility in alloplasmic *Brassica juncea* with *Erucastrum canariense* cytoplasm and development of a fertility restoration system. *Plant Breeding*120: 479–82.

Prakash, S., Bhat, S. R., Kirti, P. B., Banga, S. K., Banga, S. S. and Chopra, V. L. (2004). Oilseed-brassica Crops in India: History and Improvement. *Brassica* 6 (3–4): 1–54.

Prakash S and Chopra V L. (1988). Synthesis of alloplasmic *Brassica campestris* as a new source of cytoplasmic male sterility. *Plant Breeding* 101: 253–5.

Prakash, S. and Chopra, V.L. (1990) Male sterility caused by cytoplasm of *Brassica oxyrrina* in *B. campestris* and *B. juncea. Theor. Appl. Genet.*79: 285-287

Prakash, S., Kirti, P. B. and Chopra, V. L. (1995). Cytoplasmic malesterility systems other than *ogu* and *polima*-current status.*Proceeding. 9th International Rapeseed Congress Cambridge* **I**:44–8.

Prakash, S., Kirti, P. B. and Bhat, S. R. (1998). A *Moricandia arvensis* based cytoplasmic male sterility and fertility restoration system in *B. juncea*. *Theoretical Applied Genetics* **97**: 488–92.

Prakash, K.S. and Prakash, B.G. 1993. Yield structure analysis of oil yield in safflower (*Carethamus tinctorius* L.). *Oleagineux* **48**: 83–89.

Prasad, B.R., Khadeer, M.A., Seeta, P. and Anwar, S.Y. (1991) *In-vitro* induction of androgenic haploids in safflower (*Carthamus tinctorius* L.) *Plant Cell Reports* **10**:48.

Prasad, M.N.R. and Gowda, M.V.C. (2006). Mechanisms of resistance to tobacco cutworm (*Spodoptera litura)*and their implications is screening for resistance in groundnut. *Euphytica* **149:** 387-399.

Prasad, M.N.V. (2007) Phytremediation in India –In: *Methods and Reviews.* Niel Willey(Ed). Humana Press, Totowa, NJ USA. Pp435-454

Prasad, M.V.R. (1972a). A comparison of mutagenic effectiveness and efficiency of gamma rays, EMS, NMU and NG in *Triticum durum. Indian J. Genetics*, **32** (3):360-367.

Prasad, M.V.R. (1972b). Use of induced mutants for grain characters in *durum* wheat breeding. *Current Science*, **41** (15):570-571.

Prasad, M.V.R. (1973) Some Considerations on Plant-types for Dryland Agriculture, *Annals of Arid Zone*, **12** (3&4):124-134 September/December, 1973.

Prasad, M.V.R. (1984). Enhancing oil production through improved genotypes of castor and peanut. *Proc. III Congresso Brasileiro de Energia*, October 8, 1984, Rio de Janeiro, Brazil

Prasad, M.V.R. (1985). Aerial podding in groundnut (*Arachis hypogaea*, L.). *Indian J. Genetics.***45**(1):89-91.

Prasad, M.V.R., Mamede, F.B.F. and De Silva, F.P. (1985). Mutational improvement of peanut (*Arachis hypogaea*, L.). *Pesq. Agro. Brasileira* **20** (4):439-445.

Prasad, M.V.R. (1987). Possibilities of heterosis with new genetic variability in groundnut (*Arachis hypogaea*, L.). *Proc. Symp. On Exploitation of Heterosis Accomplishments and Prospects, Indian Soc. Genetics, Plant. Breeding.,* Oct. 15-17, 1987, Marathwada Agricultural Univ., Parbhani, India.

Prasad, M.V.R. (1988). Genetic restructuring of the groundnut plant. *Proc. Conf. on Cytology and Genetics, Indian Soc. Of Genetics and Cytology,* Bangalore. **1**:247-254.

Prasad, M.V.R. (1989). Report on the Progress of Castor Research at AICORPO Centre Similiguda and Evaluation of improved castor production technologies on Farmers' fields in Koraput district of Orissa. *Directorate of Oilseeds Research,*ICAR. Hyderabad, Nov. 1989. P.1-9

Prasad, M.V.R. (1991). Improved agro-economic and management practices for annual oilseed crops in India. *Country report presented at the FAO Regional Expert Consultation of Asian Network on Oilseed Crops, **FAO,Bangkok***, Thailand, Dec 17-20, 1991.

Prasad, M.V.R. (1992). New approaches to groundnut breeding: *Special Session on Groundnut Breeding*, Aug. 17, 1992. *Kharif Oilseeds Research Workers Group Meeting AICORPO-DOR-ICAR,* College of Agriculture, Nagpur. India.

Prasad, M.V.R. (1994a). Oilseeds: Growth should accelerate. *Survey of Indian Agriculture (The Hindu),* 1994, Sec. **3:** 46-53.

Prasad, M.V.R. (1994b) Progress in Sesame research and relevance of hybrid sesame in achieving rapid breakthrough in production and productivity. *HYBRID SESAME:Proceedings of Group Discussion.1994.* Directorate of Oilseeds Research,I.C.A.R. Hyderabad: pp5-7

Prasad, M.V.R. (1994c) Minor oil bearing species of forest origin for diversificatin of vegetable oil production. In: Prasad, M.V.R. *et al.,* (Ed) *Sustainability in Oilseeds. Indian Society of Oilseeds Research,* Dirctorate of Oilseeds Research, Hyderabad. Pp91-98

Prasad, M.V.R. (1994d) Genetic enhancement of groundnut for higher productivity. In: Prasad, M.V.R. *et al.*(Ed).*Sustainability in Oilseeds. Indian Society of Oilseeds Research,* Dirctorate of Oilseeds Research, Hyderabad Pp 17-27.

Prasad, M.V.R. (2009) Oilseed Plants for Bioenergy In:*Emerging Trends in Agriculture,* RICAREA M.S. Swaminathan Award Souvenir **2009**:87-89

Prasad, M.V.R. (2012a) Sustainable Bioenergy through *Pongamia* Vayusap. *Proc. 2nd Natl. Conf. on Recent Advnces in Bioenrgy. Ministry of New and Renewable Energy,* December 7-8, 2012, Kapurthala, (Punjab) India.

Prasad, M.V.R. (2012b) *Pongamia* Based Sustainable Bio-energy System. *J. Ecosyst Ecogr*, **2** (4): 65-66.

Prasad, M.V.R. (2012c) VAYUSAP ™ -The Elite *Pongamia* Tree. *Sustainunance,* **1**(4):2-6.

Prasad, M.V.R. (2013) Genetic Enhancement of *Pongamia pinnata* for Bio-energy. *Proc. Recent Advances in Bio-energy Research* Vol.III (3) *http://www.nire.res.in/e-Book%20 III.pdf.Ministry of New & Renewable Energy, Sardar Swaran Singh National Institute of Renewable Energy*, Govt. of India, Kapurthala, India,

Prasad, M.V.R. (2017) Production of *Moringa* leaf for better nutrtion in Africa. *Ricarea News*, RICAREA,Hyderabad, India,**16** (1&2): 2-4

Prasad, M.V.R. (2019a) Environmental Amelioration through *Pongamia pinnata* based Phytoremediation,*International Journal of Science and Research* **8** (7): 93 – 102

Prasad, M.V.R. (2019b)Some Considerations on Priorities in Breeding of Oilseed Crops. In:M. Sujatha and A. Vishnuvardhan Reddy(Eds.) *A Reorientation on Breeding Perspectives for Kharif Oilseed Crops. Resource Material, Orientation-cum-Training Programme ICAR-Indian Institute of Oilseeds Research* Hyderabad, India Pp 1-13

Prasad, M.V.R and Ankineedu, G. (1994) Proposal on Release of Crop Variety Castor (*Ricinus communis* L.) **48-1** (*Jwala*) submitted to the Central Sub-Committee on Crop Standards Notification and Release of Varieities. *Directorate of Oilseeds Research, Hyderaabad, ICAR*. P: 1-19.

Prasad, M.V.R., Kalpana Sastry, R., Raghavaiah, C.V.and Damodaram, T (Ed) (1994) Proceedings of the National Seminar on "*Sustainability in Oilseeds*" UDC 633.85(063) Indian Society of Oilseeds Research, Directorate of Oilseeds Research. August 02-05, 1993, Hyderabad, pp602

Prasad, M.V.R. and Kaul, S (1980) Problems and prospects in the genetic improvement of groundnut. *Proceedings of the National Seminar on Application of Genetics to the improvement of Groundnut. Tamilnadu Agricultural University, Coimbatore*, India,July 16-17, 1980. Pp127

Prasad, M.V.R., Langa, A. and Consolo, J.P. (2000). Selection of superior genotypes of cashew in Nampula zone of Mozambique. *The Cashew*, Jan-March, 2000: 8-23

Prasad, M.V.R., Mamede, F.B.F. and da Silva, F.P. (1985). Mutational improvement of peanut (*Arachis hypogaea*, L.).*Pesq. Agro. Brasileira*20 **(4)**:439-445.

Prasad, M.V.R. and Muralidharudu, Y. (1991)Calcium uptake in relation to pod development in aerial podding genotype of groundnut (*Arachis hypogaea*, L.). *Proceedings of the International Groundnut Workshop* (Nov. 25-30, 1991), ICRISAT, Patancheru, India.

Prasad, M.V.R. and Nagabhushanam, G.V.S. (1991). Genetic manipulation of canopy to achieve yield advancement in groundnut (*Arachis hypogaea*, L.). *Proceedings* of the *Second International Groundnut Workshop* (Nov. 25-30, 1991), ICRISAT, Patancheru, India.

Prasad, M.V.R. and Ranganatha, A.R.G. (1992). Invetsgations on Neem and Milea for Sustainable Production. *World Neem Conf. held in Bangalore*, India, Feb 24-28. Book of Abstracts: 61.

Prasad, M.V.R., Ranganatha, A.R.G. and Sudhakar Babu, S.N. (1994). Non-traditional oilseed crops. In: R.P. Singh (Ed)"*Management of Dryland Crops, Volume 1 - Oilseeds and Pulses"* Scientific Publishers, Jodhpur, India.

Prasad, M.V.R., Ranganth a, A.R.G., Sujatha, M., Dhingra, M., Ramachandram, M. and Nagaraj, G. (1993) Investigations on Neem and *Melia* for sustainable production. *Proc. World Neem Conference*, Bangalore, India.

Prasad, M.V.R. and Reddy, B.N. (1992). Advances in oilseeds production under dryland situations. In:Singh, R.P (Ed). *"Sustainable Development of Dryland Agriculture in India, Scientific Publishers*, Jodhpur, India. Pp 318-333

Prasad, M.V.R. and Sudhakar Babu, S.N. (1997). Technology for Dryland agriculture. *Monograph in Agriculture and Environment Series*, ICAR, New Delhi, India. 39pp

Prasad, M.V.R. and Swarnalatha, Kaul (1980). Induced mutants in groundnut. *Proc. of the International Groundnut Research Workshop*, October 13-17, 1980, ICRISAT, Pattancheru, India :316

Prasad, M.V.R., Swarnalatha Kaul and Jain H.K. (1984) Induced Mutations of Peanut *(Arachis hypogaea* L.) for canopy and pod characters. *Indian J. Genet and Plant Breeding*. 44: 25-34

Prasad, R andGangopadhyay, G. (2011)Phenomic analyses of Indian and exotic accessions of Sesame (*Sesamum indicum* L.)*Journal of Plant Breeding and Crop Science* **3**(13): 335-351, 15 November, 2011

Prasad, R.D. and Anjani, K. (2008) Sources of resistance to Alternaria leaf spot among *Carthamus* wild species. *Proceedings of the 7th International Safflower Conference* , Wagga Wagga, Australiahttps://safflower.wsu.edu/7th-international-safflower-conference-germplasm/

Prasad, R. D. and Suresh, M. (2012). Diseases of safflower and their maintenance. In: Murthy IYLN, Basappa, H., Varaprasad, K.S. and Padmavathi (Eds) *Safflower Research and Development in the World: Status and Strategies*. *Indian Soc Oilseeds Res*. pp. 97-106

Prasad, T.N.V.K.V., Sudhakar, P., Sreenivasulu, Y., Latha, P., Munaswamy, V., Raja Reddy, K., Sreeprasad,T.S, Sajanlal, P.R. and Pradeep, T.(2012) Effect of nanoscale zinc oxide particles on the germination, growth and yield of peanut. *J Plant Nutr***35** (2012) 905-927

Prasad, Y.G. and Anjani, K. (2001). Resistance to serpentine leafminer *Liriomyza trifolii* (Burgess) in castor (*Ricinus communis* L.). *Indian J. of Agricultural Sciences,***71**(5), 351-2.

Prasada Rao, R.D.V.J., Reddy, A.S., Chander Rao, S., Varaprasad, K.S., Thirumala Devi, K., Nagaraju,Muniyappa, V. and Reddy, D.V.R. (2000). Tobacco streak *ila virus* as a causal agent of sunflower necrosis disease in India. *J. Oilseeds Res*., **17**(2): 400-401.

Pratap, A., Gupta, S.K. and Vikas, (2007) Advances in doubled haploid technology of oilseed rape. *Indian J. Crop Sci*., 2: 267-271

Prayaga L., Lakshmamma, P. and Anjani,K (2001) Enhancement of male Sterility inSafflower by Growth Regulators and Chemicals. *Sesame and Safflower Newsletetr*, **16**: 85-87

Premnath A., Narayana, M., Ramakrsihnan, C., Kuppusamy S. and Chockalingam V. (2016). Mapping quantitative trait loci controlling oil content, oleic acid and linoleic acid content in sunflower (*Helianthus annuus* L.). *Mol. Breed.* 36:106 10.1007/s11032-016-0527-2

Preťová A., Obert B. (2003): Flax (*Linum usitatisimum* L.) – a plant system for study of embryogenesis. *Acta Biologica Cracoviensia*, **45**:15–1

Puangbut, D., Jogloy, S., Vorassot, N. *et al.*, (2009) Variability in response of Peanut (*Arachis hypogaea* L.) genotypes under early season drought. *Asian J. Plant Sciences*, **8**: 254-264.

Pustovoit, V.S. (1964). Conclusionsof work on relative to size and selection and seed production of sunflowers. *Crop Sci.* **14**:616-618.. *Agrobiologiyai*: 672-697(in Russian)

Putnam, D.H., Oplinger, E.S., Hicks, D.R., Durgan, B.R., Noetzel1, D.M., Meronuck1, R.A., Doll, J.D. and Schulte, E.E. (1990) Sunflower, *Altrnative Field Crops Manual*, University of Wisconscin & University of Minnesota, USA.

Puvvala, S.S., Muddanuru, T., Thangella, P.Kumar, A. *et al* (2019) Deciphering the transcriptomic insight during organogenesis in castor (*Ricinus communis* L.), jatropha (*Jatropha curcas* L.) and sunflower (*Helianthus annuus* L.)June 2019 Book bioRxiv 679027; doi: https://doi.org/10.1101/679027 Now published in *3Biotech* doi: 10.1007/s13205-019-1960-9

Qayum, M.A. and Sanghi, N.K. (1994) Red hairy caterpillar management through group action and non-pesticidal methods. *ASW and Oxfam (India) Trust.* Pp 118.

Quilleré, I. and Triboi, A.M., 1987. Dynamique des reserves carbonée chez le colza d'hiver: impacts sur la croissance. *7ème Congrès International sur le Colza Pologne*, 11-14 Mai 1987

Quinn, J. and Myers, R.L.(2002). Nigerseed: Specialty Grain Opportunity for Midwestern US. *ASAH Press*, Alexandria, VA.

Rabab.R and Dixit, Madhurima (1997)Protease inhibitors and *in vitro* digestability of karanj (*Pongamia glabra*) oilseed residue. A comparative study of various treatments. J. *American Oil Chemists Society*, **74**(9): 1161-1164http://doi.org/10.1007/s1174-997-0040-1

Rabedeaux, P. F.; Gaska, J. M.; Kurtzweil, N. C.; Grau, C. R. (2005). "Seasonal Progression and Agronomic Impact ofTobacco streak viruson Soybean in Wisconsin". *Plant Disease*. 89 (4): 391–396. doi:10.1094/pd-89-0391. PMID 30795455

Rabindra Prasad, R., Minz, P. K., Devendra Prasad (2010) Status of prevalence of the naturally occurring predatory fauna of insect pest complex of linseed in the agro-climatic condition of Ranchi, Jharkhand. *Uttar Pradesh Journal of Zoology* 2010 30 **(1)**: 41-45

Rachinskaia OA, Lemesh VA, Muravenko OV, Iurkevich OIu, Guzenko EV, Bol'sheva NL, Bogdanova MV, Samatadze TE, Popov KV, Malyshev SV, Shostak NG, Heller K, Khotyleva LV and Zelenin AV. (2011) Genetic polymorphism of flax *Linum usitatissimum* based on use of molecular cytogenetic markers, *Genetika.* 2011;**47**(1):65-75

Radanovic, A., Miladinovic, D., Cvejic, S., Jockovic, M. and Jocic, S. (2018)Sunflower Genetics from Ancestors to Modern Hybrids—A Review. *Genes* 2018, **9**: 528; *doi:10.3390/genes9110528*

Radovich, T. (2011). In: C.R. Elevitch (ed.). *"Farm and Forestry Production and Marketing Profile for Moringa (revised February 2011)" (P*DF)

Radhika, K., Sujatha, M. and Nageswara Rao, T. (2006) *Thidiazone* stimulates adventitiousshoot regeneration in different safflower (*Carthamus tinctorius* L). *Biologia Platarum,* **5**:174-179

Radwanski, S.A. (1977). Neem tree IV. A plantation in Nigeria, *World Crops Livestock* **29**: 222

Radwanski, S.A. and Wickens, G.E. (1981). Vegetative fallows and potential value of the neem tree (*Azadirachta indica*) in the tropics. *Economic Botany* **35**: 398-414.

Raghavaiah, C.V. and Suresh, G. (2010) Good Agronomic practices for sustaining higher production in castor. In: Hegde, D.M. (Ed) 2010. *Research and Development in Castor: Present Status and Future Strategies*. Indian Society of Oilseeds Research, Hyderabad. Pp36-55

Raghavan, R.S. and Arora, C.M. (1958) Chromosome numbers in Indian medicinal plants—II, *Proceedings of the Indian Academy of Sciences - Section B.*, June 1958, **47**(6) : 352–358

Raghuvanshi, R.K., Sinha, M.K. and Aggarwal, S.K. (1997) Effect of sulhur and zinc in ybean (*Glycine max)* – wheat (*Triticum aestivum)* cropping sequences, *Indian Journal of Agronomy,* **42** (1) : 29-32

Rahman, H. (2013) Review: Breeding spring canola (*Brassica napus* L.) by the use of exotic germplasm. *Canadian J. Plant Sci*., **93**: 363-373

Rahman, M. and Jimenez (2016) Designer Oil Crops, In: Gupta, S.K. (2016) *Breeding oilseed Crops for Sustainable Production*: *Opportunities ad Constraints*. *Academic Press* (AP)/ Elsevier. Pp.363-376

Rahman, M and Rahman, L. (1976) Correlation studies on groundnut. *Bangladeh Journal of Agricultural Sciences*. **3** (1): 23-27

Rai, B. (1995). Development of hybrid cultivars in oilseed *Brassica*: status and strategies. *Journal of Oilseed Research.* **12**(2): 239-244.

Rai, B.K. (1976) Pests of oilseed crops in India and their control. *ICAR, New Delhi.* 121pp

Raina, S.N., Sharma, S., Sasakuma, T., Kishii, M., and Vaishnavi, S. (2005). Novel repeated DNA sequences in safflower (*Carthamus tinctorius* L.) (Asteraceae): cloning, sequencing and physical mapping by fluorescence in situ hybridization. *J. Here*d. **96**: 424–429

Raj, D.N., Ramana, J.V., Rao, S.B.N., Dinesh kumar, D., Suryanarayana, M.V.A.N., Ravindra Reddy, Y. and Prasad, K.S. (2016) Effect of incorporation of detoxified karanja (*Pongamia pinnata)* and neem (*Azadirachta indica)* seed cakes in total mixed rations on milk yield, composition and efficiency in cross bred dairy cows. *Indian Journal of Animal Sciences***86 (4)**: 489–492,

Raja, A. and N. Jayabalan (2011) In vitro shoot regeneration and flowering of Sesame (*Sesamum indicum* L.) cv. SVPR - 1. *Journal of Agricultural Technology* 7(4):1089-1096.

Rajamohan, N. (1976). Pest complex of sunflower *PANS***22**: 546-563.

RajasthanDirect (2013) Global Jojoba World 2013 In Jaipur on July 30 – 31, 2013.http://news.rajasthandirect.com/global-jojoba-world-2013-in-jaipur-on-july-30-31-2013.html

Rajendiran, K. (1996) Impact of ultra violet radiation on growth and chlorophyll content in *Helianthus annuus. Environment and Ecology.* **14** (3) :731-732.

Rajpurohit,T.S. (1993) Occurrence, varietal reaction and chemical control of new powdery mildew *(Erysiphe orontii* cast) of sesame Zimin. *Indian J. Mycol. Pl. Patho.* 1993; **23**:207-307.

Raju, P.R.K. (1982) Studies on yield and resistance attributes in groundnut (*Arachis hypogaea* L.) *Ph.D. Thesis, Andhra Pradesh Agricultural University*, Hyderabad, India.

Raju, A.J.S., and Rao, S.P. (2006) Explosive pollen release and pollination as a function of nectar feeding activity of certain bees in bio-diesel plant *Pongamia pinnata* Pierre (*Fabaceae). Curr.Sci.* **90 (**7):960-967.

Raju, V.K. and Reni, M. (2001). In: K. V. Peter (Ed.) *Handbook of Herbs and Spices. Elsevier.* pp. 207–213. ISBN 978-1-85573-645-0.

Rakouský S. and Tejklová E. (1999): Methods and prospects of flax transgenesis. *Czech Journal of Genetics and Plant Breeding,* **35**: 125–129.

Rakouský S., Tejklová E., Wiesner I., Wiesnerová T., Kocábek T. and Ondřej, M. (1999): Hygromycin B – An alternative in flax transformant selection. Biologia Plantarum, **42**: 361–369

Rakouský S., Tejklová E.and Wiesner I. (2001): Recent advances in flax biotechnology in the Czech Republic. In: Kozlows-ki R., Atanasov A. (eds*): Natural Fibres: Bast Plants in the New Millennium*, June 3–6, 2001, Borovets: 151–155

Rakow, G. and D. Woods. (1987). Outcrossing in rape and mustard under Saskatchewan prairie conditions. *Can. J. Plant Sci.***67**:147-151.

Rakshit, S. (2008) Heterotic gene pool development in maize. Lecture given in the Winter School on Innovative approaches productivity in oilseed crops. Nov. 4-24, 2008, Directorate of Oilseeds Research. Hyderabad.

Rakshit S., Rakshit A. and Patil J. V. (2012) Multiparent intercross populations in analysis of quantitative traits. *J. Genet.* **91**, xx–xx

Ram, R., Catlin, Romero, D.J. and Cowley, C. (1990) Sesame: New approaches for crop improvement. p. 225-228. In: J. Janick and J.E. Simon (Eds.), *Advances in new crops. Timber Press, Portland, OR.*

Ramachandram, M. (1985). Genetic improvement of oil yield in safflower: Problems and prospects. *J. Oilseeds Res.,***2**: 1-9

Ramachandram, M. and Goud, J.V.(1981). Genetic analysis of seed yield, oil content and their components in safflower *(Carthamus tinctorius L.). Theor. Appl. Genet.* **60:**191–195.

Ramachandram, M. and Goud. J.V. (1982). Components of seed yield in safflower *(Carthamus tinctorius L.). Genetica Agraria* **36**: 211–222.

Ramachandram and Goud, (1982A). Extent of perfect and fertilized florets in safflower. Indian J. agric. Sci., **52**(2): 78-80.

Ramachandram, M. and Goud, J. V.(1983). Quantitative inheritance of spine index and its components in safflower. *Indian J. Agric. Sci.,* **53**(5): 311-313.

Ramachandram, M.and Prasad, M.V.R. (1995). In: Rai, M. and Mauria, S. (Eds.), Hybrid Castor Research and Development in India. *Hybrid Research & Development*, New Delhi, pp. 93–104.

Ramachandram, M. and Rao, V.R. (1984). Seed setting as influenced by differences in in anther dehiscence in safflower. *Indian J. Genet.*, **44**(1): 113-116)

Ramachandram, M. and Ranga Rao V (1984) Seed setting as influenced by differences in anther dehiscence in safflower. *Indian Journal of Genetics and Plant Breeding,* **44:** 113-116.

Ramachandram, M. and Ranga Rao, V. (1988) Maintenance and manipulation of varieties, parental lines. *Seed Production in Castor(Ricinus communis)* Directorate of Oilseeds Research, Hyderabad. Pp 39-45.

Ramachandram, M. and Ranga Rao, V. (1990) Seed Production in castor Technology Bulletin. *Directorate of Oilseeds Research,* Hyderabad.pp61.

Ramachandram, M. and Ranga Rao, V. (1991). Some considerations for restructuring traditional plant type of safflower.In: Ranga Rao, V. and Ramachandram, M. (Eds). *Proceedings of the Second International Safflower Conference, January 9-13, 1989.Indian Society of Oilseeds Research, ICAR-Directorate of Oilseeds Research,* Hyderabad, pp133-144.

Ramachandram, M. and Sujatha, M. (1991). Development of genetic male sterile lines in safflower. *Indian J. Genet.* **51**: 268–269.

Ramachandran, C., Peter, K. V. and Gopalakrishnan, P. K. (1980). Drumstick (*Moringa oleifera*): a multipurpose Indian vegetable. *Economic botany*, **34**(3), 276-283.

Ramachandran, S., Singh, K.S., Larroche, C., Soccol, C.R. and Pandey, A. (2007) Oil cakes and their biotechnological applications – A review. *Bioresource Technol.* **98:**2000 -2009

Ramachandran, T.K. and Menon, P.M. (1979). Pollination mechanism and inbreeding depression in niger (*Guizotia abyssinica* Cass). *Madras Agric. J.* **66**(7): 449-454.

Ramadan, M.F. (2012). Functional properties, nutritional value, and industrial applications of niger oilseeds (*Guizotia abyssinica* Cass.). *Critical Rev. in Food Sci.and Nutr.* **52**(1):1-8.

Ramaiah, M., Bhat, A.I., Jain, R.K.*et al* (2001) Isolation of an isometric virus causing sunflower necrosis disease in India. *Plant Dis.* **85**:443

Ramakrishnan, S., Gunasekaran, C.R. and Vadivelu S, 1995. Effect of various leaf extracts on root-knot nematode *Meloidogyne incognita* (Kofoid and White). *Journal of Root Crops,* **21**(1): 43-45

Raman, G.V., Rao, M.S. and Srimannarayana, G. (2000) The efficacy of botanical formulations from *Annona squamosa* L. and *Azadirachta indica,*against semilooper, *Achoea Janata* L. infesting castor in the field. *Journal of Entomological Research.* **24** (3): 235-238.

Ramanamurthy GV (1963) Relationship of cultivated safflower (*Carthamus tinctorius* L.) to the wild species *C. oxyacantha* M. B. *Ph.D. Thesis, University of California, Davis*, USA

Ramanathan, T., 1979. Mutational studies in groundnut. *Ph. D. Thesis*, Tamil Nadu Agri. University., Coimbatore, India.

Ramanjaneyulu, A., Reddy, D.K,Reddy, A.V., Kumar, N., Khayum, A.S., Shankar, V. and Neelim, T. (2014) Upscaling and outsclaing of *Rabi* castor in Andhra Pradesh-opportunities and limitations. *International Journal of Bio-resource and Stress Management,* **5** :138

Ramanujam, S. (1942) Some interspecific hybrids in sasame*(Sesamum orientale,* Retz.)*Curr. Sci.* **11**: 426-428.

RamBhajan, Chauhan, Y.S., Kumar, K., Bhajan, R. and Kumar.K. (1993). Studies on genic male sterility and its use in exploitation of heterosis in *B. campestris*. *Acta-Agronomica Hungarica.* **42**(1-2): 99-102

Rameeh, V.(2014)Cytoplasmic malessterilityand inter and intrasubgenomic heterosis studies in *Brassica* speceis: A Review. *Journal of Agricultural Sciences,* **59** (3) : 207-22

Ramesh, T., Sridevi, O., Shenoy, V.V. and Salimath, P.M. (2003)Study of fertilization barriers in crosses between *Sesamum indicum* and its wild relatives, *Indian Journal of Genetics and Plant Breeding*, **63** (2):132-13

Ramesh Kumar and Padhya, M.A. (1988). In vitro propagation of neem (*Azadirachta indica*) leaves and its callus culture. *Indian Drugs* **25:** 526-527.

Rana, R.S. and Singh, R. (1992) Present status of rapeseed-mustard germplasm in India. In: Kumar, D. and Rai, M., (Eds) *Adv. Oilseeds Res. Vol.1.***Scientific Publishers,** Jodhpur, India. 189-200pp

Rana, R.S., Arora, R.K., Lokanathan, T.R. and Patel, D.P. (1994) Sesame genetic resources in India: their diversity, utilization and conservation. Sesame Biodiversity in Asia: Conservation, Evaluation and Improvement, In: R.K. Arora and K.W. Riley(Eds) *Proceedings of the IBPGR-ICAR/NBPGR Asian Regional Workshop on 'Sesame Evaluation and Improvement' held at Nagpur & Akola in India* 28-30 September, 1993.pp109-134.

Ranganatha, A.R.G., Dudhe, M. Y., Prabhakaran, A. J and Babu, S. N. S. (2013) Sunflower In: Singhal, N.D. (Ed) *Hybrid seed Production Technology in Field Crops*, Kalyani Publishers, New Delhi, pp 125-148.

Ranganatha, A.R.G., Misra. R., Aasfa, T. and Pandey, A.K. (2011) Low cost production technologies for oilseed crops Sesame, Niger, and Linseed. *Training Manual, Directorate of Oilseeds Research,* ICAR, Hyderabad.

Ranganatha, A.R.G., Panday, A.K., Bisen, R., Jain, S. and Sharma, S. (2016) Niger. In: S.K. Gupta(Ed). '*Breeding Oilseed Crops for Sustainable Production, Opportunities and constraints'* Elsevier/ Academic Press. London/San Diego, Ca, USA/: pp 169-200

Ranganatha ARG, Rukminidevi K, and Chander Rao S (2003) Evaluation of sunflower genetic stocks for necrosis reaction. Paper presented at *ISOR National Seminar on Stress Management in Oilseeds, Directorate of Oilseed Research,* Hyderabad, 28–30 January, 2003

Ranga Rao. G., Wightman, J. and Ranga Rao D, (1993). World review of the natural enemies and diseases of *Spodoptera litura* (F.) (Lepidoptera: Noctuidae). *Insect Science and Its Application,* **14:**273–284.

Ranga Rao V (1982) Heterosis for agronomic characters in safflower. *Indian Journal of Genetics*, 42: 364–37

Ranga Rao, V. (1983) Combining ability for yield, percent oil and related components in **safflower.** *Indian J. Genet.***43**: 68–7

Ranga Rao, V. (2006) The Jatropha hype: Promise and Performance. *Proceedings of the Biodiesel Conference Towards Energy Independence - Focus on Jatropha.(9-10 June 2006) at Rashtrpati Nilayam, Bolarum, Hyderabad.* Eds. Brahma Singh *et al.*,2006. Rashtrapati Bhawan, New Delhi. P. 16 : https://www.slideshare.net/P7Y/v3m446

RangaRao, V., Ramachandram, M. and Arunachalam, V. (1977)An Analysis of Association of **Components of Yield and Oil in Safflower** (*Carthamus tinctorius* L.) *Theor. Appl. Genet.* **50**: 185-191 (1977)

Rao, G.U., Sarup, U.B., Prakash, S. and Shivanna, K. R. (1994). Developmentof a new cytoplasmic male sterility system in *B. juncea* throughwide hybridization. *Plant Breeding***112**: 171–4

Rao, K.R. and Vaidyanath, K. (1997). Induction of multiple shoots from seedling shoot tips of different varieties of Sesamum. *Indian Journal of Plant Physiology* **2: 257-261**

Rao, K.S.and Rohini, V.K. (1999) *Agrobacterium*-**mediated Transformation of Sunflower** (*Helianthus annuus* L.): A Simple Protocol. *Annals of Botany* **83**(4): 347-354

Rao, M.V.S., Rao, Y.V., Rao, V.S. and Manga, V. (1988). Induction and growth of callus in *Azadirachta indica* A. Juss. *Crop improvement* **5** (2): 203-205.

Rao, N.G.P. (1976) Groundnut in India. Present Status and Future Strategy. *Indian Agricultutal Research Institute Regional Station,* Hyderabad, India. Pp56.

Rao, P.S., Bharathi, M. and Bayyapu Reddy, K. (2013) Identification of Peanut (*Arachis hypogea* L.) varieties through chemical tests and electrophoresis of soluble proteins, *Legume Res.*, **36** (6): 475 - 483

Rao, P.V.R., Anuradha, G., Jayaprada, M., Gourishankar, V., Reddy, K.R., Reddy, N.P.E. and Siddiq, E.A. (2012) Inheritance of Powdery mildew tolerance in Sesame. *Archives of Phytopathology and Plant Protection.* **45** (4): 404-412

Rao, R.V.S. (2000) Resistance to leaf miner *Aproarema modicella.* In groundnut (*Arachis hypogaea)Indian J. Entomol.* **62** (3), 280-285

Rao, R.V.S., Sridhar, R., Singh, U. and Rao, G.V.R. (1998) Biochemical basis for groundnut (*Arachis hypogaea*) resistance to leaf mner (*Aproaerema modicella) Indian J. Agric. Sci.*, **68**(2) : 104-109

Rao, V.R. (1988) Botany In: Reddy, P.S. (Ed) *Groundnut.Indian Council of Agricultural Research.* New Delhi, India. Pp 24-64.

Rao, V. R. and Ramachandram, M. (1977). Growth and development of safflower (Carthamus tinctorius L.). *Mysore J. Agric. Sci.*, **11**: 3-11.

Rao, Y.P. (1962) Bacterial blight of Sesamum (*Sesamum orientale*, L.) *Indian Phytopath.,***15:** 297

Rapson. S., Wu M, Okada S, Das A, Shrestha P, Zhou X-R, Wood C, Green A, Singh S and Liu Q (2015).A case study on the genetic origin of the high oleic acid trait through FAD2-1 DNA sequence variation in safflower (*Carthamus tinctorius* L.). *Front. Plant Sci.* **6**:691.

Rashid, K. Y. (2003). Principal diseases of flax. In: *Flax, the genus Linum*, Muir, A. D. and Westcott, N. D. (eds.) *Taylor & Francis*, London, UK. Pp. 92–123

Rasmusson, D.C.(1991). A plant breeder's experience with ideotype breeding. *Field Crops Res.,***26**: 191-200.

Rathore, A. L., Pal A. R., and Sahu, K. K. (1999) Tillage and mulching effects on water use, root growth and yield of rainfed mustard and chickpea grown after lowland rice, *Journal of Science of Food and Agriculture*, **78**(2): 149–161, (1999). View at Google Scholar

Rathore, M., Tripathi, Y.C. and Kumar, S. (1998). Inter-dependence of seed quality, oil and protein content with reference to neem seeds from different origins. *Annals of Agri-Bio Research.,***3**: 133-138.

Rathore, S.S., Shekhawat,K., Dass, A., Kandpal, B.K. and Singh, V.K. (2017) Phytoremediation Mechanism in IndianMustard (*Brassica juncea*) and Its Enhancement Through AgronomicInterventions. *Proc. Natl. Acad. Sci., India, Sect. B Biol.Sci.* DOI 10.1007/s40011-017-0885-5ISSN 0369-8211

Rathnakumar, A.L. (2014) Groundnut (*Arachis hypogaea* L). In: D.N. Bhardwaj (Ed) *Breeding Field Crops. Agribios India,* Jodhpur, India. Pp: 340-383

Ratnabhargavi, A. R.(2013). Setting up of a Mahua oil extraction unit. Vidarbha Livelihood Forum, *Oxfam India.http://vidarbhalivelihoodforum.org/wpcontent/uploads/Draft_DPR_MahuaOilUnit.doc...*

Ratzka, A., Vogel, H., Kliebenstein, D. J., Mitchell-Olds, T. and Kroymann, J. (2002). "Disarming the mustard oil bomb". *Proceedings of the National Academy of Sciences.* **99** (17): 11223–11228.

Ravi KS, Buttgereitt A, Kitkaru AS, Deshmukh S, Lesemann DE, and Winter S (2001) Sunflower necrosis disease from India is caused by an *Ilarvirus* related to *Tobacco streak virus. New Dis Rep***3:**1–2

Ravi Kumar, Jagadeesha, B., Gajanana Kustagi, N., Lingamurthy, K.R., Ashoka, N. and Nagaraju (2019) Sap Transmission and Detection of Sunflower Necrosis virus in Sunflower,*Int.J.Curr.Microbiol.App.Sci* (2019) **8**(2): 3203-3208

Rawlings ND, Tolle DP, Barrett AJ (March 2004). "Evolutionary families of peptidase inhibitors". *Biochem. J.* **378** (Pt 3): 705–16. doi:10.1042/BJ20031825. PMC 1224039. PMID 14705960.

Rawat, D. S. and Anand, I. J. (1979). Male sterility in Indian mustard. *Indian Journal of Genetics and Plant Breeding* **39**: 412–5

Raza, G., Siddique, A., Khan, I.A., Ashraf, M.Y. and Khatri, A. (2009). Determination of essential fatty acid composition among mutant lines of canola (*Brassica napus*) through high pressure liquid chromatography. *J. integr. Plant Biol.* **51**: 1080-1085

Raza, M.A. (2019)Effect of planting patterns on yield, nutrient accumulation and distribution in maize and soybean under relay intercropping systems. *Scientific Reports***9**: Article number: 4947 (2019)

Ray, S.D. Sen, S. (1992) Heterosis in Sesame (*sesamum indicum* L.)*Tropical Agriculture* (Trinidad and Tobago) *ISSN* : 0041-3216

Reddy, A.T., Reddi, M., Sekhar., Vijayabharathi, A., Lakshmipathy, T., Lakshmikantha Reddy, G. and Jayalakshmi, V. (2017) Studies on Combining Ability and Heterosis for Yield and its Component Traits in Groundnut (*Arachis hypogaea* L.) *Int.J. Curr.Microbiol. App. Sci.,***6**(12): 551-559

Reddy, A.V. and Prabhakaran, A.J. (2010) Challenges in Maintaining Genetic Purity in Seed Production of Castor.In: Hegde, D.M. (Ed) 2010. *Research and Development in Castor: Present Status and Future Strategies. Indian Society of Oilseeds Research, Hyderabad.* Pp 91-109.

Reddy, C.D. and Haripriya, S (1993) Studies of powdery mildew reaction and yield of sesame parents and hybirds. *Journal of Research of Andhra Pradesh Agricultural University (*APAU).**21** (2):109-110.

Reddy, M. P., Arsul, B.T., Shaik, N. R. and Maheshwari, J. J. (2013). Estimation of heterosis for some traits in linseed (*Linum usitatissimum* L.). *J. Agri and Vet. Sci.* **2**(5): 11-17

Reddy, N.S., Ramesh, G. and Suryanarayana, B. (2009) Evaluation of Tree species under different land use systems for higher Carbon sequestration. *Indian J. Dryland Agri. Res. & Dev.* **24**(2):74-78.

Reddy, P.S. (1988) Genetics, breeding and varieties. In: P.S. Reddy (Ed.)*Groundnut.* Indian Council of Agricultural Research. New Delhi. Pp. 200-317

Reddy, R.P., Rao, K.V. and Rao N.G.P. (1984)Implications of variety x season interactions in groundnut breeding. *Indian Journal of Genetics and Plant Breeding,* **44**: 372-378

Reddy Prasad, D.M., Izam, A. and Khan, M.M.R. (2012) *Jatropha curcas*: Plant of medical benefits, Review. *Journal of Medicinal Plants Research* **6**(14): 2691-2699. DOI: 10.5897/ JMPR10.977 ISSN 1996-0875 ©2012 Academic Journals

Redlinger., L.M.and Davis, R. (1982) Insect control in postharvest peanuts. In: H.E. Pattee and C.T. Young(Eds). *Peanut Science and technology. American Peanut Research and Education Society,* Yoakum, Texas, USA.pp520-571

Regitano Neto A., de Oliveira Miguel A. M. R., Mourad A. L., Henriques E. A. and Alves R. M. V. (2016). Environmental effect on sunflower oil quality. *Crop Breed. Appl. Biotechnol.* **16:** 197–204. 10.1590/1984-70332016v16n3a30

Rehman, A. (1996) New Mutant Cultivar, *MBNL*, **42:** 27

Rehman, A., Das, M.L., Hawildar, M.A.R. and Mansure, M.A. (1987) Promising mutants in *Brassica campestris. MBNL*. **29** :14-15

Rehman, K.A. and Singh, S. (1944) Bees and the W effort. *Ind. Bee J.,***6:** 95-96.

Renard, M., Delourme, R., Vallée, P. and Pierre, J. (1997). Hybrid rapeseed breeding and production. *Acta Horticulturae*. 1997; **459**: 291–298.

Rengasamy, S., Kaushik, N., Kumar, J., Koul, O. and Parmar, B.S. (1993). Azadirachtin content and bioactivity of some neem ecotypes of India. In: *Proceedings of the World Neem Conference 1993, Bangalore, India.* pp. 207-217.

Rengasamy, S. and Parmar, B.S. (1994). Azadirachtin content at different states of flowering and fruiting in neem. *Pesticide Research Journal.,***6** (2): 193-194.

Rengasamy, S., Parmar, B.S. and Rengasamy, S. (1995). Azadirachtin-A content of seeds of neem ecotypes in relation to the agro-ecological regions of India. *Pesticide Research Journal***7** (2): 140-148.

Renouard S., Tribalatc M.A., Lamblin F., Mongelard G., Fliniaux O., Corbin C., Marosevic D., Pilard S., DemaillyH., Gutierrez L., Hano C., Mesnard F., Lainé E. (2014): RNAi-mediated pinoresinol lariciresinol reductase gene silencing in flax (*Linum usitatissimum* L.) seed coat: consequences on lignans and neolignans accumulation. *Journal of Plant Physiology,* **171**: 1372–1377.

Rerkasem, B.*et al.*, (1997) Relationship of seed boron concentration to germination and growth of soybean (*Glycine max). Nutrient Cycling in Agroecosystems.* **48** (3): 217-223

Rheenen, H.A.van. (1981) Genetic resources of *Sesamum* in Africa. Collection and Exploration. In *Sesame: Status and Improvement. Proceedings of the Expert Consultation.Rome, Italy* 8-12 December 1980. FAO Plant Protection and Production Division-Paper 29. Pp170-172

Ricelli, M. and Mazzani, B. (1964) Manifestations of heterosis in development of earliness and yield in diallel crosses of 32 sesame cultivars. *Agron. Trop. Venezuela.* **14**:101-125

Richharia, R.H. (1937). A note on the cytogenetics of *Ricinus communis* Linn. *J. Agric. Sci.* **7**: 707–711

Richaria, R.H. (1939) Progress Report on the Oilseeds Research Scheme. Central Province, Berar. India.

Richharia, R.H. and J.P. Kotval. (1940). Chromosome numbers in safflower. *Curr. Sci.* **9**: 73–74.

Riley, K.W. and Belayneh, H. (1989). Niger. In: G. Röbbelen, R.K. Downey and A. Ashri, (Eds.). *Oil Crops of the World.*McGraw Hill Publishing Company, New York. pp. 394-403

Riley, R. and Law, C.N. (1965) Genetic variation in chromosome pairing. *Adv. Genet.*, **13** :57-114

Rincon, C.A. and Salazar, N. (1997) Description of the stages of sesame development. *Agronomia Trop. (Maracay)* **47** (4): 475-487

Robbelen, G. (1990) Mutation breeding for quality improvement, A case study for oilseed crops. *Mutation Breeding Review No.6.*pp 1-43

Rodrigues, S.D. *et al.,* **(1997)Effect of calcium, boron and zinc deficiencies on yield of soyabean** *(Glycine max)* cultivar. *Revista Agricultural* (Piracicaba). **72** (1):131-145.

Rohilla, H.R., Chillar, B.S. and Singh, H. (2003) Assessment of yield losses caused by *Antigastra catalaunalis* (Dup.) in different genotypes of sesame under agroclimatic conditions of Haryana,*Journal of Oilseeds Research* **20** (2): 315-316

Rogers, C.E. (1992) Insect pests and strategies for their management in cultivated sunflower *Field Crops Research* **30** (3–4):301-332

Rogers, C.M. (1982) The systematics of *Linum* sect. *Linopsis* (*Linaceae*). *Pl Syst Evol* 140, 225 (1982). https://doi.org/10.1007/BF02407299

Rout Sandeep and Nayak Saswat (2015) Effect of Storage Condition and Duration on Seed Germination of *Pongamia pinnata*L. Pierre. *Research Journal of Agriculture and Forestry Sciences* ISSN 2320-6063. **3**(4), 19-24, April (2015)

Rowland, G.G.(1991). An EMS-induced low-linolenic-acid mutant in McGregor flax (*Linum usitatissimum* L.). *Can. J. Plant Sci.* **71**: 393–396

Rowland, G.G., Bhatty, R.S. (1990). Ethyl methanesulfonate induced fatty acid mutations in **flax.** *JAOCS***67**: 213–214. S

Roxburgh. W. (1832). Flora India III. *Thacker Spink and Co. Calcutta.* pp 119-125

Roy, A.K. (1965) Out breaks and new records: Occurrence of Powdery Mildew caused by *Oidiumerysiphoides. Plant Protection Bulletin* **13** :42.

Roy S.G., Mukherjee S.K. and Somnath, B. (2007). *Journal of Mycopathological Research,* 2007; **45**(1):122-128.

Rubis, D. D. (1967). Genetics of safflower seed characters related to utilization. *Conf. USDA Agri. Res. Service Publication,* Albany, CA, pp: 74-93.

Rubis, D.D. (1970). Breeding insect-pollinated crops the indispensable pollinators *Arkansas. Agric. Ext. Seil Misc. Pub.,***127**: 19-21.

Rubis, D.D. (2001). Developing new characteristics during 50 years of safflower breeding. *Vth International Safflower Conference, Williston, N.D.*, U.S.A., July 23-27,2001.

Ruchi (2008)Studies on growth and Development in Sesame {*Sesamum indicuni)* in Relation to Seed Yield. *M.Sc. Thesis, Chaudhary Sarwan Kumar Himachal Pradesh Krishi VishvaVidyalaya Palampur-176 062 (*H.P.) INDIA 70pp

Rudolphi, S., Becker, H.C. and van Witzke-Ehbrecht. (2008) Outcrossing rate of safflower (*Carthamus tinctorius* L.) under the agro-climatic conditions of North Germany.*The Biology of Carthamus tinctorius L. (Safflower). Office of the Gene-technology Regulator,* Dept. of Health. Australian Government. Pp.54.

Ruddle, H.P., Whetten, R., Cardinal, A., Upchurch, R.G. and Miranda, L. (2013) Effect of novel mutation in Delta9-stearoyl-ACP-destutrase on soybean seed and seed oil composition. *Ther. Appl. Genet.* **126**: 241-249.

Rui-López N., Haslam R.P., Venegas-Calerón M., LarsonT.R., Graham I.A., Napier J.A., and Sayanova, O. (2009): The synthesis and accumulation of stearidonic acid in transgenic plants: a novel source of heart-healthy omega-3 fatty acids. Plant Biotechnology Journal, **7**: 704–716.

Saad, A. M., Al-doori and Moaid, Y. H., (2011) **Influence of zinc fertilization levels on growth, yield and quality of some sunflower genotypes (Helianthus annuus L.).** *College of Basic Education Researchers Journal.* **11**(4)

Sabet, K.A. and Dowson, W.J. (1960) Bacterial Leaf spot of Sesame (*Sesamum orientale*, L.). *Phytopath. Z.*, **37**: 252

Sabir, A.M., Bhatti, A.H., Izhar-ul-Haq and Suhail, A.(1999). The Foraging Behaviour and Value of Pollination by Honeybees (*Apis mellifera* L.) in Linseed. *Pakistan Journal of Biological Sciences*, **2**: 645-646.DOI: 10.3923/pjbs.1999.645.646 URL: https://scialert.net/abstract/?doi=pjbs.1999.645.646

Sachan, J.N. and B. Singh. (1986). Linked epistasis for six quantitative traits in Indian mustard. *Theoretical and Applied Genetics*. **71**: 644-648.

Sachan, J.N. and B. Singh. (1988). Genetic analysis of quantitative characters in a cross of Indian mustard. *Indian Journal of Agricultural Science*. **58**(3): 176-179.

Sadakshari, Y.G., Virupakshappa, K. and Siddaramaiah, A.L. (1989) Reactions of sesame genotypes against powdery mildew and bacterial leaf spot.*Current Reseach* **18** (1): 4-5

Saeidi, K., Mirfakhraie, S.and Mehrkhou, F. (2016)Hymenopterous Parasitoids Attacking *Acanthiophilus Helianthi* **Rossi (Diptera: Tephritidae) Pupae in Kohgiluyeh Safflower** Farms of Iran,*Journal of Entomological and Acarological Research,***48** (3): 308-313

Saha, N.C. and Bhargava, S.K. (1980). Physiological studies on the growth and yield of sesame (*Sesamum indicum* L.). *Agriculture Science Digest* **6**: 143-149.

Saha, S.and Paul, A. (2017) Gamma Ray Induced Macro Mutants in Sesame *(Sesamum indicum* L.)*Int.J. Curr.Microbiol. App.Sci* (2017) **6**(10): 2429-2437

Saharan, G.S. (1984) A review of research on rapeseed-mustard pathology in India. *Paper presented at the Annual rabi Oilseeds Workshop at Jaipur.* Aug.6-10, 1984

Saharan, G.S. (1992) Management of rapeseed and mustard diseases.In: Kumar, D. and Rai, M.(Eds). *Advances in Oilseeds Res*. 1 **1992 Scientific Publisher, Jodhpur (India): pp151-18** Fabr

Saharan, G.S and Kadian, A.K. (1984) Epidemiology of *Alternaria* blight of rapeseed-mustard. *Cruciferae Newsletter* **9**: 84-86

Saharan, G.S. and Kaushik, J.C. (1981) Occurrence and epidemiology of powdery mildew of *Brassica. Indian Phytopath.* **34: 54-57**

Saharan GS, Mehta N and Sangwan MS. (2005) Diseases of Oilseeds, *Indus Publishing Company,* New Delhi, pp 475-479

Saharan, G.S. and Verma, P.R. (1989) White Rusts- *A Monograph* IDRC, Ottawa, Canada. 75pp

Said.M., Kamel.S., Bilal, A. and Mahfouz, H. (2013) Impact of insect pollinators on sesame production. ISBN-13:978-659-44290-2. Book

Saini, P., Lakshmayya, L. and Bisht,V.S.(2017) Anti-Alzheimer activity of isolatedkaranjin from*Pongamia pinnata* (L.) pierre and embelin from *Embelia ribes*Burm.f.. *Ayu*, **38**:76-81. www.ayujournal.org

Sajjanar, S.M and Eswarappa, G. (2015) Sesame (*Sesamum indicum, L.)* crop Insect Pollinators with Special Reference to Foraging Activity of Different Species of Honeybees.*IOSR Journal of Agriculture and Veterinary Science (IOSR-JAVS)* e-ISSN: 2319-2380, **8** (11): 09-14

Sakurai, H. and Pokorny, J. (2003) Development and application of novel vegetable oils t ailor-made for specific human dietary needs. Euro *J. Lipid Sci. Tech.* **105**:769-778

Salazar, B., Laurentin, H. and Davila, M. (2006) Reliability of RAPD technique for germplasm analysis of sesame (*Sesamum indicum* L.) from Venezuela. *Interciencia,* **31**: 456-460.

Salam A. L. A. and Ahmed A. A. I. (1997), Evaluation of using the extract of chinaberry fruits, *Melia azedarach* L., in the control of the cotton leafworm, *Spodoptera littoralis*, in Egypt. *Proceedings of the International Conference on Pests in Agriculture*, 6 – 8 January 1997, Le Corum, Montpellier, France, pp. 1159 – 1162

Saleh AA, Ahmed, HU, Todd, TC, Travers SE, Zeller KA, Leslie JF and Garrett, KA. (2010). Relatedness of *Macrophomina phaseolina* isolates from tall grass prairie, maize, soybean and sorghum. *Mol. Eco*., 2010; **19**:79-9

Salunkhe, D.K., Sathe, S.K.and Reddy, N.R. (1983) In: Chemistry and Biochemistry of legumes, Arrora S.K. (ed) Edward Arnold Publ. London. 1983.pp51-109

Salunkhe, D.K. Chavan, J.K., Adsule, R.N. and Kadam, S.S. (1992) *World OilSeeds: Chemistry, Technology and Utilization.* 1992. Van Nostrand Reinhold, New York.

Salvador, A. C., Nunez, I.F., Nagaraja, H., Easaw, P.T. and Vanialingam. T. (1986). Record of Egg Parasites of the Castor Semi-Looper *Achaea janata* (L.) in the Philippines. *Philippine Entomologist.***6**(6): 589-593.

Samadi A., Mahmoozadeh A., HassanzadehGorttapeh A., Torkamani M.R. (2007). Cytogenetic studies in four species of Flax (*Linum* spp.). *J. Appl. Sci.***7:** 2832–2839. https://dx.doi.org/10.3923/ jas.2007.2832.2839

Samadhiya, V.K. (2017) Effect of combined foliar application of micronutrient and nitrogen on quality parameter of safflower in Vertisol of Chhattisgarh.*International Journal of Chemical Studies* 2017; **5**(4): 149-153.

Samuel, S., Scott, P.T. and Gresshoff, P.M. (2013) Nodulation in the legume biofuel feedstock tree Pongamia pinnata. *Agricultural Research*(*Australia*) DOI: 10.1007/s40003-013-0074 https://link.springer.com/article/10.1007/s40003-013-0074-6

Sanbagavalli S., *et al*., (2016) Perspectives of Soybean Based Cropping Systems in India: A Review. *International Journal of Agriculture Sciences,* ISSN: 0975- 3710 & E-ISSN: 0975-9107, **8,** (63): 3578-3585.

Sandanasamy, J., Nour, A.H., Tajuddin, S.N.B. and Nour, A.H. (2013) Fatty Acid Composition and Antibacterial Activity of Neem (*Azadirachta indica*) Seed Oil. *The Open Conference Proceedings Journal,* 2013, **4:** (Suppl-2, M11) 43-48

Sandeep, G., Anitha, T., Vijayalatha, K.R. and Sadasakthi, A. (2019) Moringa for nutritional security (*Moringa oleifera* Lam.) *International Journal of Botany Studies* **4**(1): 21-24 www.botanyjournals.com

Sandeep, K.R. and Kubsad, V.S. (2017) Growth and yield of niger (*Guizotia abyssinica* Cass.) as influenced by planting geometry, fertilizer levels and cycocel spray, *J. Farm Sci.,***30**(2): 172-176.

Sanders, T.H. (1982) *Journal of American Oil Chem. Soc.* 1982, 59: 346-35

Sandhu, B.S, and Khera, A.S. (1977) Genoype X environment interaction in groundnut. *Crop Improvement* **4:** 225-226.

Sankara Rao, K. and Rohini, V.K. (1999). Gene transfer into Indian cultivars of safflower (*Carthamus tinctorius* L.) using *Agrobacterium tumefaciens*. *Plant Biotechnology***16**: 201-206.

Sangha, A.S. (1971) Analysis of diversity in groundnut. II. Spreading group. *Madras Agricultural Journal* **58** (12): 875-881.

Sangha, A.S. and Jaswal, S.V. (1975) Genotype X environment interactions in spreading groundnut. I. Elite strains. *Oilsseed Journal.* **5:** 9-12.

Sangha, A.S. and Sandhu, R.S. (1974) Genetic variability and correlation studies in groundnut. *Oilseeds Journal.* **4:** 22-30.

Sangwan, S., Rao. D.V.and Sharma, R.A. (2010) A review on *Pongamia pinnata*L. Pieere. Agreat versatile leguminous plant. Nature Sci., **8**(11).

Sannappa, B., Manjunath, D., Manjunath, K.G. and Prakash, B.K. (2016) Evaluation of castor hybrids / varieties employing growth and yield characters for exploitation in seed production and ERI silkworm rearing, *International Journal of Applied Research* 2016; **2**(2): 135-140

Santha Lakshmi Prasad, M., Raoof, M.A., Gayatri, B., Anjani, K. and Lavanya, C. (2019) Wilt disease of castor: an overview, *Indian Phytopathology* (2019) **72** (4):575-585

Sarbhoy, R.K. (1977) Cytogenetical Studies in *Pongamia pinnata* (L.) *Cytologia*,**42** (3-4): 415-423 *http://doi.org/10.1508/cytologia.42.415*

Sardaru, P., Viswanath, B., Johnson, A.M.A. and Golla, N. (2014) Sunflower Necrosis Disease – A Threat to Sunflower Cultivation in India. *Annals of Plant Sciences* **2**

Sarkar, M.S. and Datta, P.C. (1986) Age factor in biosynthesis of nimbin and sitoserol in the bark and callus of *Azadirachta indica. Indian Drugs***24:** 62-63.

Sarvesh, A., Reddy, T.P. and. Kavi Kishor, P.B. (1993a). Plant regeneration from cotyledons of niger. *Plant Cell Tiss. Organ Cult.* **32**: 131–13

Sarvesh, A., T.P. Reddy, T.P. and Kavi Kishor, P.B. (1993b). Embryogenesis and organogenesis in cultured anthers of an oil yielding crop niger (*Guizotia abyssinica* Cass). *Plant Cell Tiss. Organ Cult.* **35**: 75–80

Sarvesh, A., Reddy, T.P. and Kavi Kishor, P.B. (1994a). Somatic embryogenesis and organogenesis in *Guizotia abyssinica. In Vitro Cell Dev. Biol.***30**P: 104–107.

Sarvesh, A., Reddy, T.P. and KaviKishor, P.B. (1994b). Androclonal variation in niger (*Guizotia abyssinica* Cass). *Euphytica* **79**: 59–64

Sarvesh, A., Reddy, T.P. and Kavi Kishor, P.B.(1996). In vitro anther culture and flowering in *Guizotia abyssinica* Cass. *Ind. J. Exp. Biol.* **34**: 565–568

Sasanuma, T., Sehgal, D., Sasakuma, T.and Raina, S. N. (2008). Phylogenetic analysis of *Carthamus* species based on nucleotide sequence of nuclear-encoded SACPD gene and chloroplast trnL-F IGS region. *Genome*, **51**: 721–727https://doi.org/10.1139/G08-059

Sasidharan, N., Sivarajan V.V.(1996). Flowering plants of Thrissur Forests. ***Scientific Publishers,*** Jodhpur, India

Sastry, K.R. and Ramachandram, M. (1992). Differential genotypic response to progressive development of wilts of safflower. *J. Oilseeds Res.* **9**: 297–305.

Sato, S.; Hirakawa, H.; Isobe, S.; Fukai, E.; Watanabe, A.; Kato, M.; Kawashima, K.; Minami, C.; Muraki, A.; Nakazaki, N.; Takahashi, C.; Nakayama, S.; Kishida, Y.; Kohara, M.; Yamada, M.; Tsuruoka, H.; Sasamoto, S.; Tabata, S.; Aizu, T.; Toyoda, A.; Shin-i, T.; Minakuchi, Y.; Kohara, Y.; Fujiyama, A.; Tsuchimoto, S.; Kajiyama, S.; Makigano, E.; Ohmido, N.; Shibagaki, N.; *et al.* (2010-12-08). "Sequence Analysis of the Genome of an Oil-Bearing Tree, *Jatropha curcas* L". *DNA Research.* **18** (1): 65-76. doi:10.1093/dnares/dsq030. PMC 3041505. PMID 21149391.

Satyanarayana, J., Ballal. C. R., and Rao, N.S. (2005) Evaluation of egg parasitoid *Telenomus remus* Nixon and larval parasitoid *Compoletis chloridae* Uchida on *Spodoptera litura*(Fabr) in castor.*Indian Journal of Plant Protection***33** (1): 26-29

Savaliya, V.A. (2015) Management of Root-rot of Sesame (*Sesamum indicum* L.) caused by *Macrophomina phaseolina* (Tassi) Goid, *M. Sc(Ag). Thesis, JAU Junagadh*http://krishikosh.egranth.ac.in/handle/1/5810025304

Sawant, A.R. (1990). Annual Progress Report of All India Coordinated Research Project on Oilseeds (Safflower). *Jawaharlal Nehru Krishi Vishwa Vidyalaya, Indore, India, 117 pp.*

Schank, S.C. and Knowles, P.F. (1964) Cytogenetics of Hybrids of *Carthamus* species (Compositae) with Ten Pairs of Chromosomes, *American Journal of Botany* **51**(10):1093-1102

Scheiner, J.D., Gutierrez-Boem, F.H. and Lavado, R.S.(2002.) Sunflower nitrogen requirement and 15N fertilizer recovery in Western Pampas, Argentina. *Eur. J. Agron.* **17**:73–79.

Schilling, E. (1930). Bestehen Beziehungen zwischen Faserqualitat und Blattzhal des Flachsstengels ? Die Deutsche Leinen Industrielle. **48**: 254-255

Schmandke. H. (2001) Development and use of hybrid oils from transgenic plants. *Ernahrungs Umschau* 48: 153-160

Schmutz,J., Cannon,S.B. and Jackson, S.A. (2010) Genome sequence of the palaeopolyploid soybean *Nature* **463** : 178–183 (14 January 2010)

Schnable, P.S. and Wise, R.P. (1998) The molecular basis of cytoplasmic male sterility and fertility restoration. *Trends in Plant Science.* **3**(5): 175-180

Schuppert G. F., Tang S., Slabaugh M. B.and Knapp S. J. (2006). The sunflower high-oleic mutant Ol carries variable tandem repeats of *FAD2-1*, a seed-specific oleoyl-phosphatidyl choline desaturase. *Mol. Breed.* **17** :241–256. 10.1007/s11032-005-5680-y

Schwarzenbacher, H., Dolezal, M., Flisikowski, K. *et al.* (2012). Combining evidence of selection with association analysis increases power to detect regions influencing complex traits in dairy cattle. *BMC Genomics* **13,** 48 (2012). https://doi.org/10.1186/1471-2164-13-48

Sedbrook, J., Phippen, W. and Marks, M. (2014). New approaches to facilitate rapid domestication of wild plant to oilseed crop: example penny cress (*Thalapse arvense* L.,). *Plant Sci.,* **227**: 122-132

Sedgley, R.H., (1991). An appraisal of the Donald ideotype after 21 years. Field Crops Res.,**26**:93-112.

Seervi, D., Choyal, P. and Seervi, K.S. (2018)The effect of micronutrients applied as foliar spray, on yield, yield attributes and oil content of sesame crop (*Sesamum indicum* L.) *Journal of Pharmacognosy and Phytochemistry.* **7**(4): 1402-1404

Seetharam, A. (1972). Interspecific hybridization in *Linum. Euphytica* **21:** 489–495 (1972). https://doi.org/10.1007/BF00039344

Sehgal, D., Rajpal, V. R. and Raina, S. N. (2008). Chloroplast DNA diversity reveals the contribution of the two wild species in the origin and evolution of diploid safflower (*Carthamus tinctorius* L.). *Genome*, **51:** 638–643)

Sehgal D. and Raina S.N. (2011) Carthamus. In: Kole C. (Ed.) *Wild Crop Relatives: Genomic and Breeding Resources*. *Springe*r, Berlin, Heidelberg

Sehgal, D., Raina, S. N., Devarumath, R. M., Sasanuma, T., and Sasakuma, T. (2009). Nuclear DNA assay in solving issues related to ancestry of the domesticated diploid safflower (*Carthamus tinctorius* L.) and the polyploid (*Carthamus*) taxa, and phylogenetic and genomic relationships in the genus *Carthamus* L. (Asteraceae). *Molecular Phylogenetics and Evolution*, **53**: 631–644.

Sehgal, D., Rajpal, V. R., Raina, S. N., Sasanuma, T., and Sasakuma, T. (2009). Assaying polymorphism at DNA level for genetic diversity diagnostics of the safflower (*Carthamus tinctorius* L.) world germplasm resources. *Genetica* **135:** 457–470.

Sehr E. M., Okello-Anyanga W., Hasel-Hohl K., Burg A., Gaubitzer S., Rubaihayo P. R., *et al.* (2016). Assessment of genetic diversity amongst Ugandan sesame (*Sesamum indicum* L.) landraces based on agromorphological traits and genetic markers. *J. Crop Sci. Biotechnol.* **19:**117–129. 10.1007/s12892-015-0105-x

Seiler, G.J. and Gulya, T.J. (2004) Exploration for wild *Helianthus* species in North America: Challenges and Opportunies in the search for Global Treasures.*Proc. 16th International Sunflower Conference*, Fargo, ND USA

Sekhri S. (2011) "Textbook of Fabric Science: Fundamentals to Finishing". PHI Learning Private Limited, New Delhi, p. 76

Selvi, B. and Kadamban. D. (2009) Studied on the parasitic plants of Pondicherry Engineering College Campus, Puducherry. *International Journal of Plant Sciences*, (July to December, 2009) **4** (2): 547-550

Sem-Partners, (2016). Niger ASO-FIX. Coupez les rotations céréalières à bas coûts. Sem-Partners, la nouvelle dynamique des semences

Semenov, V.G. (2003)Optimization of composition of binary alternative diesel fuel. Chem. Tech. Fuel Oils. **39**: 192-196

Sengupta, A., and Gupta, M. P. (2000) Studies on the seed fat composition of Moringaceae family. *Fette, Seifen, Anstrichm.* 2000;**72**: 6–10

Sengupta, R.P. (2013) Ecological Limits and Economic Development. *Oxford University Press,* New Delhi.

Seo, H.Y., Kim, Y.J., Park, T.I., Kim, H.S., Yun, S.J., Park, K.H., Oh, M.K., Choi, M.Y., Paik, C.H., Lee, Y.S. and Choi, Y.E. (2007). High-frequency plant regeneration *via* adventitious shootformation from de-embryonated cotyledon explants of *Sesamum indicum* L. *In VitroCellular Developmental Biology - Plant.* **43**: 209-214

Serieys H., (1996). Identification, study and utilization in breeding programs of new CMS sources. FAO Progress Report. *Helia* **19**: 144–158. [Google Scholar] [Ref list]

Sethupathi Ramalingam, R., SreeRangaswamy, S.R. and Narayana, A. (1994) Identification of cytoplasmic –genic male sterile (CMS) sources in sesame through reciprocal crosses. *HYBRID SESAME: Proceedings of Group Discussion.1994.* Directorate of Oilseeds Research, I.C.A.R. Hyderabad: pp 23-26.

Severino, L.S., Costa, F.X., Beltrão, N.E.M., Lucena, A.M.A. and Guimarães, M.M.B. (2004) Mineralização da torta de mamona, esterco bovino e bagaço de cana estimada pela respiração microbiana. *Revista de Biologia Ciênc. da Terra*, **5** (2004), pp. 1-6 (in Portuguese)

Shah, S.A., Ali, I., Iqbal, M.M., Khattak, S.U. and Rahaman, K. (1999) Evolution of high yielding and early flowering variety of rape seed (*Brassica napus* L.,) through the *in-vitro* mutagenesis. In: *Proceedings of the International Symposium. New Genetical Approaches to Crop Improvement-III, Nuclear Institute of Agriculture,* Tandojam, Pakistan. Pp47-53

Shah, S.A., Ali, I. and Rhman, K. (1990) Induction and slection of superior genetic variables of oilseed rape *Brassica napus* L. *Pak. J. Bot.***7** :37-40

Shalini.Y., Sagar, B.V., Giribabu, P. and Krishna Rao, V (2014) - Management of wilt disease complex in castor. *Indian Journal of Plant Protection.***42**: 255-258

Shanthi, K., Manimuthu, L. and Singh, B.G. 1996. Genetic significance of late flowering forms in neem (*Azadirachta indica* A. Juss.) as reflected by germination studies. *Indian Forester***122** (3): 263-264.

Sharma, D. (2001) India's oilseeds revolution. *Business Line.* Oct.01, 2001.

Sharma, H.C., Pampapathy, G. and Kumar (2002) Techniqueto screenpeanutsfor resistance to tobacco army worm *Spodoptera litura* under no choince cage conditions. *Peanut Sci.,***29:** 35-40

Sharma, H.C., Pampapathy, G., Dwivedi, S.L. and Reddy, L. J. (2003) Mechanisms and Diversity of Resistance to Insect Pests in Wild Relatives of Groundnut. *Journal of Economic Entomology* **96**(6):1886-97 · January 2004 DOI: 10.1603/0022-0493-96.6.1886

Sharma, M. and Pareek, L.K. (1998). Direct shoot bud differentiation from different explants of in vitro regenerated shoots in sesame. *Journal of Phytological Research***11**: 161-163.

Sharma, N.K. and Anand, I.J. (1994) Genetic enhancement of Oilseed crops: rapeseed-mustard. In: Prasad, M.V.R., Kalpana Sastry, R., Raghavaiah, C.V. and Damodaram, T.(Eds). *Sustainability in Oilseeds. Indian Society of Oilseeds Research, Directorate of Oilseeds Research.* Hyderabad. Pp48-53

Sharma, S., Kumar, V. and Mathur, S. (2009) Comparative Analysis of RAPD and ISSR Markers for Characterization of Sesame (*Sesamum indicum* L) Genotypes. *Journal of Plant Biochemistry and Biotechnology* **18**(1) DOI: 10.1007/BF03263293

Sharma, S.K. and Gupta, J.S. (1979) Biological control of leaf blight of brown *sarson* caused by *Alternaria brassicae* and *Alternaria brassicola. Indian Phytopath.* **31:** 448-449

Sharma, S.K. and Gupta J.S. (1980) *Streptomyces rochei* in relation to *Alternaria brassicae* and *Alternria brassicola* on the surface of brown *sarson. Indian J. Bot.***59**: 161-163

Sharma, S.M. (1994) Present status of Hybrid sesame with specific reference to commercial utilization. *HYBRID SESAME: Proceedings of Group Discussion,* ICAR-Directorate of Oilseeds Research, Hyderabad, India.8-15

Sharma, R.S. and Sharma, U. (1997) Effect of sowing date and sampling time on nectar sugar content in Sunflower. *Crop Research (Hissar)* **13** (1): 151-155

Sharma, T. R. and Singh, B. M. (1992). Parameters of resistance to *Alternaria brassicae* in some *Brassica* species. *Indian Phytopathology***45**(Suppl.) P. LIX (Abst.)

Sharma, V.P. (2014) Problems and Prospects of Oilseeds Production in India. Final Report. *Centre for Management in Agriculture (CMA) Indian Institute of Management (IIM) Ahmedabad* 380 015 November 2014: Pp 1-225.

Sharman, M., Thomas, J.E. and Persley, D.M (2008) First report ofTobacco streak virus in sunflower *(Helianthus annuus*), cotton (*Gossypium hirsutum),* chickpea *(Cicer arietinum)* and mung bean *(Vigna radiata*) in Australia. December 2008.*Australasian Plant Disease Notes* **3**(1):27-29. DOI: 10.1007/BF03211228

Shchori, Y. and Ashri, A. (1970) Inheritance of several macromutations induced by diethyl sulphate in peanuts. *Arachis hypogaea* L. *Radiation Botany* **10:** 551-555

Shehu, H.E. (2014)Effects of Manganese and Zinc Fertilizers on Shoot content and Uptake of N, P and K in Sesame (*Sesamum indicum* L.) on Lithosols. *International Research Journal of Agricultural Science and Soil Science* (ISSN: 2251-0044), **4**(8) pp. 159-166.

Shekhawat, K., Rathore,S. S. , Premi,O. P., Kandpal,B. K. ,and Chauhan,J. S. (2012)Advances in Agronomic Management of Indian Mustard (*Brassica juncea* (L.) Czernj. Cosson)- An Overview. *International Journal of Agronomy*Article ID 408284, 14 pages http://dx.doi.org/10.1155/2012/408284

Shetgar, S.S., Bilapte, G.G., Puri, S.N., Patil, V.V. and Londhe, G.M., (1992) Assessment of yield loss in safflower due to aphids. Journal of *Maharashtra Agricultural Universities,* **17:** 303-304

Shifriss, O. (1960). Conventional and unconventional systems controlling sex variations in Ricinus. *J. Genet.* **57:**361

Shiga, T. and Baba, S. (1973) Cytoplasmic Male Sterility in Oil Seed Rape. *Brassica napus* L., and its Utilization to Breeding, *Japanese Journal of Breeding.* **23** (4): 187-197

Shim. Y.Y., Gui, B., Wang, Y. and Reaney, M.J.T (2015) Flax seed (*Linum usitatissimum* L) oil processing and selected Products.*Trends in Food Science and Technology* **43** (2): 162-167

Shin, D.H., Kim, J.S., Kim, I.J., Yang, J., OH, S.K., Chung, D.C. and Han, H.K. (2000) A shoot regeneration protocol effective on diverse genotypes of sunflower (*Helianthus annuus*, L.), *In Vitro Cell Dev. Biol. Plant.* **36**: 273-278

Shirazi, B.J., Mahbub, M.M., Some, T.A. and Islam, S. (2017) Effect of Nitrogen rates and foliar spray of micro nutrients on growth and yield of sesame. *American Journal of Plant Biology.***3** (2): 1-21

Shirshikar SP. 2003. Influence of different sowing dates on the incidence of sunflower necrosis disease. *Helia,* 26(39): 109-116.

Shivakumar, M., Verma, K., Talukdar, A., Lal, S.,Kumar, A. and Mukherjee, K. (2016) Introgression of null allele of Kunitz trypsin inhibitor through marker-assisted backcross breeding in soybean (Glycine max L. Merr.)BMC Genetics. **17:**10.1186/s12863-016-0413-2

Shivani, D.,Sreelakshmi, C.and Kumar, C.V.S. (2011) Inheritance of resistance to *Fusarium* wilt in safflower (*Carthamus tinctorius* L.) genotypes. *J Res ANGRU***39**:25-27

Shobha Rani, T. and Ravikumar, R.L. (2006) Sprophytic a d gametophytic recurrent selection for improvement of partial resistance to *Alternaria* leaf blight in sunflower (*Helianthus annuus*, L.) *Euphytica,***147**: 21-431.

Shou, H.X. *et al* (1998). Study on the difference of resistance to seed deterioration between cultivated and wild soybean genotypes. *Soyabean Science.* **17** (1): 59-64.

Shrinivasa, U. (2001) Honge (Pongamia) Oil proves to be a good biodiesel.*Good news India.2001 http://goodnewsindia.com/Pages/content/discovery/honge.html*

Shrivastava, P.S. and Shomwanshi, K.P.S. (1974). Investigation on the extent of cross pollination and selfing and crossing techniques in niger (*Guizotia abyssinica* Cass). *JNKV Res. J.***8**:110-112

Shukla, A.K. (1995) Effect of intercropping on major diseases and yield of soybean. *Legume Research,***18**(3/4): 217-219

Shysha E.N., Korkhovyu V.I., Bayer G.I.A., Guzenko E.V., Lemesh V.A., Kartel N.A., Yemets A.I., Blume Ya.B. (2013): Genetic transformation of flax (*Linum usitatissimum* L.) with chimeric GFP-TUA6 gene for visualisation of microtubules. *Cytology and Genetics*, **47:** 3–11

Siddiq E.A., Govila O.P., Banga S.S. (2004) Heterosis in Crop Improvement. In: Jain H.K., Kharkwal M.C. (eds) Plant Breeding. Springer, Dordrecht [https://doi.org/10.1007/978-94-007-1040-5_19]

Silva. N., Mendes-Bonato, A. B., Sales, J. G. and Pagliarini, M. S. (2011) Meiotic behavior and pollen viability in *Moringa oleifera* (Moringaceae) cultivated in southern Brazil. *Genet Mol Res.* 2011;**10**(3):1728-32.

Simpson, C.E. (1991) Pathways for introgression of pest resistance into *Arachis hypogaea* L. *Peanut Sci***18**:22-26.

Simpson, C. E. (2001). Use of wild *Arachis* species for introgresion of genes into *A. hypogaea* L. *Peanut Sci.,***28**(2): 114-116.

Simpson C.E., Krapovickas A. and Valls J.F.M. (2001) History of *Arachis* included evidence of *Arachis hypogaea* progenitors.*Peanut Sci.* 2001; **28**: 78-80

Simpson, C. E., Starr, J, L., Church, G. T., Burrow, M. D., and Paterson, A.H. (2003) Registration of NemaTAM peanut. *Crop Sci.,***43**:1561

Simopoulos, A.P. (1999). Essential fatty acids in health and chronic disease. *Am J Clin Nutr.* 1999; **70**(3): 560S–569S.

Simopoulos, A. P. (2000). Human requirement for N-3 poly-unsaturated fatty acids. *Poult Sci.*, **79**: 961–970

Sinclair, T.R. and DeWitt, C.T. (1975) Photosynthate and nitrogen requirements for seed production by various crops. *Science*, **189**: 565-567.

Sindhu Kanya, C. and Kantaraj-Urs, M (1989). "Studies on taramira (*Eruca sativa*) seed oil and meal". *Journal of the American Chemical Society.* **66** (1): 139- 140. doi:10.1007/BF02661804.

Singaravelu, B. and Ramakrishnan, N. (1998) Characterization of granulosus virus from castor semilooper *Achoea Janata* L. *Journal of Invertebrate Pathology,* **71** (3): 227-235

Singh, A. (2013) Comparative Performance and Restoration Potential of *Jatropha curcas* and *Pongamia pinnata* on Degraded Soil of North India, *Ph.D. Thesis, University of Lucknow*, National Botanical Research Institute, (CSIR)Lucknow 226 001, U.P., India.pp 198

Singh, A.K and Simpson, C.E. (1994) Biosystematics and genetic resources In: Smartt, J. (Ed) ***The Groundnut Crop: A scientific Basis for Improvement. Chapman & Hall,*** London, pp 96-137

Singh, A.L. (2004) Growth and Physiology of Groundnut. In: M.S. Basu and N.B. Singh (Eds) *Groundnut Research in India, Edition: 1National Research Center for Groundnut (ICAR), Junagadh, India.*pp 178-212.

Singh, A.L. (2011) Physiological basis for realizing yield potentials in groundnut. *Advances in plant Physiology*, **12:** 131-242

Singh, A.S., Singh, M. and Labana, K.S. (1979) Variability and correlation studies in groundnut after hybridization. *Madras Agricultural Journal***66** (9): 565-570

Singh, B., Sachan, J.N. Ram Bhajan and Singh, S.P. (1987). Inheritance of male sterility in yellow sarson *B. campestris* var. ys. *Oil Crops Newsletter*. **4**: 31-32.

Singh, B.G., Mahadevan, N.P., Shanthi, K., Geetha, S. and Manimuthu, L. (1995). Multiple seedling development in neem (*Azadirachta indica).Indian Forester***121** (11): 1049-1052.

Singh, B.G., Mahadevan, N.P., Shanthi, K., Manimuthu, L. and Geetha, S. (1997). Effect of moisture content on the viability and storability of *Azadirachta indica* A. Juss (neem) seeds. *Indian Forester***123** (7): 631-636.

Singh, B.P. (1983) Response of Taramira to Sulphur application. *Indian.J. Agric. Sci.* **53**:676-680

Singh, B.P. and Chowdhury, R.K. (1983) Correlation and path coefficient analysis of seed yield and oil content in mustard (*Brassica juncea*)**.***Canadian Journal of Genetics and Cytology*, 1983, **25**(3): 312-317, https://doi.org/10.1139/g83-049

Singh, D. and Mehta, T.R.(1954). Studies in breeding of brown sarson I. comparison of F_1's and theirs parents. *Indian Journal of Genetics and Plant Breeding.* **14**: 74-77.

Singh, H. and Singh, D. (1992) Methods of genetic improvement in oleiferous *Brassica.* In: Kumar, D. and Rai, M.(Eds) *Adv. Oilseeds Res.*Vol.**1*Scientific Publishers***, Jodhpur (India). Pp 29-58

Singh, H. C., Nagaich, V. P. and Singh, S. K. (2000) Genetic variability for dry matter production in sesame (*Sesamum indicum* L.). *Annals of Agricultural Research*, **21**(3): 323-327

Singh, I. S. (2016). Mahua. *Fruitipedia,* Indiahttp://www.fruitipedia.com/mahua.htm

Singh, J. (2004). Field evaluation of linseed, *Linum usitatissimum*L. germplasm for resistanceto *Alteranaria* and powdery mildew diseases. *J. Oilseeds Res.,***21**: 208–209

Singh KK, *et al.* **(2011) Flaxseed: a potential source of food, feed and fiber.** *Crit Rev Food Sci Nutr.* 2011 Mar;**51**(3):210-22

Singh, M.P. (1982) Studies on the damage of castor semilooper. *Achoea Janata* L. to some fruit plants and its chemical control. *Indian Journal of Plant Protection.* **10** (1/2): 37-39

Singh N B. 2003. Accomplishment and challenges in rapeseed-mustard research *Brassica***3** (3&4): 1–11

Singh, P.K. and Chopra, P. (2018) Double purpose linseed: a viable option for doubling farmers' income in the north-western Himalyan region. *Indian Farming***68**(01): 49–54

Singh, P.K., Kumar, S. and Trivedi, H.B.P. (1997) Heterosis in Niger hybrids. *J.Res.(BAU)***9**(1): 9-11

Singh, P.K. and Trivedi, H.B.P. (1993) Male sterility in niger. *J.Res.(BAU)* **5**(1): 81-82.

Singh, R. B., Pillai, R. S. N. and Kumar, H. (1981) Induced translocations in safflower. *Crop Sci* **21**: 811 815

Singh, R.B., Singh, H.K. and Parmar, A. (2013) Integrated Management of Alternaria Blight in Linseed. *Proc. Natl. Acad. Sci., India, Sect. B Biol. Sci.* **83,** 465–469 (2013). https://doi.org/10.1007/s40011-012-0152-8

Singh, R.J. (2011)Soybean genetic resources and crop improvement. *Genome* **42**(4):605-616· DOI: 10.1139/gen-42-4-605

Singh, R.J. (2017) Botany and Cytogenetics of Soybean- 10.1007/978-3-319-64198-0_2 In Book: *Soybean Genome.*pp. 11-42

Singh, R.J., and Nelson, R.L. (2015) Inter-sub-generic hybridization between *Glycine max* and *G. tomentella*: production of F1, amphidiploid, BC1, BC2, BC3, and fertile soybean plants,"*Theoretical and Applied Genetics,* **128**(6):1117–1136

Singh, R. P.; Upadhyay, S. K. (2008). The beneficial effects of feeding mahua (*Bassia latifolia* Roxb.) flower syrup to honey bee (*Apis mellifera*) colonies during periods of dearth. *J. Apicult. Res.,* **47** (4): 261-264

Singh, S.B., Kumar, P. and Prasod, K.G. (1991). Response of various tree species to irrigation regimes and leaf residue management. *Journal of the Indian Society of Soil Science* **39**: 229-232.

Singh, S. P. and Chaudhary, A. K. (2003). Selection criteria for drought tolerance in Indian mustard [*Brassica juncea* (L.) Czern & Coss. *Indian Journal of Genetics and Plant Breeding,* **63**: 263–264

Singh, V. (1996.) Inheritance of genetic male sterility in safflower. *Indian J. Genet.* **56**: 490–494.

Singh, V. (1997B) Identification of genetic linkage between male sterility and dwarfness in safflower. *Indian J. Genet.* **57**: 327–332

Singh, V. (2004.)Annual Report of Ad Hoc Project on "Biometrical Investigations of Flower Yield and Its Components and Their Maximization in Safflower.*" ICAR, New Delhi,* 35 pp.

Singh, V. *et al.,* (1995) An appraisal of 25 years of safflower research under irrigated conditions at NARI, India (1968-1993). *Sesame and Safflower newsletter,* **10**:69-79

Singh, V., Deshpande, M.B. and Nimbkar.N.(2005). Polyembryony in safflower and its role in crop improvement. In *Proceedings of the 6th International Safflower Conference, Instanbul, Turkey,* June 6–10, 2005. Esendal, E., Ed., pp. 14–20.

Singh, V, Deshpande M.B. and Nimbkar, N. (2003) NARI-NH-1: the first non-spiny hybrid safflower released in India. Sesame Safflower Newsletter, 18: 77–79

Singh, V., Galande, M.K., Deshpande, M.B. and Nimbkar, N. (2001) Inheritance of wilt (*Fusarium oxysporium* fsp. Carthami) resistance in safflower. In: Bergman, J.W. and Mundel, H.H. (Eds) *Proceedings of the fifth International Conference on Safflower, Williston, ND and Sydney.* MT, USA. 23-27 July 2001. Pp 127-131

Singh, V. and Nimbkar, N.(2007) Safflower (*Carthamus tinctorius* L.) Chapter 6, In:Ram J. Singh (Ed) "*Genetic Resources, Chromosome Engineering and Crop Improvement"* **4**.CRC Press 2007: PP 167-194

Singh, V. and Nimbkar (2016) Safflower, In: S.K. Gupta (Ed.) *Breeding Oilseed Crops for Sustainable Production* Academic Press, ElsevierInc. CA. USA. Pp 149-167

Singh, V.K., Singh, H.G. and Chauhan, N. (1983) Heterosis in Sesame. *Indian J. Agric. Sci.* **53:** 305-310

Singh, V.P., Dhillon, R.S. and Jhorar, B.S. (1996.) Floral biology, breeding behaviour and breeding strategy in neem (*Azadirachta indica* A. Juss.). In: *Proceedings of the International Neem Conference*, University of Queensland, Australia

Singhal, K. K., Chahil, S. M. and Sharma, D. D. (1986). Versatile mahua (*Madhuca indica*) seed cake - a review. *Agric. Rev.*, **7** (2): 105-123

Singhal,V. K., Gill, B.S., and Sidhu,M. S. (1990) Cytological explorations of *Indian woody legumes,Proceedings: Plant Sciences* **100** (5) : 319-331

Siring (1986) Annual Report on 'Rapeseed Research'. *Hyzazhong Agricultural University, Wuhan*, China. 69-72

Siruguri, Vasanthi; Bharatraj, Dinesh Kumar; Vankudavath, Raju Naik; Rao Mendu, Vishnu Vardhana; Gupta, Vibha; Goodman, Richard E. (2015). "Evaluation of *Bar, Barnase, and Barstar* recombinant proteins expressed in genetically engineered *Brassica juncea* (Indian mustard) for potential risks of food allergy using bioinformatics and literature searches". *Food and Chemical Toxicology*. **83**: 93–102. doi:10.1016/j.fct.2015.06.003. ISSN 0278-691

Sivagnanam, K., Vinaya Rai, R.S., Swaminathan, C. and Surendran, C. (1989). Studies on rooting response to growth regulators in *Azadirachta indica* A. Juss. In: *Proceedings of a seminar on Vegetative Propagation*, July 27-28 1989, Coimbatore, India

Sivaram, V., Jayaramappa. K.V., Menon, A. and Ceballos, R.A. (2013) Use of bee-attractants in increasing crop productivity in Niger (*Guizotia abyssinica*. L) *Braz. arch. biol. technol.*.**56** (3) Curitiba May/June 2013 http://dx.doi.org/10.1590/S1516-89132013000300003

Skoric, D. (1988) Sunflower breeding. *J. Edible Oil Ind.* **25**(1): 1-90

Skoric, D (1992)Achievements and future directions of sunflower breeding, *Field Crops Research*, **30** : 231-270

Skoric, D. (2009) Sunflower breeding for resistance to abiotic stresses. *Helia*,**32** (50): 1-16.

Skoric, D (2012) The Genetics of Sunflower, In: Kovacevic, Z., Skoric, D. and Sakac, Z.(Eds) *Sunflower Genetics and Breeding-International Monogram. Serbian Academy of Sciences, Serbia*, pp1-125

Skoropad, W. P. and Tewari, J. P. 1977. Field evaluation of theepicuIticular wax in Rapeseed and mustard in resistance to*Alternaria brassicae*. *Can Journal Plant Science***57**: 1001–1003

Smartt, J., Gregory, W.C. and Gregory, M.P. (1978) Genomes of *Arachis hypogaea*.1. Cytogenetic genetic studies of putative genome donors. *Euphytica*, **27**: 665-675.

Smith, B.D. (2014) The domestication of *Helianthus annuus* L. (sunflower). Veg. Hist. Archaeobot. 2014, **23:** 57–74. [Google Scholar] [CrossRef]

Smith, B.W. (1950) *Arachis hypogaea* L. Aerial flower and subterranean fruit. *American Journal of Botany* **37:** 802-815

Smith, D.M., Jarvis, P.G. and Odongo, J.C.W. (1997A). Sources of water used by trees and millet in Sahelian windbreak systems. Journal of Hydrology **198**: 140-153.

Smith, I.M., McNamara, D.G., Scott, P.R. and Holderness, M. (eds.), (1997). *Spodoptera litura*. In: Quarantine Pests for Europe, 2nd Edition. *CAB International*, Wallingford, Oxon, UK. pp. 518–525.

Smith Jr., J.W. and Barfield, C.S. (1982)Management of preharvest insects, In: H.E. Pattee and C.t. Young (Eds).*Peanut Science and Technology. American Peanut Research and Education Society*. Yoakum, Texas, pp 250-325.

Smýkal, P., Bačová-Kerteszová, N., Kalendar, R., Corander, J., Schulman, A.H. and Pavelek, M. (2011). Genetic diversity of cultivated flax (*Linum usitatissimum* L.) germplasm assessed by retrotransposon-based markers. *Theor Appl Genet*. **122**(7):1385-97. doi: 10.1007/s00122-011-1539-2. Epub 2011 Feb 4.

Snowdon, R., Luy, F. (2012) Potential to improve oilseed rapeand canola breeding in the genomics era. *Plant Breed*. **131**: 351-360.

Sodhi, Y. S., Chauhan, A., Verma, J. K., Arumugam, N., Mukhopadhyay, A., Gupta, V., Pental, D. and Pradhan. A. K. (2006). A new cytoplasmic male sterility system for hybrid seed production in in Indian oilseed mustard (*Brassica juncea*). *Theoretical and Applied Genetics* **114**: 93–99

Soest L. J.(2001). Status of national collections – the Netherlands. In: Maggioni L, Pavelek M, van Soest LJM, Lipman E, (Compilers) *Flax genetic resources in Europe. Ad hoc* meeting, 7–8 Dec 2001, Prague, Czech Republic. Rome: International Plant Genetic Resources Institute; 2001.

Sogbo, K. A. (2006a). "Moringa Leaf Farming Systems: Conditions for Profitability and Sustainability" (PDF). Retrieved 19 November 2013.

Sogbo, K.A. (2006b). Moringa and other highly nutritious plant resources: Strategies,standards and markets for a better impact on nutrition in Africa. In: MORINGANEWS, Réseau plantes ressources / Plant resources network. www.moringanews.org.

Sokal, R. R. and Michener, C.D. (1958). A statistical method for evaluating systematic relationships. *University of Kansas Science Bulletin.* **38**: 1409–1438.

Solanki, K.R. (1998). A decade of Neem Research 1988-1998. *National Research Centre for Agroforestry*, Jhansi, India. pp. 26-29.

Soldatov K. I. (1976). "Chemical mutagenesis in sunflower breeding," in *Proceedings of the 7th International Sunflower Conference* Krasnodar: 352–357.

Soltis, D.E and Soltis, P.S. (1998) Molecular Systematics of Plants II. Chapter from *book Molecular Systematics of Plants II: DNA Sequencing* (pp.1-42)DOI: 10.1007/978-1-4615-5419-6_1

Somasekhar, Sekharappa, Patil, B.V. and Patil, S.A. (1993) Occurrence of castor semilooper, *Achoea Janata* Linn., and its parasatoid *Microplitis maculipennis* Szepligeti., in Raichur. *Karnatak Journal of Agricultural Sciences.***6**(2): 200-202

Song K., Osborn T. C. and Williams P. H. (1990) *Brassica* taxonomy based on nuclear restriction fragment length polymorphisms (RFLPs): 3. Genome relationships in *Brassica* and related genera and the origin of *B. oleracea* and *B. rapa (syn. campestns)*. *Theor. Appl. Genet.* **79:** 497–506 (1990).

Songsri, P., Jogloy, S., Kesmala, T., Vorasoot, N., Akkasaeng, C., Patanothai, A. and Holbrook, C.C. (2008). Response of Reproductive Characters of Drought Resistant Peanut Genotypes to Drought. *Asian Journal of Plant Sciences,* **7:** 427-439.

Soni, R.P., Katoch, M., Kumar, A. and Verma, C.S.K. (2016) Flax seed composition and its health benefits. *Res. Environ. Life Sci.* **9**(3) 310-316 (2016) ISSN: 0974-4908

Sood, S., Kalia, R., Bhateria, S. and Kumar, S. (2007) Detection of genetic components of variation for some biometrical traits in *Linum usitatissimum* L. in sub-mountain Himalayan region.*Euphytica* **155**(1):107-115; DOI: 10.1007/s10681-006-9309-y

Soren, N.M. and Sastry, V.R.B. (2009) Replacement of soybean meal with processed karanj (*Pongamia glabra*) cake on the balances of karanjin and nutrients, as well as microbial protein synthesis in growing lamb, *Animal Feed Science and Technology*, **149** (1–2): 16-29https://doi.org/10.1016/j.anifeedsci.2008.04.011

Soren, N.M., Sastry, V.R.B., Saha, S.K., Wankhade, U.D., Lade, M.H. and Kumar, A. (2009) Performance of growing lambs fed processed karanj (*Pongamia glabra*) oil seed cake as partial protein supplement to soybean meal, *Journal of Animal Physiology and Animal Nutrition,* **93**(2): 237-244. https://doi.org/10.1111/j.1439-0396.2008.00811.x

Soto-Cerda, B. J., Diederichsen, A., Ragupathy, R. and Cloutier, S. (2013). Genetic characterization of a core collection of flax (*Linumusitatissimum* L.) suitable forassociation mapping studiesand evidenceof divergent selection between fibre and linseed types. *BMC Plant Biology*, **13**: 78

Soto-Cerda, B.J., Duguid, S., Booker, H., Rowland, G., Diederichsen, A. and Cloutier, S. (2014a). Genomic regions underlying agronomic traits in linseed (*Linum usitatissimum* L.) as revealed by association mapping. *J. Integr. Plant Biol.*, **56:** 75–87

Soto-Cerda, B.J., Westermeyer, F., Iñiguez-Luy, F., Muñoz, G., Montenegro, A. and Cloutier, S. (2014b) Assessing the agronomic potential of linseed genotypes by multivariate analyses and association mapping of agronomic traits, *Euphytica, **96***(1): 35-49

Spaan, W.P., Bodnar, F., Idoe, O. and de Graaff, J. (2004) Implementation of contour vegetation barriers in Burkina Faso and Mali. Quarterly *J. Int. Agric*. **43**: 21-38

Spandana, B., Reddy V.P., Prasanna, G. J., Anuradha, G. and Sivaramakrishnan, S. (2012) Development and characterization of microsatellite markers (SSR) in *Sesamum (Sesamum indicum* L.) species *Appl Biochem Biotechnol. 2012 Nov;* **168**(6):1594-607.

Spears, J.F., Tekrony, D.M. and Egli, D.B.(1997) Temperature during seed filling and soybean germination and vigour. *Seed Science and Technology.* **25** (2):233-44.

Spehar, C.R. and Galwey, N.W. (1997) Screening soybeans (*Glycine max*. L. Merrill) for calcium efficiency by root growth in low-Ca-nutrient solution. *Euphytica,* **94** (1):113-117

Sreedevi, Muniappa, T.V. and Babu, V.S. (1997) Uptake of nutrients by castor and associated weeds under different weedmanagement systems. *World Weeds.***4** (3-4):125-129

Sridharan, S., Venugopal, M.S., Dhaliwal, G.S., Arora, R., Randhawa, N.S. and Dhawan, A.K. (1998). Effect of environmental conditions on the yield of azadirachtin and oil in **neem**. In: *Ecological agriculture and sustainable development: Volume 1. Indian Ecological Society and the Centre for Research in Rural and Industrial Development, Chandigarh, India.* pp. 510-518.

Srilatha, A.N., Masthan, S.C. and Mohammad, S. (2000) Economic viability of intercropping castor with legume components. Extended Summaries (2000) National Seminar on Oilseeds and Oils: *Research and Development Needs in the Millenium. Feb.2-4, 2000.* Indian Society of Oilseeds Research Hyderabad. Pp 37-38.

Srimannarayana, B. and Raju, A.S. (1993) Direct and residual effect of applied Sulphur in sunflower based cropping systems. *Fertiliser News.***38** (9): 39-42

Srinivas and Krishna Ravi (2017), "The Regulatory Regime for Genetically Modified Crops in India", *Genetically Modified Organisms in Developing Countries*, Cambridge University Press, pp. 236–246, doi:10.1017/9781316585269.021, ISBN 9781316585269

Srinivasachar, D. and Malik., R.S (1972) An induced aphid resistant non-waxy mutant in turnip. *Brassica rapa. Curr. Sci.* **41:** 820-821.

Srinivasan, K., Viraktamath, C.A., Gupta, M. and Tewari, G.C. (1995) Geographical distribution, host range and parasitoids of serpentine leaf-miner *Liriomyzatrifolii* (Burgess) in South India. *Pest Management in Horticultural Ecosystems.* **1**(2):93-100

Srinivasan,R., Abney M. R., Culbreath, A.K., Tallury, S.and Leal-Bertioli, S.C.M. (2018) Resistance to Thrips in Peanut and Implications for Management of Thrips and Thrips-Transmitted Orthotospoviruses in Peanut, *Front. Plant Sci.,* 06 November 2018 https://doi.org/10.3389/fpls.2018.01604

Srinivasulu. (1991) Studies on *Sesamum* Phyllody disease with special reference to disease resistance. (c.f. Thangavelu, 1994).

Srinivasulu, N and Chandrasekharan, N. R. (1958) Anote on natural crossing in groundnut *Arachis hypogaea* Linn. *Science and Culture.* 23:650

Srivastava, R. and Srivastava, G.K. (2002) Autopolyploids of Helianthus annuus L. var. morden Cytologia,67 (2) :213-220

Stalker, H.T. (1981) Hybrids in the genus Arachis between sections Erectoides and Arachis. Crop Sci., 21: 359-362.

Stalker, H.T. and Campbell, W.V. (1983) Resistance of wild species of peanut insect complex. Peanut Sc., 10: 30-33.

Stalker, H.T. and Dalmicio, R.D. (1981) Chromosomes of Arachis species, Section Arachis (Leguminosae). Journal of Heredity72: 403-408

Stalker, H.T., Seitz, M.H., and Reece, P. (1987) Effect of gibberellic acid on pegging and seed set of Arachis species. Peanut Sci.,14: 17-21

Stalker, H.T. and Simpson, C.E. (1995) Genetic Resources in Arachis In: Pattee, H.E., and Stalker, H.T. (Eds). Advances inPeanut Science. American Peanut Research Education Society, Stillwater, Ok.pp.14-53.

Stefansson. B.R. (1983) High and Low Erucic Acid Rapeseed Oils. In:Kramer, J.K.G., Sauer, F.D., Pigden, W.J. (eds) Academic Press, New York. 1983. Pp143-160

Stefansson, B.R., Hougen, G.W. and Downey, R.K. (1961)Note on the isolation of rape plants with seed oil free from erucic acid. *Canadian J. Plant Sci.***41**:218-219

Stein, K., Coulibaly, D., Stenchly, K., Goetze, D., Porembski, S., Lindner, A., Konaté. S., and Linsenmair, E.K. (2017) Bee pollination increases yield quantity and quality of cash crops in Burkina Faso, West Africa. *Sci Rep*. 2017 Dec **18**:7(1):17691. doi: 10.1038/s41598-017-17970-2

Stephen, J.D.A, Gurumurthi, K. and Subramanian, K. (1998). Studies on the in vitro extraction of neem (*Azadirachta indica* A. Juss.) chemicals and propagation. *Final Report submitted to the Council of Scientific and Industrial Research,* New Delhi, India. 55 pp.

Steven, P. J., Comus and Wentworth, P. (2000). A Natural History of the Sonoran Desert. University of California Press. pp. 256–257. ISBN 0-520-21980-5.

Stevenson, P.C., Blaney, W.M., Simmonds, M.J.S. and Wightman, J.A. ((1993) The identification and characterization of resistance in wild species of *Arachis* to *Spodoptera litura. Bull. Entomol. Res.*, **83:** 421-429.

Stokes, W.E. and Hull, F.H. (1930) Peanut breeding. *Journal of the American Society of Agronomy,***22:** 1004-1009

Streit, L.G., Beach, L.R., Register, III, J.C., Jung, R. and Fehr, W.R. (2001). Asociation of the Brazilnut protein gene and Kunitz trypsin inhibitor alleles with soybean proteae inhibitor activity and agronomic traits. *Crop Sci.*,**41**: 1757-1760.

Stuber, C.W., (1992) Biochemical and molecular markers in plant breeding. *Plant Breeding Rev*. **9**: 37-61

Sturtevant, D., Lu,S., Zhou,Z.W., Shen, Y., Wang, S. *et al.,* (2020). The genome of jojoba (*Simmondsia chinensis*): A taxonomically isolated species that directs wax ester accumulation in its seeds., *Science Advances* 11 Mar 2020;**6** (11): 32-40.DOI: 10.1126/ sciadv. aay3240.

Styer, C.H., Cloe, R.J. and Hill, R.A. (1983). Inoculation and infection of peanut flowers by *Aspergillus flavus*. *Proceedings of the American Peanut Research and Education Society.***15**: 91

Stymne, S. and Appleqvist, L.A. (1980) The biosynthesis of linolenate in homogenatesfrom developing soybean cotyledons. *Plant Science Letters*, **17**: 287-294

Subba Rao, B.R., Parshad, B., Ram, A., Singh, R.P. and Srivastava, M.L. (1968) Studies on the hymenopterous egg parasites of the genus *Empoasca* (Jassidae: Homoptera). *Entomologia Exp. Appl.* **11**: 250-254.

Subba Reddy, G., Seshasailasree, P. and Maruthi, V. (1999) Contribution of production factors to yield and income of rainfed castor. *Journal of Oilseeds Research***16** (1): 57-61

Subramanian AK, Singal SK, Saxena M, and Singhal S. (2005) Utilization of liquid biofuels in automotive diesel engines: An Indian perspective. *Biomass and Bioenergy*.**9:**65–72. [Google Scholar]

Subrahmanian, M. (1995)Pollen germination studies in sesame. *Annals of Agricultural Research* **16** (2):225-226.

Subrahmnyam, P., Mehan, V.K., Nevil, D.J. and McDonald, D. (1980) Research on fungal diseases of groundnut at ICRISAT. In: *Proceedings of International Workshop on Groundnut.* ICRISAT, Patancheru, India. Pp 193-198

Sudhakaran, R. (2002) Efficacy of luferon (Match 5% EC) against *Spodoptera litura* (F) under *in vitro* condition. *Insect Environment* **8** (1): 47-48.

Sugla, T., Purukayastha, J., Singh, S.K., Solleti, S.K., Sahoo, L., (2007). Micropropagation of *Pongamia pinnata* through enhanced axillary branching. *InVitro Cell Dev.Biol.* **43:** 409–414.

Suh, M.C., Kim, M.J., Hur, C.G., Bae, J.M., Park, Y.I., Chung, C.H., Kang, C.W.and Ohlrogge, J.B.(2003). Comparative analysis of expressed sequence tags from *Sesamum indicum* and *Arabidopsis thaliana* developing seeds. *Plant Molecular Biology* **52***:* 1107-1123

Sujatha, K., Hazra, S., (2006). Invitro regenerationof *Pongamia pinnata*. Pierre. *J. Plant Biotechnol.* **33:** 263–270.

Sujatha, K. and Hazra, S., (2007). Micropropagation of mature *Pongamia pinnata* Pierre. *In Vitro Cell. Dev. Biol.Plant***43**: 608–613.

Sujatha, K., Panda, B.M. and Hazra, S., (2008). *De novo* organogenesis and plant regeneration in *Pongamia pinnata,* oil producing tree legume. *Trees***22:** 711–716

Sujatha, M. (1993). Pollen-pistil interactions and the control of self-incompatibility in niger (*Guizotia abyssinica* Cass). *J. Oilseeds Res.* **10**: 334-336

Sujatha, M. (1997) In-vitro adventitious shoot regeneration for effective maintenance of male sterile niger (*Guizotia abyssinica* L.f Cass) *Euphytica*, **93**(1):89-91

Sujatha, M (2005). Genetic improvement of *Jatropha curcas* (L.) possibilities and prospects. *Indian Journal of Agroforestry*, **8** (2): 58-65

Sujatha, M. (2006). Heterosis breeding in safflower: problems and prospects. In: G. Kalloo, M. Rai, M. Singh and S. Kumar (Eds). Heterosis in Crop Plants, Researchco Publisher, New Delhi, pp. 291-300.

Sujatha, M. (2007) Advances in **Safflower Biotechnology.** *Fumctional Plant Science and Biotechnology* **1** (1):160-170 *@Global Science Books*

Sujatha, M. and S. P. Angadi. (1989) **Refined crossing method for niger** (*Guizotia abyssinica* Cass). *J. Oilseeds Res.* **6** (1): 152 – 155.

Sujatha, M. and Dutta Gupta, S. (2013) Tissue Culture and Genetic Transformation of Safflower (*Carthamus tinctorius* L.). In: S. M. Jain and S. Dutta Gupta (Eds.), *Biotechnology of Neglected and Underutilized Crops*, pp.297-318. DOI: 10.1007/978-94-007-5500-0_12, © Springer Science+Business Media Dordrecht 2013

Sujatha M and Prabakaran AJ (2001) High frequency embryogenesis in immature zygotic **embryos of SunflowerHigh frequency embryogenesis in immature zygotic embryos of sunflower.** *Plant Cell Tiss Org Cult.* **65**:23-29

Sujatha, M. and Lakshminaryana, M. (2005)Susceptibility of castor semilooper to insecticide crystals proteins from *Bacillus thuringiensis. Indian Journal of Plant Protection,* **33** (2): 286-287.

Sujatha, M., Lakshminarayana, M., Tarakeswari, M., Singh, P.K. and Tuli, R. (2009) Expression of *Cry 1 Ec* **gene in castor conferring field resistance to tobacco caterpillar** *Spodoptera litura* and semiloopeer *Achoea janata. Plant Cell Reports.***28**: 935-940

Sujatha, M., Mukta Dhingra, Prasad, M.V.R. and Ranganatha, A.R.G. (1994). Development and evaluation of an invitro system of propagation of *Melia azedarch*, L.*Proc. of "Second **Asia Pacific Conf. on Agricultural Biotechnology" held in Madras(Chennai), India,*** March 6-10, 1994: 70.

Sujatha, M., Reddy, T.P. and Mahasic, M.J. (2008) Role of biotechnological interventions in the improvement of castor (*Ricinus communis* L.) and *Jatropha curcas* L. *Biotechnology Advances***26** (5): 424-435

Sujatha, M. and Suganya, A. (1996) In vitro organogenic comparison of different seedling **tissues of safflower** (*Carthamus tinctorius* L.). ***Sesame Safflower Newsl.*****11**: 85–90.

Sujatha M., Tarakeswari M. (2018) Biotechnological Means for Genetic Improvement in Castor Bean as a Crop of the Future. In: Kole C., Rabinowicz P. (eds) *The Castor Bean Genome. Compendium of Plant Genomes. Springer*, Cham bDOI:https://doi.org/10.1007/978-3-319-97280-0_14

Sujatha, M., Vijay, S., Vasavi, S. and Chander Rao, S. (2012). Agrobacterium-mediated **transformation of cotyledons of mature seeds of multiple genotypes of sunflower** (*Helianthus annuus L.).Plant Cell Tissue and Organ Culture* **110(**2) ·https://www.researchgate.net/publication/257668418_Agrobacterium-mediated_transformation_of_cotyledons_of_**mature_seeds_of_multiple_genotypes_of_sunflower_Helianthus_annuus_L**

Sundararaj, D.P. and Thulasidas, G.1(993). *Botany Field Crops. McMillan Ind. Ltd.*, New Delhi, Pp: 327.

Sunray International (2019) Products-Castor Oil Cake-Organic Fertilizer.https://sites.google.com/site/sunrayinternational/products/castor-oil-cake

Supriya, S.M.,Kulkarni, V.V. and Suresh, P.G. (2016) Screening of sunflower genotypes against powdery mildew. *The Bioscan* **11**: 2991-2996

Surapaneni M., Yepuri V., Vemireddy L. R., Ghanta A., Siddiq E. A. (2014). Development and characterization of microsatellite markers in Indian sesame (*Sesamum indicum* L.). *Mol. Breed.* **34** :1185–1200. 10.1007/s11032-014-0109-0

Suri, R.K. and Mathur, K.C. (1988) Oilseeds and their utilization. *International Book Distributors, Dehradun*, India

Suryawanshi, D. and Pawar, V.M. (1980) Effect on the growth and yield due to aphid *Dactynotus sonchi* in safflower. *Proc. Indian Sci.* **89:** 347-349

Sutic, D. and Dowson, W.J. (1962) Bacterial Leaf Sot of Sesame in Yugoslavia.*Phtopath. Z.* **45**: 57

Suthanthirapandian *et al* (1989) Variability in *Moringa* given in https://greenagrow.blogspot.com/2012/11/all-about-moringa.html#comment-form

Suvendu M., A. M. Badigannavar, D. M. Kale and G. S. S. Murty. (2007). Induction of genetic variability in a diseaseresistant groundnut breeding line. *BARC. Newsletter.***285**: 237-

Suzuki, M.M. and Bird, A. (2008). DNA methylation landscapes: provocative insights for epigenomics. *Nat. Rev. Genet*. **9:**65–476. T247.

Swaminathan, M.S., Chopra, V.L. and Bhaskaran, S. (1962) Chromosome aberrations and frequency and spectrum of mutations induced by EMS in barely and wheat. *Indian Journal of Genetics and Plant Breeding* **22**: 192-207

Swaminathan, M.S. and Jain, H.K. (1973) Food legumes in Indian Agriculture. Improvement of food legumes by breeding. *Proc. Symp. Protien Advisory Group.* UN Rome. July 1972: 69-82

Swarnlata, K. Prasad, M.V.R. and Rana, B.S. (1984a) Stability of pod and kernel yield of some mutants in groundnut. *Annals of Agricultural Research.***5** (1-2): 54-62

Swarnalatha, Kaul, Prasad, M.V.R. and Rana, B.S. (1984b). Genetic variability and yield component analysis in castor, *SABRAO Journal* **16**(2):*123-128.*

Syakudo, K. and Kawabata, S. (1965) Studies on peanut breeding with reference on thecombination of some main characters. II. Genotypic and phenotypic correlations between all pairs of 15 charcters in the F2 populations. *Japanese Journal of Breeding.* **15:** 167-170.

Szarejko, I. and Forster, B.P. (2007). Doubled haploidy induced mutation. *Euphytica***158**: 359-370

Szonyi, B.; Gizachew, D.; Tegegne, A.; Hanson, J. and Grace, D.(2015). Aflatoxin contamination of milk and feeds in the greater Addis Ababa milk shed in Ethiopia. *ILRI Research Brief,* **46,** *June 2015*

Tai, P.Y.P., Hammons, R.O. and Matlock, R.S. (1977). Genetic relationships among three chlorophyll-deficient mutants in peanut, *Arachis hypogaea L. Theor. Appl. Genet.* **50:**35–40. *[Crossref] [Google Scholar]*

Tajima, R., Lee, O.N., Abe, J., Lux, A. and Moritaops, S.(2007) Nitrogen-Fixing Activity of Root Nodules in Relation to Their Size in Peanut (*Arachis hypogaea L.)/Plant Production Science /* **10** (2007) 4 /

Talbot, G. (2015) Tropical exotic oils In: G. Talbot (Ed). *Specialty Oils and Fats in Food and Nutrition, 2015, Woodhead Publishing*. Pp 87-123. eBook ISBN: 9781782423973,

Tamborindeguy, C., Ben, C., Liboz, T. and Gentzbittel, L. (2004) Sequence evaluation of specific DNA libraries for development of genomics of sunflower. *Moi. Genet. Genomics.* **271:** 367-375

Tammes, T., (1928). The genetics of the genus *Linum.* In: Lotsy, J.P., Goddijn, W.A. (Eds.), *Bibliographia Genetica. Martinus Nijhoff,* The Netherlands, pp. 1–34

Tanaka D.L., Riveland N.R., Bergman J. W., Schneiter A.A. (1997). Safflower Plant Development Stages. *Proceedings of IVth International Safflower Conference, Bari (Italy),* June 2-7, 179- 180.

Tang, S. and Knapp, S.J. (2003) Microsatellites uncover extraordinary diversity in native American land races and wild populations of cultivated sunflower. *Theor. Appl. Genet.,* **106:** 990–1003.

Tanzubil, P.B. and Yhaya, B.S. (2017)Assessment of yield losses in groundnut (*Arachis hypogaea* L.) due to arthropod pests and diseases in the Sudan savanna of Ghana. *Journal of Entomology and zoology studies.* **5**(2):1561-1564

Tashiro, T., Fukuda, Y., Osawa, T. and Namiki, M. (1990) Oil and minor Components of sesame (*Sesamum indicum* L.)*Journal of the American Oil Chemists' Society* **67** (8): 508-511

Taskin, K., Ercan, A. andTurgut K (1999) *Agrobacterium tumefaciens*-mediated transformation of sesame (*Sesamum indicum* L.). *Tr J Bot***23**:291–295

Taskin, K.M. and Turgut, K. (1997*). In vitro* regeneration of sesame (*Sesamum indicum* L.). *Turkish Journal of Botany* **21***: 15-18.*

Tejklová E. (1996): Some factors affecting anther culture in *Linum usitatissimum* L. *Rostlinná výroba,***42**: 249–260. (in Czech)

Tejklová, E., Bjelková, M. and Pavelek M. (2011) Medium-linolenic linseed (*Linum usitatissimum* L.) Raciol. *Czech J Genet Plant Breed.* 2011; **47**(3):128–30.

Tejavathi, D.H., Lakshmi Sita, G. and Sunita, A. (2000). Somatic embryogenesis in Flax. *Plant Cell, Tissue and Organ Culture* **63,** 155–159 (2000). https://doi.org/10.1023/A:1006469900025

Tejovathi, G. and Anwar, S.Y. (1987). Plant regeneration from cotyledonary cultures of safflower (*Carthamus tinctorius* L.). In *Proceedings of the National Symposium "Plant Cell and Tissue Culture of Economically Important Plants," Hyderabad, India.* Reddy, G.M., Ed., pp. 347–354.

Tejovathi, G. and Anwar.S.Y.(1993). 2,4,5-trichlorophenoxy propionic acid induced rhizogenesis in *Carthamus tinctorius* L. *Proc. Ind. Natl. Sci. Acad.* **B59***: 633–636.*

Tewari, D.N. (1992), Monograph on Neem (*Azadirachta indica.*A.juss) , International Book Distributors, Dehra Dun, India

Thakur, H.L. and Bhateria. S(1993). Heterosis and inbreeding depression in Indian mustard. *Indian Journal of Genetics and Plant Breeding.* **53**(1): 60-65.

Thakur, H.L. and J.C. Sagwal. 1997. Heterosis and combining ability in rapeseed *B. napus. Indian Journal of Genetics and Plant Breeding.* **57**(2): 163-167.

Thangavelu, S. (1994) Problems and Prospects in Production of Hybrids in Sesame. *HYBRID SESAME: Proceedings of the Group Discussion,* ICAR-Directorate of Oilseeds Research, Hyderabad.Pp 20-22

Thangavelu, S. (1994a) Diversity in wild and cultivated species of *Sesamum* and its use: Sesame Biodiversity in Asia: Conservation, Evaluation and Improvement.In: (Eds). R.K. Arora and K.W. Riley,*Proceedings of the IBPGR-ICAR/NBPGR Asian Regional Workshop on 'Sesame Evaluation and Improvement' held at Nagpur & Akola in India* 28-30 September, 1993. Pp13-23

Thangavelu, S., Radhamanoharan, G. and Vindhya Varman, P.(1980) Studies on pod setting in interspecific crosses of *Arachis* spp. *Proceedings of the National Seminar on the application of Genetics to the improvement of Groundnut.Tamilnadu Agricultural University, Coimbatore*, India. July 16-17, 1980. :20-26.

Thanki, K.V., Patel, G.P. and Patel, J.R. (2001). Varietal preference in castor to *Spodoptera litura* Fbr. *Gujarat Agricultural University Journal.***26**(2): 39-43

Thanki, K.V., Patel, G.P. and Patel, J.R. (2003) Population dynamics of *Spodoptera litura* on castor *Ricinus communis. Indian Journal of Entomology.* **65** (3):347-350

Thiagarajan, M. and Murali, P.M. (1993). High frequency regeneration of neem *in vitro* (*Azadirachta indica* A. Juss.) from immature embryo explants. *World Neem Conference, Banagalore,* India, 1993.

Thobbi, V.V. (1961) Growth potential of *Prodenia litura* F. in relation to certain food plants. *Indian Journal of Entomology***23**: 262-264

Thobbi, V.V. and Srihari, T. (1968) Prospect of work done with regard to the use of biological agents at the Regional Research Centre, Hyderabad and the proposed technical programme of work. *Paper presented at the First Oilseeds Research Workshop in Hyderabad during April 22-24, 1968. Hyderabad*

Thomas, A. (2005). "Fats and Fatty Oils". *Ullmann's Encyclopedia of Industrial Chemistry.* Weinheim: Wiley-VCH. doi:10.1002/14356007.a10_173. ISBN 978-3527306732.

Thomas, C.A. 1964. Registration of 'US10' safflower. *Crop Sci.***4**: 446–447

Thomas, C.A. 1971. Registration of 'VFR-1' safflower germplasm. *Crop Sci.* **11**: 606.

Thomas, C. A. and Zimmer D. E.(1970). Resistance of Biggs safflower to *Phytophthora* root rot and its inheritance. *Phytopathol* **60**:63-64

Thompson, K.F. (1964) Triple-cross hybrid kale,*Euphytica,***13:**173-177

Thompson, K.F., (1983). Breeding winter oilseed rape *Brassica napus*. In: T.H. Coaker (Editor), *Advances in Applied Biology, Vol. VII, Academic Press, London, pp. 2-86*

Thompson K.F. and Taylor J.P. (1966) Non-linear dominance relationships between 'S' alleles. *Heredity* **21**: 345–362

Thurling, N., (1974). Morphophysiological determinants of yield in rapeseed *(Brassica camprestris and Brassica napus). 1.* Growth and morphological characters, *Aust. J. Agric. Res.,***25**:697-710.

Thurling, N., (1991). Application of the ideotype concept in breeding for higher yield in the oilseed *Brassica. Field Crops Res.*, **26**: 201-219.

Times of India (2012) *"Forest department, LIT develop new products from mahua - The Times of India"*

Tiwari, S., Kumar, S. and Gonti, I. (2011) Minireview: Biotechnological approaches for sesame (*Sesamum indicum* L.) and niger (*Guizotia abyssinica* L.f. Cass.) *AsPac J. Mol. Biol. Biotechnol.***19** (1): 2-9

TNAU (Tamil Nadu Agricultural University). (2017) Advances in Production of moringa. http://agritech.tnau.ac.in/horticulture/pdf/Moringa%20English%20book.pdf

Todd, J.W., Beach, R.M. and Branch, W.D. (1991) Resistance in eight peanut genotypes to foliar feeding of fall army worm, velvet bean caterpillar and corn earworm. *Peanut Sci.,***18:** 38-40.

Trail, F., Richards, C., and Wu, F.S. (1989) Genetic manipulation in *Brassica*. In: Bajaj, Y.P.S. (Ed.) *Biotechnology in Agriculture and Forestry. - Plant Protoplasts and GeneticEngineering – II* Springer verlag, Berlin. Vol. **9:** pp 198-216

Trifed (2009) **Product profile, Mahuwa, Trifed, Ministry of Tribal Affairs, Government** of India". *Trifed.nic.in.* Archived from the original on 2009-06-19. Retrieved 2013-11-21.

Tripathy SK, Mishra DR, Panda S, Senapati N, Nayak PK, Dash GB, Dash S, Pradhan KC, Dash **AP, Mishra D, Mohanty SK, Mohapatra PM &Lenka D (2016) Identification of heterotic** crosses for sesame breeding using diallel matting design.*Tropical Plant Research* **3**(2): 320–324

Tripathi, N.N., Saharan, G.S., Kaushik. C.D., Kaushik, J.C. and Gupta, P.P. (1987) Magnitude of losses in yield and management of *Alternaria* blight of rapeseed and mustard. *Haryana Agirc. Univ. J. Res.***17**(1): 14-

Troup, R.S. and Joshi, H. B. (1981). Troup's The Silviculture of Indian Trees., *IITDelhi,* India: Controller of Publications.

Tsaknis, J., Lalas, S., Gergis, V., and Spiliotis, V. (1998). A total characterisation of *Moringa oleifera* Malawi seed oil. *Riv. Ital. Sost. Gras.* **75**(1): 21–27.

Tse, G. and Eslick, G.D. (2014). “Cruciferous vegetables and risk of colorectal neoplasms: a systematic review and meta-analysis”. *Nutrition and Cancer*. **66** (1): 128–139.

Tu, L.W. (1993) Studies on sesame heterosis and its practical use in production., *Sesame and Safflower newsletter*,**8** :9-17

Tulod, A.M., Castillo, A.S.A., Carandang, W.M. and Pampolina, N.M. (2012) Growth performance and phytoremediation potential of *Pongamia pinnata* (L) Pierre, *Samana saman & Vitex parviflora* copper contaminated soils, amended with Zeolite and VAM. *Asia Life Sciences* **21**(2): 499-522.

Turlapati, S.A., Jami, S.K., Dutta, R. and Kirti, P.B. (2006) Genetic transformation of peanut (*Arachis hypogaea* L.) using cotyledonary node as explant and a promoterless gus: nptII fusion gene based vector. *Journal of Biosciences* **31(**2):235-246 · June 2006 DOI: 10.1007/BF02703916

Tutin, T.G., Heywood, V.H., Burges, N.A., Valentine, D.H., (1980). Flora Europaea. *Cambridge University Press, Cambridge*, UK. 480 p.

Tyagi, V., Dhillon, S.K., Kaushik, P. and Kaur, G. (2018)Characterization for Drought Tolerance and Physiological Efficiency in Novel Cytoplasmic Male Sterile Sources of Sunflower (*Helianthus annuus* L.), *Agronomy*, **8**:232; doi:10.3390/agronomy810023

Udin, M.A., Ullah, M.A., Sultana, F., Rahman, K.M. and Rahman, M.Z. (2012) Evaluation of some rapeseed mutants based on morpho-physiological, biochemical and yield attributes. *J. Environ. Sci.Nat. Resour*. **5**: 281-285.

UHS Bagalkot. (2017) Varieties developed from UHS Bagalkot. 1. Drumstick-Bhagya (KDM-01). http://uhsbagalkot.edu.in/downloads/VARIETIES%20EDITED.pdf [accessed May 30, 2017].

Umberto, Q. (2000). *CRC World Dictionary of Plant Names*. I: A-C. CRC Press. p. 664. ISBN 978-0-8493-2675-2.

Uncu A. O., Frary A., Karlovsky P., Doganlar S. (2016). High-throughput single nucleotide polymorphism (SNP) identification and mapping in the sesame (*Sesamum indicum* L.) genome with genotyping by sequencing (GBS) analysis. *Mol. Breed.* **36:**173

Uncu A. O., Gultekin V., Allmer J., Frary A., Doganlar S. (2015). Genomic simple sequence repeat markers reveal patterns of genetic relatedness and diversity in sesame. *Plant Genome* **8**:1–12. 10.3835/plantgenome2014.11.0087

Upadhyaya, H.D. (2005) Variability fordrought resistance related traits in the mini core collection of peanut. *Crop Science*, **45**: 1432-1440

USAID (2007) Vegetative propagation techniques, *Perennial Crop Support Series, Jalalabad, Afghanistan.* Publication No.2007-003-AFG. Nov.18, 2007: 38pp https://www.sas.upenn.edu/~dailey/VegetativePropagationTechniques.pdf

USDA (2017) United States Department of Agriculture. Retrieved 9 June 2017.*"Azadirachta indica"*. Germplasm Resources Information Network (GRIN). Agricultural Research Service (ARS); USDA, USA.

Urie, A. L. and Zimmer, D.E. (1970) Yield reduction in safflower hybrids caused by female selfs. *Crop Sci***10**:419-422.

Uzo, J.O. and Adedzwa, D.K. (1985a) A search for drought resistance in wild relatives of cultivates sesame (*Sesamum indicum* L.). *Paper of FAO Plant Protection and Production Department. Paper No.66*: pp163-165

Uzo, J.O., Adedzwa, D.K. and Onwukwe, R.O. (1985b) Yield and yield components and nutritional attributes of cultivated sesame (*Sesamum indicum* L.) and its endemic wild relatives in Nigeria. *Paper of FAO Plant Protection and Production Department. No. 66*: pp166-176

Uzun, B., Arslan, C. and Furat, S. (2008) Variation in Fatty Acid Compositions, Oil Content and Oil Yield in a Germplasm Collection of Sesamum (Sesamum indicum L.)*Journal of the American Oil Chemists' Society***85** (12):1135-1142

Uzun, B., Lee, D., Donini, P.and Cagirgan, M.I.(2003). Identifcation of a molecular marker linked to the closed capsule mutant trait in sesame using AFLP. *Plant Breeding***122**: 95-97

Uzun, B. and Cagirgan, M.I. (2009). Identification of molecular markers linked to determinate growth habit in sesame. *Euphytica* **166** (3): 379-384

Vaissiere, B., Moffet, J. and Loer, G. (1984) Honey bees as pollinators for hybrid cotton seed production on the Texas High Plains. *Agron. J.***76**(6).DOI: 10.2134/agronj. 1984 00021960007600060033.

Valls, J.F.M. and Simpson, C.E. (2005) Genomic characterisation of *Arachis porphyrocalyx. Comparative cytogenetics***11**(1):29-43

Vamshi, J., Uma Devi, G., Chander Rao, S. and Sridevi, G. 2018. Sesame Phyllody Disease: Symptomatology and Disease Incidence. *Int.J.Curr.Microbiol.App.Sci.* **7**(10): 2422-2437. doi: https://doi.org/10.20546/ijcmas.2018.710.281

Vance, C.P and Graham, P.H. (1995) Nitrogen fixation in Agriculture: Application and Perspectives, In: *Proceedings 10th International Congress on Nitrogen Fixation.*St. Petersburg, Russia. In *Fertilizer Research,* **42**(1/3)

van der Esch, S.A. (1999). The AZTEC project. *Paper presented at the World Neem Conference 1999, UBC,* Vancouver, Canada.

Vanisree, G. (1992) Analysis of Breeding Potential of Certain Induced Mutants of Groundnut (*Arachis hypogaea* L.) for Canopy and reproductive Attributes. *Ph.D. Thesis, Andhra Pradesh Agricultural University*, Hyderabad, India. Pp: 396

Vannozzi G. P. (2006). The perspectives of use of high oleic sunflower for oleochemistry and energy laws. *Helia* 29 1–24. 10.2298/hel0644001v

Van Soest, L. J. M. and Bas, N. (2002). Current status of the CGNLinum collection. In: Maggioni, L. M., Pavelek, M., van Soest, L. J. M., Lipman, E. (eds.) *Flax genetic resources in Europe. IPGRI*, Maccarese Rome, Pp.44-48

Vantomme, P.(2002). Non-wood forest products in 15 countries of tropical Asia: an overview. *EC-FAO partnership programme (2000-2002), FAO*, Rome, Italy

Varaprasad, K.S. (2014.) Oilseeds Scenario in India and Way Forward. *Souvenir, Indian Oilseeds Produce Export Promotion Council,* Mumbai

Varaprasad,K.S. and Sudhakara Babu, S.N. (2015). Technologies for Increasing Oilseeds Production. In*: Lead papers of Technologies for Enhancing Oilseeds Production through NMOOP, National Seminar held at Hyderabad during January* 18-19, 2015.

Varaprasad, K.S.and Swarup, G. (1986) Nematode diseases and their control in 2000 AD. *Proceedings of Indian National Science Academy* **51** (2) :66-76

Varaprasad, P.V., Kakani, V. G. and Upadyaya, H.D. (2011) Growth and production of groundnut. *http://greenplanet.eolss.net/EolssLogn/LoginFormMessage.aspx?ReturnUrl=%2fEolssLogn%2fmss%2fC10%2fE1-05A%2fE1-05A-19%2fE1- UNESCO-EOLSS eBook:*pp 1-26.

Varshney, R. K. (2016). Exciting journey of 10 years from genomes to fields and markets: Some success stories of genomics-assisted breeding in chickpea, pigeon pea and groundnut. *Plant Science,***242**: 98–107. https://doi.org/10.1016/j.plantsci.2015.09.009

Vasiljevic, L., Atlagic, J. and Skoric, D. (1991) Applicability of new biotechnology methods in sunflower breeding. *Consult. Meeting FAO-Sunflower Network, Pisa*, Italy.

Vasal, S.K., Dhillon, B.S. and Pandey, S.(2010) Recurrent Selection Methods Based on Evaluation-Cum-Recombination Block, *Plant Breeding Reviews* (June 2010) 139-163; SN 97804706 50073; DO 10. 1002/9780470650073.ch5

Vavilov, N.I. (1926) Studies on origin of cultivated plants, *Bull. Appl. Bot.* **16:** 1-248

Vavilov, N.I. (1951) The origin, variation, immunity and breeding of cultivated plants. *Chronica Botanica* **13**(1-6):59-63

VAYUGRID (2014) Annual Report 2014 of M/s Vayugrid Market Places Pvt. Ltd. https://www.zaubacorp.com/company/VAYUGRID-MARKETPLACE-SERVICES-PRIVATELIMITED/U74900KA2008PTC079566

Vear, F. (2011) *Helianthus* In: Kole(Ed.) *Crop Relatives: Genomic and Breeding Resources: Oilseeds*. Springer, New York. Pp161-170

Veerendra, H.C.S. (1995). Variation studies in provenances of *Azadirachta indica* (the neem tree). *Indian Forester* **121**: 1053- 1056.

Veitch, G.E., Boyer, A. and Ley, S.V (2008) The azadirachtin story. *Angew Chem Int Ed***47**:9402–9429

Vemana, K. and Jain, R.K. (2010) New Experimental Hosts of *Tobacco streak virus* and Absence of True Seed Transmission in Leguminous Hosts, *Indian J Virol.* 2010 Oct; **21**(2): 117–127. doi: 10.1007/s13337-010-0021-0

Venkatachalam, P., Geetha, N., Jayabalan, N. *et al.* (1998) *Agrobacterium*-mediated genetic transformation of groundnut (*arachis hypogaea* L.): An assessment of factors affecting regeneration of transgenic plants. *J. Plant Res.* **111:** 565–572 (1998) doi:10.1007/BF02507792

Venkateswaran, A.N. (1980) Discriminant function as a toolin groundnut breeding. *Proceedings of the National Seminar on the application of Genetics to the Improvement of Groundnut.* Tamilnadu Agricultural University, Coimbatore, India. July 16-17, 1980

Venkateswarlu, B., Mukhopadhyay, K., Mukhopadhyay, J. and Katyal, J.C. (1999). Selection of plus trees of neem with emphasis on azadirachtin content and development of micropropagation protocol for mass propagation. *Paper presented at the World Neem Conference 1999, UBC*, Vancouver, Canada.

Venkatraman, K. and Ashwath, N.(2009) 'Phytocapping – Importance of tree selection and soil thickness', Water, Air and Soil Pollution: *Focus***9**(5–6): 421–30.

Vera-Ruiz, E.M. and Fernandez-Martinez, J.M. (2006) Genetic mapping Tphl gene controlling beta-tocopherol accumulation in sunflower seeds. *Mol. Breed.***17**:291-296

Verma, M.L., Mehta, N. and Sangwan, M.S. (2005) Fungal and Bacterial Diseases of Sesame. Chapter 10. In: G.S. Sahran, N. Mehta, and M.S. Sangwan(Eds.) *Diseases of Oilseed Crops.* 2005. Indus Publishing Co., New Delhi. India

Verma, O.P., Singh, H.P. and Singh, P.V.(1998). Heterosis and inbreeding depression in Indian mustard. *Cruciferae Newsletter.* **20**: 75-76.

Verma P., Goyal R., Chahota R. K., Sharma T. R., Abdin M. Z., Bhatia S. (2015). Construction of a genetic linkage map and identification of QTLs for seed weight and seed size traits in Lentil (*Lens culinaris* Medik.). *PLoS ONE* 10:e0139666 10.1371/journal.pone.0139666

Verma, R. and Singh, R. R. (2007). Multiple shoot formation and *in vitro* flowering in *Brassica juncea* Var. Bhavani. *Our Nature* **5**: 21–24.

Verma, V.D. and Rai, B. (1980) Mutation in seed coat colour in Indian mustard. *Indian J. Agric. Sci.*, **50**: 545-548

Vercoe TK, 1989. Fodder value of selected Australian tree and shrub species. *ACIAR Monograph, No.* **10:**187-192;

Vic, B.A., Jan, C.C. and Miller, J. (2002) Inheritance of reduced saturated fatty acid content in sunflower. *Helia* **25**: 113-122

Vidal, R.,Gerbaud, A. and Vidal, D. (1996) Short-term effects of high light intensities on soybean nodule activity and photosynthesis. *Environmental and Experimental Botany.*, **36** (3): 394 - 357

Vikaspedia : http://vikaspedia.in/agriculture/crop-production/integrated-pest-managment/ipm-for-oilseeds/ipm-strategies-for-mustard-rapeseed/insectpests-management

Vighneswaran.V and Krishnakumar K (2010) Breeding for hybrid sesame suited for summer rice fallow. *Gregor Mendel Foundation Journal***1**: 1-4.

Vikender Kaur, Rashmi Yadav and Wankhede, D.P. (2017)Linseed (*Linum usitatissimum* L.) genetic resources for climate change intervention and its future breeding. *Journal of Applied and Natural Science* **9** (2): 1112 -1118 (2017)

Vilatersana R, Brysting AK, Brochmann C. (2007). Molecular evidence for hybrid origins of the invasive polyploids *Carthamus creticus* and *C. turkestanicus* (Cardueae, Asteraceae). *Molecular Phylogenetics and Evolution* **44:** 610-621.

Vilatersana, R., Garnatje, T., Susanna, A. and Garcia-Jacas. (2005). Taxonomic problems in *Carthamus* (Asteraceae): RAPD markers and sectional classification. *Bot. J. Linn. Soc.* **147**: 375–383.

Vindhiyavarman, P., Mothilal, A., Ganesan, K.N. and Nina Mohammed, S.E. (2000). An appraisal of triploids of amphidiploids in the genus *Arachis. Int. Arachis Newsletter.* **20** **:8-9**

Vines. R. A. (1987) Trees of Central Texas. *University of Texas Press.*

Vishnuvardhan Reddy, A. , Sudhakara Babu, S.N., Radhakrishnan, T., Bhatia, V.S. and Rai, P.K. (2020) Status and technological options for increasing oilseeds production.*Souvenir, Natioanal Seminar on "Technological Innovations in Oilseed Crops for Enhancing* ***Productivity, Profitability and Nutritional Security,*** Feb. 7-8, 2020, Indian Society of Oilseeds Research & ICAR-Indian Institute of Oilseeds Research, Hyderabad, India.pp.1-14.

Vismaya, W. Sapna Eipeson,J.R. Manjunatha,P. Srinivas, T.C. Sindhukanya (2010) Extraction of karanjin: A value addition to karanja *(Pongamia pinnata*) seed oil. *Industrial Crops and Products* **32** :118-122

Vismaya, W., Srikanta M. Belagihally2, Sindhu Rajashekhar1, Vinay B. Jayaram1, Shylaja M. Dharmesh, Sindhu Kanya C. and Thirumakudalu1 (2010A). Gastroprotective Properties of Karanjin from Karanja (*Pongamia Pinnata*) Seeds; Role as Antioxidant and H+,K+-ATPase Inhibitor)Evidence-Based Complementary and Alternative Medicine,Volume 2011 (2011), Article ID 747246, 10 pages

Vivek and Gupta, A.K. (2004) Biodiesel production from Karanj oil, ***Journal of Scientific and*** *Industrial Research*, **63**: 39-47

Vivekanandan, P. (1998). New tree, new system: neem production in south India. *Agroforestry Today***10**: 12-14

Vrbová M., Kotrba P., Horáček J., Smýkal P., Švábová L., Větrovcová M., Smýkalová I., and Griga M. (2013): Enhanced accumulation of cadmium in *Linum usitatissimum* L. plants due to overproduction of metallothionein α-domain as a fusion to β-glucuronidase protein. *Plant Cell Tissue and Organ Culture*, **112**: 321–330

Vromans, J., 2006. Molecular Genetic Studies in Flax *(Linum usitatissimum* L.) *(Ph.D. thesis). Wageningen University*, Wageningen, The Netherlands, 226 p.

Vyas, S. C. 1981. Diseases in sesamum in India and their control. *Pesticides.***15**:10. V

Vyas, S. C., Prasad, K. V. V. and Khare, M. N. 1984. Diseases of sesamum and niger in India and their control. *Department of Plant Pathology, JNKVV, Jabalpur, M. P. Bull. pp.16.*

Wakayama, S., Kausaka, K., Kanehira, T., Yamada, Y., Kawazu, K. and Kobayashi, A. (1994) *Kinobeoyn A*, a novel pigment produced in safflower in tissue culture systems. *Zeitscrift fur Naturforsch.* **49c**: 1-5

Walia, N., Kaur, A. and Babbar.S.B. (2005). In vitro regeneration of a high oil-yielding variety of safflower *(Carthamus tinctorius var. HUS-305). J. Plant Biochem. Biotech.* **14**: 1–4.

Waliar, F. (1991) Evaluation of yield losses due to groundnut diseases in West Africa. In: *Proceedings of the Second ICRISAT Regional Groundnut Meeting for West Africa.* Sept. 11-14, 1991. Niamey, Niger. ICRIST, Pattancheru, India. Pp 32-33

Waliyar, F., Vijay Krishna Kumar, K., Diallo.M., Traore, A., Mangala, U.N., Upadhyaya, H.D. and Sudini, H. (2016) Resistance to pre-harvest aflatoxin contamination in ICRISAT's groundnut mini core collection. *European Journal of Plant Pathology* (2016), **145**(4): 901–913

Wan, Z., B. Jing, J. Tu, C. Ma, J. Shen, B. Yi, J. Wen, T. Huang, X. Wnag and T. Fu (2008) Genetic characterization of a new cytoplasmic male sterility system (hau) in *Brassica juncea* and its transfer to *B. napus. Theor. Appl. Genet.* **116**: 355–362.

Wan, C.W. *et al* (1998) A study on chemical composition of soybean seeds collected from the regional trials Yellow, Huaihe and Haihe river valleys.*Soyabean Science*, **17** (1): 10-18.

Wanasundara, P.K.J.P.D., Wanasundara, U.N. and Shahidi, F.(1999) Changesin flax (*Linum usitatissimum* L.) seed lipids during germination. *Journal of American Oil Chemists' Society,***76**: 41–48 (1999)

Wang, C.T., Yue Yi Tang, Y.Y. Wang, X.Z. *et al*., (2012) Mutagenesis: A Useful Tool for the Genetic Improvement of the Cultivated Peanut (*Arachis hypogaea* L.) *Intech Open.*, DOI: 10.5772/50514. *https://www.intechopen.com/books/mutagenesis/mutagenesis-a-useful-tool-for-the-genetic-improvement-of-the-cultivated-peanut-arachis-hypogaea-l-*

Wang, D.Q. *et al* (1992) Fertilizer response of crops to urea containing borate. *Journal Shandong Agricultural University (China),* **32** (2): 108-113.

Wang *et al.,* (1995) A study of effectiveness of hybrid seed production by utilizing genic male sterility in Sesame *(Sesamum indicum* L.). *Oil Crops of China***17** (1): 12-15.

Wang, L. *et al*., (2012). Global gene expression response to waterlogging in roots of sesame. (*Sesamum indicum* L). *Acta Physiologiae Plantarum* **34:** 2241-2249

Wang L. *et al*., (2014) Genome sequencing of the high oil crop **sesame** provides insight into oil biosynthesis. *Genome Biol.* **15** (2): R39

Wang L, Yu J, Li D. andZhang X (2015) Sinbase: an integrated database to study genomics, genetics and comparative genomics in *Sesamum indicum*. *Plant Cell Physiol.* **56**(1): 2.

Wang, T., Harp, T., Hammond, E.G., Burrisaa, J.S. adnFehrr, W.R. (2001) Seed physiological performance of soybeans with altered fatty acid contents. *Seed. Sci. Res.* **11**: 93-97

Wang Z., Hobson N., Galindo L., Zhu S., Shi D., McDill J., *et al.* (2012). **The genome of flax** (*Linum usitatissimum*) assembled de novo from short shotgun sequence reads. *Plant J.* **72:** 461–473 10.1111/j.1365-313X.2012.05093.x [PubMed] [CrossRef] [Google Scholar]

Wani, S.A. and A.N. Srivastava. (1989). Combining ability analysis in Indian mustard. *Crop Improvement.* **16**(1): 72-75

Wani, S.P., Sreedevi, T.K., (2005). *Pongamia's* journey from forest to micro-enterprise for improving livelihoods. *International Crops Research Institute for the Semiarid Tropics*, Patancheru, Andhra Pradesh, India, pp. 4.

Watanabc. II., Sahunalu, and Khemnark, C. (1988). Combinations of trees and crops in the *taungya* method as applied in Thailand. *Agroforestry Systems.* **6** (2): 169-177. DOI:10.1007/BF02344754

Watling, J., Shock, M. P., Mongeló, G.Z., Almeida, F.O., Krater, T., De Oliveira, P.E. and Neves, E.G. (2018) Direct archaeological evidence for South-western Amazonia as an early plant domestication and food production centre. *PLOS ONE*, 2018; **13** (7): e0199868 DOI: 10.1371/journal.pone.0199868

Watt, G. (1966) The Commercial Products of India. Reprinted by *Today andTomorrow Printers and Publishers* (1980) New Delhi. India.

Weaver D. B. and Wilcox J. R. (1982) Heritabilities, gains from selection, and genetic correlations for characteristics of soybeans grown in two row spacings. *Crop Sci.* **22**(3):625–629. Google Scholar

Wei LB, Zhang HY, Zheng YZ, Guo WZ, Zhang TZ. (2008) Developing Est-derived microsatellites in sesame (*Sesamum indicum* L.) *Acta Agron Sin.* **34**(12):2077–2084. doi: 10.1016/S1875-2780(09)60019-5.

Wei L.B, Miao H., Li C., Duan Y., Niu J., Zhang T., *et al.* (2014). Development of SNP and InDel markers via de novo transcriptome assembly in *Sesamum indicum* L. *Mol. Breed.* **34**: 2205–2217.

Wei W, Zhang Y, Lü H, Li D, Wang L, Zhang X (2013). Association analysis for quality traits in a diverse panel of Chinese sesame (*Sesamum indicum L.)* germplasm.*J Integr Plant Biol. 2013 Aug;* **55**(8):745-58

Wei, X., Wang, L., Zhang, Y., Qi, X., Wang, X., Ding, X., Zhang, J. anf Zhang, X. (2015) Development of simple sequence repeat (SSR) markers of sesame (*Sesamum indicum*) from a genome survey. *Molecules.* 2014 Apr. 22.,**19**(4):5150-62.

Wei, X., Liu, K., [….] Zhnag, X. (2015) Genetic discovery for oil production and quality in sesame. *Nat Commun* **6:** 8609 (2015)

Wei X., Wang L., Zhang Y., Qi X., Wang X., Ding X., et al. (2014). Development of simple sequence repeat (SSR) markers of sesame (*Sesamum indicum*) from a genome survey. *Molecules* **19:** 5150–5162. 10.3390/molecules19045150

Weisker, A.C. (1995). Hybrid safflower production utilizing genetic dwarf male sterility. *United States Patent, Patent No.5,436,386.*

Weiss, E.A. (1966) Sunflower trials in Western Kenya. *East African and Forestry Journal***31** (4): 405-408

Weiss, E.A. (1971) *Castor, Sesame and Safflower.*Leonard Hill. London.

Weiss, E.A. (2000) *Oilseed Crops*, Second Edition, (ISBN 0-632-05259-7), Blackwell Science Ltd. Pp.364

Weselake, R.J., Tylor, D.c. and Rahman, M.H., Shaha, S., Laroche, A., McVetty, P.B.E. and Harwood, J.L. (2009) Increasingthe flow of Carbon into seed oil. *Biotech. Adv.***6:** 866 -878

Westcott, N. A. and Muir, A. D. (2003). Flax seed lignan in disease prevention and health promotion. *Phytochem. Rev.***, 2:** 401–417

Wheat, C. W.; Vogel, H.; Wittstock, U.; Braby, M. F.; Underwood, D.; Mitchell-Olds, T. (2007). "The genetic basis of a plant insect coevolutionary key innovation". Proceedings of the National Academy of Sciences. 104 (51): 20427–31.

Wheeler, J.R., Rye, B.L., Koch, B. L. and Wilson, A.J.G. (1992). Flora of the Kimberley region. *Western Australian Herbarium,* Perth: Department of Conservation and Land Management, 1327 pp

Wickremasinghe, I.P. and Simons, A.J. (1994). Differentiation of jak (*Artocarpus heterophyllus* Lam.) and neem (*Azadirachta indica* A. Juss.) by polyacrylamide gel electrophoresis. In: H.P.M. Gunasena(Ed.) *Proceedings of a review seminar on the UP-OFI Link Project, University of Peradeniya*, Peradeniya, Sri Lanka. pp. 77-96.

Wijayanto T., McHughen A. (1999): Genetic transformation of *Linum* by particle bombardment. *In Vitro Cellular and Development Biology*, **6:** 456–465

Williams, I.H. (1988) The pollination of linseed.*Bee World,***69**: 145-152.

Wills, D.M. and Burke, J.M. (2007) Quantitative trait locus analysis of the early domestication of sunflower. *Genetics* 2007**, 176:** 2589–2599

Wilson, I. M. (1946) Observations on wilt disease of flax, *Transactions of the British Mycological Society,* **29** (4): 221-231, IN6. https://doi.org/10.1016/S0007-1536(46)80004-4

Wilson,J.N., Baring, M.R., Mark D. Burow, M.D., Rooney, W.L. and Simpson, C.E. (2013) Diallel Analysis of Oil Production Components in Peanut (*Arachis hypogaea* L.) *International Journal of Agronomy*, 2013, 5 pages. http://dx.doi.org/10.1155/2013/975701

Wilson, R., Van Schie, B.J. and Howes, D. (1998). "Overview of the preparation, use and biological studies on polyglycerol polyricinoleate (PGPR)". *Food and Chemical Toxicology*. **36** (9–10): 711–8. doi:10.1016/S0278-6915(98)00057-X. PMID 9737417.

Witt, U. Hansen, S., Albaum, M. and Abel, W.O. (1991) Molecular analyses of the CMS in inducing *Polina* cytoplasm in *Brassica napus.* L., *Curr. Genet.***19:** 323-327

Wood, C., Liu, Q., Taylor, M., Okada, S., Zhou, X., Singh, S., Green, A.(2015). Development of super high oleic safflower oil for oleochemical applications. *Nutrients*, **7:** 9999–10019*).*

World Agroforestry (2013) https://web.archive.org/web/20131109205555/http://www.worldagroforestry.org/treedb2/AFTPDFS/Shorea_robusta.pdf

Worthington, R.E., Hammons, R.O., Allison, J.R., (1972) *J. Agr. Food Chem.* 1972. **20:** pp727-730

Wright, G. (2004). Peanuts, In: Colin Wrigley(Ed) *Encyclopedia of Grain Science, Academic Press*. Copyright © 2004 Elsevier Ltd. ISBN.978-0-12-765490-4 Pp. 1700

Wrobel-Kwiatkowska, M., J. Zebrowski, M. Starzycki, J. Oszmianski, and J. Szopa. (2007a). Engineering of PHB Synthesis Causes Improved Elastic Properties of Flax Fibers. *Biotechnol. Prog.* **23**:269–277

Wrobel-Kwiatkowska, M., M. Starzycki, J. Zebrowski, J. Oszmianski, and J. Szopa. (2007b). **Lignin defi ciency in transgenic fl ax resulted in plants with improved mechanical** properties. *J. Biotechnol.* **128**:919–934

Wu, K., Yang. M., Liu, H., Tao, Y., Mei, J. and Zhao, Y. (2014) Genetic analysis and molecular characterization of Chinese sesame (*Sesamum indicum* L.) cultivars using insertion-deletion (InDel) and simple sequence repeat (SSR) markers.*BMC Genet.* **15** (35)

Wynne, J.C. and Coffelt, T.A. (1982) Genetics of *Arachis hypogaea* L. In: *Peanut Science and Technology, pp 50-94. American Peanut Research and Education Society,* Yokum, Texas, USA

Wynne, J.C., Elkan, G.H., Isleib, T.G. and Schneeweis, T.J. (1983). Effect of host plant, Rhizobium strain and host x strain interaction on symbiotic variability in peanut, *Peanut Sci.,* **10**: 110-114.

Wynne, J.C., Emery, D.A. and Rice, P.N. (1970) Combibining ability estimates in *Arachis hypogaea* L. II. Field performance of F1 hybrids. *Crop Science* **10:** 713-715.

Wynne, J.C. and Gregory, W.C. (1981)Peanut breeding. *Advances in Agronomy*. **34:** 39-72

Xavier, A., Hall, B., Hearst, A.A., Cherkauer, K.A. and Rainey, K.M.(2017) Genetic Architecture of Phenomic-Enabled Canopy Coverage in *Glycine max*.Genetics March 2017, genetics.116.198713; DOI: https://doi.org/10.1534/genetics.116.198713

Xiaohui Zhang, Tongjin Liu, Xixiang Li, Mengmeng Duan, Jinglei Wang, Yang Qiu, Haiping **Wang, Jiangping Song and Di Shen (2016) Interspecific hybridization, polyploidization, and** backcross of *Brassica oleracea var. alboglabra* with *B. rapa* var. *purpurea* morphologically recapitulate the evolution of *Brassica* vegetables *Sci Rep.* 2016; **6:** 18618. Published online 2016 Jan 4. doi: 10.1038/srep18618 PMCID: PMC469863

Xu, B.S. *et al.,* (1992) A study of dry matter accumulation and the law of fertilizer requirement in sesame.*Acta Agricultural Universitatis Henanensis,***26** (4): 331-334

Xu, L., Najeeb, U., Tang, G.X., Gu, H.H., Zang, G.Q., He, Y. and Zhou, W.J. (2007) Haploid and double haploid technology. In: S.K. Gupta (Ed). *Rapeseed Breeding. Advances in Botnical Research.* **45.** *Academic Press/ Elsivier. San Diego, CA*. pp.1821-216

Xu, Z.Q., Jia, J.F. and Hu, Z.D. (1997). Somatic embryogenesis in *Sesamum indicum L. cv. nigrum. Journal of Plant Physiology* **150***: 755-758.*

Yadav, R.D., Jain, S.K., Alok, S., Prajapati, S.K. and Verma, A. (2011) *Pongamia pinnata. An overview. J. Pharma Sci. Res.* **2**(3): 494-500

Yadav, I.S., Yadava, T.P., Kumar, P. and Thakural, N.K.(1998). Heterosis for seed yield and its components in rapeseed under ranifed conditions. *Cruciferae Newsletter*. **20***:79.*

Yadav M, Chaudhary D, Sainger M and Jaiwal PK (2010). *Agrobacterium tumefaciens*-mediated genetic transformation of Sesame (*Sesamum indicum* L.) Plant Cell Tissue Org Cult. 2010;103:377–386. doi: 10.1007/s11240-010-9791-8. [CrossRef] [Google Scholar]

Yadav, R.D., Jain, S.K., Alok, S., Kailasiya, D., Kanaujia, V.K. and Kaur, S. (2011) A Study on Phytochemical Investigations of *Pongamia pinnata*, Linn. Leaves, *IJPSR;***2**(8): 2073-2079

Yadav.R.K., Tripathi, A.K. and Yadav, A.K. (2009) Effect of Micronutrients in Combinations with Organic Manure on Production and net Returns of Sesame (*Sesamum indicum*) in Bundelkhan Tract of Uttar Pradesh.*Ann. Agric. Res. New Series,***30** (1&2):53-58

Yadav, S.S., Jhakar, M.L. and Yadav, L.R. (2013) Response of taramira *(Eruca sativa)* to varying levels of FYM and vermicompost under rainfed conditions. *Journal of Oilseed Brassica,* **4**(1): 49-52

Yadav, Y. S., Siddiqui, A. U. and Aruna Parihar, (2005). Management of root-knot nematode *Meloidogyne incognita* infesting gram through oil cakes. *J. Phytolog. Res.,* **18** (2): 263-264

Yadava, D. K., Hossain, F. and Mohapatra, T. (2018). Nutritional security through crop biofortification in India: Status & future prospects. *The Indian journal of medical research, 148*(5), 621–631. https://doi.org/10.4103/ijmr.IJMR_1893_18

Yadava, T.P (1976) Present available status of production technology of rapeseed-mustard and future programme of work. *Paper @ International Seminar on Rapeseed-Mustard. Mysore. Nov.22-24, 1976*

Yadava, T.P., Kumar, P. and Thakral, S.K. (1984) Association of pod yield with some quantitative traits in bunch group of groundnut (*Arachis hypogaea* L.) *Journal of Research, Haryana Agricultural University,* **14**: **85-88**

Yadava, T.P., Friedt, D.W. and Gupta, S.K. (1998)Oil content and fatty acid composition of taramira (*Eruca sativa* L.) genotypes.*Journal of Food Science and Technology Mysore***35**(6): 557-558

Yakkundi, S.R., Thejavathi, R. and Ravindranath, B. (1995). Variation of azadirachtin content during growth and storage of neem (*Azadirachta indica*) seeds. *Journal of Agricultural and Food Chemistry***43** (9): 2517-2519.

Yamagishi, H. and Bhat, S.R. (2014) Cytoplasmic male sterility in *Brassicaceae* crops. *Breed Sci.,***64(**1): 38–47. doi: 10.1270/jsbbs.64.38

Yance, Donald (2020)Herbal Medicine, Healing & Cancer, Flax Seed Studies & Reviews, *OAW Health-oasis advanced wellness. © Copyright 2004 - 2020* | All Rights Reserved. *www.oawhealth.com*

Yang· G.S., Duan, Z.H.,Fu, T.D. and Wu, C.S.(1995) A Promising Alternative way of Utilizing POL CMS for Hybrd Breeding in *Brassica napus.* L

Yang, G. S. and Fu, T.D. (1991) A preliminary study on the restoring-maintaining relationship in rapeseed *(Brassica napus* and *Brassica campestris). Acta Agronomica Sinica***1991**:172151~156

Yang, X., Hu, R., Yin, H. *et al.* (2017) The *Kalanchoë* genome provides insights into convergent evolution and building blocks of crassulacean acid metabolism. *Nat Commun* **8:** 1899 (2017). https://doi.org/10.1038/s41467-017-01491-7

Yang,Y.X., Wu, W., Zheng, Y.L., Chen, L., Liu, R.J. and Huang, C.Y. (2007) Genetic diversity and relationships among safflower (*Carthamus tinctorius* L.) analyzed by inter-simple sequence repeats (ISSRs), *Genetic Resources and Crop Evolution,* **54** (5) : 1043-1051

Yarbrough, J.A. (1949) *Arachis hypogaea* L. The Seedling, Its Cotyledons, Hypocotyl and Roots. *American Journal of Botany***36** (10): 758-772

YashRoy, R.C.; Gupta, P.K. (2000). Neem-seed oil inhibits growth of termite surface-tunnels. *Indian Journal of Toxicology*. **7** (1): 49–50.

Yates, D.J. and Steven, M.D.(1987). Reflexion and absorption of solar radiation by flowering canopies of oil-seed rape (*Brassica napus* L.). *J. Agric. Sci., Camb.,* **109**: 495-502

Yermanos, D. M. (1979). "Jojoba – a crop whose time has come". *California Agriculture.*

Yermanos, D. M. (1980) Sesame. In: Fehr.W. R.; Hadley, H. H. (Ed.). ***Hybridization of crop plants***. Madson: ASA, 1980. p. 549-563.

Yermanos, D.M. *et al.,* (1967). Inheritance of quantity and quality of seed oil in safflower. *Crop Science.***7** (5): 417-422.

Yermanos, D.M. and Kostopoulos, S.S.(1970). Heterostyly and incompatibility in *Linum **grandiflorum*** Desf. in the indispensable pollinators. *Arkansas Agri. Ext. Ser. (Misc. Pub.),***127**: 50-54.

Yermanos, D.M. and Kotecha, A. (1979) Diallel analysis of four traits in sesame *(Sesamum indicum* L) Abst. In *Agron. Abs. Madison. USA. American Society of Agronomy*, **68.**

Yin, J. Z., Xiao, M., Wang, A. Q. and Xiu, Z. L. (2008). Synthesis of biodiesel from soybean oil by coupling catalysis with subcritical methanol. *Energy Conver. Manage.*,**49**:3512 - 3516

Yin Yan, Chen Zhao-bo,Yu Jian,Wang Han-zhong and Feng Zhong-chao (2010) Analysis of Potential for Rapeseed Production in China,*Journal of Agricultural Science and Technology* 2010-**03** (http://en.cnki.com.cn/Article_en/CJFDTOTAL-NKDB201003004.htm)

Yirgu, D. 1964. Some diseases of *Guizotia abyssinica* in Ethiopia. Plant Disease Reporter **48**:672.

Yogesh, S, *et al.*, (2014) Light use efficiency, productivity and profitability of maize and soybean intercropping as influenced by planting geometry and row proportion.*Karnataka J. Agric. Sci.*, **27** (1): 1-4.

Yol, E·, Ustun., R ·, Golukcu, M., Uzun, B. (2017) Oil Content, Oil Yield and Fatty Acid Profile of Groundnut Germplasm in Mediterranean Climates, *J. Am. Oil Chem. Soc.*, (2017) **94**:787–804. DOI 10.1007/s11746-017-2981-3.

Yoopraset, S and Sooksathan, I. (1992) Allelopathic effect of sesame *(Sesamum indicum L.)* on some field crops. *Sesame Safflower Newsletter*, **7** :111

Younghee K (2001). Effects of BA, NAA, 2,4-D and AgNO3 treatments on the callus induction and shoot regeneration from hypocotyl and cotyledon of sesame (*Sesamum indicum* L.). *Journal of the Korean Society for Horticultural Science* **42**: 70-74.

Yu J, Dossa K, Wang L, Zhang Y, Wei X, Liao B, Zhang X (2017) PMDBase: a database for studying microsatellite DNA and marker development in plants. *Nucleic Acids Res. 2017 Jan 4;* **45**(D1): D1046-D1053.

Zaveri, P.P. (2009) Castor Hybrid Seed Production. Availability and Economics. *Paper presented at 'Seminar on Castor Hybrid Seed Production' by Gujarat State Seed Producers Association.* Ahmedabad. 2009. Pp: 31-38

Zeven, A.C. and P.M. Zhukovsky. (1975). Dictionary of Cultivated Plants and their Centres of Diversity. *Centre for Agric. Publ. and Document. (Pudoc),* Wageningen, Netherlands. Pp219

Zhan, X-c., Jones, D. A., and Kerr, A. (1988) Regeneration of flax plants transformed by *Agrobacterium rhizogenes*. *Plant Mol Biol***11**(5):551–559

Zhang, G. Q., Zhang, D.Q., Tang, G.X., He, Y. and Zhou, W.J. (2006) Plant development from microspore-derived embryos in oilseed rape as affected by chilling, desiccation and cotyledon excision. Biol. Plantarum, **50**: 180-186.

Zhang H, Wei L, Miao H, Zhang T, Wang C (2012) Development and validation of genic-SSR markers in sesame by RNA-seq. *BMC Genomics. 2012 Jul 16*; 13():316.

Zhang H, Miao H, Li C, Wei L, Duan Y, Ma Q, Kong J, Xu F, Chang S (2016) Ultra-dense SNP genetic map construction and identification of SiDt gene controlling the determinate growth habit in *Sesamum indicum* L. *Sci Rep. 2016 Aug* **16** (6):31556.

Zhang, O.H. (1982) Oil Crops of China No.**1**:12-18

Zhang, Y. *et al.*, (2015) Row Ratios of Intercropping Maize and Soybean Can Affect Agronomic Efficiency of the System and Subsequent Wheat.PLOS/ONEhttps://doi.org/10.1371/journal.pone.0129245

Zhang, Z. and Johnson R.C. (1999). Safflower germplasm collection directory. IPGRI Office for East Asia, Beijing

Zhao LJ, Peralta-Videa JR, Rico CM, Hernandez-Viezcas et al., (2014). CeO2 and ZnO nanoparticles change the nutritional quality of cucumber (*Cucumis sativus*). *J Agric Food Chem***62** (2014) 2752-2759

Zhou, J., Raman, H. *et al.*, (2014) Constructing a dense linkage map and mapping QTL for traits of flower development in *Brassica carinata. Theor. Appl. Genet.***127**: 1593-1605

Zhou, W.J., Tang, G.X. and Hagberg, P. (2002) Efficient production of double haploid plants by immediate cochicine treatment of isolated microspores in winter *Brassica napus. Plant Growth Regul.* **37**: 185-192.

Zhu, J., Struss, D. and Robbelen, G. (1993) Studies on resistance to *Phoma lingam* in *Brassica napus-Brassica nigra* addition lines. *Plant Breed.* **111**: 192-197

Zhuchenko, A. A. and Rozhmina, T. A. (2000). Mobilizacija genetičeskich resursov lna [Mobilization of flax genetic resources], VILAR AND VNIIL, *Starica*. Pp. 224

Zimmer, D.E. (1965) Sources of rust resistance in safflower. *Plant Disease Report* **49**:440-442.

Zimmer D.E. and Urie, A.L. (1968) Inheritance of rust resistance in crosses between cultivated safflower, *Carthamus tinctorius* L., and wild safflower, *Carthamus oxyacantha* M.B. *Phytopathology,* **58**:1340–1342).

Zimmer, D.E. and Urie, A.L. (1970) Registration of rust resistant safflower breeding lines PCA, PCM-1, PCM-2, PCN1, and PCOy. *Crop Science* **10**:1463

Zimmerman, L.H. and Smith, J.D. (1966) Production of F1 seed in castor beans by use of sex genes sensitive to environment. *Crop Science.* **6**: 406-409

Zohary, D., Hopf, M. (2000) Domestication of Plants in the Old Word: The Origin and Spread of Cultivated Plants in West Asia, Europe and the Nile Valley. *Oxford University Press,* Oxford, UK. 316 p

Zomlefer, Wendy B. (1994). Guide to flowering plant families. Chapel Hill, NC: *The University of Carolina Press*. 430 p. [72836]

Zope, R.E., Katule, B.K. and Ghorpade, D.S. (1994). Seed filling duration and yield in safflower. *Sesame Safflower Newsletter* **9**: 39-42

Zuk, M., Rihter, D., Matula, J., Szopa, J. (2015) Linseed, the multipurpose plant. *Industrial Crops and Products 75* · May 2015 DOI: 10.1016/j.indcrop.2015.05.005